Table 2 Units of quantities in electromagnetics

Quantity	Symbol	SI unit and abbreviation	SI unit expressed in terms of SI base units
Length	l	meter (m)	m
Mass	m	kilogram (kg)	kg
Time	t	second (s)	s
Electric current	I	ampere (A)	A
Temperature	$\mathscr{T}$	kelvin (K)	K
Luminous intensity	$\mathscr{I}$	candela (cd)	cd
Velocity	v	meter/second (m/s)	$m \cdot s^{-1}$
Acceleration	a	meter/second^2 (m/s)^2	$m \cdot s^{-2}$
Area	A	meter^2 (m^2)	m^2
Volume	v	meter^3 (m^3)	m^3
Energy, work	W	joule (J)	$kg\ m^2\ s^{-2}$
Energy density	w	joule/meter^3 (J/m^3)	$kg\ m^{-1}\ s^{-2}$
Force	$\mathbf{F}$	newton (N)	$kg\ m\ s^{-2}$
Momentum	$m\mathbf{v}$	newton-second (N · s)	$kg\ m\ s^{-1}$
Torque	$\mathbf{T}$	newton-meter (N · m)	$kg\ m^2\ s^{-2}$
Power	P	watt (W)	$kg\ m^2\ s^{-3}$
Electric charge	Q, q	coulomb (C)	$A \cdot s$
Electric field strength	$\mathbf{E}$	volt/meter (V/m)	$m\ kg\ s^{-3}\ A^{-1}$
Potential, emf	$V, \phi, \mathscr{V}$	volt (V)	$kg\ m^2\ A^{-1}\ s^{-3}$
Electric flux	Ψ	coulomb (C)	$A \cdot s$
Electric flux density	$\mathbf{D}$	coulomb/meter^2 (C/m^2)	$A\ s\ m^{-2}$
Volume charge density	ρ	coulomb/meter^3 (C/m^3)	$A\ s\ m^{-3}$
Surface charge density	ρ_s	coulomb/meter^2 (C/m^2)	$A\ s\ m^{-2}$
Line charge density	ρ_L	coulomb/meter (C/m)	$A\ s\ m^{-1}$
Sheet-current density	$\mathbf{K}$	ampere/meter (A/m)	$A \cdot m^{-1}$
Current density	$\mathbf{J}$	ampere/meter^2 (A/m^2)	$A \cdot m^{-2}$
Capacitance	$\mathbf{C}$	farad (F)	$A^2\ s^4\ kg^{-1}\ m^{-2}$
Polarization	$\mathbf{P}$	coulomb/meter^2 (C/m^2)	$A\ s\ m^{-2}$
Dipole moment	$\mathbf{p} = q\mathbf{l}$	coulomb-meter (C · m)	$m\ A\ s$
Impedance	Z	ohm (Ω)	$kg\ m^2\ A^{-2}\ s^{-3}$
Resistance	R	ohm (Ω)	$kg\ m^2\ A^{-2}\ s^{-3}$
Resistivity	ρ	ohm-meter (Ω · m)	$kg\ m^3\ A^{-2}\ s^{-3}$
Conductivity	σ	siemens/meter (S/m)	$A^2\ s^3\ kg^{-1}\ m^{-3}$
Permittivity	ε	farad/meter (F/m)	$A^2\ s^4\ kg^{-1}\ m^{-3}$
Permeability	μ	henry/meter (H/m)	$m\ kg\ s^{-2}\ A^{-2}$

Quantity	Symbol	SI unit and abbreviation	SI unit expressed in terms of SI base units
Inductance	L	henry (H)	$kg\ m^2\ A^{-2}\ s^{-2}$
Magnetic flux	ϕ	weber (Wb)	$kg\ m^2\ A^{-1}\ s^{-2}$
Magnetic flux density	$\mathbf{B}$	tesla (T) = weber/meter^2 (Wb/m^2)	$kg\ A^{-1}\ s^{-2}$
Magnetic field strength	$\mathbf{H}$	ampere/meter (A · m^{-1})	$A \cdot m^{-1}$
Magnetic dipole moment	$\mathbf{m} = I\mathbf{A}$	ampere-meter^2 (A · m^2)	$A \cdot m^2$
Flux linkage	Λ	weber-turn (Wb)	$kg\ m^2\ A^{-1}\ s^{-2}$
Magnetization	$\mathbf{M}$	ampere/meter (A · m^{-1})	$A \cdot m^{-1}$
Magnetomotive force (mmf)	$\mathscr{F}$	ampere-turn (A)	A
Magnetic (scalar) potential	ϕ, Φ	ampere (A)	A
Magnetic vector potential	$\mathbf{A}$	weber/meter (Wb/m)	$kg\ m\ A^{-1}\ s^{-2}$
Magnetic pole strength	Q_m	ampere-meter (A · m)	$A \cdot m$
Magnetic pole density	ρ_m	ampere/meter^2 (A/m^2)	$A \cdot m^{-2}$
Magnetic surface pole density	ρ_{sm}	ampere/meter (A/m)	$A \cdot m^{-1}$
Reluctance	$\mathscr{R}$	1/henry (H^{-1})	$A^2\ s^2\ kg^{-1}\ m^{-2}$
Poynting vector	$\mathscr{P}$	watt/meter^2 (W/m^2)	$kg \cdot s^{-3}$
Attenuation constant	α	neper/meter (Np/m)	m^{-1}
Phase constant	β	radian/meter (rad/m)	m^{-1}
Frequency	f	hertz (Hz) = cycle/s	s^{-1}
Wavelength	λ	meter (m)	m
Intrinsic impedance	η	ohm (Ω)	$kg\ m^2\ A^{-2}\ s^{-3}$
Skin depth	δ	meter (m)	m

Physical constants	
Permittivity of vacuum	$\varepsilon_0 = 8.854 \times 10^{-12}$ F/m
Permeability of vacuum	$\mu_0 = 4\pi \times 10^{-7}$ H/m
Electronic charge	q_e or $e = 1.602 \times 10^{-19}$ C
Speed of light in vacuum	$c = 2.9979 \times 10^8$ m/s
Electron rest mass	$m_e = 9.11 \times 10^{-31}$ kg
Planck's constant	$h = 6.626 \times 10^{-34}$ J · s
Boltzmann's constant	$k = 1.38 \times 10^{-23}$ J/K
Free-space wave impedance	$\eta_0 = 376.7\ (\cong 120\pi)\ \Omega$
Standard gravitational acceleration	$g = 9.807$ m/s^2

APPLIED ELECTROMAGNETICS

APPLIED ELECTROMAGNETICS

MARTIN A. PLONUS
Professor of Electrical Engineering
and Computer Science
Northwestern University

McGRAW-HILL BOOK COMPANY
New York St. Louis San Francisco Auckland Bogotá Düsseldorf Johannesburg London Madrid Mexico Montreal New Delhi Panama Paris São Paulo Singapore Sydney Tokyo Toronto

APPLIED ELECTROMAGNETICS

4567890 DODO 8321

This book was set in Times Roman. The editors were Peter D. Nalle and Michael Gardner; the cover was designed by Scott Chelius; the production supervisor was Leroy A. Young. The drawings were done by J & R Services, Inc.
R. R. Donnelley & Sons Company was printer and binder.

Library of Congress Cataloging in Publication Data

Plonus, Martin A
Applied Electromagnetics.

Bibliography: p.
Includes index.
1. Electromagnetic theory. 2. Electric engineering.
I. Title.
QC670.P73 621.3'01 77-11986
ISBN 0-07-050345-1

To the memory of my father,
Christopher Plonus

Contents

Chapter Two

CONDUCTORS AND CHARGES 53

Chapter Three

SOURCES OF VOLTAGE (emf) AND STEADY ELECTRIC CURRENT 101

Chapter Four

DIELECTRICS AND POLARIZATION 123

Chapter Twelve
RELATIVITY AND MAXWELL'S EQUATIONS 458

Chapter Thirteen
APPLICATION OF MAXWELL'S EQUATIONS: EM WAVES AND PROPAGATION OF ENERGY 471

Chapter Fourteen
APPLICATIONS OF MAXWELL'S EQUATIONS: REFLECTION OF EM WAVES 509

Chapter Fifteen
TRANSMISSION LINES 552

Preface

This book is designed to be used in a first undergraduate course in electromagnetics (EM) which follows the usual introductory physics courses. It is different in two respects. After teaching a first EM course several times, it was observed that some topics present special difficulties for students. These are developed in more detail, either by adding sections which are more rigorous, or by adding sections which cover more of the characteristics of the subject. This should aid students in mastering the subject since they will not have to search in reference books. It should also help instructors in tailoring the material to a class since they can easily omit or use the material in these extra sections. The second characteristic of the book is the heavy emphasis on the practical side of EM. Sections have been added on devices and phenomena from many diverse fields which have EM principles as an underlying base. The intention here is to convince students that their understanding of many areas, such as solid state, physical electronics, linear and rotating machines, microwaves, superconductivity, etc., depends on EM. This and the sections on devices should also stimulate students' interest, for they often tend to view a course in EM as a dry experience which does not go beyond mathematical manipulations.

For an understanding of EM, in addition to the sections on applications and uses of EM in other fields, a minimum understanding of microscopic or atomic behavior is necessary. Hence the sprinkling of such sections throughout the book (for example, Secs. 2.1 and 4.2). Having presented basic electromagnetics in a rather thorough fashion, it is now apparent that to cover the entire book in a span of a quarter or even a semester would be impractical.

If we view a first course as consisting of three parts, electrostatics, magnetostatics, and a unification of these two in a theory of EM as given by Maxwell's equations, including elementary applications of Maxwell's equations (such as plane waves and the propagation of energy), then an outline of sections suitable for a single course in basic EM is suggested below.

Chapter 1: All sections, except perhaps 1.19 and 1.20
Chapter 2: 2.1, 2.2, 2.3, 2.4, 2.6, 2.8, 2.9, 2.10, plus 1 or 2 application sections
Chapter 3: 3.1, 3.2, 3.3, 3.5, 3.6, 3.7
Chapter 4: 4.1, 4.2, 4.3, 4.4, 4.5, 4.9, 4.10, 4.11
Chapter 5: All sections, except perhaps 5.12, 5.15, 5.16, 5.17
Chapter 6: 6.1 to 6.11, plus 1 or 2 application sections
Chapter 7: All sections
Chapter 8: All sections, except perhaps 8.4, 8.5
Chapter 9: 9.1 to 9.5, 9.9
Chapter 10: 10.1 to 10.7, plus some application sections
Chapter 11: All sections, except perhaps 7.4 and parts of 7.8 and 7.9
Chapter 12: 12.1, 12.2, 12.3
Chapter 13: 13.1 to 13.4, plus remaining sections if time permits
Chapters 14, 15: Optional

It should be emphasized that this is only a suggested outline. For a one-quarter course it still might be too ambitious, whereas additional sections can be included if the course is longer or more advanced. In a typical electrical engineering curriculum such a course is then followed by a second course in EM which begins with Maxwell's equations and continues with the application of these equations to transmission lines, waveguides, propagation, reflection of waves, radiation problems, etc.

The motivation for this book was to close the gap in EM texts between sound treatment of theory and inadequate or absent coverage of applications of such theory. Such a gap, if not removed in the classroom, results in the usual emphasis on derivations and mathematics, and has given beginning EM courses the dubious status of being too tough and not very interesting. This state is not helped given today's students, with their continual questioning of the relevance of a professor's teaching as well as their emphasis on practical rather than theoretical material. The penalty in a text taking a practical approach is in the added length of the application sections. On the other hand, a book of this type should lend itself easily to a balance of material; one can make a course as practical or as theoretical as desired by the choice of sections.

The organization of the book is traditional: electrostatics, magnetostatics, followed by electromagnetics. Thus the simpler engineering applications are presented at the beginning of the book. Other organizations, some that start with Maxwell's equations or with relativity, appear to be an attempt to present dynamics early in the course, in the hope of stimulating and making the course more exciting for students. As there is little hope that students can grasp advanced concepts adequately without mastering fundamentals first, the frequent backtracking to cover static material proves to be disruptive. The more logical presentation of the traditional approach can be made sufficiently exciting to engineering students by relating the theory to real-world problems which are covered in the application sections.

ACKNOWLEDGMENT

I would like to thank Professor R. F. Frerichs, the inventor of the photoconductive CdS cell, for many discussions on microscopic phenomena. My thanks are also extended to Professor R. E. Beam for a careful reading of the manuscript, to S. C. H. Wang for helping with the Problems, and to my wife, Tina, for her assistance on the manuscript. The typing was ably done by Mrs. Leslie Rindler, Miss Joanna Hague, and Paul Burczyk.

MARTIN A. PLONUS

Bibliography

BEGINNING

Boast, W. B.: "Vector Fields," Harper & Row, Publishers, Incorporated, New York, 1964

Bradshaw, M. D. and W. J. Byatt: "Introductory Engineering Field Theory," Prentice-Hall, Inc., Englewood Cliffs, N.J., 1967.

Carter, G. W.: "The Electromagnetic Field in Its Engineering Aspects," American Elsevier Publishing Company, Inc., New York, 1967.

Cheston, W. B.: "Elementary Theory of Electric and Magnetic Fields," John Wiley & Sons, Inc., New York, 1964

Hayt, W. H.: "Engineering Electromagnetics," 3d ed., McGraw-Hill Book Company, New York, 1974.

Holt, C. A.: "Introduction to Electromagnetic Fields and Waves," John Wiley & Sons, Inc., New York, 1966.

Kraus, J. D. and K. R. Carver: "Electromagnetics," 2d ed., McGraw-Hill Book Company, New York, 1973.

Magid, L. M.: "Electromagnetic Fields, Energy and Waves," John Wiley & Sons, Inc., New York, 1972.

Paris, D. T. and F. K. Hurd: "Basic Electromagnetic Theory," McGraw-Hill Book Company, New York, 1969.

Skilling, H. H.: "Fundamentals of Electric Waves," John Wiley & Sons, Inc., New York, 1964.

Silvester, P.: "Modern Electromagnetic Fields," Prentice-Hall, Inc., Englewood Cliffs, N.J., 1968.

Tomboulian, D. H.: "Electric and Magnetic Fields," Harcourt, Brace & World, Inc., New York, 1965.

Walsh, J. B.: "Electromagnetic Theory and Engineering Applications," The Ronald Press Company, New York, 1960.

Whitmer, R. M.: "Electromagnetics," 2d ed., Prentice-Hall, Inc., Englewood Cliffs, N.J., 1962.

INTERMEDIATE

Bohn, E. V.: "Introduction to Electromagnetic Fields and Waves," Addison-Wesley Publishing Company, Inc., Reading, Mass., 1968.

Della Torre, E. and C. V. Longo: "The Electromagnetic Field," Allyn and Bacon, Inc., Boston, 1969.

Dearholt, D. W. and W. R. McSpadden: "Electromagnetic Wave Propagation," McGraw-Hill Book Company, New York, 1973.

Durney, C. H. and C. C. Johnson: "Introduction to Modern Electromagnetics," McGraw-Hill Book Company, New York, 1973.

Fano, R. M., L. J. Chu, and R. B. Adler: "Electromagnetic Fields, Energy and Forces," John Wiley & Sons, Inc., New York, 1960.

Javid, M. and P. M. Brown: "Field Analysis and Electromagnetics," McGraw-Hill Book Company, New York, 1963.

Johnk, C. T.: "Engineering Electromagnetic Fields and Waves," John Wiley & Sons, Inc., New York, 1975.

Jordan, E. C. and K. G. Balmain: "Electromagnetic Waves and Radiating Systems," 2d ed., Prentice-Hall, Inc., Englewood Cliffs, N.J., 1968.

Lorrain, P. and D. R. Corson: "Electromagnetic Fields and Waves," 2d ed., W. H. Freeman and Company, San Francisco, 1970.

Moon, P. and D. E. Spencer: "Foundations of Electrodynamics," D. Van Nostrand Company, Inc., Princeton, New Jersey, 1960.

Nussbaum, A.: "Electromagnetic Theory for Engineers and Scientists," Prentice-Hall, Inc., Englewood Cliffs, N.J., 1965.

Owen, G. E.: "Introduction to Electromagnetic Theory," Allyn and Bacon, Inc., Boston, 1963.

Paris, D. T. and F. H. Hurd: "Basic Electromagnetic Theory," McGraw-Hill Book Company, New York, 1969.

Plonsey, R. and R. E. Collin: "Principles and Applications of Electromagnetic Fields," McGraw-Hill Book Company, New York, 1961.

Popovic, B. D.: "Introductory Engineering Electromagnetics," Addison-Wesley Publishing Company, Inc., Reading, Mass., 1971.

Ramo, S., J. R. Whinnery, and T. Van Duzer: "Fields and Waves in Communication Electronics," John Wiley & Sons, Inc., New York, 1965.

Rao, N. N.: "Elements of Engineering Electromagnetics," Prentice-Hall, Inc., Englewood Cliffs, N. J., 1977.

Reitz, R. and F. J. Milford: "Foundations of Electromagnetic Theory," Addison-Wesley Publishing Company, Inc., Reading, Mass., 1960.

Schelkunoff, S. A.: "Electromagnetic Fields," Blaisdell Publishing Company, New York, 1963.

Scott, W. T.: "The Physics of Electricity and Magnetism," 2d ed., John Wiley & Sons, Inc., New York, 1966.

Shadowitz, A.: "The Electromagnetic Field," McGraw-Hill Book Company, New York, 1975.

ADVANCED

Abraham, M. and R. Becker: "The Classical Theory of Electricity and Magnetism," Hafner Publishing Company, Inc., New York, 1951.

Clemmow, P. C.: "An Introduction to Electromagnetic Theory," Cambridge University Press, New York, 1973.

Elliot, R. S.: "Electromagnetics," McGraw-Hill Book Company, New York, 1966.

Harrington, R. F.: "Time-harmonic Electromagnetic Fields," McGraw-Hill Book Company, New York, 1961.

Jackson, J. D.: "Classical Electrodynamics," 2d ed., John Wiley & Sons, Inc., New York, 1975.

Johnson, C. C.: "Field and Wave Electrodynamics," McGraw-Hill Book Company, New York, 1965.

Kong, J. A.: "Theory of Electromagnetic Waves," John Wiley & Sons, Inc., New York, 1975.

Langmuir, R. V.: "Electromagnetic Fields and Waves," McGraw-Hill Book Company, New York, 1961.

Maxwell, J. C.: "A Treatise on Electricity and Magnetism," vol. 1 and 2, Dover Publications, Inc., New York, 1954.

Panofsky, W. K. H. and M. Phillips: "Classical Electricity and Magnetism," Addision-Wesley Publishing Company, Inc., Reading, Mass., 1956.

Stratton, J. A.: "Electromagnetic Theory," McGraw-Hill Book Company, New York, 1952.

Van Bladel, J.: "Electromagnetic Fields," McGraw-Hill Book Company, New York, 1964.

Weeks, W. L.: "Electromagnetic Theory for Engineering Applications," John Wiley & Sons, Inc., New York, 1964.

APPLIED ELECTROMAGNETICS

CHAPTER
ONE

STATIC ELECTRIC FIELDS

1.1 DIMENSIONS AND UNITS

Throughout this book the mksa (meter-kilogram-second-ampere) system of units, now a subsystem of the SI units, is used. A dimensional analysis should always be the first step in checking the correctness of an equation.† A surprising number of errors can be detected at an early stage simply by checking that both sides of an equation balance dimensionally in terms of the four basic dimensions. For example, Newton's second law gives the force in newtons as

$$F = ma \qquad \text{mass(length)/(time)}^2$$

An element of work in joules is given by

$$dW = \mathbf{F} \cdot \mathbf{dl} \qquad \text{mass(length)}^2\text{/(time)}^2$$

Power, which is the time rate of doing work in joules per second, is

$$P = \frac{dW}{dt} \qquad \text{mass(length)}^2\text{/(time)}^3$$

The fourth fundamental quantity is the electric current, which is the rate of flow of electric charge. In the older mksq system of units electric charge was the fourth fundamental quantity. The smallest naturally occurring charge e is that possessed by an electron and is equal to -1.6×10^{-19} C. Any charged object is a collection of elementary particles, usually electrons. The possible values of total charge Q of such an object are given by

$$Q = \pm ne \qquad \text{where } n = 0, 1, 2, \ldots$$

† A dimension defines a physical characteristic. A unit is a standard by which a dimension is expressed numerically. For example, a second is a unit in terms of which the dimension time is expressed. One should not confuse the name of a physical quantity with its units of measure. For example, power should not be expressed as work per second, but as work per unit time.

Electric charge is quantized and appears in positive and negative integral multiples of the charge of the electron. The discreteness of electric charge is not evident, simply because most charged objects have a charge that is much larger than e. Besides the law of charge quantization, we should observe that a law of charge conservation also exists. It says that the net charge in any isolated system remains constant.

1.2 COULOMB'S FORCE LAW

Experimentally one can show that a force exists between two charges. Mathematically the force relationship between two point charges Q_1 and Q_2 can be stated as†

$$F = k\frac{Q_1 Q_2}{R^2}$$

where R is the distance between the Q's and k is a constant of proportionality which depends on the system of units used. In the SI system of units the constant of proportionality is given by $1/4\pi\varepsilon$. The symbol ε is the permittivity of the medium in which the charges are situated. For vacuum, which is a medium characterized by the absence of all material, the permittivity in the SI system of units is given by

$$\varepsilon_0 = 8.854 \times 10^{-12}\ \text{F/m} \cong \frac{1}{36\pi}\,10^{-9}\ \text{F/m}$$

Note that the permittivity of air, a very tenuous material, is substantially the same as for vacuum. Hence, the symbol ε_0 will be used to denote the permittivity of air, free space, atmosphere, as well as vacuum. For material media it is convenient to express the permittivity of the media normalized with respect to vacuum. It is then known as the relative permittivity or dielectric constant $\varepsilon_r = \varepsilon/\varepsilon_0$ and is commonly listed in tables for dielectric constants (see Table 1.1). Dielectric materials will be considered in detail in Chap. 4. It is sufficient for now to note that the presence of polarizable atoms or molecules in a material is expressed by $\varepsilon_r > 1$. A dimensional check shows that ε has the dimensions of capacitance per length, or $(\text{time})^4 \times (\text{current})^2/\text{mass}(\text{length})^3$. Since force is a vector, we can write Coulomb's law for two point charges immersed in a uniform material media of infinite extent as

$$\boxed{\mathbf{F} = \frac{Q_1 Q_2}{4\pi\varepsilon R^2}\hat{\mathbf{R}}} \tag{1.1}$$

where $\hat{\mathbf{R}} = \mathbf{R}/R$ is a unit vector along a line connecting the two charges. If the medium is air, as is usually the case, then $\varepsilon = \varepsilon_0$.

† When a charge is confined to a region whose dimensions are small compared with the distance to other charges, we can consider it as a *point charge*.

Table 1.1 Dielectric constants of dielectric materials†

Material	Relative permittivity ε_r
Vacuum	1
Air (atmospheric pressure)	1.0006
Polyethylene	2.2
Polystyrene	2.7
Rubber	3
Paper (impregnated)	3
Bakelite	5
Quartz	5
Lead glass	6
Mica	6
Flint glass	10
Glycerin	50
Water (distilled)	81
Barium titanate	1200
Barium strontium titanate	10,000

† For additional values see A. R. von Hippel, ed., "Dielectric Materials and Applications," John Wiley & Sons, Inc., New York, 1954.

It should be emphasized that we have chosen to introduce dielectric material in the first discussion of Coulomb's law. The force between two charges is inversely dependent on the permittivity ε of the material in which the charges are immersed. This is a simplified "practical" approach. This approach can be justified by pointing out that a large number of practical problems are concerned with linear, isotropic, homogeneous dielectrics for which the permittivity ε is a constant, which implies that the same equations that apply to vacuum apply also to dielectric media. The parameter ε for dielectric media simply has a higher value than ε_0 for vacuum. This book will be mainly concerned with such materials. There are materials (primarily those with a crystalline structure) which have a permittivity which is more complicated than a scalar constant. Because of this, the approach that we are taking has been criticized.† Although criticism of this nature is valid, the advantage of our approach lies in its simplicity. In Chap. 4 a more careful look is given to dielectric material from the microscopic point of view.

Homogeneity, Linearity, and Isotropy

Before closing this section, let us define homogeneity, linearity, and isotropy. A medium is *homogeneous* if its physical characteristics do not vary from point to

† The Teaching of Electricity and Magnetism at the College Level, Coulomb's Law Committee, *Am. J. Phys.*, pp. 1–25, January 1950.

point in the medium. A medium is *linear* in permittivity, if ε remains constant when the magnitude of the charges in Coulomb's law are varied. A medium is *isotropic* if its properties are independent of direction.

1.3 THE ELECTRIC FIELD

Coulomb's law gives the force that will be exerted on a point charge Q_2 when it is placed in the vicinity of another point charge Q_1. If we now remove Q_2 but retain Q_1 in a fixed position, we can say that an electric field remains in all space about Q_1. The magnitude of the electric field at a point is the force per unit charge on a positive test charge at the point, provided the test charge, call it ΔQ_2, is small enough so as not to disturb the field that we are testing. The electric field is a vector because the force on ΔQ_2 has direction as well as magnitude. We can now define the *electric field* at the point where ΔQ_2 is located as

$$\boxed{\mathbf{E} = \frac{\mathbf{F}}{\Delta Q_2} = \frac{Q_1}{4\pi\varepsilon R^2}\hat{\mathbf{R}} \qquad \text{N/C}} \tag{1.2}$$

If the medium is free space, as is usually the case, the substitution of ε_0 for ε should be made. To aid in visualizing the **E** field, we introduce the concept of field lines which give the direction of the force on a positive test charge. If the test charge were released, it would move in the direction of the field lines. From now on we shall call the point where an effect is observed the *observation or field point* and denote it by **r**, and the point where the source or cause is located the *source point* and denote it by **r**′. Figure 1.1 shows the field lines of a stationary point charge Q_1 as measured by a test charge ΔQ_2 as well as the source and observation point coordinates. In this new notation we can write the electric field of a point charge as

$$\mathbf{E}(\mathbf{r}) = \frac{Q_1(\mathbf{r}')}{4\pi\varepsilon|\mathbf{r}-\mathbf{r}'|^2}\hat{\mathbf{R}} = \frac{Q_1(\mathbf{r}')}{4\pi\varepsilon|\mathbf{r}-\mathbf{r}'|^2}\frac{\mathbf{r}-\mathbf{r}'}{|\mathbf{r}-\mathbf{r}'|} \tag{1.3}$$

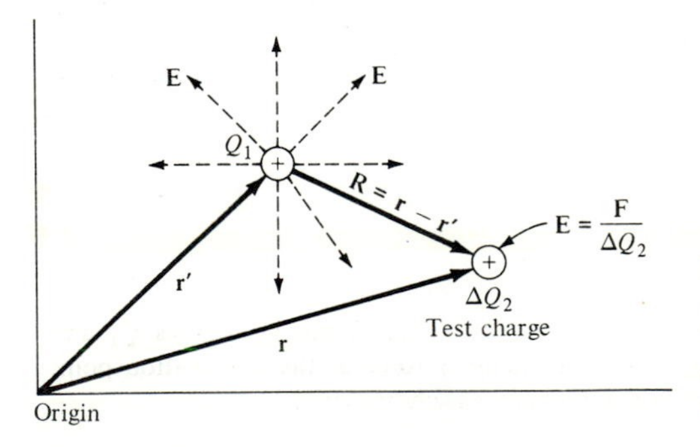

Figure 1.1 The dashed lines show the radially outward force field from the point charge Q_1. The solid lines relate to the notation for source and observation points, with $\mathbf{R} = \mathbf{r} - \mathbf{r}'$.

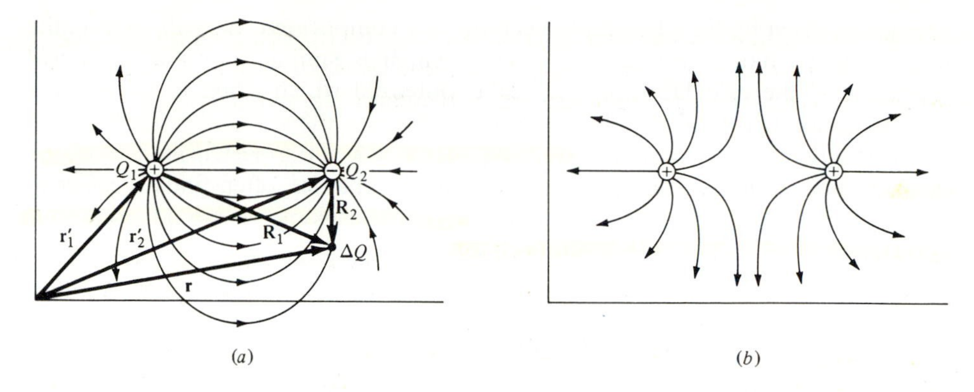

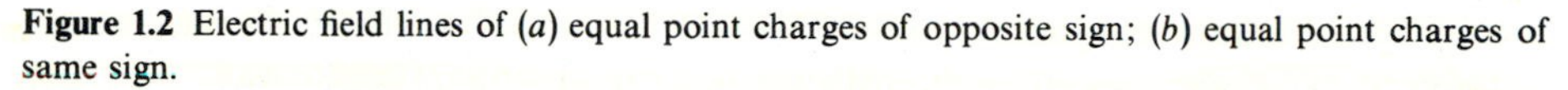

Figure 1.2 Electric field lines of (*a*) equal point charges of opposite sign; (*b*) equal point charges of same sign.

Figure 1.2 shows the electric field configurations for two point charges. The shape of the field lines is obtained by moving our infinitesimally small test charge ΔQ about and noting the direction of the force on ΔQ. Mathematically the electric field at the position $\mathbf{r}$ of the test charge ΔQ is given by

$$\mathbf{E}(\mathbf{r}) = \frac{Q_1(\mathbf{r}'_1)}{4\pi\varepsilon|\mathbf{r} - \mathbf{r}'_1|^2}\hat{\mathbf{R}}_1 + \frac{-Q_2(\mathbf{r}'_2)}{4\pi\varepsilon|\mathbf{r} - \mathbf{r}'_2|^2}\hat{\mathbf{R}}_2 \tag{1.4}$$

which gives the electric field configuration of Fig. 1.2*a* as $\mathbf{r}$ is varied.

The above expression was obtained by applying the superposition principle. Because the $\mathbf{E}$ field of a point charge is a linear function of charge, it follows that the fields of more than one point charge are linearly superposable. This principle is simple, but not self-evident. It says that the force between any two charges is independent of the presence of other charges. To find the resultant force we simply add the individual forces vectorially.

It is a simple matter to extend this to the case when N charges are present. The resultant force per unit charge at the observation point $\mathbf{r}$ is then

$$\mathbf{E}(\mathbf{r}) = \frac{1}{4\pi\varepsilon}\sum_{n=1}^{N}\frac{Q_n}{|\mathbf{r} - \mathbf{r}'_n|^2}\hat{\mathbf{R}}_n \tag{1.5}$$

To perform the vector addition in the presence of more than one charge is usually tedious. For example, the force vectors for the point in Fig. 1.2*a* at which the test charge is located are shown in Fig. 1.3. Needless to say, a vector addition for N

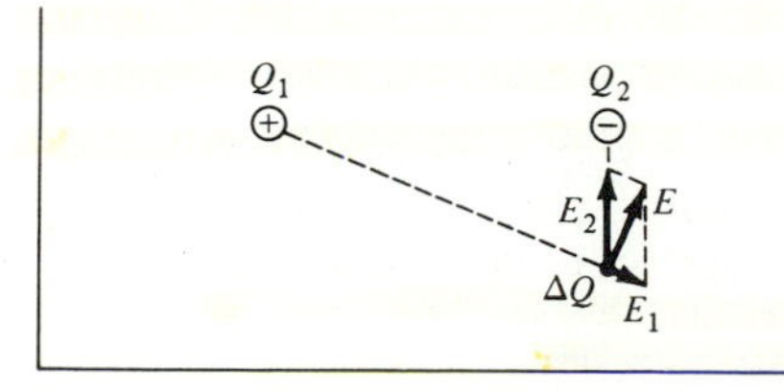

Figure 1.3 The individual forces of charges Q_1 and Q_2, and the resultant force at the observation point (the point where test charge ΔQ is).

charges as given by Eq. (1.5) could become very complicated. We will show in the next sections that an alternate approach which is simpler is possible. It will involve first the calculation of the scalar potential which when followed by a differentiation will yield the electric field.

It should be pointed out that it is helpful to think of the electric field as a force field. The beginning student, having had some mechanics, often finds it easier to think in terms of force fields, even though strictly speaking we should refer to an electric field as a force field per unit charge.

1.4 THE ELECTRIC POTENTIAL

Whenever we have a force field, we can associate a work field with it. If we divide the expression for an element of work† in a force field $\Delta W = \mathbf{F} \cdot \mathbf{dl}$ by our test charge ΔQ, we obtain

$$\frac{\Delta W}{\Delta Q} = \frac{\mathbf{F}}{\Delta Q} \cdot \mathbf{dl} \tag{1.6}$$

In the previous section we defined force per unit charge as the electric field. Defining work per unit charge as a scalar potential V, Eq. (1.6) can be written as

$$\boxed{\Delta V = -\mathbf{E} \cdot \mathbf{dl} \qquad \text{J/C}} \tag{1.7}$$

where the reason for the minus sign is given in the next paragraph. It should be noted that a work field, in contrast to a force field, is always scalar. The work per unit charge required to transport the test charge from l_1 to l_2 is called the *difference in electric potential* of the two points.‡ The unit of electric potential is the *volt* and is equal to 1 J/C. If the work field is given in volts, the unit for the electric field is either newtons per coulomb or volts per meter.

In general, electric force fields are nonuniform. The energy per unit charge required to move a test charge from l_1 to l_2 must then be expressed as an integral; i.e., the potential difference is

$$V_2 - V_1 = -\int_{l_1}^{l_2} \mathbf{E} \cdot \mathbf{dl} \tag{1.8}$$

This is called a line integral and means that the component of electric field in the direction of the path is to be multiplied by an element of distance along the path, and the sum is to be taken by integration as one moves along the path. The negative sign is introduced so that when a positive test charge is moved against the force field of a positive charge, the work will be given as a positive quantity.

† The dot product in ΔW implies a projection of the line of motion on the force lines.

‡ Potential difference and voltage are synonymous in the static case.

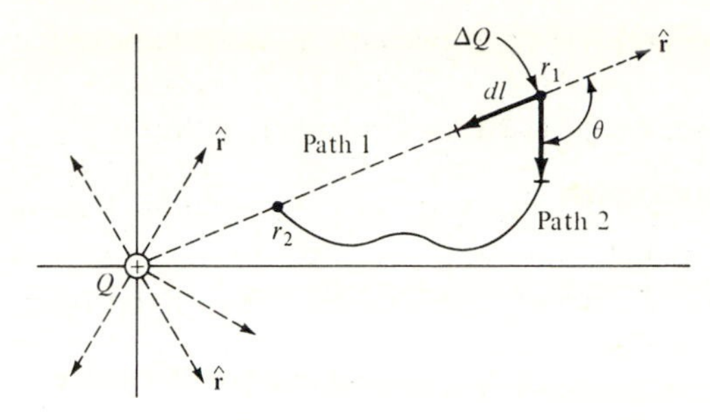

Figure 1.4 A positive charge located at the origin, and its force field shown by the dashed lines. A small charge ΔQ is moved from r_1 to r_2.

Let us now consider a simple example. We would like to find the work which is required to move a test charge toward a point charge located at the origin, as shown in Fig. 1.4. The work per unit charge or the potential difference in volts between r_1 and r_2 $(r_1 > r_2)$ is then

$$V_{r_1 \text{ to } r_2} = V_2 - V_1 = -\int_{r_1}^{r_2} \frac{Q}{4\pi\varepsilon r^2} \hat{\mathbf{r}} \cdot \mathbf{dl} \tag{1.9}$$

where $\hat{\mathbf{r}}$ is a unit vector given by $\hat{\mathbf{r}} = \mathbf{r}/r$. If we move the test charge ΔQ along path 1 which is in the direction of the force lines of Q, we see that $\hat{\mathbf{r}} \cdot \mathbf{dl} = -dl = dr$. The negative sign accounts for the fact that the path of motion is opposite to the direction of the field. However, as we move a distance dl toward the charge at the origin, we are moving in the direction of decreasing r because r is measured from the origin; hence $dl = -dr$. The work along path 1 is then

$$V_2 - V_1 = -\int_{r_1}^{r_2} \frac{Q}{4\pi\varepsilon r^2}\, dr = \frac{Q}{4\pi\varepsilon}\left(\frac{1}{r_2} - \frac{1}{r_1}\right) \tag{1.10}$$

The integrated result depends only on the starting and final positions of the test charge.

Now let us make one important observation which will be relevant time after time in the material which we will consider in the remainder of the book. If we elect to move the test charge along path 2 (or, for that matter, any arbitrary path between the points r_1 and r_2), we find that the dot product in integral (1.9) gives the same result as for path 1; that is,

$$\hat{\mathbf{r}} \cdot \mathbf{dl} = \cos\theta\, dl = dr \tag{1.11}$$

This happens because $\hat{\mathbf{r}} \cdot \mathbf{dl}$ is the projection of an elemental line of motion on the force lines of Q which are radially outward in the direction given by the unit vector $\hat{\mathbf{r}}$. Hence, the potential difference between r_1 and r_2 is independent of the shape of the path and is always given by Eq. (1.10). Another way of saying it is that the scalar potential depends on the initial and final positions; it is single-valued regardless of the path taken from r_1 to r_2. This conclusion leads us immediately to another which can be stated as follows: If a charge is taken around any closed path, the final and initial positions coincide and no net work is done. Mathematically, this is expressed as

$$\oint \mathbf{E} \cdot d\mathbf{l} = 0 \tag{1.12}$$

where the circle denotes that a closed-path-line integral is considered. In returning a charge to its starting position we restore the system to its initial state, doing no net work. A force field such as the electrostatic field for which the closed-loop integral is zero is referred to as a *conservative field*, meaning that since no mechanism for friction is provided, the conservation of energy principle applies. Conservative fields will be discussed again in Sec. 1.6.

Before concluding this section, let us introduce the absolute potential. Equation (1.10) gives the difference in potential between two points. It would be convenient if one of the points could be chosen such that the potential is zero at that point. Observing that the interaction force between two charges is zero when they are infinitely separated, we can say that $V_1 = 0$ when $r_1 = \infty$. The *absolute potential* at the point r due to charge Q located at the origin is then defined as

$$V = -\int_{\infty}^{r} \mathbf{E} \cdot d\mathbf{l} = \frac{Q}{4\pi\varepsilon r} \tag{1.13}$$

and is the work per charge required to bring a test charge from infinity to the point r. Whenever a potential at a point is given, it is understood that it is the absolute potential.

The absolute potential due to N charges located at the source points r_i', as shown in Fig. 1.5, can now be written using our more precise notation introduced in Fig. 1.1 as

$$V(\mathbf{r}) = \sum_{i=1}^{N} \frac{Q_i}{4\pi\varepsilon|\mathbf{r} - \mathbf{r}_i'|} = \sum_{i=1}^{N} \frac{Q_i}{4\pi\varepsilon R_i} \tag{1.14}$$

Although the absolute potential is useful, strictly speaking only potential differences have meaning. This is because the point of zero potential is arbitrary. In some applications the point at infinity is chosen as the zero reference potential, whereas in applications that involve the potential between the two conducting surfaces of a capacitor, one of the surfaces can be chosen as zero. The absolute potential is then the potential difference $V(\mathbf{r}) - \mathbf{V}(\infty)$. For completeness we can give the potential difference between two observation points r_1 and r_2 for the charges of Fig. 1.5 as

$$V(\mathbf{r}_1) - V(\mathbf{r}_2) = \sum_{i=1}^{N} \frac{Q_i}{4\pi\varepsilon}\left(\frac{1}{R_{1i}} - \frac{1}{R_{2i}}\right) \tag{1.15}$$

where R_{1i} and R_{2i} are the scalar distances between the ith charge and observation points r_1 and r_2, respectively. The absolute potential is obtained by moving one of the two observation points to infinity.

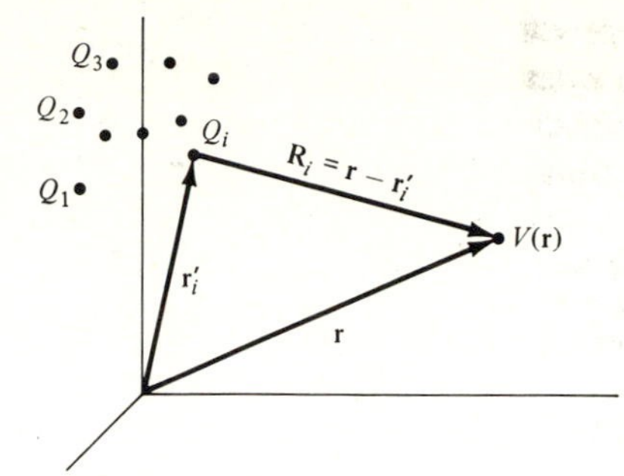

Figure 1.5 The absolute potential at the observation point **r** for N charges is defined as the work per unit charge when a test charge ΔQ is moved from ∞ to the observation point **r**.

An exercise: Figure 1.6 shows three point charges Q_1, Q_2, and Q_3. The absolute potential at the observation point is readily stated as

$$V(\mathbf{r}) = \sum_{n=1}^{3} \frac{Q_n}{4\pi\varepsilon|\mathbf{r} - \mathbf{r}'_n|}$$

The mathematical method that we have used thus far will be summarized well if we calculate the above potential function for the three charges by integrating the elements of work as we move the test charge from infinity to **r**. This is, of course, doing it the hard way, but it will bring out the key points. Thus

$$V = -\int_{\infty}^{\mathbf{r}} \mathbf{E} \cdot \mathbf{dl} = -\int_{\infty}^{\mathbf{r}} \sum_{n=1}^{3} \frac{Q_n}{4\pi\varepsilon R_n^2} \hat{\mathbf{R}}_n \cdot \mathbf{dl}$$

where $R_n = |\mathbf{r}_{dl} - \mathbf{r}'_n|$ and is the distance between the nth charge and the position of the test charge at the point dl, as shown in Fig. 1.6. Note that the position of the test charge ΔQ and the position of dl coincide. We again observe that [see Eq. (1.11)]

$$\hat{\mathbf{R}} \cdot \mathbf{dl} = \cos\theta \, dl = dR$$

where θ is the angle between the direction of motion of the test charge and the direction of the line connecting the position of ΔQ with that of one of the charges Q. Furthermore, since the summation operator and the integral operator in the expression for V commute, we can write

$$V = -\sum_{n=1}^{3} \frac{Q_n}{4\pi\varepsilon} \int_{\infty}^{\mathbf{r}} \frac{dR_n}{R_n^2} = \sum \frac{Q_n}{4\pi\varepsilon} \frac{1}{R_n}\bigg|_{\infty}^{\mathbf{r}}$$

After integration and substitution of limits, we obtain

$$V = \sum \frac{Q_n}{4\pi\varepsilon}\left(\frac{1}{R_n(\mathbf{r})} - \frac{1}{R_n(\infty)}\right) = \sum \frac{Q_n}{4\pi\varepsilon}\left(\frac{1}{|\mathbf{r} - \mathbf{r}'_n|} - \frac{1}{|\infty - \mathbf{r}'_n|}\right)$$

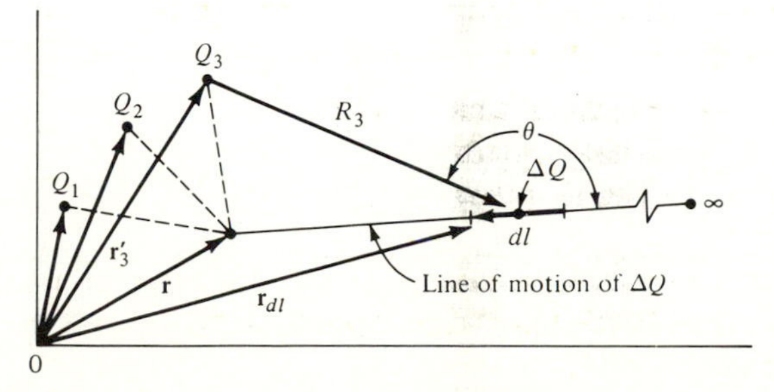

Figure 1.6 The absolute potential at the point **r** is obtained by calculating the work performed as a test charge ΔQ is moved from infinity to **r**.

Since the $1/\infty$ term is zero, we obtain for the potential

$$V = \sum \frac{Q_n}{4\pi\varepsilon} \frac{1}{|\mathbf{r} - \mathbf{r}'_n|}$$

which is the same expression with which we started.

1.5 FIELDS OF CONTINUOUS CHARGE DISTRIBUTIONS

Equations (1.14) and (1.15) give the potential for N discrete charges. We can use these expressions to derive the work or potential function for a continuous distribution such as a charged gas or plasma "cloud." Let the charges be continuously distributed throughout a volume v such that the charge density in coulombs per meter cubed at any point is given by ρ. Let us further divide the volume v into small elemental volumes Δv, as shown in Fig. 1.7. The total charge contained in an elemental volume Δv located at $\mathbf{r}'$ is then given by $\rho(\mathbf{r}')\,\Delta v$ and is treated as a point charge. In order to obtain the V field of the continuous distribution of charge, we use Eq. (1.14) to add the contribution of all point charges $\rho(\mathbf{r}'_n)\,\Delta v_n$; then, by using the definition of an integral as shown below:

$$V(\mathbf{r}) = \lim_{\substack{\Delta v_n \to 0 \\ N \to \infty}} \frac{1}{4\pi\varepsilon} \sum_{n=1}^{N} \frac{\rho(\mathbf{r}'_n)\,\Delta v_n}{|\mathbf{r} - \mathbf{r}'_n|} = \frac{1}{4\pi\varepsilon} \iiint_v \frac{\rho(\mathbf{r}')\,dv}{|\mathbf{r} - \mathbf{r}'|} \tag{1.16}$$

we obtain the potential field of a continuous charge distribution. Similarly, by use of Eq. (1.5), which gives the electric field due to N point charges, we can obtain the electric field due to a continuous charge distribution confined to volume v as

$$\mathbf{E}(\mathbf{r}) = \lim_{\substack{\Delta v_n \to 0 \\ N \to \infty}} \frac{1}{4\pi\varepsilon} \sum_{n=1}^{N} \frac{\rho(\mathbf{r}'_n)\,\Delta v_n}{|\mathbf{r} - \mathbf{r}'_n|^2} \hat{\mathbf{R}}_n = \frac{1}{4\pi\varepsilon} \iiint_v \frac{\rho(\mathbf{r}')}{|\mathbf{r} - \mathbf{r}'|^2} \hat{\mathbf{R}}\,dv \tag{1.17}$$

where $\mathbf{R} = \mathbf{r} - \mathbf{r}'$, and $\hat{\mathbf{R}}$ is the unit vector $\hat{\mathbf{R}} = \mathbf{R}/R$.

In many applications involving conductors we find that a charge Q which is placed on a conducting object will distribute itself in a thin layer on the surface of the object such that

$$Q = \iint_A \rho_S\,dA \tag{1.18}$$

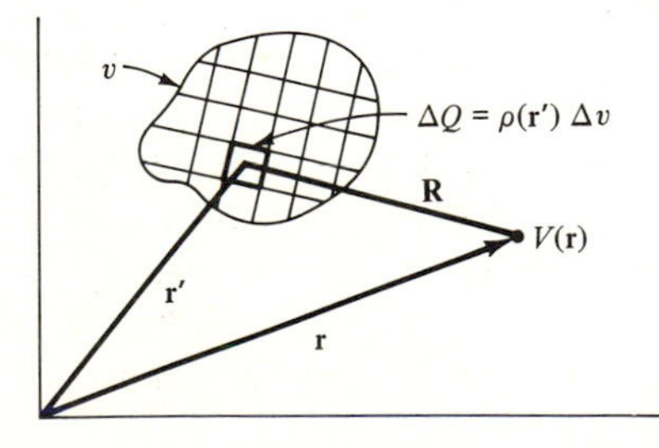

Figure 1.7 A continuous charge cloud confined to the volume v. By subdividing v into elemental volumes Δv, the potential at $\mathbf{r}$ can be obtained as a summation over all elemental volumes.

where ρ_s is a surface charge density in coulombs per meter squared. When the object is a thin wire, the charge will assume a one-dimensional distribution such that

$$Q = \int_l \rho_L \, dl \tag{1.19}$$

where ρ_L is a line charge density in coulombs per meter. The potential from these two charge distributions then becomes

$$V(\mathbf{r}) = \begin{cases} \dfrac{1}{4\pi\varepsilon} \displaystyle\iint_A \frac{\rho_s(r')\, dA}{R} \\ \\ \dfrac{1}{4\pi\varepsilon} \displaystyle\int_l \frac{\rho_L(r')\, dl}{|\mathbf{r} - \mathbf{r}'|} \end{cases} \tag{1.20}$$

The electric field can be obtained in a similar manner.

1.6 EQUIPOTENTIAL SURFACES

In Sec. 1.4 it was shown that the electrostatic field is conservative. A mathematical statement of that property as given by Eq. (1.12) is that the closed-loop integral of the field is zero. It means that any work that is done on the test charge in moving it about the electrostatic field will be returned by the field as the test charge returns to its starting position. This must be so since the electrostatic field has no mechanism for friction losses. Out of the infinitely many closed-loop paths that a test charge can take, one in particular merits special attention. In the expression for the work function (1.9) we observe that the direction $\hat{\mathbf{r}}$ of the force field and the direction $\mathbf{dl}$ of motion of our test charge are related by

$$\hat{\mathbf{r}} \cdot \mathbf{dl} = \cos\theta \, dl$$

If the test charge is moved perpendicular to the direction of the field, $\cos 90° = 0$, and no work is performed at any point along the path of motion; that is,

$$\int \mathbf{E} \cdot \mathbf{dl} = 0$$

Such a path is called an *equipotential*, and all points of a field having the same potential may be thought of as connected by equipotential surfaces. An important property of fields is that the electric field vectors must be perpendicular to these surfaces at every point. For example, the equipotential surfaces for a point charge are concentric spherical shells, as shown in Fig. 1.8*a*; for a uniform field, shown in Fig. 1.8*b*, such as that which exists in the space between the conducting plates of a parallel plate capacitor, the equipotentials are plane surfaces. The work that would be done in moving a test charge from point *a* to point *b* in Fig. 1.8*a* would

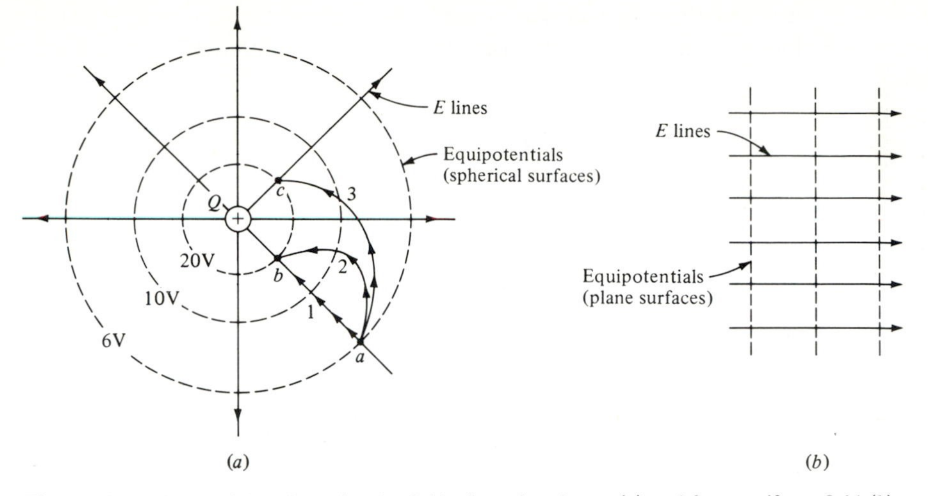

Figure 1.8 Equipotential surfaces for the field of a point charge (a) and for a uniform field (b).

be 14 voltcoulomb (VC) regardless of whether path 1 or 2 was chosen. As a matter of fact, since points b and c are on the same equipotential surface, no potential difference exists between b and c, and the motion of a test charge along path 3 involves the same amount of work as that between a and b. Therefore, the work involved is always 14 VC when a test charge is moved from anywhere on the 6-V equipotential surface to anywhere on the 20-V equipotential.

The notion of an equipotential surface will be particularly useful when considering metallic objects since the surfaces of such objects are equipotential surfaces.

1.7 FIELDS DUE TO LINE AND SURFACE CHARGES

We have found that the electric and potential fields of a point charge vary with distance R from the charge as

$$E \propto \frac{1}{R^2} \quad \text{and} \quad V \propto \frac{1}{R} \tag{1.21}$$

In the following sections we will calculate the fields due to specific line and surface charge distributions. We will treat these distributions as assemblages of point charges as was done in Sec. 1.5.

Example: Fields due to a finite and infinite line charge Figure 1.9 shows a finite line charge of length $2l$ along which is distributed a charge Q such that the charge density is $\rho_L = Q/2l$. If we situate the line charge along the z axis in a cylindrical coordinate system, we obtain for the E field due to ΔQ

$$dE = \frac{\Delta Q}{4\pi\varepsilon R^2} = \frac{\rho_L\, dz}{4\pi\varepsilon(r^2 + z^2)} \tag{1.22}$$

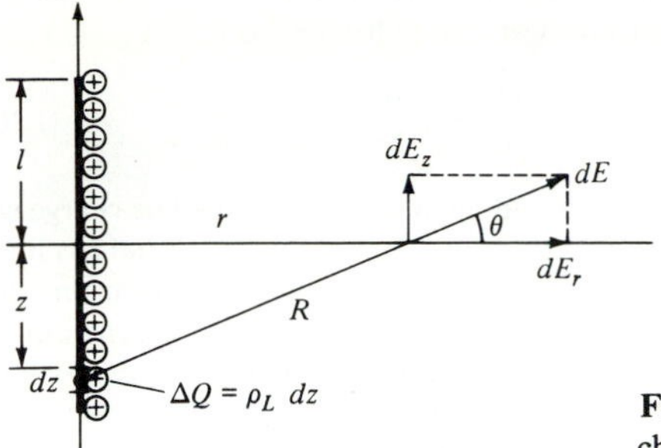

Figure 1.9 The electric field due to an element of a line charge which has a line charge density ρ_L, C/m.

This can be resolved into two components E_r and E_z by multiplying by cos θ and sin θ, respectively. Along the r axis the E_z components from the upper and lower half of the line charge cancel each other, leaving only the E_r component. In the $z = 0$ plane, we have then

$$dE_r = \frac{\Delta Q}{4\pi\varepsilon R^2}\cos\theta = \frac{\rho_L\,dz}{4\pi\varepsilon(r^2+z^2)}\,\frac{r}{(r^2+z^2)^{1/2}} \tag{1.23}$$

To obtain the total component E_r, we integrate over the entire length of charge to get

$$E_r = \int_{-l}^{l} \frac{\rho_L r}{4\pi\varepsilon(r^2+z^2)^{3/2}}\,dz = \frac{\rho_L l}{2\pi\varepsilon r(r^2+l^2)^{1/2}} \tag{1.24}$$

Note that r is treated as a constant in the above integration. Equation (1.24) gives the total E field in a plane which bisects and is perpendicular to the line charge. In any other perpendicular plane we can obtain the components of the E field simply by changing the limits of the integral appropriately. For example, in the $z = l$ plane, E_r can be obtained by using (1.24) with a change of limits from $-2l$ to 0.

The field of an infinite line charge is now a simple extension of (1.24), obtained by letting $l \to \infty$; that is,

$$E_r = \lim_{l\to\infty} \frac{\rho_L l}{2\pi\varepsilon r(r^2+l^2)^{1/2}} = \frac{\rho_L}{2\pi\varepsilon r} \tag{1.25}$$

We now observe that the direction of the E field produced by an infinite line charge is perpendicular to the line charge; that is, $\mathbf{E} = E_r\hat{\mathbf{r}}$. Also, as one moves away from the line charge, the E field goes to zero as $1/r$ in contrast to the E field of a point charge which decays as $1/r^2$. It is interesting to observe that the electric field due to an infinite line charge remains finite despite the fact that the total charge on the wire must be infinite.

Another point to consider here is the usefulness of solving problems which involve structures of infinite extent. As these structures are not found in real life, such solutions, although simple, should be of academic interest only. However, one should realize that many real-life engineering problems cannot be solved rigorously. They are usually too complicated to be even expressed in mathematical language. Should a solution be available, most likely it will be so complex as to make it useless for engineering applications. Such problems are usually approached by making some idealizations and approximations. The ability to make the appropriate approximations in complicated problems and still obtain useful solutions is a characteristic of an accomplished engineer and applied scientist. It should be developed at the earliest stage in a student's education. In our case, for example, for points near a finite charged rod but away from the ends, the simple expression (1.25) yields results so close to the exact values that the difference can be ignored for most practical purposes. On the other hand, for points close to the ends of the rod, a point charge approximation might give useful results.

The potential between two points r_0 and r, which lie along any normal to the infinite line charge, is the work per unit charge required to transport a positive test charge from r_0 to r:

$$V = -\int_{r_0}^{r} \mathbf{E} \cdot d\mathbf{l} = -\int_{r_0}^{r} E_r\, dr = -\int_{r_0}^{r} \frac{\rho_L\, dr}{2\pi\varepsilon r} = -\frac{\rho_L}{2\pi\varepsilon} \ln \frac{r}{r_0} \tag{1.26}$$

where Eq. (1.11) was used to show that $\mathbf{E} \cdot d\mathbf{l} = E_r\, dr$. It should be noted that for the line charge we cannot select infinity as the reference of zero potential, and therefore the absolute potential has little meaning in this case. If we take $V = 0$ for $r_0 = \infty$, the above equation would lead to an infinite potential for all finite r's. The reason for this difficulty is that it is meaningless to say that one can be infinitely far from an infinite structure. On the other hand, for localized structures the assumption of zero fields at infinity is usually valid (localized meaning that the structure can be confined inside a fictitious sphere of large but finite radius). In the case of the infinite line charge the difficulty at infinity is not bothersome since it is the potential difference between two points that is of interest.

Example: Fields due to a ring charge Figure 1.10 shows a charged ring of radius r. For points on the axis of the ring which are a distance z from the center of the ring, the normal component of the electric field from a small portion ΔQ of the ring charge is

$$\Delta E_n = \frac{\Delta Q}{4\pi\varepsilon R^2} \cos\theta \tag{1.27}$$

Adding the contributions of all the ΔQ's, we obtain

$$E_n = \frac{Q_{\text{ring}}}{4\pi\varepsilon R^2} \frac{z}{R} \tag{1.28}$$

where $\cos\theta = z/R$. From symmetry we see that on the z axis only the normal component of the E field remains. At the center of the loop, given by $z = 0$, the electric field is therefore zero. On the other hand, for $z \gg r$, Eq. (1.28) gives

$$E_n = \left.\frac{Q_r z}{4\pi\varepsilon(r^2 + z^2)^{3/2}}\right|_{z \gg r} \cong \frac{Q_r}{4\pi\varepsilon z^2} \tag{1.29}$$

which is as expected because at large distances from the charged ring the ring looks like a point charge. Hence the fields must decay to zero as given by Eq. (1.21).

Example: Fields due to a charged disk The above example can be immediately used to obtain the axial field from a charged disk by integrating the contributions from all rings that constitute the disk. If a charge Q_r is placed on a thin ring and if this charge distributes itself as a uniform line charge, the line charge density ρ_L and total charge are related by $Q_r = \rho_L 2\pi r$. On the other hand, if we think of the ring as having a washer shape of width dr, then the total charge and the surface charge density ρ_s will be related by

$$Q_r = \rho_s 2\pi r\, dr$$

We now assume that the total charge Q_r will distribute itself on the disk such that the surface charge density ρ_s is constant. The normal component of the electric field along the z axis is then, using Eq. (1.28),

$$E_n = \int_0^a \frac{\rho_s 2\pi r\, dr}{4\pi\varepsilon R^3} z \tag{1.30}$$

where a is the radius of the charged disk and $R^2 = z^2 + r^2$. Integrating this, we obtain

$$E_n = \frac{\rho_s z}{2\varepsilon} \int_0^a \frac{r\, dr}{(r^2 + z^2)^{3/2}} = -\frac{\rho_s z}{2\varepsilon} \left.\frac{1}{(r^2 + z^2)^{1/2}}\right|_0^a$$

$$= -\frac{\rho_s}{2\varepsilon}\left[\frac{z}{(a^2 + z^2)^{1/2}} - 1\right] \tag{1.31}$$

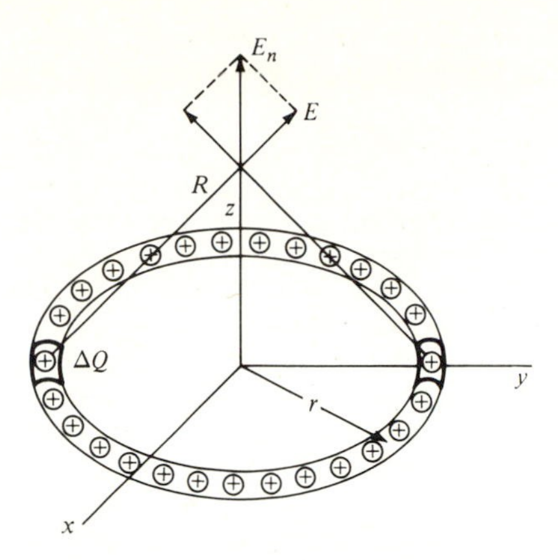

Figure 1.10 A charged ring of radius r, located in the xy plane where $r = (x^2 + y^2)^{1/2}$ and $R = (r^2 + z^2)^{1/2}$. The ring carries a total charge Q_r. The normal component is along the z axis.

In the special case when the observation point is far from the disk, $z \gg a$, Eq. (1.31) reduces to

$$E_n \cong \frac{\rho_s a^2}{4\varepsilon z^2} = \frac{Q_d}{4\pi\varepsilon z^2} \tag{1.32}$$

Note that the total charge on the disk is given by $Q_d = \rho_s \pi a^2$. This limiting result should be expected because at large distances the disk will behave like a point charge. In (1.31) the square root term was approximated using the binomial expansion

$$(1 + \Delta)^n \cong 1 + n\Delta \qquad \Delta \ll 1 \tag{1.33}$$

The electric potential for points on the axis of the uniformly charged circular disk can be obtained by integration, using (1.31), as

$$V = -\int_{\infty}^{z} E_n \, dz = \frac{\rho_s}{2\varepsilon}[(a^2 + z^2)^{1/2} - z] \tag{1.34}$$

Example: Fields due to an infinite surface of charge As a further extension of the fields from a charged ring, we can obtain the electric field from an infinite sheet of charge by letting $a \to \infty$ in (1.31), which gives

$$E_n = \frac{\rho_s}{2\varepsilon} \tag{1.35}$$

This is an interesting result. It says that the electric field remains constant in magnitude with distance from the surface and is always normal to the surface. If we have a thin sheet of charge as shown in Fig. 1.11, then a uniform electric field of magnitude $E = \rho_s/2\varepsilon$ is produced on both sides of the charged sheet. It should be noted at this time that a uniformly charged sheet as shown in Fig. 1.11 and a surface charge that exists on the surface of an infinite conductor are different situations [see (Eq. 2.44)].

We can now summarize the characteristics of various charge distributions by saying that for

Point charge:	$E \propto \dfrac{1}{r^2}$	$V \propto \dfrac{1}{r}$	
Line charge:	$E \propto \dfrac{1}{r}$	$V \propto \ln r$	(1.36)
Surface charge:	$E \propto$ constant	$V \propto r$	

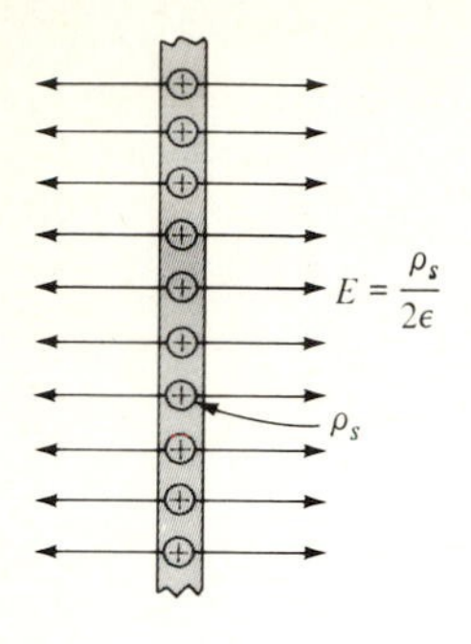

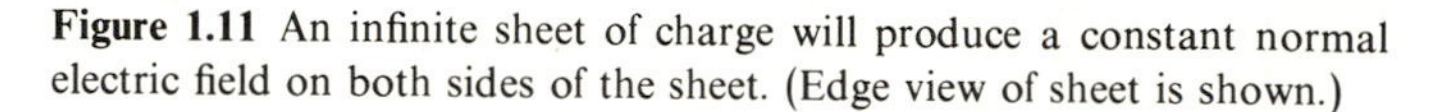
Figure 1.11 An infinite sheet of charge will produce a constant normal electric field on both sides of the sheet. (Edge view of sheet is shown.)

This list should be used with caution for charged rods of finite length and charged surfaces of finite extent. For example, when very close to the charged surface of a disk, the fields should look like those of an infinite sheet of charge. Taking Eq. (1.31) and letting $z \to 0$, we obtain $E_n = \rho_s/2\varepsilon$, which is a constant electric field and agrees with the table. On the other hand, when the observation point is very far from a finite disk, the electric field will look like that from a point charge.

Example: Electrostatic deflection in an oscilloscope The principle of an oscilloscope is illustrated in Fig. 1.12. A set of accelerating anodes being at a higher potential than the cathode accelerate electrons from the cathode. The acceleration is given by

$$a = \frac{F}{m} = \frac{eE_a}{m} \tag{1.37}$$

where E_a is the electric field created by the accelerating voltage V_a. As the electron drifts through this potential difference, it will acquire kinetic energy given by

$$W = eV_a \tag{1.38}$$

which will be equal to $\frac{1}{2}mv^2$ if the electron starts from rest. It therefore will enter the deflecting plates with a velocity v given by

$$v_x = \left(\frac{2eV_a}{m}\right)^{1/2} \tag{1.39}$$

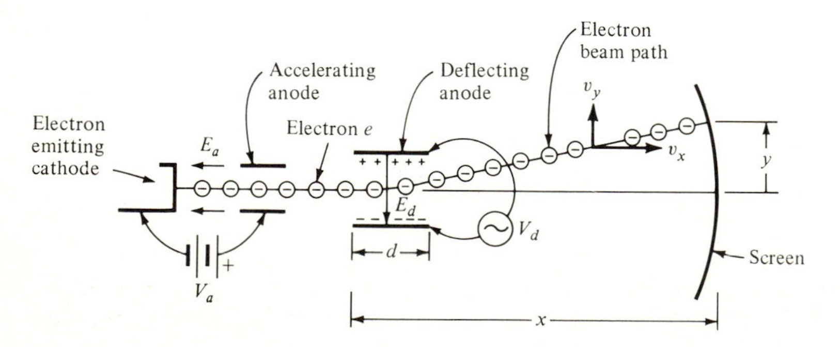

Figure 1.12 The elements of an oscilloscope are: a source of electrons, accelerating and deflecting anodes, and an electroluminescent screen.

As the electron e enters the deflecting plates, it will also enter an electric field E_d which is produced by the deflecting voltage V_d. The electron will be accelerated upward during the time t_d, given by d/v_x, that it is between the plates. The vertical component of velocity that it will acquire goes from zero to

$$v_y = a_y t_d = \frac{eE_d}{m} t_d = \frac{eE_d}{m} \frac{d}{v_x} \tag{1.40}$$

when the electron leaves the plates. It will then follow a deflected straight line path until it impinges on the fluorescent screen. At the time when it just leaves the deflecting plates, it will have traveled a distance upward given by

$$y = \int_0^{t_d} v_y \, dt = \frac{1}{2} a_y t_d^2 = \frac{1}{2} \frac{eE_d}{m} t_d^2 = \frac{1}{2} \frac{eE_d}{m} \left(\frac{d}{v_x}\right)^2 \tag{1.41}$$

which shows that while the electron is under the influence of the deflecting voltage V_d, the equation for the trajectory is a parabola ($y \propto d^2$). Notice that during deflection the horizontal component v_x of the electron velocity which was imparted by the accelerating plates is preserved. After leaving the deflecting plates, the electron will proceed in a straight line path to the screen. The deflection angle α can be given by

$$\alpha = \arctan \frac{v_y}{v_x} = \arctan \frac{eE_d d}{mv_x^2} \cong \arctan \frac{y}{x} \tag{1.42}$$

where the approximation $v_y/v_x = y/x$ was made since $x \gg d$. Equating the arguments in the last two terms gives us the position y on the screen that the electron will strike

$$y = \frac{eE_d xd}{mv_x^2} \tag{1.43}$$

If a variable voltage V_d is placed on the plates, the electron, because it has such a small inertia, will be able to follow rapid fluctuations of V_d.

1.8 THE GRADIENT AND ITS USE IN OBTAINING THE FORCE FIELD FROM THE WORK FIELD

We have shown that the work field can be obtained from a force field by

$$V = -\int \mathbf{E} \cdot \mathbf{dl}$$

Symbolically we can write this as

$$V = \mathscr{L}(\mathbf{E}) \tag{1.44}$$

where the operator $\mathscr{L}$ is given by

$$\mathscr{L} = -\int (\ ) \cdot \mathbf{dl} \tag{1.45}$$

It is simple to show using Eq. (1.44) that the force function (which is a vector field) can be obtained from the work function (a scalar) by the inverse operation

$$\mathbf{E} = \mathscr{L}^{-1} V \tag{1.46}$$

It is clear that since $\mathscr{L}$ is an *integral operator*, $\mathscr{L}^{-1}$ must be a *differential operator*.

We will soon find that $\mathscr{L}^{-1} = \nabla$ and is called the *gradient operator*. Whereas $\mathscr{L}$ operates on a vector field and reduces it to a scalar, $\mathscr{L}^{-1}$ operates on a scalar to produce a vector.

In order to find the explicit form of $\mathscr{L}^{-1}$, let us examine the expression for an element of work (strictly speaking, a differential amount of work per unit charge)

$$dV = -\mathbf{E} \cdot \mathbf{dl} = -E\,dl \cos\theta \tag{1.7}$$

where θ is the angle between **E** and the direction of motion **dl**. The last term in the above expression is particularly useful since it gives us the magnitude of the E field, if we write it as

$$E \cos\theta = -\frac{dV}{dl} \tag{1.47}$$

The meaning of this equation becomes clearer with the aid of Fig. 1.13. If we are given V as a function of position, we can draw the equipotential surfaces. The lines of force can then be found by drawing lines perpendicular to the equipotentials thus determining **E**. It is the mathematical equivalent of this graphical process that we seek to find here. Looking at Fig. 1.13, we can say that the left side of Eq. (1.47) is a component of E obtained by projecting E onto the path of motion and is equal to the right side which is the directional derivative of V along the path l. There will be one direction for which the directional derivative dV/dl is a maximum, and for this direction the left side of Eq. (1.47) will also be a maximum, and in fact will be E itself. That is,

$$E = -\left(\frac{dV}{dl}\right)_{\max} \tag{1.48}$$

Since a maximum, as used here, implies a direction (that direction for which $\theta = 0$), we have in fact determined not just the magnitude of E but also its direction. We shall call the maximum value of dV/dl at a given point the *gradient* of potential V at that point and use the symbol ∇, called *del*, to write **E** as

$$\boxed{\mathbf{E} = -\nabla V} \tag{1.49}$$

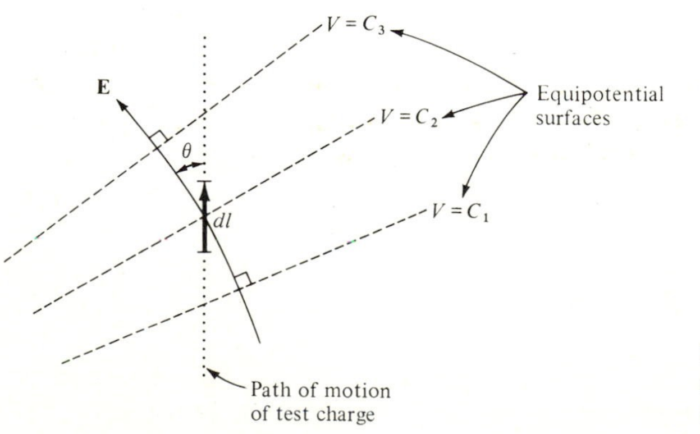

Figure 1.13 A set of equipotential surfaces given by the constants $C_1, C_2, \ldots$ are shown.

The explicit form for operator ∇ is found as follows: Let us first express an element of work by expanding the dot product in Eq. (1.7) as

$$-\Delta V(x, y, z) = \mathbf{E} \cdot \mathbf{dl} = E_x\, dx + E_y\, dy + E_z\, dz \tag{1.50}$$

where an elemental distance **dl** is written in terms of its components and unit vectors as

$$\mathbf{dl} = dx\, \hat{\mathbf{x}} + dy\, \hat{\mathbf{y}} + dz\, \hat{\mathbf{z}} \tag{1.51}$$

Treating ΔV as a total differential and using the chain rule for partial derivatives, we can also express ΔV as

$$\Delta V(x, y, z) = \frac{\partial V}{\partial x} dx + \frac{\partial V}{\partial y} dy + \frac{\partial V}{\partial z} dz \tag{1.52}$$

From a comparison of expressions (1.50) and (1.52), we find that the **E** field is

$$-\mathbf{E} = \frac{\partial V}{\partial x} \hat{\mathbf{x}} + \frac{\partial V}{\partial y} \hat{\mathbf{y}} + \frac{\partial V}{\partial z} \hat{\mathbf{z}} \tag{1.53}$$

which determines the del operator in rectangular coordinates; that is,

$$\boxed{\mathscr{L}^{-1} = \nabla = \frac{\partial}{\partial x} \hat{\mathbf{x}} + \frac{\partial}{\partial y} \hat{\mathbf{y}} + \frac{\partial}{\partial z} \hat{\mathbf{z}}} \tag{1.54}$$

We could have just as well used the spherical or cylindrical coordinate system to find the explicit form for the gradient operator. However, the rectangular coordinate system is the most familiar and most widely used coordinate system. The gradient in spherical and cylindrical coordinates is given in the back cover.

1.9 RELATIONSHIP BETWEEN GRADIENT AND DIRECTIONAL DERIVATIVE

Much can be learned about the gradient operation by finding its relationship to the directional derivative. As we have seen in Eqs. (1.48) and (1.49), the gradient is the maximum directional derivative. A graphical interpretation of this statement using Fig. 1.13 is that the gradient gives the maximum rate of change across the equipotential surfaces; hence it must be normal to them. If the path of motion in Fig. 1.13 is a curve whose parametric equations are $x(l)$, $y(l)$, and $z(l)$, we can write for the unit tangent at an arbitrary point along the curve

$$\widehat{\mathbf{dl}} = \frac{dx}{dl} \hat{\mathbf{x}} + \frac{dy}{dl} \hat{\mathbf{y}} + \frac{dz}{dl} \hat{\mathbf{z}} \tag{1.55}$$

where $\widehat{\mathbf{dl}} = \mathbf{dl}/dl$. Let us suppose that $V(x, y, z)$ is an arbitrary scalar function of position (like potential, temperature, density, etc.). The directional derivative of V

along this curve is given by

$$\frac{dV}{dl} = \frac{\partial V}{\partial x}\frac{dx}{dl} + \frac{\partial V}{\partial y}\frac{dy}{dl} + \frac{\partial V}{\partial z}\frac{dz}{dl} \tag{1.56}$$

We know already that the gradient of V is given by

$$\nabla V = \left(\frac{\partial}{\partial x}\hat{\mathbf{x}} + \frac{\partial}{\partial y}\hat{\mathbf{y}} + \frac{\partial}{\partial z}\hat{\mathbf{z}}\right) V \tag{1.57}$$

Comparing the above three expressions, we see that the directional derivative is the dot product of the gradient of V and the unit tangent along the direction of differentiation; that is,

$$\boxed{\frac{dV}{dl} = \nabla V \cdot \hat{\mathbf{dl}}} \tag{1.58}$$

At the risk of boring the reader, we can say that the directional derivative is the maximum rate of change of a function, projected on the direction of differentiation. For example, if the curve is along the surfaces of constant V, then $dV/dl = 0$. Whereas, if the curve is perpendicular to the surfaces, the derivative in that direction is a maximum and is given by

$$\frac{dV}{dl} = |\nabla V|$$

which is the same as Eqs. (1.47) or (1.48) when $\theta = 0$. Therefore, the maximum directional derivative can be represented by the operation

$$\max \frac{d}{dl}(\ ) = |\nabla(\ )| \tag{1.59}$$

It should be clear that the normal to a family of surfaces, represented by a function $V(x, y, z)$, is related to the gradient of V. The gradient when operating on a scalar function yields a vector function which is normal to the surfaces. The unit normal $\hat{\mathbf{n}}$ to the surfaces V is then

$$\hat{\mathbf{n}} = \frac{\nabla V}{|\nabla V|} \tag{1.60}$$

Example: Since the equipotential surfaces for a point charge located at the origin are spherical surfaces concentric with the origin, the normal to these surfaces has the same direction that the E field has; that is, $\hat{\mathbf{n}} = \nabla V/|\nabla V| = -\hat{\mathbf{E}}$.

We will now introduce an important characteristic of a gradient. We want to show that a closed-loop integral of a gradient of a scalar function is always zero; that is,

$$\boxed{\oint \nabla\phi \cdot \mathbf{dl} = 0} \tag{1.61}$$

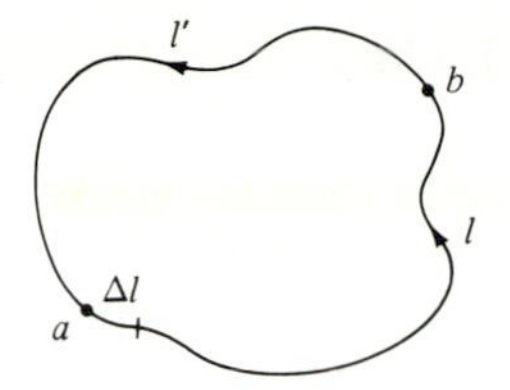

Figure 1.14 Path l and l' are arbitrary paths from a to b to a, respectively. Path $l + l'$ forms a closed loop.

where ϕ is a scalar function of position such as potential, temperature, density, etc. Consider Fig. 1.14. Expand the function ϕ about the point a by writing the first two terms in a Taylor expansion; that is,

$$\phi(a + \Delta l) = \phi(a) + \frac{\partial \phi}{\partial l} dl + \cdots \tag{1.62}$$

Using Eq. (1.58), the directional derivative term in the above expression can be written as

$$\phi(a + \Delta l) = \phi(a) + \nabla\phi \cdot \mathbf{dl} + \cdots \tag{1.63}$$

The function at b can be obtained by integration

$$\phi(b) = \phi(a) + \int_a^b \nabla\phi \cdot \mathbf{dl} \tag{1.64}$$

which is independent of the path followed from a to b. We can now construct a closed loop by integrating from b to a along path l'. This gives us

$$\phi(a) = \phi(b) + \int_b^a \nabla\phi \cdot \mathbf{dl} \tag{1.65}$$

Adding Eqs. (1.64) and (1.65) and noting that $l + l'$ is a closed path, we obtain

$$\oint \nabla\phi \cdot \mathbf{dl} = 0 \tag{1.61}$$

We can now restate the conservative property of electrostatic fields, which is given by Eq. (1.12) as

$$\oint \mathbf{E} \cdot \mathbf{dl} = 0 \tag{1.12}$$

in a more elegant way. We can simply say that since a (vector) force field can be derived from a (scalar) work field by operating on the work field with the gradient operator, as given by (1.49),

$$\mathbf{E} = -\nabla V \tag{1.49}$$

it follows immediately from (1.61) that (1.12) holds. An electrostatic **E** field is conservative because it is a gradient of a scalar field and (1.61) holds for all such fields.

1.10 ELECTRIC FLUX DENSITY AND GAUSS' LAW

The electric field **E** depends in general upon the permittivity ε and thus upon the medium. For example, the electric field from a point charge Q is given by $E = Q/4\pi\varepsilon r^2$. A brief list of dielectric constants $\varepsilon_r = \varepsilon/\varepsilon_0$, where the free-space permittivity $\varepsilon_0 \cong 10^{-9}/36\pi$ F/m, is given in Table 1.1. We see now that if the intervening medium between the point charge and the observation point is bakelite instead of air, the electric field will, be less by a factor of 5. To remove this dependence upon the medium, let us define an electric flux density **D** by multiplying E by ε

$$\boxed{\mathbf{D} = \varepsilon \mathbf{E} \qquad \text{C/m}^2} \tag{1.66}$$

If the medium is isotropic, ε will be a scalar, and the direction of **E** and **D** will be the same. Since **D** is a flux density, the total flux Ψ streaming through an elemental area ΔA can be obtained by forming the dot product of **D** with that area; that is,

$$\Delta\Psi = \mathbf{D} \cdot \Delta\mathbf{A} \qquad \text{C} \tag{1.67}$$

Figure 1.15 shows this graphically. In this figure $\hat{\mathbf{n}}$ is the unit vector denoting the normal to the elemental surface which permits us to express an elemental area in vector form as $\Delta\mathbf{A} = \hat{\mathbf{n}}A$. Equation (1.67) can then also be written as

$$\Delta\Psi = \mathbf{D} \cdot \hat{\mathbf{n}}\, \Delta A$$

Of course, the total flux Ψ streaming through a large area can be obtained by integrating $\Delta\Psi$. The total flux coming out of a closed surface, for example, a balloon surface, is simply denoted by

$$\Psi = \oint\!\!\oint \mathbf{D} \cdot \mathbf{dA} \tag{1.68}$$

If such a closed-surface integral is finite, Ψ is finite, implying that net flux must come out of the surface. Furthermore, since such net flux has the dimensions of charge, it is not difficult to imagine that a net amount of charge must reside inside the closed surface. Putting it another way, the sources of the flux must be charges.

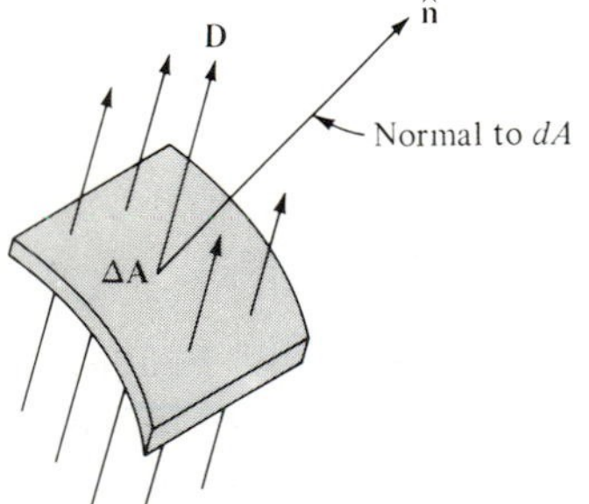

Figure 1.15 The total flux streaming through the area ΔA is given by the dot product of **D** and $\Delta\mathbf{A}$.

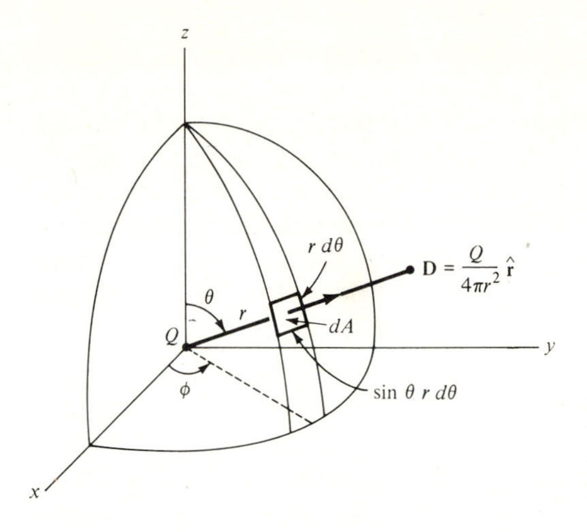

Figure 1.16 The spherical coordinate system, showing a point charge Q at the origin and an elemental surface area $dA = (r\,d\theta)(\sin\theta\; r\,d\phi)$.

This conjecture can be put on a firm basis and is known as Gauss' law, which we will consider next.

Let us take a simple case first. A point charge Q is located at the origin. Let us surround this charge with a fictitious spherical surface with its center at the origin. The total flux Ψ streaming through this balloon-type surface is then given by

$$\Psi = \oiint \frac{Q}{4\pi r^2}\hat{\mathbf{r}} \cdot \mathbf{dA} \tag{1.69}$$

Since the normal to this surface is along the radial direction, that is, $\mathbf{dA} = \hat{\mathbf{r}}\, dA$, we have

$$\Psi = \oiint \frac{Q}{4\pi r^2}\, dA = \frac{Q}{4\pi r^2} \oiint dA \tag{1.70}$$

where the flux-density term was taken outside the integral since over a spherical surface of constant radius r, the flux density is uniform everywhere on that surface. The remaining surface integral is the surface area of a balloon-type surface of radius r and is equal to $4\pi r^2$. We will find it useful to calculate this result in detail, using the spherical coordinate system. Referring to Fig. 1.16, which shows a portion of a balloon surface and an elemental surface area dA, we obtain

$$\oiint dA = \int_0^{2\pi}\int_0^{\pi} (r\,d\theta)(\sin\theta\; r\,d\phi) = 2\pi r^2 \int_0^{\pi} \sin\theta\, d\theta = 4\pi r^2$$

We can now make the important observation that the total flux Ψ emanating from the closed surface is equal to the charge enclosed; that is, $\Psi = Q$. This is Gauss' law.

This statement will now be shown to be true for an arbitrarily shaped closed surface. Figure 1.17 shows again a point charge Q located at the origin, but this time it is surrounded by a nonspherical surface. The total flux leaving this surface

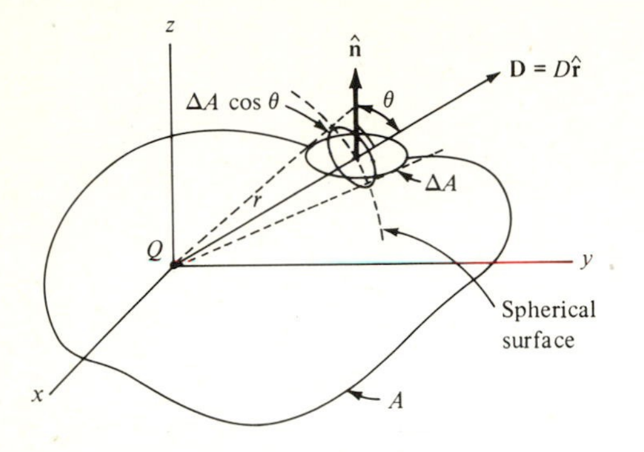

Figure 1.17 An arbitrarily shaped closed surface surrounding a point charge Q is used to show Gauss' law $\Psi = Q$.

is given by Eq. (1.68), which can be written as

$$\Psi = \oiint D\hat{\mathbf{r}} \cdot \hat{\mathbf{n}}\, dA = \oiint D \cos\theta\, dA = \oiint \frac{Q}{4\pi r^2} \cos\theta\, dA \tag{1.71}$$

where θ is the angle between the normal $\hat{\mathbf{n}}$ to a surface element and the radial direction r of the flux density. The key point here is the term $\hat{\mathbf{r}} \cdot \hat{\mathbf{n}}\, dA = \cos\theta\, dA$; it is the projection of the surface element $\hat{\mathbf{n}}\, dA$ on a spherical surface that cuts the arbitrarily shaped surface at the point r. Therefore, $\cos\theta\, dA$ is just an elemental area on a spherical surface and is given by

$$\cos\theta\, dA = r\, d\theta \sin\theta\, r\, d\phi$$

as shown in Fig. 1.16. The integral in Eq. (1.71) can then be written

$$\Psi = \frac{Q}{4\pi} \oiint \frac{\cos\theta\, dA}{r^2} = \frac{Q}{4\pi} \oiint \frac{r^2 \sin\theta\, d\theta\, d\phi}{r^2} = Q \tag{1.72}$$

It should be noted that the r^2 term in the denominator which belongs to the flux density **D** always cancels the r^2 term in the numerator due to the area. In other words, the proof of Gauss' law is based upon the assumption that the electric force from a charge varies as $1/r^2$. Gauss' law can now be stated as

$$\boxed{\oiint \mathbf{D} \cdot \mathbf{dA} = Q} \tag{1.73}$$

that is, the total flux out of a closed surface equals the charge enclosed. If the charge is distributed, either as many discrete charges or as a continuous charge distribution, throughout the volume $v(A)$ which is bounded by the surface A, Gauss' law can be written as

$$\oiint_{A(v)} \mathbf{D} \cdot \mathbf{dA} = \sum_n Q_n = \iiint_{v(A)} \rho\, dv \tag{1.73a}$$

where $A(v)$ is the surface area that bounds volume v. We should carefully take note of the fact that only the net charge that is enclosed by the surface A contributes to

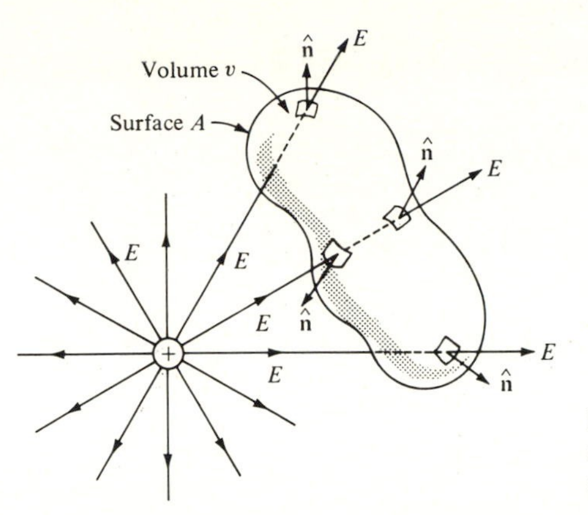

Figure 1.18 A closed surface with a point charge outside the closed surface. The outward normal at each point to the surface is shown as $\hat{\mathbf{n}}$.

the net flux at the boundaries of A. Any charge residing outside of A has no effect on the net flux out of the volume.

Figure 1.18 shows a closed surface which does not contain any charge inside but is immersed in an electric field. We readily see that by dotting the flux density $\mathbf{D}$ with all elemental areas $\hat{\mathbf{n}}\, dA$ and adding, the resultant net flux will be zero, because as much flux penetrates into the surface as emerges outward from it. Another way of saying it is as follows: If no net flux "sticks" out of a closed surface, no net charge can be enclosed.

1.11 RELATIONSHIP BETWEEN GAUSS' LAW AND COULOMB'S LAW

We can show that Gauss' law is an alternative statement of Coulomb's law. To deduce Coulomb's law from Gauss' law, let us apply Gauss' law to a point charge located at the origin as shown in Fig. 1.16. If we consider a spherical surface centered about the origin, the normal to that surface will be radial everywhere, and we can write for Gauss' law

$$\oiint \varepsilon \mathbf{E} \cdot \mathbf{dA} = \oiint \varepsilon E_r \, dA = Q \tag{1.74}$$

where E_r is the radial component of the E field. However, from the symmetry of the problem we readily deduce that the total E field can only be in the radial direction and that it must have the same value everywhere on the spherically symmetric surface. Therefore, Eq. (1.74) becomes

$$\varepsilon E_r \oiint dA = Q \tag{1.74a}$$

which reduces to

$$\varepsilon E_r 4\pi r^2 = Q \tag{1.74b}$$

or

$$\mathbf{E} = \frac{Q}{4\pi\varepsilon r^2}\hat{\mathbf{r}} \tag{1.74c}$$

since the surface area of a closed spherical surface is $4\pi r^2$ and the direction of the E field has been established from symmetry considerations to be radial. If we now place a second charge Q' at the point where we have calculated the E field, the force that Q' will experience is given by

$$\mathbf{F} = Q'\mathbf{E} \tag{1.75}$$

Using Eq. (1.74*c*), we can write

$$\mathbf{F} = \frac{QQ'}{4\pi\varepsilon r^2}\hat{\mathbf{r}} \tag{1.76}$$

which is Coulomb's law. We have thus obtained Coulomb's law from Gauss' law. The reason why Coulomb's law need not be listed in a table of equations which describe the electromagnetic field and which are known as Maxwell's equations is precisely because Gauss' law will be shown to be one of Maxwell's equations. Therefore, Coulomb's law is implicit in Maxwell's equations.

Let us repeat again that the validity of Gauss' law is dependent upon the inverse square law of Coulomb. For example, if the force law were not exactly $1/r^2$, the field inside a uniformly charged spherical shell could not be exactly zero—a result that is shown in (1.80*b*).

1.12 SOME APPLICATIONS OF GAUSS' LAW

Gauss' law allows us to find the net charges within a closed surface if the electric flux emerging from that surface is known. But perhaps the greatest use of Gauss' law is in calculating the electric field from symmetric charge distributions. The simplicity of this approach, whenever it is applicable, is startling especially when contrasted with the amount of work needed if Coulomb's law is used.

Electric Field from an Infinite Line Charge

This problem was already solved in Sec. 1.7 using Coulomb's law. The electric field from a uniformly charged infinite wire is given by Eq. (1.25). Let us choose a barrel-shaped surface which surrounds the line charge, as shown in Fig. 1.19, for our gaussian surface. Gauss' law can then be written as

$$\oiint \mathbf{D} \cdot d\mathbf{A} = Q = \int \rho_L \, dl \tag{1.77}$$

The closed-surface integral can be divided into two parts

$$\iint_{\text{ends}} \mathbf{D} \cdot \hat{\mathbf{z}} \, dA + \iint_{\substack{\text{cylindrical}\\\text{surface}}} \mathbf{D} \cdot \hat{\mathbf{r}} \, dA = \int \rho_L \, dl \tag{1.77a}$$

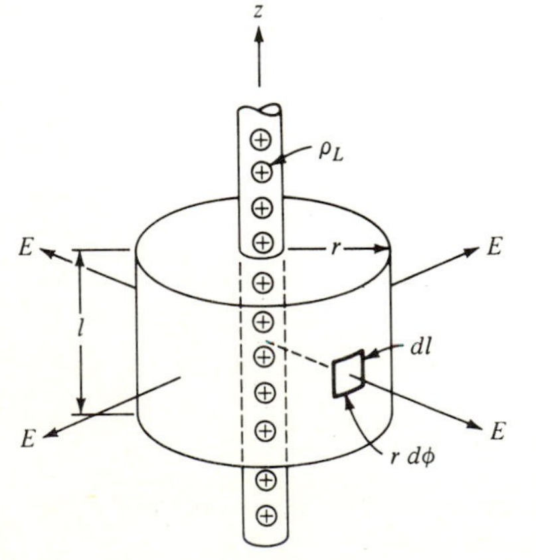

Figure 1.19 A line charge of uniform linear charge density ρ_L is surrounded by a cylindrical surface of radius r and length l.

where the first integral gives the flux streaming through the planar end caps and the second integral is the flux through the cylindrical surface. The normal to the end caps is given by $\hat{\mathbf{z}}$, the normal to the cylindrical surface by $\hat{\mathbf{r}}$. From the symmetry of the problem and because of the uniformness of the line charge, the E field can only be in the radial direction. Therefore, no flux streams through the ends of the barrel-shaped surface, leaving only the cylindrical surface to be considered. Equation (1.77a) becomes

$$\int_0^l \int_0^{2\pi} \varepsilon E r \, d\phi \, dl = \int_0^l \rho_L \, dl \tag{1.77b}$$

Performing the l integration and taking the constant E outside the remaining integral, we have

$$l\varepsilon E \int_0^{2\pi} r \, d\phi = \rho_L l \tag{1.77c}$$

Canceling l and completing the ϕ integration, we obtain for the electric field from a uniform line charge

$$\boxed{E = \frac{\rho_L}{2\pi\varepsilon r}} \tag{1.77d}$$

It should be noted that the procedure for the solution using Gauss' law is much simpler than that using Coulomb's law.

Electric Field from a Uniformly Charged Infinite Plane Sheet

This problem was also considered in Sec. 1.7. Figure 1.20 shows a portion of a nonconducting infinite plane sheet of uniform charge. The surface charge is ρ_s in coulombs per meter squared. A convenient gaussian surface is a "pill box" oriented perpendicular to the surface, of height $2l$ and radius r, with the infinite sheet of charge dividing the "pill box" symmetrically. Gauss' law (1.73) can be written as

$$\iint_{\text{ends}} \mathbf{D} \cdot \mathbf{dA} + \iint_{\substack{\text{cylindrical} \\ \text{surfaces}}} \mathbf{D} \cdot \mathbf{dA} = \iint \rho_s \, dA \tag{1.78}$$

Considering the symmetry of the sheet, we can immediately deduce that the E field is normal to the sheet. Hence, the electric field pierces only the end caps of the "pill box" and no E field or flux goes through the curved cylindrical portion; thus

$$\iint_{\text{ends}} \mathbf{D} \cdot \mathbf{dA} = \iint \rho_s \, dA \tag{1.78a}$$

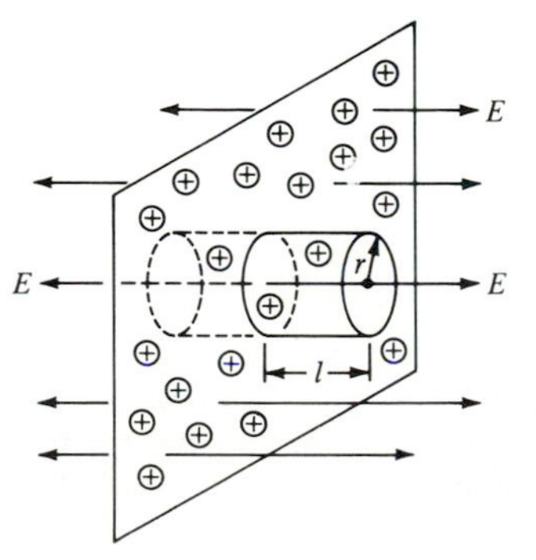

Figure 1.20 A cylindrical gaussian surface extends on both sides of the surface charge. See also Fig. 1.11.

Notice that the "pill box" has two ends, with the normal for each end in the same direction as the E field that pierces the ends. Therefore, we can write

$$2D \iint dA = \rho_s \iint dA \tag{1.78b}$$

or

$$\boxed{E = \frac{\rho_s}{2\varepsilon}} \tag{1.78c}$$

which is the same E field as obtained in Eq. (1.35). The E field has the same magnitude no matter how far the observation point from the infinite sheet is.

Spherical Shell of Charge

Gauss' law is particularly useful when the charge distributions have spherical symmetry. Let us take a charge Q_T and assume it to be distributed over an imaginary spherical shell of radius a, as shown in Fig. 1.21. Since the charge will spread itself uniformly over the spherical surface, the E field outside the shell must be radial from the charge center. The total charge will distribute itself such that

$$Q_T = \oint\!\!\oint \rho_s \, dA = \rho_s 4\pi a^2 \tag{1.79}$$

To find the field inside the shell, we apply Gauss' law to a spherical gaussian surface of radius $r = a - \Delta r$, which is just inside the spherical shell, and obtain

$$\oint\!\!\oint \mathbf{D} \cdot dA = 0 \tag{1.80}$$

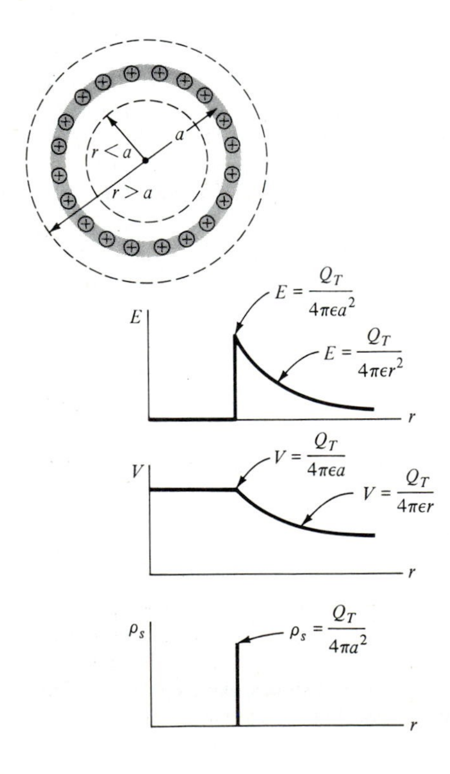

Figure 1.21 A total charge Q_T is distributed uniformly over a spherical shell. The variations of the E and V fields are shown inside and outside the shell.

because the net amount of charge enclosed inside the surface of radius $r = a - \Delta r$ is zero. Furthermore, from symmetry the E field on the surface $a - \Delta r$ must be uniform, which permits us to take D outside the integral

$$D \oiint dA = 0 \tag{1.80a}$$

Since the surface area in general is not zero, we conclude that D must be zero; that is,

$$\boxed{D = 0 \qquad r < a} \tag{1.80b}$$

By similar reasoning, we conclude that this holds for any interior surface, which is indicated by $r < a$ in the above equation. A test charge placed anywhere within the interior of the shell would experience no force; hence, from the point of view of the test charge, the spherical shell of charge is nonexistent.

To find the external field, we can apply Gauss' law to a spherical gaussian surface of radius $r > a$ and obtain

$$\begin{aligned}\oiint \varepsilon \mathbf{E} \cdot \mathbf{dA} &= \oiint \rho_s \, dA \\ &= Q_T\end{aligned} \tag{1.81}$$

Using again the spherical symmetry condition, the above equation simplifies to

$$\varepsilon E 4\pi r^2 = Q_T \tag{1.81a}$$

Since the electric field is radial, we can write it now as

$$\boxed{\mathbf{E} = \frac{Q_T}{4\pi\varepsilon r^2}\hat{\mathbf{r}} \qquad r > a} \tag{1.81b}$$

This equation is identical to that for the electric field from a point charge. Therefore, for points outside of the shell it appears as if the distributed charge is concentrated at the origin. The force that a test charge would experience at points $r > a$ would be indistinguishable from that of a point charge Q_T at the origin. At this point, we should observe again the elegance and simplicity of the gaussian approach. An approach to this problem by Coulomb's law would have resulted in considerably more detail. The behavior of the E field is shown graphically in Fig. 1.21.

The absolute potential for points outside the positively charged shell, which is the work required to move a positive test charge from infinity to those points, is given by

$$\begin{aligned}V &= -\int_\infty^r \mathbf{E} \cdot \mathbf{dl} = -\int_\infty^r E_r \, dr \\ &= -\int_\infty^r \frac{Q_T}{4\pi\varepsilon r^2} dr = \frac{Q_T}{4\pi\varepsilon r} \qquad r > a\end{aligned} \tag{1.82}$$

where we have used expression (1.81b) for the E field. If we bring the test charge right up to the shell, the amount of work done is

$$V = \frac{Q_T}{4\pi\varepsilon a} = \frac{\rho_s a}{\varepsilon} \tag{1.83}$$

If we move the test charge past the shell into the interior, the additional work that would have to be done would be zero since the electric field is zero within the shell. The work function, unlike the force function, is therefore continuous, as is shown graphically in Fig. 1.21. We should always expect a continuous work function, for if it had a discontinuity at some point, the corresponding force field at that point would have to be infinite, and infinite force fields are nonphysical. This comes about because the force field is derived from the gradient of the work field, and a derivative of a function at a point at which it is discontinuous is infinite.

It should be observed here that the charged shell was assumed to be infinitesimally thin. Hence the electric field will show an abrupt change from a finite value just outside the shell to zero value just inside. If the charged shell is assumed to have finite thickness, the electric field will show a continuous variation. The charge density ρ_s versus radial distance r is shown as the last graph in Fig. 1.21. It is zero everywhere except at the point $r = a$ where it has a finite value given by Eq. (1.79).

Three Spherical Shells of Charge

As the next example, let us consider three spherical and concentric shells each carrying the same total charge which is distributed uniformly over each shell, as shown in Fig. 1.22. Since Gauss' law is so

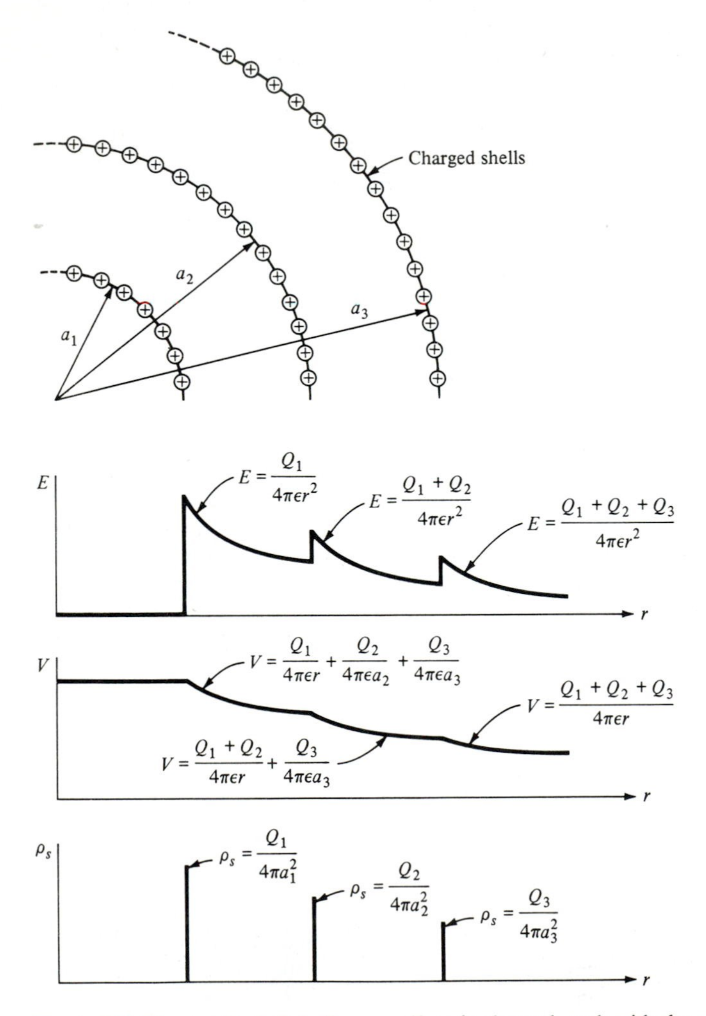

Figure 1.22 Three spherical shells are uniformly charged, each with the same total charge. The radii are $a_3 = 3a_1$ and $a_2 = 2a_1$.

elegant and simple, we can predict the electric and potential field variations without any further calculations if we understand the previous example well. Figure 1.22 shows all necessary equations which should be self-explanatory.

A Uniformly Charged Solid Sphere

If we take a total charge Q_T and distribute it uniformly throughout the interior of a sphere of radius a, the volume charge density ρ (in coulombs per cubic meter) is related by

$$Q_T = \rho \tfrac{4}{3}\pi a^3 \tag{1.84}$$

A practical example of such a situation might be a charged dielectric sphere or a spherical cloud of charged plasma or gas. We can again use the results obtained for a single shell simply by imagining the solid sphere to be composed of many closely spaced shells. For points outside the sphere, $r > a$, the electric and potential fields are those of a point charge of total charge Q_T located at the center. For points inside the sphere, $r < a$, we know from Gauss' law that only those shells matter for which $r < a$. That is, only the charge contained within a sphere of radius r will contribute to the field. That part of the total charge Q_T which lies outside r makes no contribution to **E** at radius r. In summary the fields outside the sphere are

$$\mathbf{E} = \frac{Q_T}{4\pi\varepsilon r^2}\hat{\mathbf{r}} \qquad r \geq a \tag{1.84a}$$

$$V = \frac{Q_T}{4\pi\varepsilon r} \qquad r \geq a \tag{1.84b}$$

To obtain the fields inside the uniformly charged sphere, we observe first that the volume of a sphere varies as the cube of the radius. Therefore, the remaining charge inside a sphere of radius r, where $r < a$, is

$$Q_r = Q_T\left(\frac{r}{a}\right)^3 = \rho \tfrac{4}{3}\pi r^3 \tag{1.84c}$$

The electric field is then

$$\mathbf{E} = \frac{Q_r}{4\pi\varepsilon r^2}\hat{\mathbf{r}} = \rho\frac{r}{3\varepsilon}\hat{\mathbf{r}} \qquad r \leq a \tag{1.84d}$$

and as shown graphically in Fig. 1.23, the E field approaches zero uniformly. To repeat, the electric field at a point r inside the sphere is that of a point charge Q_r located at the origin with the charge outside the sphere of radius r making no contribution.

The absolute potential for points inside is the sum of the work required to bring a test charge from infinity to the edge of the sphere [Eq. (1.84b)] and the additional work required to move the test charge against the force field of the remaining charge Q_r from a to r. Mathematically we have

$$\begin{aligned} V &= -\int_\infty^a \mathbf{E}_{\text{out}} \cdot d\mathbf{l} - \int_a^r \mathbf{E}_{\text{in}} \cdot d\mathbf{l} \\ &= \frac{Q_T}{4\pi\varepsilon a} - \int_a^r \frac{Q_r}{4\pi\varepsilon r^2}\,dr \\ &= \frac{Q_T}{4\pi\varepsilon a} + \frac{Q_T}{8\pi\varepsilon a^3}(a^2 - r^2) \qquad r \leq a \end{aligned} \tag{1.84e}$$

Note that in the second integral Q_r is a function of r and cannot be taken outside the integral. We have considered a sphere with a constant charge density. Should the charge density be a function of r, then the total charge Q_T in (1.84) will be related to the charge density by an integral; that is,

$$Q_T = \iiint \rho(r)\,dv$$

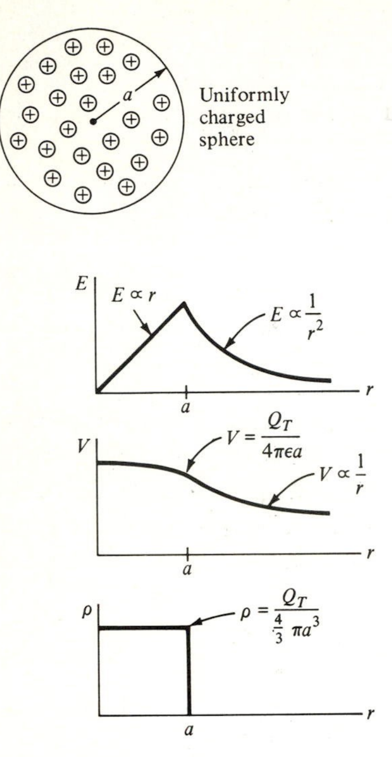

Figure 1.23 Variation of the fields and charge density for a uniformly charged sphere.

1.13 BOUNDARY CONDITIONS FOR THE ELECTRIC FIELD

Many practical problems in electromagnetics involve situations that have materials of different properties next to each other. It is, therefore, of great interest to know the behavior of the fields at the interface. How the electric field changes across an interface between, let us say, water and air, two different dielectrics, or a dielectric and a metal is very useful to know. The apparent shortening of your feet when you stand in water is an example that the electric field of the light rays is discontinuous when crossing the water-air interface. To show this and boundary conditions in general, we will use the integral field equations

$$\oint \mathbf{E} \cdot \mathbf{dl} = 0 \tag{1.12}$$

and

$$\oiint \mathbf{D} \cdot \mathbf{dA} = \iiint \rho \, dv \tag{1.73}$$

to find the relations between the electrostatic field vectors on both sides of a discontinuity. We can also point out at this time that even though the behavior of the electric field at a boundary will be derived from electrostatic equations, the relations obtained will be valid even when the electric field is time-varying. Even at the very high frequencies of light these relations will be applicable.

There is another reason why knowledge of boundary conditions is needed. We will show in the next few sections that the behavior of the electric and potential fields is determined by partial differential equations. Solutions to differential equations always involve arbitrary constants. A first-order differential equation has one arbitrary constant, a second-order equation has two arbitrary constants, and so on. These arbitrary constants are determined by the application of boundary conditions.

In order to find the behavior at the boundary, we will resolve the electric field along the boundary into tangential and normal components and analyze these two components separately.

Boundary Condition for Tangential Components of Electric Field at the Interface of Two Dielectrics

Consider rectangle *abcd* which lies with its longest sides parallel to the surface separating the two dielectric media. Let sides *ab* and *cd* become vanishingly small, but in such a way that the boundary is always between the two longer sides Δl, as shown in Fig. 1.24. An electric field $\mathbf{E}_2$ upon crossing the boundary will kink and become $\mathbf{E}_1$. The closed-line integral (1.12) which expresses the conservative property of electrostatic fields when applied to contour *abcd* gives

$$\lim_{ab,cd \to 0} \oint_{abcd} \mathbf{E} \cdot \mathbf{dl} = 0$$

$$\int_b^c \mathbf{E}_2 \cdot \mathbf{dl} + \int_d^a \mathbf{E}_1 \cdot \mathbf{dl} = 0$$

$$(\mathbf{E}_2 - \mathbf{E}_1) \cdot \mathbf{\Delta l} = 0 \tag{1.85}$$

The longer sides of length Δl are small enough that E can be considered a constant over Δl, yielding the last equation in (1.85). The minus sign was introduced because the vectors $\mathbf{\Delta l}_1$ and $\mathbf{\Delta l}_2$ which are on opposite sides of the boundary are opposite to each other. Since $\mathbf{\Delta l}$ in general is finite and can be written as $\mathbf{\Delta l} = \hat{\mathbf{t}}\,\Delta l$ where $\hat{\mathbf{t}}$ is a unit vector tangential to the surface, we obtain from (1.85)

$$\boxed{E_{t2} = E_{t1}} \tag{1.86}$$

This says that the tangential components of an E field are continuous across an interface.

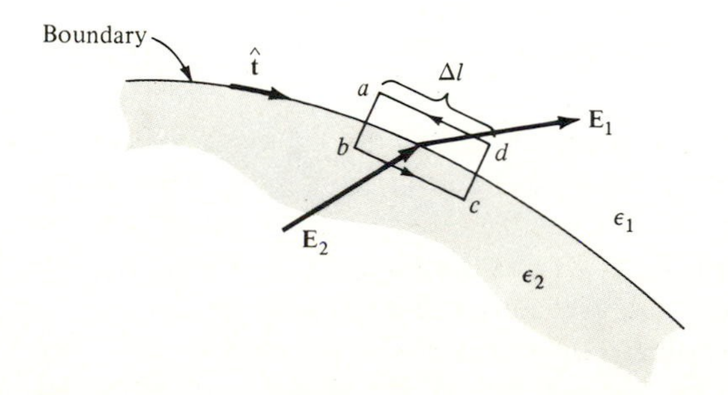

Figure 1.24 Contour *abcd* used in obtaining boundary conditions for *E*.

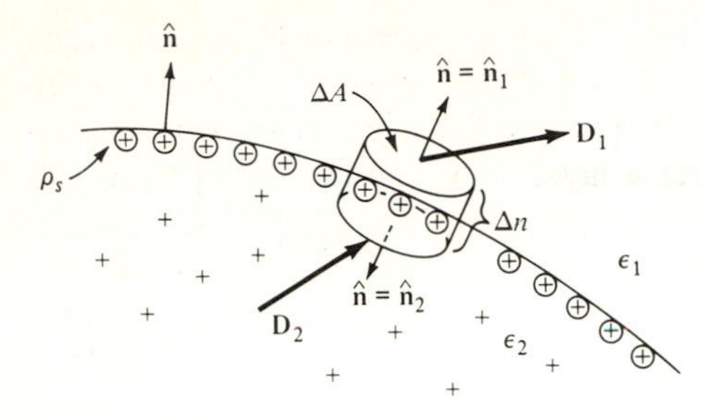

Figure 1.25 Pill box of height Δn and surface ΔA contains the interface with surface charge ρ_s. The outward normal to the pill box surface is $\hat{\mathbf{n}}$. Note that $\hat{\mathbf{n}}_1 = -\hat{\mathbf{n}}_2$. A surface charge layer ρ_s is shown to exist as well as some volume charge density throughout region 2.

Boundary Condition for Normal Components of Electric Field at an Interface

To obtain the condition on the normal components of the electric field, we can use Gauss' law (1.73) and apply it to a small pill box constructed to lie in the interface of the two dielectrics, as shown in Fig. 1.25. If the height Δn becomes vanishingly small, no flux will pass through the sides of the pill box. Gauss' law for this situation can be written as

$$\lim_{\Delta n \to 0} \left(\oiint \mathbf{D} \cdot \hat{\mathbf{n}}\, dA = \iiint \rho\, dv \right)$$

$$\iint (\mathbf{D}_1 - \mathbf{D}_2) \cdot \hat{\mathbf{n}}\, dA = \lim_{\Delta n \to 0} \iiint \rho\, dn\, dA = \iint \rho_s\, dA \tag{1.87}$$

where we have assumed that a surface charge which is

$$\rho_s = \lim_{\Delta n \to 0} \int_0^{\Delta n} \rho\, dn = \lim_{\Delta n \to 0} \rho\, \Delta n \tag{1.87a}$$

exists in an infinitesimally thin layer on the boundary. Since ΔA is small, we can remove the integral signs in (1.87) and obtain

$$(\mathbf{D}_1 - \mathbf{D}_2) \cdot \hat{\mathbf{n}}\, \Delta A = \rho_s\, \Delta A$$

or

$$\boxed{(\mathbf{D}_1 - \mathbf{D}_2) \cdot \hat{\mathbf{n}} = D_{n_1} - D_{n_2} = \rho_s} \tag{1.87b}$$

where D_n are the normal components. Thus, when a surface charge exists, the normal components of electric flux density, unlike the tangential components of electric field, are discontinuous by the amount of surface charge present. The interface between two dielectrics is usually charge-free. Hence for a charge-free boundary

$$D_{n_1} = D_{n_2} \qquad \text{or} \qquad \varepsilon_1 E_{n_1} = \varepsilon_2 E_{n_2} \tag{1.87c}$$

Equation (1.87*a*) merits further discussion. It implies that at points where a surface charge density ρ_s exists, the volume charge density ρ must necessarily be infinite. Charged dielectric bodies have a finite volume charge density distributed throughout the interior of the body. Hence ρ_s is zero everywhere. For conductors, on the other hand, a charge placed on them will distribute itself in an infinitesimally thin layer on the surface, yielding a finite ρ_s. (This will be considered in the next chapter.) Hence, unless a layer of real surface charge exists at the interface between two dielectrics,† the appropriate boundary condition for dielectrics is (1.87*c*), which states that the normal components of flux density are continuous.

It is now apparent that the normal components of the electric field of light must contribute to the bending of light rays when light crosses the air-water interface. The bending can be appreciable since the dielectric constant of water is 81 times larger than that of air (see Table 1.1).

1.14 DIVERGENCE AND THE DIFFERENTIAL FORM OF GAUSS' LAW

A second form of Gauss' law is expressed in terms of derivatives. If you recall, the integral form of Gauss' law relates the net flux out of a finite volume to the net amount of charge enclosed in that volume. The differential form of Gauss' law can be found by applying Gauss' integral law to an infinitesimally small volume surrounding a point. As we shall see, the integrals will then transform to differentials in the limit as the volume goes to zero and we will obtain a point relationship of Gauss' law involving derivatives only. Since derivatives, as compared to integrals, are easier to calculate, the differential form of Gauss' law will be very useful and, in a sense, more general.

Let us consider a small volume $\Delta v = \Delta x \, \Delta y \, \Delta z$ which is permeated by a flux density **D** as shown in Fig. 1.26. Applying Gauss' law to this elemental volume and dividing by Δv, we obtain

$$\lim_{\Delta v \to 0} \frac{\oint\!\!\oint \mathbf{D} \cdot \mathbf{dA}}{\Delta v} = \lim_{\Delta v \to 0} \frac{\iiint_{\Delta v} \rho \, dv}{\Delta v} = \rho \tag{1.88}$$

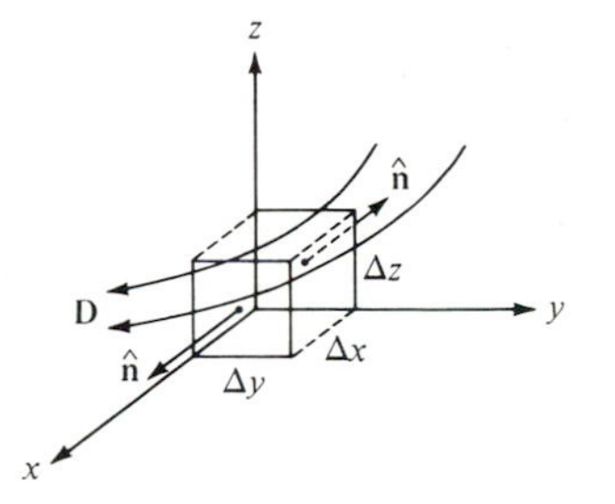

Figure 1.26 A small cubical volume in a vector field **D** is used to calculate the divergence of the vector field.

† A real surface charge could be obtained by tearing away surface electrons, as when glass and silk are rubbed together.

where the volume integral over the elemental volume Δv is simply $\rho\ \Delta v$, which when divided by Δv, gives ρ. Note that the surface integral on the left-hand side cannot be approximated by its integrand even though the integration is over the small surface which bounds the infinitesimal Δv. Unlike an ordinary area integral, a closed-surface integral has special properties which do not permit this relaxation. For example, if a point charge is located in Δv, then the total flux coming out of an enclosing surface, be it small or large, is always the same.

A simpler way to obtain Eq. (1.88) is to assume that a charge is distributed throughout a volume v with charge density ρ. Then the total charge enclosed in Δv is simply $\Delta Q = \rho\ \Delta v$. Gauss' law applied to Δv becomes

$$\oint\!\!\oint \mathbf{D} \cdot \mathbf{dA} = \Delta Q = \rho\ \Delta v \tag{1.88a}$$

which when divided by Δv gives (1.88).

The limit operation on $\mathbf{D}$ expressed by Eq. (1.88) is called the divergence of $\mathbf{D}$; that is,

$$\boxed{\operatorname{div} \mathbf{D} = \lim_{\Delta v \to 0} \frac{\oint\!\!\oint \mathbf{D} \cdot \mathbf{dA}}{\Delta v}} \tag{1.89}$$

where "div" is an abbreviation for divergence. The *divergence* of a vector flux density is therefore a measure of the outward flow of the flux per unit volume. From Gauss' law we also know that the right-hand side of (1.89) is equal to the volume charge density; that is,

$$\boxed{\operatorname{div} \mathbf{D} = \rho} \tag{1.90}$$

This expression will be shortly identified as the differential form of Gauss' law. For a field at a point to have a finite divergence means that an equivalent to a source must exist there. If the divergence of a field is zero for a small volume, this can be interpreted to mean that all the flux that enters the volume also leaves it; this region of space is then source-free.

To relate the divergence to derivatives, we will calculate the flux entering and leaving the six faces of the cubical volume of Fig. 1.26. For convenience we will choose the rectangular coordinate system and list the divergence operations for other coordinate systems on the back cover. It should be clear that divergence is an intrinsic operator which does not depend on the choice of a particular coordinate system.

Consider the pair of faces $\Delta z\ \Delta y$ in Fig. 1.26 first and calculate the net flux through these two faces. The flux entering through the face in the zy plane is given by

$$\mathbf{D} \cdot \hat{\mathbf{n}}\ \Delta A = -D_x\ \Delta y\ \Delta z \tag{1.91}$$

where the minus sign comes about because D_x and the outward normal $\hat{\mathbf{n}}$ to face $\Delta z\,\Delta y$ are opposite to each other. The flux leaving through the adjacent face is

$$\mathbf{D}\cdot\hat{\mathbf{n}}\,\Delta A = \left(D_x + \frac{\partial D_x}{\partial x}\Delta x\right)\Delta y\,\Delta z \tag{1.91a}$$

where the outward flux density is approximated by the flux density entering plus the rate of change of D between the two faces times the distance between the faces. This should be recognized as the first two terms in the Taylor expression of D about the origin. If the distance Δx between faces is small, the approximation by two terms in the expansion is valid since the higher-order terms involving $(\Delta x)^n$ will be negligible. The net flux out of the two faces can now be obtained by adding the flux entering and leaving, which gives

$$\frac{\partial D_x}{\partial x}\Delta x\,\Delta y\,\Delta z \tag{1.91b}$$

If the same procedure is repeated for the remaining two sets of faces, we obtain for the net flux out of volume $\Delta x\,\Delta y\,\Delta z$

$$\oint\!\!\oint \mathbf{D}\cdot d\mathbf{A} = \left(\frac{\partial D_x}{\partial x} + \frac{\partial D_y}{\partial y} + \frac{\partial D_z}{\partial z}\right)\Delta x\,\Delta y\,\Delta z \tag{1.91c}$$

Dividing this expression by the volume Δv, as in Eq. (1.89), gives us the expression for divergence in terms of derivatives; that is,

$$\text{div}\,\mathbf{D} = \frac{\partial D_x}{\partial x} + \frac{\partial D_y}{\partial y} + \frac{\partial D_z}{\partial z} \tag{1.91d}$$

The above can be written in simple form by employing the del operator which was first used in the gradient operation. Equation (1.91d) can be seen to be the dot product of the operator ∇ with $\mathbf{D}$; that is,

$$\text{div}\,\mathbf{D} = \nabla\cdot\mathbf{D} = \left(\frac{\partial}{\partial x}\hat{\mathbf{x}} + \frac{\partial}{\partial y}\hat{\mathbf{y}} + \frac{\partial}{\partial z}\hat{\mathbf{z}}\right)\cdot(D_x \quad + D_y \quad + D_z \quad) \tag{1.91e}$$

This completes the derivation of the divergence operator $\nabla\cdot$ in the rectangular coordinate system. Because divergence involves the dot product, the quantity that it operates on must be a vector. The resultant of this operation is then a scalar quantity. Note that this is different from the gradient which operates on a scalar and yields a vector.

Finally, we can write the differential form of Gauss' law, which is a point relationship expressed in terms of space derivatives, as

$$\nabla\cdot\mathbf{D} = \rho \tag{1.92}$$

It will be shown in Chap. 11, that the differential and integral forms of Gauss' law constitute Maxwell's second equation in differential and integral form, respectively. In summary, if a charge density exists at a point in space, then the flux density must have a divergence at this point, which means that if the point is surrounded by a small volume, a net flux must emerge from it. Conversely, if the flux density **D** shows a divergence at a point, a charge density must exist at such a point.

1.15 GAUSS' DIVERGENCE THEOREM

The discussion of the previous section can now be easily extended to derive a theorem which will be found useful in manipulating vector equations in order to reduce them to their most convenient forms. This theorem is known as Gauss' divergence theorem.

Using Eq. (1.89), we can say that the outward flux from the surface of an infinitesimal cube is equal to the divergence of the flux-density vector multiplied by the volume of the cube; that is,

$$\oint\!\!\oint \mathbf{F} \cdot \hat{\mathbf{n}}\, dA = (\nabla \cdot \mathbf{F})\, \Delta v \tag{1.93}$$

where **F** stands for an arbitrary vector field. For a finite volume we can, after subdividing it first into elemental volumes Δv, obtain the net flux by integrating the divergence over the entire volume. Doing this, we obtain Gauss' divergence theorem

$$\boxed{\oint\!\!\oint_{A(v)} \mathbf{F} \cdot \mathbf{dA} = \iiint_{v(A)} \nabla \cdot \mathbf{F}\, dv} \tag{1.94}$$

where $A(v)$ is the surface area bounding volume v, and $v(A)$ is the volume v bounded by area A.

In the previous section we have shown that the divergence of a vector field can be considered to be a measure of scalar sources of the field. For example, we know that charge is the source for the electrostatic field. This is expressed by Eqs. (1.90) or (1.92) which state that a finite divergence of a field at a point implies the presence of a source density there. *Gauss' divergence theorem* should then be read as follows: The net flux through a closed surface is equal to the summation of the scalar sources inside the surface.

1.16 THE LAPLACIAN OPERATOR AND LAPLACE'S EQUATION

At this time perhaps the student is beginning to feel that there will be no end to the introduction of new operators. At this stage the student might also feel that these operators are abstract and not very useful. The usefulness of these operators will be in the ease with which electromagnetic phenomena can be described. This will become more apparent as we progress in our studies and solve more problems.

Besides the two differential operators, the gradient and divergence, we will have to become familiar with two more. These will also be intimately connected with the gradient and are the curl and the laplacian.

The laplacian operator comes about when the differential relation of Eq. (1.92) between the force field and charge density

$$\varepsilon\nabla \cdot \mathbf{E} = \rho \tag{1.95}$$

is considered. We have assumed the permittivity ε in the relation $\mathbf{D} = \varepsilon\mathbf{E}$ to be a constant (appropriate for linear, homogeneous, and isotropic medium). Often it is more convenient to use the work field instead of force field. From (1.49) we know that the electric field and potential are related by

$$\mathbf{E} = -\nabla V \tag{1.49}$$

Substituting this in (1.95), we obtain

$$\nabla \cdot \nabla V = -\frac{\rho}{\varepsilon} \tag{1.96}$$

Here we have a new operation, namely, the divergence of a gradient which in rectangular coordinates can be written as

$$\left(\frac{\partial}{\partial x}\hat{\mathbf{x}} + \frac{\partial}{\partial y}\hat{\mathbf{y}} + \frac{\partial}{\partial z}\hat{\mathbf{z}}\right) \cdot \left(\frac{\partial}{\partial x}\hat{\mathbf{x}} + \frac{\partial}{\partial y}\hat{\mathbf{y}} + \frac{\partial}{\partial z}\hat{\mathbf{z}}\right) V = -\frac{\rho}{\varepsilon} \tag{1.96a}$$

After performing the dot product, the above expression becomes

$$\left(\frac{\partial^2}{\partial x^2} + \frac{\partial^2}{\partial y^2} + \frac{\partial^2}{\partial z^2}\right) V = -\frac{\rho}{\varepsilon} \tag{1.96b}$$

where the operator in parentheses is called the *laplacian operator* in rectangular coordinates. This is a new differential equation which relates the work or potential variation at any point to the charge density at that point. It is known as *Poisson's equation* and is written using the ∇^2 notation for the laplacian operator as

$$\boxed{\nabla^2 V(x, y, z) = -\frac{\rho(x, y, z)}{\varepsilon}} \tag{1.96c}$$

The independent variables are the coordinates x, y, z of an arbitrary point. They have been explicitly written to emphasize that this is a point relationship.

At most points in space, charge does not exist. Hence, the right-hand side of Poisson's equation is zero, and (1.96c) for this case is

$$\boxed{\nabla^2 V(x, y, z) = 0} \tag{1.97}$$

This is the equation for charge-free regions, known as *Laplace's equation.* At all points in space not occupied by charge, the potential variation is restricted or confined to that given by Laplace's equation. This important equation not only describes the potential variation in electrostatics, but also describes many phenomena in engineering, chemistry, and physics. A great deal of literature is devoted to the solution of Laplace's equation in the various coordinate systems and for various boundary conditions.

Example: Heat flow A fine example of the use of the laplacian which is both simple and familiar to everybody is the description of heat flow. We know that if the surfaces of equal temperature are given by $T(x, y, z)$, the flow of heat will be at right angles to the surfaces; i.e., heat will flow along a normal from a high-temperature surface to a lower one. Heat flow H is then given by

$$\mathbf{H} = -k \nabla T \tag{1.98}$$

where k is a constant. Let us take the common situation where heat Q is conserved. If we consider a closed surface and a net amount of heat comes out of the surface, then the heat flow through that surface must be equal to the rate of decrease of heat inside; that is,

$$\oint\!\!\oint \mathbf{H} \cdot d\mathbf{A} = -\frac{dQ}{dt} = -\frac{d}{dt}\iiint \rho \, dv \tag{1.99}$$

where Q is the heat inside the surface and can be assumed to be distributed throughout the region, with ρ as heat per unit volume. Applying Gauss' divergence theorem to the left side of the above equation, we have

$$\iiint \nabla \cdot \mathbf{H} \, dv = -\frac{d}{dt}\iiint \rho \, dv \tag{1.99a}$$

Since this statement is true for volumes of any shape, we can equate the integrands and obtain

$$\nabla \cdot \mathbf{H} = -\frac{\partial \rho}{\partial t} \tag{1.99b}$$

Substituting for heat flow from Eq. (1.98), the above equation becomes

$$\nabla^2 T = \frac{1}{k}\frac{\partial \rho}{\partial t} \tag{1.99c}$$

Temperature variation is therefore described by a partial differential equation, where the left side is the laplacian of temperature and where the inhomogeneous, or source, term is proportional to the time rate of increase of heat density. The similarities between the equations of heat flow and electrostatics are striking. From a mathematical point of view we can say that electrostatics is a study of Poisson's and Laplace's equations and their solution. Once the solution to these equations is obtained, the electric field (or heat-flow vector) can be obtained simply by taking the gradient of the potential (or temperature). In the next few sections we will concentrate on solutions of such partial differential equations in various coordinate systems.

1.17 SOLUTION TO POISSON'S AND LAPLACE'S EQUATIONS

If all space would be empty except for the presence of some charge distributed over some finite volume v, the potential variation for points inside and outside the charged region would be given by Poisson's and Laplace's equations, respectively

$$\nabla^2 V(x, y, z) \begin{cases} = -\dfrac{\rho(x, y, z)}{\varepsilon} & \text{points } x, y, z \text{ inside } v \\ = 0 & \text{points } x, y, z \text{ outside } v \end{cases} \tag{1.100}$$

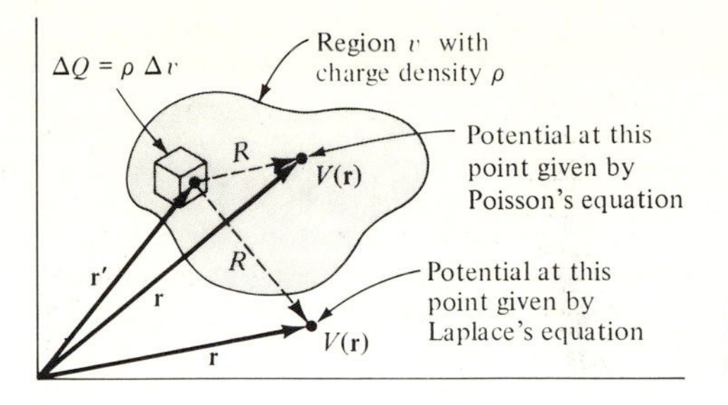

Figure 1.27 A charged region with charge density ρ surrounded by free space.

The solution to this equation is already known and was given in Eq. (1.16); that is,

$$\boxed{V(\mathbf{r}) = \frac{1}{4\pi\varepsilon} \iiint \frac{\rho(\mathbf{r}')\, dv}{|\mathbf{r} - \mathbf{r}'|}} \tag{1.101}$$

where $\mathbf{r}$ is the distance vector to the observation points x, y, z and $\mathbf{r}'$ is the distance vector to the source points x', y', z'. To verify that this is a solution to (1.100), we can substitute (1.101) into (1.100) and see if (1.101) will satisfy it. Doing this, we obtain

$$\frac{1}{4\pi\varepsilon} \iiint \rho(\mathbf{r}')\, \nabla^2 \left(\frac{1}{R}\right) dv \begin{cases} = -\dfrac{\rho(\mathbf{r})}{\varepsilon} & \text{inside } v \\ = 0 & \text{outside } v \end{cases} \tag{1.102}$$

where $R = |\mathbf{r} - \mathbf{r}'|$, and the laplacian has been moved inside the integral because it operates only on the observation point $\mathbf{r}$. Figure 1.27 shows some of the detail graphically. The laplacian of $1/R$ with $\mathbf{r}$ outside v, when integrated over v, will give zero. With $\mathbf{r}$ inside v, R will become zero when the observation and source point coincide giving a singularity which, when integrated over v, will give $-\rho/\varepsilon$ for the left side of (1.102).†

Another example of the solution to Poisson's equation, which is somewhat easier, is the case of a point charge Q located at the origin. We know the solution to this configuration to be

$$V(\mathbf{r}) = \frac{Q}{4\pi\varepsilon r}$$

To verify that this is a solution to Poisson's equation, let us integrate Poisson's equation over a volume containing the origin. Doing this, we obtain

$$\iiint \nabla \cdot \nabla V\, dv = -\frac{1}{\varepsilon} \iiint \rho\, dv = -\frac{Q}{\varepsilon}$$

† The student familiar with delta functions should note that

$$\nabla^2 \frac{1}{|\mathbf{r} - \mathbf{r}'|} = -4\pi\, \delta(\mathbf{r} - \mathbf{r}')$$

where the integral in the middle is just the total charge Q of the point charge divided by ε. Applying Gauss' divergence theorem to the left side, we can convert the volume integral to a surface integral

$$\oiint \nabla V \cdot \mathbf{dA} = -\frac{Q}{\varepsilon}$$

Considering the symmetry of this problem, we know that the electric field ($\mathbf{E} = -\nabla V$) of a point charge located at the origin must be radial. Therefore, the gradient can have only a radial component, and the above integration reduces to

$$\frac{\partial V}{\partial r} 4\pi r^2 = -\frac{Q}{\varepsilon} \quad \text{or} \quad \frac{\partial V}{\partial r} = -\frac{Q}{4\pi\varepsilon r^2}$$

which is the electric field from a point charge. Integrating, we obtain

$$V = \frac{Q}{4\pi\varepsilon r}$$

which is the same expression with which we started. Note that the method used here is very similar to that in Sec. 1.11.

The solutions that we have obtained thus far are known as free-space solutions. They are not realistic since in real-life situations charges or charged regions are usually near boundaries, such as conducting planes, dielectric regions, etc. Nevertheless, we need to know the nature of free-space solutions before we can tackle solutions to Poisson's or Laplace's equation with boundary conditions. Free-space solutions are practical since they can serve as approximations to real-life problems.

1.18 SOLUTION OF LAPLACE'S EQUATION WITH SIMPLE BOUNDARY CONDITIONS

Let us now consider three problems involving the solution of Laplace's equation in the rectangular, cylindrical, and spherical coordinate systems. Notice that in this kind of problem the region of interest is free of charge. The problem is specified strictly in terms of the potentials which exist on the surfaces (usually conductors) that bound the region of interest. Once the solution to this problem is found, the electric field can be obtained from $\mathbf{E} = -\nabla V$. The charge distribution on the bounding surfaces is then given by the boundary condition $\varepsilon E_n = \rho_s$. Since Laplace's equation is a second-order partial differential equation, the solution will have two arbitrary coefficients which must be determined from given boundary conditions of the problem.

Example: Potential variation between parallel plates Let us consider two parallel and infinite plates composed of conducting material as shown in Fig. 1.28. It is desired to find the potential variation in the space between the plates when the bottom plate located in the xy plane is at zero potential and the top plate located in the $z = z_1$ plane is at potential V_1. It will be shown in the next chapter that the surfaces of conducting bodies are equipotential surfaces so that assumptions of the

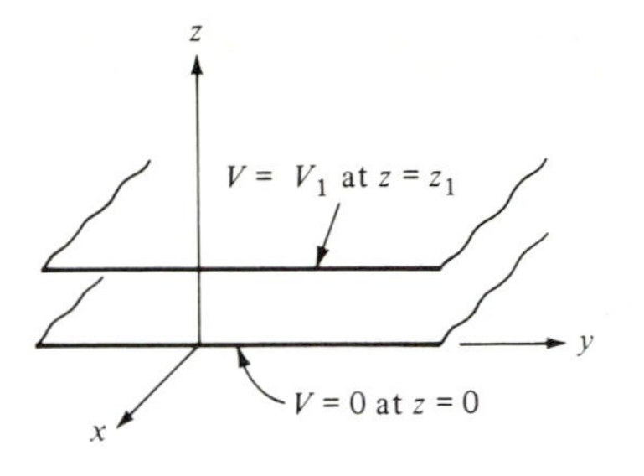

Figure 1.28 Two parallel conducting plates of infinite extent with a potential difference of V_1 volts between them.

same potentials all along the conducting planes are realistic. The region between the plates contains no charge. Hence the potential must satisfy Laplace's equation. Since the rectangular coordinate system is most suitable to describe this problem (a coordinate system is suitable to describe an object if it takes only one coordinate to do so), Laplace's equation can be written as

$$\frac{\partial^2 V}{\partial x^2} + \frac{\partial^2 V}{\partial y^2} + \frac{\partial^2 V}{\partial z^2} = 0 \tag{1.103}$$

The plates, being of infinite extent in the x and y directions, have no rate of change along these directions; that is, $\partial/\partial x = \partial/\partial y = 0$. This leaves

$$\frac{\partial^2 V}{\partial z^2} = 0 \tag{1.103a}$$

which when integrated twice gives

$$\boxed{V = C_1 z + C_2} \tag{1.103b}$$

where C_1 and C_2 are constants of integration. These must be determined from boundary conditions which are given as

$$V(z = 0) = 0 \qquad \text{and} \qquad V(z = z_1) = V_1 \tag{1.103c}$$

Applying the first boundary condition to solution (1.103*b*), we obtain

$$0 = C_1 0 + C_2 \tag{1.103d}$$

which determines constant C_2 as $C_2 = 0$. Our solution to Laplace's equation after one arbitrary constant has been determined is therefore

$$V = C_1 z \tag{1.103e}$$

Applying the second boundary condition† to determine the remaining unknown constant, we obtain $V_1 = C_1 z_1$. The solution for the potential variation between the plates, when one plate is at a higher potential than the other one, is now complete and reads as

$$V = \frac{V_1}{z_1} z \tag{1.103f}$$

The potential varies linearly from $V = 0$ to $V = V_1$ in the region between the plates. The electric field can be immediately obtained either from $dV = -\mathbf{E} \cdot \mathbf{dl}$, which gives $dV = (V_1/z_1)\, dz$ and the electric

† Notice that a sufficient number of boundary conditions are needed to determine the solution uniquely. Because a second-order differential equation has two unknown constants in its solution, two independent boundary conditions are needed.

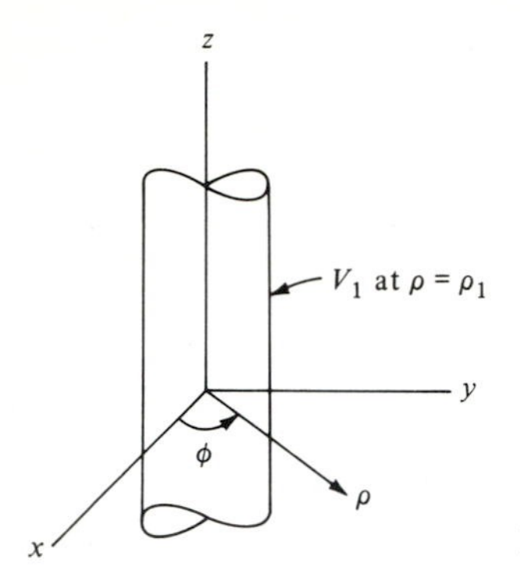

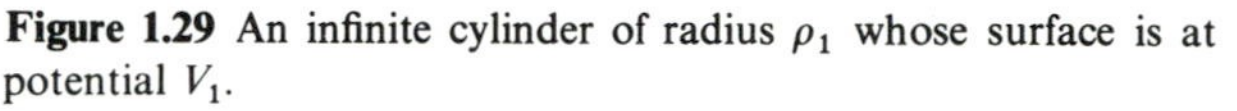

Figure 1.29 An infinite cylinder of radius ρ_1 whose surface is at potential V_1.

field as $E = V_1/z_1$, or from $\mathbf{E} = -\nabla V$, which gives $\mathbf{E} = (-\partial/\partial z)V\hat{\mathbf{k}} = (-V_1/z_1)\hat{\mathbf{k}}$. The electric field is therefore normal to the plates and has a constant value V_1/z_1.

Example: Potential variation inside and outside a hollow cylinder Let us now consider a solution to Laplace's equation of a problem which is best described in the cylindrical coordinate system. Let us consider an infinitely long, metallic cylindrical shell of radius ρ_1 which is at potential V_1 as shown in Fig. 1.29. Note that we have only specified one boundary condition

$$V(\rho = \rho_1) = V_1 \tag{1.104}$$

If a second boundary condition cannot be obtained, we will not be able to obtain a unique solution to this problem. The cylindrical coordinate system is most suitable here since we can describe mathematically the surface of the cylinder with one coordinate. Laplace's equation in cylindrical coordinates (see back cover) reads as follows:

$$\nabla^2 V = \frac{1}{\rho}\frac{\partial}{\partial \rho}\left(\rho \frac{\partial V}{\partial \rho}\right) + \frac{1}{\rho^2}\frac{\partial^2 V}{\partial \phi^2} + \frac{\partial^2 V}{\partial z^2} = 0 \tag{1.104a}$$

ρ as used here is a radial distance from the origin in the xy plane and should not be confused with the charge density ρ. Since the cylinder is infinite in the z direction, $\partial/\partial z = 0$. Also, since the cylinder is circular, $\partial/\partial\phi = 0$. Laplace's equation for this problem simplifies to

$$\frac{1}{\rho}\frac{\partial}{\partial \rho}\rho\frac{\partial V}{\partial \rho} = 0 \tag{1.104b}$$

Discarding the $1/\rho$ term and integrating once, we obtain

$$\rho\frac{\partial V}{\partial \rho} = C_1 \tag{1.104c}$$

Dividing by ρ and integrating once more, we obtain

$$\boxed{V = C_1 \ln \rho + C_2} \tag{1.104d}$$

as the general solution to Laplace's equation in cylindrical coordinates for problems with circular symmetry which are independent of the z direction. Compare this solution to the generic solutions given in Eq. (1.36).

INSIDE What is the potential variation inside a hollow cylinder when the conducting walls are maintained at a constant potential? Applying boundary condition (1.104), we can determine one of the constants in terms of the other, say C_2 in terms of C_1, and obtain for (1.104*d*)

$$V = C_1 \ln \frac{\rho}{\rho_1} + V_1 \tag{1.104e}$$

How do we determine the other unknown constant C_1? Whenever we encounter a situation such as this, where an insufficient number of boundary conditions for the number of unknown constants are given, we should take a critical look at the problem and see whether some additional boundary conditions might not be hidden in the stated problem. In this case, if we examine the solution (1.104*e*), we find that the potential V becomes infinite at $\rho = 0$. Since we have no reason to expect the potential of a hollow cylinder to be suddenly infinite along its axis, it is only realistic to *force* a solution to be finite along the axis. Therefore, we can construct a second boundary condition by saying that

$$V(\rho = 0) = \text{a finite value} \tag{1.104f}$$

Applying this to (1.104*e*), we have no choice other than to force the constant C_1 to be zero which will remove the infinity of ln 0. Our solution for the potential inside the cylinder becomes then

$$V = V_1 \tag{1.104g}$$

It is a constant value. This solution is expected from Gauss' law since in Sec. 1.12 it was shown that the potential is constant inside a spherical shell.

OUTSIDE The potential outside with boundary condition (1.104) applied is that given by (1.104*e*). As this solution still contains the arbitrary constant C_1, it is natural to attempt to construct a second boundary condition by saying that $V = 0$ at $\rho = \infty$. However, as the cylinder is infinitely long, it is meaningless to speak of being infinitely far from an infinite structure. This was discussed in Sec. 1.7. Therefore, either we must accept solution (1.104*e*) as is, or if we can have V specified at a second surface ρ_2, then we can determine C_1. Should we be given that $V(\rho = \rho_2) = V_2$, then our solution to this problem can be stated as

$$V = \frac{V_2 - V_1}{\ln \rho_2/\rho_1} \ln \frac{\rho}{\rho_1} + V_1 \tag{1.104h}$$

Example: Potential variation from a sphere As a final example, let us consider the variation of potential in a region outside a sphere when the spherical surface is at a constant potential. Since the sphere is an object of finite extent, it represents a realistic situation unlike that of an infinite cylinder. A practical example of this problem would be a metallic sphere held at a constant potential by a battery connected between ground (earth) and the sphere, as shown in Fig. 1.30. Since the region outside this isolated sphere is assumed to be charge-free, the potential variation will be given by a solution of Laplace's equation. For a sphere the most natural coordinates to use are the spherical coordinates

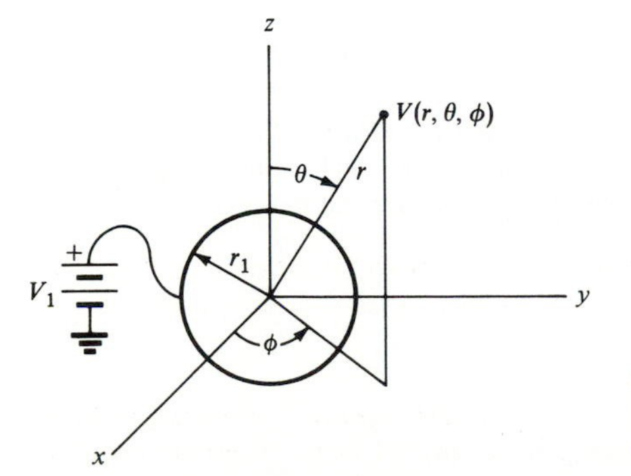

Figure 1.30 A sphere of radius r_1 held at constant potential V_1 with respect to earth. The spherical coordinates (r, θ, ϕ) are also shown.

since one coordinate, r = constant, describes mathematically the surface of a sphere. Laplace's equation in spherical coordinates (see back cover) is

$$\nabla^2 V = \frac{1}{r^2}\frac{\partial}{\partial r}\left(r^2 \frac{\partial V}{\partial r}\right) + \frac{1}{r^2 \sin\theta}\frac{\partial}{\partial \theta}\left(\sin\theta \frac{\partial V}{\partial \theta}\right) + \frac{1}{r^2 \sin^2\theta}\frac{\partial^2 V}{\partial \phi^2} = 0 \tag{1.105}$$

From the symmetry of the sphere we have $\partial/\partial\phi = \partial/\partial\theta = 0$. Laplace's equation reduces to

$$\frac{1}{r^2}\frac{\partial}{\partial r}\left(r^2 \frac{\partial V}{\partial r}\right) = 0 \tag{1.105a}$$

We can delete the $1/r^2$ term as an arbitrary multiplier and integrate the remaining expression with respect to r to obtain

$$r^2 \frac{\partial V}{\partial r} = C_1 \tag{1.105b}$$

where C_1 is a constant of integration. Dividing by r^2 and integrating once more, we have, for the general solution of the potential outside a sphere†

$$\boxed{V = -\frac{C_1}{r} + C_2} \tag{1.105c}$$

The given boundary condition here is that

$$V(r = r_1) = V_1 \tag{1.105d}$$

Applying it to (1.105*c*), we obtain

$$V = \frac{V_1 r_1}{r} + C_2\left(1 - \frac{r_1}{r}\right) \tag{1.105e}$$

To determine the second constant C_2, we have to construct a second boundary condition. As we are considering an isolated sphere (we can consider earth to which the other side of the battery V_1 is connected to be infinitely far away), a realistic assumption for a finite body such as the sphere is that the potential V must vanish as $r \to \infty$. Therefore, our second boundary condition can be stated as

$$V(r = \infty) = 0 \tag{1.105f}$$

Applying this to (1.105*e*), constant C_2 is determined as zero, and we obtain for the potential outside the sphere

$$V = \frac{V_1 r_1}{r} \tag{1.105g}$$

This solution should be compared to that obtained for a charged spherical shell (1.82). In a sense the problems are identical. The battery V_1 connected to the metallic sphere will hold the spherical surface at V_1 and will deposit a net amount of charge Q_1 on the sphere. If we look at this problem in this way, solution (1.82) applies and we can say that

$$V = \frac{V_1 r_1}{r} = \frac{Q_1}{4\pi\varepsilon r} \qquad r > r_1 \tag{1.105h}$$

† This solution should be compared to those stated in Eq. (1.36).

Therefore, the solution to Laplace's equation permits us to relate the potential to the charge on the sphere; that is,

$$V_1 = \frac{Q_1}{4\pi\varepsilon r_1} \tag{1.105i}$$

We will find this a useful relation when the capacitance of a sphere is to be determined.

The potential inside the sphere, if the sphere is hollow, can be obtained from (1.105*c*) or (1.105*e*) as $V = V_1$.

1.19 POTENTIAL MAXIMA AND MINIMA AND SOLUTIONS TO LAPLACE'S EQUATION

If we examine the solutions to Laplace's equation obtained thus far, we will notice that they are given by monotonic functions. In other words, where Laplace's equation describes the variation of the potential, the solutions contain no maxima nor minima at any points in space. That this must be the case can be easily seen if we observe that a local maxima or minima for the potential would require the presence of a negative or positive charge at such points. Mathematically local maxima and minima have the common property that at such points the first derivatives are zero. Furthermore, if the second derivative is negative (positive) at such a point, the point is a maximum (minimum) point. Since the second partials in rectangular coordinates must sum to zero in order to satisfy Laplace's equation, it follows that neither local maxima nor minima can exist unless such points in space are occupied by electric charge.

From the definition of potential it follows that for a positive point charge to be in stable equilibrium, it must be at a point of minimum potential, and a negative point charge must be at a point of maximum potential. We can now make the following important observation: A charge which is acted on by electric forces only, cannot rest in stable equilibrium in an electric field. This is known as Earnshaw's theorem. For example, if we have an arrangement of charges in space, which are not allowed to move, the potential between these charges is such that a free charge "tossed" into this space cannot find a stable point in the space between charges. The only stable points are those occupied by the fixed charges. Summarizing, we can say that there is no stable spot or point in the field of a system of fixed charges.

1.20 SUMMARY OF METHODS OF SOLUTIONS TO ELECTROSTATIC FIELD PROBLEMS

Let us examine some possible ways in which a problem in electrostatics can be formulated. If the distribution of charges is specified and it is desired to determine the electric field, we can use Eq. (1.14) or (1.16)

$$V(\mathbf{r}) = \frac{1}{4\pi\varepsilon} \iiint \frac{\rho(\mathbf{r}')\, dv}{|\mathbf{r} - \mathbf{r}'|} \tag{1.16}$$

to find the potential first and then apply the gradient operation to find the electric field; that is,

$$\mathbf{E} = -\nabla V \tag{1.49}$$

This type of problem can be classified as one for which the solution to Poisson's equation is sought (see Sec. 1.17).

If, on the other hand, the electric field is specified everywhere and the charge distribution is to be determined, we can use Gauss' law in differential form

$$\nabla \cdot \mathbf{D} = \rho \tag{1.92}$$

and the boundary condition

$$D_{n_1} - D_{n_2} = \rho_s \tag{1.87b}$$

to find the charge distribution.

Perhaps the largest class of problems falls in the category for which the solution to Laplace's equation is to be obtained. Several such problems were illustrated in Sec. 1.18. In this kind of problem the surfaces and the potentials on these surfaces are prescribed. If the objects are conductors, as is usually the case, the surfaces are equipotentials. The problem is then one of finding the potential distribution V in the free space between the conductors. Since the potential in free space obeys Laplace's equation

$$\nabla^2 V = 0 \tag{1.97}$$

this problem is simply stated as one of finding the solution to Laplace's equation subject to boundary conditions. Once the solution is found, the electric field can then be obtained from the potential by $\mathbf{E} = -\nabla V$. For completeness, let us try to formulate this problem from another point of view. Suppose we are trying to find the electric fields for the given boundary conditions. The field equations are

$$\oint \mathbf{E} \cdot d\mathbf{l} = 0 \tag{1.12}$$

and

$$\nabla \cdot \mathbf{E} = 0 \tag{1.92}$$

But since, in view of (1.61) and (1.49), these two equations are equivalent to Laplace's equation, we are back to solving Laplace's equation subject to boundary conditions.

In Sec. 1.18, solutions to Laplace's equation for simple problems were obtained. The symmetry in these problems was such that Laplace's equation simplified to a differential equation in one dimension which we were able to integrate. In the majority of problems such a simplification cannot be expected, and we have to consider solving a multidimensional partial differential equation—usually very difficult. Many mathematicians have devoted their lives to the study of such equations with the result that solutions to many special situations exist, but a general method for its solution does not exist. Perhaps one of the most general methods, in the sense that many practical problems can be solved by it, is the separation of variables technique. In this method we assume that the potential function $V(x, y, z)$ can be "separated" into a product of three functions

$$V(x, y, z) = X(x)Y(y)Z(z) \tag{1.106}$$

where X, Y, and Z are functions of only x, y, and z, respectively. The assumption of a product solution often reduces the partial differential equation (1.97) to three independent ordinary differential equations for which solutions might be readily available.†,‡,§

Another technique for solving Laplace's equation is the method of images (Sec. 2.11). For two-dimensional problems, conformal mapping, which is borrowed from the theory of complex variables, is a very effective technique to solve Laplace's equation.§,‡ Field mapping (Sec. 3.7) and analog techniques*,§ are also very effective means of obtaining solutions to Laplace's equation. Since there are so many specialized ways to obtain solutions to Laplace's equation, how can we be sure that the solution is unique? Any way that we obtain a solution, be it by guessing or by a more formalized technique, if that solution satisfies all the boundary conditions, it is the only solution possible. The uniqueness of the solution to Laplace's equation is usually proven by assuming a second solution and then showing that the difference between the first and second solutions must be zero.§

PROBLEMS

1.1 Check the dimensional correctness of

$$W = \int \mathbf{F} \cdot d\mathbf{s} \qquad E = \frac{Q}{4\pi\varepsilon r^2} \qquad V = -\int \mathbf{E} \cdot d\mathbf{s}$$

1.2 For any vector $\mathbf{A} = A_x \hat{\mathbf{x}} + A_y \hat{\mathbf{y}} + A_z \hat{\mathbf{z}}$ show that

$$A_x = \mathbf{A} \cdot \hat{\mathbf{x}} \qquad A_y = \mathbf{A} \cdot \hat{\mathbf{y}} \qquad A_z = \mathbf{A} \cdot \hat{\mathbf{z}}$$

where $\hat{\mathbf{x}}$, $\hat{\mathbf{y}}$, and $\hat{\mathbf{z}}$ are unit vectors along the x, y, and z directions, respectively, and hence that

$$\mathbf{A} = (\mathbf{A} \cdot \hat{\mathbf{x}})\hat{\mathbf{x}} + (\mathbf{A} \cdot \hat{\mathbf{y}})\hat{\mathbf{y}} + (\mathbf{A} \cdot \hat{\mathbf{z}})\hat{\mathbf{z}}$$

1.3 If vectors **A** and **B** are given by

$$\mathbf{A} = -3\hat{\mathbf{x}} + 2\hat{\mathbf{y}} \qquad \text{and} \qquad \mathbf{B} = \hat{\mathbf{x}} - 4\hat{\mathbf{y}}$$

determine $\mathbf{A} + \mathbf{B}$, $\mathbf{A} - \mathbf{B}$, $\mathbf{A} + 3\mathbf{B}$, and $2\mathbf{A}$ analytically and graphically.

† R. M. Whitmer, "Electromagnetics," 2d ed., sec. 5.2, Prentice-Hall, Inc., Englewood Cliffs, N.J., 1962.

‡ R. Plonsey and R. E. Collin, "Principles and Applications of Electromagnetic Fields," chap. 4, McGraw-Hill Book Company, New York, 1961.

§ S. Ramo, J. R. Whinnery, and T. Van Duzer, "Fields and Waves in Communication Electronics," chap. 3, John Wiley & Sons, Inc., New York, 1965.

* W. H. Hayt, "Engineering Electromagnetics," 3d ed., chap. 6, McGraw-Hill Book Company, New York, 1974.

1.4 A force is radial from the origin with a magnitude which is inversely proportional to the square of the distance from the origin (that is, $\mathbf{F} = \hat{\mathbf{r}}/r^2$). Express this force at a position (x, y) in terms of its x and y components.

1.5 A point charge $Q = 10^{-9}$ C is located at $(-0.5, -1, 2)$ in air.

(*a*) What is the magnitude of the electric field intensity at a distance of 1 m from the charge?

(*b*) Find $\mathbf{E}$ at $(0.9, 1.2, -2.4)$.

1.6 A point charge of -10^{-6} C is located at the origin (0, 0, 0) of a rectangular coordinate system. Another point charge of -10^4 C is located on the positive x axis at (1, 0, 0), that is, at a distance of 1 m from the origin. What is the force on the second charge if the intervening medium is air? What is the force if both charges are positive?

1.7 What is the force of attraction between the electron and the nucleus of the hydrogen atom, which are spaced at approximately 10^{-10} m (= 0.1 nm)? The hydrogen atom has one electron, and the nucleus has a charge equal but opposite in sign to that of the electron.

1.8 Calculate the magnitude of the repulsive force between two point charges of 1 C each and separated by a distance of 1 mm.

Answer: 9×10^{15} N. Since 10^4 N $\approx$ 1 U.S. ton, this is an extremely large force.

1.9 Two identical point charges are placed 50 mm apart in free space and exhibit a force of repulsion of 0.161 N. What is the magnitude of each charge?

1.10 Two point charges, $Q_1 = 2 \times 10^{-4}$ C located at (1, 2, 4) and $Q_2 = -10^{-4}$ C located at (2, 0, 6), are situated in air. Find the vector force $\mathbf{F}_2$ on charge Q_2. Note that $\mathbf{F}_2 = -\mathbf{F}_1$.

1.11 Two charges are arranged in the xy plane as follows: $Q_1 = 10^{-9}$ C at (0, 0), and $Q_2 = 4 \times 10^{-9}$ C at (3, 0).

(*a*) Determine the electric field $\mathbf{E}$ at the points (1, 0) and (1, 2) by determining the $\mathbf{E}$ field due to each charge and adding the results vectorially.

(*b*) Determine the potential V at (1, 0) and (1, 2). Explain why V can have a finite value at a point at which $\mathbf{E} = 0$.

1.12 Two charges are arranged in the xy plane as follows: $Q_1 = 10^{-9}$ C at (0, 1), and $Q_2 = -10^{-9}$ C at (0, −1).

(*a*) By vector addition find the $\mathbf{E}$ field at the points (1, 0) and (1, 1).

(*b*) Find the potential V at (1, 0) and (1, 1).

(*c*) Explain why $\mathbf{E}$ can have a finite value at a point at which $V = 0$.

(*d*) Find $\mathbf{E}$ and V anywhere along the x axis.

(*e*) Find $\mathbf{E}$ and V anywhere along the y axis.

1.13 Two conducting spheres of negligible diameter have masses of 0.2 g each. Two nonconducting threads, each 1-m long and of negligible mass, are used to suspend the two spheres from a common support. After placing an equal charge on the spheres, it is found that they separate with an angle of 45° between the threads.

(*a*) If the gravitational force is 980×10^{-5} N/g, find the charge on each sphere.

(*b*) Find the angle between threads if the charge on each sphere is 0.5 μC.

1.14 Four charges of 1 μC each are located in free space in a plane at (0, 0), (8, 0), (8, 6), and (0, 6). What is the potential rise from point (4, 3) to point (0, 3)?

1.15 Four charges of 1 μC each are located in free space in a plane at $(\pm 1, \pm 1)$. Find $\mathbf{E}$ at (3, 0).

1.16 A uniform line charge of ρ_L C/m is lying on the z axis and extending from −1 to +1. Determine the potential at a point (x, y, z).

1.17 A piece of straight-line charge, 1.5 m long, possesses a uniform linear charge density of $\rho_L = 2\ \mu$C/m. Find the $\mathbf{E}$ field at a point P located 1 m directly away from one end and in line with the line charge.

1.18 What are the equipotential surfaces between two concentric spheres of radius a and b, maintained at a potential difference V_0?

1.19 Two point charges, 1 μC and 2 μC, are 1 m apart in free space. Find the work required to move a third charge of 4 μC from infinity to a point midway between the two charges. Avoid a path that passes through either of the two charges.

1.20 Two charges, $+2Q$ and $-Q$, are located at (0, 1) and (0, −1), respectively. Find the line in the xy plane for which $V = 0$ and draw a graph.

1.21 If a charge of 1 μC is uniformly distributed throughout a spherical volume of radius $r = 10$ mm, what is **E** and V everywhere?

1.22 Consider an infinitely long cylindrical volume of radius a. If the charge density ρ is constant inside this volume and is zero outside, what is **E** and V at all points?

1.23 In rectangular coordinates, the volume from $z = 0$ to $z = a$ has a uniform charge density $\rho = c$. Outside this volume, $\rho = 0$. Find **E** at all points inside and outside this volume.

1.24 The charge density inside a sphere of radius a is given by $\rho = kr^2$. Find **E** and V inside and outside the sphere.

1.25 The charge density in a shell of thickness $b - a$ is given by $\rho = kr$, where k is a constant and b and a are the radii of two concentric spheres.

(a) Determine **E** and V for all values of r. Six answers will be needed.

(b) Sketch a graph of **E** and V as a function of r.

1.26 Two parallel conducting plates are spaced 1-mm apart in air. If a potential of 100 V applied across them causes a transfer of 10^{-8} C of charge, what is the area of the plates?

1.27 A hemispherical shell is uniformly charged with a surface density ρ_s.

(a) Determine the **E** field at the center of the sphere by direct integration from Coulomb's law.

(b) Determine the **E** field at the center by first calculating the potential V anywhere on the axis of symmetry and then differentiating with respect to the distance z along this axis.

1.28 A circular washer of outer radius b and inner radius a is charged to a uniform surface density ρ_s. Find the electric field on the axis of the washer at a distance z from the center.

1.29 Determine the magnitude of an **E** field which would be sufficiently strong to balance the gravitational force on an electron ($m_e = 9.1 \times 10^{-31}$ kg; $q_e = -1.6 \times 10^{-19}$ C). Find the distance a second electron would have to be positioned below the first one in order to produce the same **E** field.

1.30 Potential distributions are given by $V = 3x + 1$ V, $V = 2y^{1/2}$ V, $V = 5y^2 + 10x$ V, and $V = 4/(x^2 + y + z^2)$ V. In each case find the expressions for **E**.

1.31 What is the value of the **E** field at the surface of a flat conducting sheet which has placed on it a surface charge density of $\rho_s = 10^{-2}$ C/m^2.

1.32 An infinitely long cylindrical conductor of radius a has a charge of ρ_L C/m distributed along its length. Determine the **E** field for $r > a$

(a) by applying Gauss' law,

(b) by finding the potential V first and deriving E from $\mathbf{E} = -\nabla V$.

1.33 Derive Coulomb's law from Gauss' law making any reasonable assumption you deem necessary.

1.34 Show that Gauss' law depends on the inverse-square law.

1.35 An infinite plane sheet of charge has a surface charge density of ρ_s C/m^2. The magnitude of the E field is uniform and is $\rho_s/(2\varepsilon_0)$ on both sides of the sheet, directed normally away from the sheet. Show that Gauss' law is satisfied for any spherical surface whether or not it cuts the sheet.

1.36 A charge Q is uniformly distributed in a half-circular ring of radius a. Determine E and V at the origin.

1.37 A charge Q is distributed along a finite line whose end points are located at (0, 0) and (l, 0) in the xy plane. Find the variation of V along the x axis when $x > 0$, for the case when the linear charge density is given by $\rho_L = x^2$.

1.38 A charge Q is uniformly distributed over a plane square whose sides have length l. Find the expressions for E and V as a function of distance along the normal axis of symmetry to the square.

1.39 Three cylindrical and concentric shells of charge have a line charge uniformly distributed on each cylinder. If the inner cylinder of radius $r = 2$ m has a line charge $\rho_L = \rho_s 2\pi r = 10\pi$ C/m, the middle cylinder has $r = 4$ m and $\rho_L = -4\pi$ C/m, and the outer cylinder has $r = 6$ m and $\rho_L = -6\pi$ C/m, find **D** at $r = 1$ m, 3 m, 5 m, and 7 m.

1.40 Find the gradient and laplacian of a scalar field which varies as

(a) $1/r$ in two dimensions.

(b) $1/r$ in three dimensions.

1.41 Find the divergence of the following vector functions:

(*a*) $\mathbf{A} = 2y\,\hat{\mathbf{x}} + z\,\hat{\mathbf{y}} + xy\,\hat{\mathbf{z}}$

(*b*) $\mathbf{B} = \sin \alpha z\,\hat{\mathbf{x}} + \cos \alpha x\,\hat{\mathbf{y}} + y\,\hat{\mathbf{z}}$

(*c*) $\mathbf{C} = r^{-1}\,\hat{\mathbf{r}} + 2 \sin\theta \sin\phi\,\hat{\boldsymbol{\theta}} - \sin\theta \sin\phi\,\hat{\boldsymbol{\phi}}$

(*d*) $\mathbf{D} = r^{1/2}\,\hat{\boldsymbol{\phi}} + \cos\theta \sin\phi\,\hat{\boldsymbol{\theta}} + r^2\,\hat{\mathbf{r}}$

(*e*) $\mathbf{E} = \sin\phi\,\hat{\boldsymbol{\rho}} + \cos^2\phi\,\hat{\boldsymbol{\phi}} + z^2\,\hat{\mathbf{z}}$

1.42 If the potential is expressed as

$$V = k \sin ax \sin by\, e^{\alpha z} \qquad \alpha = (a^2 + b^2)^{1/2}$$

where a, b, and k are constants, find the $\mathbf{E}$ field and charge density ρ as a function of x, y, and z.

1.43 If a sphere of radius a has

(*a*) a uniform charge density $\rho = k$

(*b*) a charge density $\rho = kr$

(*c*) a charge density $\rho = kr^3$

find $\mathbf{D}$ and $\nabla \cdot \mathbf{D}$ as a function of radius and sketch your results. k is a constant.

1.44 Determine $\mathbf{D}$ at (3, 0, 0) if there is a point charge $Q = 2$ C at (2, 0, 0), and a line charge with $\rho_L = 2$ C/m along the z axis.

1.45 To Prob. 1.44 add a surface charge $\rho_s = 10$ C/m^2 in the plane $x = 9$.

1.46 If a charge distribution is given by $\rho = \rho_0(r/a)^{1/2}$ in the spherical coordinate system, find the total charge that is contained in a sphere of radius a, and find $\mathbf{D}$ at $r = a$.

1.47 Referring to Fig. 1.21, what value of point charge at the center of the concentric shell arrangement would cause $\mathbf{D}$ to be zero for $r > a_3$ if the charge density of shell 1 is 5 C/m^2, of shell 2 is 2 C/m^2, and of shell 3 is -1 C/m^2?

1.48 A sphere of radius a has a charge density which varies as $\rho = \rho_0 r^2$ C/m^3. What value of point charge at the center of the sphere would cause $\mathbf{D}$ to be zero for $r > a$?

1.49 An infinitely long, cylindrical charge distribution of radius a has a uniform charge density ρ_0. Using Gauss' law find the $\mathbf{E}$ field and potential inside.

1.50 Show that $r^n \sin n\theta$ and $r^n \cos n\theta$ and the sum of such terms is a solution to Laplace's equation in two dimensions. n is a positive or negative integer.

1.51 (*a*) Use Poisson's equation to find the potential V in the region between two infinite parallel plates separated by a distance l. The potential on the two plates is 0 and V_0, and the space between the plates contains a space charge $\rho = kx$, where k is a constant and x is the distance measured from the grounded plate.

(*b*) Find the surface charge density on each plate.

1.52 Using Laplace's equation find the potential distribution between two concentric spherical conductors separated by free space. The inner conductor of radius a is at potential V_0, and the outer conductor of radius b is at $V = 0$.

1.53 Repeat Prob. 1.52 except that now the outer sphere is at potential V_0 and the inner sphere is at zero potential. Find $\mathbf{E}$ and V between the spheres.

1.54 Given two concentric cylinders with radii a and b ($b > a$), solve Laplace's equation and find the potential V and electric field $\mathbf{E}$ between the two cylinders if the inner cylinder is kept at potential V_0 and the outer cylinder is at zero potential.

1.55 Two infinite conducting shells that are cylindrical and concentric and that have inner and outer radii a and b, respectively, have a battery of potential V_0 connected between them (a coaxial transmission line would have such a configuration). If the space between the shells is filled with charge which distributes itself with density $\rho = k/r$, where k is a constant, find the potential distribution in the space between the cylindrical conductors.

CHAPTER

TWO

CONDUCTORS AND CHARGES

Guide to the Chapter

This chapter brings together material which at first glance appears disconnected and even out of place in a text on electromagnetics. The intent here is to strengthen the few electrostatic principles learned thus far by showing how these form an underlying base for diverse topics and applications.

A major objective of this chapter is to study current flow through a conducting media. As many dissimilar materials are conductors, we have to look at the properties of metals, dielectrics, gases, and plasmas, particularly at their behavior on the atomic scale when a current passes through. Confining the study of conductors to the metallic variety would be a poor preparation for later material on dielectric and magnetic substances. Difficulties in understanding bound versus free charge, why an electron does not leave a metal surface, why a current-carrying wire remains uncharged, polarization, magnetization, etc., can usually be traced to an inadequate preparation during the study of conduction.

The chapter can be divided into three parts. The first part deals with the usual macroscopic properties of conductors and is expected to be covered in a basic EM course. It provides a working knowledge. The second part is a study of the microscopic behavior of solid and gaseous conduction. It provides an understanding of the physical processes behind conduction. The last four sections form the last part. As is the emphasis throughout the book, the fundamentals just learned are applied to real-life situations. They are used to explain phenomena of general interest, such as thunderstorms, and to solve practical problems, such as electrostatic precipitators.

2.1 GENERAL PROPERTIES OF MATERIALS

Solids, as compared with most liquids and gases, have their molecules or atoms arranged much closer to each other. The arrangement of the atoms in some solids, such as diamond or rock salt, is highly ordered with a regularly repeating lattice pattern throughout the solid. The atomic structure of metals is in such a regular pattern, which is called a *crystalline state*. Some other solids, on the other hand, such as glass or pitch, have their atoms arranged more or less randomly. But in all solids, the atoms are tightly bound to one another. It is the interaction between the atoms which leads to the formation of bands of energy levels. These bands can be used to obtain the electrical properties which distinguish conductors, insulators, and semiconductors. In a book on semiconductors we would have to proceed at this point with a detailed study of band formation and other topics of solid-state theory. However, when the emphasis is on the macroscopic, as compared with microscopic, study of metals and dielectrics, band theory is not particularly relevant.

The atoms which occupy the lattice points in a material are neutral because the positively charged protons of the nucleus delicately balance the negative charges of the electrons surrounding the nucleus. The electrons surround the nucleus in a shell-like arrangement. It is the electrons in the outermost shell—usually referred to as valence electrons—which are of practical significance. These electrons enter into ordinary chemical reactions in materials, such as burning, rusting, melting, etc., and in conduction of current. The electrons in the innermost shells of an atom are tightly bound to the nucleus and cannot enter into ordinary chemical reactions because such reactions do not involve enough energy to overcome the strong attractive forces these electrons experience. It is only at temperatures in the millions of degrees that the innermost electrons are stripped (ionized) from their nucleus.

The outermost electrons of the atoms of some materials can form complete shells, others form incomplete shells. A completed shell with all its electrons in place is chemically inert. For example, rare gases such as helium and neon are inert because their atoms have only completed shells. Most materials have atoms in which the outermost shell is incompletely filled with electrons. The electrons in this shell are called *valence electrons* because they participate in the formation of chemical compounds. Materials which have one or two valence electrons missing in their outermost shell (which would otherwise be complete) will usually show an affinity to complete the shell. Since materials with almost completed shells show no interest in giving up electrons, such materials do not conduct electricity and are known as *insulators* or *dielectrics*. This includes the inert materials which have their outermost shells completed and hence cannot conduct current. On the other hand there are substances like the metal group which have their outermost shells highly incomplete. The one or two valence electrons making up the last shell are very loosely bound to the atoms and are readily available for conduction of current. The slightest force (i.e., electric field) will dislodge them making them available to participate in the conduction process. For example, sodium, a good conductor, has only one of eight possible electron states filled with an electron in the valence shell. The good electrical properties of metals, many electrons (at least one per atom) plus the ease of movement of each electron, result in the high conductivity of metallic substances. It should be pointed out at this time that only the negative charge of the atom is free to move. The remaining positively charged atom is immobile in crystalline structured metals or dielectrics as well as in amorphous materials such as glass. The actual charge carriers are the free electrons.

We must now consider an important mechanism which dislodges bound electrons and makes them available for conduction. We know that in any material, the atoms, molecules, and electrons are in a continual state of vibration. The energy associated with this vibration is less in cold substances than in hot ones. Temperature can be regarded as a measure of this motion. A *thermometer* is a device that interacts with the vibrational energy of the substance and measures this energy on some scale, such as Celsius, Fahrenheit, or Kelvin. If each atom in the lattice structure of a material is considered as a mathematical point, then the atom has three degrees of freedom (corresponding to the x, y, z axes) in which to vibrate. According to Boltzmann's law, each degree of freedom has a vibrational energy W proportional to the absolute temperature of the substance; that is,

$$W = \frac{kT}{2} \tag{2.1}$$

where k is the Boltzmann constant and is equal to $k = 8.62 \times 10^{-5}$ electron volts/degree on the Kelvin scale (eV/K), or $k = 1.38 \times 10^{-23}$ J/K. Thus, for a point particle with mass m moving in space with velocity v, the total energy per particle, also known as the *equipartition energy* or the *mean thermal energy*, is

$$W = \frac{mv^2}{2} = \frac{3kT}{2} \tag{2.2}$$

This law is strictly true for gas molecules only but will serve as an approximation for molecules in denser media. One can now visualize the atomic lattice of a substance as being in a continual state of vibration. The individual lattice points are "jiggling" about their center position in a random manner as shown graphically in Fig. 2.1. Some bound electrons, if the temperature is high enough and

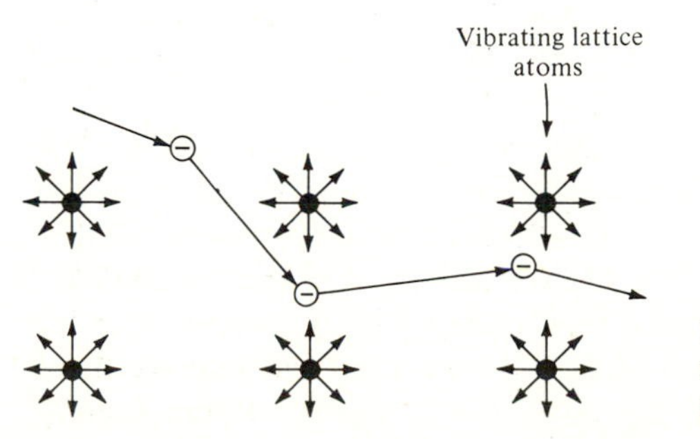

Figure 2.1 Graphical representation of electron motion through a solid. The lattice atoms are shown as vibrating about their center positions.

depending on the substance, will be knocked loose and become available for conduction. This is the reason why dielectrics, having a great deficiency of free charge carriers to begin with, become better conductors at higher temperatures. On the other hand it is well known that metallic conductors become poorer conductors of electricity as the temperature rises. The reason for this can be seen when we consider what happens to an electron in a metal as it drifts through the lattice in response to an applied electric field. At the higher temperatures the lattice points will vibrate about their center positions with larger excursions. An electron drifting through the maze of the lattice will, in effect, encounter what appear to be larger atoms. The chances for a collision are increased with a resulting decrease in the effectiveness of the electron as a current carrier. The fact that the increased vibrational energy at the higher temperature in metal might free more electrons is of no importance since in a metallic substance all valence electrons are considered free. That is, for monovalent metallic conductors the density of conduction electrons is nearly the same as the density of atoms and nearly independent of temperature. For copper the density n of conduction electrons is $n = 8.4 \times 10^{28}$ electrons/m^3. Incidentally, the density n is approximately the same for all solids.

2.2 ELECTRIC CURRENT

Before we continue with the study of conductors, let us define current. We know that if we touch a charged conductor to an uncharged one, charges will flow to the latter with the result that it will also become charged. The same thing will occur if the charged body is connected to the uncharged one by a conducting wire. Charges will now flow through the wire until the potential at the two ends of the wire becomes equalized, at which point charge motion seizes. Before the potential between the two bodies became equalized, a force field (electric field) existed on the free charges in the wire which caused the free charges to drift in the direction of this field. The number of charges which pass a given cross section of the conducting wire per unit time is called the *electric current* and is defined by

$$I = \frac{dQ}{dt} \tag{2.3}$$

The current I is the same for all cross sections of the conducting wire, even though the cross-sectional area may be different at different positions. If the rate of flow of charge with time is constant, the current does not vary with time and is simply given by $I = Q/t$. The current I is in amperes, if Q is in coulombs and t is in seconds.

Electric current, a macroscopic quantity, is the result of motion of many microscopic charges. The flow of charge at each point in the material has a direction which is that of the electric field **E**. The contribution of the individual charges to the total current can be better expressed by the current density **J**. The vector **J** at any point in the material is in the direction that a positive charge would move at that point. If there are positive and negative charge carriers in the material, they will move opposite to each other but will contribute to the current in the *same* direction. We can express J at any point in a conducting material if we consider an area element dA whose normal is in the direction of charge motion; the current density at this point is

$$J = \frac{dI}{dA} \qquad \text{A/m}^2 \tag{2.4}$$

where dI is the current that flows through dA. The current flowing through an elemental area $\mathbf{dA}$ whose normal is not necessarily parallel to the current density vector $\mathbf{J}$ is

$$dI = \mathbf{J} \cdot \mathbf{dA} \tag{2.5}$$

The total current I that flows through any macroscopic cross section is then given by the integral $I = \iint \mathbf{J} \cdot \mathbf{dA}$.

In the case of good conducting materials such as metals, we consider the conduction process to be a drifting of the cloud of free electrons inside the material, while the material remains pointwise neutral and neutral as a whole (for conduction in gases and in electrolytes, the picture of a moving charge cloud is literally true). We can then express current density in terms of the velocity of motion of the charged particles in the cloud. Let us consider Fig. 2.2 which shows an elemental volume $\Delta_{\text{vol}} = dA(v\ dt)$ of a material with a density of n electrons per cubic meter. Each electron has a charge e. If the cross-sectional area of the material is dA and the electron velocity due to the field $\mathbf{E}$ is v, then in time dt, a block of charge $v\ dt$ long and dA in area flows past a given point. The total charge in this block is $dQ = ne(dA\ v\ dt)$. The current from Eq. (2.3) is then $dQ/dt = ne\ dA\ v$, and the current density $\mathbf{J}$ can be written with the help of Eq. (2.4) as

$$\mathbf{J} = ne\,\mathbf{v} \qquad \text{A/m}^2 \tag{2.6}$$

Since $ne = \rho$ is charge density in coulombs per cubic meter, the current density can also be written as $\mathbf{J} = \rho\mathbf{v}$.

In metallic media, the slowly drifting free electrons give rise to the current density which is expressed by Eq. (2.6). The metallic medium itself remains at rest.

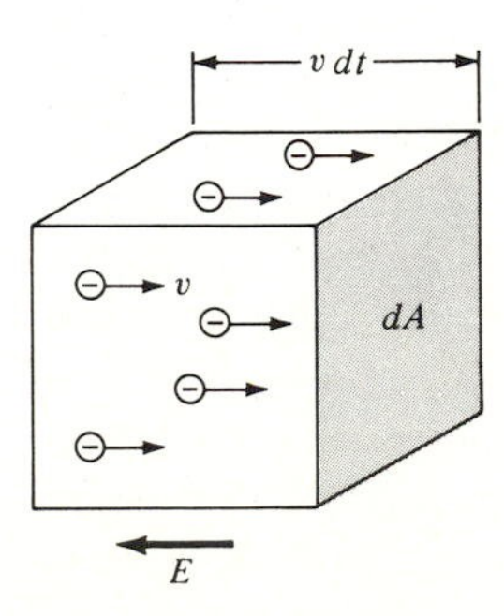

Figure 2.2 Under an applied **E** field the free electrons drift with velocity v through a block of material.

Such a current is known as a *conduction current*. The **J** of Eq. (2.6) is known as a *convection current* when it represents the motion of a charged medium as a whole. This occurs when current flows in a liquid, gas, or vacuum. For example, in highly evacuated regions such as those which exist within vacuum tubes and television picture tubes, the convection current is a beam of electrons. In gaseous discharges of high-power rectifier tubes and flourescent lamps, moving electrons and ions enter into the convection process, whereas in electrolytes, slowly moving positive and negative ions are involved.

Unfortunately, before it was known that a current in a metal is due to the motion of negatively charged electrons, the convention was adopted which gave current the same direction that positive-charge motion would possess. Therefore, the direction of current and electron motion are opposite.

Example: Let us consider a typical situation of a copper wire with a cross-sectional area A of 1 mm^2 carrying a current of 1 A. Let us calculate the average drift velocity of the conduction electrons. If we multiply the current density in Eq. (2.6) by the cross section, we obtain for the drift velocity

$$v = \frac{I}{Ane} \tag{2.7}$$

Since we know that in copper the valence-electron density is the same as the number of atoms per unit volume, which is $n = 8.4 \times 10^{28}$ free electrons/m^3, we obtain for the drift velocity

$$v = 7.4 \times 10^{-5} \text{ m/s} \tag{2.8}$$

It takes 135 s for the electrons in this wire to drift 1.0 cm. This is a very slow velocity, considering that a signal on a wire propagates with the speed of light.

Let us take an example of a wire which is 1000 km long. It will take a signal (a current pulse or a change of current on the wire) 3.3 ms to arrive at the other end, whereas if we could tag an electron at one end of the wire, it would take on the average 10^{10} s, or approximately 400 years, for the tagged electron to arrive at the other end. The explanation for the great difference in speeds is given by the fact that a signal on a wire travels as an electromagnetic wave just outside the wire.

2.3 CHARGE CONSERVATION AND THE CONTINUITY EQUATION

Since charge like matter is conserved, we must have a build-up of charge within a region if there is a net current flow into it. Similarly, if there is a net current flow out of a region, charge within the region decreases. Figure 2.3 shows a current flowing out of a region; this current must equal the time rate of decrease of positive charge within that region. Since the net outward current through the enclosing surface of a volume can be given in terms of current density **J** on the surface as $\oiint_{A(v)} \mathbf{J} \cdot \mathbf{dA}$ and since total charge inside a volume can be given in terms of charge density ρ as $Q_{\text{in}} = \iiint_{v(A)} \rho \, dv$, then

$$\oiint_{A(v)} \mathbf{J} \cdot \mathbf{dA} = -\frac{\partial}{\partial t} \iiint_{v(A)} \rho \, dv \tag{2.9}$$

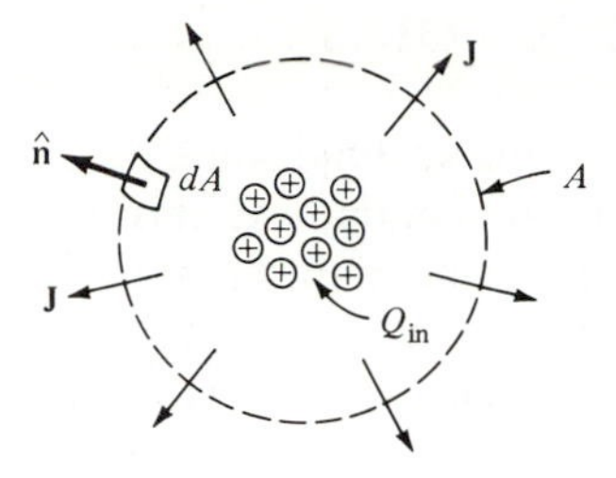

Figure 2.3 Charges inside a volume bounded by surface A can leave or enter this volume only by current flow **J** across the bounding surface A.

expresses the principle of conservation of charge. By applying the divergence theorem (1.94) to the left side of the above equation, we can write

$$\iiint_{v(A)} \nabla \cdot \mathbf{J}\, dv = -\frac{\partial}{\partial t} \iiint_{v(A)} \rho\, dv \tag{2.10}$$

Since this statement must be true for any shape of volume, the integrands can be equated, yielding the well-known continuity equation

$$\boxed{\nabla \cdot \mathbf{J} = -\frac{\partial \rho}{\partial t}} \tag{2.11}$$

which gives the relation between current density J and charge density ρ at a point. This result implies that the amount of current diverging from an infinitesimal volume element is equal to the time rate of decrease of charge contained within.

If we consider a junction or node of n wires, each carrying a steady current I_n, we can say that at the junction

$$\sum I_n = 0 \tag{2.12}$$

since a junction of wires is neither a source nor a sink for currents. This is *Kirchhoff's current law* which states that the algebraic sum of the currents at a junction is zero. If the currents are distributed, Kirchhoff's current law can be written as

$$\oint\!\!\oint \mathbf{J} \cdot \mathbf{dA} = 0 \tag{2.12a}$$

where the closed surface surrounds the junction. Applying the definition of divergence (1.89) to the above statement, we obtain the point relation

$$\nabla \cdot \mathbf{J} = 0 \tag{2.12b}$$

for any point where a steady current flows. For example, (2.12*b*) applies to any point in a conductor where current is flowing. This relation can be obtained immediately from the continuity Eq. (2.11) since for steady currents as much charge enters a volume as leaves it, so that $\partial \rho / \partial t = 0$ and (2.11) reduces to (2.12*b*).

2.4 CONDUCTIVITY AND OHM'S LAW AT A POINT

Two phenomena can be associated with the transport of charge. One is the heating of the conductor, and the other is the existence of a magnetic field about a current. The latter will be explored in Chap. 6, while the heating of the conductor will be considered next.

If the resistance between two points of a conductor is R, Ohm's law relates the current I flowing between two points to the voltage V between the points as

$$V = IR \tag{2.13}$$

where the units of R are ohms (*resistance* is one ohm if a voltage of one volt can maintain a current of one ampere). Since V is work per unit charge and I is charge per unit time, we immediately find that power is equal to $P = VI$. Power delivered and dissipated in a circuit element with resistance R is given by

$$P = IV = I^2R \tag{2.14}$$

The electric energy is dissipated and appears as thermal energy. The concept of resistance can be associated with any current-carrying device in which electric energy is converted to thermal energy. Since power P is related to energy by $P = dW/dt$, we have, for the thermal energy W dissipated in a resistor over a time interval T,

$$W = I^2RT \tag{2.15}$$

if the current and resistance remain constant over the time interval. This is known as *Joule's heating law*. Joule's experiments showed that in the presence of an applied electric field, the free electrons in a conductor travel on the average with a constant drift velocity. They accelerate between lattice centers, collide with the lattice atoms, and thereby transfer electric energy into lattice vibrational energy, which by Eq. (2.2) can be expressed as thermal kinetic energy.

Ohm's law in the form $V = IR$ cannot be used when distributed quantities such as electric fields $\mathbf{E}$ and current densities $\mathbf{J}$ are considered. However, with the concept of resistivity ρ we are able to express Ohm's law for a field point. *Resistivity* ρ of a material is given by

$$\rho = \frac{RA}{l} \quad \Omega\text{-m} \tag{2.16}$$

where ρ is the constant of proportionality relating the length l and cross-sectional area A of a piece of material to its resistance R, as shown in Fig. 2.4. Ohm's law for

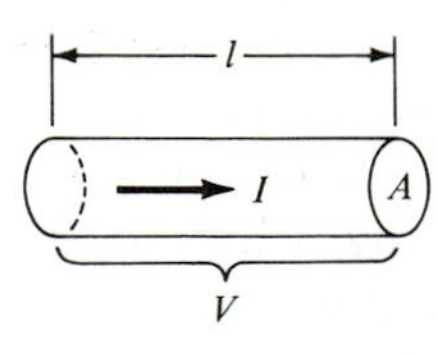

Figure 2.4 A cylindrical piece of material of cross section A and length l with potential V across it.

such a piece of material can then be written as

$$V = IR = I\frac{\rho l}{A} \tag{2.17}$$

If we rearrange this as

$$\frac{V}{l} = \rho\frac{I}{A} \tag{2.17a}$$

and observe that for small dimensions of the material, the electric field E is given by $E = V/l$ and the current density by $J = I/A$, we obtain Ohm's law as

$$E = \rho J \tag{2.17b}$$

Usually it is more convenient to use electric conductivity σ, which is the reciprocal of ρ; that is, $\sigma = 1/\rho$. Ohm's law in vector form can then be expressed as

$$\mathbf{J} = \sigma\mathbf{E} \tag{2.17c}$$

since for homogeneous and isotropic materials (which most practical materials are) the drift velocity of the charge carriers is in the same direction as that of the applied field. In the next section Ohm's law is derived rigorously from a microscopic view. The units of σ in the older system are in mhos/meter, with mho = 1/ohm. In the newer SI system of units, σ is in siemens/meter (S/m).

Values of conductivity for some common materials at a temperature of 20°C are shown in Table 2.1. We can make use of the difference in conductivity to distinguish between insulators and conductors. Although there are no perfect insulators, the insulating ability of fused quartz is about 10^{25} times as great as that of copper.

Joule's law (2.14) can now be expressed at a point by considering a differential element of cross section ΔA and length Δl, as shown in Fig. 2.4. The potential drop ΔV across the element in terms of the field is $\Delta V = \mathbf{E} \cdot \Delta\mathbf{l} = \mathbf{E} \cdot \mathbf{J}\ \Delta l/J$, where the element Δl is in the direction of the current. The power dissipated in this element is then, according to Eq. (2.14),

$$\Delta P = \Delta I\ \Delta V = J\ \Delta A\ \mathbf{E} \cdot \mathbf{J}\frac{\Delta l}{J} = \mathbf{E} \cdot \mathbf{J}\ \Delta v \tag{2.18}$$

where Δv is the volume of the differential element. The power density in watts per unit volume is then

$$\frac{\Delta P}{\Delta v} = \mathbf{E} \cdot \mathbf{J} \tag{2.18a}$$

The electric field $\mathbf{E}$ therefore gives up $\mathbf{E} \cdot \mathbf{J}$ watts of power per unit volume to the conduction current $\mathbf{J}$. Since this current flows in the conductor, the energy of the drifting electron will be transferred to the lattice by collisions which in turn will be converted into heat.

Most practical materials are isotropic and homogeneous. For those the $\mathbf{E} \cdot \mathbf{J}$ term can be expressed as σE^2 or J^2/σ because $\mathbf{E}$ and $\mathbf{J}$ have the same direction for isotropic materials.

Table 2.1 Table of conductivities

Material	Conductivity, S/m
Conductors	
Silver	6.1×10^7
Copper	5.7×10^7
Gold	4.1×10^7
Aluminum	3.5×10^7
Tungsten	1.8×10^7
Brass	1.1×10^7
Iron	10^7*
Nichrome	10^6
Mercury	10^6
Graphite	10^5
Carbon	3×10^4
Germanium	2.3
Seawater	4
Insulators	
Wet earth	10^{-3}*
Silicon	3.9×10^{-4}
Distilled water	10^{-4}*
Dry earth	10^{-5}*
Rock	10^{-6}*
Bakelite	10^{-9}*
Glass	10^{-12}*
Barium titanate	10^{-12}–10^{-13}
Rubber	10^{-15}*
Mica	10^{-15}*
Wax	10^{-17}*
Quartz (fused)	10^{-17}*

* Approximate values.

2.5 CONDUCTIVITY AND THE FREE-ELECTRON GAS MODEL OF METALS

In this section we will derive Ohm's law from a microscopic or atomic point of view. We have stated before that the valence electrons of metals are very loosely bound to their atoms and are essentially free to wander about the lattice structure of the material. For example, copper atoms have 29 electrons with 28 in tightly bound shells. The outer, or last, electron exists alone outside the closed shells and is essentially free to move about the crystalline structure of copper. Without an external field applied, these electrons are in thermal motion. They drift randomly in all directions through the conductor, with an average drift velocity of zero, making collisions with the atoms of the lattice and with one another. We may regard the free electrons as constituting a kind of electron gas, with the conductor acting as a container enclosing the gas. Many properties of interest, including

Ohm's law which relates the conduction current to the applied electric field, can be obtained from the free-electron model of metals.

As pointed out above, the motion of the conduction charges in the absence of an applied field is random, with an average velocity of zero. However, between collisions the electrons in a metal move with rather high speed. We can obtain this speed from Eq. (2.2) as 10^5 m/s at room temperature (300 K or 27°C). Actually, quantum mechanical considerations modify (2.2) so that the average speed of the conduction electrons between collisions with the lattice ions is on the order of 10^6 m/s at room temperature. This speed is much higher than a typical drift velocity of 10^{-4} m/s in conducting materials [see Eq. (2.8)]. The result of this random high-speed motion between collisions is a fluctuating current known as *Johnson noise.* Such a current can be measured between the terminals of any resistor.

Now let us see what happens when we apply a voltage across the conductor which will give rise to an electric field E within the conductor. All electrons are accelerated opposite to the direction of the field, with the acceleration given by Newton's second law as

$$\frac{dv}{dt} = \frac{F}{m} = \frac{eE}{m} \tag{2.19}$$

where e and m are the charge and mass of the electron, respectively. Integrating, we obtain for the velocity

$$v = \frac{eE}{m} t \tag{2.20}$$

This says that the velocity of the electrons increases linearly with time toward infinity, which when substituted for v in Eq. (2.6) also implies that the current in the conductor will keep on increasing indefinitely with time. Nothing could be further from the truth. Such a result violates Ohm's law which predicts a constant current. Besides, we already know that the drift velocity in conductors is an extremely small value ($\sim 10^{-4}$ m/s). What have we done wrong in the above two equations? We have ignored the additional damping force which results when electrons after being accelerated collide with the lattice. Such collisions will result in energy transfer to the lattice which can be observed as a warming of the current-carrying conductor. Let us therefore introduce the concept of *relaxation time*, denoted by τ, which is the mean free time between collisions. In other words, τ is the time required for the electron velocity to be randomized by electron-lattice collisions when electrons starting from rest acquire a drift velocity. The average drift velocity due to the applied electric field can then be written as

$$v_d = \frac{eE}{m} \tau \tag{2.21}$$

The above equation was obtained by substituting τ for t in (2.20).

The velocity given above is also the solution to an equation of motion of the type shown in Eq. (2.19) which includes a damping term; that is,

$$m\frac{dv_d}{dt}+\frac{mv_d}{\tau}=eE \tag{2.22}$$

where the term m/τ is the coefficient of friction. It should be pointed out that the mean free time τ between collisions remains unchanged after an external field is applied and the free electrons begin to drift in the direction of the external field. This is easily understood when we consider that the average thermal speed (no external field present) is on the order of 10^6 m/s and the average drift velocity along the direction of an applied field is only 10^{-4} m/s. Hence the thermal speed and τ remain essentially unchanged in the presence of an applied field.

Substituting the drift velocity (2.21) in the expression for current density (2.6), we have

$$\boxed{\mathbf{J}=\frac{ne^2\tau}{m}\mathbf{E}} \tag{2.23}$$

This shows that the current density is directly proportional to the electric field, which is Ohm's law. Comparing this to $\mathbf{J}=\sigma\mathbf{E}$ of Eq. (2.17), we obtain for the conductivity

$$\sigma=\frac{ne^2\tau}{m} \tag{2.24}$$

which is independent of the applied electric field E because the relaxation time τ is independent of $\mathbf{E}$.

From the atomic point of view, we now have a picture of a great number of electrons darting about haphazardly in all directions with great speed, but still managing to advance slowly in the direction of an applied field. The additional energy that the electrons acquire from the external field and which sets them to drift slowly goes into raising the temperature of the conductor as the electrons collide with the lattice. The mean free time τ for a typical conductor such as copper can now be computed since all the other parameters in (2.24) are known. We obtain for the relaxation time τ

$$\tau=2.4\times10^{-14}\text{ s} \tag{2.25}$$

where we have used $\sigma=5.7\times10^7$ S/m, $n=8.4\times10^{28}$ particles/m^3, the mass of the electron $m=9\times10^{-31}$ kg, and $e=-1.6\times10^{-19}$ C. Thus, the mean free time between collisions is on the order of 10^{-14} s for a conduction electron. This gives a mean free path of approximately 10^{-8} m when the average thermal speed of 10^6 m/s is used. Since an average atomic diameter is on the order of several angstroms (1 Å $=10^{-8}$ cm), the mean free path turns out to be in excess of a hundred atomic diameters. This is puzzling and indicates a difficulty with the classical free-electron gas model. In classical terms one would expect a collision to occur every few atomic diameters. The long mean free paths can be understood only in terms of quantum mechanics. There, one considers a free electron as a wave traveling through the lattice structure of the material. The lattice structure can be considered as a filter that passes the wave freely if the lattice is arranged in a perfectly regular array. As the wave picture is not based upon a lattice-electron collision mechanism, it makes the large mean free path more plausible.

2.6 THE TEMPERATURE DEPENDENCE OF CONDUCTIVITY

As was pointed out at the end of Sec. 2.1, the resistance of metallic materials increases, whereas the resistance of insulating substances decreases with tempera-

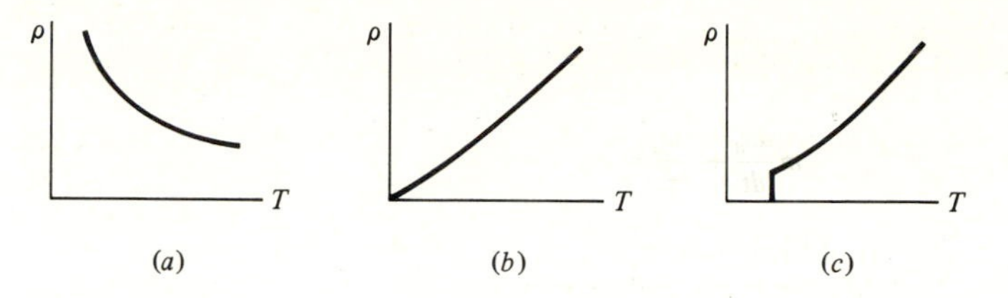

Figure 2.5 Variations of resistivity ρ for a dielectric (*a*), conductor (*b*), and superconductor (*c*).

ture.† Figure 2.5 shows the variation of the resistivity ρ versus temperature T for some typical materials. We can define a *temperature coefficient of resistivity* α as

$$\alpha = \frac{1}{\rho}\frac{\Delta\rho}{\Delta T} \tag{2.26}$$

where we have approximated the temperature variation of materials by a linear curve. The resistivity ρ is then approximately given by

$$\rho = \rho_0[1 + \alpha(T - T_0)] \tag{2.27}$$

and the temperature coefficient, which is positive (negative) for conductors (insulators), is tabulated in many handbooks ("Handbook of Chemistry and Physics," for example). The variation in conductivity can be expressed by using (2.27) and noting that $\sigma = 1/\rho$ as

$$\sigma = \frac{\sigma_0}{1 + \alpha(T - T_0)} \tag{2.28}$$

The zero subscripts indicate a temperature T_0 at which σ_0 is known. For example, copper, silver, and carbon have α in $(°\mathrm{C})^{-1}$ given by 0.0039, 0.0038, and -0.0005, respectively, at room temperature.

The curve for a conductor in Fig. 2.5*b* is shown as going to zero at absolute zero. This is only an approximation. In fact for copper a residual resistivity of 0.02×10^{-8} Ω-m exists at this temperature. However, there are materials known as superconductors which do in fact have zero resistance at temperatures close to absolute zero. Figure 2.5*c* shows the variation of ρ for such materials as, for

† In metallic conductors the density of conduction electrons is essentially independent of temperature, and the resistance increases because at the higher temperatures the lattice atoms undergo larger vibrations thus increasingly impeding the progress of the drifting conduction electrons. Any other lattice imperfections, such as impurities, will also increase the resistance. Dielectrics and some semiconducting materials have much larger resistances than metallic conductors because they have, relatively speaking, few free electrons available for conduction. When the temperature rises, the lattice vibrations will become stronger as the mean thermal energy [see Eq. (2.2)] increases. This will free more electrons and the resistivity will decrease to such a degree that it will overcome the increase in resistivity resulting from increased thermal vibrations.

example, mercury and lead whose resistivity drops very abruptly to zero at temperatures 6 and 7.2 K, respectively. Since the resistance of superconducting materials is exactly zero, currents once established in closed superconducting circuits persist indefinitely without diminution. The absence of resistance to the flow of charges also implies that superconductors cannot dissipate energy and therefore show no heat loss. This latter property is of great interest for industrial applications.

2.7 CONDUCTORS, PERFECT CONDUCTORS, AND SUPERCONDUCTORS

We have seen that the electrical conductivity depends on a supply of free electrons in a material and on the mean free time τ of the electrons. This is expressed by Eq. (2.24) which gives the conductivity as $\sigma = ne^2\tau/m$. Since metals have an abundance of free electrons, they make up the group of good conductors. When τ is large, the conduction electrons can travel large average distances before making collisions; as a result, the conductivity is large. Copper, a widely used conducting material, has a conductivity $\sigma = 5.7 \times 10^7$ $(\Omega\text{-m})^{-1}$ and a relaxation time $\tau = 10^{-14}$ s.

The picture that we now have of the metallic conductors is one in which the scattering of the free electrons by the lattice atoms, which are themselves randomly displaced by thermal agitation, brings about the definite conductivity. Theoretically speaking, if we had a regularly arranged nonvibrating lattice array of atoms, no scattering would take place and the resistance would be zero. In other words, we would have a perfect conductor. We can now see that any irregularities in the lattice, not just those due to temperature agitation, will contribute to the resistance. These include distortion of the lattice, presence of impurities, cold work, annealing, ducting, alloying, etc., all of which will increase the scattering probability.

A perfect conductor can now be defined as a material for which $\sigma \to \infty$ and $\tau \to \infty$. For a material like copper these two parameters differ greatly from infinity, yet copper and silver are treated by many as though they were perfect conductors. In a perfect conductor, the equation of motion for an electron of mass m and charge e in the presence of an electric field E does not contain a retarding term due to collisions and, hence, is given simply by Newton's second law

$$m\frac{dv}{dt} = eE \tag{2.29}$$

This equation was already discussed as Eq. (2.19). In a situation such as this, the velocity will increase indefinitely, which in turn means that the conductivity will increase indefinitely; that is, $\sigma \to \infty$. At the same time, since collisions are absent, the collision damping term mv/τ in (2.22) must vanish, which implies that $\tau \to \infty$. A material for which this happens cannot be realized physically since it would have to have completely motionless lattice atoms arranged in perfect symmetry. The uncertainty principle alone does not allow such a material to exist.

To show how impractical the dream of obtaining a perfect conductor is, we can perform the following experiment. Let us take a pure sample of our best conductors, such as silver or copper. One would expect that if the sample were chilled to absolute zero, the resistance of the sample would be reduced to zero. Ignoring the formidable task of reducing the temperature of the sample to absolute zero, we would find that the resistance is not zero but that a residual resistivity of about 0.02×10^{-8} Ω-m still exists. From quantum mechanics it is known that absolute zero temperature does not imply the absence of all motion. A motion of the lattice at this temperature still exists and is called a zero-point energy vibration. This motion—in addition to imperfections—accounts for the residual resistance. A search for perfectly conducting materials would then appear quite hopeless until superconducting materials were discovered. These materials in a sense even outdo the desirable properties of perfect conductors. They have absolute zero resistance for a range of temperatures, as shown in

Fig. 2.5c, and they expel all magnetic fields from their interior. We will not pursue superconductivity now but will leave it until we have studied magnetism and then take it up again in Sec. 9.9.

At this point we can say that a *perfect conductor* is a material that has infinite conductivity and allows static magnetic fields to exist within the material. A *superconducting material* is a perfect diamagnetic material which expels all magnetic fields from its interior and also has zero resistance.

If we consider Ohm's law

$$\mathbf{J} = \sigma \mathbf{E} \tag{2.30}$$

and define a perfect conductor as a material for which $\sigma \to \infty$, then to maintain a finite current J, we see that the electric field inside the material must vanish; that is, $E \to 0$. Since the resistance of a perfect conductor or superconductor is zero, no power is dissipated in the material. Consequently, the temperature of such a material will not change when a current flows in it. The property of superconductors that no power is lost or converted into heat has great implications in power-distribution systems because it makes lossless transmission lines possible.

2.8 THE REARRANGEMENT TIME OF FREE CHARGE IN A CONDUCTING MATERIAL

We will now derive an important property of conducting materials: A charge when placed at a point inside a conducting body will vanish from that point with a time which is inversely proportional to the conductivity σ of the material. Since the charge must reappear somewhere else, we will find that it moves to the surface and distributes itself in such a way as to produce zero field inside the conducting body. For purposes of discussion, we will consider any substance a conducting material if it has a nonzero conductivity. (In practice, materials which have a conductivity of less than 1 S/m are considered to be insulators.)

Since we are interested in what happens to a charge which by some means is placed inside a conducting body, let us begin with the continuity relation (2.11) which relates the time rate of change of charge to the resulting current; that is,

$$\nabla \cdot \mathbf{J} = -\frac{\partial \rho}{\partial t}$$

Any current that will flow as a result of a change in charge density ρ can be expressed using Ohm's law $\mathbf{J} = \sigma \mathbf{E}$. The above expression can then be written as

$$\nabla \cdot \mathbf{E} = -\frac{1}{\sigma}\frac{\partial \rho}{\partial t} \tag{2.31}$$

The left side is the divergence of the electric field which can be directly related to charge density by Gauss' law (1.92) as $\nabla \cdot \mathbf{E} = \rho/\varepsilon$. Substituting this in Eq. (2.31), we obtain a simple differential equation

$$\frac{\partial \rho}{\partial t} + \frac{\sigma}{\varepsilon}\rho = 0 \tag{2.32}$$

for the time variation of the charge density $\rho(t)$ at an arbitrary point within the conducting material. The solution of this equation is by inspection

$$\rho(t) = \rho_0 e^{-t/T} \tag{2.33}$$

where ρ_0 is the initial charge density (i.e., at time $t = 0$, a charge $\Delta Q_0 = \rho_0 \, \Delta v$ is placed at a point located inside the volume element Δv). T is the free-charge rearrangement time

$$\boxed{T = \frac{\varepsilon}{\sigma}} \tag{2.34}$$

and is sometimes referred to as the relaxation time. We have chosen rearrangement time so as not to confuse it with the relaxation time τ of (2.21) which is the mean free time of a free electron between collisions in a metal. The term "rearrangement time" is particularly suitable since as the charge vanishes from a point, it must reappear somewhere else.

Whenever a process decays exponentially, as in Eq. (2.33), we can associate with it a time constant T, which is the time it takes a process to decay to $1/e$ of its initial value. Here we shall call T the free-charge rearrangement-time constant. For example, if we consider copper or silver which have $\sigma \cong 6 \times 10^7$ S/m and a permittivity ε which is that of free space, or $\varepsilon_0 = 10^{-9}/36\pi$ F/m, we obtain for the time required for an initial charge distribution ρ_0 to decay to 36.8 percent of its initial value

$$T_{\text{Cu,Ag}} = \frac{\varepsilon}{\sigma} \cong 10^{-19} \text{ s} \tag{2.35}$$

This is an extremely short time, and we can consider that in the interval of a few time constants, i.e., about 10^{-18} s for the conducting metals, most of the charge will have vanished from any interior point of the body and will have appeared at the surface. On the other hand if we consider a dielectric such as mica, the rearrangement time is

$$T_{\text{mica}} = 10 \text{ h}$$

where we have used $\sigma \cong 10^{-15}$ as the conductivity for mica (see Table 2.1) and $\varepsilon_r = 6$ as the relative permittivity (see Table 1.1). It takes a long time for the charge to decay. For a good dielectric or insulator the rearrangement is very long, and for many practical purposes one can consider a placed charge to remain at the point of placement.

We have now another convenient way to differentiate between insulators and conductors. If a material substance has a very short rearrangement time, which means that a free charge is extremely mobile in it, the material is considered to be a conductor; whereas, if T is very long, the material behaves like an insulator. We can get an idea of the meaning of short and long by comparing the values for T in Table 2.2 for materials known to be conductors or insulators.

Should the materials be used in the presence of signals that vary with time, then the definition of a good conductor would be that the rearrangement time T must be very short in comparison to the period of the signal. We can now see that some materials which are considered to be good conductors at some frequency,

Table 2.2 Free-charge rearrangement times for some common materials

Material	Rearrangement time $T = \varepsilon/\sigma$
Copper	1.4×10^{-19} s
Carbon	3×10^{-16} s
Seawater	2×10^{-10} s
Distilled water	10^{-6} s
Mica	$\frac{1}{2}$ day
Fused quartz	50 days

tend to become insulators if the frequency is raised sufficiently. We can express this quantitatively by saying that if

$$\frac{T_{\text{rearrangement}}}{T_{\text{signal}}} \ll 1 \tag{2.36}$$

the material acts like a conductor, and if

$$\frac{T_{\text{rearrangement}}}{T_{\text{signal}}} \gg 1 \tag{2.37}$$

the material acts like an insulator, where T_{signal} is the period of the harmonic signal. For example, seawater is considered to be a conductor for frequencies below 1 GHz and an insulator for frequencies above 1 GHz.

2.9 INTERIOR AND EXTERIOR FIELDS OF CONDUCTORS AND BOUNDARY CONDITIONS

We will primarily consider metallic conductors here. From the previous section we know that if we place a charge ΔQ at some point within a conductor, it will vanish from that point with a time behavior given by (2.33)

$$\Delta Q(t) = \rho_0 \, \Delta v \, e^{-t/T} \tag{2.38}$$

where Δv is an elemental volume surrounding the point and the initially placed charge is $\Delta Q_0 = \rho_0 \, \Delta v$. For metallic conductors the decay time constant T is on the order of 10^{-18} s, which is an extremely short time. Since the charge has no place to go but to the surface, it will accumulate there in a very thin layer of surface charge ρ_s. From solid-state physics we know that the excess charge will accumulate within one or two atomic layers of the surface. The relationship between ρ_s and volume charge density ρ was already derived and given by Eq. (1.87*a*). Strong forces keep the surface charges from leaving the surface, and only in that sense are the charges inside the conductor not completely free. The picture that we now have of a conductor with some charge placed within it is one of the charge "rushing" to the surface. The inside of a metallic conductor for any time after the charge has been placed (denoted by $t > 0$, or t^+) is, therefore, charge-free; i.e., from (2.38) we can say that

$$\Delta Q(t^+) = 0 \qquad \text{within conductor} \tag{2.39}$$

From the above discussion it also follows that no electrostatic field can exist inside a conductor. Electrons will move in response to any remaining field until they have arranged themselves to produce a zero field everywhere inside the conductor; that is,†

$$\boxed{E = 0 \qquad \text{within conductor}} \tag{2.40}$$

We can ask ourselves now what the mechanism is that makes the excess charge rush to the surface. We know that a metal has so many free electrons that even a vanishingly small electric field will set large numbers of electrons in motion. Once such an electron motion is started, it must be sustained by some external source, or the motion of electrons will cease. We will see later that chemical action in a battery or magnetic fields can maintain an electric current. In our case, the charge placed within the conductor can be considered to be an external source which creates a temporary radial electric field from the point where the charge was placed. The charge is composed of many discrete particles (see Sec. 1.1) which are electrons. These electrons will respond to the field and essentially fly apart, leaving the interior charge-free and with zero electric field.

Since the electric field is zero inside, the gradient of the potential V must also be zero; that is, $\mathbf{E} = -\nabla V = 0$. This implies that the potential does not vary inside and is a constant; that is,

$$\boxed{V = \text{constant} \qquad \text{inside and on surface of conductor}} \tag{2.41}$$

where the constant is proportional to the charge on the body. The interior of the conductor is, therefore, an equipotential region, and its surface is an equipotential surface. Another way to arrive at the same conclusion is to observe that the arrangement of the excess electrons on the surface of the conductor must be such as to produce a zero tangential E field on the surface. Any other arrangement which does not result in a zero tangential field would require that the free charges on the surface move around in response to the nonzero tangential E field. An arrangement of free charge is depicted in Fig. 2.6.

Gauss' law tells us that an electric field must emanate from any region containing charge. Therefore we can conclude that an electric flux or field must leave a charged conducting surface normally, because it has been shown that the tangential electric field is zero. If we state Gauss' law for the conductor shown in Fig. 2.6 as

$$\oint\!\!\oint \mathbf{D} \cdot \hat{\mathbf{n}}\, dA = \oint\!\!\oint \rho_s\, dA \tag{2.42}$$

† Note that $E = 0$ within a conductor is a property of a conducting material and as such cannot be derived. For example, the fact that $\rho = 0$ inside a conductor does not imply $E = 0$ from Gauss' law [see relevant discussion preceding Eq. (2.50).]

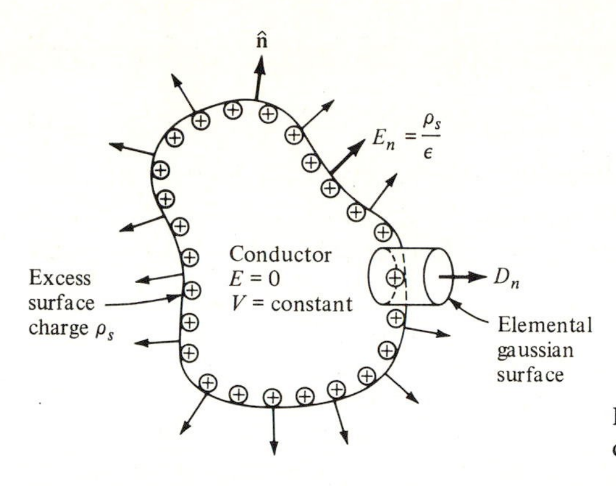

Figure 2.6 A conductor with the excess charge distributed on the surface.

then for an elemental gaussian surface we have that

$$D_{\text{normal}} = \rho_s$$

$$E_{\text{tangential}} = 0 \tag{2.43}$$

because the D field inside is zero. These are the boundary conditions for electric fields at a conducting surface. Sometimes they are written, using the normal $\hat{\mathbf{n}}$ to the surface, as $\hat{\mathbf{n}} \cdot \mathbf{D} = \rho_s$ and $\hat{\mathbf{n}} \times \mathbf{E} = 0$ (the $\hat{\mathbf{n}} \times$ operation picks out the tangential component).

Example: Let us now examine the electric field due to a charged conducting surface and due to an isolated charge layer. These are two different situations, even though at first glance it does not appear so. In the case of the former, the electric field is given from (2.43) as

$$E_n = \frac{\rho_s}{\varepsilon} \tag{2.44}$$

and in the case of the surface charge layer ρ_s, the electric field is given from (1.35) or (1.78*c*) as

$$E_n = \frac{\rho_s}{2\varepsilon}$$

Why the difference of the factor of 2? Gauss' law will be most helpful here. If we examine the derivation of Eq. (1.78*c*) and Fig. 1.20, we find that an infinite sheet of surface charge has a normal electric field on both sides of the sheet, whereas a conducting surface with a surface charge has a field in the free-space region only. Comparing the electric flux through the elemental gaussian surfaces of Figs. 2.6 and 1.20, the difference of the factor of 2 is readily apparent. Since this usually presents a perplexing situation to the student, let us examine it another way. We know that the general boundary condition for the normal components at an interface are, from (1.87*b*),

$$D_{n_1} - D_{n_2} = \rho_s \tag{1.87b}$$

where the normal points from region 2 into region 1, as shown in Fig. 1.25. Assuming region 2 is the conductor, we let D_{n_2} be zero and obtain (2.44). On the other hand, in the case of a sheet of surface charge with free space on both sides, D_{n_1} points in the direction of the normal, but D_{n_2} is opposite to $\hat{\mathbf{n}}$. Observing that the magnitudes of the flux on both sides of the sheet are the same, we obtain $D_n = \rho_s/2$. If we were to account for the remaining charges on the surface of a

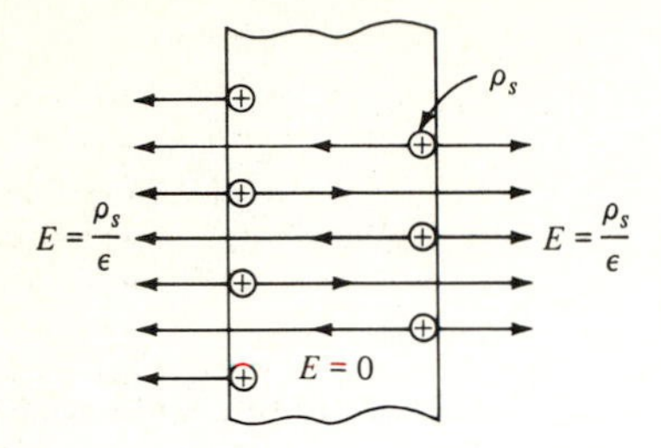

Figure 2.7 A piece of a large charged conducting sheet. The excess charge is distributed on the surface as surface charge ρ_s.

conducting body, we would see that their effect is to produce zero E field inside and ρ_s/ε field outside a metallic conductor. A simple case of this is shown in Fig. 2.7. We see that the field from one side of the surface charge will cancel the field from the charges on the other side if the observation point is within the metal, but the fields will add to provide an electric field $E = \rho_s/\varepsilon$ outside.

To complete this section we can now state Gauss' law in differential form and the continuity equation for the interior of a conductor as

$$\boxed{\nabla \cdot \mathbf{D} = 0} \tag{2.45}$$

and

$$\boxed{\nabla \cdot \mathbf{J} = 0} \tag{2.46}$$

The first equation merely states that the interior of a metallic body is pointwise neutral, which by now has been well established. Nevertheless, if we start with Gauss' law, we know that at the instant we place an excess charge, we have

$$\nabla \cdot \mathbf{D} = \rho_0|_{t=0} \tag{2.45a}$$

From the time $t = 0$ to, let us say, t^+ (where $t^+ > T$), the charges will be rearranging themselves, and so Gauss' law for this time interval reads

$$\nabla \cdot \mathbf{D} = \rho_0 e^{-t/T} \tag{2.45b}$$

It is only after $t > t^+$ that Eq. (2.45) holds and we can say the metal is pointwise neutral. Similarly, the continuity equation reduces to the special form $\nabla \cdot \mathbf{J} = 0$ only after $t > t^+$; that is,

$$\nabla \cdot \mathbf{J} = -\frac{\partial \rho}{\partial t} = \frac{\rho_0}{T} e^{-t/T} \to 0 \qquad \text{for } t > t^+ \tag{2.46a}$$

2.10 INDUCED CHARGES ON CONDUCTORS AND ELECTROSTATIC SHIELDING

The boundary condition $D_n = \rho_s$ for the surface of a conductor states that if a surface charge exists, that surface charge gives rise to an electric field on the free-space side of the boundary. The converse must also hold: If an electric field impinges on a conductor, the normal component of it will induce a surface charge

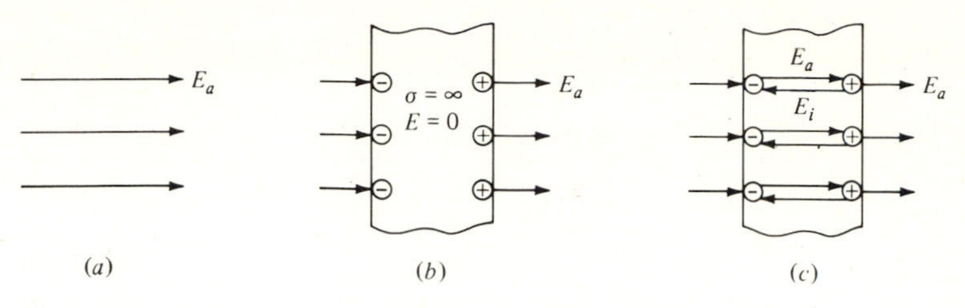

Figure 2.8 (*a*) An electric field E_a in space before anything is introduced; (*b*) a metallic slab is introduced, showing the induced charges; (*c*) the zero E field inside the conductor is shown to be composed of two fields E_a and E_i which cancel each other.

on the conductor side of the boundary. Figure 2.8 shows this graphically. A plane slab of a conducting material is introduced at right angles in an external field E_a. The applied field is apparently unaffected by the slab because on both sides of the slab the field remains the same. It acts as if the field enters the slab and then reemerges on the other side. How can this be, since no electric field can exist inside the slab? What happens is the following: The field that impinges on the left side of the slab induces negative charges there according to $D_n = \rho_s$. These charges come from the abundant supply of free electrons that exist within any metallic object. Since the slab was neutral before it was introduced in the external field, a migration of negative charges to the left side of the slab will leave many positively charged points. The only stable position for the positive charges is on the right side of the slab because we know that electric fields at interior points must be zero. Any positive charge at the interior would be neutralized by the free electrons within the rearrangement time T appropriate for that material. The positive surface charges on the right side will now be balanced by the electric field that continues to the right of the slab, as given by $\rho_s = D_n$.

Figure 2.8*c* depicts another way of looking at this. Once the negative and positive surface layers are induced, we can assume that an induced electric field E_i between these surface layers exists which balances the applied field E_a such that the total electric field in the metal is zero; that is,

$$E_{\text{metal}} = E_a + E_i = 0 \tag{2.47}$$

Therefore, in a metal the induced field is

$$E_i = -E_a \tag{2.47a}$$

The picture that we now have of a conductor introduced in an external electric field is one in which the free charges within the metal rush to the surfaces to establish an internal electric field that will exactly cancel the applied electric field and produce a zero field inside the conductor.

The Spherical Shell

Another succinct example of this is a thick conducting spherical shell that has a charge introduced at the shell center, as shown in Fig. 2.9. The charge might be placed through a small hole in the shell. If we draw a spherical gaussian surface S_i just inside the spherical shell, we will

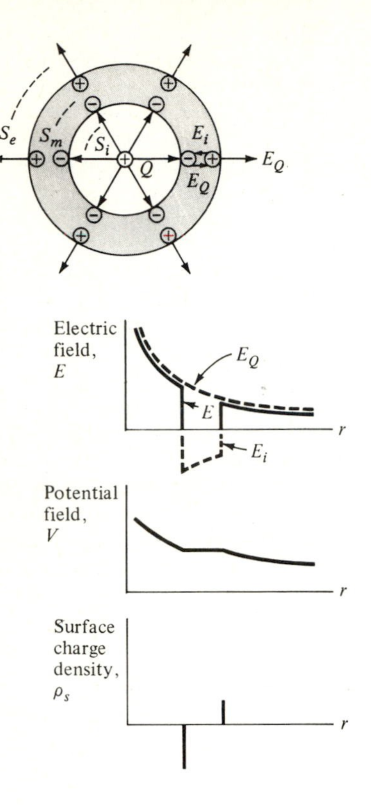

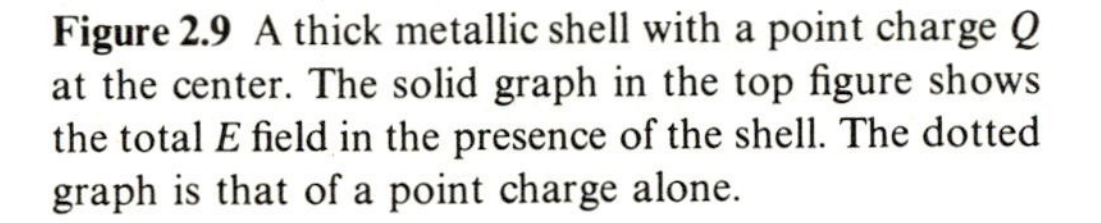
Figure 2.9 A thick metallic shell with a point charge Q at the center. The solid graph in the top figure shows the total E field in the presence of the shell. The dotted graph is that of a point charge alone.

find that the electric field is that from a point charge. This is shown in Fig. 2.9 as E_Q. The electric field E_Q will induce a negative surface layer ρ_s on the inner wall of the shell given by

$$\oiint \rho_s \, dA = |Q| \tag{2.48}$$

For a gaussian surface S_m drawn inside the metal

$$\oiint \mathbf{D} \cdot \mathbf{dA} = Q - \oiint \rho_s \, dA = 0 \tag{2.48a}$$

as by (2.40), D inside the metal is zero. The amount of charge enclosed within S_m is therefore zero. Similarly, if we consider a gaussian surface S_e exterior to the shell, we obtain

$$\oiint \mathbf{D} \cdot \mathbf{dA} = Q = \oiint \rho_s \, dA \tag{2.48b}$$

The net amount of charge enclosed is Q, which is distributed over the outer wall of the shell, and the external field is therefore that of a point charge

$$\mathbf{D} = \frac{Q}{4\pi r^2} \hat{\mathbf{r}} \tag{2.48c}$$

An interesting variation of the above problem is the case when the positive point charge Q in Fig. 2.9 is moved off center. The induced negative charge layer on the interior surface changes to a nonuniform arrangement, being denser at points of the surface that are nearest to the point charge. The nonuniform distribution can be easily depicted by a sketch in which the E lines from the point charge terminate normally on the interior surface of the conducting shell. The positive charge layer on the outside wall of the conducting shell, however, remains uniformly distributed. Therefore, the exterior field appears to originate from a point charge located at the center of the shell. What is happening here is that the induced charge on the inner surface exactly cancels the effects of the point charge for all radii greater than the inner wall radius. The positive charge on the outside of the conducting shell distributes itself uniformly over the surface as it is unaffected by the combination of point charge and inner wall charge.

We can now perform an experiment which is similar to that of Faraday's famous ice-pail experiment. If we touch the interior point charge Q to the inner wall or connect a conducting wire between Q and the inner wall, a current will flow for a brief moment until the charges on the inner wall and the point charge will be reduced to zero. Now we are left with only the outer surface layer ρ_s on the spherical shell. As far as the field external to the spherical shell is concerned, no change has taken place and the variation of the field shown in Fig. 2.9 external to the shell remains the same. This example shows that the charge that was originally on the point charge has been transferred entirely to the exterior of a conductor. If we were to connect a conducting wire between the outer shell and ground, electrons would flow from ground to the outer surface discharging it. We would be left with an uncharged structure, the point charge having been effectively transferred to ground. Note that *ground* (usually denoted by the symbol ⏚) can be considered to be an uncharged object at zero potential that has an infinite number of positive and negative charges and will always accept the excess charges of any object that is connected to it.

The Electric Field Inside a Metallic Enclosure

If we consider an empty metallic enclosure of arbitrary shape as shown in Fig. 2.10, we can show that the electric field is zero inside. Let us take a closed surface S_i just inside the inner walls of the metallic shell. Since no charge is

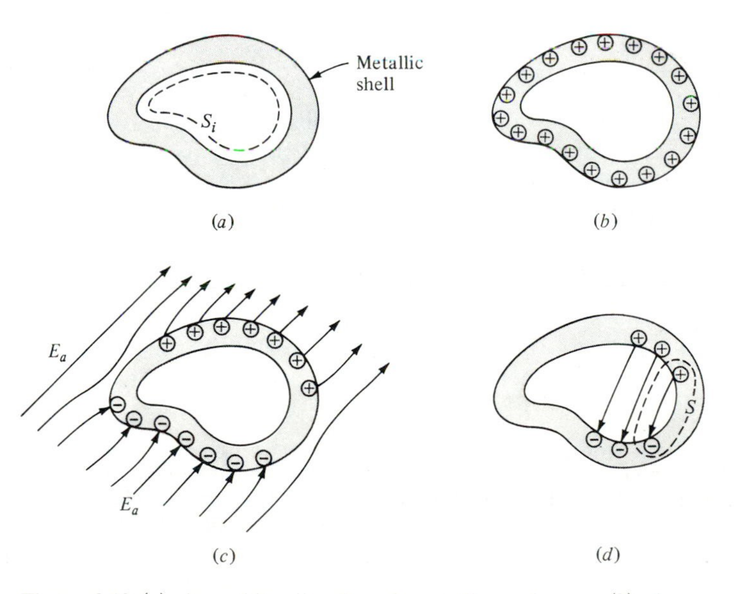

Figure 2.10 (*a*) An arbitrarily shaped metallic enclosure; (*b*) the same enclosure but now charged, with the excess charge appearing on the outside surface; (*c*) enclosure immersed in an external field; (*d*) a closed contour S partly in the metal and partly inside the cavity.

enclosed, Gauss' law for S_i becomes

$$\oiint \mathbf{D}_i \cdot \mathbf{dA} = 0 \tag{2.49}$$

and we can argue that $\mathbf{D}_i = 0$ inside the shell. It will be shown that the interior field vanishes even if the enclosure is charged, as shown in Fig. 2.10*b*, with the excess charges distributed on the exterior surface; it is also valid when the enclosure is immersed in an externally applied field E_a, as shown in Fig. 2.10*c*, where the induced charges will distribute themselves on the outside surface. We can conclude that a closed conductor, even one with small holes, such as a mesh or grid enclosure, acts as an electrical shield for apparatus that is inside the enclosure. Any external electric fields are screened out and cannot influence the zero E field in the interior of the cavity.

Let us now take a closer look at our conclusion of zero electric fields within the cavity of a metallic enclosure. If we examine Eq. (2.49), could we not say that a static field exists which is caused by positive and negative charges of like amounts which are distributed in some manner on the interior surface, for example, as shown in Fig. 2.10*d*? We would then have an electric field in the interior which starts on the positive charges and ends on the negative ones, as shown in Fig. 2.10*d*. This picture, even though it looks plausible, cannot be. It implies that the inner surface of the enclosure is not an equipotential. Since we know that conducting surfaces must be equipotential surfaces, the opposite charges will move toward each other, canceling each other and restoring the inner surface to an equipotential. Mathematically this can be shown by using Eq. (1.12) or (1.61) which state that a closed-line integral in an electrostatic field must be zero. Applying (1.12) to contour S in Fig. 2.10*d*, we find that

$$\oint \mathbf{E} \cdot \mathbf{dl} \neq 0 \tag{2.50}$$

since the portion of the contour within the metallic shell does not contribute (E is zero in a metal), but the contour along the E field going from the assumed positive charges to the negative ones does. Either we must now assume that an electric field inside the metal exists, which will cancel the electric field inside the cavity so as to reduce (2.50) to zero, or the electric field inside the cavity must vanish. From Eq. (2.40) we know that electric fields within metals must vanish. Therefore, charges on the inner walls of the empty cavity cannot exist, and we conclude that a static electric field inside a cavity cannot exist.† Now we can also understand why it is safe to be inside a metallic enclosure during a lightning storm.

Example of Two Large Conducting Plates

Let us consider two large conducting plates as shown in Fig. 2.11 and see how the induced charge distribution will change as we make some changes in the applied field and in the plates.

† Conducting shells, even if very thin, will give complete electrostatic shielding. This is not the case for time-varying fields; there, the interior of a shell is shielded only if the thickness of the shell is larger than the skin depth [see Eq. (8.76)].

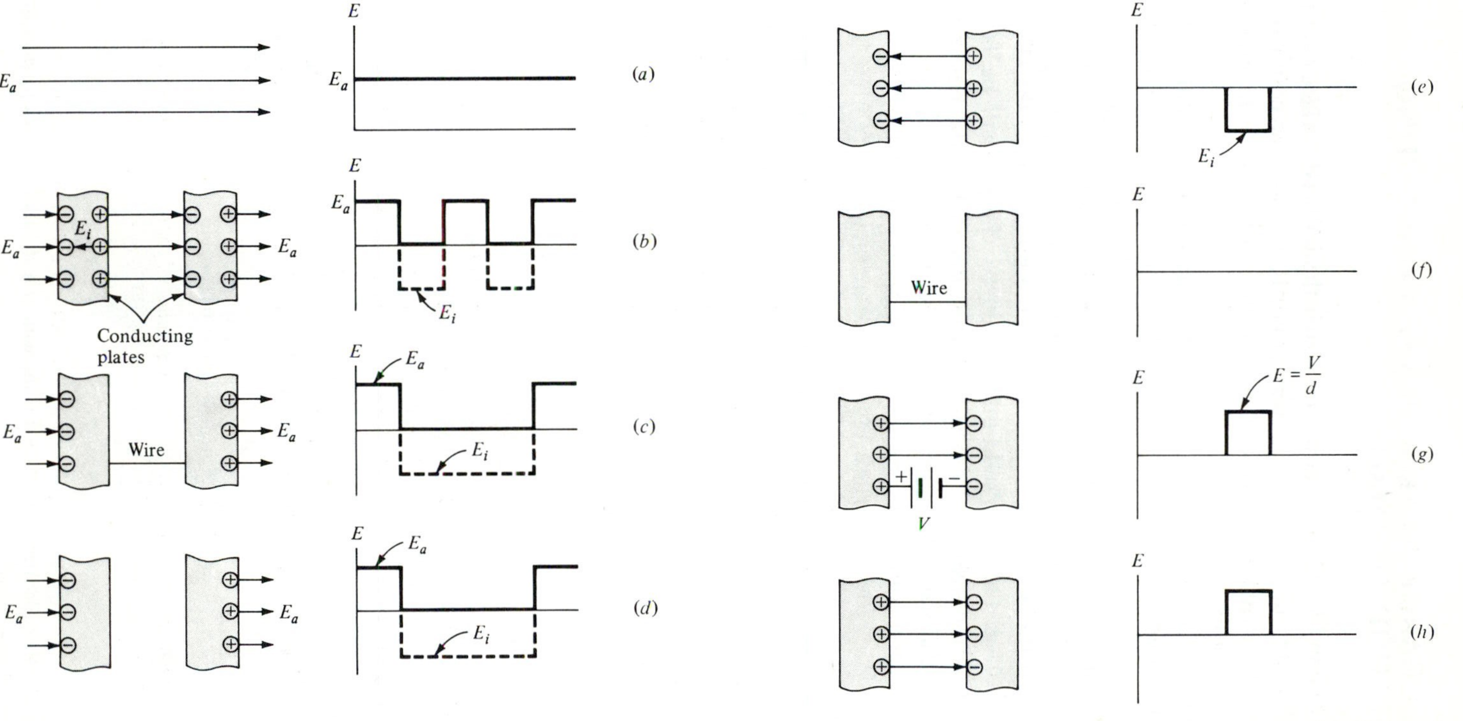

Figure 2.11 (*a*) A uniform electric field E_a in free space; (*b*) two parallel plates are introduced into the E_a field. The induced charges, the induced field E_i, and total field E are shown; (*c*) a conducting wire is connected between the inside faces, discharging them and leaving only the surface charge on the outer faces; (*d*) disconnecting the wire produces no change; (*e*) if the applied field E_a is removed, the surface charge will migrate to the inner faces of the plates, and the remaining electric field E_i is that between the surface charges; (*f*) connecting a wire again between the inner faces will discharge the surface layer of charges; (*g*) removing the wire and connecting a battery of voltage V between the plates (of separation d) will charge the plates, producing an electric field between the plates; (*h*) removing the battery, we find that the electric field between the plates remains. The plates act like a charged capacitor.

2.11 A POINT CHARGE NEAR A PLANE CONDUCTING SURFACE—METHOD OF IMAGES

Until now, we have considered conductors for which the induced surface charge distribution was known; in most cases it was uniform. Let us now consider a problem for which the induced surface charge distribution is unknown and will turn out to be nonuniform.

Figure 2.12 shows a point charge Q located a distance d above an infinite initially uncharged flat conducting plane, usually referred to as a *ground plane*. We wish to find the fields in the space above the plane as well as the charge distribution on it. Since the flat conducting plane is an equipotential surface, the electric field from the point charge must terminate on the plane normally, inducing negative surface charges at the points of termination. At the points where the charge Q is close to the surface, the magnitude of the induced charges will be large, tapering off to zero as the distance from Q increases. Recall that the boundary conditions for metallic surfaces are: Tangential E is zero, the surface is an equipotential, and the normal component of the flux is related to the induced surface charge by $D_n = \rho_s$. The induced surface charges are attracted by Q from large distances within the conductor and are held in position by the normal electric field from the point charge. No repulsive forces along the plane are experienced by the induced charges since $E_{\text{tangential}}$ is zero. Of course if the point charge Q were to be removed, the conducting surface would cease to be an equipotential, the surface charges would "fly apart" leaving the surface again charge-free and thus reestablishing the surface as an equipotential.

Figure 2.12 shows a sketch of the force lines between Q and the induced charges. To obtain a rigorous solution to this problem, we would have to solve

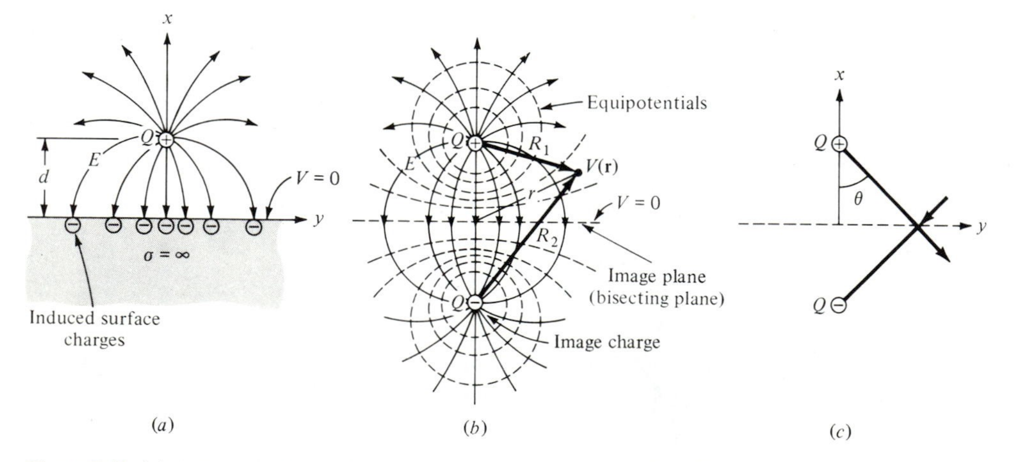

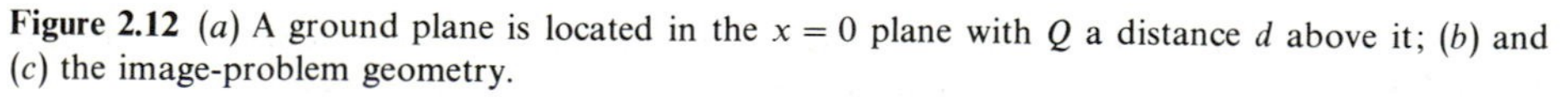
Figure 2.12 (*a*) A ground plane is located in the $x = 0$ plane with Q a distance d above it; (*b*) and (*c*) the image-problem geometry.

Poisson's equation subject to boundary conditions on the plane, i.e., solve

$$\nabla^2 V = -\frac{\rho}{\varepsilon}$$

with boundary condition $V = 0$ for $x = 0$. The electric field could then be obtained from $\mathbf{E} = -\nabla V$, and the induced surface charge distribution from $\rho_s = D_n$. The explicit expression for the induced surface charge in terms of the potential would be

$$\rho_s = -\varepsilon \left.\frac{\partial V}{\partial x}\right|_{x=0} \tag{2.51}$$

Solving the problem this way could turn out to be a formidable task. Fortunately, there is a class of problems which can be solved by a simple trick called the *method of images*, which applies to problems with plane and spherical boundaries.

If we examine Fig. 2.12*b* which shows the electric field arising from two point charges of equal magnitude but opposite sign, we will notice that the electric field is normal to the perpendicular bisecting plane. The bisecting plane is also at zero potential because it is halfway between the two charges. This means that we could insert a metallic sheet along this plane without disturbing the electric field since the boundary conditions for the conducting sheet would be automatically satisfied. As a matter of fact, we could insert metallic sheets, appropriately curved and at the appropriate potential, along any of the equipotential surfaces shown by the dashed lines in Fig. 2.12*b* without the electric field sensing their presence. It is clear now that Fig. 2.12*a* and the upper portion of Fig. 2.12*b* are alike; both have the same equipotentials and the same electric field lines. Therefore, we can replace the problem of Fig. 2.12*a* with its equivalent shown in Fig. 2.12*b*, which is much simpler. Herein lies the value of the image method: The $+Q$ charge and the $-Q$ "image charge" form an arrangement whose solution can be written down without difficulty. The potential at a point $\mathbf{r}$ is simply given by

$$V(\mathbf{r}) = \frac{Q}{4\pi\varepsilon R_1} + \frac{-Q}{4\pi\varepsilon R_2} \tag{2.52}$$

where $R_{1,2} = [(x \mp d)^2 + y^2]^{1/2}$. This expression and Eq. (2.51) can now be used to obtain the induced surface charge. Since performing the derivative operations could still become tedious, let us employ another trick to obtain the surface charge in a "gentlemanly" manner.† To calculate the surface charge, we need to know the electric field along the $x = 0$ symmetry plane only. Referring to Fig. 2.12*c*, we observe that the normal component of the electric field at a point y in the symmetry plane is twice that of the normal component of a single point charge; that is,

$$E_n = -2\left(\frac{Q}{4\pi\varepsilon R^2}\right)\cos\theta = \frac{-Q}{2\pi\varepsilon R^2}\frac{d}{R} \tag{2.53}$$

† It is said by some knowledgeable people that some parts of applied mathematics are like a bag of tricks. A skilled mathematician is one who knows how and when to use the right trick.

where $R = (d^2 + y^2)^{1/2}$. The induced surface charge is then

$$\rho_s(y) = \varepsilon E_n(y) = -\frac{Qd}{2\pi(d^2 + y^2)^{3/2}} \tag{2.54}$$

Of course, the induced charge distribution is circularly symmetric about the x axis, which can be explicitly expressed by replacing y by $r = (y^2 + z^2)^{1/2}$.

We can now perform an interesting check on our work by integrating the induced charge distribution over the entire plane; that is,

$$\iint \rho_s \, dA = -\frac{Qd}{2\pi}\int_0^\infty \int_0^{2\pi} \frac{r\,d\phi\,dr}{(d^2 + r^2)^{3/2}} = -Qd\int_0^\infty \frac{r\,dr}{(d^2 + r^2)^{3/2}}$$

$$= \left.\frac{Qd}{(d^2 + r^2)^{1/2}}\right|_0^\infty = -Q \tag{2.55}$$

where we have used polar coordinates r and ϕ, for which a surface area element in the zy plane is given by $dA = dz\,dy = r\,d\phi\,dr$. We find that the total induced charge is equal to $-Q$, which is as expected since the total induced charge should equal the image charge. As pointed out before, the combination of the fictitious image charge and the real charge Q gives the correct field in the upper half space. The contribution of the image charge to the field is equivalent to the contribution of the surface charge distribution ρ_s, given by Eq. (2.54).

2.12 EXAMPLE: FORCES THAT KEEP AN ELECTRON FROM ESCAPING A METAL SURFACE

We can apply the principles of electrostatics to answer the often-asked question: Why don't charges simply leave a charged conducting surface? We have observed in the last section that a point charge brought near a conducting surface will induce a surface charge of opposite polarity. The question that now arises is whether the point charge experiences an attractive force toward the conducting surface. The answer to this question is yes, because, as was shown in the previous section, the problem of a point charge Q located a distance d above a conducting plane is equivalent to the problem of a point charge Q and an image charge $-Q$ located a distance d below the surface. From Coulomb's law we know that a force of attraction of

$$F = \frac{Q^2}{4\pi\varepsilon(2d)^2} \tag{2.56}$$

will exist between these two charges. Since the force on the point charge Q is given by $F = QE$ and since the E field from the image charge in the upper half plane is indistinguishable from that produced by the induced surface charge, Eq. (2.56) gives the attractive force toward the plane which the point charge experiences.

An alternative method for obtaining the force is to calculate the total force acting on the surface of the plane.† The force on each charged element of area is $dF = (\rho_s \, dA)E$, where ρ_s for the case of an infinite plane is given by Eq. (2.54). The E field, which at a conducting surface can only be normal to the surface, can be obtained from the boundary condition $D_n = \rho_s$. The force on each surface element of charge can then be written as

$$dF = \frac{1}{2} \frac{\rho_s^2}{\varepsilon} dA \tag{2.57}$$

Note that a factor of one-half was used in the above expression. This is because the average field that acts on the surface charge is one-half that given by $E_n = \rho_s/\varepsilon$; on one side of the charge the field is E_n, whereas on the other side (the metal) it is zero. The total force acting on the plane which is equal and opposite to the force acting on Q is then given by an integration over the entire surface. The result of such an integration would be Eq. (2.56). Notice that (2.57) also implies that any charged surface is under the influence of a pulling force which acts normal to the surface. Because ρ_s is squared in (2.57), an induced charge of any polarity, or equivalently any impinging electric field which has a normal component to the surface, will result in a pull on the surface.

The results obtained above can now be used to explain why an excess of charge on a surface does not immediately leave the surface, as would be expected because of the repulsion that exists between the individual charges. If an element of charge is removed from the surface, we see that the image force attracts it back to the surface. Similarly, for an uncharged metallic surface, removing an electron will leave a positive image charge which tends to attract the electron back to the surface. However, if sufficient energy is supplied, the charge can of course be removed from the surface. We will call this the *work function* or *barrier energy* of a metal. It consists largely of the work done against the attractive image force in removing an electron from the metal surface.

Let us explore the mechanism of how an electron can be removed from a surface. In our picture, the metallic solid consists of atoms surrounded by a cloud of free electrons. The cloud of electrons and the positively ionized atoms taken together are electrically neutral overall and pointwise.‡ Pointwise neutrality is

† This approach will be carried out in detail in Secs. 5.14 and 5.15.

‡ "Pointwise" as used here is in a macroscopic sense, as opposed to microscopic. By a macroscopic point we mean a small point in the real-world sense. Such a point, even though small (as, for example, the tip of a very sharp needle), still contains many billions of atoms. On the other hand, a microscopic point is of small enough measure so we can talk about points between the individual atoms. In any case, it does not make much sense to talk about points mathematically. One must talk about a neighborhood or an elemental volume which when shrunk to zero localizes a point. In the macroscopic, or large-scale, sense, the neighborhood of the point is very small in terms of physical dimensions (let us say, a cube with sides on the order of 0.01 mm), whereas a point in the microscopic sense could have atomic dimensions (10^{-8} cm) for the enclosing volume. Throughout this book we will always talk in the large-scale, or macroscopic, sense unless stated otherwise.

implied by Eq. (2.45) which states that for a metal, $\nabla \cdot \mathbf{D} = 0$. We know that the electrons in the cloud have energy of thermal vibration. Some of the electrons at any given time have more energy than others. The energy which is close to the maximum is known as the *Fermi energy*. The electrons which have an energy close to the Fermi energy have the best chance to escape the surface. But since a barrier energy exists for electrons at the surface, for an electron of energy equal to the Fermi level to escape from the metal surface, it has to gain additional energy, at least equal to the barrier energy. Of course electrons lying below the Fermi level have to gain even more energy in order to escape. The escape energy can be supplied by photons (photoelectric effect), by an electric field (Schottky effect), by heating, or by high electric fields (high-field emission).

The Energy Barrier

Let us now examine the energy barrier at a metallic surface in more detail.‡ We know that the energy barrier is largely made up of the macroscopic image force, which was calculated under the assumption that the metallic surface is an equipotential and is perfectly plane and smooth. But an electron is a microscopic particle, and when close to the surface, it does not see a surface that is plane and smooth. As a matter of fact, microscopically the surface is pretty difficult to locate exactly. We can say it is the last row of atoms, but since lattice distortions and irregularities (not to mention surface roughness) will exist, an uncertainty in surface location of the order of atomic diameters (10^{-8} cm) or more will exist. The macroscopic assumption of a continuum is not valid microscopically. An electron about to leave the surface will first experience the quite erratic forces of the different atoms when it is only a small number of atomic diameters from the surface. It must first overcome the irregular surface forces. Only then, when it is far from the surface in terms of atomic diameters, will the surface begin to look smooth, and the electron will then be in the region of the image forces (the term "macroscopic" which precedes image force should be clear now). Let us refer back to Fig. 2.12 and define a distance x_0, which is at least a few atomic diameters from the surface, at which an escaping electron has overcome the surface forces and such that for $x > x_0$ the surface may be considered a true equipotential plane. To remove an electron e entirely from the surface to infinity, the work necessary will be

$$W_i = -\int_{x_0}^{\infty} \frac{e^2}{16\pi\varepsilon x^2}\,dx = -\frac{e^2}{16\pi\varepsilon x_0} \tag{2.58}$$

where we have used Eq. (2.56) for the image force. The barrier energy W_B will then consist of the surface energy W_s and the image energy W_i:

$$W_B = W_i + W_s \tag{2.59}$$

Equation (2.59) can be represented graphically by sketching the potential variation inside a metal. Figure 2.13*a* shows the potential energy of an electron as a function of radial distance from an isolated nucleus. Since an isolated nucleus is a positive ion which can be treated as a point charge e, the work function of a free electron e in the vicinity of a nucleus is given by Coulomb's law as

$$W = -eV = -\frac{e^2}{4\pi\varepsilon r} \tag{2.60}$$

which is plotted in Fig. 2.13*a*. If we now consider two nuclei, the potential variation for a free electron in the vicinity of two positive ions has the form of the solid curve in Fig. 2.13*b*. The dashed curves

‡ Forces on charged conducting surfaces are again considered in Sec. 5.15.

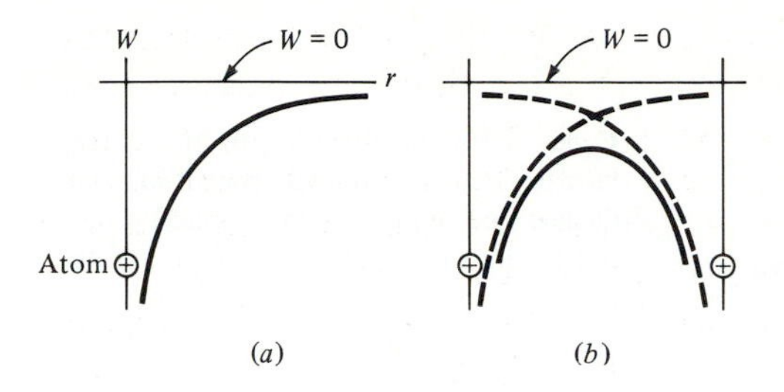

Figure 2.13 (*a*) The potential energy in the vicinity of an isolated nucleus; (*b*) the solid curve represents the potential energy variation near two nuclei. It is the combination of two dashed curves, each representing the variation of a single nucleus.

represent the variation of a single ion; combining two such curves we obtain the solid curve for two ions. Notice that the zero reference level for potential is taken as the horizontal axis in the sketches for potential. Since zero potential corresponds to an electron infinitely far from the nucleus, an electron which has reached the horizontal axis can be considered removed from the metal.†

The potential energy variation along a lattice axis inside a metal is that of an entire row of nuclei. A free electron moving along this line from one nucleus to another until it reaches the surface of the metal will experience a potential energy variation as sketched in Fig. 2.14. Within the metal, the potential energy varies in a series of humps between the nuclei and rises at the boundary. The rise comes about because there are no more nuclei to the right of the last nucleus (if we use the sketch in Fig. 2.14) to pull the potential curve down such as would happen if we were inside the metal. The last nucleus gives a potential energy variation which is that of an isolated nucleus of Fig. 2.13*a*. An electron which moves along the row of atoms inside the metal will experience the additional barrier at the surface. In order to escape the surface, it must surmount the barrier. If no additional energy is provided, the electron will bounce off the barrier and return to the interior of the metal. The picture that we now have of the electrons in a metal is that of a gas of particles free to move about inside the metal but confined to the interior by the barrier walls that exist at the metal surfaces.

Bound and Free Electrons

We can observe now that the microscopic picture of a metal is vastly different from the macroscopic one. Whereas a metal is represented as an equipotential in the large-scale, or macroscopic, sense, the atomic view shows that the potential varies greatly in the vicinities of the nuclei, approaching a value of $-\infty$ close to each nucleus. Lower energy electrons, represented by a in Fig. 2.14, are essentially

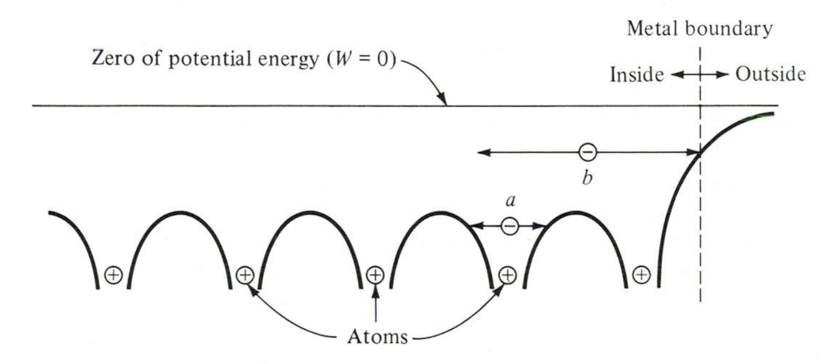

Figure 2.14 Potential energy of an electron—inside and at the boundary of a metal.

† If this sounds confusing, we should note that it really does not matter where we choose the reference level, because it is only differences of potential that have any physical significance or meaning.

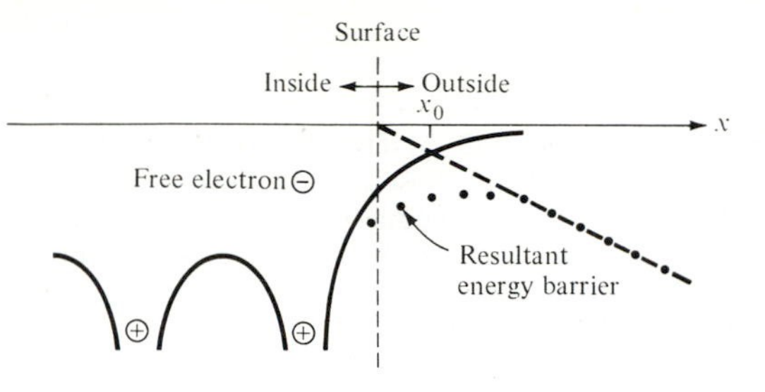

Figure 2.15 The lowering of the energy barrier by an applied potential. The dashed line represents the applied potential to the surface. The dotted line is the resultant potential energy barrier with the applied potential. The distance from the surface is given by x_0.

trapped between the humps of adjacent nuclei and are "bound." That is, electrons with energies well below the tops of the humps can only move within the "well." They cannot leave their parent atom and are called *bound electrons*. *Free electrons*, represented by b in Fig. 2.14, are those with a potential energy higher than the tops of the humps. They move freely through the lattice and contribute to the current when an external voltage is applied. When they reach the surface, they collide with the potential energy barrier and are returned to the interior. Their motion is such that they maintain charge neutrality at the interior of the metal. It should be also pointed out that the free electrons during their motion within the metal are influenced primarily by the tops of the potential energy humps, which are constant from atom to atom. These tops also occupy most of the interatomic distance. In that sense we can say that the electron is moving in a constant potential field, which can be considered as a crude approximation to the macroscopic assumption that the interior and the surface of a metal are equipotentials.

For a free electron to escape the surface, we can lower the barrier potential by applying a positive voltage to the surface, as shown in Fig. 2.15. The applied voltage tends to accelerate the electrons away from the surface. Since this opposes the retarding energy barrier, the resultant energy barrier will have a maximum after which it will decline and merge with the applied potential energy curve. The lowering of the potential energy barrier at x_0 due to an applied field is known as the *Schottky effect*. If an electron within the metal has sufficient energy to reach the point x_0, it will be emitted, because beyond x_0 it is under accelerating forces away from the surface. For ordinary applied potentials the critical distance x_0 is very large relative to interatomic distances and is well within the range of the image force law (2.58) or (2.56). The critical distance x_0 can be calculated by equating the accelerating force to the image force (2.56)

$$Ee = \frac{e^2}{16\pi\varepsilon x_0^2} \tag{2.61}$$

and solving for x_0. Thus,

$$x_0 = \frac{1}{4}\left(\frac{e}{\pi\varepsilon E}\right)^{1/2} \tag{2.62}$$

where E is the applied electric field. A typical distance is 8×10^{-8} m with an applied electric field $E = 50{,}000$ V/m.

Thermionic Emission

If the surface is heated to a high temperature, many free electrons will have energies above the barrier. An applied electric field can then draw these from the surface. The resulting current so obtained is called a *thermionic emission current*. Many practical devices are based on this principle, such as the thermionic cathode in vacuum tubes, carbon-arc lamps, and the tungsten electrode sun lamps.

High-Field Emission

On the other hand we have high-field emission upon which the cold-cathode devices are based. If we were to increase the applied accelerating field at the metal surface (in this case, usually referred to as the cathode), a value will be reached at which the barrier potential is reduced sufficiently by the Schottky effect, for the material to start emitting electrons. For tungsten this occurs at fields on the order of 10^{10} V/m. However, we should notice that as the energy barrier is lowered, the distance to point x_0 is also reduced. For a sufficiently small value of x_0, it becomes possible for electrons to "tunnel" through the barrier, with a resulting increase of emission at fields as low as 10^8 V/m. Examples of this type of emission are the mercury pool rectifiers and arc-controlled circuit breakers.

2.13 GAS BREAKDOWN AND CORONA DISCHARGE AT SHARP POINTS

Arcs and coronas are common occurrences in practical engineering whenever high electric fields are involved. As electrostatics is fundamental in an understanding of such discharges, it follows naturally that discharges be introduced at this time.

Let us first consider some macroscopic properties of conductors which are intimately related to emission of electrons at metallic surfaces. We can begin by observing that charge tends to accumulate at sharp points on a conducting body with a resulting high surface charge density at such points. Since the normal electric field E_n at a surface is related to the surface charge density ρ_s by the boundary condition $\varepsilon E_n = \rho_s$, we see that the electric field strength E will also reach high values near sharp points. If the radius of curvature of such points is sufficiently small, a glow discharge will occur there as the surrounding air is ionized by the high field. Familiar examples are glow discharges at sharp points during thunderstorms and at sharp points of a television receiver chassis. The often heard cracking sound when a television receiver is turned on is discharge between some point on the chassis and the television tube or a high-voltage power supply which carries voltages up to 30 kV. Either the arc discharge will blunt the offending point sufficiently and the arc will extinguish itself, or the power supply will be short-circuited by the very low resistance of the arc.

The reason for charge accumulation at sharp points is that the surface charge tries to spread out as much as possible over the surface of the charged conductor. A protruding sharp point, by its very nature, is farthest from most of the surface, and so charges are pushed toward it and, once there, are trapped. We can show that the electric field at points which have a small radius of curvature is high by considering a combination of a small and a large sphere, as shown in Fig. 2.16*a*. The combination is held at the same potential by a wire connection between the two spheres. Figure 2.16*a* shows a small sphere of radius a_1 carrying charge Q_1 and a larger sphere of radius a_2 with charge Q_2. The two-sphere combination can be thought of as an approximation to a conducting body of the shape shown in Fig. 2.16*b*. If the external field from the smaller sphere is found to be larger, then we can say that the field at the pointed end of the conductor in Fig. 2.16*b* will also be larger when compared with the field at the wider end. To proceed, let us find

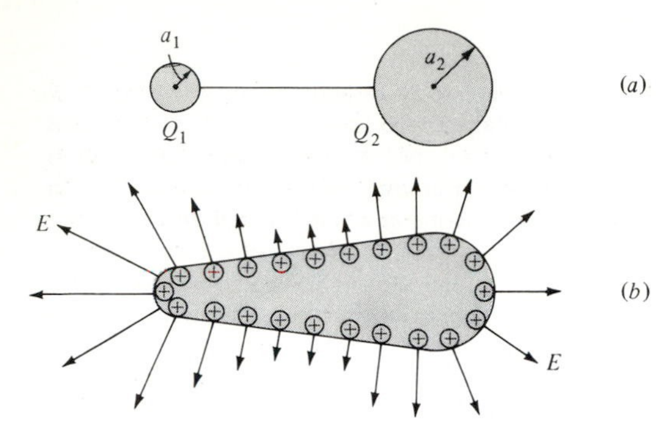

Figure 2.16 (*a*) Two spheres kept at the same potential by a conducting wire; (*b*) the electric field near sharp points on a charged conductor is very strong.

the charge densities ρ_{s1} and ρ_{s2} for the two spheres. Since the spheres are connected by a wire, the surfaces of both will be at the same potential; that is, $V_1 = V_2$. The potential in free space just outside each sphere is then given from Eq. (1.83) or (1.84*b*) as

$$V_1 = \frac{1}{4\pi\varepsilon_0}\frac{Q_1}{a_1} = \frac{1}{4\pi\varepsilon_0}\frac{Q_2}{a_2} = V_2 \tag{2.63}$$

which gives us

$$\frac{Q_1}{Q_2} = \frac{a_1}{a_2}$$

where we have assumed that the field of the wire has little influence (its primary purpose is to maintain the spheres at the same potential) and the charge distribution on both spheres is uniform (even though the presence of a charged sphere will disturb the uniform charge distribution on the other sphere somewhat). Now the surface charge density ρ_s is related to the total charge and the surface area by

$$\rho_{s1} = \frac{Q_1}{4\pi a_1^2} \qquad \text{and} \qquad \rho_{s2} = \frac{Q_2}{4\pi a_2^2} \tag{2.63a}$$

The ratio of the surface charges is then

$$\frac{\rho_{s1}}{\rho_{s2}} = \frac{Q_1}{Q_2}\left(\frac{a_2}{a_1}\right)^2 = \frac{a_2}{a_1} \tag{2.64}$$

where Eq. (2.63) was substituted in the center term. Since the free-space external field for each sphere is normal and is related to the surface charge by the boundary condition $\varepsilon_0 E_n = \rho_s$, we have for the ratio of the fields

$$\boxed{\frac{E_{n1}}{E_{n2}} = \frac{a_2}{a_1}} \tag{2.65}$$

The electric field at the smaller sphere will therefore be larger than that on the bigger sphere by the ratio a_2/a_1. We can now conclude that charge will concentrate at sharp points with the result that the electric fields will be very strong in the neighborhood of such points.

Corona Discharge†

A result of the very high electric fields near sharp points of a charged conductor will be a breakdown of the air in the immediate vicinity of the point. Let us see how this comes about. If we take a conducting body with a sharp point and begin to raise it to a high potential with respect to ground,‡ a value of potential will be reached at which the object will begin to discharge itself to the surrounding air via the point. This will be accompanied by emission of light (glow or corona) and a hissing sound. We soon will find that we cannot raise the potential of the object any further since charges from the object are "conducted" to the atmosphere as fast as they are deposited on the object. Similarly, we will find that a charged object when left alone will eventually discharge itself by leaking its charge via sharp points to the surrounding air. Now, how can air conduct charges away, since it is normally thought of as an insulator? What happens is that, at any given time, there are always some electrons and a few ions of both polarities in the atmosphere (produced by cosmic rays) which are attracted to the charged conductor, hence partially neutralizing it.

To illustrate this process, let us assume that we have a highly charged body which has a protruding tip. As atmospheric ions or electrons enter the strong field of the tip, they are accelerated to high velocities. As inevitable collisions take place between neutral air molecules and the high-speed particles, electrons will be knocked out of the neutral air molecules. As a result of this process vast numbers of additional ions and electrons are produced. Such a process is sometimes referred to as an *avalanche breakdown.* The surrounding air becomes much more conducting, with the result that the charged object will soon quickly lose most of its charge. The air in the neighborhood of the tip might even glow because of light emitted from the air molecules during these collisions.

We can now define a *corona discharge* as the leakage of charge brought about by the ionization of the air surrounding a charged conductor. The electric breakdown of the air which occurs in the corona discharge will happen at an electric field strength of about 3×10^6 V/m. Other gases have a breakdown strength either larger or smaller than that of air. Table 2.3 gives some representative values. There are some gases such as freon and nitrogen in which the free electrons can lose their energy in inelastic collisions. It is thus more difficult for the electrons to acquire enough energy to ionize the surrounding molecules. In high-voltage devices it is quite common to use freon and nitrogen to take advantage of this effect to reduce the corona discharge.

Table 2.3 Dielectric constant and dielectric breakdown strength

Material	Dielectric constant	Breakdown strength, V/cm
Air	1	3×10^4
Barium titanate	1200	7.5×10^4
Freon	1	8×10^4
Oil	2.3	1.5×10^5
Paper	3	2×10^5
Porcelain	7	2×10^5
Glass	6	3×10^5
Paraffin	2	3×10^5
Quartz (fused)	4	4×10^5
Polystyrene	2.6	5×10^5
Mica	6	2×10^6

† L. B. Loeb, "Electrical Coronas: Their Basic Physical Mechanisms," University of California Press, Berkeley, 1965.

‡ This can be done simply by connecting a high-voltage battery between ground and the object. In practice this could be done by a Van de Graaff generator which would continuously deposit charges on the object.

Example: The variation in breakdown strength for different gases is used in a simple device for checking the presence of gases other than air. For example, freon is a commonly used refrigerant in air-conditioning systems, which are commonly known to leak. Since freon has a higher breakdown strength than air, a leak detector for freon is simply a spark gap which is adjusted so that it just starts a breakdown arc in air. When this spark-gap device is brought in a region which contains freon molecules, the spark-gap arc will extinguish quickly.

Corona discharge is also very important in high-power lines. There it represents a power loss and limits the maximum voltage which can be used on power lines. The primary power loss in transmission lines is due to I^2R heating, which decreases as the voltage is increased. It is therefore desirable to use the maximum possible voltage which is corona-discharge limited.

We can conclude this section by saying that if we desire to charge an object to a high potential and not have it discharge itself in air, we must make sure that all surfaces are smooth. Only then can we be sure that at no place on the object will the field be abnormally high.

Arc Discharge†

A glow or a corona discharge which surrounds an object having a sharp point is characterized by a low current and a high potential from the point to the atmosphere. If the potential increases further, complete breakdown takes place, and a continuously ionized path (an arc or a spark) is formed to the nearest point of opposite polarity. The result is a quick discharge of the initially charged object. We can define an *arc discharge* as a type of electric conduction in gases, characterized by high currents (from amperes to thousands of amperes) and a low potential drop (a few tens of volts). An arc can, therefore, be considered essentially a short circuit. Except for the lower potential and greater current density, the ionization mechanism is similar to that in a glow discharge.

To clarify the arc mechanism further, let us assume that we have two pointed metallic objects, as shown in Fig. 2.17, which are maintained at some potential by an external battery. If the battery voltage is just high enough to cause a corona discharge, we will find that a small current flows in the closed circuit formed by the connecting wires and the region between the tips. As stated before, the ionization is started by electron impact in the high field surrounding the tip of the positive anode. The released electrons and positive ions go into forming a neutral plasma‡ which occupies the region between the tips, as suggested in Fig. 2.17. Positive ions collect at the negative cathode, and electrons at the anode. Since a current flows in the circuit, the question now arises as to where all the electrons come from to maintain the current in the region between the tips. It can be shown that they are impact-liberated electrons from fresh gas molecules streaming into the area as well as liberated by ion

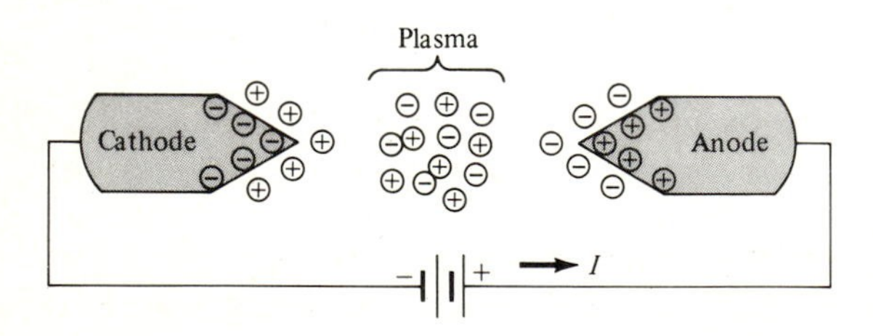

Figure 2.17 Two pointed metallic objects connected to a battery by wires. A discharge between the tips will occur for sufficiently high potentials.

† J. M. Meek and J. D. Craggs, "Electrical Breakdown of Gases," Oxford University Press, Fair Lawn, N.J., 1953.

‡ A plasma is a conducting gas which contains approximately equal concentrations of positive and negative charge plus an abundance of neutral molecules.

bombardment of the cathode which will free electrons, imparting sufficient energy to permit them to climb the potential surface barrier of the cathode material (as shown in Fig. 2.15).

Now let us see what happens as the current is increased by an increase in the battery potential until a spark jumps between the two points. First the cathode surface will become very hot, and parts of the cathode surface will be sputtered off by ion bombardment. At such a time the temperature will rise so steeply that the metal will evaporate creating ideal conditions for initiation of a plasma arc. As shown in Fig. 2.17, we can divide the resulting arc into three main parts: the regions immediately adjacent to the positive and negative electrodes, and the plasma region, which is the brightly luminous connecting arc column. Let us consider the three regions separately.

Looking at the cathode region will help answer the question of where the electrons that maintain the high-current-density arc originate. As in the glow discharge case, part of the electrons are released by positive ion bombardment of the cathode. But a more important mechanism in the release of electrons from the surface is thermionic emission (see Fig. 2.15). What happens is that the excess of positive ions in the cathode region, by continual bombardment, keep the cathode sufficiently hot for electrons to be emitted thermionically.

In the plasma region the arc column is found to be an equipotential, since the voltage gradient is uniform along the column. This indicates that the plasma region is electrically neutral. A plasma arc is, therefore, similar in many ways to a metal, since both are visualized as having a structure of heavy ions surrounded by rapidly moving electrons. In a metal the ions are in a stationary lattice structure, whereas in a plasma the heavy ions move about slowly. The plasma column that forms is created by ionization of the surrounding air, which is otherwise an excellent insulator at room temperature. That is, air becomes conducting when thermal collisions remove electrons from the outer shells of nitrogen and oxygen molecules. An air gap of 1 cm can withstand about 30,000 V at normal pressure and temperature (see Table 2.3). When air is converted into a plasma at around 5000 K, its conductivity increases by more than 13 orders of magnitude, or by a factor of more than 10 trillion. It then approaches the conductivity of carbon (see Table 2.1). This tremendous range from excellent insulator to efficient conductor cannot be reproduced by matter in any other form; hence, the plasma arc is the most effective switching medium known for circuit breakers and many forms of high-power switches.† Once an arc has been initiated, the high temperature needed to sustain the arc is provided by the current passing through the arc itself.

The anode region is one of negative space charge where electrons are accelerated in order to provide, through ionization, a supply of positive ions for the arc column. As the current increases, the anode temperature will rise which will cause vaporization of the anode material. This is sometimes accompanied by a hissing sound.

2.14 EXAMPLE: THUNDERSTORMS AND THE PRINCIPLE OF THE LIGHTNING ROD‡

Having described in detail the mechanisms of coronas and arcs, let us see how we can protect ourselves and electric equipment from damage that high-current discharges can cause. The most powerful arc is that caused by lightning. It is a discharge across a potential of up to 100 million volts which can exist between the bottom of a highly charged cloud and earth.

Let us first describe briefly some of the electrical phenomena that exist in our environment. The earth's surface, usually referred to here as ground, carries a

† W. Rieder, Circuit Breakers, *Sci. Am.*, January 1971.

‡ M. A. Uman, "Lightning," McGraw-Hill Book Company, New York, 1969. J. Alan, "Atmospheric Electricity," Pergamon Press, London, 1975. A. D. Moore, Electrostatics, *Sci. Am.*, pp. 46–58, March 1972. R. H. Golde, "Lightning Protection," Chemical Rubber Co., Inc., New York, 1975.

negative surface charge. The surface charge is induced by the electric field that exists between the top of the atmosphere and the earth surface. The potential difference between the upper part of the atmosphere and ground is about 400,000 V with an electric field of approximately 100 V/m near ground, as shown in Fig. 2.18. The reason why we do not get a shock from this potential is that the human body is a relatively good conductor. As shown in Fig. 2.18*b* where a grounded metallic rod picks up some of the negative charge of the earth and thereby raises the potential distribution, the human body acts in a similar fashion by being its own "umbrella" against the atmospheric potential. Without going into reasons why a 0.4 million volt potential between the top of the atmosphere and earth exists, let us note that in a fully developed thundercloud, the charge on the bottom of the cloud is so great as to cause a potential of up to 100 million volts to be developed between the bottom of the cloud and earth. This will give a much bigger electric field than the 0.4 million volts from the top of the atmosphere (approximately 60,000 m) to earth, considering that the bottom part of a mature thunderstorm has only an altitude of approximately 2000 m. The large negative charge in the thundercloud will induce an equally large positive surface charge in the ground directly below the cloud. Taking into account that air during conditions of a thunderstorm breaks down easier, we can readily see that such large voltage can create a giant arc of lightning between cloud and earth which transfers negative charges to the ground.

The mechanism of the lightning stroke is quite complex. During a full thunderstorm, lightning discharges occur between clouds as well as between clouds and earth with the bolts striking downward and upward as well as sideways between the clouds. If we examine a cloud-to-earth discharge, we find that it begins with a "feeler" or "leader" from the cloud which darts downward in a series of steps and is followed by a main flash called the *return stroke* from ground to cloud. In the literature the weak leader is referred to as the "step leader." The leader being full of negative charges from the cloud seeks the most easily ionized path toward earth. As it approaches ground level, it will seek out high points for several reasons. One is that, for high points, which are grounded, the cloud-

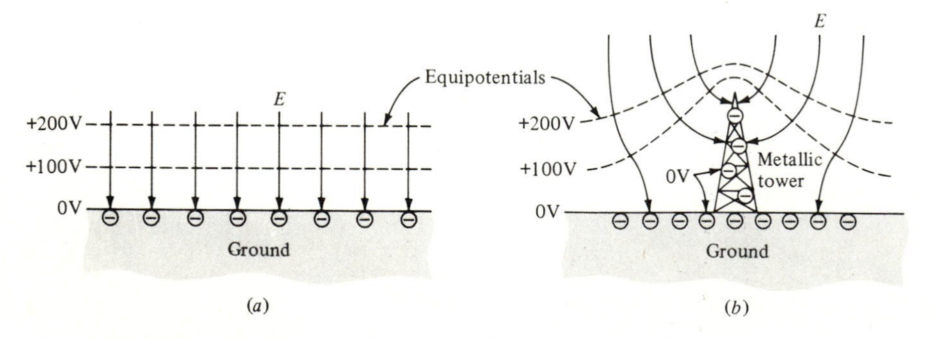

Figure 2.18 (*a*) The atmospheric electric field and the resultant surface charge along the ground; (*b*) the equipotential distribution around a protruding grounded object.

to-earth path is shorter and therefore easier to complete. The other is related to the potential distribution around grounded metallic towers. Since a metallic surface is an equipotential, the top of a grounded tower will be at zero potential, as shown in Fig. 2.18*b*. A lightning bolt "feeling" its way down will naturally seek out the nearest zero potential point.

There is another phenomenon that must be considered when a lightning arc seeks a high point. As the leader comes down, the fields near to it will become so large that a corona discharge starts from grounded sharp points and actually reaches up to the leader as shown in Fig. 2.19. Lightning therefore tends to strike such points. The old folklore saying that lightning never strikes the same place twice, apparently is not true. The Empire State Building, for example, gets repeatedly struck during each thunderstorm passing above it. Perhaps what the old folks are trying to tell us is that lightning does not strike twice because the structure that attracts the lightning in the first place will be destroyed during the strike. This brings us to the real topic of this section, and that is protection from lightning.

As we have seen before, the electric fields inside a metallic enclosure are zero. The best protection, therefore, that we could have for a structure is to surround it with a metallic shell and ground it. If lightning should strike the shell, the current will be harmlessly conducted to ground. Needless to say, protection of this kind would be rather costly and often impractical. What is usually done is to use the crudest approximation to an enclosed shell which is a slice of the shell in the form of a lightning rod on top of the structure to be protected and a thick enough ground cable between rod and earth capable of carrying the surges, which can be as high as 20,000 A in a typical lightning bolt. The theory behind this arrangement

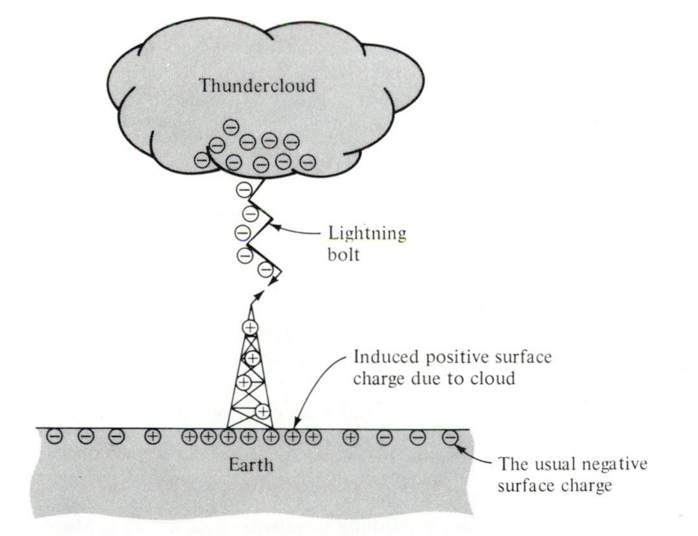

Figure 2.19 A lightning bolt reaching down from a heavily charged cloud will attract a feeler from a sharp point going upward.

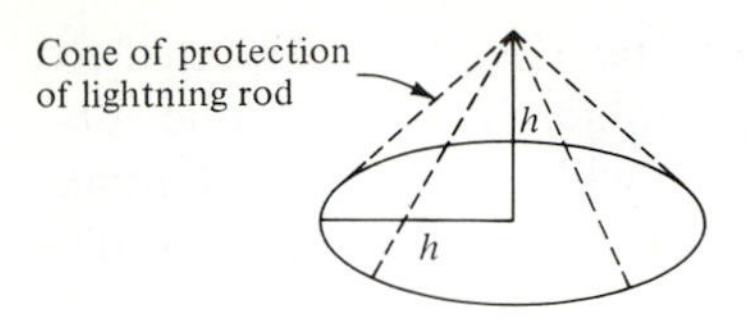

Figure 2.20 A lightning rod and its cone of protection.

is that the top of the lightning rod will attract the lightning since we know that lightning tends to strike the highest objects on the horizon. If the rod is struck, it will conduct the current surge to ground via the well-grounded cable, giving a zone of protection directly beneath the rod. Tests and experiments have shown that the protected space inside of which the probability of a strike is very low is a cone, with the top of the lightning rod as the apex of the cone and a base of radius equal to the cone height h (essentially rod height above ground) as shown in Fig. 2.20. If the base radius is increased to twice the top-of-the-rod height, a cone of protection within which an object will be struck occasionally is obtained.

2.15 EXAMPLE: ELECTROSTATIC PRECIPITATORS IN AIR POLLUTION CONTROL†

An example of a commercial device that is based on electrostatic principles is the *electrostatic precipitator.* It is a device which uses an electric discharge to remove liquid droplets or solid particles from a gas, usually air, in which they are suspended. The electrostatic precipitator separates the suspended particles from the gas by first passing the suspension through a corona-discharge area where charging and ionization of the objectionable particles takes place. The charged pollutants are then attracted to a

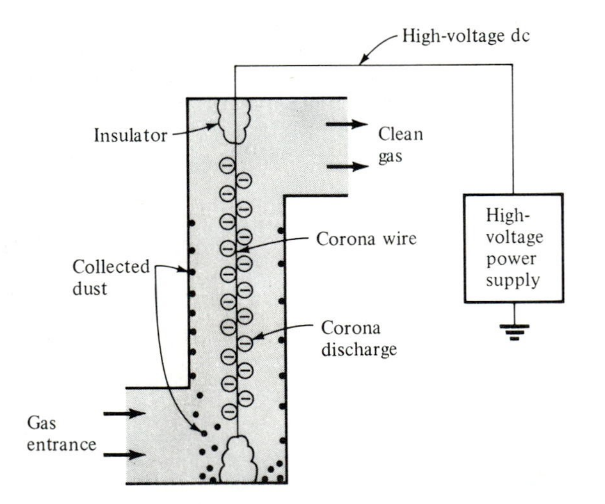

Figure 2.21 An electrostatic precipitator showing the corona wire and corona discharge. Dust particles will be charged by ions from the corona and collected on the precipitator walls.

† W. L. Faith, "Air Pollution Control," John Wiley & Sons, Inc., New York, 1959. H. J. White, "Industrial Electrostatic Precipitation," Addison-Wesley Publishing Company, Inc., Reading, Mass., 1963. A. D. Moore, Electrostatics, *Sci. Am.*, pp. 46–58, March 1972. H. C. Perkins, "Air Pollution," McGraw-Hill Book Company, New York, 1974.

collecting anode where they can be removed by jarring the collecting anode (or by some other mechanical means, such as scraping, etc.). A typical precipitator cross section is shown in Fig. 2.21.

Characteristics of Pollutants

Let us examine some properties of pollutants. Particle suspensions in gases are referred to in the literature as aerosols, particulate clouds, or dispersoids. The size of the pollutant particles is usually in the range of 0.1 to 200 μm, where 1 $\mu m = 10^{-6}$ m. For comparison, the molecules of a gas are on the order of 10^{-3} to 10^{-4} μm, and the wavelength of visible light falls in the range of 0.4 to 0.8 μm. Particles larger than 100 μm may be excluded from the category of dispersoids because they settle too rapidly. Dust is formed by pulverization or mechanical disintegration of solid matter into particles of size ranging from 1 μm up to about 100 or 200 μm. Since it contains a relatively high proportion of large particles, dust settles rather fast in a calm atmosphere. Smoke derived from burning of organic materials, such as wood, coal, and tobacco, has particles which are very fine and range in size from less than 0.01 μm up to 1 μm. Particle-size distribution of fumes is similar to that of smoke. Mists or fogs are formed by the condensation of vapors upon suitable nuclei to give a suspension of small liquid droplets. Atomization of liquids is another means of producing mist. Particle sizes lie between 5 and 100 μm.

The most important electrical property of suspended particles, natural or synthetic, is that they are charged to an appreciable degree. The fractions of positively and negatively charged particles present are generally equal, so that any given suspension is, as a whole, electrically neutral. For example, the highly charged clouds accompanying lightning storms are evidence of a natural charging process. Measurements of the free electric charge on rain droplets show an average charge of 10^{10} to 10^{11} electron charges (one electron charge $= 1.6 \times 10^{-19}$ C) depending on altitude and the intensity of the storm. Severe radio interference is experienced by airplanes in rain or snow storms and is due to the charged rain or snow drops striking the surface of the airplane. This is usually referred to as precipitation static. Dust storms will produce the same kind of static. Fumes and smoke are similarly charged. Freshly formed mists or fogs are found to be initially uncharged but tend to absorb charge from the atmosphere. We can summarize and say that particle charging occurs naturally during the formation or life of a particle by various mechanisms, such as flame ionization, friction, grinding, dispersion from the bulk state, cosmic radiation, etc. As a matter of fact, most electrostatic precipitators for home use dispense with a charging mechanism and simply use the natural charges on dust to collect the particles by passing the suspension through a set of charged plates or grids such as shown in Fig. 2.22. Alternate sets of grids are connected together, and one of these sets is maintained at a high potential with respect to the other set. The applied potential is below the corona-producing level since for home use the high potentials needed for corona discharges and the accompanying high levels of ozone released are undesirable.

Using the natural charges of pollutants to collect them is an inefficient procedure. The pollutants are attracted to the collecting plates because a Coulomb's force exists between the plates and the particles. Since the Coulomb's force increases in proportion to particle charge, we can readily see that doubling the charge will double the separation force and, other factors being equal, will reduce precipitator size by a factor of 2. In large precipitators, for industrial applications, economic considerations require that particle charge be made as large as possible. From experience we know that the

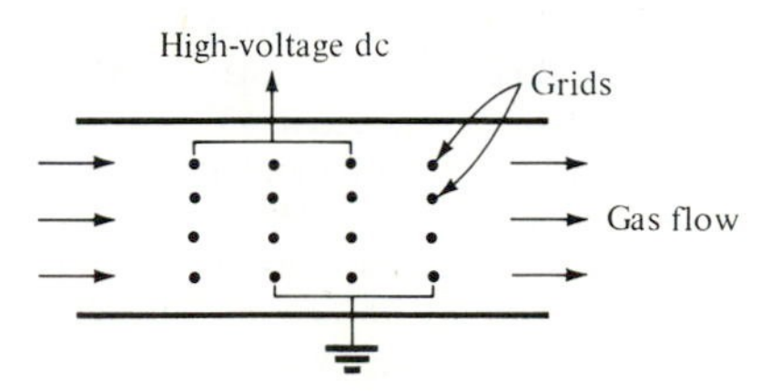

Figure 2.22 A simple electrostatic precipitator that uses the natural charges of pollutants to collect them on the grids.

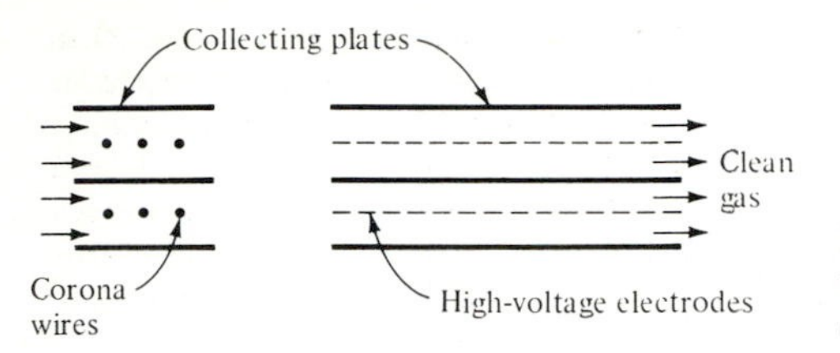

Figure 2.23 A two-stage precipitator showing the corona and collecting mechanisms in stages one and two, respectively.

corona-discharge method is by far the best means for achieving high particle charge for gas-cleaning purposes. As a matter of fact, if we take dust, for example, we find that the natural charges that exist on the pollutants are found to be of the order of 5×10^{13} electronic charges per gram, which is only 5 to 10 percent of corona-charged dusts in precipitators. Fumes usually have even lower charges. On the other hand, in precipitators that pass the polluted gas through a corona discharge first, we find that the dispersoids become charged to the degree of 5×10^{14} to 5×10^{15} electronic charges per gram.

For industrial applications where the high degree of ozone produced and the high voltages needed are not bothersome, corona discharge precipitators are used. They are of two types. The single-stage precipitator, as shown in Fig. 2.21, that charges the particles and collects them at the same time, and two-stage precipitators, as shown in Fig. 2.23. For industrial-furnace cleaning the single-stage precipitators have been found to be superior. A high corona can be produced by the use of a single fine wire which is maintained at a negative high potential. The entering dust particles become highly charged in the first few inches of travel and are then driven to the wall of the precipitator tube by the intense electric field of the corona. The cleaning action is very fast. Since virtually all particles are collected in less than 1 s (usually 0.1 s), which is much shorter than the time a particle would spend traveling down the precipitator tube, recycling is rarely used. Note that migration velocities of suspended particles are usually around 0.03 m/s for very fine particles of 0.5 μm, up to 1 or 1.3 m/s for 20 or 30-μm particles. Two-stage precipitators are used principally for cleaning air since they tend to produce less ozone than the single-stage precipitator. The second stage is essentially the same in principle as that shown in Fig. 2.22 with a purely electrostatic field between smooth, nondischarging electrodes.

Charging the Pollutants

Let us now examine the mechanism by which the pollutants become charged. First, it is relatively immaterial whether a positive or negative charge is transferred to the pollutant particle since both are equally effective for the same degree of charge. For industrial cleaning, negative polarity is preferred because of its greater stability and the possibility of higher operating voltages. For air cleaning in general, positive polarity is preferred because ozone production is lower. We have shown in the previous sections how ionization by electron impact in a high-voltage field leads to a chain reaction or electron avalanche in which each new electron produced generates new electrons by ionization in ever-increasing numbers, which finally leads to the glow of the corona discharge. In a precipitator such as that shown in Fig. 2.21, with the wire at negative polarity, positive ions are attracted to the wire with the negative ones attracted to the grounded pipe. Although both positive and negative ions are formed in equal numbers in the corona-glow region near the wire, over 99 percent of the gas space between the wire and the grounded outer wall contains a space charge of only negative ions. Thus, we have a picture of most of the precipitator space occupied by the relatively heavy and slow-moving negative ions, with the region immediately next to the wire in which the corona glow is visible primarily occupied with positive ions and free electrons. In the corona field there are about 10^8 negative ions/cm^3. On the other hand, even a dense aerosol fume contains no more than about 10^6 to 10^7 particles/cm^3. In the precipitation process the particles, in passing through the corona field, are therefore subjected to intense bombardment by the negative ions with a resultant transfer of charge from ions to pollutant particles. The pollutant particles are therefore charged to one polarity. The charging process is very quick. Usually in 0.01 s or less, the particles have achieved 300 electron charges for particles in the 1-μm range and 30,000 electron charges for 10-μm particles in a corona field of about 6000 V/cm. These values are close to the maximum predicted theoretical values. The curve for particle charge versus particle diameter is a uniform line, so that charges for particles other than the two mentioned above can be predicted. The speed with which the charging process is completed implies that the pollutants become fully charged during the first few centimeters of their travel through the corona field of the precipitator (using an average gas velocity of 2 m/s). Since an ion usually carries a single electron charge, we can see now that at least 300 ions must collide with a single particle and transfer their charges to it. The charging process in which ions attach themselves to particles in the presence of an electric field is referred to as *field or impact charging*. It is the predominant charging mechanism for particles which are larger than about 0.5 μm in diameter. Typical values for the corona-discharge fields range from several to 10 kV/cm. For particles smaller than this, charging by ion current in an electric field becomes less effective. A second process called *ion diffusion charging* becomes the important mechanism for particles smaller than about 0.2 μm. Since the ions present in a gas share the thermal energy of the gas molecules, their thermal motion will cause them to diffuse through the gas without the aid of an externally applied field. As the ions diffuse, they will collide with any particles which may be present and adhere to the particles because of the attractive electric image forces which develop when the ions approach the particles. At the same time as the charge accumulates on the particle, it will give rise to a repelling field which will prevent additional ions from reaching the particle. Since large particles can hold an appreciable amount of charge, to charge such particles fully requires the help of an externally applied field as in field or impact charging. It should be clear now that diffusion charging is of some importance only for the smaller particles. Of course for particles in the range of 0.2 to 0.5 μm both charging methods are effective.

The last stage in the electrical precipitation of pollutants is their removal from the collecting anodes. Efficient collection is very important because the particles may reenter the gas stream while still inside the precipitator. If the gas velocity is very high, reentering particles can present a difficult technical problem.

Summary

To summarize this section, we can say that a single-wire negative-corona electrostatic precipitator combines the two functions of charging and collecting pollutant particles in one unit. The negative corona can be divided into two primary zones. The first is the glow zone around the discharge wire, also called the *active zone*. It contains a mixture of negative and positive ions, free electrons, excited and normal molecules, all interacting with the corona wire and with each other. The second zone is the *passive zone*. It occupies the region between the corona glow and the positive collecting electrode which is the grounded outer wall of the precipitator tube. In this zone we find neutral molecules and a fraction of negative ions and electrons which were created in the active zone around the corona wire. They move under the influence of the electric field toward the collecting anode. The active zone is primarily an electron source for the corona. It is a thin tubular volume around the discharge wire containing an essentially neutral plasma. The passive region, on the other hand, is occupied by a large electric-space charge of negative polarity. The fundamental mechanisms of particle charging and collection take place in this region. A polluted gas which is forced to pass through this region will have the pollutant particles which are larger than the gas molecules precipitated out and collected at the passive anode.

2.16 EXAMPLE: ELECTROSTATIC PHOTOGRAPHY—XEROGRAPHY

As the final illustration of the application of the principles of electrostatics, let us discuss electrostatic photography. The photocopy field has been revolutionized in recent years by the introduction of dry photoelectric processes which produce positive copies without a (wet) negative intermediary. Unlike ordinary photography in which the silver halide emulsion must be developed by a wet process and is good only for one picture, the electrostatic "negative" can be used repeatedly. Many electrostatic methods exist and are usually classified under the general heading of electrophotography.†

We will consider one process in particular which has been highly developed by the Xerox Corp. and is known as xerography. It utilizes a photoconductive plate backed by a conducting plate. The photoconductive plate is a vitreous selenium coating which is evaporated in a thin layer (20 to 100 μm) onto the metallic plate. To prepare the plate for copying, it is precharged to a 6000- to 7000-V potential by a corona discharge which imparts a uniform electrostatic surface charge on the selenium coating thereby making it sensitive to light. Segments of the surface charge are conducted or dissipated to the grounded metallic plate by light rays reflected from the white parts of a document being copied. We now have an electrostatic image pattern in the form of the remaining surface charges on the selenium plate which correspond to the printing on the document being copied. A powder of opposite charge, usually known as "dry ink," is sprayed on the plate and is attracted to the surface charge pattern. Ordinary paper is now placed over the sprayed selenium plate. Passing a high voltage over the paper will attract some of the dry ink to it. The paper is then peeled off the selenium plate, and the stuck-on powder is fused to the paper by heat. As a final step the selenium plate is brush cleaned and is ready for another copy. The life of the selenium plate is ordinarily limited by the number of brush cleanings that it can withstand. In the commercial machines the copying process is very fast (several seconds). The selenium plate is in the form of a rotating drum, with the various functions of charging, exposing, and cleaning arranged around the exterior surface of the drum.

The xerographic process can be divided into six distinct steps which will now be considered one at a time.

Charging

Figure 2.24*a* shows schematically the charging of the amorphous selenium coating by moving a corona wire across the surface. The single fine corona wire is enclosed by a shielded housing which aids in distributing the corona spray uniformly over the surface. If the potential between ground and the corona wire is high, air molecules near the wire will be ionized. Positive ions will then be attracted and will stick to the surface. If the selenium coating is an insulating material, the surface charge will remain an appreciable amount of time on the surface. The photoconductive selenium coating will have a resistivity ρ of approximately 10^{12} to 10^{14} Ω-m in the dark. This gives for the charge rearrangement time T (see Sec. 2.8)

$$T = \varepsilon\rho = 50 \text{ s} \tag{2.66}$$

where we have used $7\varepsilon_0$ for the permittivity and 10^{12} Ω-m for the resistivity of selenium. Therefore, for times less than 50 s most of the surface charge remains on the surface provided the plate remains in the

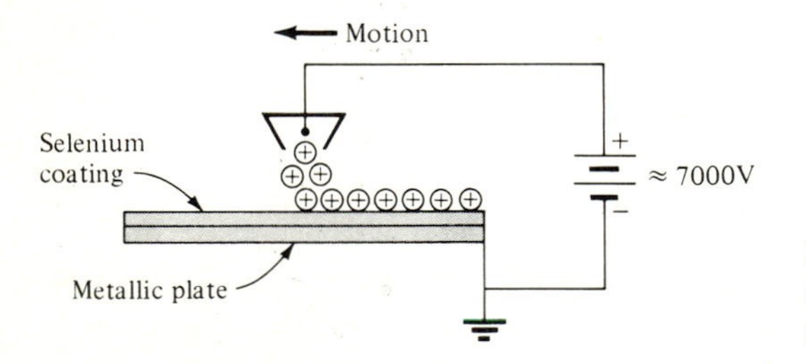

Figure 2.24a Charging of the selenium coating by moving a corona discharge grid over its surface.

† R. M. Schaffert, "Electrophotography," 2d ed., Halsted Press, 1975. R. B. Comizzoli et al., Electrophotography—A Review, *Proc. IEEE*, pp. 348–369, April 1972. A. D. Moore, Electrostatics, *Sci. Am.*, pp. 46–58, March 1972.

dark. The resistivity will be reduced several orders of magnitude when the selenium plate is exposed to light. The coating must be thin enough, usually on the order of one-tenth of a millimeter, such that the exposing light will penetrate most of the material.

Regarding the photoconductivity of amorphous selenium, much need not be said, except that amorphous selenium happens to be a semiconductor with an absorption band in the range of wavelength corresponding to the visible spectrum. When such a material is exposed to light, electrons will jump from the valence band across the relatively narrow gap, which is characteristic of semiconductors, into the conduction band, thereby making the material more conducting.

Exposing

Figure 2.24*b* shows light being reflected from a document and focused by a lens onto the selenium plate. Since the plate until now was kept in the dark, we can assume that all the uniform surface charge is still on the selenium coating. At the points on the plate corresponding to reflected light from the white areas of the document, the emulsion becomes suddenly more conductive with the result that the surface charges become more mobile [T in Eq. (2.66) decreases to perhaps 0.1 to 1 s] and begin the leak away from the surface. Since the positive surface charge is attracted to the opposite potential of the conducting plate, the exposed points will become quickly discharged to ground, leaving on the surface a charge pattern which corresponds to the dark areas of the document. As the quantity of charge remaining in any particular area is inversely related to the intensity of illumination, some grays will also be reproduced. For continuous-tone work, the exposure latitude is relatively narrow. Xerography is best suited for line-copy reproduction. We can now see that in overexposure, the images will appear light and "washed out," with fine lines missing. In overexposure the surface charge has been abnormally decreased with the result that it will not be able to attract a sufficient amount of developing powder. In underexposure, on the other hand, not enough surface charge was allowed to leak off. The result will be an abnormally high attraction of developing powder in the exposed areas yielding a heavy image and speckle in the nonimage areas. The resolution of the xerographic plates should be on the order of the emulsion thickness of the selenium coating.

Developing

Figure 2.24*c* shows a bin of dry ink moving across the latent image of surface charge on the plate. Dry ink is a powder whose particles are charged with a polarity opposite to that of the latent image charges. The number of negatively charged developer particles attracted to charged areas will be proportional

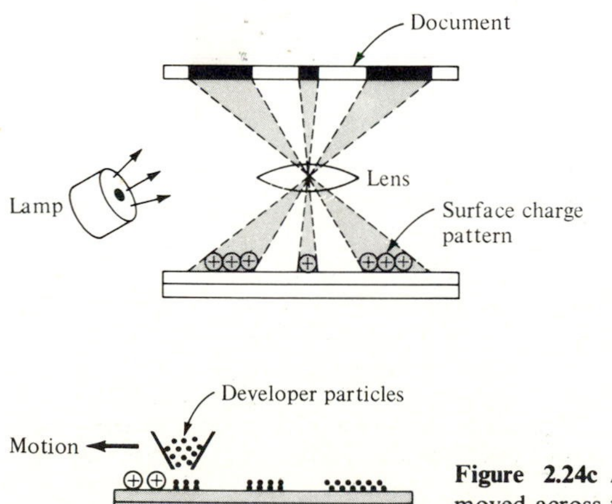

Figure 2.24b Light reflected from a document is focused by a lens onto the precharged selenium plate.

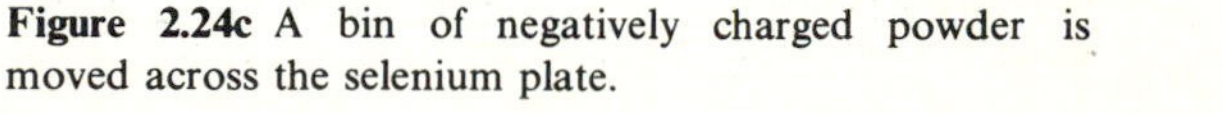

Figure 2.24c A bin of negatively charged powder is moved across the selenium plate.

to the intensity of electric fields associated with the charge densities of the electrostatic image. Uncharged areas will be free of developed particles. There are many ways in which the developer can be applied to the plate, including brushing, spraying, and rolling.

Another reason why xerography is best suited for line-copy work is that the attraction of particles is greatest near the edges separating charged from uncharged areas. Since the field intensity or voltage contrast is greatest there, more powder will be deposited at such places. The result is a solid black line around extended objects. This of course is ideal for line-copy work where the large contrast between white and dark areas is desirable.

Transferring the Image to Paper

The image on the selenium plate which has been developed with dry ink is now ready to be transferred to paper. There are two commonly used methods, electrostatic transfer and adhesive transfer. Since adhesive transfer requires specially coated paper, the electrostatic-transfer method is used mostly in xerography machines. It is illustrated in Fig. 2.24*d*. In the electrostatic method the transfer is accomplished by placing ordinary paper in contact with the image side of the plate. The paper is then charged by moving a corona charging grid similar to that used in precharging the selenium plate over the paper. The positive charge so placed on the paper will attract the top layer of developer particles which are held by attraction to the latent image and pull them onto the paper. The paper is then stripped from the plate, carrying the transferred image with it.

Fixing the Transferred Image

The dry ink used in xerography is a mixture of a carrier and a resinous pigment. The resin is fusible so the image can be permanently fixed by heating.

Cleaning the Selenium Plate

Since the charge applied to the paper cannot entirely overcome the attraction of the developer particles by the latent image, a residual particle distribution will remain on the plate after the paper is peeled off. The residual powder must first be cleaned off the plate before it can be reused. In the commercial machines a rotating drum with fur brushes is moved over the plate, thus gently removing the remaining particles. Eventually after repeated reuse a developer film will form over the entire plate. This must be removed by periodic cleaning with solvent, usually isopropyl alcohol, which will not dissolve the photoconductive emulsion.

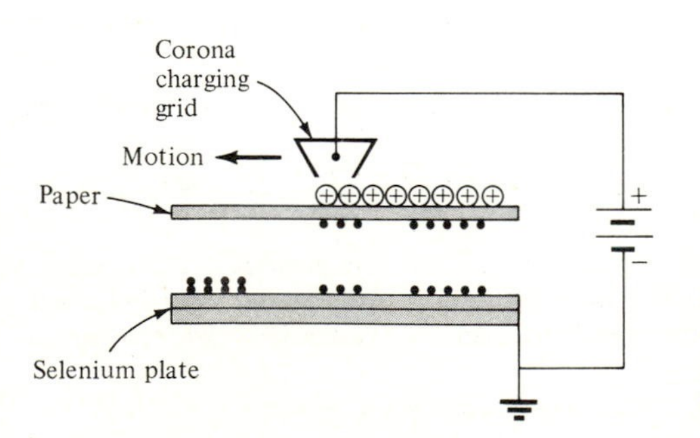

Figure 2.24d A paper which is placed over the selenium plate is charged positively and will attract the negatively charged developer particles from the latent image.

PROBLEMS

2.1 What is the velocity of an electron in free space at room temperature (20°C = 293 K), at liquid-nitrogen temperature (−196°C), and at liquid-helium temperature (−268.8°C)?

2.2 What is the velocity of a hydrogen atom (which has about 1860 times the mass of an electron) in free space at room temperature?

2.3 The resistivity of a germanium bar 2 mm by 3 mm by 5 mm long is 5 Ω-cm. Find the resistance between electric contacts placed at the ends of the long dimension.

2.4 A bar has dimensions 1 mm by 2 mm by 2 cm long, with electric contacts at the ends of the long dimension. If the bar material is glass, copper, and 10 Ω-cm germanium, determine the resistance in each case.

2.5 A field of 200 V/m is applied to a wire with conductivity $\sigma = 10^5$ S/m. Find the current density in the wire. What is the total current flowing if the wire has a diameter of 1 mm? Find the total charge that passes any point in the wire during a 1-s interval.

2.6 If a current of 1 A is flowing in a copper wire, find the number of electrons that pass a given cross section in the wire each second.

2.7 A potential difference of 200 V is applied across two electrodes in a vacuum tube which results in a current flow of 1 mA. What power is dissipated, in watts?

2.8 If a germanium bar contains 10^{18} electrons/cm^3, determine the drift velocity of electrons in the bar across which a field of 500 V/m is applied. How long does it take an electron to move 1 m?

2.9 Five wires meet at a junction point. If two wires carry a current toward the junction of 1 and 2 A, respectively, and two wires carry a current away from the junction of 5 and 6 A, respectively, find the current and its direction in the fifth wire.

2.10 Derive Eq. (2.12*b*) by starting with (2.12*a*) and applying the divergence theorem.

2.11 Find the **E** field in a wire which carries a current of 1 A, has a resistance of 1 Ω/10 m, and is 1 mm in diameter.

2.12 Find the maximum current allowed through a resistor which is rated at 100 Ω, 2 W.

2.13 If an electric heater raises the temperature of 1 kg of water 100°C when turned on for 1 h, find the heater resistance if a current of 5 A passes through it.

2.14 A vacuum-tube cathode emits electrons at the rate of 10^{18} electrons/s. If these are collected by an anode, 100 V above that of the cathode, find the current at the anode and the power dissipated at the anode.

2.15 The concentration of free electrons in copper is 8.4×10^{28} electrons/m^3. A No. 18 copper wire (diameter = 1.03 mm), which has a resistance of 6.5 Ω/1000 ft, carries a current of 3 A. Find the drift velocity and conductivity.

2.16 A copper wire with cross-sectional area of 1 mm^2 carries a current of 1 A. What is the average drift speed of conduction electrons in the wire?

2.17 The resistance of the copper windings in a motor is 100 Ω at 20°C when the motor is not in use. The resistance rises to 120 Ω after the motor has been running for some time. Find the temperature of the windings.

2.18 Brass, which is a conducting material, has a conductivity of $\sigma = 1.6 \times 10^7$ S/m. What is the rearrangement time for electrons in such a material?

2.19 Using the concept of the rearrangement time T, find the range of frequencies when brass, dry earth, and seawater act as conducting materials.

2.20 A conducting sphere of radius a and material parameters ε, μ, and σ is immersed in free space (ε_0, μ_0, $\sigma = 0$). If at time $t = 0$, a charge is distributed uniformly throughout the conducting sphere, such that the charge density is ρ_0 C/m^3, find the current distribution and the electric field inside the sphere and the electric field outside the sphere.

Hint: Use the result of Eq. (1.84*d*) to find the initial fields inside the sphere.

2.21 Show that the absolute potential at a distance a from a plane of finite thickness with uniform surface charge ρ_s is $V = \rho_s a/\varepsilon_0$ if the surface of the plane is taken as the reference level.

2.22 Show that no net electric force can exist on an isolated charged body due to its own charge.

2.23 Show that a charged conducting surface experiences a normal outward force, $\rho_s^2/2\varepsilon_0\varepsilon_r$, per unit area, where $\varepsilon_r = \varepsilon/\varepsilon_0$.

2.24 If a small charged object is placed near a large uncharged conducting body, is there a force on the small charged object?

2.25 A line charge ρ_L C/m lies inside an uncharged conducting shell which is infinitely long and cylindrical. The outer radius of the shell is b, and the inner radius is a. The line charge ρ_L is parallel, but is displaced from the axis by a distance c as shown in the figure. Calculate the electric field outside the shell and sketch the field inside. By using + and − signs, represent the approximate distribution on the inner and outer surfaces of the conducting shell.

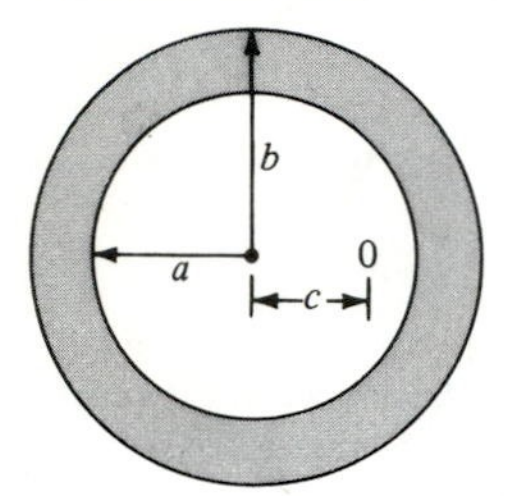

2.26 Discuss why the exterior field in the previous problem appears to come from a line charge which is situated at the center of the tube.

2.27 A point charge Q is inside an uncharged conducting spherical shell, with outer and inner radii b and a, respectively. The point charge Q is displaced from the center by a distance c. Calculate the electric field at points outside the sphere and sketch the field inside.

2.28 A space is enclosed with conducting walls. If a charge Q is moved about slowly inside the enclosed space, is it possible to detect this motion outside the space by using an electric field meter?

2.29 A point charge Q is situated at a distance d above a large conducting plane which is grounded. Find the force on charge Q.

2.30 A point charge Q is situated at a distance d above an infinite conducting plane. What is the surface charge density on the plane?

2.31 A point charge Q is placed near two conducting half-planes which are perpendicular to each other, as shown in the figure. Find the potential in the quarter space between the two conducting planes by replacing the conducting planes by three image charges.

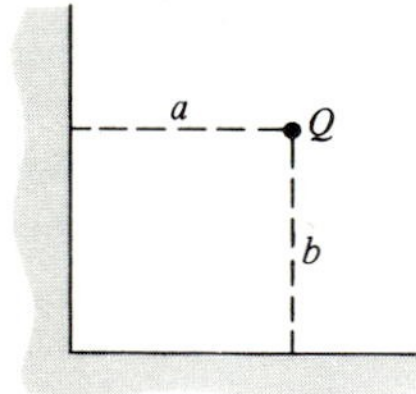

2.32 Find the force per unit length F_L on two parallel, infinitely long line charges ρ_{La} and ρ_{Lb} which are separated by a distance l.

2.33 An infinitely long, thin conductor is charged with ρ_L coulombs per unit length (C/m) and is situated parallel to and at a height h above the plane surface of the earth. Determine the force per unit length on the line charge ρ_L.

2.34 Find the work necessary to remove an electron from a metal across its boundary into free space. If the atomic radius of metals is on the order of $x_0 = 1.6 \times 10^{-10}$ m, assume that the work involved will be the work from x_0 to infinity.

CHAPTER

THREE

SOURCES OF VOLTAGE (EMF) AND STEADY ELECTRIC CURRENT

3.1 THE EMF OF A BATTERY AND NONCONSERVATIVE ELECTRIC FIELDS

In the previous chapter Ohm's law for a distributed current $\mathbf{J}$ was obtained as

$$\mathbf{J} = \sigma \mathbf{E} \tag{3.1}$$

This equation implies that an electric field E is needed to maintain a current. Since a current is a flow of charges, and since charges assume a stationary distribution very quickly on conducting surfaces that are immersed in an electrostatic field, an electrostatic field alone cannot sustain a steady current around a closed circuit. We can express this in a different way by writing (3.1) as

$$\mathbf{E} = \mathbf{J}\rho = \mathbf{J}\frac{RA}{L} \tag{3.2}$$

where $\rho =$ resistivity, $\rho = 1/\sigma$

$R =$ resistance of conducting path [see (Eq. 2.17)]

$A =$ cross section of path

$L =$ length of path

Since currents flow in closed loops, we can integrate Eq. (3.2) around a closed path of a simple circuit (Fig. 3.2) and obtain

$$\oint \mathbf{E} \cdot \mathbf{dl} = I\frac{R}{L}\oint dl$$

$$= IR \tag{3.3}$$

where $JA = I$ and the line integral of dl is the path length L of the resistor R (the resistance of the connecting wire and that of the battery are assumed to be zero). One of the properties of an electrostatic field is that it is conservative [see Eq. (1.12)]. The closed-line integral on the left side of (3.3) is therefore zero if E is an electrostatic field. Expression (3.3) then becomes

$$0 = IR$$

We conclude that an E due to static charges cannot maintain a constant current in a closed circuit since Eq. (3.3) reduces to $I = 0$.

The property that an electrostatic field cannot maintain a constant current can also be demonstrated by using Fig. 3.1 where we have taken a capacitor, charged it, and then inserted a piece of conducting wire as shown. Initially a current will flow in the wire. But in a very brief time (the rearrangement time is 10^{-18} s if the wire is copper), induced charges will appear at the ends of the wire. Within the wire, the induced field will cancel the applied field due to the capacitor with the result that current in the wire for time $t > 10^{-18}$ s is zero.

In order to sustain a steady current, we must have a source of energy. A steady current in a closed loop is the continuous motion of electrons around the circuit. The moving electrons are impeded by the resistance of the conducting medium and give up energy to the conductor in the form of heat [see Joule's law, Eq. (2.15)]. This energy must come from a nonconservative electric force field since in a conservative field an electron returning to the same point in the circuit has neither gained nor lost energy. Chemical action in a battery can produce such a nonconservative force which will push the electrons around a circuit until the chemical energy stored in the battery is depleted.

Let us consider a simple electric circuit which contains a source, shown schematically in Fig. 3.2. Ohm's law as given by Eq. (3.1) is valid at any point in the conducting material of the circuit. Since we have an additional force due to chemical action present, let us write the total field as

$$\mathbf{E}_t = \mathbf{E}_s + \mathbf{E}_e \tag{3.4}$$

where the subscript s denotes an electrostatic field derivable from charges and subscript e denotes a source field, usually referred to as the emf field. In this case

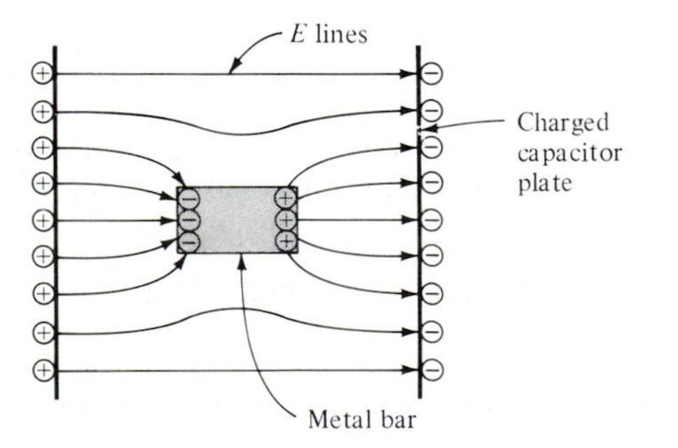

Figure 3.1 A metallic bar immersed in an electrostatic field will show a current until the induced field is established.

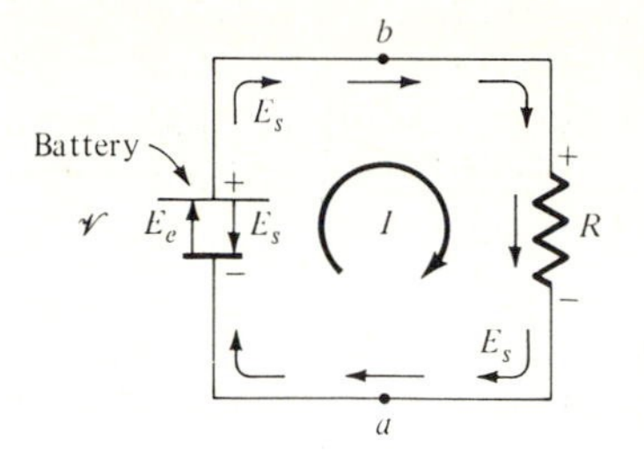

Figure 3.2 A series circuit composed of a battery and a resistance R with current I flowing.

the emf is due to battery action. Integrating (3.2) around the closed circuit gives

$$\oint \mathbf{E}_t \cdot \mathbf{dl} = I \frac{R}{L} \oint dl \tag{3.5}$$

Equation (3.5) can be written as

$$-\oint \nabla V \cdot \mathbf{dl} + \oint \mathbf{E}_e \cdot \mathbf{dl} = IR \tag{3.6}$$

noting that the electrostatic field $\mathbf{E}_s$ can be derived from a gradient of a scalar potential [see Eq. (1.49)]. Using the fact that a closed-line integral of a gradient of a scalar field is zero, (3.6) simplifies to

$$\oint \mathbf{E}_e \cdot \mathbf{dl} = IR$$

$$\mathscr{V} = IR \tag{3.7}$$

where the symbol $\mathscr{V}$ is used to denote the electromotive force† (emf) of the battery. Like potential difference V, $\mathscr{V}$ is energy per unit charge and is expressed in volts. It occurs only across the battery terminal, because as we traverse the circuit as indicated by the closed-loop integral in (3.7), the emf is zero everywhere along the circuit except across the battery terminals; that is, $\oint \mathbf{E}_e \cdot \mathbf{dl} = \int_a^b \mathbf{E}_e \cdot \mathbf{dl} = \mathscr{V}$. The presence and direction of the fields around the circuit are indicated by arrows in Fig. 3.2. Note that $\mathbf{E}_e$ and $\mathbf{E}_s$ have opposite directions in the battery.

Kirchhoff's voltage law expresses the property that sources are needed to maintain the flow of current. We can write Kirchhoffs voltage law for a closed circuit containing sources of emf and resistors as

$$\sum \mathscr{V} = \sum IR \tag{3.8}$$

Usually, emf which is related to the nonconservative E_e is referred to as a *voltage rise*; and the IR's outside the battery terminals which are related to the conservative E_s, as *voltage drops* (potential difference across R is $V_{ab} = E_s L = JL/\sigma = IR$). In a closed loop, therefore, the voltage rises must equal the voltage drops.

† Note that the use of electromotive force is actually a misnomer since $\mathscr{V}$ is not a force but work per unit charge. (See also footnote p. 443.)

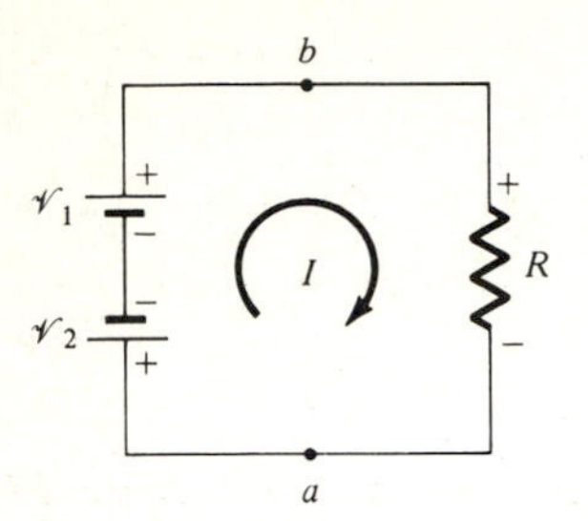

Figure 3.3 A single-loop dc circuit with two voltage sources.

To illustrate Kirchhoff's law [Eq. (3.8)], let us consider a single-loop dc circuit as shown in Fig. 3.3. Two sources of emf are connected in series. If we traverse the circuit in the direction of the indicated current flow, we see that the two emf's oppose each other. Equation (3.8) can then be written as

$$\mathscr{V}_1 - \mathscr{V}_2 = IR$$

Such a circuit is of interest if a low voltage at the terminals *ab* is desired. For example, if a dc voltage of 0.1 V is needed, a mercury cell with emf of 1.4 V and a dry cell with emf of 1.5 V can be connected in series. A setup such as this is more efficient than one using the emf of a single battery, connecting a potentiometer across it, and obtaining the low emf this way.

3.2 THE NATURE OF EMF

The convention for direction of current is the direction in which positive charges would flow. Unfortunately, this convention was adopted before it was known that the charge carriers in a current-carrying conductor are electrons. In what follows, we shall use the historical view in which current is the flow of positive charges.

Let us consider a battery in more detail. A conservative or electrostatic field can be associated with a battery. This can be seen if the circuit of Fig. 3.2 is opened. The open-circuit voltage which appears across the battery terminals is a result of the charges which are accumulated there and are symbolized by the + and − signs. A popular explanation of the workings of a battery when a load is connected to it is to say that the positive charges are attracted to the negatively charged terminal and move there through the load resistor thereby doing work. If in this view no other mechanism is present in the battery, then the positive charges which arrive at the negative terminal would discharge it until neutrality was obtained, at which point the current through the load resistor would cease. Certainly the positive charges which arrive at the negative terminal have no tendency to go through either the battery or load resistor back to the positive terminal of the battery. This kind of battery is then essentially a charged capacitor which is being discharged through a load resistor. Using such a popular explanation one would infer that only an electrostatic field is associated with a battery. However, we have already shown that an electrostatic field alone cannot maintain a steady current.

In a real battery, on the other hand, the chemical action produces a force (the electromotive force) which transports the positive charges back to the positive terminal. The chemical action force transports the positive charges through the battery against the opposing electric potential of the battery. The positive charges are now again available to travel through the load resistor and give their newly acquired energy to I^2R heating of the load resistor R. The charges will continue going around the circuit until the chemical energy which is stored inside the battery is expended or the battery is disconnected. We can now speak of a battery as the "seat" of the nonconservative emf voltage. As a matter of fact, any device† that is capable of driving charges around a circuit against an opposing potential can now be defined as a source of emf or simply voltage source. (See footnote p. 443.)

Before we go on and consider other sources of emf, let us observe that every source of emf has associated with it an internal resistance. Figure 3.4 shows a battery in a circuit with its internal resistance explicitly shown. Some energy will therefore be dissipated within the battery. The internal loss is given by I^2R_i and shows up as heat. The rate at which chemical energy is converted into electric energy by the battery is the power associated with the emf and is given by $\mathscr{V}I$ in watts. A portion of this power is dissipated in the internal resistance, while the remainder is available to the load. The energy balance per unit time (power in watts = energy in joules per second) for the above circuit is then

$$\mathscr{V}I = I^2R_i + I^2R_L \tag{3.9}$$

Canceling I on both sides of the equation, we obtain Kirchhoff's voltage law for the circuit

$$\mathscr{V} = IR_i + IR_L \tag{3.10}$$

The potential difference across the load resistor, which is also the available voltage across the battery terminals, is then

$$V_L = \mathscr{V} - IR_i \tag{3.11}$$

The available voltage across a load is therefore the emf of the battery minus the internal voltage drop of the battery. The current that flows in the series circuit is

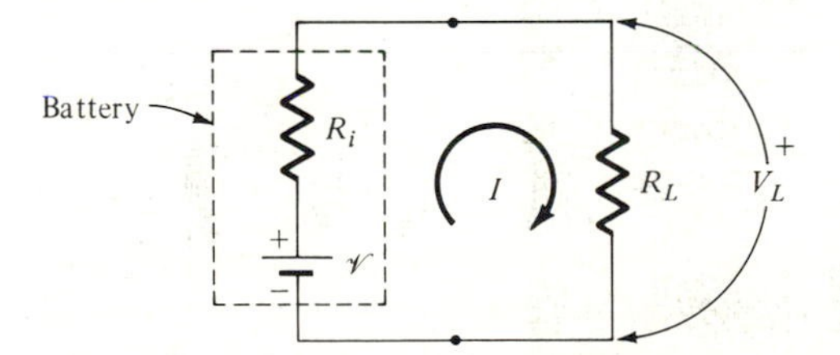

Figure 3.4 A battery is represented by its emf and internal resistance R_i.

† A particularly good example of an emf-producing device is the Van de Graaff generator. There, a moving belt transports charges continuously from one terminal and deposits them at the other terminal, thereby creating a large potential difference. The $\mathbf{E}_e$ field, whose direction (Fig. 3.2) is the direction of charge motion within the emf device, therefore represents the effects of the moving, charged belt and $\mathscr{V} = \oint \mathbf{E}_e \cdot \mathbf{dl} = \int_{\text{belt}} \mathbf{E}_e \cdot \mathbf{dl}$ represents the conversion of mechanical to electric energy.

given by

$$I = \frac{\mathscr{V}}{R_i + R_L} \tag{3.12}$$

For practical purposes one can say that the reason that current goes to zero as the battery is discharged is not that emf, whose magnitude is given by $\mathscr{V}$, goes to zero, but that the internal resistance R_i goes to infinity. In other words, as the chemical energy which is stored inside the battery is depleted, R_i becomes increasingly large. For practical purposes a discharged battery can be assumed to still have its emf but with an internal resistance which is infinitely large. The internal resistance of a battery is therefore a variable depending on the state of the charge and the age (shelf life) of the battery.

To measure the emf of a battery, we can remove the load at which point the current I will vanish. From Eq. (3.11), $V_L = \mathscr{V}$; that is, the potential difference appearing across the battery terminals on open circuit is the battery's emf. To measure the emf, even of an almost completely discharged battery, one can connect a high-resistance voltmeter (of 10^7 Ω or larger) across the battery terminals. Such a voltmeter approximates an open-circuit load and requires only the tiniest trickle of charge flow to give a reading. If the input resistance of the meter is much larger than R_i, the reading will be a measure of the $\mathscr{V}$ of the battery. We can conclude this section by saying that batteries which are expected to deliver large currents, such as the lead-acid automobile batteries, must have small internal resistances.

3.3 SOURCES OF EMF

There are other devices, besides batteries, which are sources of emf. All of them convert nonelectric energy into electric energy which is then used to drive charges around a circuit. The emf-producing arrangements are as follows:

1. *Fuel cells.* These devices convert chemical energy released by oxidation of liquid fuels directly into electricity. In one type of cell we have methyl alcohol as the fuel, which is mixed with potassium hydroxide. Two metal electrodes are inserted into the mixture. When air is bubbled over one electrode, the oxidation of the fuel results in a voltage of 0.5 V between the electrodes. One of the great advantages of fuel cells is that the fuel and oxidant can be fed to the cell continuously while it is operating. Fuel cells have been developed that can "burn" hydrogen, hydrocarbons, or alcohols with an efficiency of 60 percent. Hydrogen-oxygen fuel cells have been successfully used in space missions.
2. *Thermocouples.*† A thermocouple converts thermal energy into electric energy

† W. T. Scott, "The Physics of Electricity and Magnetism," John Wiley & Sons, Inc., New York, sec. 6.3, 1966.

with an efficiency of about 6 percent. It consists of a junction of two dissimilar metals which when maintained at different temperatures, produces an emf, usually referred to as a thermoelectric emf.

3. *Photovoltaic cells.* These contain photosensitive silicon, selenium, or cadmium sulfide which will generate a voltage when exposed to light and other radiant energy. A typical photo emf is 0.26 V per cell. Examples are exposure meters for cameras and solar cells which collect radiant energy and convert it to electricity with an efficiency of 15 to 20 percent.
4. *Electric generators.* Here mechanical energy is converted to electric energy with an efficiency of up to 99 percent for the larger generators. The production of motional emf is therefore very efficient. However, the overall efficiency for power-generating plants from fossil fuels to electricity is less than 40 percent. This is because the conversion efficiency in a steam turbine power plant from chemical to thermal to mechanical energy is at best 40 percent. If the mechanical energy were produced by diesel engines, the efficiency would drop to 38 percent, and if automobile-type engines were used, the efficiency would drop further to 25 percent. The efficiency of nuclear power plants as compared to steam power plants is even lower (approximately 30 percent) because nuclear reactors cannot be run as hot as boilers burning fossil fuel. This means that 70 percent of the energy in the fuel used in a nuclear power plant appears as waste heat, which is released either into a body of water or if cooling towers are used, into the surrounding air. Fossil-fuel plants waste 60 percent of the energy in the fuel in this way.
5. *Electric batteries.* A battery is a device for converting chemical energy into electric energy. It consists of series or parallel arrangements of voltaic cells which are composed of two different conducting materials immersed in an electrolyte; the chemical reaction of forming a new solution results in the separation of charges. The potential difference so obtained enables the cell to function as a source of emf with the immersed conductors serving as the electrodes for connection to an external circuit. The output is between 1 to 2 V per cell.

   There are two types of cells. *Primary cells* are those which cannot be recharged after all the chemical energy has been converted into electric energy. Examples of irreversible cells are the zinc-carbon dry cells, such as used in flashlights. The two electrodes in dry cells are carbon and zinc with ammonium chloride solution as the electrolyte. Charge separation occurs when zinc dissolves in the electrolyte, but the process cannot be reversed to form the zinc electrode from the solution. *Secondary cells*, on the other hand, are reversible. They act as sources for electric energy when the electrodes are being dissolved in the electrolyte solution. The cell is then discharging as the current through the external load circuit neutralizes the separated charges. The cell can be recharged by reversing the current to re-form the electrodes. The charging current is supplied by an external voltage source, with the cell to be recharged then serving as a load resistance. The reversible secondary cells are

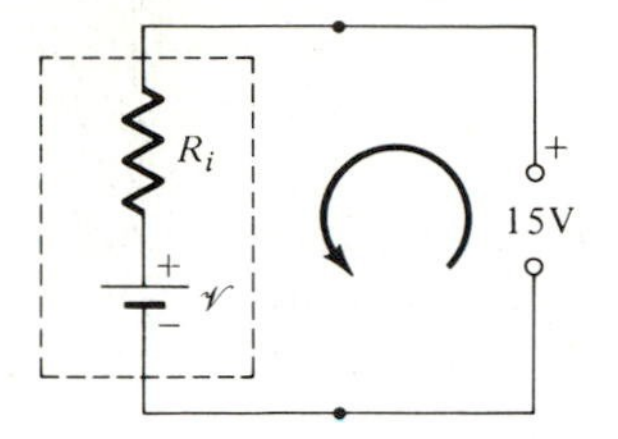

Figure 3.5 A 12-V lead-acid battery being charged by an external source of 15 V. Note that charging current flows from positive to negative terminal.

also referred to as *storage cells* for obvious reasons. The most common form of storage cell is the lead-acid wet cell used in automobile batteries.

Figure 3.5 is a schematic representation of a 12-V storage battery being charged from a 15-V source. We observe now that electric energy is converted into another form of energy (in this case chemical) only when current flows against an emf.

Example: The lead-sulfuric acid battery This wet-cell-type battery is commonly used in automobiles. It can deliver high values of load current which can go higher than 300 A during starting. Large currents are possible because of a low internal resistance which is on the order of 0.01 Ω for a 12-V battery. A single lead-acid cell has an output of 2 to 2.2 V requiring a series arrangement of six cells for a 12-V battery. The electrolyte in the lead-acid battery is a combination of concentrated sulfuric acid (H_2SO_4, with a specific gravity of 1.835) and water (H_2O, with a specific gravity of 1). In a fully charged cell the dilute solution of sulfuric acid has a specific gravity of 1.280 at room temperature. During discharge the water formed dilutes the acid further. When the specific gravity decreases to about 1.150, the cell is considered completely discharged. The positive and negative electrodes are a set of interconnected plates of lead peroxide (PbO_2) and spongy lead (Pb). On discharge, the lead and lead peroxide electrodes supply lead ions that combine with the sulfate ions (SO_4) in the sulfuric acid to form a lead sulfate ($PbSO_4$) coating on both electrodes (lead sulfate is the whitish powder often found on the outside of batteries). Also the lead peroxide of the positive electrode unites with hydrogen ions to form water. Note that the sulfuric acid electrolyte is a combination of hydrogen and sulfate ions. During the charging phase, the reversed direction of the ions flowing in the electrolyte results in a reversal of the chemical reactions. The Pb ions from the lead sulfate re-form the lead peroxide electrode, while the SO_4 ions from the lead sulfate combine with H_2 ions to produce more sulfuric acid. At the negative electrode the lead sulfate reacts with hydrogen ions to produce sulfuric acid while re-forming lead on the electrode. The chemical reaction for the lead-acid cell can be given by

Negative electrode		Positive electrode		Electrolyte		Lead sulfate on electrodes		Water
Pb	+	PbO_2	+	$2H_2SO_4$	$\underset{\text{discharge}}{\overset{\text{charge}}{\rightleftharpoons}}$	$2PbSO_4$	+	$2H_2O$

Lead-acid batteries are rated in terms of the discharge current that can be maintained continuously for a specified period of time, without the output voltage dropping below a minimum level somewhere between 1.5 to 1.8 V per cell. A commonly used rating for batteries is ampere-hours, based on an 8-h discharge. For example, a 160 ampere-hour-rated battery is capable of supplying a load current of 160/8 A, or 20 A, for a period of 8 h. It can supply more current for a shorter time or less current for a longer time. Present storage batteries have an overall conversion efficiency from chemical to electric energy of 70 to 75 percent. By comparison to other processes, this is a high conversion efficiency which leaves little room for improvement on this point in storage batteries.

3.4 BATTERIES FOR ELECTRIC CARS*

At the beginning of this century more electric automobiles were on the roads than internal-combustion cars. By the late 1920s the electric car had all but disappeared. The demise of the electric car proved to be the low-energy storage per pound of the batteries, especially when compared to the energy stored in fossil fuel. As a matter of fact even today the low capacity of batteries is the greatest stumbling block in the revival of the electric car. Let us see why this is so.

The heat of combustion of gasoline is 6100 watt-hours (W-h)† per pound of fuel. This is available as thermal energy. A piston engine converts this to mechanical energy with an efficiency of approximately 25 percent. On the other hand the energy output of lead-acid storage batteries is 17 W-h per pound of battery. The electric output of batteries can be converted to mechanical energy by electric motors with an efficiency of about 90 percent. Allowing for the higher conversion efficiency to mechanical energy in electric motors, the energy concentration by weight in fossil fuels is at least higher by a factor of 100 when compared with that of lead-acid batteries. A figure often used for the energy needed to move an average-sized car under normal conditions a distance of 1 mi is 1 kW-h/mi. Using lead-acid batteries this gives a figure of 65 lb of batteries per mile,‡ whereas applying this criterion to piston engines using fossil fuel, we would need 0.65 lb of gasoline per mile. Since gasoline has a weight of 6.15 lb/gal, this gives us a figure of 10.5 mi/gal which is a realistic figure. Clearly, the high energy concentration in hydrocarbon fuels makes them much superior to batteries.

The reason storage batteries are so heavy is that they use chemical elements at the heavy end of the periodic table, notably lead, and produce only about as much energy as would result from an equal number of atoms at the light end of the series. What is needed is a battery in which, for example, lithium is oxidized and reduced instead of lead. On the other hand, one should not expect a breakthrough in the voltages obtained per battery cell, since all chemical reactions produce low potentials.

Another advantage that fossil-fuel power plants have is that part of the fuel needed in combustion is air which need not be transported with the vehicle. The metal-air battery is a device which avails itself of that advantage. Zinc is usually the fuel, and air serves as the oxidizing agent. The output of the zinc-air battery is

* C. L. Mantell, "Batteries and Energy Systems," McGraw-Hill Book Company, New York, 1970. G. J. Murphy, "Considerations in the Design of Drive Systems for On-the-Road Electric Vehicles," *Proc. IEEE*, December 1972.

† Conversion between commonly used units of energy:

$$1 \text{ W-h} = 3.6 \times 10^3 \text{ J} = 3.413 \text{ Btu} = 8.604 \times 10^2 \text{ cal} = 2.655 \times 10^3 \text{ ft-lb} = 2.25 \times 10^{22} \text{ eV}$$

‡ 65 lb of fully charged batteries would be discharged after propelling an average-sized car a distance of 1 mi.

Table 3.1 Energy output per pound of battery

Battery	Output, W-h/lb
Lead-acid	17
Nickel-cadmium	20
Silver-zinc	30
Zinc-nickel oxide	36
Zinc-air	50
Lithium-metal sulfide	60 (operating temp. 425–500°C)
Fuel cell	80
Fossil fuel	6100

about 50 W-h of electricity per pound of battery. For comparison, the outputs of other batteries are listed in Table 3.1.

3.5 BOUNDARY CONDITIONS AT A CONDUCTOR-DIELECTRIC INTERFACE IN THE PRESENCE OF CURRENTS

In Sec. 2.9 of the previous chapter it was shown that the surface of a conductor in the static case with no current present is an equipotential surface. This means that the tangential electric field along the surface vanishes [the boundary condition for a metallic surface is given by Eq. (2.43) which is $E_{\text{tangential}} = 0$]. On the other hand, if the conductor carries a current, the relationship between current density **J** and electric field **E** in a conducting medium which is

$$\mathbf{J} = \sigma \mathbf{E} \tag{3.13}$$

specifies that a finite electric field **E** must be present in the conductor. Only if the conductor is a perfect conductor with $\sigma = \infty$ can we assume that the electric field is zero. For metals like copper and silver, a small but finite field will exist in the conductor that carries a current.

For example, if we connect a wire across a battery (usually not a good thing to do), as shown in Fig. 3.6*a*, a portion of the battery's emf will appear across the wire with the result that the equipotential surfaces coincide with the cross sections of the wire. Note that the electric field is along the wire whereas the equipotentials are across the wire. The orthogonality between the E field and the equipotential is thus maintained. Since E_{tan} just inside the wire is not zero, the boundary condition (1.86), which specifies that the tangential electric field across a boundary must be continuous, will require that a tangential electric field appear just outside the wire, that is,

$$E_{\text{tan}}^{\text{dielectric}} = E_{\text{tan}}^{\text{metal}} \tag{3.14}$$

where $E_{\text{tan}} = J_{\text{tan}}/\sigma$. We can now conclude that the tangential electric field just outside a current-carrying wire of finite conductivity cannot be zero, nor can the electric field be entirely normal to the surface as it can be in the electrostatic case.

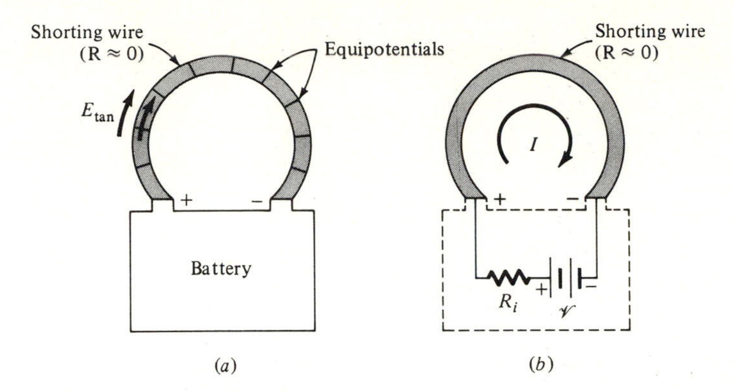

Figure 3.6 (*a*) The surface of a metallic wire is not an equipotential surface when a current flows; (*b*) when a battery is shorted, most of the emf is "dropped" across the internal resistance.

Figure 3.6*b* shows more clearly what happens when an almost zero resistance wire is connected across a battery. Most of the emf is then dropped across the internal resistance R_i of the battery with the result that the power $\mathscr{V}^2/R_i$, which is delivered by the battery, must be dissipated internally. The rapid generation of internal heat could cause the battery housing to burst.

Let us now consider two conducting objects which are separated by a dielectric such as free space. If the two objects are maintained at different potentials, a surface charge distribution ρ_s will build up on the conducting surfaces. This, in view of the boundary condition (2.43), which is

$$\varepsilon E_n = \rho_s \tag{3.15}$$

requires that a normal electric field exist between the conductors. For example, if we consider two long metal strips and connect a battery between them, as shown in Fig. 3.7*a*, the field will be normal to the conductors. Inside the metal, the field is everywhere zero. If a current is made to flow in the conducting strips by connecting a resistance R between them, as shown in Fig. 3.7*b*, we will find that the E lines

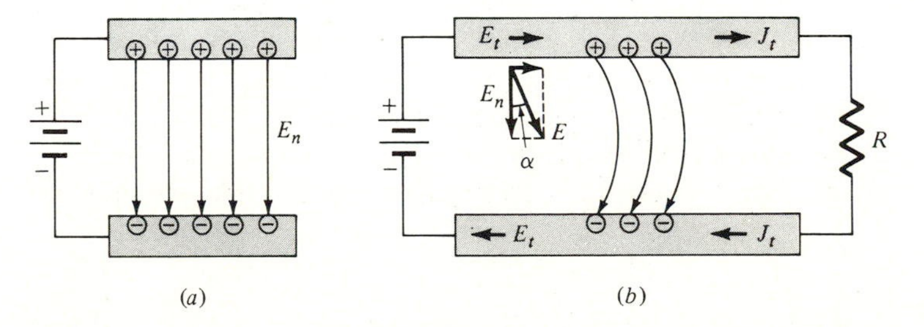

Figure 3.7 (*a*) Two strips of metal held at different potentials by a battery; (*b*) by connecting a resistor R, a current is made to flow in the strips.

do not terminate normally on the strips but make an angle α, given by

$$\alpha = \tan^{-1} \frac{E_t}{E_n} \tag{3.16}$$

with the normal to the surface. The curving of the lines is greatly exaggerated since the E_t is usually much smaller than E_n. Let us illustrate this by an example.

Example: Let us assume that the strips shown in Fig. 3.7*b* are 1-mm thick, 1-cm wide, and are of silver. Let us also assume that the battery has an emf of $\mathscr{V} = 1$ V and $R = 1\ \Omega$ such that a current of 1 A flows in the circuit. The tangential E field is then given by

$$E_t = \frac{\mathscr{V}}{RA\sigma} = 0.162 \text{ V/m} \tag{3.17}$$

where the cross-sectional area $A = 10^{-5}$ m^2, and $\sigma = 6.17 \times 10^7$ S/m for silver. In the calculation of (3.17) it was assumed that current in the circuit is approximately given by $I = \mathscr{V}/R = J_t A$; that is, the resistance of the strips is much less than R and is, therefore, ignored. The normal electric field is given by

$$E_n = \frac{\mathscr{V}}{d} \tag{3.18}$$

where d is the separation of the strips. Choosing a value for the separation as $d = 1$ cm, E_n becomes 100 V/m. The angle $\alpha = \tan^{-1} 0.00162 \approx 0.08°$. The angle by which the field lines deviate from normal is therefore extremely small in the metallic conductors, even for a large current such as 1 A.

Another example where a slight curving of the E lines occurs is between the current-carrying conductors of transmission lines. A longitudinal cross section of a coaxial transmission line is shown in Fig. 3.8, again with the curvature of the E lines greatly exaggerated.

To conclude this section let us consider a conductor consisting of two dissimilar metals strapped together, with the current flowing parallel to the boundary. Figure 3.9 shows a section of such a conductor with a copper bar strapped to a

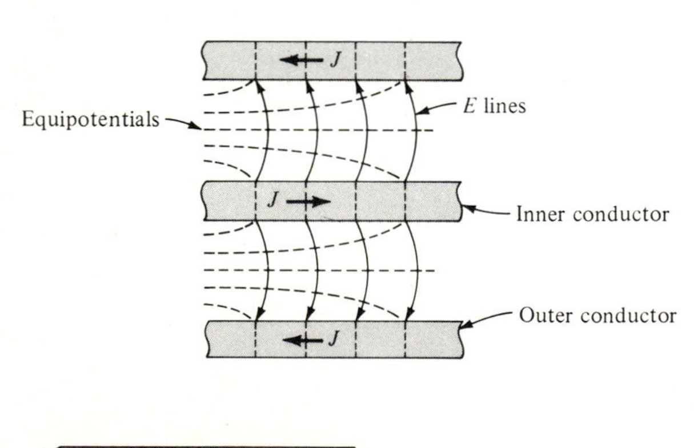

Figure 3.8 A cross section along the axis of a coaxial transmission line, showing the slight curving of the E lines when the conductors carry a current.

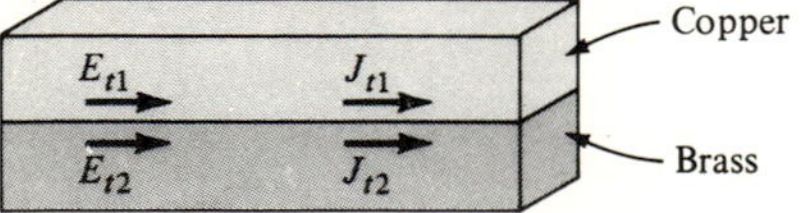

Figure 3.9 A section of a conducting wire constructed by parallel bars of copper and brass.

brass bar. The tangential components of the electric field across the boundary must be continuous [see Eq. (1.86)]; that is,

$$E_{t1} = E_{t2} \tag{3.19}$$

This condition can be derived very simply by noting that the potential imposed across the end faces of the copper bar is the same as that across the brass bar. It therefore follows that the electric field or potential per unit length must be the same for both bars. The current densities have different magnitudes because the conductivities are different. Using Eq. (3.19), we find that the current densities are related by

$$\frac{J_{t1}}{\sigma_1} = \frac{J_{t2}}{\sigma_2} \tag{3.20}$$

3.6 REFRACTION OF CURRENT AT A CONDUCTOR-CONDUCTOR BOUNDARY

When a current crosses an interface between two media of conductivities, σ_1 and σ_2, respectively, it is found that the current-flow lines are refracted. Let us consider a conductor-conductor boundary between two media of constants σ_1, ε_1 and σ_2, ε_2 as shown in Fig. 3.10. The current lines are shown to have kinked in crossing the boundary. Using result (3.20) of the previous section, we know the behavior of the tangential components in crossing the boundary. If we can find an analogous relationship for the normal components, then the behavior of the total current at the interface can be determined. From the continuity equation for steady currents, Eq. (2.12), we can write the boundary condition for the normal components as

$$J_{n1} = J_{n2} \tag{3.21}$$

or

$$\sigma_1 E_{n1} = \sigma_2 E_{n2}$$

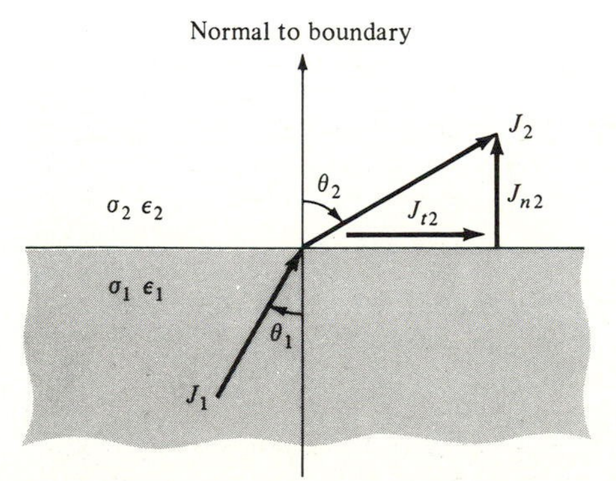

Figure 3.10 In crossing a boundary between different conducting media, current will change direction.

Equation (3.21) can be derived from (2.12*a*) in the same way the boundary condition (1.87*b*) for the normal components of the electric flux density was derived from Gauss' law (see Fig. 1.24 for details). Dividing (3.20) by (3.21), we obtain

$$\frac{1}{\sigma_1}\frac{J_{t1}}{J_{n1}} = \frac{1}{\sigma_2}\frac{J_{t2}}{J_{n2}} \tag{3.22}$$

which can be rewritten by using $\tan\theta = J_t/J_n$ as

$$\frac{\tan\theta_1}{\tan\theta_2} = \frac{\sigma_1}{\sigma_2} \tag{3.23}$$

where the angles θ are shown in Fig. 3.10. The current lines upon crossing a boundary will therefore bend an amount which is proportional to the ratio of the conductivities. It is now important to observe that when current leaves a very good conductor to enter a poor one, its direction is nearly always perpendicular to the boundary in the poorly conducting material. For example, if region 1 is a good conductor and region 2 an insulator, we have $\sigma_1/\sigma_2 \gg 1$. Then, for practically all angles of incidence from medium 1, the current will enter medium 2 at right angles to the surface. This phenomenon is observed in the leakage of current from a conductor into imperfect insulation.

Since the current upon crossing a boundary between two conducting media has a normal component, we can assume that a normal electric field given by Eq. (3.21) also exists at the boundary. Because the conductivities in the two media are different, this normal component is discontinuous across the interface. Using the boundary condition (1.87*b*) for the normal components of the electric field, which is

$$\varepsilon_1 E_{n1} - \varepsilon_2 E_{n2} = \rho_s \tag{3.24}$$

we find that discontinuous normal electric fields induce a surface charge layer ρ_s on the boundary. Re-expressing (3.24) as

$$\varepsilon_1\frac{J_{n1}}{\sigma_1} - \varepsilon_2\frac{J_{n2}}{\sigma_2} = \rho_s \tag{3.25}$$

and using $J_{n1} = J_{n2}$, the surface charge at the interface is given by

$$\rho_s = J_n\left(\frac{\varepsilon_1}{\sigma_1} - \frac{\varepsilon_2}{\sigma_2}\right) \tag{3.26}$$

This surface charge which accumulates on the boundary between two conductors across which current is flowing can be stated in terms of the free-charge rearrangement times (2.34) for the two media as

$$\rho_s = J_n(T_1 - T_2) \tag{3.27}$$

If both media are metallic conductors, the permittivities are related as $\varepsilon_1 \simeq \varepsilon_0 \simeq \varepsilon_2$ and the surface charge can then be written as

$$\rho_s = \varepsilon_0 J_n \left(\frac{1}{\sigma_1} - \frac{1}{\sigma_2} \right) \tag{3.28}$$

From the above remarks we see that the boundary condition at the interface between two dielectrics, given by Eq. (1.87*c*) as $D_{n1} = D_{n2}$, is not strictly true unless $\sigma = 0$ for the dielectrics. Since most dielectrics have a finite conductivity, a small leakage current can exist. The leakage current can give rise to a surface charge density (3.26). For most practical purposes, a surface charge density created in this manner is small and can be ignored. However, in large condensers which employ layers of different dielectrics between their plates, this effect can be significant.†

3.7 LAPLACE'S EQUATION FOR POTENTIAL DISTRIBUTION IN CONDUCTING MEDIA

In the last section we observed that the refraction of current at the interface between two conducting media is analogous to the refraction of electric flux at the interface between two dielectrics. This comes about because the current density **J** and electric flux density **D** are linearly related to the electric field **E** ($\mathbf{J} = \sigma\mathbf{E}$, and $\mathbf{D} = \varepsilon\mathbf{E}$) in most materials. A duality between **J** and **D** and between σ and ε exists therefore in linear, isotropic media. The potential distribution in conducting media can be obtained by starting with the continuity relationship for steady currents (2.12*b*) or (2.46) which is

$$\nabla \cdot \mathbf{J} = 0 \tag{3.29}$$

Introducing Ohm's law $\mathbf{J} = \sigma\mathbf{E}$, we have

$$\sigma\nabla \cdot \mathbf{E} = 0 \tag{3.30}$$

The electric field **E** in a region not containing emf's (that is, $\oint \mathbf{E} \cdot d\mathbf{l} = 0$) can be derived from a scalar potential as shown in Eq. (1.49)

$$\mathbf{E} = -\nabla V \tag{3.31}$$

Substituting this in (3.30), we obtain

$$\nabla^2 V = 0 \tag{3.32}$$

† G. P. Harnwell, "Principles of Electricity and Electromagnetism," 2d ed., p. 107, McGraw-Hill Book Company, New York, 1949.

which is Laplace's equation for potential distribution in a conducting medium. This means that distributions of steady currents in conducting media are analogous to field distributions in dielectric media. If the boundary conditions are equivalent in both cases, solution to problems in current flow can be obtained from the solution to the corresponding problems of the flux density in an insulating media. Once the solution to Laplace's equation for a boundary-value problem is obtained, the static electric field can be obtained from the gradient relationship (3.31). Multiplying $\mathbf{E}$ by the permittivity ε, we obtain the flux density $\mathbf{D} = \varepsilon\mathbf{E}$ in a dielectric medium. On the other hand, multiplying by σ, we obtain the current density $\mathbf{J} = \sigma\mathbf{E}$ for a conducting medium. If the solution to Laplace's equation cannot be obtained directly, various approximate techniques can be employed, such as graphical field mapping,† the finite-difference solution method‡ of Laplace's equation, relaxation methods,§ etc. Numerical techniques using the digital computer are very versatile and are a fast means of solving potential problems.¶

The electrolytic tank* shown in Fig. 3.11 is a good example of some of the ideas presented above. A number of metallic conductors with proper shapes and arrangement are placed in a conducting liquid medium of moderate conductivity, such as a salt solution. The conductors are then connected to external sources of potential with the result that a small current will flow between the conductors. A small probe introduced in the conduction field and connected to a high-input-

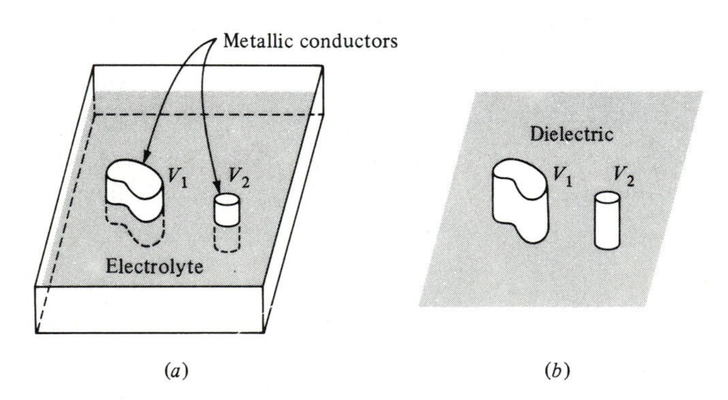

Figure 3.11 (*a*) An electrolytic tank showing two metallic conductors at potentials V_1 and V_2; (*b*) the solution to the conduction problem is also the solution to the equivalent electrostatic problem.

† W. B. Boast, "Vector Fields," Harper & Row, Publishers, Incorporated, New York, 1964.

‡ G. E. Forsythe and W. R. Wasow, "Finite Difference Methods for Partial Differential Equations," John Wiley & Sons, Inc., New York, 1960.

§ V. R. Southwell, "Relaxation Methods in Theoretical Physics," Oxford University Press, Fair Lawn, N.J., 1946.

¶ R. F. Harrington, Matrix Methods for Field Problems, *Proc. IEEE*, February 1967.

* D. L. Amort, The Electrolytic Tank Analog, *Electro-Technol.*, vol. 70, July 1962.

impedance voltmeter can explore the potential distribution in the electrolyte solution. Equipotential lines are established as the locus of all points in the solution for which the voltmeter maintains a constant reading. The flow lines, corresponding to **J** lines and **E** lines, can then be drawn normal to the equipotential lines. Since the conductivity of the electrolyte is much smaller than that of the metallic conductors, the electric field in the metal is much smaller than that in the conductive solution. This allows us to assume that the surfaces of the metallic conductors are equipotentials. Thus we can obtain solutions to Laplace's equation for geometrically complicated conductors, which might be very difficult to do analytically.

The solution found for the conduction problem can now be applied to find the field lines between the conductors in an equivalent electrostatic problem in which the same metallic conductors are surrounded by a dielectric medium, as shown in Fig. 3.11*b*.

In the next section it will be shown that the distribution of electric field and current density must be known before solutions for the resistance of arbitrarily shaped conductors can be calculated.

3.8 RESISTANCE OF CONDUCTORS OF ANY SHAPE

Ohm's law gives us the resistance R of a two-terminal device in terms of voltage V and current I between the two terminals as $R = V/I$. We can use this simple formula to obtain the resistance of an arbitrarily shaped conductor in which the voltage and current are distributed throughout the conductor by simply substituting the integral expressions for V and I into Ohm's law. Let us consider Fig. 3.12 which shows a curved piece of conductor with a varying cross section. We will assume that the end surfaces are coated with a material whose conductivity is much higher than that of the wire. The assumption of equipotentials for the end surfaces can then be made. The total resistance between the terminal surfaces a and b is then

$$R_{ab} = \frac{V_{ab}}{I_{ab}} = \frac{\int_a^b \mathbf{E} \cdot d\mathbf{l}}{\iint \mathbf{J} \cdot d\mathbf{A}} = \frac{\int_a^b \mathbf{E} \cdot d\mathbf{l}}{\iint \sigma\mathbf{E} \cdot d\mathbf{A}} \tag{3.33}$$

where ab is any path from surface a to surface b. The denominator represents the total current flowing between the end surfaces and is, therefore, an integration over any cross-sectional surface of the conductor. The above formula, although simple in concept, is usually impossible to evaluate for an arbitrarily shaped

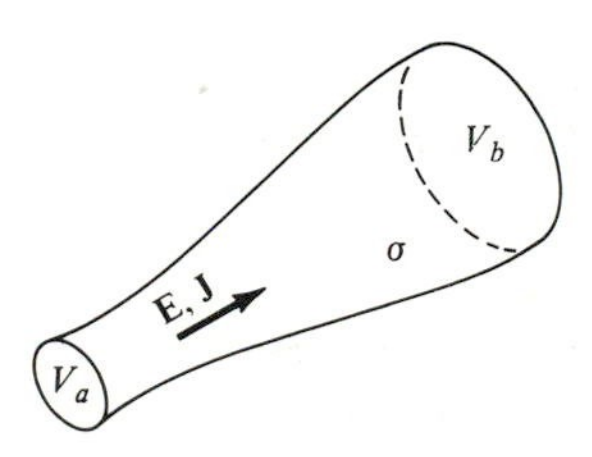

Figure 3.12 An arbitrarily shaped conductor. The total resistance between the end faces which are held at constant potentials V_a and V_b is desired.

conductor. When the total resistance between two bodies which are immersed in a conductive medium is needed, one usually resorts to graphical mapping techniques or electrolytic tank methods. On the other hand if the conducting medium has a degree of symmetry, Eq. (3.33) can be very useful in obtaining the total resistance. Let us consider a few examples.

Example: Resistance of a rectangular bar If we wish to find the end-to-end resistance of a bar of isotropic, homogeneous material whose conductivity σ is a constant, Eq. (3.33) is easily applicable. Figure 3.13 shows the geometry. For simplicity let us also assume that the connecting leads have zero resistance. The voltage between the end faces is simply

$$V = \int \mathbf{E} \cdot \mathbf{dl} = EL \tag{3.34a}$$

and the total current is

$$I = \sigma \iint \mathbf{E} \cdot \mathbf{dA} = \sigma EA \tag{3.34b}$$

which gives for the end-to-end resistance

$$R = \frac{L}{\sigma A} \quad \Omega \tag{3.34c}$$

When σ and A vary along a wire or bar, the resistance can be expressed as

$$R = \int \frac{dl}{\sigma A} \tag{3.34d}$$

Example: Resistance of a curved rectangular bar An example of a more complicated resistor is obtained by bending the bar of conducting material as shown in Fig. 3.14. If the curved sides are

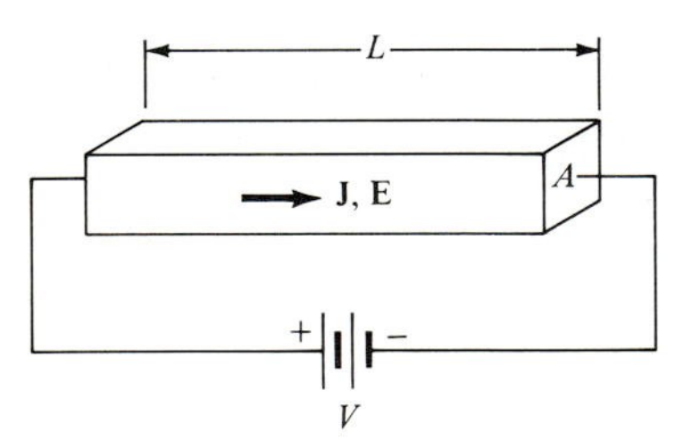

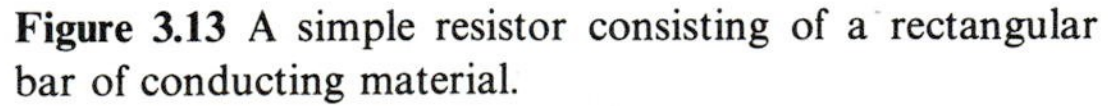
Figure 3.13 A simple resistor consisting of a rectangular bar of conducting material.

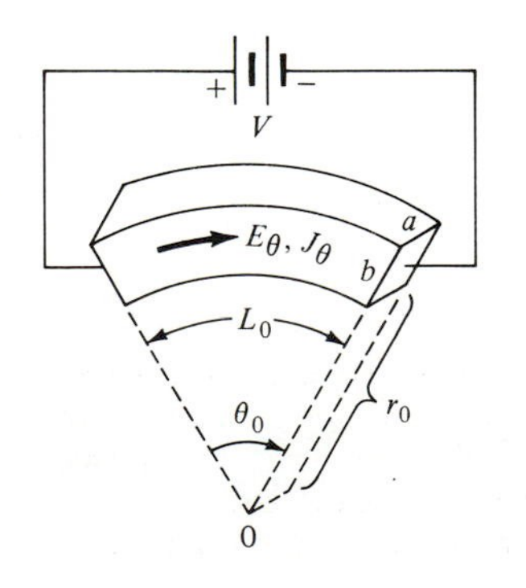

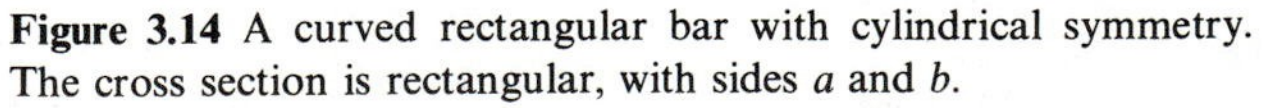
Figure 3.14 A curved rectangular bar with cylindrical symmetry. The cross section is rectangular, with sides a and b.

circular arcs whose center is 0, the equipotential surfaces are planes at right angles to the curved sides and passing through 0. The current-flow lines and E lines are then circular arcs also. The electric field E must vary with r across the curved bar, since the lower portions of the end faces are closer to each other than the top portions. The variation is given by $E_\theta = K/r$, where K is some constant, because the electric field for this problem is $E_\theta = V/L = V/(r\theta)$. The voltage between the end faces is then

$$V = \int \mathbf{E} \cdot d\mathbf{l} = \int_0^{\theta_0} \frac{K}{r} r \, d\theta = K\theta_0 = K \frac{L_0}{r_0} \tag{3.35a}$$

The total current between the end faces is given by

$$I = \iint \sigma \mathbf{E} \cdot d\mathbf{A} = \sigma \int_0^a \int_{r_0}^{r_0+b} \frac{K}{r} \, dr \, da$$

$$= \sigma a \int_{r_0}^{r_0+b} \frac{K}{r} \, dr = \sigma a K \ln \left(1 + \frac{b}{r_0}\right) \tag{3.35b}$$

which gives for the end-to-end resistance

$$R = \frac{1}{\sigma} \frac{L_0}{a r_0 \ln (1 + b/r_0)} \tag{3.35c}$$

Example: Resistance of a cylindrical shell A section of a coaxial cable, shown in Fig. 3.15, consists of a round, solid inner conductor of radius r_a surrounded by a tubular shell of radius r_b. When the cable is used as a transmission line, current flows along the inner conductor and returns in the outer concentric shell. However, when the space between r_a and r_b is filled with an imperfect dielectric, a radial leakage current tends to flow between the inner and outer conductors. The magnitude of this current depends on the conductivity σ of the dielectric. Let us assume that the inner and outer cylindrical surfaces are maintained at the constant potentials V_a and V_b, respectively. We would like to find what the total resistance of the space between r_a and r_b is. Since the total radial current

$$I_r = \iint J_r \, dA = J_r 2\pi r L \tag{3.36a}$$

that flows through any cylindrical section of radius $r(r_a < r < r_b)$ and length L is the same, we conclude that the radial electric field and current density must have an inverse radial dependence; that is,

$$J_r = \frac{K}{r} \tag{3.36b}$$

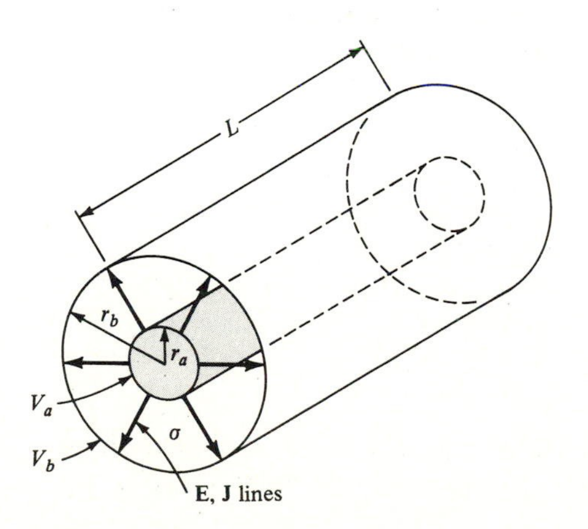

Figure 3.15 Leakage current that flows between the inner and outer conductors of a section of a coaxial line is shown by the radial arrows. The "leaky" space between r_a and r_b can be considered as a cylindrical shell resistor.

where K is some constant. The voltage between the inner and outer conductors $V_{ab} = V_a - V_b$ is then given by

$$V_{ab} = \int_{r_a}^{r_b} E_r\, dr = \int_{r_a}^{r_b} \frac{K}{\sigma r}\, dr = \frac{K}{\sigma} \ln \frac{r_b}{r_a} \tag{3.36c}$$

The total leakage resistance is then

$$R = \frac{V_{ab}}{I_r} = \frac{1}{\sigma 2\pi L} \ln \frac{r_b}{r_a} \tag{3.36d}$$

3.9 ENERGY ASSOCIATED WITH CURRENT FLOW JOULE'S LAW

Joule's heating law, or the energy needed to maintain a steady current through a conducting medium of total resistance R, was derived in Sec. 2.4 of the previous chapter. We found that when a conductor carries a current, the work done by the forces of the field is dissipated as heat because of the electron collisions with the lattice. When a small charge dQ moves through a potential difference, the work done by the forces of the field is

$$dW = V\, dQ \tag{3.37}$$

Since power P is the rate of work done by the field, we have that

$$P = \frac{dW}{dt} = V \frac{dQ}{dt} = VI \tag{3.38}$$

which by use of Ohm's law, becomes Joule's law

$$P = I^2 R \tag{3.39}$$

This expression cannot be applied directly to a field point. On the other hand, power per unit volume can be defined at a point and is given by the limit as the unit volume shrinks to zero. It is then related to the field strength and current density at that point. This differential form of Joule's law is given from Eq. (2.18) as

$$\frac{\Delta P}{\Delta v} = \mathbf{E} \cdot \mathbf{J} \qquad \text{W/m}^3 \tag{3.40}$$

It states that the electric field $\mathbf{E}$ loses $\mathbf{E} \cdot \mathbf{J}$ watts per unit volume to the steady current flow $\mathbf{J}$. For conducting media which is isotropic, that is, media with the same properties in all directions, the conductivity σ is a scalar and the above expression simplifies to

$$\frac{\Delta P}{\Delta v} = EJ = \sigma E^2 \tag{3.41}$$

The same expression (3.40) can be used when a convection current instead of a conduction current is involved. A *convection current* is represented by a stream of charged particles such as electrons in space. The current density at each point in such a stream of particles is given by Eq. (2.6) as

$$\mathbf{J} = \rho \mathbf{v}' \tag{3.42}$$

where $\mathbf{v}'$ is the velocity of the charged particles at any point. If the charged particles are under the influence of an electric field, the energy per unit time and volume transferred to the particles from the electric field is then given by Eq. (3.40) as

$$\frac{\Delta P}{\Delta v} = \rho \mathbf{E} \cdot \mathbf{v}' \tag{3.43}$$

Note that the energy absorbed by the charged-particle stream is not converted into heat but into additional acceleration of the particles. The kinetic-energy increase of the particles is then given by (3.43).

We can now define the total resistance of a conducting medium in terms of energy. Referring to Fig. 3.12 which shows an arbitrarily shaped conductor, the power dissipated in the conductor as a whole is given by I^2R. Using the differential form (3.40), we can write for total power dissipation

$$P = I^2R = \iiint \mathbf{E} \cdot \mathbf{J}\, dv \tag{3.44}$$

which gives for the total resistance R between the terminal faces

$$R = \frac{\iiint \mathbf{E} \cdot \mathbf{J}\, dv}{I^2} \tag{3.45}$$

To show that the above expression is equivalent to (3.33) is left as a problem.

PROBLEMS

3.1 Electric energy is paid for by the kilowatthour. What is the cost of operating a 110-V heater which has an internal resistance of 10 Ω for 8 h at the rate of 4 cents/kWh?

3.2 A 12-V battery has an initial charge of 200 A · h. Assuming the potential difference across its terminals remains constant until the battery is discharged, for how many hours can the battery deliver 200 W?

3.3 An electric car uses fuel cells (80 Wh/lb). If it is a small-sized electric car, we can use the figure of 0.5 kWh/mi for normal driving conditions. How many pounds of fuel-cell batteries per mile are needed?

3.4 (*a*) Calculate the power loss in a 50-m long No. 10 copper wire (2.59 mm diameter) if the wire is at a temperature of 20°C and a voltage of 100 V is applied across the ends.

(*b*) If the temperature is raised to 50°C, what is the power loss?

(*c*) Find the energy lost in heat in the wire in 10 h.

3.5 A 100-m long Nichrome heating wire has a resistance of 100 Ω. Is it possible to obtain more heat by winding one coil or by cutting the wire in half and winding two separate coils? Assume that in each case the coils are connected across a 110-V line.

3.6 A circular 1-m long copper rod has a diameter of 4 mm.

(*a*) Find the resistance between its ends.

(*b*) What must the cross-sectional edge size of a square 1-m long aluminum rod be if its resistance is to be the same?

3.7 How long will it take for a 1000-W immersion heater to bring a 1-l pot of water initially at 20°C to boiling temperature? Assume that 70 percent of the available energy is absorbed by the water.

3.8 It is desired to generate heat at the rate of 100 W by connecting a 1-Ω resistor across a battery whose emf is 12 V.

(*a*) Find the potential difference across the resistor.

(*b*) Find the internal resistance of the battery.

3.9 Referring to Fig. 3.4, for what value of R_L is the power delivered to R_L a maximum? What is the maximum power?

3.10 (*a*) Find the **E** field in a 1-mm diameter wire if it has a resistance of 1 Ω/km and carries a current of 1 A.

(*b*) If, in addition, a uniform static surface charge of 10^{-11} C/m^2 is placed on the wire, find the magnitude and direction of the **E** field that now exists just outside the surface of the wire.

3.11 A current flows across the flat boundary between two media.

(*a*) If the current direction in medium 1 ($\sigma = 10$ S/m, $\varepsilon = \varepsilon_0$) is at 40° to the normal, find the angle of the current in medium 2 ($\sigma = 1$ S/m, $\varepsilon = 4\varepsilon_0$).

(*b*) If the current density in medium 1 is 1 A/m^2, calculate the surface charge density at the boundary.

3.12 Find the resistance between the ends of a solid, round aluminum wire of radius a and length l if surrounding the aluminum wire is a copper sleeve of thickness t. Does it matter whether an infinitesimally thin insulating sleeve separates the aluminum and copper material or whether the copper material is in direct contact with the aluminum?

3.13 A material of conductivity σ is used to construct a flat washer of thickness t and inner and outer radii r_a and r_b, respectively. Find the resistance between the inner and outer edges as well as between the flat surfaces.

3.14 A resistor is in the form of a spherical shell with inner and outer radii r_a and r_b, respectively. The conductivity of the shell is σ. Find the resistance between the inner and outer surfaces.

3.15 A resistor is in the form of a truncated right circular cone. The length of the cone is h, and the radii of the two ends are a and b, respectively. Find the resistance between the two ends in terms of a, b, h, and the conductivity σ.

CHAPTER
FOUR

DIELECTRICS AND POLARIZATION

Guide to the Chapter

The main ideas of the chapter are developed in Secs. 4.4 and 4.5. Preceding the formal development is Sec. 4.1 which describes, at the molecular level, the behavior of a dielectric in the presence of an electric field and Sec. 4.2 which shows that macroscopic fields are averages of microscopic fields. A section on fields of electric dipoles, Sec. 4.3, is next because polarization is a dipole phenomena. The formal part on dielectrics can be considered complete with Sec. 4.7. In this section polarization notions are applied to conducting materials; doing so should build a better understanding of polarization. The remaining sections are applications of polarization. The examples in those sections are developed in detail. A study of some fully developed examples should give students confidence to work problems on their own.

4.1 POLARIZABILITY OF DIELECTRICS

It was shown in Chap. 1 that the electric field is a function of the permittivity ε of the medium. For example, the field of a point charge Q is given by $E = Q/4\pi\varepsilon r^2$. In Sec. 1.10 a flux density $\mathbf{D}$, given by $\mathbf{D} = \varepsilon\mathbf{E}$, which is independent of the medium, was introduced. Therefore, the understanding of force and work functions such as electric and potential fields is intimately related to an understanding of the permittivity ε of a medium. As in the study of conductivity σ for conducting materials (Sec. 2.5), where we had to examine the material microscopically, we will again have to resort to a microscopic examination of dielectric materials in order to understand ε.

In Sec. 2.1, we discussed briefly the differences between conductors and dielectrics. We pointed out that atoms of dielectric materials have their outermost electron shell almost completely filled. A characteristic of complete shells is that it is relatively difficult to dislodge an electron from the shell (it is a bound electron). The result is that dielectric materials have few electrons available for conduction of current and hence are classified as insulators, in contrast to metals that have an abundance of free electrons which can travel over large macroscopic distances inside the metal. The same conclusion can be obtained by examining the rearrangement time T (Sec. 2.8), which is the time it takes for a free charge when deposited at a point inside a material to vanish from that point. For metals, T was approximately 10^{-18} s, or an extremely brief time, after which the free charge will appear as a distribution on the surface of the metal. For a dielectric, such as mica, for example, T is approximately one-half of a day. Thus, a free charge deposited inside a mica object will, for practical purposes, stay there and can be considered as a bound charge. Free-versus-bound charges are then the essential difference between metals and dielectrics, with semiconductors falling somewhere in between.

Let us examine the effect that a charge Q placed in the interior of a dielectric medium has on the surrounding dielectric. Figure 4.1 shows a positive charge Q and its tendency to polarize the surrounding atoms. The radially outward force field of Q displaces the negative electron clouds and the positive nuclei of each atom in opposite directions creating or inducing dipoles throughout the dielectric. A *dipole* is defined as a pair of oppositely charged charges q separated by a small distance d from each other. The direction of a dipole is along the axis connecting the two charges. The *dipole moment* **p** can then be defined as

$$\mathbf{p} = q\mathbf{d} \tag{4.1}$$

Note that the dipole as a whole is electrically neutral since it is made up of opposite charges of equal strength. As a measure of intensity of polarization of a

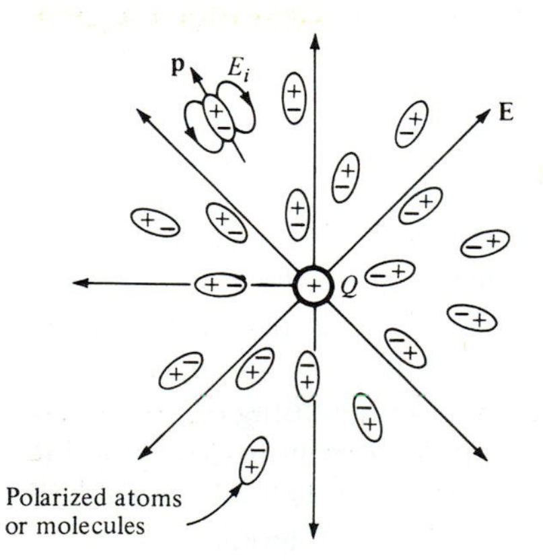

Figure 4.1 A free charge Q introduced inside an insulating medium will polarize the surrounding atoms as shown. One atom is shown with its dipole moment p and surrounding electric field E_i.

material, the total dipole moment **p** per unit volume v will be used; that is,

$$\mathbf{P} = \frac{\mathbf{p}}{v} \tag{4.2}$$

Dipoles are created in the dielectric because the electrons cannot be detached from their atoms. The electrons respond by shifting with respect to the nuclei along the direction of the applied field. If, on the other hand, the dielectric were replaced by a metal, the electrons in Q would displace the electrons on the adjacent atoms, which in turn would displace the electron on the atoms next to them until the charge Q finally appeared on the surface of the metal.

Perhaps it is clearer now that the field of a point charge which is placed in a dielectric medium of permittivity ε, that is, $E = Q/4\pi\varepsilon r^2$, should be dependent on the medium. After all, the atoms after being polarized create their own field called the induced field $\mathbf{E}_i$, which opposes the field of the point charge. The resultant field ($E = Q/4\pi\varepsilon r^2$) is different from the field ($E = Q/4\pi\varepsilon_0 r^2$) of a point charge Q in empty space. We will show that the permittivity ε can be considered to be a measure of polarizability of the dielectric.

Polar and Nonpolar Molecules

Before we continue, we must distinguish between dielectrics which have two types of molecules, polar and nonpolar molecules. *Polar molecules* have a permanent dipole moment. They are electrically neutral, but have their centers of positive and negative charge permanently displaced. For example, NaCl is a polar molecule. One end of the dipole has the negative chlorine atom, while the other is occupied by the positive sodium atom. Figure 4.2 shows polar molecules in the absence and presence of an external field. We see that while in the absence of an orienting field, the thermal agitation of the molecules randomizes their direction. They tend to align along the direction of an applied field when such a field is present. *Nonpolar molecules*, on the other hand, have no electric dipole moment in the absence of an applied field, as shown in Fig. 4.3*a*. An external field polarizes such nonpolar molecules along the external electric field **E**, as shown in Fig. 4.3*b*. Even though random thermal motion tends to randomize this alignment, each dipole experiences a torque [see Eq. (4.3)].

$$\mathbf{T} = \mathbf{p} \times \mathbf{E}$$

which realigns the dipoles with the applied field.

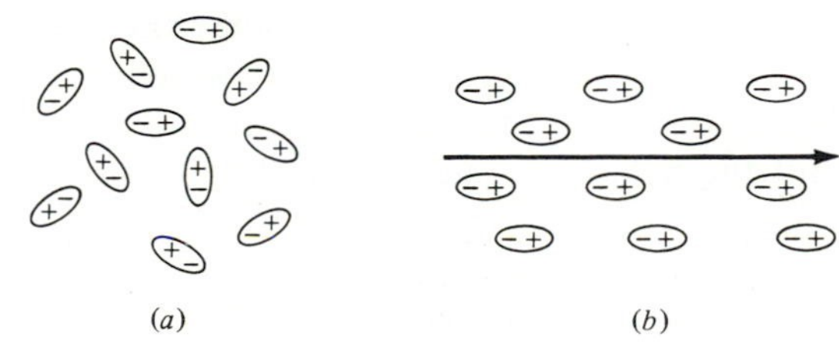

Figure 4.2 (*a*) Thermal agitation randomizes polar molecules; (*b*) a field applied to polar molecules aligns their dipoles.

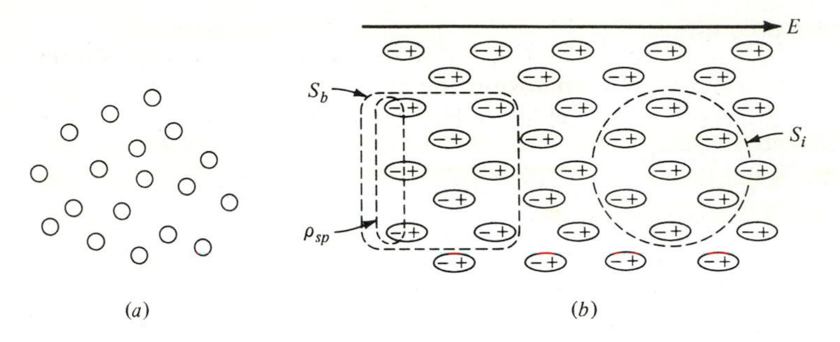

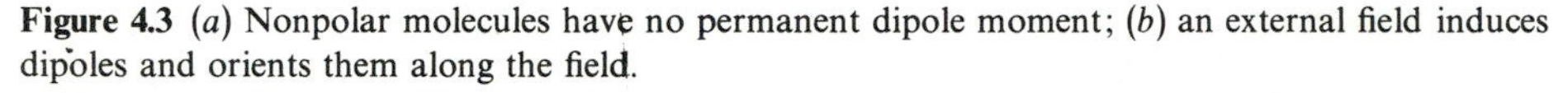

Figure 4.3 (*a*) Nonpolar molecules have no permanent dipole moment; (*b*) an external field induces dipoles and orients them along the field.

A third type of dielectric exhibits a permanent polarization in the absence of an applied field, and so can be thought of in the same sense as a permanent magnet. Some waxes, for example, when molten and placed in a strong electric field will retain some of the induced polarization after cooling even after the externally applied field is removed. This special case of aligned polar molecules occurs in ferroelectric materials below a certain temperature called the *Curie point*. For example, barium titanate has a permanent polarization below 120°C. These materials are analogous to permanent magnets and are known as *ferroelectrics*. The description of this class of dielectrics is beyond the scope of this book and is found in solid-state physics texts.† However, in Secs. 4.8 and 4.9 we will calculate the external field produced by a permanently polarized rod and sphere.

We should note that polarization does not give rise to a net charge inside a dielectric. If we refer to the interior volume S_i shown in the above figure, we see that an equal amount of negative and positive charge is enclosed by it. A net amount of polarization charge does appear on the surface of a polarized dielectric. If we examine the volume enclosed by S_b, we can see that the layer of charge nearest the boundary of the material remains uncanceled and appears as a surface polarization charge ρ_{sp}. The degree of polarization will be measured in terms of the induced surface charge, which for small external fields, will be proportional to the magnitude of the field.

4.2 MICROSCOPIC AND MACROSCOPIC FIELDS

At this point we must distinguish between microscopic and macroscopic concepts.‡ For example, the polarization **P** is a macroscopic field which was introduced to account for the change of atoms to atomic dipoles in dielectric media exposed to an electric field. But the field from these induced dipoles is a

† C. Kittel, "Solid State Physics," John Wiley & Sons, Inc., New York, chap. 8, 1956.

‡ See also footnote on page 81.

microscopic field, hence defining **P** as $\mathbf{P} = \mathbf{p}/\Delta v$, where p is the atomic dipole moment and Δv is a small volume enclosing the atom, makes **P** a microscopic field. We have now a contradiction which is best resolved by examining how **P** varies throughout the dielectric.

Matter consists principally of empty space interspersed with atoms and molecules. In solids these atoms are more densely packed than in gases. In turn, an atom is mostly empty space with the nucleus and electron occupying a tiny fraction of its volume. To get a fairly good idea of the relative emptiness of the "atomic world," we can look at our own sun and planets which also occupy only a small fraction of space. More quantitatively, we picture an atom consisting of a heavy positively charged nucleus of approximate radius of 10^{-15} m. Negatively charged electrons travel around the nucleus in elliptical orbits. These electrons are about the size of the nucleus; i.e., the radius of the electrons is approximately 10^{-15} m. The radius of the electron orbits about the nucleus is about 10^{-10} m. The spacing between the centers of adjacent atoms (at normal temperature and pressure) varies between 10^{-10} m in solids to about 10^{-8} m in gases. It is now clear that **P** plotted on a microscopic scale would look like a series of uneven spikes with zero value in between. Such a detailed interatomic description of **P** or the **E** field, besides being too complicated, is of no interest to us; we are primarily concerned with large-scale or macroscopic fields which are continuous in nature and are obtained by averaging over many atoms. For example, we do not view the density of a slab of dielectric such as wax as being a discontinuous function of space. Macroscopically, it is continuous for as small a piece of wax as we can obtain. What is macroscopically small? As already pointed out in Sec. 2.12, a cube with sides of 1 micron ($1\ \mu\text{m} = 10^{-6}$ m) would have a volume $\Delta v = (\mu\text{m})^3 = 10^{-18}\ \text{m}^3$, which is certainly physically small, and we could call it a point. Yet, such a volume would still contain 10^{11} atoms (i.e., microscopically speaking it is large). As a **P** or **E** field of a dielectric material measured over a series of adjacent cubes, each the size of a cubic micron, is a continuum field which does not fluctuate from cube to cube, we call **P** or **E** a macroscopically smooth field. The value at each point is obtained by averaging over a cell or cube, macroscopically small but containing many billions of atoms.

In the light of this discussion, let us reexamine the definition of polarization **P**, i.e., the dipole moment per unit volume as given by Eq. (4.2). Since we want it to be a continuous macroscopic field, it is obtained by averaging over a physically small volume element Δv as†

$$\mathbf{P} = \frac{\sum \mathbf{p}_i}{\Delta v} = \frac{\mathbf{p}}{\Delta v} \tag{4.2a}$$

where p_i is the dipole moment of each atomic dipole and the summation is carried out over all atomic dipoles in Δv. Therefore **p** is an average dipole moment that exists in an elemental volume Δv (on the order of, say, a cubic micron). Each

† Frequently **P** is expressed as $\mathbf{P} = N\mathbf{p}_i$, where N is the number of atomic dipoles per unit volume. This expression follows directly from (4.2a).

elemental volume Δv of polarized dielectric possesses a small dipole with moment $\mathbf{p} = \mathbf{P}\Delta v$. In the macroscopic sense, the value of $\mathbf{P}$ throughout the interior of a dielectric is now a smoothly varying function. Later in the chapter we will use $\mathbf{P}$ to determine other macroscopic quantities such as $\mathbf{D}$ and $\mathbf{E}$ for a dielectric. Because it is obtained by averaging, $\mathbf{P}$ provides for us the link between the atomic world and such measurable† quantities as $\mathbf{E}$ and $\mathbf{D}$.

4.3 THE FIELDS OF AN ELECTRIC DIPOLE

Torque and Work Associated with an Electric Dipole

The electrostatic dipole illustrated in Fig. 4.4 consists of two opposite charges separated by a distance $\mathbf{d}$. The distance $\mathbf{d}$ is a vector because it also specifies the orientation of the dipole. The strength of the dipole is given by the dipole moment $\mathbf{p} = q\mathbf{d}$, because the torque, tending to align the dipole axis along the field E, is given by

$$\mathbf{T} = \mathbf{p} \times \mathbf{E} \tag{4.3}$$

This is easily derived using the definition of *torque* which is force times moment arm. Referring to Fig. 4.5, the torque on the dipole is

$$T = Fd \sin \theta = qdE \sin \theta$$

Equation (4.3) is a vector expression of this result.

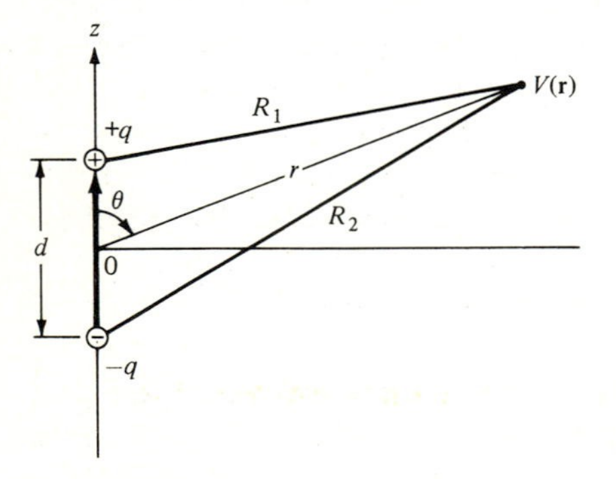

Figure 4.4 An electric dipole located at the origin and oriented along the z axis. The fields at $\mathbf{r}$ are to be obtained.

† In a sense, all macroscopic fields (frequently referred to as observables or measureables) are averaged quantities. Temperature, pressure, etc., all depend on the action of countless atoms and molecules. An observable is not only a space average but also a time average, because the individual atoms move about constantly while the electrons whirl in orbits about their nuclei (see Sec. 2.1). Thus, a space-time macroscopic description is a space-and-time average of the fields over small volumes Δv and small intervals of time Δt. The intervals are usually much smaller than the resolution capabilities of our measuring instruments. A cubic micron is very small. When differentiating variables with respect to time, Δt can be as small as 10^{-12} s. Microscopically speaking, this still is a long time, as an electron may complete thousands of orbits in that interval of time.

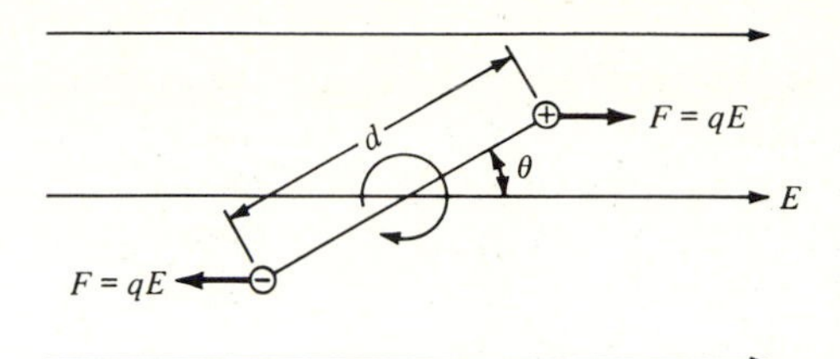

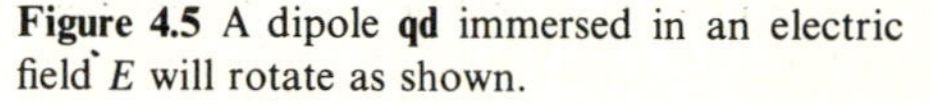

Figure 4.5 A dipole **qd** immersed in an electric field E will rotate as shown.

The dipole when aligned with the field ($\theta = 0°$) experiences no torque, while maximum torque is obtained at right angles to the field. To rotate the dipole in an electric field, work must be done. We can associate a potential energy W with reorienting a dipole in an external field. This energy is given by

$$W = \int T \, d\theta = \int_{\pi/2}^{\theta} pE \sin\theta \, d\theta = -pE\cos\theta \tag{4.4}$$

where for simplicity the zero potential energy position was chosen to coincide with the angle $\theta = \pi/2$. Using vector notation, Eq. (4.4) can be expressed as a dot product

$$W = -\mathbf{p} \cdot \mathbf{E} \tag{4.5}$$

which gives a minimum energy $W = -pE$ when the dipole is aligned with the field ($\theta = 0$) and a maximum $W = +pE$ when $\theta = \pi$, as shown in Fig. 4.6. The work involved when a dipole "flips" by 180° is, therefore,

$$W = 2pE \tag{4.6}$$

It takes work of $2pE$ J to rotate a dipole from the minimum energy configuration to the maximum energy position. This amount of work is released when the dipole flips back to its minimum energy position. Equation (4.6) can be easily derived another way by noting that "flipping" a dipole means that the $+q$ must be transported a distance d to the position where $-q$ is, and $-q$ must be transported a distance d to where $+q$ is, which involves an amount of work equal to $W = 2dF = 2dqE = 2pE$.

At this point, we should note that the net force on a dipole which is immersed in a uniform field is zero. This is because the two forces acting on the two charges of the dipole are opposite to each other. Therefore, in a uniform field the dipole does not move translationally; it only rotates until it is aligned with the field. In general, since work and force are related by $\mathbf{F} = -\nabla W$, the force on a dipole is given by use of Eq. (4.5) as

$$\mathbf{F} = \nabla(\mathbf{p} \cdot \mathbf{E}) \tag{4.7}$$

Figure 4.6 (a) Minimum energy position of a dipole $\mathbf{p} = q\mathbf{d}$ in an external field $\mathbf{E}$; (b) maximum energy position of a dipole.

For fixed **p** we can move ∇ past it so that it operates on **E** directly. With the help of vector identities† in the appendix, the force can be written as $\mathbf{F} = (\mathbf{p} \cdot \nabla)\mathbf{E}$, which shows that the force is zero in a uniform field because the derivative of a constant **E** is zero. In a nonuniform field, however, the dipole will move rectilinearly because one charge of the dipole is in a slightly stronger field than its companion charge.

The Ammonia Maser—A Simple View

In the following chapters the behavior of a dipole in a field can be used to explain the principle of a maser or laser. For example, in the ammonia maser, the ammonia molecules NH_3 have a pyramidal configuration where the three hydrogen atoms are at the corners of a triangle with the nitrogen atom positioned either above or below the plane of the triangle, as shown in Fig. 4.7. The ammonia molecule has an electric dipole moment because the centroid of all its negative electric charges does not coincide with the centroid of all its positive electric charges. The laws of quantum mechanics allow the nitrogen atom to "flip" through the plane of the hydrogen atoms. This inversion of the pyramid gives the ammonia molecule a dipole moment with two energy states, a high and a low one. The high one is usually referred to as the excited state of the NH_3 molecule.

In the ammonia maser, ammonia gas is first heated and the excited molecules are collected by passing the ammonia gas through a nonuniform electric field which deflects electric dipoles of opposite moments in different directions. The excited molecules are passed through a cavity where they give their energy to the cavity by "flipping" to their lower energy state. The radiation frequency f of the molecules, to which also the cavity is tuned, is given by Planck's law

$$\Delta W = hf \tag{4.8}$$

where h is Planck's constant and ΔW is the energy between the high and low energy states of the ammonia dipoles. The radiation frequency of the ammonia maser or oscillator can, therefore, be related to the dipole moment of the molecules by

$$f = \frac{2pE}{h} \tag{4.9}$$

Unfortunately, the exact expressions are more complicated than the above equations, since quantum mechanical considerations such as transition probabilities must be considered.‡ Nevertheless, the above does present the main principles behind the maser.

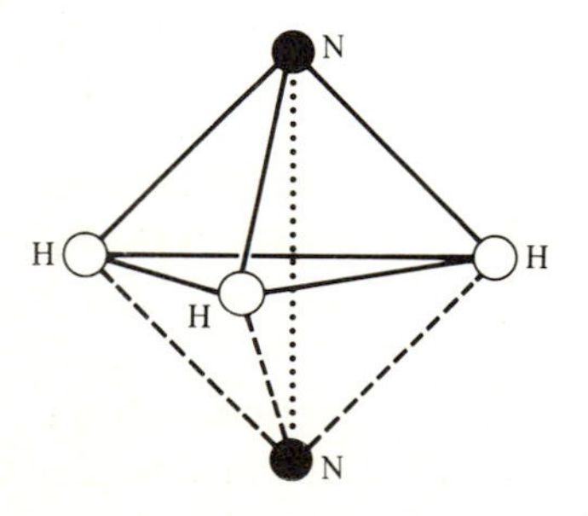

Figure 4.7 The ammonia molecule (NH_3). Since the nitrogen atom can exist on both sides of the plane formed by the three hydrogen atoms, NH_3 shows an inversion, or flipping, frequency of 23.87 MHz.

† $\nabla(\mathbf{p} \cdot \mathbf{E}) - (\mathbf{p} \cdot \nabla)\mathbf{E} = \mathbf{p} \times (\nabla \times \mathbf{E}) = 0$ since for static fields curl $\mathbf{E} = 0$.

‡ J. R. Singer, "Masers," John Wiley & Sons, Inc., New York, 1959.

Electric and Potential Fields of an Electric Dipole

The potential at $\mathbf{r}$ in Fig. 4.4 is obtained by adding the potentials of each charge. Thus

$$V(\mathbf{r}) = \frac{1}{4\pi\varepsilon}\left(\frac{q}{R_1} - \frac{q}{R_2}\right) \tag{4.10}$$

Applying the law of cosines, we have

$$R_1^2 = r^2 + \left(\frac{d}{2}\right)^2 - 2r\frac{d}{2}\cos\theta \tag{4.11}$$

Since the dipole is considered to be small, it is valid to assume that the observation distance will always be many dipole lengths; that is, $r \gg d$. With this approximation we can write

$$\begin{aligned} R_1 &\cong r - \frac{d}{2}\cos\theta \\ R_2 &\cong r + \frac{d}{2}\cos\theta \end{aligned} \tag{4.11a}$$

Note that this approximation implies that R_1 and R_2 are parallel. Equation (4.10) can then be written as

$$\begin{aligned} V(\mathbf{r}) &\cong \frac{q}{4\pi\varepsilon}\,\frac{R_2 - R_1}{R_1 R_2} \\ &= \frac{q}{4\pi\varepsilon}\,\frac{d\cos\theta}{r^2[1 - (d/2r)\cos\theta][1 + (d/2r)\cos\theta]} \\ &\cong \frac{q}{4\pi\varepsilon}\,\frac{d\cos\theta}{r^2} \end{aligned} \tag{4.12}$$

The last term was obtained by again invoking the inequality $d/r \ll 1$ and discarding the $(d/2r)\cos\theta$ factor as being small when compared with 1. Using the electric dipole moment notation and the fact that $\hat{\mathbf{z}} \cdot \hat{\mathbf{r}} = \cos\theta$, we can write for (4.12)

$$\boxed{V(\mathbf{r}) = \frac{1}{4\pi\varepsilon}\,\frac{\mathbf{p}\cdot\hat{\mathbf{r}}}{r^2}} \tag{4.12a}$$

The electric field is obtained by operating on the potential function with the gradient. Spherical coordinates are convenient here especially if we consider the dipole to be located along the z axis which makes the dipole symmetric about the z axis; that is, $\partial/\partial\phi \equiv 0$. The $\mathbf{E}$ field is then

$$\mathbf{E} = -\nabla V(\mathbf{r}) = -\left(\frac{\partial V}{\partial r}\hat{\mathbf{r}} + \frac{1}{r}\frac{\partial V}{\partial\theta}\hat{\boldsymbol{\theta}}\right) \tag{4.13}$$

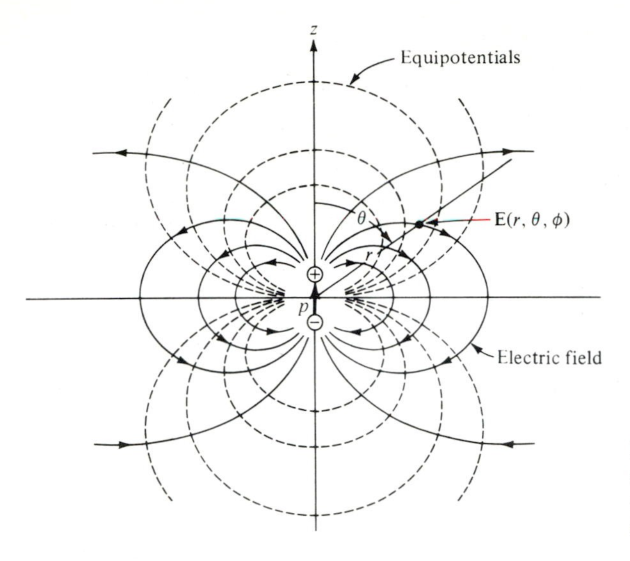

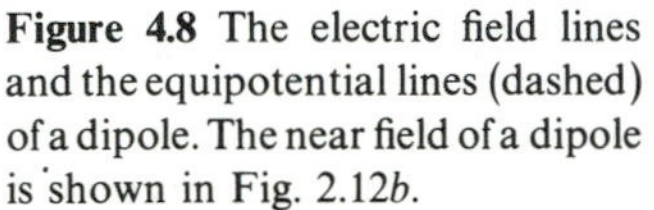
Figure 4.8 The electric field lines and the equipotential lines (dashed) of a dipole. The near field of a dipole is shown in Fig. 2.12*b*.

where the ∇ operator in spherical coordinates is given on the inside of the back cover. Performing the indicated derivatives, we obtain

$$\mathbf{E}(\mathbf{r}) = \frac{p \cos \theta}{2\pi\varepsilon r^3}\hat{\mathbf{r}} + \frac{p \sin \theta}{4\pi\varepsilon r^3}\hat{\boldsymbol{\theta}} \tag{4.14}$$

It should be pointed out again that the results for V and $\mathbf{E}$ are valid far from the dipole in terms of dipole dimensions. The fields of the dipoles also decay faster to zero, $V \propto 1/r^2$ and $E \propto 1/r^3$ when compared to those of a single charge. This can be expected since a "partial cancellation" takes place when two opposite charges are brought close to each other. The cancellation is better in some directions than in others, which is indicated by the presence of the cosine and sine terms in the expressions for V and $\mathbf{E}$. The force lines ($\mathbf{E}$ lines) and the equipotential lines, which are perpendicular to the lines of force, are shown in Fig. 4.8.

Finally, we observe that even though an $\mathbf{E}$ field exists in the space surrounding the dipole, Gauss' law is not violated. Recall that Gauss' law specifies that if no net charge is enclosed in a surface, no net flux can come out of that surface. Clearly, a surface that surrounds a dipole encloses no net charge; hence $\oiint \mathbf{D} \cdot \mathbf{d}A = 0$. Integrating the dipole fields (4.14) over a closed surface gives no net flux; therefore, Gauss' law is not violated.

4.4 POLARIZATION AND THE DIELECTRIC CONSTANT

The *relative permittivity* (*dielectric constant*) of a material is defined as $\varepsilon_r = \varepsilon/\varepsilon_0$, where ε is the permittivity of the material and $\varepsilon_0 = 10^{-9}/36\pi$ and is the permittivity of free space. Let us see why for some materials the dielectric constant is different from that for free space. If we consider a slab of material as shown in Fig. 4.9 immersed in an external field $\mathbf{E}_0$, the molecules will be polarized

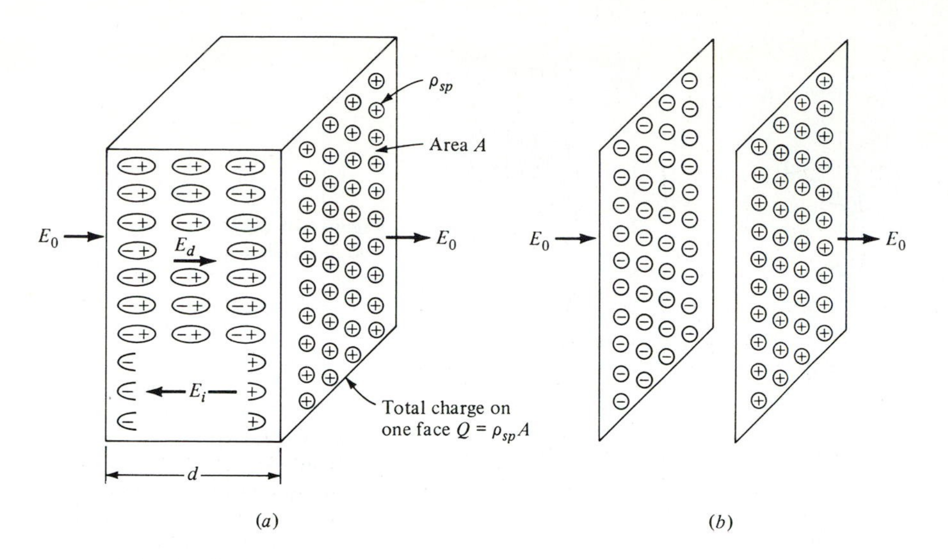

Figure 4.9 (*a*) A slab of dielectric showing polarization by an external field E_0; (*b*) since the interior charges cancel, the only charges that remain are the surface charges ρ_{sp}. Hence, for fields exterior to the slab, a slab of polarized material is equivalent to two sheets of charge.

as shown. We will assume that the dipoles line up regularly. The dipole moment of each molecule can be written as $\Delta\mathbf{p} = q\,\Delta\mathbf{d}$. Inside the material, the adjacent charges of the dipoles will annul each other, leaving, in effect, only the negative charges on the left side of the slab and the positive charges on the right side. If we integrate a line of dipoles, we obtain

$$\mathbf{p} = \int_0^d q\,\Delta\mathbf{d} = q\mathbf{d} \tag{4.15}$$

The total dipole moment of the entire rectangular slab is then

$$\mathbf{p} = Q\mathbf{d} \tag{4.16}$$

where $Q = \sum q$ and is the entire surface charge due to polarization.

The polarization $\mathbf{P}$ or dipole moment per unit volume for the slab can now be written as

$$\mathbf{P} = \frac{\mathbf{p}}{v} = \frac{Q\mathbf{d}}{Ad} = \frac{Q}{A}\hat{\mathbf{d}} = \rho_{sp}\hat{\mathbf{d}} \tag{4.17}$$

The polarization is thus equal to a surface charge density ρ_{sp} of polarization charge. Even though the polarization surface charge is a real accumulation of charge and will contribute to the electric field and potential inside and outside the dielectric, it will not enter in the boundary condition (1.87*b*)

$$D_{n_1} - D_{n_2} = \rho_s$$

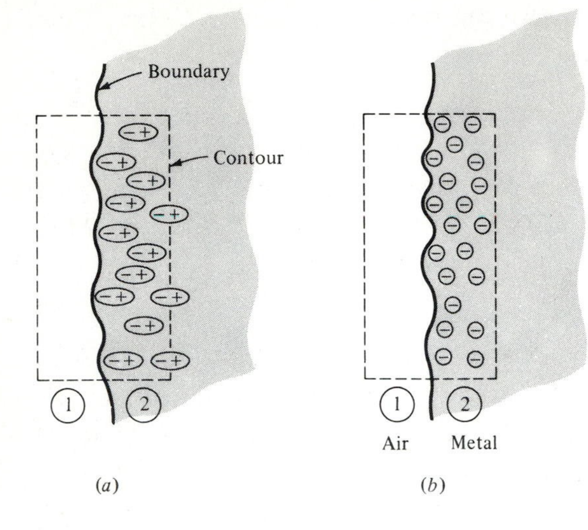

Figure 4.10 (*a*) The boundary between two dielectrics showing that a macroscopically small contour encloses zero bound charge; (*b*) a contour between air and a charged metal encloses a net amount of charge. Note that we are not allowed to draw a microscopically small contour which could go between the atoms and split the dipoles (see p. 81 and Sec. 4.2 for a discussion of macroscopic versus microscopic).

In other words ρ_s in (1.87*b*) must be a free charge of accumulation. This point was already discussed following Eq. (1.87*b*). Bound surface charge ρ_{sp} does not enter in the discontinuity of the normal components of electric flux density *D*.

The difference between bound and free charge can be clarified with the aid of Fig. 4.10*a* which shows a macroscopically small contour enclosing a boundary between free space and a polarized dielectric. Similar contours are used when deriving boundary conditions. The boundary is shown jagged to suggest that a macroscopically smooth boundary is not necessarily microscopically smooth. We see here that no matter how small we make the contour, we will still include many molecules (or dipoles) within the contour. The adjacent charges will cancel each other and leave zero charge inside the contour. The appropriate boundary condition for a dielectric is therefore†

$$D_{n_1} - D_{n_2} = 0 \tag{4.18}$$

Applying this boundary condition to the slab in Fig. 4.9 where the fields are assumed to be normal to the slab faces, we have

$$D_0 = D_d \tag{4.18a}$$

The flux density is the same in free space as in the dielectric. For comparison, the charged surface of a metal with an accumulation of free charge is shown in Fig. 4.10*b*. It is is clear that a net amount of charge is enclosed by the contour.

Since a polarized dielectric slab, as shown in Fig. 4.9*b*, is equivalent to two

† We can place a real surface charge at certain dielectric-dielectric boundaries by mechanical means, for example, by rubbing, in which case the normal components of *D* become discontinuous by that surface layer. Also, when a multilayered imperfect dielectric is used in a condenser, a free charge can appear. This effect is considered in Eqs. (3.24) to (3.28).

layers of bound surface charge ρ_{sp}, we can write for the induced electric flux density $\mathbf{D}_i$ between the two charge layers

$$D_i = -\rho_{sp} \tag{4.19}$$

$\mathbf{D}_i$ is related to the induced electric field $\mathbf{E}_i$ by $D_i = \varepsilon_0 E_i$, as shown in Fig. 4.9. Note that we have used the free-space permittivity ε_0 to multiply E_i, because we are treating the dielectric as free space to which an array of dipoles is added. Comparing this expression to (4.17), we can relate $\mathbf{P}$ to $\mathbf{D}_i$ by

$$\mathbf{D}_i = -\mathbf{P} \tag{4.20}$$

The field $\mathbf{E}_d$ inside the dielectric is seen from Fig. 4.9 to be the applied field $\mathbf{E}_0$ reduced by the induced field $\mathbf{E}_i$ (which opposes $\mathbf{E}_0$), that is,

$$\mathbf{E}_d = \mathbf{E}_0 + \mathbf{E}_i \tag{4.21}$$

Since $\mathbf{P} = -\mathbf{D}_i$, we can write using Eq. (4.21)

$$\mathbf{P} = -\varepsilon_0 \mathbf{E}_i = -\varepsilon_0(\mathbf{E}_d - \mathbf{E}_0) \tag{4.22}$$

or

$$\varepsilon_0 \mathbf{E}_0 = \varepsilon_0 \mathbf{E}_d + \mathbf{P} \tag{4.22a}$$

If we compare this expression to the boundary condition (4.18*a*), we conclude that the electric flux density $\mathbf{D}_d$ inside the dielectric is equal to

$$\boxed{\mathbf{D}_d = \varepsilon_0 \mathbf{E}_d + \mathbf{P}} \tag{4.23}$$

The dielectric constant ε_d can now be finally obtained from $\mathbf{D}_d = \varepsilon_d \mathbf{E}_d$ as

$$\varepsilon_d = \varepsilon_0 + \frac{P}{E_d} \tag{4.24}$$

for the case when the dielectric is isotropic ($\mathbf{P}$ and $\mathbf{E}_d$ are in the same direction) and ε_d is a scalar. This is often written as $\varepsilon_d = \varepsilon_0(1 + \chi)$ where χ is called the *electric susceptibility*.

The physical picture that we have now developed for a dielectric is the following: If the medium is free space, we have $\varepsilon_d = \varepsilon_0$. If dipoles of moment p are present throughout the medium, the polarization $\mathbf{P}$ of the medium enters in the dielectric constant as $\varepsilon_d = \varepsilon_0 + P/E_d$. In practice when considering simple dielectrics, we usually do not work with polarization $\mathbf{P}$ but work directly with the permittivity ε_d and electric flux density $\mathbf{D}$; that is, we work with $\mathbf{D}_d = \varepsilon_d \mathbf{E}_d$. The permittivity, which can be easily measured, accounts for all polarization effects.

4.5 FIELDS EXTERIOR TO A DIELECTRIC OBJECT

If a dielectric object is polarized throughout its interior, as shown in Fig. 4.11, the polarization $\mathbf{P}$ will give rise to a field outside the object. Let us calculate the potential and electric fields due to source $\mathbf{P}$. The polarization $\mathbf{P}$ can be either

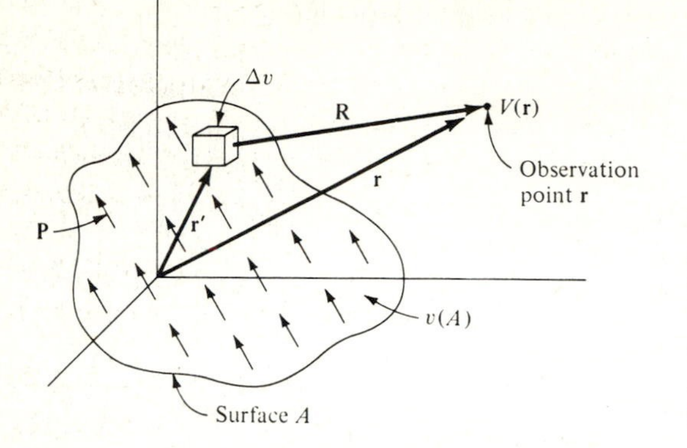

Figure 4.11 A dielectric object with dipole moment per unit volume **P**. The polarization **P** within the object produces potential $V(\mathbf{r})$. $v(A)$ is the volume of the object which is bounded by the surface area A; $R = |\mathbf{r}' - \mathbf{r}|$ is the distance from the dipole located at $\mathbf{r}'$ to the observation point $\mathbf{r}$.

induced by an applied field, in which case we must add the applied field to that due to **P** to obtain the total field (superposition), or the object can be ferroelectric, in which case it is permanently polarized (like a permanent magnet) and the permanent **P** is the only source for the exterior field. A further objective will be to show that accumulated bound charge gives rise to fields just like ordinary charge and that the definition for the interior field $\mathbf{D}_d = \varepsilon_0 \mathbf{E}_d + \mathbf{P}$ is consistent with Gauss' law.

We have shown in Eq. (4.12*a*) that the potential due to a dipole $\mathbf{p}'$ is given by

$$V(\mathbf{r}) = \frac{1}{4\pi\varepsilon_0} \frac{\mathbf{p}' \cdot \hat{\mathbf{R}}}{R^2} \tag{4.25}$$

where the dipole is located at $\mathbf{r}'$ and the distance to the observation point $\mathbf{r}$ is $R = |\mathbf{r} - \mathbf{r}'|$. Assuming that the dipoles in Fig. 4.11 are located in elemental volumes Δv within the dielectric, the contribution to potential due to each dipole moment $\mathbf{p}'(\mathbf{r}') = \mathbf{P}'(\mathbf{r}')\,\Delta v$ can be written as

$$V(\mathbf{r}) = \frac{1}{4\pi\varepsilon_0} \mathbf{P}' \cdot \nabla' \left(\frac{1}{R}\right) \Delta v \tag{4.25a}$$

where we have used the relationship $\hat{\mathbf{R}}/R^2 = \nabla'(1/R)$ [see Eq. (8.44)]. Integrating over the entire volume v of the dielectric, the potential exterior to the dielectric object is given by

$$V = \frac{1}{4\pi\varepsilon_0} \iiint_v \mathbf{P}' \cdot \nabla' \left(\frac{1}{R}\right) dv \tag{4.26}$$

This integral, with the aid of the vector identity

$$\nabla \cdot (\Psi \mathbf{A}) = \Psi \nabla \cdot \mathbf{A} + \mathbf{A} \cdot \nabla \Psi \tag{4.27}$$

can be rearranged in a form which lends itself to a simple physical interpretation. Thus, letting the scalar Ψ be $1/R$, we obtain

$$V = \frac{1}{4\pi\varepsilon_0} \iiint_v \left(\nabla' \cdot \frac{\mathbf{P}'}{R} - \frac{\nabla' \cdot \mathbf{P}'}{R}\right) dv \tag{4.28}$$

The first term can be converted to a surface integral with the aid of Gauss' theorem (1.94). This gives

$$V(\mathbf{r}) = \frac{1}{4\pi\varepsilon_0} \oiint_A \frac{\mathbf{P}' \cdot \mathbf{dA}}{R} - \frac{1}{4\pi\varepsilon_0} \iiint_v \frac{\nabla' \cdot \mathbf{P}'}{R} dv \tag{4.29}$$

where the surface integration is over the surface of the dielectric object and $\mathbf{dA} = \hat{\mathbf{n}}\, dA$. Comparing this expression to that of a potential due to a volume charge density (1.16) and a surface charge density (1.20), we can write Eq. (4.29) as

$$V(\mathbf{r}) = \frac{1}{4\pi\varepsilon_0} \oiint_A \frac{\rho_s'\, dA}{R} + \frac{1}{4\pi\varepsilon_0} \iiint_v \frac{\rho'}{R} dv \tag{4.30}$$

We can now identify $\mathbf{P} \cdot \hat{\mathbf{n}} = P_n$ (which is the normal polarization component to the surface of the dielectric) with a surface charge density. Similarly we can identify $-\nabla \cdot \mathbf{P}$ with a volume charge density. That is,

$$\rho_{sp} = \mathbf{P} \cdot \hat{\mathbf{n}} \qquad \rho_p = -\nabla \cdot \mathbf{P} \tag{4.31}$$

where the subscript p stands for the bound polarization charge. The conclusion is that the dielectric may be replaced by the charge distributions ρ_{sp} and ρ_p and the problem treated as an ordinary potential problem. If there are any other charges present, the potential due to such charges can be added to that due to the polarization charge to give the total potential. The electric fields can be calculated from the gradient of the potential as usual; that is,

$$\mathbf{E}(\mathbf{r}) = -\nabla V(\mathbf{r}) = \frac{\hat{\mathbf{R}}}{4\pi\varepsilon_0} \oiint \frac{\rho_s'\, dA}{R^2} + \frac{\hat{\mathbf{R}}}{4\pi\varepsilon_0} \iiint \frac{\rho'\, dv}{R^2} \tag{4.32}$$

noting that $\nabla(1/\mathrm{R}) = -\hat{\mathbf{R}}/R^2$ as given by Eq. (8.44). A dielectric with a uniform polarization throughout has only a layer of polarization charge ρ_{sp} on the surface. In the interior, $\rho_p = 0$ since the divergence of a constant polarization is zero. On the other hand, if the dipole density varies with position, we have, in addition to ρ_{sp}, an uncanceled bound volume charge of polarization ρ_p throughout the dielectric.

We have shown that ρ_{sp} and ρ_p represent actual accumulations of charge. These charges, even though bound, produce fields according to Coulomb's law, just as any other charges. If free charges are also present, we may simply add their charge density ρ to that of the polarization charge, that is, $\rho_p + \rho$, and calculate the fields using this sum of charges as the new source for the fields. Since Gauss' law relates the electric field through a closed surface to the charges enclosed, for points inside the dielectric, we can use the differential form of Gauss' law (1.92) and write

$$\nabla \cdot \varepsilon_0 \mathbf{E}_d = \rho_p + \rho \tag{4.33}$$

If we substitute for ρ_p from (4.31), we obtain

$$\nabla \cdot (\varepsilon_0 \mathbf{E}_d + \mathbf{P}) = \rho \tag{4.34}$$

Comparing this to Gauss' law $\nabla \cdot \mathbf{D} = \rho$, we again have

$$\mathbf{D} = \varepsilon_0 \mathbf{E}_d + \mathbf{P} \tag{4.35}$$

which is consistent with our derivation (4.23) for electric flux density inside the dielectric. The implication of (4.34) is that the divergence of vector $\mathbf{D} = \varepsilon_0 \mathbf{E}_d + \mathbf{P}$ depends only on the free-charge density ρ.

4.6 ARTIFICIAL DIELECTRICS

An *artificial dielectric* is a distribution of metallic particles which act individually as dipoles and collectively as a dielectric. Spherical conductors are often used as the dipole-acting particles which are held in a fixed arrangement by a low-density plastic foam such as styrofoam (the dielectric constant of styrofoams is usually between 1.03 and 1.10). Figure 4.12 shows a lattice of metal spheres. The spheres must be small and separated a sufficient distance from each other so that the local field is essentially uniform. Artificial dielectrics† were developed for use as dielectric material in lenses for focusing microwaves (where the artificial dielectric can be made much lighter than real dielectrics by the use of small hollow spheres). The restriction on separation, therefore, when applied to time-varying fields implies that the particle dimensions must be small compared with the wavelength of the signal.

From Sec. 2.10 we know that a metallic sphere which is introduced in an electric field $\mathbf{E}_d$ will experience a separation of the free charges in the metal as shown in Fig. 4.13. This separation of charge taken as a whole acts as a dipole. The polarization $\mathbf{P}$ of the artificial dielectric using (4.2*a*) can be written as

$$\mathbf{P} = N\mathbf{p} \tag{4.36}$$

where N is the number of spheres per unit volume and $\mathbf{p} = q\mathbf{d}$ is the equivalent dipole moment of the sphere. The permittivity of such a material can then be written as

$$\varepsilon_d = \varepsilon_0 + \frac{P}{E_d} = \varepsilon_0 + \frac{Np}{E_d} \tag{4.37}$$

where the polarization is in the same direction as the internal field E_d. We must now determine the dipole moment of the sphere.

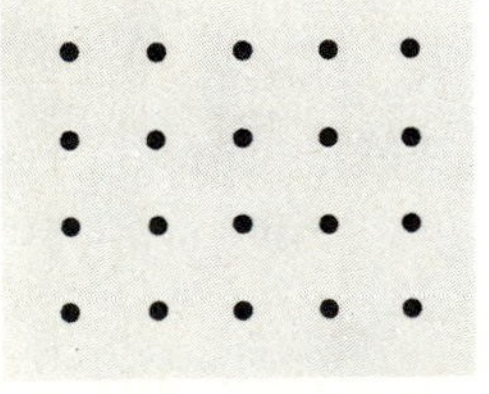

Figure 4.12 A lattice arrangement of metal spheres in an artificial dielectric.

† W. E. Kock, "Sound Waves and Light Waves," pp. 112–121, Doubleday & Company, Inc., Garden City, N.Y., 1965.

Assuming the internal field E_d to be uniform, the potential for a uniform field can be written in general as

$$V = -\int_0^r \mathbf{E}_d \cdot d\mathbf{r} = -\int_0^r E_d \cos\theta \, dr = -E_d r \cos\theta \tag{4.38}$$

where the origin is taken to be at the center of the sphere and θ is the angle between an observation point r and the axis of the induced dipole (Fig. 4.13). The potential of a dipole with moment p was derived in Eq. (4.12*a*) as

$$V = \frac{1}{4\pi\varepsilon_0} \frac{p\cos\theta}{r^2} \tag{4.39}$$

The total potential inside the artificial dielectric can be written as the sum of the uniform field plus the dipole field†

$$V = -E_d r \cos\theta + \frac{p\cos\theta}{4\pi\varepsilon_0 r^2} \tag{4.40}$$

This relationship can now be used to obtain the dipole moment of the sphere by forcing the boundary conditions on the surface of the metallic sphere which is an equipotential surface. Thus for $r = a$, we have

$$0 = -E_d a \cos\theta + \frac{p\cos\theta}{4\pi\varepsilon_0 a^2} \tag{4.41}$$

which permits us to solve for

$$\frac{p}{E_d} = 4\pi\varepsilon_0 a^3 \tag{4.42}$$

The dielectric constant of the artificial dielectric is thus

$$\varepsilon_r = \frac{\varepsilon_d}{\varepsilon_0} = 1 + 4\pi N a^3 \tag{4.43}$$

The dielectric constant is seen to depend on the volume concentration of spheres N and on the volume of each sphere. In practice it is found that dielectric constants ranging from 1 to 7 can be produced by an array of metallic particles.

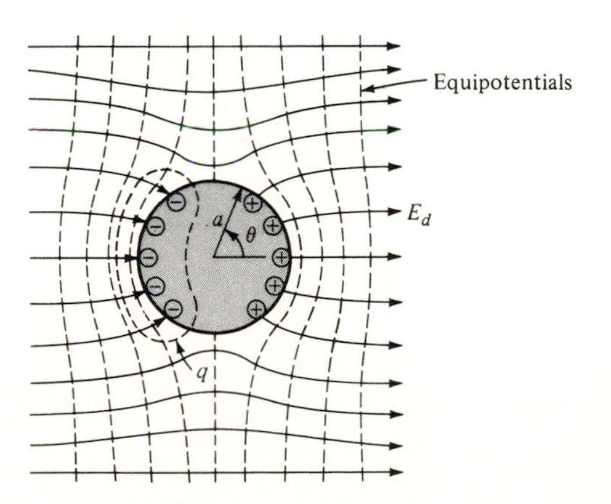

Figure 4.13 A metal sphere immersed in an external field E_d will be polarized as shown. The induced charges form a dipole with moment $p = 4\pi\varepsilon_0 a^3 E_d$.

† For a more rigorous derivation of a metallic sphere in a uniform field, see Sec. 4.12.

4.7 COMPARISON OF POLARIZATION IN A DIELECTRIC AND CONDUCTOR

There are certain similarities in polarization when a slab of dielectric and a slab of metal are introduced in an electric field. Both form layers of surface charge on their exterior surfaces. The potential and electric fields external to these polarized slabs can be correctly calculated by replacing the slabs with layers of induced surface charge densities as shown in Figs. 4.9 and 4.14. The potential at exterior points due to the induced surface charge is given by Eq. (4.30) as

$$V = \frac{1}{4\pi\varepsilon_0} \oiint \frac{\rho_s \, dA}{R} \tag{4.44}$$

For the dielectric we have

$$\rho_s = \rho_{sp} = P = (\varepsilon_d - \varepsilon_0)E_d = \frac{\varepsilon_0 E_0(\varepsilon_r - 1)}{\varepsilon_r} \tag{4.45}$$

where ε_r is the relative permittivity equal to $\varepsilon_d/\varepsilon_0$. The above expression was obtained from (4.31) and by use of (4.23) and (4.18*a*). For the metallic disc we have for the induced surface charge density

$$\rho_s = \varepsilon_0 E_0 \tag{4.46}$$

which was obtained from the boundary condition for a metal [Eq. (2.44)]. The difference between the two cases is related to the way the induced field, which points from the positive to the negative surface charges, reduces the applied field E_0 in the interior of the material. For a metal the free conduction charges are mobile and can move over large distances if an E field is present. The free charges accumulate on the metal surface. The surface charge builds up until the induced

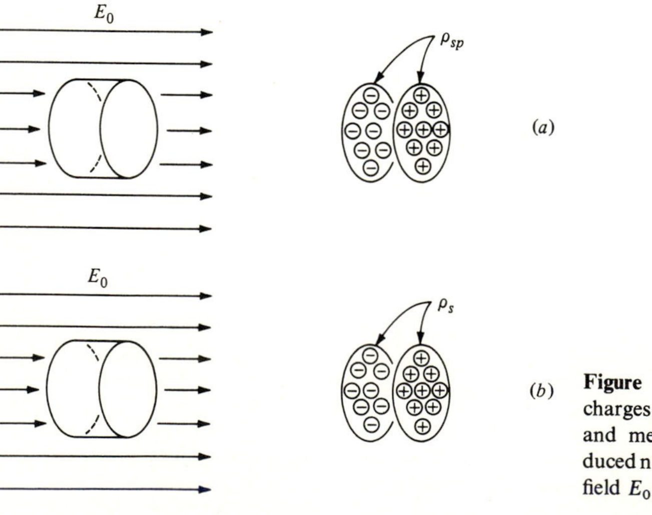

Figure 4.14 The induced surface charges when a disc of dielectric (*a*) and metallic (*b*) material is introduced normally to an external electric field E_0.

field that it produces cancels the applied field, at which point the charge motion ceases. The result is that total $E = 0$ in the interior of the metal. One can say, the polarization of the conductor by the external electric field is complete.

For dielectrics, on the other hand, the charges are bound and can travel no more than a small fraction of an atomic diameter. The induced electric field due to the polarization surface charges partially cancels the applied field, leaving a net electric field E_d inside the dielectric which points in the same direction as that of the applied field. By the same token, the induced charges appearing on the dielectric surfaces are not equal in magnitude to the induced charges on the metallic surfaces. This can be easily seen when comparing (4.46) to (4.45) where we note that

$$\rho_s\bigg|_{\text{metal}} = \rho_{sp}\bigg|_{\varepsilon_r \to \infty} \tag{4.47}$$

That is, the polarization surface charge of a dielectric will equal that of a metal only when the permittivity of the dielectric becomes infinite (dielectrics with infinite ε_r do not exist). For an infinite ε_r, the interior field of a dielectric would be zero.† In short, a dielectric introduced in an external field will become polarized but never to such an extent that the interior fields will go to zero.

The other differences between conductors and dielectrics are internal. Whereas dielectrics have a polarization $\mathbf{P}$ and electric field $\mathbf{E}_d$ throughout the interior of the material and a dielectric constant larger than that of free space, metals have a zero $\mathbf{P}$ and zero $\mathbf{E}$ throughout and a dielectric constant which is that of free space; that is, $\varepsilon_{\text{metal}} = \varepsilon_0$. To have dipole moments throughout a material requires bound charge. The free charge of a conductor, which is due to the electrons, will simply move to the surface in response to an applied field, leaving a layer of positive charge on the opposing surface. No array of dipoles can exist in the interior of a conductor; hence, $P = 0$ and the dielectric constant is that of free space. From (4.22*a*) or (4.23) we can write for P

$$P = -\varepsilon_0(E_d - E_0) = (\varepsilon_d - \varepsilon_0)E_d \tag{4.48}$$

which shows that as $\varepsilon_d \to \varepsilon_0$, P vanishes inside. On the other hand, a dielectric can be considered completely polarized only when the interior field E_d goes to zero. Rearranging (4.48) as

$$E_d = \frac{P}{\varepsilon_d - \varepsilon_0} \tag{4.49}$$

we see that for finite P, this can occur only when $\varepsilon_d \to \infty$.

† Note that in many situations a problem in dielectrics can be reduced to the equivalent problem when the medium is metal by letting $\varepsilon_r \to \infty$.

Example of Polarization by a Point Charge Inside Two Spherical Shells, One Dielectric and One Metallic

Figure 4.15 shows the fields produced by a point charge Q located at the center of two concentric shells. The inside shell is of a dielectric material; the outside shell is metallic. We can write Gauss' law for this problem as

$$\oiint \mathbf{D} \cdot d\mathbf{A} = \iiint \rho \, dV = Q \qquad r < c; r > d \tag{4.50}$$

Since the D field is symmetric about the origin, where Q is located, we have

$$\mathbf{D} = \frac{Q}{4\pi r^2} \hat{\mathbf{r}} \tag{4.51}$$

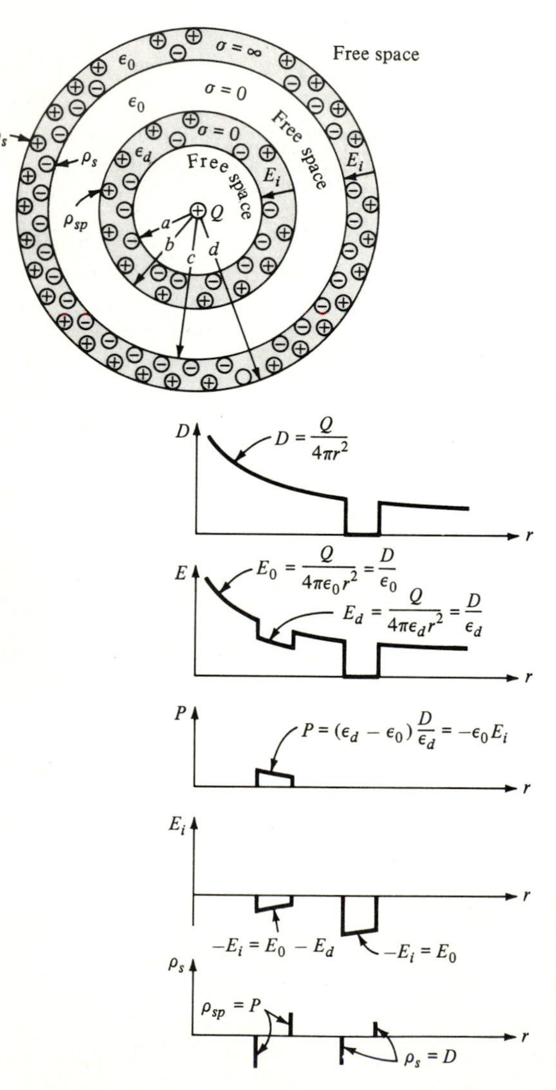

Figure 4.15 A point charge Q located at the center of two concentric spherical shells and the resultant fields. The inner shell is of dielectric material; the outer is metallic.

This D is valid both in free space ($r < a$; $b < r < c$) and inside the dielectric shell ($a < r < b$) because in Gauss' law (4.50), ρ and Q are free charges. The D field is therefore unaffected by the dielectric. Polarization effects enter through D as $D = \varepsilon_0 E + P$. The layers of surface charge are given by $P = \rho_{sp}$ and $D = \rho_s$. The verification of the remaining results is left as a problem.

4.8 POLARIZATION CURRENT

Wherever the polarization in matter changes with time, (as, for example, during the interval that it takes to establish the polarization in a dielectric), we have a flow of bound charge which constitutes a polarization or bound-charge current. This current is just as effective in producing a magnetic field as is a motion of free charge. To find the polarization current, we can apply the equation of continuity (2.9) to J_p and ρ_p; that is,

$$\oiint \mathbf{J}_p \cdot \mathbf{dA} = -\frac{\partial}{\partial t} \iiint \rho_p \, dv \tag{4.52}$$

where J_p is the polarization current due to variation of ρ_p with time. For example, when a sinusoidally varying field is applied to a dielectric, the polarization tries to keep up with the changes in the field. As the external field reverses periodically, the dipoles in the dielectric material "flip" in step with the field. Since a reversal of a dipole implies a flow of charge (remember a reversal of a dipole can be considered as the flow of $+q$ to the position of $-q$ and the flow of $-q$ to the position of $+q$), we have an oscillating polarization current J_p. This should also explain why the dielectric constant of materials decreases at very high frequencies. At these frequencies the molecular dipoles are simply not able to follow in step with the applied ac field.

We can replace ρ_p in Eq. (4.52) by $-\nabla \cdot \mathbf{P}$ according to (4.31). We can then write for the right-hand integral of (4.52)

$$\iiint \rho_p \, dv = -\iiint \nabla \cdot \mathbf{P} \, dv = -\oiint \mathbf{P} \cdot \mathbf{dA} \tag{4.53}$$

where the last step was obtained by an application of Gauss' theorem (1.94). Therefore

$$\oiint \mathbf{J}_p \cdot \mathbf{dA} = \oiint \frac{\partial \mathbf{P}}{\partial t} \cdot \mathbf{dA} \tag{4.54}$$

Since this statement is true for an arbitrary closed surface area, we can equate the integrands and obtain for the polarization current

$$\mathbf{J}_p = \frac{\partial \mathbf{P}}{\partial t} \tag{4.55}$$

Note that we cannot have a steady bound-charge current, that is, J_p is zero for dc polarization since under dc polarization, the bound charges separate and there is no further motion of charge and, hence, no current. The differential form of the continuity equation can be obtained by taking the divergence of the above expression, that is,

$$\nabla \cdot \mathbf{J}_p = \frac{\partial}{\partial t} \nabla \cdot \mathbf{P} = -\frac{\partial \rho_p}{\partial t} \tag{4.56}$$

4.9 EXAMPLE: FIELDS OF A PERMANENTLY POLARIZED ROD

Suppose we have a rod of ferroelectric material which has a uniform and permanent polarization P as shown in Fig. 4.16. The polarization will give rise to equivalent charge densities which, according to (4.31), are

$$\rho_{sp} = \mathbf{P} \cdot \hat{\mathbf{n}} = \pm P \qquad \rho_p = -\nabla \cdot \mathbf{P} = 0 \tag{4.57}$$

The assumption of uniform polarization leads to a layer of surface charge on the top and bottom of the rod and zero volume charge density. For fields outside and far from the rod, the rod can be approximated by two lumped charges, $Q = \pm PA$, separated by the height d of the rod. This combination will act as a dipole. Hence, the exterior field is a dipole field which is given by Eq. (4.12a) as

$$V = \frac{1}{4\pi\varepsilon_0} \frac{p \cos\theta}{r^2} \tag{4.58}$$

with the equivalent dipole moment of the rod as $p = PAd = Pv$ where v is the volume of the rod.

Unfortunately, when a rod of dielectric material is introduced in a uniform electric field, the induced polarization in the rod is not uniform, even for the case when the axis of the rod lies along the electric field. Therefore, we cannot say that

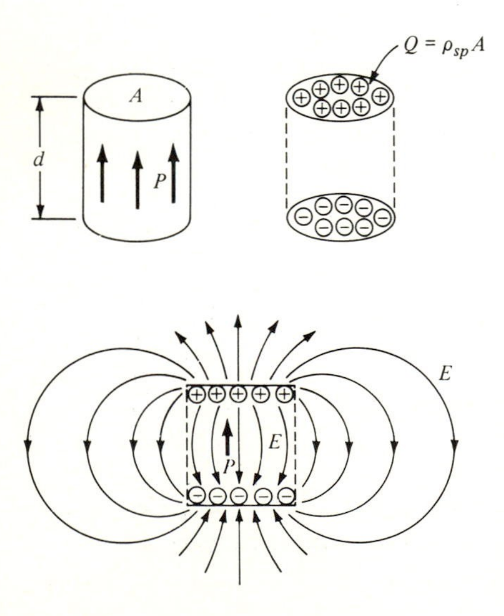

Figure 4.16 A rod of height d with end surfaces A having a uniform polarization P throughout is equivalent to two layers of surface charge as shown.

the resultant field is simply the combination of the uniform field and a dipole field (4.58). The problem of a dielectric rod in a uniform field must be solved as a regular boundary-value problem, i.e., solving Laplace's equation with the proper boundary conditions on the surface.

On the other hand, the sphere and other ellipsoidal bodies become uniformly polarized when introduced in a uniform electric field. Therefore, knowing the external field of a uniformly polarized sphere, the problem of a dielectric sphere in a uniform field can also be solved and is simply the external field of the polarized sphere superposed on the uniform field.

Fields on the Axis. For the uniformly polarized rod shown in Fig. 4.16, the fields on the axis of the cylinder of radius a and height d can be derived rigorously by using the results for the field of a charged disc, Eqs. (1.31) or (1.34). Positioning the cylinder so its center coincides with the origin of a coordinate system, as shown in Fig. 1.10, reduces the problem to two discs, one positively charged and located at $z = d/2$, the other negatively charged and located at $z = -d/2$. The potential of the ferroelectric rod, using Eq. (1.34), is then given by

$$V = \frac{\rho_{sp}}{2\varepsilon_0}\left(\left\{\left[a^2 + \left(z - \frac{d}{2}\right)^2\right]^{1/2} - \left(z - \frac{d}{2}\right)\right\} - \left\{\left[a^2 + \left(z + \frac{d}{2}\right)^2\right]^{1/2} - \left(z + \frac{d}{2}\right)\right\}\right) \tag{4.58a}$$

where ρ_{sp} is the surface polarization charge density, and from Eq. (4.57) is equal to polarization P. This expression is valid anywhere along the axis of the cylinder.

If we restrict ourselves to fields far from the cylinder, i.e., when $z \gg a$ or d, we can use the binomial series to make certain approximations in (4.58a). Thus, in the far field of the cylinder the potential simplifies to

$$V = \frac{\rho_{sp}\pi a^2 d}{4\pi\varepsilon_0 z^2}\left(1 + \frac{d^2}{4a^2}\right) \tag{4.58b}$$

Since the area of the disc is πa^2, the total polarization charge on each disc is $Q = \rho_{sp}\pi a^2$. The dipole moment should then simply be $p = Qd = Pv$ as given in (4.58). However, since the charge is distributed over a disc instead of being concentrated at a point, we have a correction term which from (4.58b) is seen to be $1 + d^2/4a^2$. The correct dipole moment is thus

$$p = Pv\left(1 + \frac{d^2}{4a^2}\right) \tag{4.58c}$$

and should be used in (4.58). The **E** field is given by Eq. (4.14).

4.10 EXAMPLE: FIELDS OF A PERMANENTLY POLARIZED SPHERE

A sphere with uniform polarization is shown in Fig. 4.17. The permanent polarization P throughout the interior of the sphere is equivalent to charge densities which according to Eq. (4.31) are

$$\rho_{sp} = \mathbf{P} \cdot \hat{\mathbf{n}} = P\cos\theta' \qquad \rho_p = \nabla \cdot \mathbf{P} = 0 \tag{4.59}$$

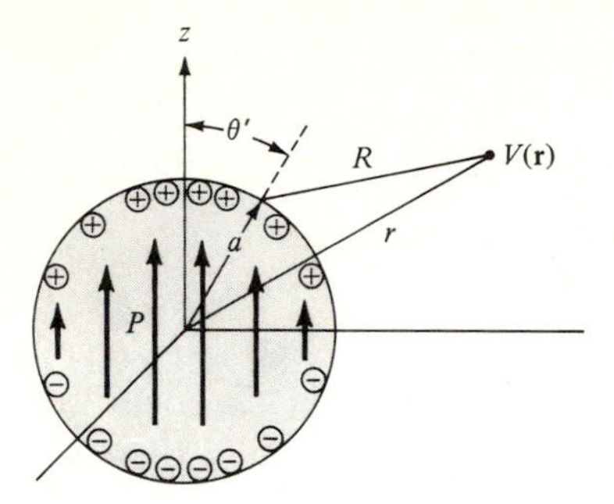

Figure 4.17 A ferroelectric sphere of radius a. The uniform polarization P is equivalent to a surface layer $P \cos \theta'$ of charge.

where θ' is measured from the z axis. The equivalent surface charge density ρ_{sp} is shown in Fig. 4.17 and gives rise to a potential [see Eq. (4.30)]

$$V(\mathbf{r}) = \frac{1}{4\pi\varepsilon_0} \oiint \frac{\rho'_{sp}\, dA'}{R} = \frac{1}{4\pi\varepsilon_0} \int_0^{2\pi} \int_0^{\pi} \frac{(P \cos \theta')a^2 \sin \theta'\, d\theta'\, d\phi'}{|\mathbf{a} - \mathbf{r}|} \tag{4.60}$$

The distance from each element of surface charge to the observation point $\mathbf{r}$ can be expressed as

$$|\mathbf{a} - \mathbf{r}| = (a^2 + r^2 - 2ar \cos \Psi)^{1/2} \approx r - a \cos \Psi \tag{4.61}$$

where Ψ is the angle between the two radial vectors $\mathbf{a}$ and $\mathbf{r}$ and the last term was obtained by applying the far-field approximation $r/a \gg 1$. The cosine of the angle between two radial vectors is given by the well-known expression

$$\cos \Psi = \cos \theta \cos \theta' + \sin \theta \sin \theta' \cos (\phi - \phi') \tag{4.62}$$

where the primed quantities are the direction angles of $\mathbf{a}$ and the unprimed are those of the observation point $\mathbf{r}$. After some additional algebra and noting that $(1 + \Delta)^{-1} \approx 1 - \Delta$ when $\Delta \ll 1$, the result of the integration of (4.53) is

$$V(\mathbf{r}) = \frac{Pa^2 \cos \theta}{3\varepsilon_0 r^2} \tag{4.63}$$

which gives the potential outside the uniformly polarized sphere. By comparing this expression to that of a dipole (4.12a), the equivalent dipole moment of the sphere can be expressed as

$$p = \frac{4\pi}{3} a^3 P = vP \tag{4.64}$$

where v is just the volume of a sphere. Thus, a simple relationship between dipole moment p and polarization P can be obtained for a sphere.

Strictly speaking, $V(\mathbf{r})$ is valid only far from the sphere since the far field approximation $r/a \gg 1$ was used. However, we will show in the next section that (4.63) is valid not only close to the sphere but also on the surface of it. This unexpected result can be explained as follows: The equivalent dipole for a uniformly polarized sphere is much smaller in extent than the sphere. Therefore, if the equivalent dipole is located at the center of the sphere, one can say that the dipole field at a distance equal to the radius of the sphere is in the far field of the dipole. This is shown pictorially in Fig. 4.18. A particularly lucid explanation of this is given by Purcell.† He shows that a uniformly polarized sphere can be considered as the superposition of two spheres, one with a uniform negative volume charge throughout, the other with a uniform positive charge. The superposition is not concentric, as the centers of the two spheres are slightly displaced which leaves a net surface charge as shown in Fig. 4.17.

† E. M. Purcell, "Electricity and Magnetism," McGraw-Hill Book Company, New York, 1963.

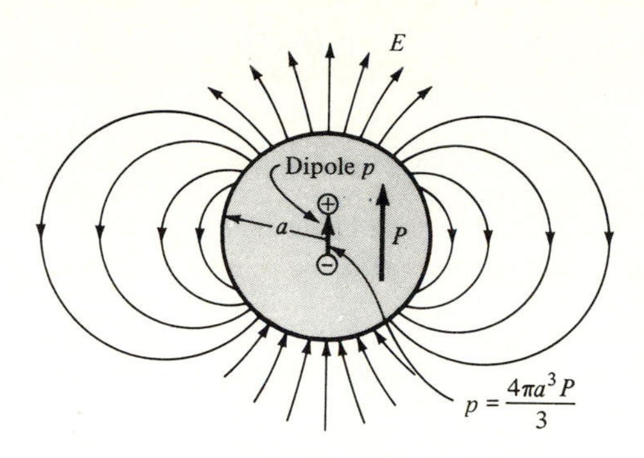

Figure 4.18 The fields outside a uniformly polarized sphere are identical to those of a dipole p located at the center of the sphere. The electric field lines are shown, which are given by Eq. (4.14).

Of course, the volume charge in the interior of the two spheres cancels. The small displacement of the two spheres is an indication of the size of the equivalent dipole p.

If (4.63) is valid on the surface of the sphere, then the potential on the surface $r = a$ is

$$V = \frac{Pa \cos \theta}{3\varepsilon_0} = \frac{P}{3\varepsilon_0} z \tag{4.65}$$

and is seen to depend on the z coordinate only. This suggests that the electric field inside the uniformly polarized sphere is uniform and is given by

$$\mathbf{E} = -\nabla V = -\frac{\partial V}{\partial z}\hat{\mathbf{k}} = -\frac{P}{3\varepsilon_0}\hat{\mathbf{k}} = -\frac{\mathbf{P}}{3\varepsilon_0} \tag{4.66}$$

where $\hat{\mathbf{k}}$ is a unit vector in the z direction and P is the uniform polarization of the sphere. This gives us the important relationship between internal field and polarization and shows, which is rather remarkable, that the field inside a uniformly polarized sphere is constant throughout its interior. Figure 4.19 is a sketch of the fields inside and outside the uniformly polarized sphere. Comparing the normal component of the E field outside the sphere as given by Eq. (4.14) to the normal component inside (4.66), we find that the E field is discontinuous by the surface layer of polarization charge. This layer acts as a source for the fields inside and outside, and one can say that the E lines inside and outside begin and end on this surface charge layer. The tangential components of the E field, on the other hand, are continuous across the boundary, as required by the boundary conditions.

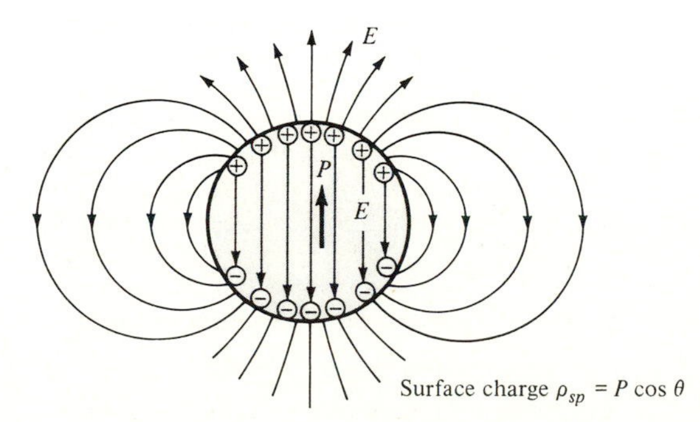

Figure 4.19 A uniformly polarized sphere, showing the electric fields inside and outside.

4.11 EXAMPLE: DIELECTRIC SPHERE IN A UNIFORM ELECTRIC FIELD

A dielectric sphere in a uniform field presents a particularly interesting problem at this time. It can be solved as a regular boundary-value problem with comparative ease. The results will show that the sphere acts as a dipole, and the total field will be the sum of the uniform field and a dipole field.

Let a dielectric sphere of radius a and dielectric constant $\varepsilon/\varepsilon_0$ be placed in an initially uniform electric field† E_0 as indicated in Fig. 4.20. The relationship of the applied field E_0 to its potential V_0 is $V_0 = -E_0 z = -E_0 r \cos\theta$. Since there are no free charges inside or outside the sphere, the problem is one of solving Laplace's equation inside and outside with the proper boundary conditions at $r = a$. An additional boundary condition can be immediately constructed by observing that far from the sphere the field must remain essentially unperturbed, i.e., for $r \gg a$, $V = -E_0 z$. For the potential external to the sphere we can write

$$V_e = Ar\cos\theta + B\frac{\cos\theta}{r^2} \tag{4.67}$$

where the first term is a uniform field and the second term is a dipole field $\cos\theta/r^2$ with an arbitrary coefficient. The coefficients A and B are to be determined from the boundary conditions. A $1/r$ term is not needed in the solution since such terms in a potential function imply the presence of a net charge on the sphere. The need for a dipole term is obvious since the induced charge distribution on the sphere, positive for one-half and negative for the other, looks like a dipole. Quadrupole and other charge distributions are not induced. Therefore, terms higher than $1/r^2$, such as $1/r^3$, are not needed in the solution for V_e.

Let us assume a solution of the same form for the interior of the sphere; that is,

$$V_i = Cr\cos\theta + D\frac{\cos\theta}{r^2} \tag{4.68}$$

where C and D are arbitrary constants. At distances far from the sphere the external field must be uniform and equal to the original field $-E_0 r\cos\theta$; that is, $V_e \approx Ar\cos\theta \to -E_0 r\cos\theta$ for $r \gg a$. Hence, $A = -E_0$. Another condition for determining the coefficients is obtained by noting that in our assumed solution for the internal field (4.68), the field becomes infinite as $r \to 0$. Since such infinities imply the presence of sources at $r = 0$ (in our case the $1/r^2$ term implies a presence of a dipole at the origin), we must force the coefficient D to be zero. Said more simply, we have no reason for believing that the potential or macroscopic electric field at the center of a dielectric sphere must go to infinity when the sphere is introduced in a uniform electric field; therefore, $D = 0$.

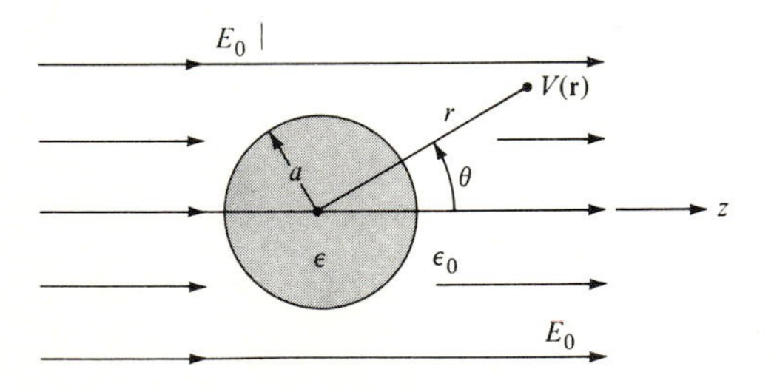

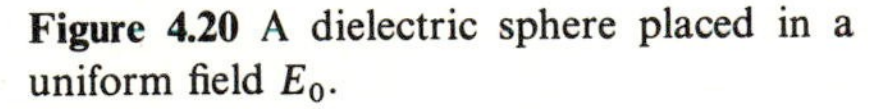
Figure 4.20 A dielectric sphere placed in a uniform field E_0.

† A uniform electric field can be thought of as being produced in the space between the charged parallel plates of a very large capacitor.

Our solutions for the exterior and interior fields are then

$$V_e = -E_0 r \cos\theta + B\frac{\cos\theta}{r^2} \tag{4.69}$$

$$V_i = Cr\cos\theta \tag{4.70}$$

The remaining coefficients must be determined from boundary conditions: E_t and D_n are continuous at $r = a$. Since the normal to a sphere is the radial unit vector $\hat{\mathbf{n}} = \hat{\mathbf{r}}$ and since $\mathbf{D} = \varepsilon\mathbf{E}$, we can re-express the continuity of the normal components $D_{ne} = D_{ni}$ at $r = a$ as $\varepsilon_0 E_{re} = \varepsilon E_{ri}$ at $r = a$. The external permittivity is that of free space (ε_0) and that of the sphere is ε. Using the gradient relationship between potential and electric fields in spherical coordinates

$$\mathbf{E} = -\nabla V = -\left(\frac{\partial}{\partial r}\hat{\mathbf{r}} + \frac{1}{r}\frac{\partial}{\partial\theta}\hat{\boldsymbol{\theta}} + \frac{1}{r\sin\theta}\frac{\partial}{\partial\phi}\hat{\boldsymbol{\phi}}\right)V \tag{4.71}$$

the above boundary condition for the radial components becomes

$$\varepsilon_0\frac{\partial V_e}{\partial r} = \varepsilon\frac{\partial V_i}{\partial r} \qquad \text{for } r = a \tag{4.72}$$

Substituting V_e and V_i we obtain

$$\varepsilon_0\left(-E_0 - \frac{2B}{a^3}\right) = \varepsilon C \tag{4.73}$$

The remaining boundary condition $E_{te} = E_{ti}$, which expresses the continuity of the tangential components across the boundary $r = a$, can be written as

$$\frac{1}{r}\frac{\partial V_e}{\partial\theta} = \frac{1}{r}\frac{\partial V_i}{\partial\theta} \qquad \text{for } r = a \tag{4.74}$$

Note that the tangential components to a sphere immersed in a z-directed field are the θ components. Equation (4.74) leads to

$$E_0 - \frac{B}{a^3} = -C \tag{4.75}$$

We now have two equations and two unknowns. Solving for B and C we obtain

$$B = a^3 E_0\frac{\varepsilon - \varepsilon_0}{\varepsilon + 2\varepsilon_0} = a^3 E_0\frac{\varepsilon_r - 1}{\varepsilon_r + 2} \tag{4.76}$$

$$C = -E_0\frac{3\varepsilon_0}{\varepsilon + 2\varepsilon_0} = -E_0\frac{3}{\varepsilon_r + 2} \tag{4.77}$$

where $\varepsilon_r = \varepsilon/\varepsilon_0$ is the dielectric constant of the material of the sphere. Thus, the problem is solved, and the potential is

$$V_e = -E_0 z + a^3 E_0\frac{\varepsilon_r - 1}{\varepsilon_r + 2}\frac{\cos\theta}{r^2} \qquad r \geq a \tag{4.78}$$

$$V_i = -\frac{3E_0}{\varepsilon_r + 2}z \qquad r \leq a \tag{4.79}$$

Figure 4.21 shows how an originally uniform electric field is distorted by the presence of a dielectric sphere.

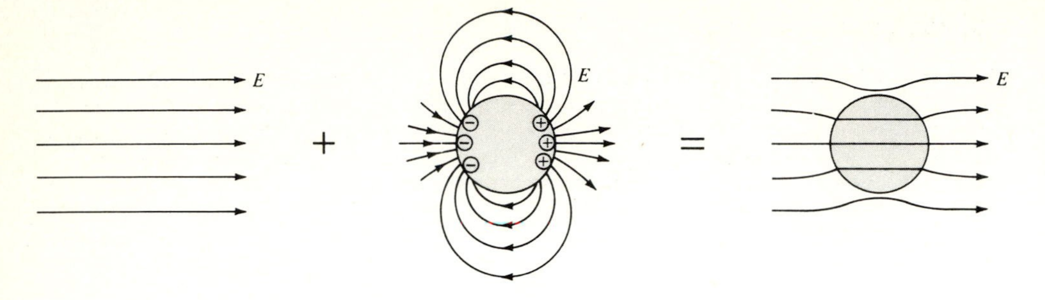

Figure 4.21 The resultant field of a dielectric sphere immersed in a uniform field is the sum of the uniform field plus a dipole field.

Relation of Internal and External Fields to Induced Polarization

Inside the sphere, V_i gives rise to a uniform field†

$$\mathbf{E}_i = -\nabla V_i = -\frac{\partial V_i}{\partial z}\hat{\mathbf{z}} = \frac{3}{\varepsilon_r + 2}\mathbf{E}_0 \tag{4.80}$$

which is seen to be in the same direction but smaller than the applied field E_0. Outside the sphere, the potential V_e is given by the original unperturbed field $-E_0 r \cos\theta$ plus the field of an equivalent electric dipole at the origin, with dipole moment

$$p = 4\pi\varepsilon_0 a^3 E_0 \frac{\varepsilon_r - 1}{\varepsilon_r + 2} \tag{4.81}$$

This result follows by comparing (4.78) to the expression (4.12*a*)

$$V = \frac{p}{4\pi\varepsilon_0}\frac{\cos\theta}{r^2} \tag{4.82}$$

which is the potential due to a dipole of moment p. This dipole moment can be related to the volume integral of polarization P by

$$p = \int P\,dv = P\frac{4\pi}{3}a^3 \tag{4.83}$$

P is constant because E_i is constant. The polarization can also be obtained from (4.22) or (4.45) as

$$P = (\varepsilon - \varepsilon_0)E_i = (\varepsilon_r - 1)\varepsilon_0 E_i = 3E_0\varepsilon_0\frac{\varepsilon_r - 1}{\varepsilon_r + 2} \tag{4.84}$$

† The internal field can also be expressed in terms of the induced polarization P with the aid of (4.84); i.e.,

$$\mathbf{E}_i = \frac{3}{\varepsilon_r + 2}\mathbf{E}_0 = \mathbf{E}_0 - \mathbf{E}_0\frac{\varepsilon_r - 1}{\varepsilon_r + 2} = \mathbf{E}_0 - \frac{\mathbf{P}}{3\varepsilon_0}$$

The Depolarization Field

Equation (4.80) implies that for the interior of the sphere the original uniform electric field E_0 is reduced by the factor† $3/(\varepsilon_r + 2)$. This reduction of the field inside the sphere is due to the field of the polarization surface charge $\rho_{\text{pol}} = P \cos \theta$ acting in opposition to the original field and is known as *depolarization*. The opposing field is called the *depolarization field* $\mathbf{E}_{\text{dep}}$. For the dielectric sphere, it is the difference between the original field and the internal field; that is,

$$\mathbf{E}_{\text{dep}} = \mathbf{E}_i - \mathbf{E}_0 = -\frac{\varepsilon_r - 1}{\varepsilon_r + 2}\mathbf{E}_0 \tag{4.85}$$

For the case of a permanently polarized sphere, $\mathbf{E}_{\text{dep}}$ must be equal to the internal electric field of Eq. (4.66). That is, E_{dep} and the E of (4.66) have as their source the surface polarization layer $\rho_{\text{pol}} = \mathbf{P} \cdot \hat{\mathbf{n}} = P \cos \theta$. To confirm this we can substitute E_0 from (4.84) in (4.85). We obtain

$$\mathbf{E}_{\text{dep}} = -\frac{\mathbf{P}}{3\varepsilon_0} \tag{4.86}$$

which is the same as (4.66).

The total electric field is shown in Fig. 4.22*a*. The smaller internal field is denoted by the thinner lines. The depolarization field can be imagined as pointing from the positive to the negative charges, i.e., opposite to the resultant field shown.

The Electric Flux

The electric flux density $\mathbf{D}_e$ external to the sphere can be obtained by operating with the gradient on (4.78). In spherical coordinates we obtain

$$\begin{aligned}\mathbf{D}_e &= -\varepsilon_0 \nabla V_e \\ &= \varepsilon_0 E_0 \hat{\mathbf{z}} + a^3 E_0 \frac{\varepsilon_0}{r^3}\frac{\varepsilon_r - 1}{\varepsilon_r + 2}(2 \cos \theta \, \hat{\mathbf{r}} + \sin \theta \, \hat{\boldsymbol{\theta}})\end{aligned} \tag{4.87}$$

which is just a constant field plus a dipole field as given by Eq. (4.14). Note that the unit vector $\hat{\mathbf{k}}$ in the z direction is equal to $\hat{\mathbf{k}} = \cos \theta \, \hat{\mathbf{r}} - \sin \theta \, \hat{\boldsymbol{\theta}}$. The normal components of $\mathbf{D}$ are continuous across the

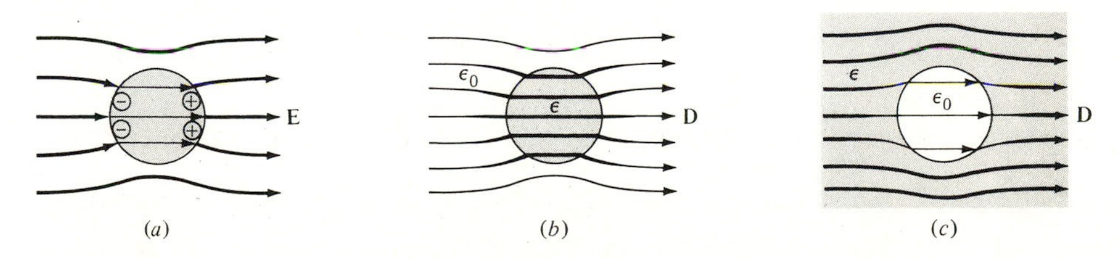

Figure 4.22 (*a*) The electric field for a dielectric sphere. The thinner lines inside the sphere denote that the field is reduced by the polarization charges; (*b*) lines of electric flux density D. The thicker lines represent larger values of D; (*c*) flux density inside and outside a spherical cavity in dielectric material.

† For dielectric materials $\varepsilon_r > 1$.

boundary, while lines of **E** are not. Hence, the external flux density close to the boundary is the same as that inside, which is constant and is given by†

$$\mathbf{D}_i = \varepsilon \mathbf{E}_i = \varepsilon_0 \varepsilon_r \mathbf{E}_i = \frac{3\varepsilon_r}{\varepsilon_r + 2} \mathbf{D}_0 \tag{4.88}$$

We see therefore that the flux density inside the dielectric sphere is increased over the value of the original field $\mathbf{D}_0$, whereas $\mathbf{E}_i$ is decreased as a result of the depolarization field. As a matter of fact, $E_i/E_0 \to 3/\varepsilon_r \to 0$, and $D_i/D_0 \to 3$ as ε_r becomes large. This increase in D is shown pictorially in Fig. 4.22*b* by the thicker lines of D inside.

A Spherical Cavity in a Dielectric Medium

Let us consider a spherical hole, or cavity, in a dielectric medium of permittivity ε, as shown in Fig. 4.22*c*. If the applied electric field E_0 in the dielectric before the introduction of the cavity is uniform and along the z axis, the solution to this problem is the same as that for a dielectric sphere in a uniform field. However, since the interior of the cavity is free space with permittivity ε_0 and the exterior with ε, these permittivities must be exchanged in the constants B and C given by (4.76) and (4.77); that is,

$$B = a^3 E_0 \frac{\varepsilon_0 - \varepsilon}{\varepsilon_0 + 2\varepsilon_0} = a^3 E_0 \frac{1 - \varepsilon_r}{1 + 2\varepsilon_r} \tag{4.89}$$

$$C = -E_0 \frac{3\varepsilon}{\varepsilon_0 + 2\varepsilon} = -E_0 \frac{3\varepsilon_r}{1 + 2\varepsilon_r} \tag{4.90}$$

The electric field inside the cavity is thus

$$E_i = \frac{3\varepsilon_r}{2\varepsilon_r + 1} E_0 \tag{4.91}$$

It is parallel to E_0 and is larger than E_0. The external field is the applied field E_0 plus the field of a dipole located at the center of the cavity and oriented opposite to the applied field with dipole moment

$$p = 4\pi\varepsilon a^3 E_0 \frac{1 - \varepsilon_r}{1 + 2\varepsilon_r} \tag{4.92}$$

The original flux density in the dielectric, before the introduction of the cavity, is $\mathbf{D}_0 = \varepsilon E_0$. The flux density inside the cavity is given by

$$\mathbf{D}_i = \varepsilon_0 \mathbf{E}_i = \frac{3}{2\varepsilon_r + 1} \mathbf{D}_0 \tag{4.93}$$

which is less than the original D_0 because the factor $3/(2\varepsilon_r + 1)$ is less than 1 for $\varepsilon_r > 1$. Figure 4.22*c* shows the lines of electric flux density inside and outside the cavity.

We can observe now that a less dense medium‡ tends to diverge a field, whereas a more dense medium tends to converge or concentrate it. This observation will be important when we come to the study of forces on dielectric objects where it will be shown that materials which concentrate flux density will be drawn into regions of stronger electric fields, whereas materials that diverge D will be

† To check (4.88), derive it by using the general relationship $\mathbf{D} = \varepsilon_0 \mathbf{E} + \mathbf{P}$.

‡ In a first approximation, the dielectric constant of a medium is proportional to the density of the medium.

expelled. The same observation will be made in the study of magnetic fields. Paramagnetic materials which have a relative permeability $\mu_r > 1$ will also concentrate and be drawn into regions of higher magnetic flux. Diamagnetic materials, such as superconductors, for which $\mu_r < 1$, on the other hand, will expel magnetic fields.

4.12 CAVITIES IN DIELECTRIC MATERIALS

A novel way to measure the electric field **E** and flux density **D** in a dielectric is to cut needle-shaped and disc-shaped cavities in the material and measure the fields in these. If a needle-shaped cavity, as shown in Fig. 4.23*a*, is cut or drilled parallel to the electric field, we have from the boundary condition on the tangential components $E_{t1} = E_{t2}$ that

$$E_{\text{cav}} = E \tag{4.94}$$

The polarization charges at the ends of the long, slender cavity can be ignored as being small, especially if the ends are rounded. No polarization charge exists on the sides since the polarization **P** is parallel to the electric field **E** in the dielectric. If we confine ourselves to the center of the cavity and stay away from the edges of the cavity where the field fringes, we can define the field in a needlelike cavity parallel to **P** as the average field in the dielectric medium. The same reasoning applies to a long, thin rod of dielectric material which is placed with its axis parallel to an electric field E_0 in free space; that is, $E_{\text{rod}} = E_0$.

The electric flux density is determined by cutting a disc-shaped cavity normal to D in the dielectric material, as shown in Fig. 4.23*b*. In this case the field E_{cav} is not the same as E in the dielectric because polarization charges appear on the surfaces. P, D, and E in the dielectric are related by $D = \varepsilon E = \varepsilon_0 E + P$. The polarization charge layers on top and bottom are then given by

$$\rho_{\text{pol}} = P = E(\varepsilon - \varepsilon_0) \tag{4.95}$$

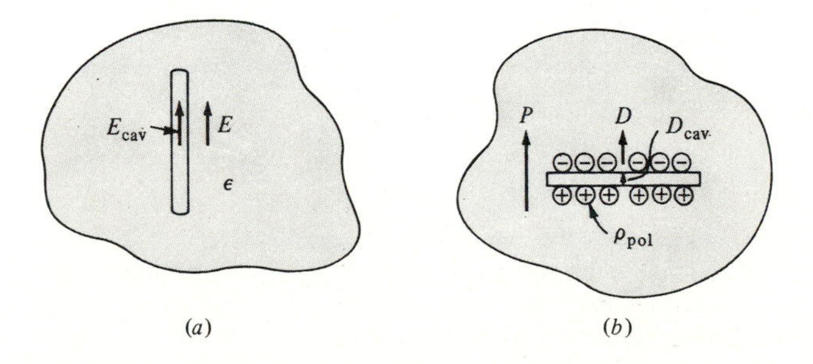

Figure 4.23 Cavities for measuring internal fields in a dielectric.

and will produce an electric field E_{pol} inside the cavity, parallel to P and of magnitude $E_{pol} = \rho_{pol}/\varepsilon_0$. The total field strength in the cavity is then

$$\mathbf{E}_{cav} = \mathbf{E} + \frac{\mathbf{P}}{\varepsilon_0} \tag{4.96}$$

We should point out again that this is valid far from the edges of the cavity. The vector **D** is simply obtained by applying the boundary condition for the normal components of flux density: $D_{n1} = D_{n2}$. Therefore,

$$D_{cav} = D \tag{4.97}$$

that is, the flux density at the center of a disc-shaped cavity oriented normally to D gives the average D in the dielectric material. The cavity field can also be expressed as $E_{cav} = D/\varepsilon_0 = \varepsilon_r E$. The boundary conditions on the normal components of D can also be used to obtain the fields in a dielectric plate or a disc which is immersed in an external field E_0. If the plane of the plate is perpendicular to E_0, $D_{n1} = D_{n2}$ gives

$$E_{plate} = \frac{\varepsilon_0}{\varepsilon} E_0 = \frac{E_0}{\varepsilon_r} \tag{4.98}$$

where ε is the permittivity of the plate material.

4.13 EXAMPLE: METALLIC SPHERE IN A UNIFORM ELECTRIC FIELD

The solution to the problem of a conducting sphere immersed in a uniform electric field can be obtained from Laplace's equation and the appropriate boundary conditions. In that sense this problem could have been considered at the end of Chap. 1. The reason for considering it in a chapter on polarization is that it follows naturally the problem of a dielectric sphere in a uniform field. This problem will show the difference between a partially (dielectric) and completely (metallic) polarized sphere. Another reason is that metallic spheres are used as the elements in artificial dielectrics, as was shown in Sec. 4.5. In that section the solution to this problem was already obtained, but with a lack of rigor.

Let an initially uncharged metal sphere of radius a be placed in a uniform electric field E_0. The spherical coordinate system can be readily used if we let the applied E_0 be along the z direction and the sphere be centered at the origin, as shown in Fig. 4.24. The introduction of the metal sphere will alter the field lines of E_0. The reason for this is that field lines must strike a metallic surface, which is an equipotential surface, normally. If we had no special ways of guessing the form of the solution, we could proceed strictly formally. That is, we know that in the region outside the sphere the potential is given by Laplace's equation, which from Eq. (1.105) or the back cover is

$$\nabla^2 V(r, \theta) = \frac{\partial V}{\partial r}\left(r^2 \frac{\partial V}{\partial r}\right) + \frac{1}{\sin\theta}\frac{\partial}{\partial\theta}\left(\sin\theta\,\frac{\partial V}{\partial\theta}\right) = 0 \tag{4.99}$$

where the derivatives with respect to ϕ vanish because the combination of the sphere and the z-directed E field is symmetric about the z axis. A solution to this equation can be obtained by the separation-of-variables technique for partial differential equations (Sec. 1.20). However, whenever possible one should try to construct the solution from the given problem. This is usually not only simpler, but it also brings the physical aspects of the problem into play at an early part in the solution. Anyway, according to the uniqueness theorem, which is shown in more advanced texts, we have a guarantee that there is

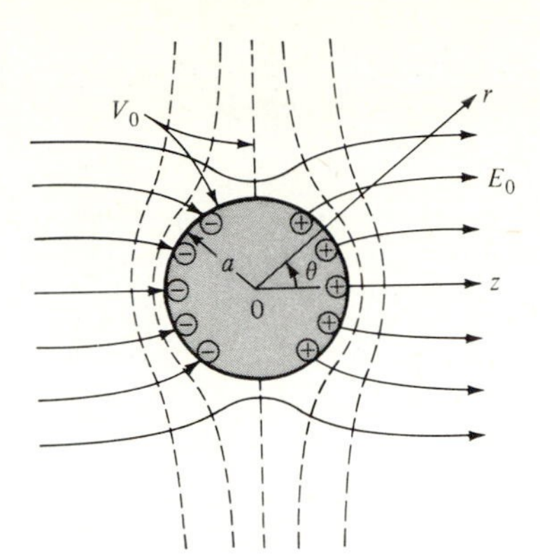

Figure 4.24 The electric field and equipotentials (dashed) for a metallic sphere in a uniform electric field. The E lines are always normal to the equipotentials. Note that the field is little disturbed at distances of a few radii from the sphere.

only one solution to Laplace's equation which satisfies a given set of boundary conditions. Therefore, any way or any method that we use to obtain a solution, if it satisfies Laplace's equation and the boundary conditions, it is unique.

The solution that we are looking for must contain a dipole term (since the induced charges on the surface of the sphere are in effect a large dipole) and far from the sphere the field must look like the original field (since far from the sphere the field is only slightly perturbed from its initial form); that is,

$$\mathbf{E}(r, \theta)\Big|_{r=\infty} = \mathbf{E}_0 = E_0\hat{\mathbf{z}} \tag{4.100}$$

$$V(r, \theta)\Big|_{r=\infty} = -E_0 z = -E_0 r \cos\theta \tag{4.101}$$

Thus, we can write for the solution

$$\begin{aligned} V(r, \theta) &= \text{initial field} + \text{perturbation due to sphere} \\ &= (-E_0 r \cos\theta) + (A + Br^{-1} + Cr^{-2}\cos\theta + \cdots) \end{aligned} \tag{4.102}$$

The terms in the perturbation field correspond to a constant potential, a point charge at the origin, and a dipole term. Higher order terms are not needed since the initially uniform field does not induce a more complex charge distribution than that corresponding to a dipole term. Terms with positive powers in r are not allowed since then the perturbation field, instead of vanishing far from the sphere, would in fact increase. Also the sphere was assumed initially uncharged, which forces $B \equiv 0$.

Suppose the surface of the sphere is at the constant potential V_0. Our solution (4.102) must then satisfy

$$V(r, \theta)\Big|_{r=a} = V_0 = -E_0 a \cos\theta + A + Ca^{-2}\cos\theta \tag{4.103}$$

The only way this can be satisfied for all θ is for $A = V_0$ and for the coefficients of $\cos\theta$ to be equal; that is, $-E_0 a + Ca^{-2} = 0$. The solution for a metallic sphere at constant potential V_0 is therefore

$$\boxed{V(r, \theta) = V_0 - E_0 r \cos\theta + E_0 a^3 \frac{\cos\theta}{r^2}} \tag{4.104}$$

It should be pointed out that the choice of reference plane for potential is arbitrary. Choosing the plane through the origin of the coordinate system of Fig. 4.24 as the $V = 0$ plane is convenient since then the surface of the sphere is at zero potential. For this case V_0 would be zero. However, for the sake of generality we include an arbitrary reference potential V_0.

The last term in the above solution for the potential is the perturbation of the original uniform field by the conducting sphere. It corresponds to that of a dipole of moment p given by

$$p = 4\pi\varepsilon_0 E_0 a^3 \tag{4.105}$$

and located at the center of the sphere with its axis along z. When we compare this dipole moment to that of a dielectric sphere in a uniform field (4.81), we see that we can obtain p for a conducting sphere by letting $\varepsilon_r \to \infty$ in p for the dielectric sphere. The depolarization field can be obtained from (4.85) by substituting $\varepsilon_r \to \infty$. We obtain that $E_{\text{dep}} = -E_0$. That is, the interior field E_i for a metallic sphere is zero, a condition we derived already in Sec. 1.12.

We can make use now of the expression for the potential to calculate the electric field at all points in space exterior to the sphere

$$E_r = -\frac{\partial V}{\partial r} = E_0\left(1 + \frac{2a^3}{r^3}\right)\cos\theta \tag{4.106}$$

$$E_\theta = -\frac{1}{r}\frac{\partial V}{\partial \theta} = -E_0\left(1 - \frac{a^3}{r^3}\right)\sin\theta \tag{4.107}$$

The induced surface charge on the sphere can be obtained from the boundary condition on the surface of a conductor. Thus for the normal component of D we have at $r = a$ that $\rho_s = D_n = \varepsilon_0 E_n = -\varepsilon_0\, \partial V/\partial n$ or

$$\rho_s(\theta) = \varepsilon_0 E_r\Big|_{r=a} = 3\varepsilon_0 E_0 \cos\theta \tag{4.108}$$

The total charge on the sphere

$$Q = \oint\!\!\oint \rho_s\, dA = 2\pi a^2 \int_0^\pi \rho_s(\theta)\sin\theta\, d\theta \tag{4.109}$$

is zero, which agrees with our initial assumption. Of course, an initially uncharged sphere experiences no force when introduced in a uniform field. The torque on the sphere, which is given by $\mathbf{T} = \mathbf{E}_0 \times \mathbf{p}$, is also zero since the induced dipole moment $\mathbf{p}$ is in the direction of the applied field $\mathbf{E}_0$. The potential at the interior of the sphere is the same as its value at the surface of the sphere, that is, V_0, since $E = 0$ for the interior.

If the sphere were given a charge Q, the total potential would be simply that given by (4.104) with the addition of a $Q/4\pi\varepsilon_0 r$ term.

PROBLEMS

4.1 Inside a uniformly polarized dielectric, the number of dipoles per unit volume is N and their moment is given by $\mathbf{p}$. What is the polarization vector?

4.2 Show that one can expect the permanent electric dipole moment of polar molecules to be on the order of 10^{-29} C · m.

4.3 The polarizability α is defined by $\mathbf{p} = \alpha\mathbf{E}$ because an atom acquires a dipole moment proportional to the external electric field E.

(*a*) What is the total force that acts on an atom when it is introduced in a uniform **E** field?

(*b*) What is the force that acts on an atom when it is introduced into the **E** field of a point charge Q? Assume the distance r between the atom and the point charge Q is much greater than the diameter d of the atom; that is, $r \gg d$.

4.4 An electric field of 10^3 V/m is applied to a helium gas whose relative permittivity is given as 1.00007. If the concentration is such that there are 10^{25} atoms/m^3, find

(*a*) Polarization **P**.

(*b*) Dipole moment **p**.

(*c*) The shift between the electron cloud and nucleus, i.e., the separation between the negative and positive charge.

4.5 The electric flux density in Bakelite (a plastic material) is given as 4×10^{-6} C/m^2. Find the magnitude of the polarization vector **P**.

4.6 A uniform electric field with a flux density of 1 C/m^2 is normal to a plane slab of glass which has a dielectric constant $\varepsilon_r = 6$. If this results in uniform polarization of the material and if the volume of the slab is 0.5 m^3, find the polarization P and the total dipole moment p of the slab.

4.7 A uniform electric field in air is normal to a plane slab of polystyrene for which $\varepsilon_r = 2.56$. If this induces a polarization-surface charge density $\rho_{sp} = 0.1$ C/m^2 on the slab surfaces, find the

(*a*) Polarization P, flux density D, and electric field E in the slab.

(*b*) Flux density D_{air} and electric field E_{air} outside the slab.

4.8 At a dielectric-air interface the electric field in the dielectric ($\varepsilon_r = 10$) is at an angle of 75° to the normal. What is the angle of the free-space **E** field with respect to the normal?

4.9 The electric field in air above a dielectric slab is at a 30° angle to the normal of the air-dielectric boundary. If the corresponding angle in the dielectric is 71°, what is the dielectric constant ε_r of the dielectric?

4.10 A plane dielectric slab consists of many layers of dielectric materials sandwiched together as shown. If this slab is immersed in a uniform electric field in free space, show that the entrance angle θ_i is the same as the exit angle θ_e.

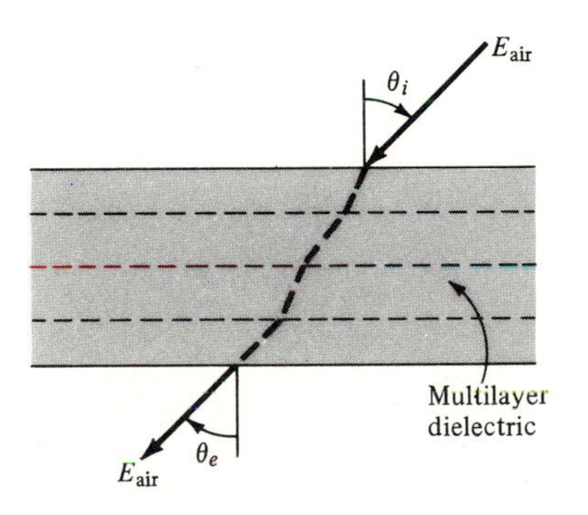

4.11 An electric dipole is originally aligned with an electric **E** field. It takes 0.05 J of work to turn it by 180°. Find the torque required to hold it at right angles to the **E** field.

4.12 A linear quadrupole is an arrangement of four charges as shown. If the separation of charges d is small compared to the observation distance r, find the potential and electric field of the quadrupole.

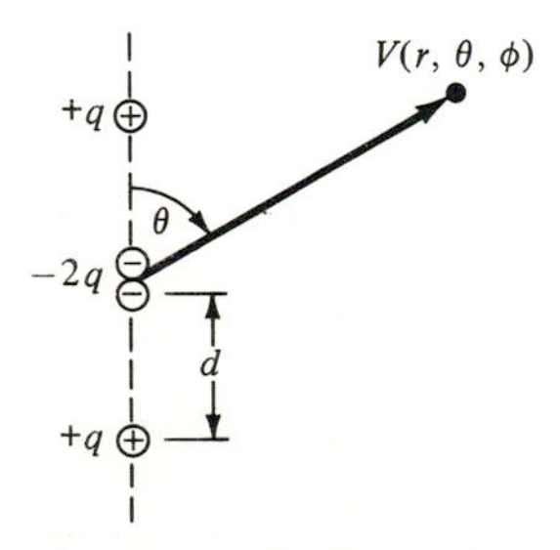

4.13 A quadrupole is an arrangement of four charges as shown. If the separation distance d is small compared to the observation distance r, find the potential and electric field (in the plane of the paper) of the quadrupole.

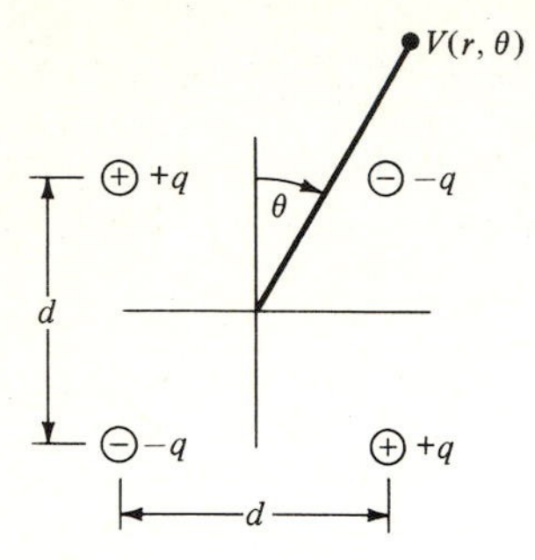

4.14 Referring to Fig. 4.8 and Eq. (4.14), determine the equation for the **E** field lines of a dipole.

4.15 Determine the equation for the equipotential surfaces of a dipole.

4.16 A dipole of moment **p** is situated at the origin of an xy coordinate system and is parallel to the x axis. What is the force on a second dipole, also of moment **p**, that is located at (b, b) and points toward the origin?

4.17 What is the volume density ρ' of a bound charge in

(*a*) A homogeneous dielectric?

(*b*) A nonhomogeneous dielectric?

4.18 Demonstrate that in the time-varying case the polarization current $\partial \mathbf{P}/\partial t$ has dimensions of current density.

4.19 Polarization current results from the relative motion of a bound positive charge and a bound negative charge. Using the definition of **P**, show that $\mathbf{J}_p = \partial \mathbf{P}/\partial t$ is consistent with the expression $\mathbf{J}_p = \rho_+ \mathbf{v}_+ + \rho_- \mathbf{v}_-$, where ρ and v *are the polarization charge density and velocity, respectively.*

4.20 A spherical shell of thickness $b - a$, where a and b are the inner and outer radii of the shell, respectively, surrounds a point charge Q located at the center of the shell. If the shell is of a dielectric material with dielectric constant ε_r, find **E** and **D** for the three regions $r < a$, $a < r < b$, $r > b$. Assume the material outside the shell is free space with permittivity ε_0.

4.21 Find the polarization **P**, volume polarization charge density ρ_p, and surface polarization charge density ρ_{sp} in the shell material of Prob. 4.20.

4.22 Sketch the fields produced by Q, ρ_{sp} at $r = a$, and ρ_{sp} at $r = b$ of the previous problem. Then show that

(*a*) The **E** field produced by Q and the two ρ_{sp}'s of Prob. 4.21 is the same as the **E** field obtained in Prob. 4.20.

(*b*) The **E** field produced by ρ_{sp} at $r = a$ and ρ_{sp} at $r = b$ is zero for $r < a$ and $r > b$.

4.23 Three coaxial cylinders separated by dielectrics of permittivities ε_1 and ε_2 are shown in the figure. A charge ρ_{L1} C/m is placed on the inner conductor, and a charge $-\rho_{L2}$ C/m is placed on the outer conductor. The middle conductor is connected to ground. Give an expression for the potential variation and sketch a curve showing the potential distribution as a function of the radial distance r.

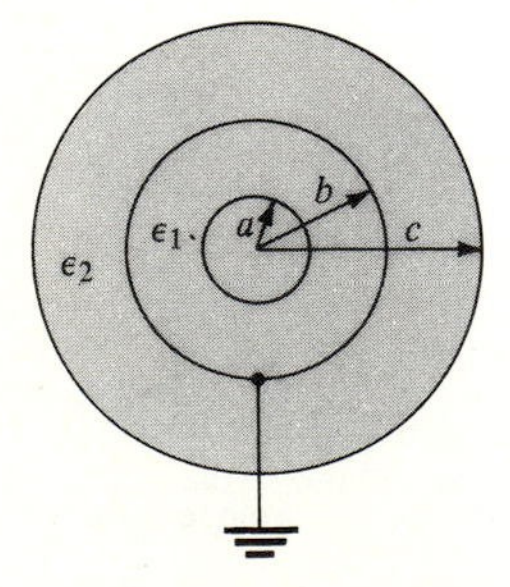

4.24 At the center of a dielectric sphere of permittivity ε and radius a is placed a point charge Q. Find V, **E**, **P**, ρ_p, and ρ_{sp} inside and outside the sphere. Assume the sphere is immersed in free space (ε_0).

4.25 Dielectric breakdown was discussed in Sec. 2.13. Do you expect the introduction of a dielectric material in a space formerly occupied by free space would increase or decrease the breakdown strength of that space?

4.26 Discuss why the **E** field in a space is less when a dielectric is present than when it is not.

4.27 An artificial dielectric material consists of a number of brass spheres of diameter d, spaced $3d$ apart, in a regular lattice. If each sphere is influenced only by the external electric field, determine the dielectric constant for this material.

4.28 A dielectric shell has an inner and outer radius of a and $2a$, respectively. The shell material has a dielectric constant $\varepsilon_r = 3$. If the shell is placed in an initially uniform electric field $\mathbf{E}_0$, calculate the **E** field inside the spherical cavity $r < a$.

4.29 A thin dielectric disk of radius a and thickness t has a permanent polarization with **P** parallel to the axis of the disk.

(*a*) What are **E** and **D** inside the disk?

(*b*) Calculate the **E** field on the axis of the disk.

4.30 In Sec. 4.7 a comparison of polarization in dielectrics and conductors was made. Consider and discuss a metal as a polarizable body. Determine the value of polarization **P** and susceptibility χ.

4.31 A conducting sphere of radius a has a charge Q distributed over its surface. If the sphere is covered with a dielectric layer with inner and outer radii a and b, respectively, calculate

(*a*) The polarization surface charge on the inside and outside of the dielectric.

(*b*) The volume density of polarization charge inside the dielectric.

4.32 A dielectric cylinder of radius a and relative permittivity ε_r, is immersed in a uniform electric field E_0 which is at right angles to the axis of the cylinder. Find the internal and external potential. Check your answer by observing the behavior when $\varepsilon_r = 1$ and $\varepsilon_r = \infty$.

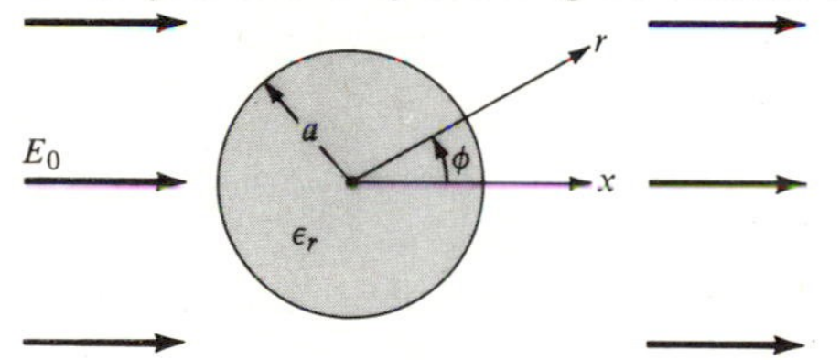

4.33 In the previous problem, classify the external field as a uniform field plus a linear dipole field. What is the dipole moment?

4.34 A grounded (zero potential) conducting cylinder of radius a is perpendicular to a uniform electric field E_0. Find the potential at all points outside the cylinder.

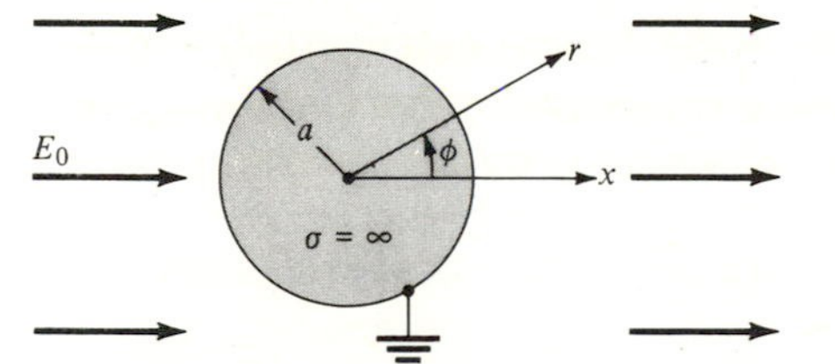

4.35 A large block of dielectric material is polarized uniformly with a dipole moment per unit volume **P**.

(*a*) What is the electric field inside a small spherical cavity which is cut into the material?

(*b*) What is the electric field on the axis of a small cylindrical cavity which is cut into the material with its axis parallel to **P**?

4.36 Referring to Fig. 4.23, show that a charge Q in the needle-shaped cavity will experience a force $\mathbf{F} = Q\mathbf{E}$, and in the disk-shaped cavity the force on Q will be $\mathbf{F} = Q\mathbf{D}/\varepsilon_0$.

CHAPTER

FIVE

CAPACITANCE, ENERGY, AND FORCES

Guide to the Chapter

The organization of the chapter is as follows: First, the concept of capacitance is developed. Capacitance for several configurations is then studied. Second, it is shown that capacitors can store energy. Again, energy storage for several configurations is studied. Finally, since forces must be involved in situations where energy is stored, transferred, etc., we develop the force expressions for the elements involved in energy storage. The last four sections give examples of forces exerted by electric fields.

5.1 DEFINITION OF CAPACITANCE

Capacitance is a property of a geometric configuration, usually of two conducting objects separated by an insulating medium. It is a measure of how much charge a particular configuration is able to retain when a battery of V volts is connected and then removed. For example, Fig. 5.1 shows two initially uncharged conducting bodies to which a battery had been connected. It shows that an equal but opposite charge of magnitude Q was deposited on the two conductors. After the battery is removed, these charges hold the conductors at a potential difference of V volts. An electric field E is associated with the charges† which in turn is related to the potential V by $V = -\int \mathbf{E} \cdot d\mathbf{l}$. Note that if the objects are conductors, the surface of each one is an equipotential. Therefore, the potential between any two points, one on one conductor and the other on the other conductor, is given by V.

† The boundary condition on the conducting surface is $\varepsilon_0 E_n = \rho_s$.

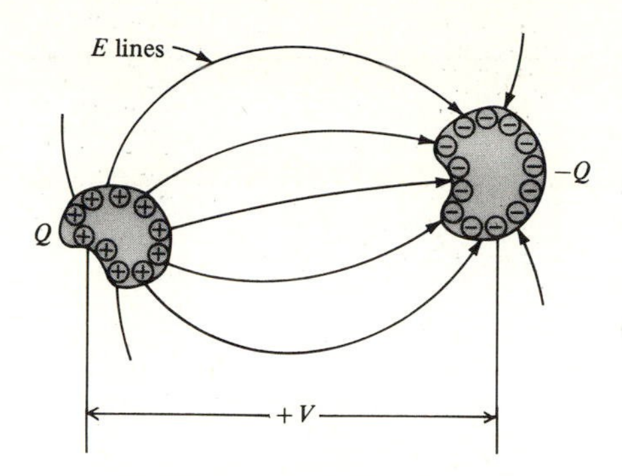

Figure 5.1 A capacitor formed by two conductors carrying equal but opposite charges. The potential difference between the conductors is V.

The amount of charge Q deposited on each conductor will be proportional to the voltage V of the battery and some constant C, called the capacitance; that is,

$$\boxed{Q = CV} \tag{5.1}$$

The determination of this constant for various arrangements of conductors is one of the objectives of this chapter. Although capacitance will be derived using electrostatics, the values obtained are valid up to very high frequencies.

The question of why charge is transferred to the two capacitor conductors when a battery is connected can be answered as follows: The potential between the two initially uncharged conductors is zero (a condition that can be obtained by first touching and then separating the conductors). When a battery is connected, the potential difference across the capacitor becomes that of the battery, because battery terminal, connecting wire, and capacitor conductor are made of metallic material and thus form an equipotential surface. Each capacitor conductor becomes an extended battery terminal. At the instant of connection, electrons flow until the conductor is at the same potential as the corresponding battery terminal. Then the flow of charge stops. Equal and opposite charges are transferred to the capacitor conductors because the current that the battery emf "forces" to flow is made up of electrons leaving one conductor and "flowing" to the other. The electrons do not come from the battery but from the metallic material connected to the battery. Overall charge neutrality is preserved because the deficiency of electrons on one conductor shows up as an excess of equal magnitude on the other conductor.

The proportionality of Q to V was previously demonstrated when the potential V from a point charge Q was obtained as $V = Q/4\pi\varepsilon r$. Comparing this to (5.1) we can say that the spherical surface of radius r which surrounds the point charge has a capacitance of $C = 4\pi\varepsilon r$ with respect to infinity. To aid in visualizing this, consider an uncharged small metal sphere of radius r, isolated in space and centered at the origin, as shown in Fig. 5.3*a*. If a battery of potential V could be connected from infinity to such a sphere, an amount of charge Q would be transferred to the sphere. The sphere capacitance (between it and infinity) is therefore $C = 4\pi\varepsilon r$. We note that the capacitance of a point charge is zero because $C \to 0$ as $r \to 0$. The reason for this is that the potential from infinity to the point $r = 0$

(which is the work required to transfer a charge from infinity to $r = 0$) is infinite. From this we can observe that if the voltage required to transfer a given amount of charge to a capacitor is high, the capacitance must be low.

The unit for capacitance is the coulomb per volt (C/V). This ratio was given the name farad (F). Since a capacitance of 1 F is very large, more commonly used units are the microfarad ($1\ \mu F = 10^{-6}$ F) and the picofarad ($1\ pF = 10^{-12}$ F).

5.2 CHARGING A CAPACITOR: CHARGING ENERGY

If a charge Q is transferred to an initially uncharged capacitor or, what amounts to saying the same thing, if the voltage on the capacitor is "built-up" to V, the capacitor now stores potential energy. Note that V is a measure of potential energy since it is work per unit charge. On the other hand, to charge a capacitor requires a finite amount of time, and so we must speak of power or a time rate of doing work. This is easily seen since the transfer of charge is a time rate known as the current:

$$I = \frac{dQ}{dt} = C\frac{dV}{dt} \tag{5.2}$$

and current flowing across a potential V is power $P = IV$. The picture of the charging process is as follows: A current flows through the initially uncharged capacitor (initially uncharged also implies that the initial voltage across the capacitor is zero), depositing charges on the capacitor. Because charges are continually being deposited, the voltage across the capacitor rises from the initially zero voltage. This process will go on until the voltage across the capacitor reaches that of the source battery. At this point the voltage difference between battery and capacitor becomes zero and current flow stops. The capacitor is considered charged. When the battery is disconnected, a charge Q given by

$$Q = CV_{\text{battery}}$$

will remain on the capacitor.

The actual, rather than the conceptual, derivation of the charging process can be easily given. Figure 5.2*a* shows a circuit composed of a battery V_b, a resistance R representing the connecting resistance of the wires and the internal resistance of the battery, and a capacitor C assumed to be initially uncharged. If at time $t = 0$, the switch is closed, a current $I(t)$ will flow in the circuit. Kirchhoff's voltage law (3.8) for the circuit in Fig. 5.2 can be written as

$$V_b = RI + \frac{1}{C}\int_0^t I\, dt \tag{5.3}$$

Note that the voltage V_c across the capacitor which is given by the integral term above can be obtained either by integrating Eq. (5.2) or by substituting $Q = \int_0^t I\, dt$ in Eq. (5.1).

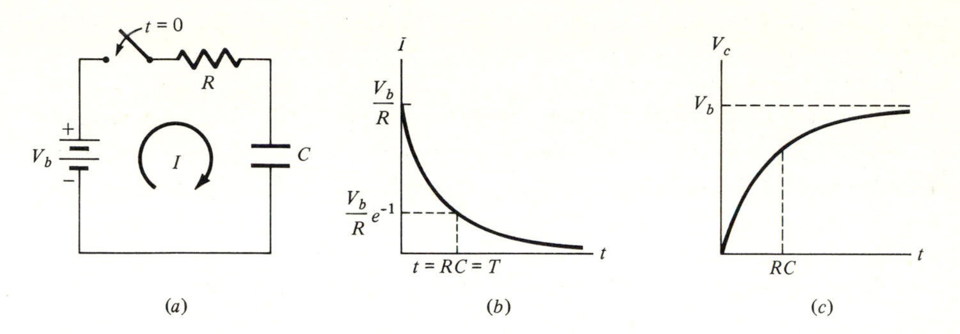

Figure 5.2 (*a*) A charging circuit for a capacitor in which energy is transferred from the battery to C. (*b*) and (*c*) show the charging current and voltage, respectively.

There are many ways to solve (5.3) for I. Let us proceed by differentiating with respect to t. We obtain

$$0 = \frac{dI}{dt} + \frac{1}{RC} I \tag{5.4}$$

By inspection a solution to this equation is

$$I = Ae^{-t/RC} \tag{5.5}$$

The constant A can be evaluated from the initial condition that $V_c|_{t=0} = 0$. Since voltage across a capacitor cannot change instantaneously, V_c will remain zero for an instant after the switch is closed (i.e., a capacitor has inertia for voltage but not for current). From (5.2) we see that for an instantaneous change in voltage across a capacitor to occur, the charging current would have to be infinite. Since this is impossible (finite jumps in current are possible, but not infinite ones), we conclude that voltage cannot change instantaneously across a capacitor. The current at $t = 0$ is then $I(t = 0) = V_b/R = A$. Thus, the charging current as a function of time is given by

$$I = \frac{V_b}{R} e^{-t/RC} \tag{5.6}$$

The charging current, shown graphically in Fig. 5.2*b*, decreases exponentially to zero. The time during which the current decreases to $1/e$ (or 37 percent) of its initial value is known as the time constant T and is given by $T = RC$. Reducing the circuit resistance R will thus decrease the time that it takes to charge the capacitor.

The capacitor voltage V_c is given by

$$V_c = \frac{1}{C}\int_0^t I\,dt = -V_b e^{-t/RC}\Big|_0^t = V_b(1 - e^{-t/RC}) \tag{5.7}$$

and is shown graphically in Fig. 5.2*c*. Theoretically it will take an infinite time for the capacitor to reach the voltage of the battery since $V_c(t = \infty) = V_b$. For practical purposes it is assumed that the charging process is complete when a time of several time constants has elapsed.

It is tempting to say that since $V = W/Q$, the energy transferred to the capacitor from the battery is simply $W = QV_b$. During charging, capacitor voltage and current are functions of time. Therefore the correct way to obtain energy is by integrating the instantaneous power, $P = IV_c$. The energy stored in C after the charging process is complete ($t \to \infty$) is then given by

$$\boxed{\begin{aligned} W &= \int_0^\infty IV_c\,dt = \frac{V_b^2}{C}\int_0^\infty e^{-t/RC}(1 - e^{-t/RC})\,dt \\ &= \frac{CV_b^2}{2} \end{aligned}} \tag{5.8}$$

This can also be expressed as $W = CV_b^2/2 = Q^2/2C = QV_b/2$ by the use of $Q = CV$. Thus the energy transferred is less by a factor of one-half when compared with $W = QV_b$. The reason for this is that initially little energy is needed to transfer charge to the capacitor because the initial voltage V_c is zero. As the capacitor voltage increases, it takes more work to deposit additional charges. The factor of one-half is therefore an averaging factor by which QV_b is multiplied. This averaging effect can also be seen from a graph of power $P = IV_c$ versus time. The power graph begins at zero, reaches a maximum, and then decays to zero.

The energy given by (5.8) expresses the amount of chemical potential energy which comes from the battery and is transferred to the capacitor. This amount of energy is available from the capacitor during discharge.

5.3 EXAMPLE: SPHERICAL-SHELL CAPACITOR

A useful capacitor configuration consists of two concentric spherical shells, as shown in Fig. 5.3*b*. If the shells are of a conducting material, the surfaces are equipotentials. Connecting a battery V in between will deposit a $+Q$ $(-Q)$ on the inner (outer) shell. As was shown in Sec. 2.10, Q will distribute itself on each surface in such a manner that the surface charge density is uniform. The field in the space between the shells is then like that from a point charge Q located at the origin, or

$$E_r = \frac{Q}{4\pi\varepsilon r^2} \qquad a \le r \le b \tag{5.9}$$

The potential difference V between the shells, or the work per unit charge to move a positive test charge from radius b to radius a in the field of a point charge Q

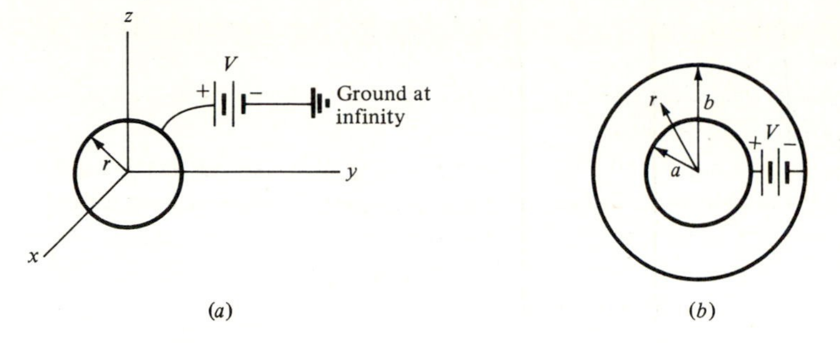

Figure 5.3 Capacitance of a spherical structure. (*a*) Capacitance of an isolated sphere with respect to infinity; (*b*) capacitance of the space between two spherical shells.

located at the origin, is given by

$$V = -\int_b^a E_r \, dr = \frac{Q}{4\pi\varepsilon} \int_b^a \frac{dr}{r^2} = \frac{Q}{4\pi\varepsilon} \frac{b-a}{ba} \tag{5.10}$$

The capacitance is then

$$C = \frac{Q}{V} = 4\pi\varepsilon \frac{ab}{b-a} \tag{5.11}$$

Decreasing the spacing between the shells increases the capacitance, whereas increasing the space, for example, by letting $b \to \infty$, will reduce the capacitance to that of an isolated sphere, $C = 4\pi\varepsilon a$.

5.4 PARALLEL-PLATE CAPACITOR AND DIELECTRIC STRENGTH†

Figure 5.4 shows a capacitor in which the two conductors that carry the opposite charges are parallel plates, each of area A and separated by the distance d. Since the two oppositely charged plates attract each other, they must be held apart by some insulating medium, usually some dielectric sandwiched between the plates. The force between capacitor plates will be calculated in Sec. 5.13.

In order to calculate C we need to know the charge Q on either plate in terms of the potential difference V. For large plates and small separation, the electric field E between the plates will be uniform over most of the area of the plates and is given by $E = V/d$.‡ The charges are confined to the interior surfaces of the two plates by mutual attraction. The surface layer of charge ρ_s is determined from boundary conditions as $D_n = \rho_s$. The total charge on each plate is therefore

† See Table 2.3 for values of breakdown strength.

‡ The field distribution for a similar problem was given in Sec. 1.18

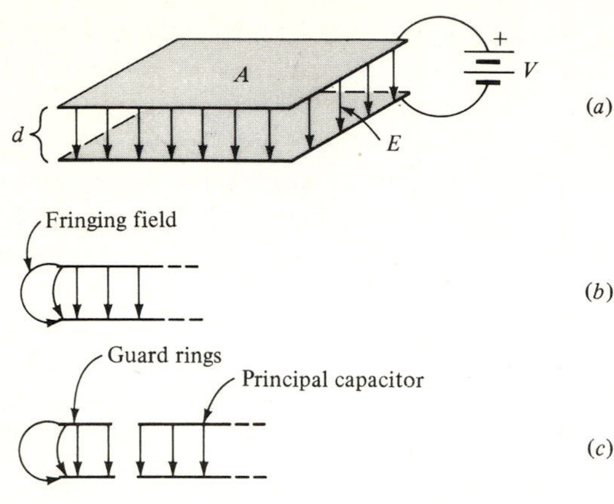

Figure 5.4 (*a*) Two plates of area A separated by a distance d form a parallel-plate capacitor. The fringing electric field at the edges (*b*) can be minimized in the principal capacitor by surrounding it with guard rings (*c*).

$Q = \rho_s A$, and the capacitance is

$$C = \frac{Q}{V} = \frac{\rho_s A}{Ed} = \frac{\varepsilon E A}{Ed} = \frac{\varepsilon A}{d} \tag{5.12}$$

The above formula is strictly valid only if the fringing field near the edges of the parallel plates is negligible. Some fringing, which is a continuation of the field beyond the volume defined by the two plates, is always present, for an electric field cannot end abruptly. If the field were to go to zero abruptly outside the plates, it would violate the boundary condition which imposes continuity on the tangential electric field. The fringing field, therefore, provides a gradual transition between the internal field of the capacitor and the zero field outside. In practice, A is sufficiently large such that a uniform E exists for most of A and (5.12) can be used with good accuracy. For example, high-quality capacitors are manufactured by rolling two strips of aluminum foil which are separated by a dielectric, usually mica or paper. Two wire leads are soldered, one to each strip of foil, and the "log" is then sealed in wax or some other sealer. The capacity can be increased by increasing the length of the aluminum strips, decreasing their spacing, or using an insulating material with a high dielectric constant $\varepsilon_r = \varepsilon/\varepsilon_0$. Note that capacitance can be written as

$$C = \varepsilon_r C_0$$

where C_0 is a capacitor having free space as the insulating medium.

At a first glance it might appear that the easiest way to increase C is to decrease d. However, minimum d is determined by the maximum voltage V that will be applied to the capacitor in use. For example, if the insulating medium is air with a dielectric breakdown strength of 30,000 V/cm and if the circuit in which C is to be used is such that the voltage across C does not exceed 300 V, then $d_{\min}$ must be

$$d_{\min} > 10^{-2} \text{ cm}$$

If a d lower than d_{min} is used, sparking or arcing between the two plates can occur which can destroy the capacitor. Table 2.3 gives values of dielectric strength for other materials. The most widely used dielectric medium for high-quality capacitors is mica. It has a very high value for dielectric strength, and its low value for conductivity (see Table 2.1) makes the leakage current across the capacitor low. In places where leakage current is not critical, barium titanate capacitors are used. The high value for ε_r of titanate dielectrics makes these capacitors very small in size.

As an example, we can calculate the capacitance of a pair of metal plates with $A = 1\ \text{cm}^2$, $d = 1$ mm, and $\varepsilon = \varepsilon_0 = 8.85 \times 10^{-12}$ F/m. We obtain $C \approx 10^{-12}$ F = 1 pF. Since free space with a breakdown strength of 30 kV/cm is used as the insulating medium, the operating voltage on the capacitor cannot exceed 30 kV/cm × 1 mm = 3000 V.

5.5 EXAMPLE: CAPACITANCE OF COAXIAL TRANSMISSION LINE

A coaxial transmission line consisting of two concentric cylindrical conductors of radii a and b is shown in Fig. 5.5. Since this and the parallel-wire transmission line are the most commonly used transmission lines, let us determine the capacitance C and, in a later chapter, the inductance L per unit length for both types of transmission cables. Knowing these two parameters, the characteristic impedance Z_0 and velocity of propagation v for a transmission line can then be found from $Z_0 = \sqrt{L/C}$ and $v = 1/\sqrt{LC}$.

In practice, a potential difference $V_a - V_b$ is applied between the inner and outer conductors. As a result current flows in the center and outer wire in opposite directions. Therefore, during any short interval of time, a charge ρ_L coulombs per meter exists on one conductor and an equal and opposite line charge exists on the other conductor. This line charge will distribute itself equally on the outside of the inner conductor and on the inside of the outer conductor such that $\varepsilon E_r = \rho_s$, where E_r is the radial field shown in Fig. 5.5 and ρ_s is the surface charge. To derive the capacitance of the coaxial line, we observe now that the

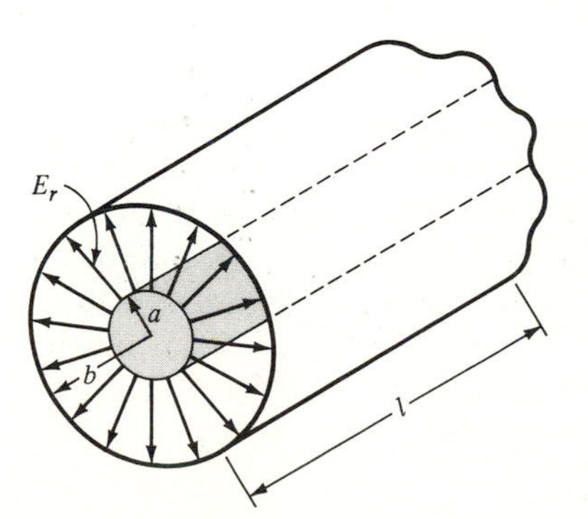

Figure 5.5 Cross section of a coaxial transmission line. For the direction of E field shown, a + charge must exist on the inner and a − charge on the outer conductor.

inner and outer conductors are equipotential surfaces. In Eq. (1.26) it was shown that the equipotential surfaces of a line charge ρ_L are also coaxial cylinders.† The potential difference between two coaxial cylinders was found to be

$$V_a - V_b = \frac{\rho_L}{2\pi\varepsilon} \ln \frac{b}{a} \tag{5.13}$$

If we now take ρ_L to represent the charge per unit length l on each of the two coaxial capacitor surfaces, the capacitance C of a coaxial cable of length l is then

$$\boxed{C = \frac{\rho_L l}{V_1 - V_2} = \frac{2\pi\varepsilon l}{\ln (b/a)}} \tag{5.14}$$

This can be re-expressed in terms of the dielectric constant ε_r for the medium between the coaxial cylinders as

$$\frac{C}{l} = \varepsilon_r \frac{55.6}{\ln (b/a)} \qquad \frac{pF}{m} \tag{5.15}$$

where $\varepsilon_r = \varepsilon/\varepsilon_0$ and $\varepsilon_0 = 8.85 \times 10^{-12}$ F/m.

5.6 EXAMPLES: CAPACITANCE OF TWO PARALLEL-WIRE TRANSMISSION LINES

In this section we will find the capacitance of two parallel conducting cylinders which are assumed to be infinitely long. Although this problem can be solved for two cylinders of unequal radii, we will confine ourselves to the case in which the cylinders are of the same size. The results can be used to find the capacitance per unit length for parallel-wire transmission lines such as, for example, the 300-Ω TV twin-lead cable.

This problem is more difficult than the corresponding one for the coaxial line. Hence, it is important that we understand the method of solution before we begin with the details. As in the coaxial cable we know that in use, a current flows in one wire and an equal and opposite one flows in the other wire. At any instant of time we can assume therefore that the two lines are equally and oppositely charged. The problem now is reduced to that of solving for the potential-field distribution in the region between two conducting cylinders which have a surface charge. Unfortunately, because the distance between the cylinders is finite, the charge distribution on each cylinder will not be uniform. As shown in Fig. 5.6, the surface charge density is greater for points on the cylinder which are nearest to the other

† A similar problem was considered in Eqs. (1.77*d*) and (1.104*h*).

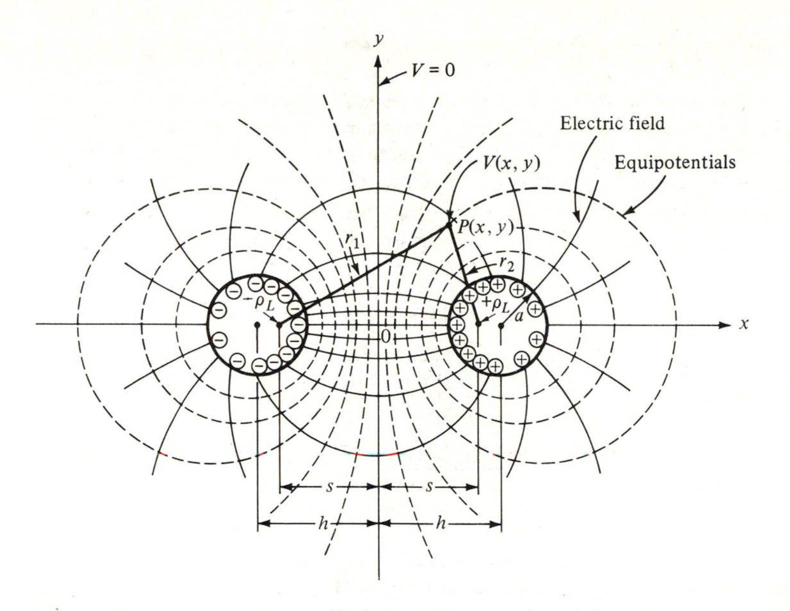

Figure 5.6 The electric field and equipotential lines (dashed) about two line charges ρ_L and $-\rho_L$. The surfaces of two conducting cylinders of radius a can be imagined to coincide with the heavy circles without disturbing the fields of the line charges outside the circles of radius a.

cylinder. Each charged cylinder can again be replaced by an equivalent line charge ρ_L, but because of the nonuniform charge distribution, the location of the equivalent line charge is not at the center of the cylinder but is displaced from the center toward the other cylinder, as shown in Fig. 5.6.

The solution to the two-wire problem therefore reduces to the solution for the potential distribution between two line charges. We will find that the equipotential surfaces will be cylinders about the line charges, as shown in Fig. 5.6. This fact can be used in the solution of the parallel-wire transmission line as follows: The conducting surfaces of a transmission line with a given spacing $2h$ and wire radius a can always be made to coincide with some equipotential cylinders of the two line charges by varying the spacing $2s$ between the line charges. If, in fact, two conducting cylinders (whose surfaces are equipotentials) were so placed, it would not disturb the field distribution of Fig. 5.6. Hence, we can use the potential distribution between two line charges to calculate the capacitance between two parallel conducting wires.

Field Distribution of Two Line Charges

Two line charges $+\rho_L$ and $-\rho_L$ are shown positioned along the x axis a distance $\pm s$ from the origin. The perpendicular bisector, which is the y axis, is seen to be a line of $V = 0$ potential. Along it a test charge can be moved without being affected by the force fields of ρ_L and $-\rho_L$. Let us pick the origin 0 as a convenient reference point from which to measure potentials.

The potential $V(x, y)$ at point $P(x, y)$ can be obtained by adding the potential from $+\rho_L$ when $-\rho_L$ is not present to the potential from $-\rho_L$ when $+\rho_L$ is not present. For an isolated line charge, the equipotentials are circles about the line charge. Let us take the line charge $+\rho_L$ first. The potential difference between a circle about $+\rho_L$ that goes through P and one that goes through 0 is, from (5.13), given by $V_{0P}^+ = (\rho_L/2\pi\varepsilon) \ln (s/r_2)$, where r_2 and s are radii measured from $+\rho_L$. Similarly the potential from the $-\rho_L$ line charge is given by $V_{0P}^- = (-\rho_L/2\pi\varepsilon) \ln (s/r_1)$. By the superposition theorem the potential difference between 0 and P in the presence of both line charges is therefore equal to the sum†

$$V_{0P} = V_{0P}^+ + V_{0P}^- = \frac{\rho_L}{2\pi\varepsilon} \ln \frac{r_1}{r_2} \tag{5.16}$$

From this equation, we observe that $V = 0$ along the y axis, because $r_1/r_2 = 1$ along $x = 0$. We are now looking for the family of equipotentials in the xy plane. These curves, for which $V =$ constant, are defined by setting

$$\frac{r_1}{r_2} = k \qquad k = \text{constant} \tag{5.17}$$

in (5.16). Expressing r_1 and r_2 in terms of the x-y coordinates as $r_1 = [(s + x)^2 + y^2]^{1/2}$ and $r_2 = [(s - x)^2 + y^2]^{1/2}$, we can write (5.17) as

$$(s + x)^2 + y^2 = k^2[(s - x)^2 + y^2] \tag{5.18}$$

or

$$x^2 - 2xs\frac{k^2 + 1}{k^2 - 1} + s^2 + y^2 = 0 \tag{5.19}$$

To put this equation in a more recognizable form, we can add $s^2(k^2 + 1)/(k^2 - 1)$ to both sides of (5.19). This completes the square on the left side, and (5.19) becomes

$$\left(x - s\frac{k^2 + 1}{k^2 - 1}\right)^2 + y^2 = \left(\frac{2ks}{k^2 - 1}\right)^2 \tag{5.20}$$

This is the equation of a family of circles (the dotted lines in Fig. 5.6) having the form

$$(x - h)^2 + y^2 = a^2 \tag{5.21}$$

where the radii are given by

$$a = \frac{2ks}{k^2 - 1} \tag{5.22}$$

and the centers of these circles lie at the point $(h, 0)$ on the x axis, where h is given by

$$h = s\frac{k^2 + 1}{k^2 - 1} \tag{5.23}$$

Note that h can be positive or negative, depending on whether the constant k is greater or less than unity. For $k > 1$ we have a circle to the right of the y axis. A corresponding circle to the left of the y axis exists and is given by $k' = 1/k$ and shown in Fig. 5.6.

Let us examine the family of equipotential circles of radius a and center $(\pm h, 0)$. As k increases, corresponding to larger potentials V_{0P} in (5.16), the radii $a \to 0$ and $h \to s$. In the limit as $V \to \infty$ and

† Notice that when we are ready to calculate the capacitance of the parallel-wire line, we will need the potential between the conducting cylinders which is $V_{P'P}$, where $P' = P(-x, y)$ and is the mirror-image point of P. Since by symmetry $V_{P'P} = 2V_{0P}$, $V_{P'P}$ is readily obtained once V_{0P} is known. V_{0P} is used because the writing of the equations for the equipotentials is greatly simplified when 0 is the origin of the x-y coordinate system.

$k \to \infty$, $a = 0$ and $h = s$, so that the equipotentials are infinitely small circles with their centers at the line of charge. On the other hand as $k \to 1$, we have that $V_{0P} \to 0$, $a \to \infty$, and $h \to \infty$, but in such a way that $h/a \geq 1$. This means that the y axis which is the $V = 0$ line is also the circumference of an infinitely large circle with center at infinity. Since the distance from the origin 0 to the center of this circle is always larger than the radius of this circle, that is, $h/a \geq 1$, the circles never cross the y axis. To summarize, as the potential goes through its entire range from $V = \infty$ to $V = 0$,

$$V_{0P} = \infty \Rightarrow k = \infty,\ h = s,\ a = 0$$

$$V_{0P} = 0 \Rightarrow k = 1,\ h = \infty,\ a = \infty$$

the corresponding equipotential lines vary from an infinitesimally small circle located at $h = s$, to an infinitely large circle with center at infinity.

The potentials to the right of the y axis, which is the $V = 0$ line, are positive. To the left of the y axis, the potentials are negative corresponding to values of $k < 1$ which make the ln term in (5.16) negative. Therefore, the potential between point P and its mirror image P' is given by $V_{PP'} = 2V_{0P}$.

Two-Conductor Transmission Line

The previous results can now be used to calculate the capacitance of a parallel-wire transmission line by simply fitting the outside surfaces of the two conductors which are equipotentials to a pair of equipotential circles in Fig. 5.6. For example, a particular transmission line can be represented by the two heavy-lined circles of radius a and center-to-center spacing $2h$. If a potential $V_{PP'}$ is applied between the two conductors, an equivalent line charge density ρ_L can be determined from the equations. The capacitance per unit length is then the ratio of ρ_L to the potential difference $V_{PP'}$.

In order to use the previous results, we have to express the potential between two cylinders in terms of the parameters a and h of the transmission line. If $V_{PP'}$ is applied between the two cylinders, one conductor will be at $+V_{PP'}/2 = V_{0P}$ and the other at $-V_{PP'}/2$. From (5.16) we can then write

$$V_{PP'} = \frac{\rho_L}{\pi\varepsilon} \ln k \tag{5.24}$$

for the equipotentials. We can express k in terms of a and h by eliminating s from (5.22) and (5.23), giving

$$k^2 - \frac{2kh}{a} + 1 = 0 \tag{5.25}$$

The two solutions to this equation are

$$k = \frac{h}{a} \pm \sqrt{\left(\frac{h}{a}\right)^2 - 1} \tag{5.26}$$

If we let k correspond to the $+$ sign root in (5.26) and k' to the $-$ sign, we can readily show that $kk' = 1$. Because the spacing h between the wires must be larger than the wire radius a, that is, $h/a > 1$, we have that $k > 1$ and $k' < 1$. The first root k gives the equipotentials in the right half plane, and k' gives the equipotentials in the left half plane. The potential difference between the two conducting cylinders can now be written as

$$V_{PP'} = V_P - V_{P'} = \frac{\rho_L}{2\pi\varepsilon} \ln k - \frac{\rho_L}{2\pi\varepsilon} \ln k' = \frac{\rho_L}{\pi\varepsilon} \ln k$$

or

$$V_{PP'} = \frac{\rho_L}{\pi\varepsilon} \ln \left[\frac{h}{a} + \sqrt{\left(\frac{h}{a}\right)^2 - 1}\right] \tag{5.27}$$

Finally, the capacitance per unit length, C/l, in farads per meter is given by

$$\frac{C}{l} = \frac{\rho_L}{V_{PP}} = \frac{\pi\varepsilon}{\ln\left[\dfrac{h}{a} + \sqrt{\left(\dfrac{h}{a}\right)^2 - 1}\right]} \tag{5.28}$$

We can express this in terms of the dielectric constant ε_r of the medium between the wires which gives

$$\frac{C}{l} = \frac{27.8\varepsilon_r}{\ln\left[\dfrac{h}{a} + \sqrt{\left(\dfrac{h}{a}\right)^2 - 1}\right]} \quad pF/\mathrm{m} \tag{5.29}$$

where $\varepsilon = \varepsilon_0 \varepsilon_r$ and $\varepsilon_0 = 8.85 \times 10^{-12}$ F/m. For many practical transmission lines the ratio of center-to-center spacing and wire radius is such that $2h/a > 10$. Using $a/h \ll 1$ in (5.29), we obtain the simple expression

$$\frac{C}{l} \approx \frac{27.8\varepsilon_r}{\ln(2h/a)} \quad \mathrm{pF/m} \tag{5.30}$$

We can now observe that the capacitance per unit length for practical parallel-wire transmission lines is approximately half of that for coaxial lines [compare Eqs. (5.30) and (5.15)].

An afterthought to this section: If one compares the length of the derivation of the formula for capacitance for the parallel line to that for the coaxial line, one cannot help but notice that the derivation is significantly longer for the parallel line. The explanation for this difference also serves to demonstrate the power of Gauss' law. Because of the symmetry of the coaxial line, Gauss' law was applicable and the results came quickly. No such circular symmetry is present for the parallel line; hence Gauss' law could not be used. The potential had to be calculated by adding the separate potentials of the two line charges, a considerably longer procedure.

5.7 CAPACITORS WITH A DIELECTRIC MEDIUM

Considering our previous discussions, not much needs to be said about capacitors which have a dielectric material as the insulating medium between the capacitor plates. As was pointed out in Sec. 5.4, comparing these capacitors to those having air as the insulating medium, we found that capacitance C is larger than the corresponding air capacitance C_0 by the factor ε_r; that is,

$$C = \varepsilon_r C_0 \tag{5.31}$$

where $\varepsilon_r = \varepsilon/\varepsilon_0$ is the relative permittivity of the insulating medium. Relation (5.31) is obtained from (5.12) by writing $C = \varepsilon_r \varepsilon_0 A/d = \varepsilon_r C_0$. As a matter of fact, in practice (5.31) is often used to determine the dielectric constant.

The mechanism which increases the capacitance is that of polarization.† This can be simply demonstrated as follows:

$$C = \frac{Q}{V} = \frac{\rho_s A}{V} = \frac{DA}{V} = \frac{(\varepsilon_0 E + P)A}{V} \tag{5.31a}$$

where the boundary condition at the metal-dielectric interface, $\rho_s = D$, was used. It helps if we refer to Fig. 5.7 which shows a parallel-plate capacitor in which the voltage V is related to the electric field by $V = Ed$. Using this relationship in the equation above, we obtain

$$C = \frac{\varepsilon_0 A}{d} + \frac{PA}{Ed} \tag{5.31b}$$

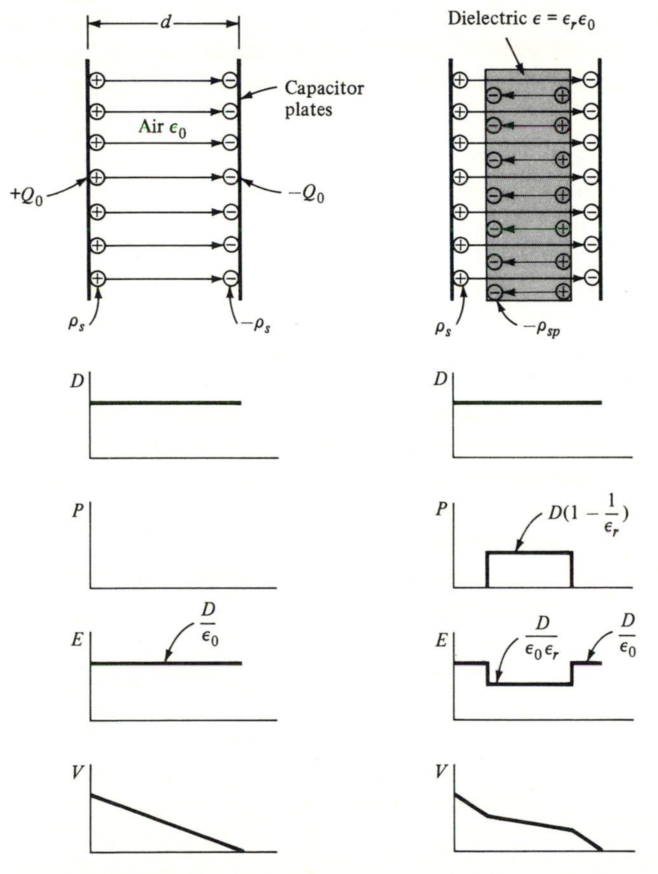

Figure 5.7 The fields in an air- and dielectric-filled capacitor. The electric field inside the dielectric is reduced by the polarization field. The polarization is obtained from (4.48) as $P = D - \varepsilon_0 E = D(1 - 1/\varepsilon_r)$.

† In Sec. 5.9 we will show that it takes energy to polarize a medium.

which shows that C consists of a free-space capacitor plus a polarization capacitance. We can again use $D = \varepsilon_0 E + P$ to express polarization as $P = \varepsilon_0(\varepsilon_r - 1)E$ which gives

$$\begin{aligned} C &= \frac{\varepsilon_0 A}{d} + (\varepsilon_r - 1)\frac{\varepsilon_0 A}{d} \\ &= C_0 + (\varepsilon_r - 1)C_0 \\ &= \varepsilon_r C_0 \end{aligned} \tag{5.31c}$$

The above equation shows that a dielectric inside a capacitor boosts its capacitance by a factor of ε_r. But perhaps more interesting, it tells us that we can consider a dielectric-filled capacitor as two capacitors in parallel, one simply a free-space capacitor C_0, the other a free-space capacitor multiplied by $\varepsilon_r - 1$. This is particularly helpful when we consider partially filled capacitors. For example, a capacitor half-filled [volume is $v = (A/2)d$, not $v = A(d/2)$] with a dielectric would have a capacitance $C = C_0 + \frac{1}{2}(\varepsilon_r - 1)C_0$.

To further aid us in the understanding of the increase in capacitance, let us look at a capacitor at constant V and constant Q and see what happens as a dielectric is introduced.

Constant Q

Let us consider first the case when the charging battery deposits an amount of charge Q_0 on C_0 and is then disconnected from the free-space capacitor C_0. As is shown in Fig. 5.7, when a dielectric is introduced, the free charges on the capacitor plates polarize the dielectric which leads to a decrease in the electric field between the plates and, in turn, to a decrease in the potential difference $V = Ed$. Since $C = Q_0/V$, and Q_0 remains constant, the capacitance increases because V decreases. In this case, Q_0 remains constant and V decreases as the dielectric is introduced. The increase in capacitance can be shown explicitly as follows:

$$C = \frac{Q_0}{V} = \frac{Q_0}{Ed} = \frac{Q_0}{(D/\varepsilon)d} = \frac{Q_0}{\varepsilon_0 E_0 d/\varepsilon}\bigg|_{D=D_0} = \varepsilon_r C_0 \tag{5.32}$$

where $C_0 = Q_0/V_0$ is the free-space capacitance with battery V_0 connected. Note the use of the boundary condition $\varepsilon E = \varepsilon_0 E_0$ at the dielectric-air interface.

Constant V

Figure 5.8 shows a second case in which the battery V_0 remains connected to the capacitor as the dielectric is introduced. The voltage across the capacitor, instead of decreasing when the dielectric is inserted, must now remain constant. Therefore

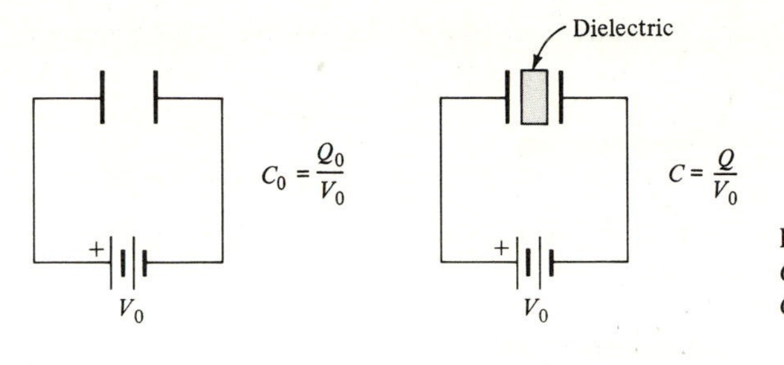

Figure 5.8 A free-space capacitor C_0 and a dielectric-filled capacitor C.

an additional amount of charge will be added to the capacitor such that the new capacitance is again $C = \varepsilon_r C_0$. The increased charge which accounts for the increase in capacitance by the factor ε_r in (5.31) can be given by

$$Q = CV_0 = C\frac{Q_0}{C_0} = \varepsilon_r Q_0 \tag{5.33}$$

The increase in capacitance can also be shown as follows:

$$C = \frac{Q}{V_0} = \frac{\rho_s A}{V_0} = \frac{\varepsilon EA}{V_0} = \frac{\varepsilon_r \varepsilon_0 EA}{V_0} = \varepsilon_r \frac{\varepsilon_0 E_0 A}{V_0}\bigg|_{E=E_0} = \varepsilon_r \frac{Q_0}{V_0} = \varepsilon_r C_0 \tag{5.34}$$

Note that the electric fields in the air and dielectric-filled capacitors are the same when the battery V_0 remains connected; that is, $E_0 = E = V_0/d$. In this case V remains constant and Q increases as the dielectric is introduced.

5.8 SERIES AND PARALLEL COMBINATIONS OF CAPACITORS

Connecting two capacitors in parallel will add their capacitances. This can be shown by use of Fig. 5.9. Applying Kirchhoff's current law (2.12), we can determine the condition under which the networks of Fig. 9a and b are equivalent. Thus, for part (a) we can write

$$I_a = I_1 + I_2 = C_1\frac{dV_a}{dt} + C_2\frac{dV_a}{dt} \tag{5.35}$$

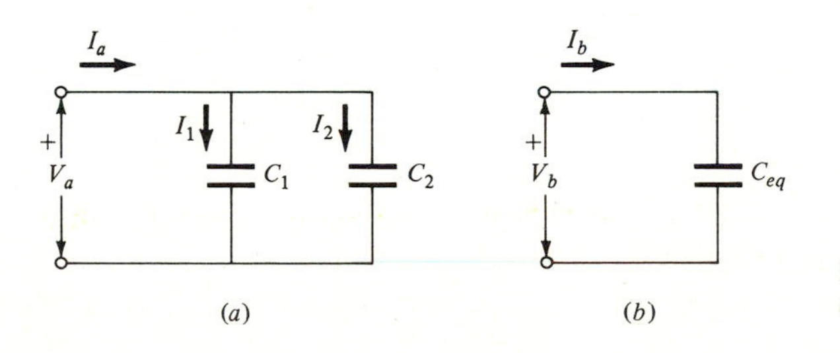

Figure 5.9 Two networks which are equivalent when $C_{eq} = C_1 + C_2$.

where the relation between current and voltage in a capacitor is given by (5.2). Similarly, for part (*b*), we have

$$I_b = C_{eq} \frac{dV_b}{dt} \tag{5.36}$$

The conditions for the networks to be equivalent are $I_a = I_b$ and $V_a = V_b$. Therefore the equivalent capacitance C_{eq} is

$$\boxed{C_{eq} = C_1 + C_2} \tag{5.37}$$

which implies also that $Q_{eq} = Q_1 + Q_2$. The equivalent capacitance is therefore the sum of the capacitances. This can also be seen from an examination of the parallel-plate capacitance formula $C = \varepsilon A/d$. If the separation d in C_1 and C_2 is the same, paralleling C_1 and C_2 results in a capacitor with a plate area A equal to the sum of the plate areas of C_1 and C_2.

The equivalent capacitance for two capacitors in series can be obtained from Fig. 5.10. Kirchhoff's voltage law can be applied as

$$V_a = V_1 + V_2 = \frac{1}{C_1} \int I_a\, dt + \frac{1}{C_2} \int I_a\, dt \tag{5.38}$$

Similarly, for network (*b*),

$$V_b = \frac{1}{C_{eq}} \int I_b\, dt \tag{5.39}$$

For the two networks to be equivalent, we must have $I_a = I_b$ and $V_a = V_b$. Therefore, we conclude that

$$\boxed{\frac{1}{C_{eq}} = \frac{1}{C_1} + \frac{1}{C_2}} \tag{5.40}$$

For only two capacitors in series, (5.40) can be simply written as $C_{eq} = C_1 C_2/(C_1 + C_2)$. The equivalent capacitance is therefore always less than the smallest of the series capacitors. We should observe that the charge deposited on each capacitor in a series combination is the same; that is, $Q_1 = Q_2$. This is because in a series circuit the same current flows through each capacitor for the same length of time. Therefore also $Q_{eq} = Q_1 = Q_2$.

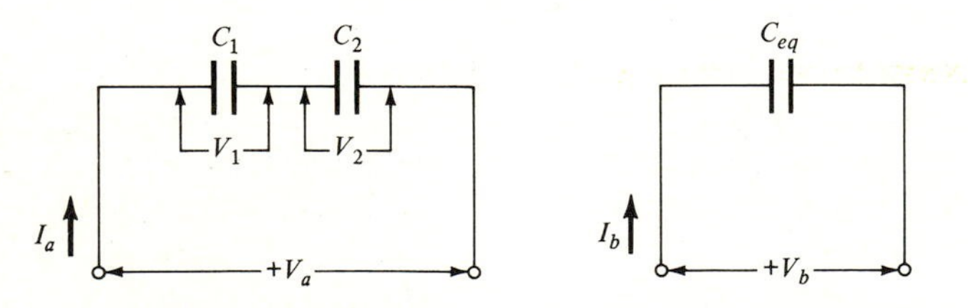

Figure 5.10 Two networks which are equivalent when $C_{eq} = C_1C_2/(C_1 + C_2)$.

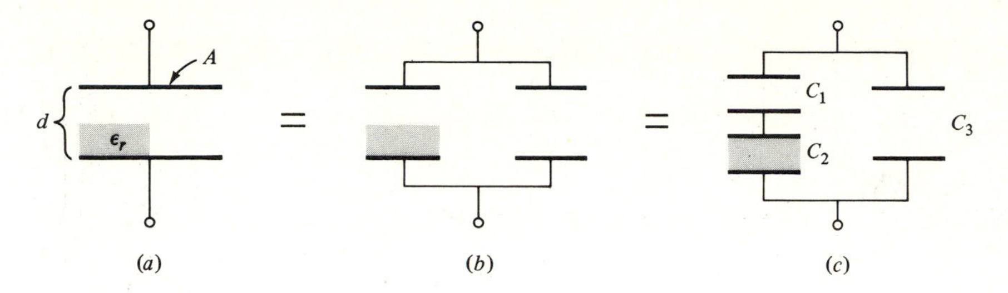

Figure 5.11 (*a*) Parallel-plate capacitor quarter-filled with dielectric. Equivalent circuits of (*a*) are shown in (*b*) and (*c*).

Example A parallel-plate capacitor is quarter-filled with a dielectric of dielectric constant ε_r, as shown in Fig. 5.11. Find the equivalent capacitance.

The capacitor shown in part (*a*) is equivalent to two capacitors in parallel, shown in part (*b*), because the potential difference across both is the same. The capacitor in part (*b*) which is half-filled with a dielectric is equivalent to two capacitors in series, as shown in part (*c*). The capacitances of the three capacitors can be written as

$$C_1 = \varepsilon_0 \frac{A/2}{d/2} \qquad C_2 = \varepsilon_r C_1 \qquad C_3 = \varepsilon_0 \frac{A/2}{d}$$

where A represents the plate area of the whole capacitor. Using (5.37) and (5.40), the equivalent capacitance can be written as

$$C = \frac{C_1 C_2}{C_1 + C_2} + C_3 = \frac{\varepsilon_0 A}{d}\left(\frac{\varepsilon_r}{\varepsilon_r + 1} + \frac{1}{2}\right)$$

The above expression can be checked by letting $\varepsilon_r \to 1$, which gives $C = \varepsilon_0 A/d$.

It should be pointed out that we could have also used the approach indicated in (5.31*c*) to solve the above problem.

5.9 ENERGY STORED IN A CAPACITOR

In Sec. 5.2 we have shown that in charging a capacitor an amount of energy equal to†

$$W = \frac{CV_b^2}{2} = \frac{Q^2}{2C} = \frac{QV_b}{2} \tag{5.41}$$

is transferred from the battery to the capacitor. In that section we examined the charging process closely. Had we been interested only in the energy stored in a charged capacitor, we could have obtained (5.41) simply by noting that charge q on a capacitor which is at voltage V is given by $q = CV$. The work done in

† These three expressions are related by $Q = CV$.

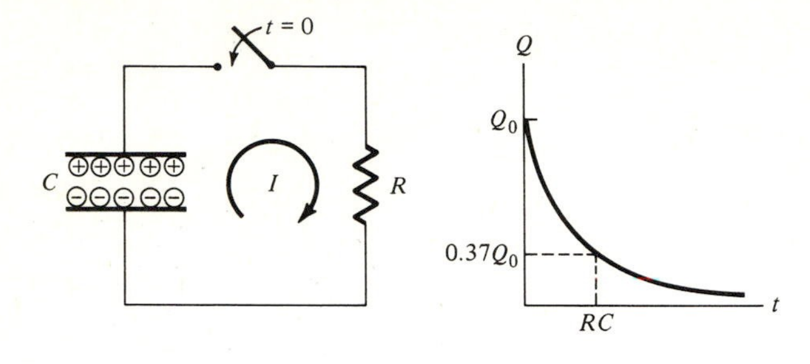

Figure 5.12 (*a*) A charged capacitor will discharge through R; (*b*) charge variation during discharge.

increasing the charge by an amount dq is $dW = V\,dq = q\,dq/C$. The total work in charging from zero to Q coulombs is then

$$W = \int_0^Q \frac{1}{C} q\,dq = \frac{1}{2}\frac{Q^2}{C} \tag{5.42}$$

A charged capacitor has a voltage V between its plates or, alternatively, a charge Q on its plates. The stored energy can be converted into heat by discharging C through a resistance R, as shown in Fig. 5.12*a*. Since the equation for discharge current I in Fig. 5.12*a* is the same as (5.4), the discharge current is given by Eq. (5.6); i.e., I decays exponentially from the initial value V_b/R to zero with a time constant $T = RC$. The voltage across the resistance V_R is equal to $V_R = IR$ and is therefore similar in shape to the discharge current. The shape of the charge variation during discharge is given by $Q = V_R/C$ and is shown in Fig. 5.12*b*.

Example: An apparent paradox Figure 5.13*a* shows a circuit for charging a capacitor by a battery. The energy transferred to C during the charging process is

$$W_a = \tfrac{1}{2}CV_a^2 = \frac{1}{2}\frac{Q_a^2}{C}$$

Figure 5.13*b* shows an arrangement where the charged capacitor is connected to an identical but uncharged capacitor. The charged capacitor will charge the uncharged one until the charges on both capacitors are equal. Since charge must be conserved, the charge on the capacitor in part (*a*) must be

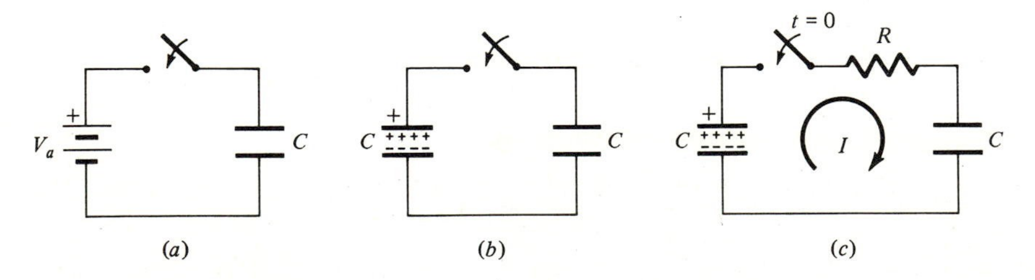

Figure 5.13 (*a*) Battery charges capacitor; (*b*) charged capacitor charges an identical uncharged capacitor; (*c*) same circuit as (*b*) except for resistance R which has been added.

equal to the charges on the two capacitors in part (*b*); that is,

$$Q_a = Q_b$$

$$CV_a = (C + C)V_b \qquad \text{or} \qquad V_b = \tfrac{1}{2}V_a$$

The total energy in part (*b*) is

$$W_b = \tfrac{1}{2}C_b V_b^2 = \tfrac{1}{2}(2C)(\tfrac{1}{2}V_a)^2 = \tfrac{1}{4}CV_a^2$$

We see now that $W_b = \frac{1}{2}W_a$, or only one-half the original energy is left. What happened to the remaining $\frac{1}{2}W_a$ energy? One can speculate that it was dissipated during the charging process in the wire resistance. But what about if the connecting wires are superconductors with zero resistance?

To answer this question let us include a resistance R, as shown in Fig. 5.13*c*, calculate the energies involved, and let R go to zero. Kirchhoff's voltage equation for Fig. 5.13*c* can be written as

$$RI + \frac{2}{C}\int I\,dt = 0$$

with the initial condition that the current at the time the switch is closed is $I|_{t=0} = V_a/R$ which accounts for the charged state of one of the capacitors and the uncharged state of the remaining one. Differentiating, we obtain

$$\frac{dI}{dt} + \frac{2}{RC}I = 0$$

which with the above initial condition has the solution

$$I = \frac{V_a}{R}e^{-(2/RC)t}$$

The final voltage across the initially uncharged capacitor is

$$V_c = \frac{1}{C}\int_0^\infty I\,dt = \frac{V_a}{2}$$

which by symmetry will also be that across the initially charged capacitor. The energies in Fig. 5.13*c* can be written as

$$\begin{aligned} W_c &= W_{\text{res}} + W_{\text{cap}} \\ &= \int_0^\infty RI^2\,dt + \tfrac{1}{2}(C + C)\left(\frac{V_a}{2}\right)^2 \\ &= \frac{CV_a^2}{4} + \frac{CV_a^2}{4} \\ &= W_a \end{aligned}$$

where the integration was performed by using the expression for I above. The energies now balance since $W_c = W_a$. Thus we see that the energy in R which is converted to heat during the charging process is exactly equal to $CV_a^2/4$ and is independent of the value of R. In the limit as $R \to 0$, an infinitesimally short $(T = RC \to 0)$ but infinitely large current pulse transfers $Q_a/2$ charge from the charged to the uncharged capacitor.

5.10 ENERGY STORED IN AN ELECTRIC FIELD

Let us answer a question that is often asked about charged capacitors. Is the energy stored associated with the charge on the conductor plates, or is it associated with the electric field in the space between the conducting plates (that space may be vacuum or a dielectric)? As it turns out, this question does not have much meaning because the electric field E between the capacitor plates is always related to the voltage and charge on the plates. For example, let us consider a parallel-plate capacitor which is held at constant potential. Ignoring fringing of the electric field we can substitute $V = Ed$ and capacitance $C = \varepsilon A/d$ into the expression for work $W = CV^2/2$ and obtain $W = \frac{1}{2}\varepsilon E^2 v$, where v is the volume, $v = Ad$, between the plates. The energy density w at each point in the electric field is then $w = W/v = \frac{1}{2}\varepsilon E^2$.

Similarly if we begin with a capacitor in which the charge is constant, we can use $W = Q^2/2C$ and substitute the expression $Q = \rho_s A$ for charge Q, where ρ_s is the surface charge density on the plates. We thus obtain $W = \rho_s^2 v/2\varepsilon$. From the boundary condition for a metal surface (2.43) we know that $\varepsilon E_n = \rho_s$. Substituting for ρ_s we obtain once again for the energy stored, $W = \frac{1}{2}\varepsilon E^2 v$. This last formula lends itself to the interpretation that the energy stored is in the electric field E that exists in the volume v between the plates, whereas $W = Q^2/2C$ can be identified with the energy needed to deposit charge Q on the plates. Since charge density and electric field are related by the boundary condition $D_n = \rho_s$, these two expressions represent two different points of view of a single experimental fact.

We can derive the energy density at points in an electric field more precisely if we consider a small volume element $\Delta v = \Delta d\, \Delta b^2$ between the plates of a parallel-plate capacitor, as shown in Fig. 5.14. The top and bottom faces of area Δb^2 are parallel to the capacitor plates and normal to the **E** field. As a matter of fact, if we imagine thin sheets of metal foil of area Δb^2 placed coincident with the top and bottom faces of the small cube Δv, a small parallel-plate capacitor of capacitance

$$\Delta C = \varepsilon \frac{\Delta b^2}{\Delta d} \tag{5.43}$$

is formed.† The potential difference between the top and bottom faces is

$$\Delta V = E\, \Delta d \tag{5.44}$$

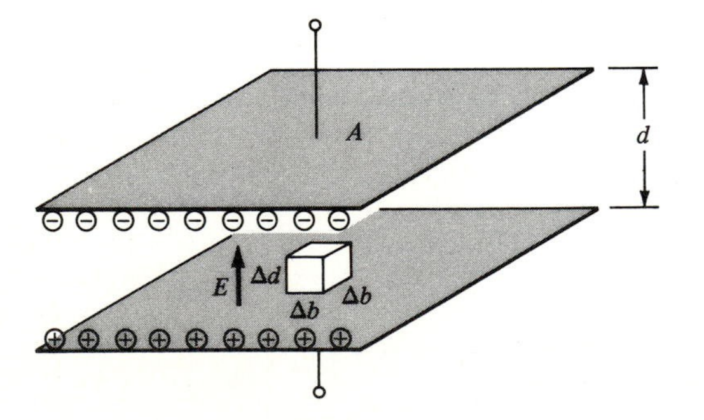

Figure 5.14 A small capacitor of volume $\Delta v = \Delta d\, \Delta b^2$ inside a larger one of volume $v = Ad$.

† Note that an introduction of a thin metal sheet perpendicular to the E field does not disturb the E field.

The energy stored in volume Δv is then, from $W = \frac{1}{2}CV^2$, given by

$$\Delta W = \tfrac{1}{2}\,\Delta C\,\Delta V^2 = \tfrac{1}{2}\varepsilon E^2\,\Delta v \tag{5.45}$$

As the volume Δv shrinks to zero, it defines a point. The energy density at this point is the limit of the ratio $\Delta W/\Delta v$ as Δv approaches zero, or

$$\boxed{w = \lim_{\Delta v \to 0} \frac{\Delta W}{\Delta v} = \frac{1}{2}\varepsilon E^2} \tag{5.46}$$

Therefore, if we have a space in which an electric field E exists, any point of that space (ε_0 if vacuum; ε if dielectric) has associated with it an energy density† of $\frac{1}{2}\varepsilon E^2$ J/m³. Thus the energy stored in a volume depends on E^2 in that volume. An expression for stored energy in an electric field is therefore

$$W = \tfrac{1}{2}\varepsilon \iiint E^2\,dv \tag{5.47}$$

We can easily demonstrate that by integrating w throughout the volume of the capacitor in Fig. 5.14, we obtain $W = \frac{1}{2}CV^2 = \frac{1}{2}QV$ as the energy stored in C. If we assume that no fringing of the electric field occurs at the edges of the capacitor, then E is constant throughout the space between the plates. The integration is then simply the product of the constant energy density and the volume, or $W = wAd = \frac{1}{2}\varepsilon E^2 Ad = \frac{1}{2}(DA)(Ed) = \frac{1}{2}QV$ J.

Although w was derived for the special case of a parallel-plate capacitor, it gives the electric energy density for any electric field. For example, the electric field of an electromagnetic wave in free space exists without being attached to electric charges. The energy per unit volume in the electric field of such a wave is correctly given by w of (5.46).

5.11 FIELD CELLS AND THE MEANING OF ε

Among the concepts that seem to baffle many students is the permittivity ε and permeability μ of a medium. The fact that the velocity of light is given by $c = 1/\sqrt{\mu\varepsilon}$ (which for free space is $c = 1/\sqrt{\mu_0\varepsilon_0} = 3 \times 10^8$ m/s), or that the force between point charges is inversely proportional to ε of the medium in which the charges are embedded does not seem to aid in the understanding. However, the formula for a parallel-plate capacitor gives ε as $\varepsilon = Cd/A$, where A is the area of the plates and d the separation between them. Dimensionally this says that ε is capacitance per length. Since this approach seems promising, we will explore it further.

The significance of ε becomes clear when we divide a space in which an elecric field E exists into many small volumes, as was done in Fig. 5.14. The field E in each small volume could be reproduced by placing thin metal sheets at the top and bottom faces, as shown in Fig. 5.15, and by applying a potential difference of $\Delta V = E\,\Delta d$ between the metal sheets. The small cubical volume Δv shown with metal sheets is called a *field-cell capacitor.* The side walls of the field cell are parallel to the

† A more general analysis valid also for anisotropic fields would give $w = \frac{1}{2}\mathbf{D}\cdot\mathbf{E}$ which reduces to (5.46) for isotropic dielectrics.

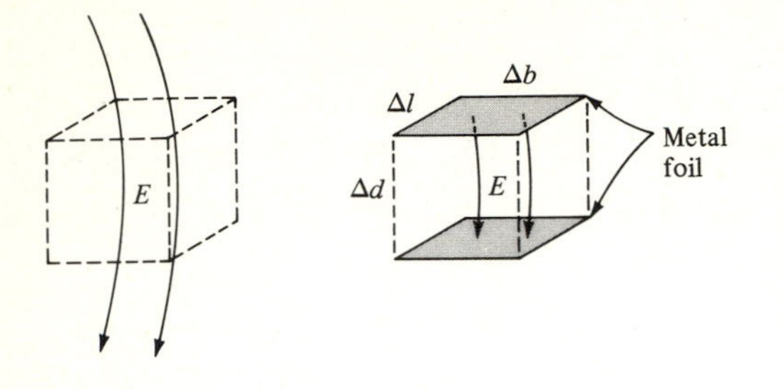

Figure 5.15 (*a*) An electric field **E** in space; (*b*) a field cell produces the original **E** field inside the cell if the applied potential between the metal foils is $\Delta V = E\,\Delta d$.

E field, while the top and bottom coincide with the equipotential surfaces of the original *E* field. The capacitance of such a field cell is (using the parallel-plate capacitor formula)

$$\Delta C = \frac{\varepsilon\,\Delta l\,\Delta b}{\Delta d} \tag{5.48}$$

Let us choose now one of the metal-foil sides to be equal to the height of the field-cell capacitor; that is, $\Delta d = \Delta b$ (we could have chosen just as well $\Delta l = \Delta d$). With this choice we obtain for ε:

$$\varepsilon = \frac{\Delta C}{\Delta l} \qquad \text{F/m} \tag{5.49}$$

We can now identify the permittivity ε as the capacitance per unit depth of a field-cell capacitor imagined to be at the point at which ε is given. In vacuum, ε_0 is given as 8.85×10^{-12} F/m or 8.85 pF/m. A field-cell capacitor per meter depth in vacuum has a capacitance of 8.85 pF. Two such field-cell capacitors are shown in Fig. 5.16. Part (*a*) shows a cubical volume with all sides 1-m long, whereas part (*b*) shows a volume which is $\Delta d \times \Delta d \times 1$ m. Both have the same capacitance of 8.85 pF (it is assumed that the electric field is uniform throughout the volume of each capacitor).

The concept of field cells is used in graphical mapping of electric fields in dielectrics. One begins at the conducting surfaces which bound a dielectric by drawing field and potential lines perpendicular and parallel to the conducting surfaces. This divides the space into squares which can be identified with field cells. Since the capacitance per unit depth of any field cell, large or small, is equal to ε, a field at places of great variation can be divided into finer cells to obtain more accuracy in the field map.

5.12 ENERGY STORED IN AN ASSEMBLY OF CHARGES

A charged capacitor consists of an assembly of positive charges on one plate and an assembly of negative charges on the other. To accumulate these two charges in the respective locations requires an amount of work from the charging battery equal to $W = \frac{1}{2}QV$.

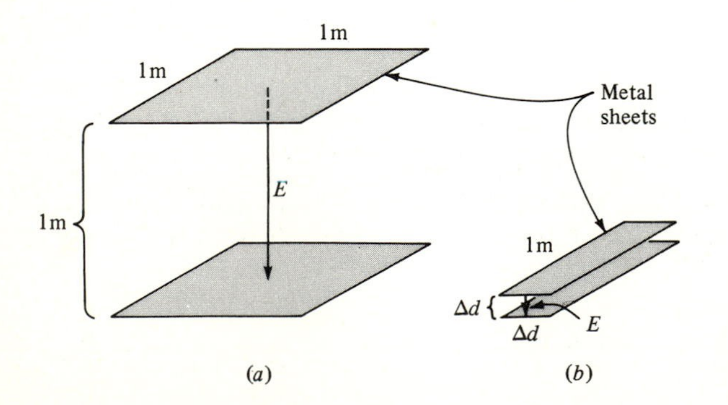

Figure 5.16 Two field-cell capacitors of different dimensions but with the same capacitance.

Let us now calculate the work that is needed to assemble a "pile" of charge by bringing in discrete charges from infinity and adding them to the "pile." Clearly the energy that is stored in such an assembly is the work that is done by some external force in assembling the system of charges. The stored energy can be released at a later time by allowing the charges to fly apart. The energy available when the system disintegrates in that fashion is also called the *potential, or free, energy* of the system.

If we assume empty space, no work will be required to bring the first charge Q_1 into position, say at $\mathbf{r}_1$. However, to bring a second charge Q_2 from ∞ to point $\mathbf{r}_2$ will require work, as Q_2 now has to be moved against the force field of the charge already in position. The energy that is stored in a system of two point charges is therefore, from Coulomb's law or Eq. (1.14), given by

$$W_2 = Q_2 \frac{Q_1}{4\pi\varepsilon_0 |\mathbf{r}_1 - \mathbf{r}_2|} \tag{5.50}$$

To move a third charge into position will require work to be done against the combined force field of Q_1 and Q_2. It is

$$W_3 = Q_3 \left(\frac{Q_1}{4\pi\varepsilon_0 |\mathbf{r}_1 - \mathbf{r}_2|} + \frac{Q_2}{4\pi\varepsilon_0 |\mathbf{r}_2 - \mathbf{r}_3|} \right) \tag{5.51}$$

We can now continue to move additional charges into position. Each time a new charge is brought in, the total energy stored increases by the amount it took to position the new charge against the force field of all charges already present. Adding, the total potential energy to position N point charges is

$$\boxed{W = \frac{1}{2} \sum_{i=1}^{N} Q_i V_i} \tag{5.52}$$

where V_i is the potential [given by Eq. (1.14)] due to all charges except Q_i at the point at which Q_i is located.

In terms of the field point of view developed in Sec. 5.10, we can say that bringing in additional charges increases the energy stored because the electric field increases. Since energy density w is proportional to the square of the field, an increased electric field implies an increased energy stored.

Charge Distribution

To assemble a charge with continuous volume charge density $\rho(x, y, z)$ will require an amount of energy which can be calculated by a procedure in which all parts of the system are brought simultaneously to their final charge values. Suppose the system has been built up to charge density ρ and potential V. To add a small amount of charge δQ which is brought from infinity and added to the system will require work equal to

$$\delta W_1 = V\,\delta Q = \iiint V\,\delta\rho\,dv \tag{5.53}$$

Simultaneously with the addition of δQ, the potential V of the system will increase by δV. The increase in energy due to δV can be given by

$$\delta W_2 = Q\,\delta V = \iiint \rho\,\delta V\,dv \tag{5.54}$$

Since δW_1 must equal δW_2, we can write for the increase of stored potential energy

$$\begin{aligned}\delta W &= \tfrac{1}{2}(\delta W_1 + \delta W_2)\\ &= \frac{1}{2}\iiint (V\,\delta\rho + \rho\,\delta V)\,dv\end{aligned} \tag{5.55}$$

The integrand is an exact differential, which allows us to write

$$W = \frac{1}{2}\iiint \rho V\,dv \tag{5.56}$$

where W is the complete energy stored in a system and is equal to the work done by an external force in assembling the charge distribution. Notice that the factor of one-half is an averaging factor which expresses the fact that it requires less energy to position δQ at the beginning of the assembly process.

If the charges are placed on a conductor, we know that they will distribute themselves in a thin layer on the conducting surface. Equation (5.56) can then be written as

$$W = \frac{1}{2}\oiint \rho_s V\,dA \tag{5.57}$$

and if we have N conductors,

$$\begin{aligned}W &= \frac{1}{2}\sum_{n=1}^{N}\oiint_{S_n} V_n\rho_{sn}\,dA\\ &= \frac{1}{2}\sum_{n=1}^{N} V_n\oiint \rho_{sn}\,dA\\ &= \frac{1}{2}\sum_{n=1}^{N} V_n Q_n\end{aligned} \tag{5.58}$$

where S_n is the surface of the nth conductor. Notice that V_n can be taken outside the integral because the surface of each conductor is an equipotential. The last statement is in terms of the equipotential and total charge Q_n on each conductor.

It is interesting now to relate (5.58) to the energy stored in a capacitor which is a system of two conductors; that is, $N = 2$. The charges on the capacitor plates are equal, but the voltages are different. This gives

$$\begin{aligned}W &= \tfrac{1}{2}(V_1 Q_1 + V_2 Q_2)\\ &= \tfrac{1}{2}Q(V_1 - V_2)\end{aligned} \tag{5.59}$$

where $Q_1 = -Q_2$ and $V_1 - V_2$ is the potential difference across the charged capacitor. Thus (5.59) is equal to (5.41).

5.13 ENERGY STORED IN POLARIZATION

We have shown in Sec. 5.7 that adding a dielectric to a free-space capacitor will increase its capacitance. The increased capacitance C is given by

$$C = \varepsilon_r C_0 \tag{5.60}$$

where C_0 is the free-space capacitance and ε_r is the dielectric constant. Again, we must distinguish between two cases. The constant-V case is one in which the external source (the charging battery) remains connected to the system (the capacitor). The constant-Q case is one in which the external source is disconnected after initially charging the system.

It was pointed out that the mechanism which increases the capacitance is that of polarization in the dielectric material. Since the energy stored in a capacitor is

$$W = \tfrac{1}{2}CV^2 = \tfrac{1}{2}\varepsilon_r C_0 V^2 = \varepsilon_r W_0 \tag{5.61}$$

we see that the energy stored in a dielectric-filled capacitor increases by the factor ε_r over that of a free-space capacitor, where we have assumed that V remains constant when the dielectric is inserted. The additional energy supplied by the battery in charging the dielectric-filled capacitor is therefore the energy required to polarize the dielectric. Hence, the polarization energy is given by

$$\boxed{W_{\text{pol}} = W - W_0 = (\varepsilon_r - 1)W_0} \tag{5.62}$$

As the dielectric is removed, $\varepsilon_r \to 1$ and the polarization energy goes to zero (it is returned to the system). Note that from (4.24) or (4.45) the factor $\varepsilon_r - 1$ in the above equation can be related explicitly to polarization P by

$$\varepsilon_r - 1 = \frac{P}{\varepsilon_0 E} \tag{5.63}$$

For the constant-Q case, $W = W_0/\varepsilon_r$, $W_{\text{pol}} = -W_0(\varepsilon_r - 1)/\varepsilon_r$, and $W_0 = Q^2/2C_0$. Introducing the dielectric decreases the total energy of the system because the energy to polarize the dielectric can only come from the system, as all external sources are assumed to be disconnected. Because a dielectric medium is under an attractive force toward the capacitor, removing the dielectric from the capacitor will restore energy W_{pol} to the system.

Polarization Energy from the Field Point of View

It is instructive to reexamine the polarization energy from the field point of view. If we begin with a generalized expression of (5.47), we have for the energy stored in a dielectric medium

$$W = \frac{1}{2}\iiint \mathbf{D}\cdot\mathbf{E}\,dv = \frac{1}{2}\iiint(\varepsilon_0\mathbf{E} + \mathbf{P})\cdot\mathbf{E}\,dv$$

$$= \frac{1}{2}\iiint \varepsilon_0 E^2\,dv + \frac{1}{2}\iiint \mathbf{P}\cdot\mathbf{E}\,dv \tag{5.64}$$

where $\mathbf{D}$ is given by (4.23). Specializing this to an isotropic dielectric in the volume v between the capacitor plates, we obtain

$$W = \tfrac{1}{2}\varepsilon_0 E^2 v + \tfrac{1}{2}PEv \qquad (5.65)$$

The first term on the right is the energy stored in the electric field, whereas the second is the energy in the polarization. Substituting for P from (5.63), we have

$$W = \tfrac{1}{2}\varepsilon_0 E^2 v + \tfrac{1}{2}(\varepsilon_r - 1)\varepsilon_0 E^2 v \qquad (5.66)$$

As before, if we assume that the battery remains connected (V does not change) as the medium changes from free space to dielectric, the electric field will remain constant during and after the change; that is, $E_0 = E = V/d$. The above expression can then be written as

$$W = \frac{1}{2}\varepsilon_0 \left(\frac{V}{d}\right)^2 v + \frac{1}{2}(\varepsilon_r - 1)\varepsilon_0 \left(\frac{V}{d}\right)^2 v \qquad (5.67)$$

The first term can now be identified with W_0, which allows us to write

$$W = W_0 + (\varepsilon_r - 1)W_0 \qquad (5.68)$$

The polarization energy is therefore the second term:

$$W_{\text{pol}} = (\varepsilon_r - 1)W_0 \qquad (5.69)$$

which agrees with (5.62). The above is often written in terms of the electric susceptibility χ which is defined by (4.24) as $\varepsilon_r - 1 = \chi$. For the case when the sources of the electric field are fixed (Q constant), the polarization energy is obtained in a similar way as $W_{\text{pol}} = W_0(1 - \varepsilon_r)/\varepsilon_r$, where $W_0 = \tfrac{1}{2}\varepsilon_0 E_0^2 v$.

5.14 FORCE BETWEEN CAPACITOR PLATES

The plates in a charged capacitor attract each other, for the simple reason that one plate carries a negative while the other carries a positive charge. Considering a free-space capacitor as shown in Fig. 5.17, we see that a mechanical force F_m must exist that will counterbalance the attractive electric force F_e. If the plates are allowed to move together by the small distance Δx, the mechanical work performed by the electric forces is

$$\Delta W_m = F_x \, \Delta x \qquad (5.70)$$

Capacitor Charge Q Constant

If the battery is removed after charging C, the charge Q on the plates will remain constant as the plates are allowed to move. Considering an isolated system, we observe that if any mechanical work is done by the system, the electrostatic energy W_e of the system must decrease. That is, the energy changes

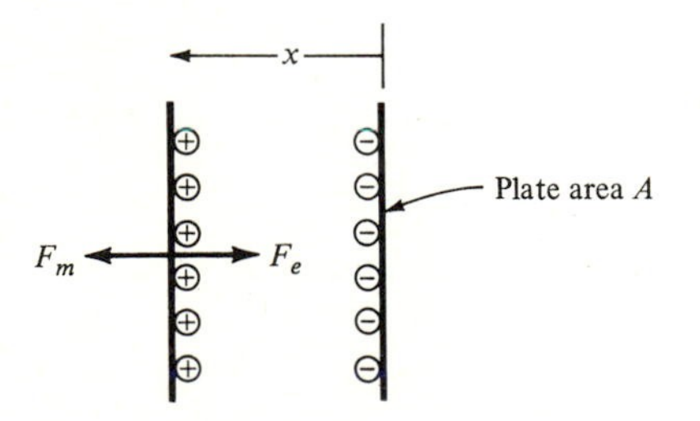

Figure 5.17 The charged conductors of a free-space parallel-plate capacitor are held apart by mechanical force F_m.

must balance as

$$\Delta W_m + \Delta W_e = 0 \tag{5.71}$$

Substituting from (5.70) and solving for the force acting on the plates, we obtain

$$\boxed{F_x = -\frac{\partial W}{\partial x}\bigg|_Q} \tag{5.72}$$

where the subscript e is deleted from F and W. The above is a general relation between force and energy for an isolated system in which charge Q remains constant during displacement [denoted by the subscript Q in Eq. (5.72)].† The minus sign in Eq. (5.62) means that the force is in the direction that will decrease the stored energy.

To find the attractive force between the plates we can use for energy stored in a capacitor, $W = \frac{1}{2}Q^2/C$, which is a convenient expression when Q remains constant. Thus, in Fig. 5.17, the force on any plate is

$$F = -\frac{\partial}{\partial x}\left(\frac{1}{2}\frac{Q^2}{C}\right) = -\frac{\partial}{\partial x}\left(\frac{1}{2}\frac{Q^2}{\varepsilon_0 A}x\right) = -\frac{1}{2}\frac{Q^2}{\varepsilon_0 A} \tag{5.73}$$

where the expression for parallel-plate capacitance, $C = \varepsilon_0 A/x$, was used. The minus sign means that F is in such a direction that the distance x between the plates is decreased. Reversing the plates does not change the results since Q in (5.73) is squared. The above force can be written in terms of the electric field E as $F = -\frac{1}{2}\rho_s^2 A/\varepsilon_0 = -\frac{1}{2}\varepsilon_0 E^2 A$, where Q is expressed as surface charge ρ_s multiplied by plate area A; that is, $Q = \rho_s A$. The last term was obtained by use of the boundary condition for a charged metallic surface, $D_n = \varepsilon_0 E_n = \rho_s$.

We derived (5.73) using the expression for energy stored in a capacitor in terms of charge Q. We could have also started with the expression (5.46) which gives the energy stored in a capacitor in terms of the electric field as $W = \frac{1}{2}\varepsilon_0 E^2 Ax$, and obtained $F = -\frac{1}{2}\varepsilon_0 E^2 A$ by differentiating W with respect to x. In relation to the force expression, we observe that when the plates move, the electric field (but not the potential) inside the capacitor remains constant when the charge Q on the plates remains constant.

An examination of the energy expression reveals that the energy stored decreases as x becomes smaller. In an isolated system, with the charges fixed, there is no external source of energy. The decrease in energy $\Delta W = \frac{1}{2}\varepsilon_0 E^2 A\ \Delta x$ as the plates move together by an amount Δx is accounted for by the increase in mechanical energy $F_m\ \Delta x$, where F_m is the mechanical force holding the plates apart.

If an isotropic dielectric occupies the capacitor, the force on the plates is obtained by replacing ε_0 by ε in (5.73). The new force is therefore F/ε_r and is smaller than that for the free-space capacitor.

Capacitor Voltage V Constant

The problem changes substantially when the charging battery remains connected to C and the plates are allowed to move. The system (consisting here of the capacitor) now has an external source of energy W_{batt}. If the plates of the parallel-plate capacitor are allowed to move under the influence of the electric forces, the mechanical work that will be done, this time by the system and the batteries, is again given by (5.70). Conservation of energy for the constant V case can be stated as

$$\Delta W_m + \Delta W_e = \Delta W_b \tag{5.74}$$

where the subscripts refer to mechanical, electric, and battery energies, respectively.

† If an object rotates about an axis, the proper relationship between torque T and angular displacement ϕ would be

$$T = -\frac{\partial W}{\partial \phi}\bigg|_Q$$

If the plates move closer by an infinitesimal distance Δx, the capacitance will increase by an amount

$$\Delta C = \varepsilon_0 A \left(\frac{1}{x - \Delta x} - \frac{1}{x} \right) = \varepsilon_0 A \frac{\Delta x}{x^2} \tag{5.75}$$

Additional charge equal to $\Delta Q = (\Delta C)V$ will be deposited on C. The energy stored in the capacitor, given by $W = \frac{1}{2}QV$, will then increase by an amount

$$\Delta W = \tfrac{1}{2}(\Delta Q)V \tag{5.76}$$

On the other hand, the battery had to supply an amount of energy equal to

$$\Delta W_b = (\Delta Q)V \tag{5.77}$$

as the battery in effect had to move electrons from the positive capacitor plate, through the constant potential V, and deposit them on the negative plate. Thus

$$\Delta W_b = 2\,\Delta W \tag{5.78}$$

Only half the energy supplied by the battery shows up as an increase in the electric energy; the other half is given to the mechanical field in changing the capacitance by ΔC (either by moving the plates closer to each other, or by inserting a dielectric with $\varepsilon_r > 1$). Eliminating ΔW_b from (5.74) and using (5.70), we can write for the force†

$$\boxed{F_x = \left.\frac{\partial W}{\partial x}\right|_V} \tag{5.79}$$

This is different in sign from (5.72). The force on the plates is now in the direction that will increase the stored electric energy. The two cases demonstrate the importance of properly accounting for all energy changes. Simply equating mechanical work to changes in electric field energy would have resulted in the wrong sign for the force in (5.79).

To find the force, the proper expression to use for W when V is held constant is $W = \frac{1}{2}CV^2$. The force is then, using the parallel-plate capacitor of Fig. 5.17,

$$F = \frac{\partial}{\partial x}\left(\frac{1}{2}CV^2\right) = \frac{\partial}{\partial x}\left(\frac{1}{2}\frac{\varepsilon_0 A}{x}V^2\right) = -\frac{1}{2}\frac{\varepsilon_0 A}{x^2}V^2 \tag{5.80}$$

The last term in this expression can be rewritten as $F = -\frac{1}{2}\varepsilon_0 E^2 A$ since $V = Ex$. Note that for constant V the electric field E (and therefore the charge Q on the plates) must change when the spacing x varies. The minus sign indicates again that the force on the plates is in such a direction as to decrease the distance x between plates.

In the constant-V case, the force on the plates would increase by the factor ε_r if a dielectric of permittivity $\varepsilon = \varepsilon_r \varepsilon_0$ were placed between the plates.

The difference between this case and the case before is that now the stored energy $W_e = \frac{1}{2}(\varepsilon_0 A/x)V^2$ increases in the capacitor as the plates are allowed to move closer together. The increase in energy is, from (5.76) and (5.75),

$$\Delta W = \frac{1}{2}\Delta C\ V^2 = \frac{1}{2}\frac{\varepsilon_0 A}{x^2}\Delta x\ V^2 \tag{5.81}$$

† The corresponding torque on an object would be

$$T = \left.\frac{\partial W}{\partial \phi}\right|_V$$

In addition, an equal amount of energy is given by the battery to the mechanical field as $F_m \, \Delta x$. For both cases the force is such as to increase the capacitance [see Eqs. (5.87) and (5.88)].

5.15 FORCES ON CONDUCTING SURFACES WHICH ARE CHARGED

In the last section it was shown that a force exists on each element of surface bearing a surface charge ρ_s. The force per unit area ΔA, from (5.73) or (5.80), is given by

$$\frac{F}{\Delta A} = -\frac{1}{2}\varepsilon_0 E^2 = -\frac{1}{2}\frac{\rho_s^2}{\varepsilon_0} \tag{5.82}$$

where the minus sign gives the direction of the force which is from the conductor to free space, and the boundary condition for a conducting surface is $\varepsilon_0 E_n = \rho_s$. Therefore, any surface charge, or any **E** field terminating on the conducting surface, will exert a pull on the surface. It does not matter whether the surface is at constant-V or constant-Q; the force is the same. In Sec. 2.12 it was shown that the surface charge ρ_s is bound inside the conductor by internal forces. Hence, the force acting on each element of charge $\rho_s \, \Delta A$ is transmitted to the conductor itself which gives rise to the pull on its surface.

It is instructive to derive the force from fundamentals. The surface can be considered as a continuum of patches of charge; each $\Delta Q = \rho_s \, \Delta A$. We also know that an electric field given by $E = \rho_s/\varepsilon_0$ exists normal to the charged conducting surface. It is now tempting to use Coulomb's law [as was done in Eq. (2.57)] and say that each ΔQ experiences a force $F = \Delta Q \, E$. Re-expressing ΔQ in terms of surface charge ρ_s, we obtain for the force per unit area $F/\Delta A = \varepsilon_0 E^2 = \rho_s^2/\varepsilon_0$, which is exactly twice the magnitude of the value given in (5.82). This is easily diagnosed as an incorrect result since we have used the total electric field outside the conductor, $E = \rho_s/\varepsilon_0$, which includes the self field of the charge ΔQ upon which we want to find the force. The portion of the field that is subjecting the charge $\Delta Q = \rho_s \, \Delta A$ to an outward force can only be the field from all other charges on the surface. The self field of charge $\Delta Q = \rho_s \, \Delta A$, or that portion of the total field on a conducting surface that comes from only $\rho_s \, \Delta A$, is given by Gauss' law as $E = \rho_s/2\varepsilon_0$, a result which was derived in (1.78*c*). Hence, the self field in this case is exactly one-half of the total field. Figure 5.18 shows that the field $E = \rho_s/\varepsilon_0$ outside a charged conductor is composed of the self field of the local charge and the field due to remote charges on the remainder of the surface. The above figure is similar to Fig. 2.7. For clarity, Fig. 2.7 should be referred to at this time since it shows the remote charges explicitly.

The force on an entire object having a surface charge $\rho_s(x, y, z)$ is obtained by

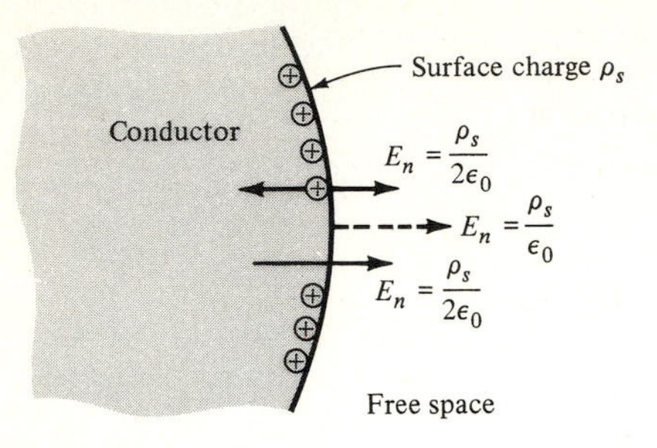

Figure 5.18 Components of the surface field due to a charged conductor with surface charge ρ_s. The upper vector shows the electric field from the surface charge alone. The lower vector is the contribution due to remote charges. The dashed middle vector is the total field outside a charged conductor.

integrating (5.82) over the surface of the object. Thus

$$F = \frac{1}{2} \iint \rho_s E \, dA$$
$$= \frac{1}{2\varepsilon_0} \iint \rho_s^2 \, dA \tag{5.83}$$

Two examples To illustrate the use of Eq. (5.83), let us apply it to two cases. One is a metallic sphere that is charged; the other is a metallic sphere immersed in an external electric field.

In the first case, the radial electric field which is produced by the surface charges on the metallic sphere will create a radial force field. If the sphere were of expandable material, the outward radial force would increase the size of the sphere.† For example, if a charge Q is placed on a shell of radius a, the charge would distribute itself uniformly with $\rho_s = Q/4\pi a^2$. The total radial force would be, using the spherical coordinate system,

$$F = \frac{1}{2\varepsilon_0} \int_0^{2\pi} \int_0^{\pi} \left(\frac{Q}{4\pi a^2} \right)^2 a^2 \sin \theta \, d\theta \, d\phi$$
$$= \frac{Q^2}{8\pi\varepsilon_0 a^2} = 2\varepsilon_0 E^2 \pi a^2 \tag{5.84}$$

where the boundary condition $\rho_s = \varepsilon_0 E$ was used to obtain the last expression. Of course no net translational force exists on the sphere since the force components on opposite symmetric points are equal and opposite.

The second case, that of an uncharged metallic sphere of radius a immersed in a uniform electric field E_0, was considered in Sec. 4.13. Referring to Fig. 4.24, we see that the left half of the sphere carries a negative surface charge, while the right half is positively charged. The induced surface charge is given by (4.108) and is

$$\rho_s = \varepsilon_0 E_r = 3\varepsilon_0 E_0 \cos \theta \tag{5.85}$$

Since ρ_s appears squared in the integrand of the force expression, a net outward force exists on the sphere trying to pull it apart.‡ The total radial force is, from (5.83), given by

$$F = \frac{1}{2\varepsilon_0} \int_0^{2\pi} \int_0^{\pi} (3\varepsilon_0 E_0 \cos \theta)^2 a^2 \sin \theta \, d\theta \, d\phi$$
$$= 6\varepsilon_0 E_0^2 \pi a^2 \tag{5.86}$$

† The tendency to increase the size of the charged sphere can also be concluded from the results of the previous section where it was observed that the force is always in the direction as to increase the capacitance. From (5.11), the capacitance of an isolated sphere is proportional to radius a.

‡ Notice how easy it is to conclude wrongly that the sphere should contract, because one expects the oppositely charged halves of the sphere to attract each other.

Again no net translational force exists on the sphere since on opposite points the force components are equal and opposite. This could have also been predicted from the fact that the external field induces a dipole moment in the sphere, and we know that a dipole in a uniform field experiences no net translational force [see Eq. (4.7)].

5.16 WHY DIELECTRIC OBJECTS MOVE TOWARD STRONGER ELECTRIC FIELDS

This question can be answered very simply with the aid of Fig. 5.19. The fringing electric field of a parallel-plate capacitor is a nonuniform field which becomes stronger as the capacitor is approached. A dielectric object brought into this fringing field will have charges induced in it, as shown in Fig. 5.19*a*. Since the induced negative (positive) charge is closer to the positive (negative) charge on the capacitor plate than is the induced positive (negative) charge, a net force exists on the dielectric object which draws the object toward the capacitor. This attractive phenomena exists irrespective of whether the capacitor is held at a fixed potential V (battery remains connected) or at a constant charge Q (battery charges capacitor and is then disconnected). To see this, we can examine the expressions derived in Sec. 5.14. At fixed V we have $W = \frac{1}{2}CV^2$ for energy. The force is then, from (5.79),

$$F_x = \left.\frac{\partial W}{\partial x}\right|_V = \frac{1}{2}V^2\frac{\partial C}{\partial x} \tag{5.87}$$

and thus acts in the direction of increasing capacitance (a dielectric object introduced into a capacitor increases its capacitance). We did not even have to

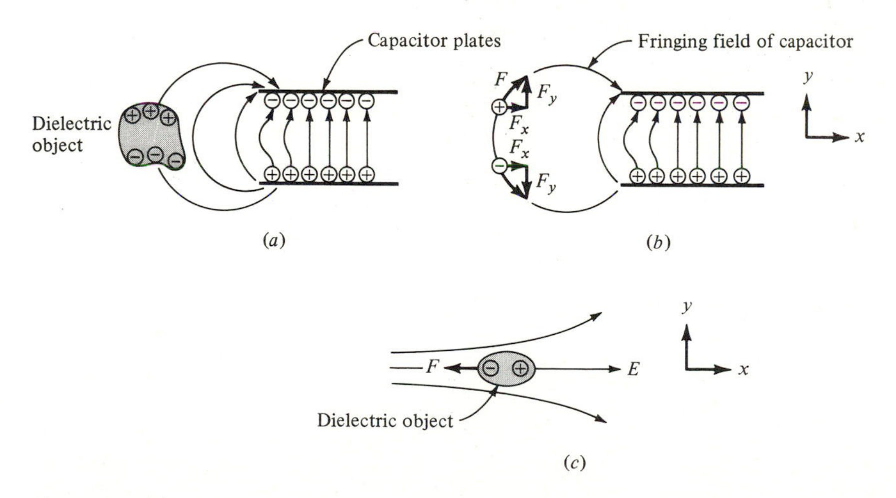

Figure 5.19 (*a*) A dielectric object in a nonuniform field. The induced dipole term shows that the object will be attracted toward the stronger electric field; (*b*) the force is at right angles to the field; (*c*) the force is parallel to the field.

differentiate W to arrive at this conclusion. We can simply observe that the plus sign in $F = \partial W/\partial x$ means that the force acts in the direction that increases the stored energy, and a dielectric that moves inside a capacitor increases stored energy. On the other hand for fixed Q we use $W = Q^2/2C$, which by use of (5.72) gives for the force

$$F_x = -\left.\frac{\partial W}{\partial x}\right|_Q = \frac{Q^2}{2C^2}\frac{dC}{dx} \tag{5.88}$$

Again, dC/dx shows that the force is in the direction of increasing capacitance. The negative sign in $F = -\partial W/\partial x$ implies that the force is in the direction that decreases stored energy, which is the case since increasing C decreases W at constant Q.

The induced charge distribution in a dielectric object is in effect a large dipole. The force on such an object is thus given by a dipole term. Figure 5.19*b* shows the forces acting on a single dipole. Since the force on the individual charges of the dipole acts along the electric field lines as shown, we see that the dipole as a whole will experience a force component F_x toward the interior of the capacitor. For purposes of calculating force, we can approximate the induced charges on the dielectric object by a constant dipole moment $\mathbf{p} = p_y\hat{\mathbf{y}}$. Since from (4.7), the force on an electric dipole is given by

$$\mathbf{F} = \nabla(\mathbf{p} \cdot \mathbf{E})$$

we have

$$\begin{aligned} \mathbf{F} &= \nabla(p_y E_y) \\ &= \hat{\mathbf{x}} p_y \frac{\partial E_y}{\partial x} \end{aligned} \tag{5.89}$$

which shows that the force on the dielectric slab has an x component if E_y has a rate of change in the x direction. Note that E_y is at a maximum in the median plane; hence $\partial E_y/\partial y = 0$.

Figure 5.19*c* shows an electric field that is allowed to spread out along the x direction. The field strength thus decreases along the positive x direction. A dielectric object introduced in such a field will again move toward the high-field region which is in the direction of the negative axis. The electric field E_x will induce a dipole moment p_x in the object with a resultant force

$$\mathbf{F} = \nabla(\mathbf{p} \cdot \mathbf{E}) = \hat{\mathbf{x}} p_x \frac{\partial E_x}{\partial x} \tag{5.90}$$

In contrast to Fig. 5.19*b* which shows a force at right angles to the field, the dielectric object now moves in a direction which is parallel to the electric field. The physical reason for the net force on the object is that the negative induced charge is in a stronger field than is the positive charge.

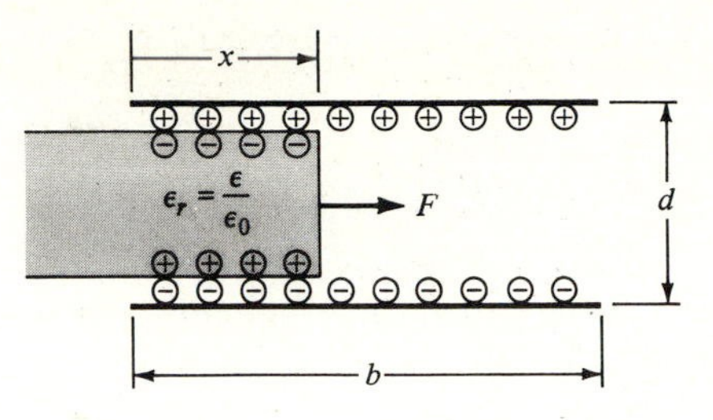

Figure 5.20 A dielectric slab partially inside the capacitor. The area of the capacitor plates is $A = bl$.

Force on a Dielectric Slab Between Capacitor Plates

Figure 5.20 shows a slab of dielectric partly inside the capacitor. We can calculate the force tending to pull the dielectric slab of dielectric constant ε_r completely inside the dielectric by several methods. For example, the capacitance for the capacitor of Fig. 5.20 is given by (5.31c) as

$$C = \frac{\varepsilon_0 lb}{d} + (\varepsilon_r - 1)\varepsilon_0 \frac{lx}{d} \tag{5.91}$$

where the area of the capacitor plates is $A = lb$. The force is then, from (5.87),

$$\begin{aligned} F_x &= \frac{1}{2} V^2(\varepsilon_r - 1)\varepsilon_0 \frac{l}{d} \\ &= \tfrac{1}{2}E^2(\varepsilon_r - 1)\varepsilon_0 ld \end{aligned} \tag{5.92}$$

when the capacitor is held at constant V [for constant Q use Eq. (5.88)].† We observe that the force trying to pull the slab inside is proportional to $\varepsilon_r - 1$. Thus, denser dielectrics (in general, ε_r is proportional to density) will experience a larger force. Of course as $\varepsilon_r \to 1$, $F_x \to 0$, as expected.

We can now make an interesting observation. If we calculate the work required to place the dielectric slab completely inside the capacitor, we find that

$$\begin{aligned} W &= \int_0^b F_x\, dx = F_x b = \frac{1}{2} V^2(\varepsilon_r - 1)\varepsilon_0 \frac{lb}{d} \\ &= \tfrac{1}{2}V^2(\varepsilon_r - 1)C_0 \\ &= (\varepsilon_r - 1)W_0 \end{aligned} \tag{5.93}$$

is the energy needed to polarize the inserted dielectric and is equal to the energy of polarization (5.69).

Another way to calculate the force on the slab is to begin with the energy in the capacitor of Fig. 5.20 which is

$$W = \iiint \tfrac{1}{2}\varepsilon E^2\, dv = \tfrac{1}{2}\varepsilon E^2\, xld + \tfrac{1}{2}\varepsilon_0 E_0^2(b - x)\, ld \tag{5.94}$$

The force given in (5.92) is again obtained by use of (5.87).

† In the case when the charge Q on the capacitor plates is constant, V changes as the capacitance changes with position x of the slab. However, for any given V, (5.92) gives the correct force on the slab. As $Q = CV$, the constant-Q case can be obtained simply by replacing V by Q/C in (5.92).

5.17 EXAMPLE: PRESSURE RISE IN A LIQUID SURFACE DUE TO AN ELECTRIC FIELD

We can make use of the previous results to show that a liquid surface which is exposed to an electric field will rise. Let us consider first an electric field E_t that is tangential to the surface, as shown in Fig. 5.21. The capacitor is held at constant V by the battery. Treating the dielectric liquid inside the capacitor as a dielectric slab, we can use (5.92) to obtain the force trying to draw the liquid into the capacitor as

$$\frac{F}{A} = p = \frac{1}{2} E_t^2 (\varepsilon_r - 1)\varepsilon_0 \tag{5.95}$$

where p is the pressure (defined as force per unit area) and $A = ld$ is the liquid-air surface inside the capacitor. In general, if we have two media of dielectric constants ε_2 and ε_1, the pressure on the boundary between the media will be given by $p = \frac{1}{2}E_t^2(\varepsilon_2 - \varepsilon_1)$, where the direction of the pressure is from medium 2 to 1 if $\varepsilon_2 > \varepsilon_1$. It is interesting to observe that the origin for the force that lifts the liquid is not at the surface of the liquid but is in the liquid dipoles which are in the fringing field of the capacitor. They experience a force which pushes them into the region of stronger electric field with the result that the pressure inside the capacitor is higher than in the rest of the medium. This in turn results in a small rise of the dielectric liquid surface. The height h of the rise is obtained by equating the weight of the raised liquid to the upward pressure on the liquid column inside the capacitor plates. The weight of the raised liquid is $F = g\rho lhd$, where g is the gravitational constant (9.81 m/s^2), ρ is the mass density of the liquid, and lhd is the raised volume. Equating this force to (5.95), we can solve for h and obtain

$$h = \frac{E_t^2(\varepsilon_r - 1)\varepsilon_0}{2g\rho} \tag{5.96}$$

where the electric field is related to the applied voltage across the capacitor plates by $E = V/d$.

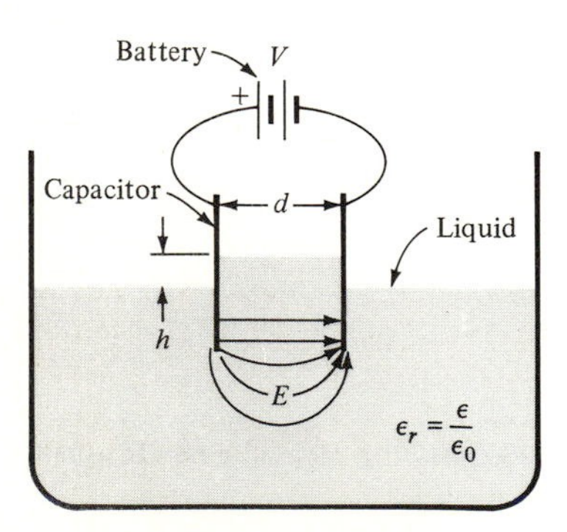

Figure 5.21 A capacitor partially immersed in a dielectric liquid of dielectric constant ε_r. The liquid inside the capacitor will rise to a height h above the liquid outside the capacitor.

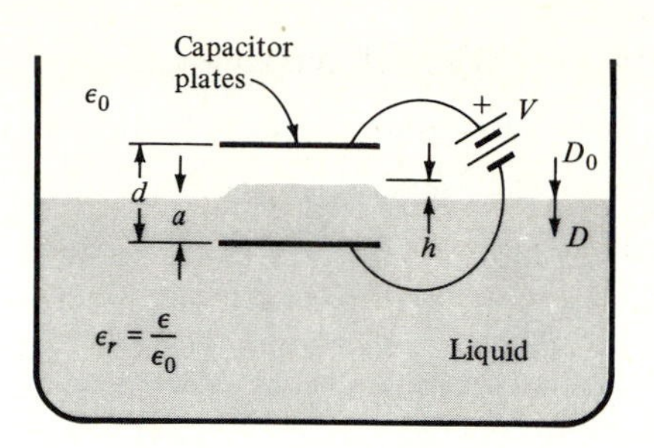

Figure 5.22 An electric field normal to the liquid surface will exert an upward pressure.

Another example of a liquid surface rise is obtained when the electric field is normal to the dielectric liquid-air interface. Such a case is shown in Fig. 5.22 where a capacitor is partly submerged. In this case, unlike the previous case, the liquid surface can be visualized as directly lifted because the top layer of polarization charge experiences the stronger electric field E_0 in the free-space portion of the capacitor, whereas the next polarization layer down is already in the weaker electric field of the dielectric. Recall that the boundary condition on the normal components of a dielectric-air interface is $D = D_0$, which shows that the E field inside the dielectric is less by a factor of ε_r; that is, $E = E_0/\varepsilon_r$. To calculate the force on the liquid surface, we can again use (5.87) or (5.88). The capacitance† of the partially filled capacitor is needed and is readily found as

$$C = \frac{\varepsilon A}{\varepsilon_r d + a(1 - \varepsilon_r)} \tag{5.97}$$

The boundary condition for the normal components, $\varepsilon_0 E_0 = \varepsilon E$, and the potential across the capacitor, $V = E_0(d - a) + Ea$, were used in calculating C. We obtain for the force

$$F = \frac{Q^2}{2C^2}\frac{dC}{da} = \frac{AE^2}{2}\varepsilon(\varepsilon_r - 1) = \frac{AD^2(\varepsilon_r - 1)}{2\varepsilon} \tag{5.98}$$

when the capacitor is held at constant Q and

$$F = \frac{1}{2}V^2\frac{dC}{da} = \frac{1}{2}V^2\frac{\varepsilon A(\varepsilon_r - 1)}{[a(1 - \varepsilon_r) + \varepsilon_r d]^2} \tag{5.99}$$

for constant potential V.

Perhaps an easier way to obtain the force is to calculate the energy stored in the capacitor and differentiate using (5.87) or (5.88). The energy stored in the capacitor of Fig. 5.22 is

$$\begin{aligned} W &= \iiint \frac{1}{2}\varepsilon E^2\,dv = \iiint \frac{1}{2}\frac{D^2}{\varepsilon}\,dv \\ &= \frac{1}{2}D^2 A\left(\frac{a}{\varepsilon} + \frac{d - a}{\varepsilon_0}\right) \end{aligned} \tag{5.100}$$

where A is the capacitor-plate area. Differentiating, we obtain for the upward electrostatic pressure

$$p = \frac{1}{A}\frac{dW}{da} = \frac{1}{2}D^2\left(\frac{1}{\varepsilon_0} - \frac{1}{\varepsilon}\right) = \frac{D^2}{2\varepsilon_0}\left(1 - \frac{1}{\varepsilon_r}\right) \tag{5.101}$$

This expression is the same as (5.98) because for the case when the charge is held constant, D also remains constant as per the boundary condition at the dielectric-metal interface, $D = \rho_s$. Incidentally

† To include the he:ght h of the lifted surface explicitly, change a to $a + h$. As the rise is usually very small, $a + h \approx a$.

(5.101) can also be used to obtain the pressure on an air-metal boundary by letting $\varepsilon_r \to \infty$ which gives $p = D^2/2\varepsilon_0$ for pressure, which is the same as (5.82). To use (5.101) when the potential V on the capacitor is held constant, one first applies the dielectric-air boundary condition $\varepsilon_0 E_0 = \varepsilon E$ to it. After this substitution we have

$$p = \frac{\varepsilon_0 E_0^2}{2}\left(1 - \frac{1}{\varepsilon_r}\right) = \frac{\varepsilon_0 E^2}{2}\varepsilon_r(\varepsilon_r - 1) \tag{5.102}$$

which is the appropriate expression since it now can be related to the constant potential V by $V = E_0(d - a) + Ea = E[\varepsilon_r(d - a) + a]$. The height h of the lifted surface is obtained by equating the weight of the lifted volume $F = g\rho Ah$ to one of the appropriate expressions for the upward electrostatic force.

As a generalization we can now give the total electrostatic pressure on a dielectric-air boundary by combining the force terms when the electric field is parallel and normal to the boundary. Thus

$$p = \tfrac{1}{2}\varepsilon_0(\varepsilon_r - 1)(E_t^2 + \varepsilon_r E_n^2) \tag{5.103}$$

where E_t and E_n are the tangential and normal components, respectively, just inside the dielectric. Since the total field at a boundary can always be decomposed into a normal and tangential field, (5.103) gives the pressure for an arbitrarily oriented field. The direction of force is always toward the dielectric of smaller permittivity; in our examples the force is from a dielectric to free space.

PROBLEMS

5.1 The radius of earth is 6400 km. If we regard it as a single spherical capacitor, what is the capacitance of the earth?

5.2 A charged conducting sphere of radius a carries a charge Q. What is the total electric potential energy of the charged sphere?

5.3 A conducting object is given a positive charge. Does its mass increase, decrease, or remain the same?

5.4 An isolated conducting sphere with a diameter of 5 cm is situated in air.

(*a*) What is the maximum voltage to which the sphere can be charged before breakdown of the surrounding air occurs?

(*b*) What is the electric energy density at the surface of the sphere maintained at that potential?

(*c*) What is the largest charge density that can be retained by the conducting sphere?

5.5 Determine the largest charge density that can be retained on the surface of any conductor in air.

5.6 A 12-V battery is rated as having a chemical potential energy of 10^6 J. How many times can the battery be used to charge a 50-μF capacitor?

5.7 A typical automotive storage battery can store 1 kW · h of energy.

(*a*) If a parallel-plate capacitor with free space between the plates is to store this amount of energy, what volume must the capacitor have?

(*b*) If a typical storage battery has a volume of 0.01 m^3, calculate the ratio of the two volumes. What can you say about the usefulness of such capacitors to store electric energy?

5.8 Referring to the previous problem, if a typical electrolytic capacitor rated at 100 μF and 500 V occupying a volume of 3×10^{-5} m^3 is used, calculate the ratio of battery volume to capacitor volume. How many capacitors would be needed? Is this a practical way to store electric energy?

Hint: It would be a practical way as long as the ratio of capacitor volume to battery volume is not greater than, let us say, 10.

5.9 Show that the expression (5.11) for capacitance between two concentric spherical shells reduces to the parallel-plate capacitor formula (5.12) when the separation between the shells becomes very small.

5.10 Show that the capacitance between two concentric spherical shells reduces to the capacitance of an isolated sphere ($C = 4\pi\varepsilon_0 a$) when the separation $b - a$ between the shells becomes very large.

5.11 Derive the spherical capacitor formula (5.11) by first dividing the spherical capacitor into many parallel-plate capacitors, applying the parallel-plate capacitor formula (5.12) to each elemental capacitor, and integrating the resulting expression over the surface of the sphere.

5.12 Two spheres, both of radii a, are separated by a distance r such that $r \gg a$. Show that the capacitance between the spheres is given by $C \cong 2\pi\varepsilon_0 a$.

5.13 Calculate the number of time constants it takes for a capacitor in an RC circuit to be charged to half its final value.

5.14 The battery in the circuit of Fig. 5.2 has been connected a long time to the circuit. If the battery is now disconnected and replaced by a short, prove that all energy stored in the capacitor C is transformed into Joule heat in the resistor R.

5.15 A bank of 100 parallel-connected 10-μF capacitors is used to store electric energy. If it is desired to charge this bank to 10,000 V, calculate the cost using a rate of 4 cents/kW · h.

5.16 After charging a 1-μF capacitor to a potential difference of 100 V, it is disconnected from the charging circuit and connected to a second, initially uncharged capacitance. If the potential difference drops to 60 V, what is the capacitance of the second capacitor?

5.17 The construction of a paper capacitor is as follows: Aluminum foil of 100-cm^2 area is placed on both sides of paper of thickness 0.03 mm. If the dielectric constant of paper (from Table 2.3) is given as 3, and its dielectric breakdown strength as 2×10^5 V/cm, what is the rating of the capacitor? (Find capacitance and working voltage; assume that working voltage is one-half the maximum voltage which can be applied across the capacitor.)

5.18 What is the capacitance of the capacitor shown? A dielectric slab of thickness a and dielectric constant ε_r is positioned as shown in the free-space capacitor.

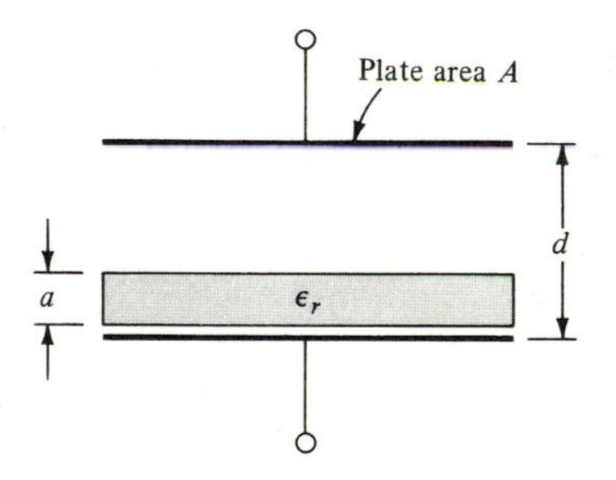

5.19 What is the capacitance of the capacitor shown?

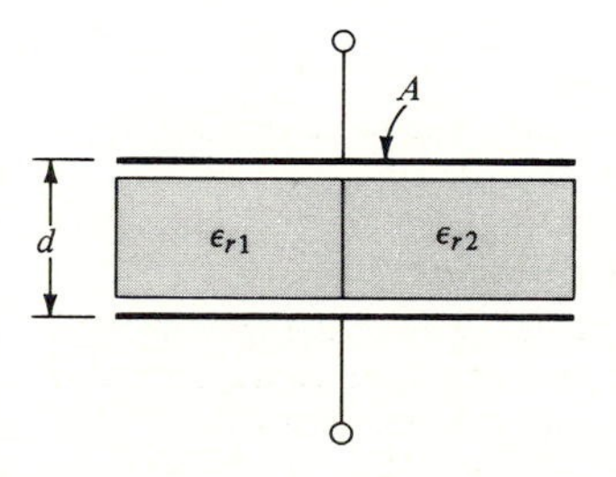

5.20 What is the capacitance of the capacitor shown?

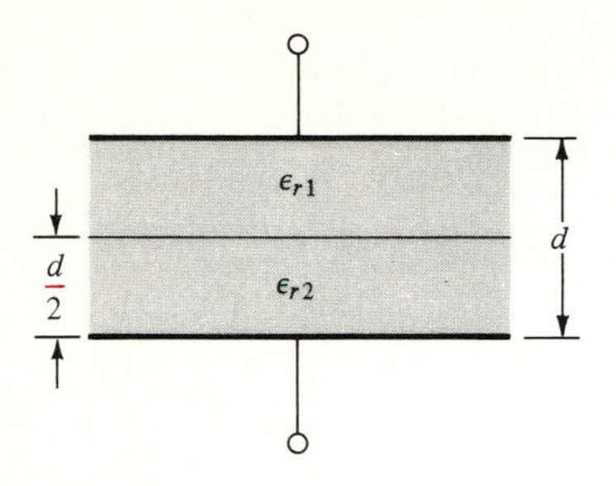

5.21 What is the capacitance of the capacitor shown?

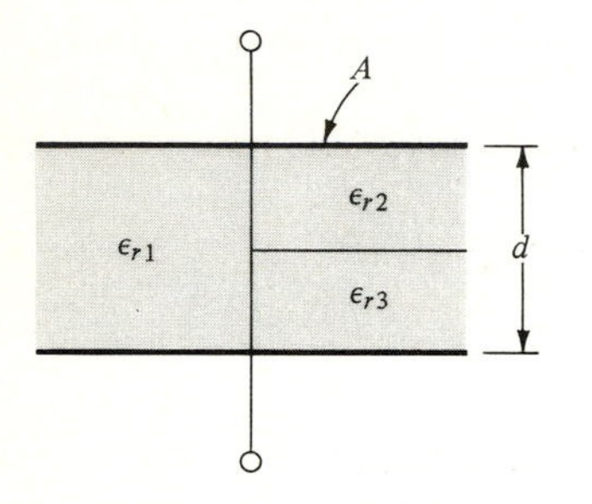

5.22 A metal foil of negligible thickness is introduced in a capacitor as shown. What effect does it have on the capacitance?

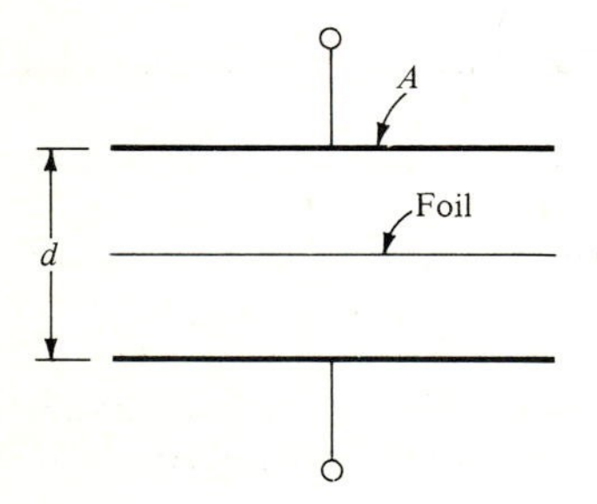

5.23 A slab of brass centered between the plates of a parallel-plate capacitor is shown. What is the effect of the slab on the capacitance?

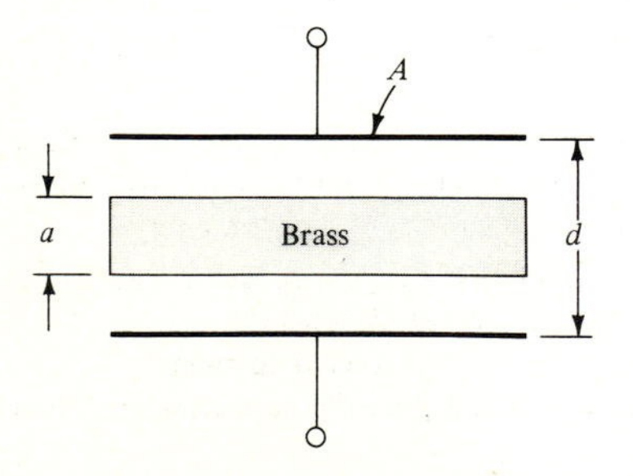

5.24 One plate of a parallel-plate capacitor is tilted at an angle θ as shown. If the angle θ is small, find the capacitance.

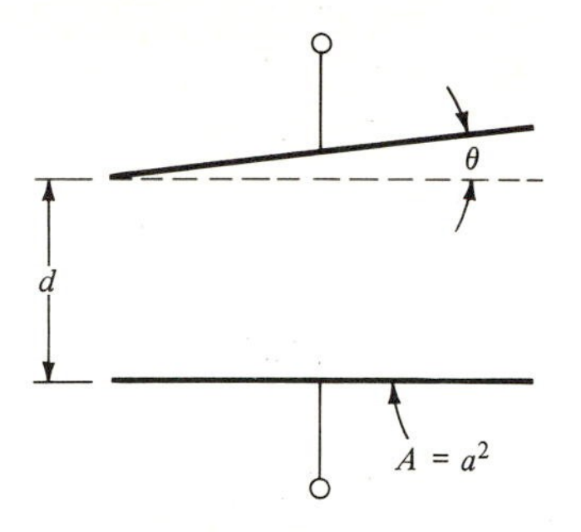

5.25 Show that $\mathbf{P} = (\varepsilon - \varepsilon_0)\mathbf{E}$ and that $\nabla \cdot \mathbf{P} = (\varepsilon_r - 1)\varepsilon_0 \nabla \cdot \mathbf{E}$.

5.26 The permittivity of the dielectric material between the plates of a parallel-plate capacitor varies linearly from ε_1 at one plate to ε_2 at the other plate.

(*a*) Show that the capacitance is given by

$$C = \frac{A(\varepsilon_2 - \varepsilon_1)}{d \ln (\varepsilon_2/\varepsilon_1)}$$

where A is the area and d is the separation between the plates.

(*b*) What is the volume density of polarization charge ρ_p in the dielectric?

5.27 If the voltage between the inner and outer conductors of a coaxial cable is increased continuously, dielectric breakdown of the air will eventually occur. Will the breakdown start at the inner or at the outer conductor?

5.28 A coaxial cable has inner and outer radii of a and b, respectively.

(*a*) Show that the relationship between the electric field E_a at a and the potential V between a and b is given by $V = aE_a \ln (b/a)$.

(*b*) If the radius of the outer conductor is fixed, but the inner one assumed to be variable, determine radius a such that E_a is a minimum for a fixed V.

5.29 Using the results of Prob. 5.28*b*, show that the maximum potential difference between the conductors of a coaxial cable, before a breakdown of the dielectric occurs, is given by $V_{\max} = aE_a^{\text{breakd}}$, where E_a^{breakd} is the breakdown strength of the dielectric between the radii.

5.30 Referring to Fig. 5.6, determine the approximate maximum allowable value of the potential difference between the conductors of a parallel-wire transmission line, such that no breakdown of the dielectric between the conductors takes place.

5.31 Show that the plates of a parallel-plate capacitor attract each other with a force per unit area given by $F = \rho_s^2/2\varepsilon_0$, where ρ_s is the surface charge density on either plate.

5.32 Two parallel metal plates separated by a distance of d and each of area A are charged by a battery of V volts. The battery is then disconnected, and the plates are pulled apart to a distance of $2d$. Neglecting fringing, find the work done in separating the plates.

5.33 An air parallel-plate capacitor of capacitance C is charged until the energy stored is $Q^2/2C$. A sheet of mica ($\varepsilon_r = 6$) is then placed between the plates increasing the capacitance by a factor of 6. Since the charge is unchanged, the stored energy is now $Q^2/12C$. Account for the rest of the energy.

5.34 A dielectric slab (refer to Fig. 5.20) is withdrawn from a parallel-plate capacitor, such that part of the slab remains between the plates. Calculate the force which acts on the slab when

(*a*) The capacitor plates are held at a fixed potential by a battery connected to them.

(*b*) The capacitor plates have a constant charge (obtained by charging the capacitor and then disconnecting the battery from the plates).

5.35 Referring to the previous problem, is there a difference between the two cases? Explain your answer.

5.36 Calculate the work necessary to pull a mica sheet ($\varepsilon_r = 6$) out from between the plates of a parallel-plate capacitor. The area of each plate is 1 m^2, and the thickness of the mica sheet is 2 mm.

(a) The capacitor is connected to a battery of 500 V.

(b) The capacitor is charged by a battery of 500 V which is subsequently disconnected.

5.37 Determine an expression for the force tending to pull the dielectric slab of thickness t, where $t < d$, into the space between the plates of the capacitor. Consider both cases of constant-V and constant-Q.

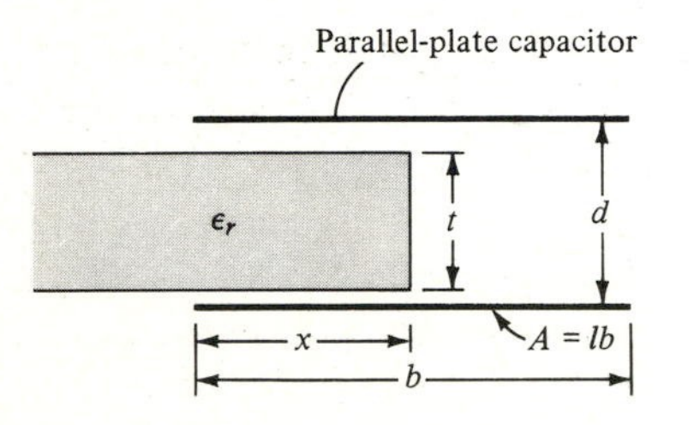

5.38 Calculate the potential V necessary to float the lower plate of a parallel-plate capacitor if the upper plate is fixed in a horizontal plane. The weight W of the lower plate is 10^{-3} N, its area is 25 cm^2, and it is desired to suspend the lower plate at a distance d of 1 mm from the fixed upper plate. Neglect fringing.

5.39 The surface of an initially uncharged soap bubble of radius r_0 acquires a charge Q. The radius will now increase to r because of the mutual repulsion by the charges of the charged surface. Show that by equating the work done by the bubble in pushing back the atmosphere to the decrease in the stored electric energy that follows the expansion, we obtain

$$Q = \left[\frac{32}{3}\pi^2\varepsilon_0 p r_0 r(r_0^2 + r_0 r + r^2)\right]^{1/2}$$

where p is the atmospheric pressure.

5.40 A positive point charge near an air-liquid dielectric will deform the liquid surface. Will the surface rise or sink? What about if a negative point charge is substituted?

5.41 What is the capacitance of a coaxial cable of length L which has a dielectric with permittivity ε that occupies a length x of the cable?

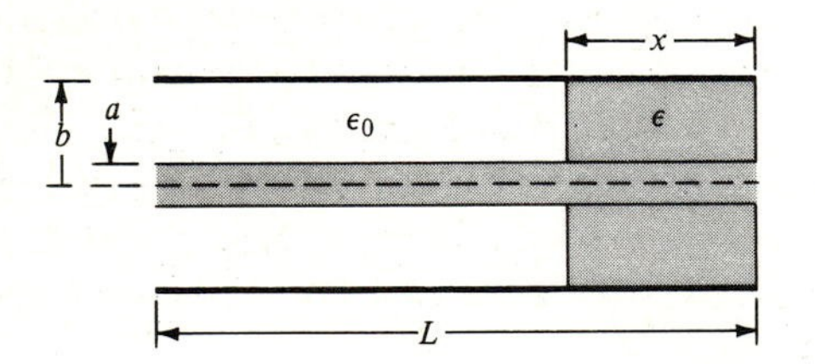

5.42 A coaxial cable of inner and outer radii a and b, respectively, is lowered vertically in a bath of liquid dielectric. A voltage V is applied to the cable. Show that the liquid between the inner and outer conductors rises a height

$$h = \frac{\varepsilon_0(\varepsilon_r - 1)V^2}{\rho g(b^2 - a^2)\ln(b/a)}$$

above the liquid surface. ε_r and ρ are the relative permittivity and the mass density of the liquid, respectively.

5.43 A parallel-plate capacitor which has a constant charge Q on its plates is completely immersed in a liquid dielectric.

(*a*) Show that the force on the plates will decrease to $F = F_0/\varepsilon_r$, where F_0 is the force on the plates of a free-space capacitor and ε_r is the dielectric constant of the liquid.

(*b*) Show that F is the difference between F_0 and the force of the liquid surface that pushes against the plates.

5.44 A parallel-plate capacitor which has a constant voltage V between its plates is completely immersed in a liquid dielectric.

(*a*) Show that the force on the plates will increase to $F = \varepsilon_r F_0$, where F_0 is the force on the plates of a free-space capacitor and ε_r is the dielectric constant of the liquid.

(*b*) Why is F in this case not equal to the difference in F_0 and the force of the liquid surface against the plates?

CHAPTER

SIX

THE MAGNETIC FIELD AND MAGNETIC FORCE IN FREE SPACE

6.1 INTRODUCTION

In our study of electricity we have shown that stationary charges produce an electric field. If the charges are moving with uniform velocity, a secondary effect takes place, which is the phenomenon of magnetism. If we accelerate charges, we have an additional effect; the accelerated charges now produce a radiating electromagnetic field, i.e., a field that can transport energy. Thus any radiating system such as an antenna must somehow accelerate charges in order to produce such a field. In that sense, magnetism and electromagnetic fields are special cases of electricity. It can be shown that Coulomb's law, if modified to include charges in motion, will give terms which can be identified with a magnetic field B. Since motion is relative, a given physical experiment which is purely electrostatic in one coordinate system can appear as electromagnetic in another coordinate system that is moving with respect to the first. Magnetic fields seem to appear and vanish merely by a change in the motion of the observer. Hence the subject of relativity plays a fundamental role in electromagnetics. We will show in Chap. 12, "Relativity and Maxwell's Equations," that all the laws of electromagnetic fields can be derived by applying the relativistic transformation to Coulomb's law. The above ideas can be summarized by showing a list of the velocity v_q of a charge q and the fields that the charge produces:

$$v_q = 0: \qquad E \neq 0,\ B = 0$$

$$v_q \neq 0: \qquad E \neq 0,\ B \neq 0$$

$$dv_q/dt \neq 0: \qquad E \neq 0,\ B \neq 0, \text{ radiation fields}$$

A magnetic field is thus associated with moving charges. One can say therefore that the sources of magnetic field are currents.

Units of the Magnetic Field

In the SI system of units the magnetic field B is given in teslas [1 tesla (T) = 1 weber/meter^2 (1 Wb/m^2)]. Since a tesla is relatively large, the magnetic field is usually given in the cgs units of gauss (G), where

$$1 \text{ T} = 10^4 \text{ G}$$

For reference, the earth magnetic field is about 0.5 G, that of a small permanent magnet is about 100 G, that of large electromagnets is up to 20,000 G, and that of magnets of some large particle accelerators is in the range of 60,000 G.

The magnetic field vector, here designated as $\mathbf{B}$, is also referred to as the *magnetic induction*, or the *magnetic flux density*. It should be distinguished from the magnetic field strength $\mathbf{H}$ (where $\mathbf{B} = \mu\mathbf{H}$), which is quite different but is sometimes also referred to as the magnetic field. In the SI system the unit for H is the ampere-turn per meter. The cgs unit of oersted (Oe) is often used for H, where

$$1 \text{ ampere-turn/m} = 4\pi \times 10^{-3} \text{ Oe}$$

The *magnetic flux* ϕ through an area A which is normal to the B lines is defined by $\phi = BA$. In the SI system the unit of ϕ is the weber (Wb), and in the cgs system it is the maxwell (Mx), where

$$1 \text{ Wb} = 10^8 \text{ Mx}$$

The above definition of flux shows that B is a flux density.

6.2 TWO POSTULATES FOR THE MAGNETIC FIELD

We shall introduce the results of the experimental work of Ampère and Biot-Savart as two postulates:

Postulate 1 A current element $\mathbf{I}\,dl$ immersed in a magnetic field $\mathbf{B}$ will experience a force $\mathbf{dF}$ given by

$$\boxed{\mathbf{dF} = \mathbf{I} \times \mathbf{B}\,dl} \tag{6.1}$$

The force is in newtons (N).

This is shown pictorially in Fig. 6.1. Note that $\mathbf{I}\,dl$ cannot exist by itself as it always must be part of a complete loop or circuit. However, the concept of a current element greatly simplifies the development of the magnetic field and its effects. In a physical situation the force on the entire loop can be obtained by integrating the current elements; that is,

$$\mathbf{F} = \oint I\,\mathbf{dl} \times \mathbf{B} \tag{6.2}$$

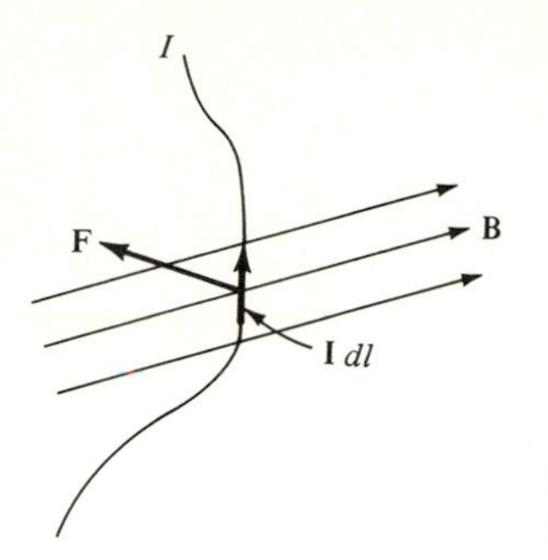

Figure 6.1 The force on a current element $\mathbf{I}\,dl$ immersed in a magnetic field **B** is at right angles to both $\mathbf{I}\,dl$ and **B**.

It should now be noted† that since current flow is restricted to the path of the circuit, usually a conducting wire, the direction of the current element can be either specified by the direction of the current **I** or the direction of the circuit element **dl**; that is, $\mathbf{I}\,dl = I\,\mathbf{dl}$.

A current element experiences a force which is at right angles to the plane formed by the current element and the applied magnetic field **B**. The force is of magnitude

$$dF = IB\,dl\sin\theta \tag{6.3}$$

where θ is the angle between the direction of the current **I** and the magnetic field **B**, as shown in Fig. 6.2*a*. This can be easily remembered by the *FBI* rule which uses the left hand to relate the directions of *F*, *B*, and *I*, as shown in Fig. 6.2*c*.

If a straight wire of length l is oriented at right angles to a magnetic field B, as shown in Fig. 6.3*a*, it will be acted on by a force $F = IBl$. This expression can be used to explain motor and relay action (see Sec. 6.11).

Example Let us calculate the total force on a wire bent into a shape of a straight piece and a semicircle, as shown in Fig. 6.3*b*. A current I flows along the wire, and the entire structure is immersed in an external magnetic field B which is normal to the plane of the circle. The sense of B

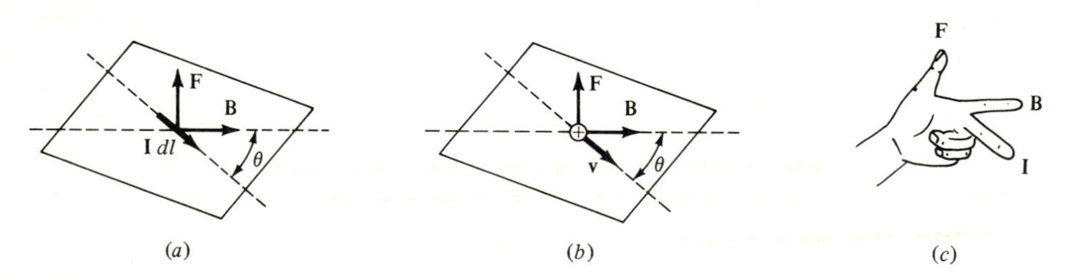

Figure 6.2 (*a*) The magnetic force is perpendicular to the plane of **I** and **B**; (*b*) of **v** and **B**; (*c*) the left hand *FBI* rule gives the directions between *F*, *B*, and *I*.

† The advantage of making a filamentary current a vector quantity is that $\mathbf{I} \times \mathbf{B}$ is emphasized in the expression for magnetic force. The form of Eq. (6.1) is now closer in appearance to the magnetic force on a moving charge, $\mathbf{F} = q(\mathbf{v} \times \mathbf{B})$, and to $\mathbf{J} \times \mathbf{B}$ of (6.5) which is known in general as the magnetic force term.

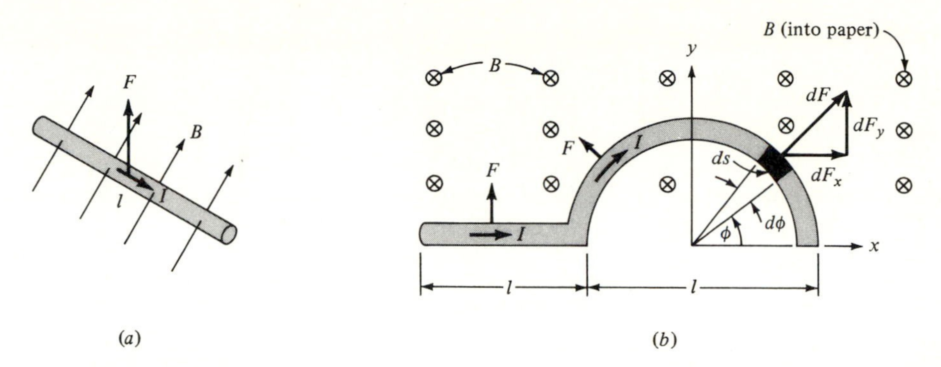

Figure 6.3 (*a*) Magnetic force F on a length l of a current-carrying wire; (*b*) magnetic force on a sickle-shaped wire.

is indicated by the cross symbols, which denote that the B field is into the page (a dot symbol denotes a B field out of the page). The force dF on a current element $I\ ds$ is as shown and is given by $dF = IB\ ds = IB(l/2)\ d\phi$. Since the x components of force cancel, the total upward force on the sickle-shaped structure is

$$F = F_y = IBl + \int_0^\pi \sin\phi\ dF$$

$$= IBl + IBl$$

Hence the upward force on the semicircle is the same as the force on a straight wire of length equal to the diameter of the semicircle.

Current is a time rate of charge; that is, $I = dq/dt$. The current I that we have talked about thus far is due to moving electrons in the wire and is known as a *conduction current.* We can modify postulate 1 to apply explicitly to moving charges in free space which would constitute a *convection current.* Such a convection current has a current density $\mathbf{J} = \rho\mathbf{v}$ [see Eq. (2.6)], where $\mathbf{v}$ is the velocity of the charges and ρ is their volume charge density. Since current I in a thin wire of cross section dA is related to current density J by $I = J\ dA$, the expression for a current element can be written as $\mathbf{I}\ dl = \mathbf{J}\ dA\ dl$. We can now substitute the current density of a convection current for J. This gives $\mathbf{I}\ dl = \mathbf{J}\ dA\ dl = \rho\mathbf{v}\ dA\ dl = q\mathbf{v}$, where $dA\ dl$ is an elemental volume enclosing the charge q; that is, $\rho\ dA\ dl = q$. Therefore a charge q moving with velocity $\mathbf{v}$ is equivalent to a current element $\mathbf{I}\ dl = q\mathbf{v}$. Substituting this in Eq. (6.1), we can state postulate 1 explicitly for a moving charge q in free space as

$$\boxed{\mathbf{F} = q\mathbf{v} \times \mathbf{B}} \tag{6.4}$$

which is illustrated in Fig. 6.2*b*. This force is called the *Lorentz force* and is sometimes used to define the magnetic field as $B = F/qv$. To illustrate the Lorentz force, consider an electron e moving at right angles to a magnetic field. The electron will experience a force $F = Bev$, which is a well-known relation in elementary physics. If, on the other hand, a conducting wire of length l is moved at right angles to the field, as shown in Fig. 6.3*a*, the free electrons will experience a force $F = Bev$ which is parallel to the wire. This force divided by charge is equivalent to an electric field $E = F/e = Bv$. Hence the induced voltage in a rod of length l moving with velocity v is $V = El = Bvl$.

Perhaps even a more general expression than (6.4) is an expression that gives the force per unit volume, F', on a current density **J** in an external field **B.** Since $\mathbf{I}\,dl = \mathbf{J}\,dv_{ol}$, where v_{ol} stands for volume, we can write

$$\boxed{\mathbf{F}' = \frac{d\mathbf{F}}{dv_{ol}} = \mathbf{J} \times \mathbf{B}} \tag{6.5}$$

where J is any current density, conduction or convection. Equation (6.5) shows clearly that a force $\mathbf{J} \times \mathbf{B}$ exists on each elemental volume dv_{ol} of current density. The $\mathbf{J} \times \mathbf{B}$ force is predominant in causing the twisting and bending of plasma columns or a moving cloud of charged particles in a magnetic field.

The magnetic force is very different from the electric force. For one thing, the electric force acting on a charge immersed in an electric field is along the direction of the electric field lines and is not dependent on the velocity of the charge. The magnetic force, on the other hand, depends on the velocity of the charge. As a matter of fact, only the velocity component which is perpendicular to the magnetic field, $v_\perp = v \sin\theta$, contributes to the force.† Therefore a charged particle shot into a constant B field at right angles to B will experience a force which will change the particle motion to a uniform circular motion about the B lines, as shown in Fig. 6.4. From Newton's second law, the radial force on the particle is $F_r = ma_r = qv_\perp B$, with the radial or centripetal acceleration given by $a_r = v_\perp^2/r = \omega^2 r$. Therefore, $qv_\perp B = m\omega^2 r$, where r is the radius of the circle as shown in Fig. 6.4 and ω is the angular velocity of the orbiting charged particle of mass m. On the other hand, the parallel component of velocity, $v_\parallel = v\cos\theta$, is unaffected by the presence of B. A charged particle injected into a B field along the B lines will continue to coast along the direction of magnetic field with unchanged speed. Thus a particle entering a magnetic field at some angle to the

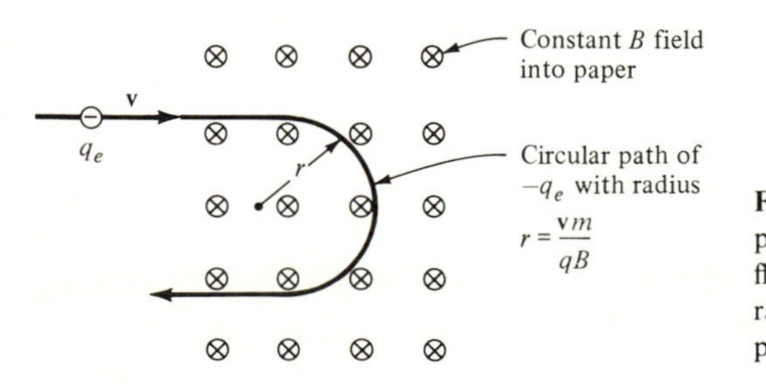

Figure 6.4 Circular path of electron $-q_e$ in a perpendicular magnetic field. Magnetic deflection of an electron beam in a cathode-ray tube makes use of the same principle.

† Refer to Fig. 6.2*b* and note that the magnitude of Eq. (6.4) is $F = qvB \sin\theta$.

B lines will follow a helical path about the *B* lines, with $v_{\parallel}$ unaffected and $v_{\perp}$ causing the circular motion.

Another difference between the electric and magnetic forces is that the magnetic force can do no work; i.e., the kinetic energy of a charged particle does not change as it moves through a uniform magnetic field. The reason is that the magnetic force on a charged particle is at right angles to its direction of motion. For example, if we let the displacement **dl** of a moving charge be $\mathbf{dl} = \mathbf{v}\, dt$, where dt is a small interval of time, we obtain for the work $dW = \mathbf{F} \cdot \mathbf{dl} = q(\mathbf{v} \times \mathbf{B}) \cdot \mathbf{v}\, dt = 0$. A charged particle moving through a uniform magnetic field will experience a magnetic force which will change its velocity but not its speed. It will be deflected without a loss or gain in energy.

Postulate 2 A current element $\mathbf{I}\, dl$ produces a magnetic field **B** which at distance R from the element is given by

$$\mathbf{dB} = \frac{\mu_0}{4\pi} \frac{\mathbf{I} \times \hat{\mathbf{R}}}{R^2} dl \tag{6.6}$$

The geometry involved is shown in Fig. 6.5. The constant $\mu_0 = 4\pi \times 10^{-7}$ H/m or Wb/A · m; μ_0 is called the *permeability of free space* and has the same significance in magnetics as ε_0 had for electric fields. Postulate 2 implies that the magnetic field is everywhere normal to the element of length dl; that is, **dB** is normal to the plane formed by the unit vectors $\hat{\mathbf{I}}$ and $\hat{\mathbf{R}}$, where $\hat{\mathbf{R}} = \mathbf{R}/R$. Except for the vectors involved, (6.6) is similar to Coulomb's law for electrostatics. If we write the magnitude of (6.6) as

$$dB = \frac{\mu_0}{4\pi} \frac{I\, dl}{R^2} \sin\theta \tag{6.7}$$

where θ is the angle between **I** and **R**, the inverse square nature of this law with distance is clearly revealed. Comparing this to Coulomb's law, magnetic charge Q_m (if there would be such a thing) would have to be equal to $Q_m = I\, dl$.

As before we note that a current element cannot exist by itself. Thus (6.6) is complete only if we integrate the expression over the entire circuit of which $I\, dl$ is only a differential element. A complete current loop, as shown in Fig. 6.6, will

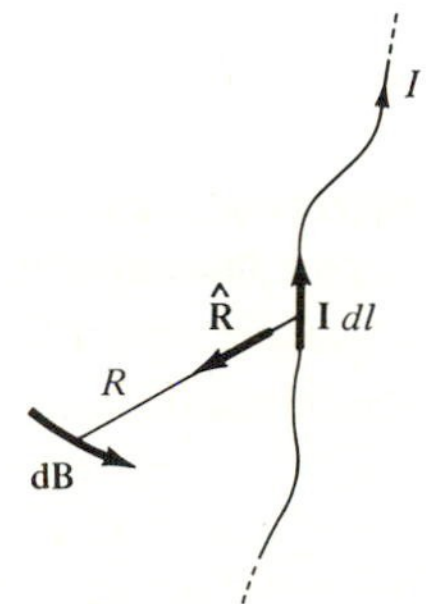

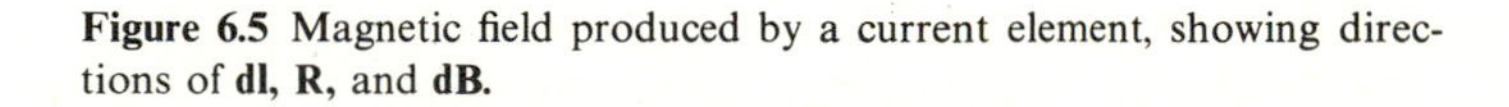

Figure 6.5 Magnetic field produced by a current element, showing directions of **dl**, **R**, and **dB**.

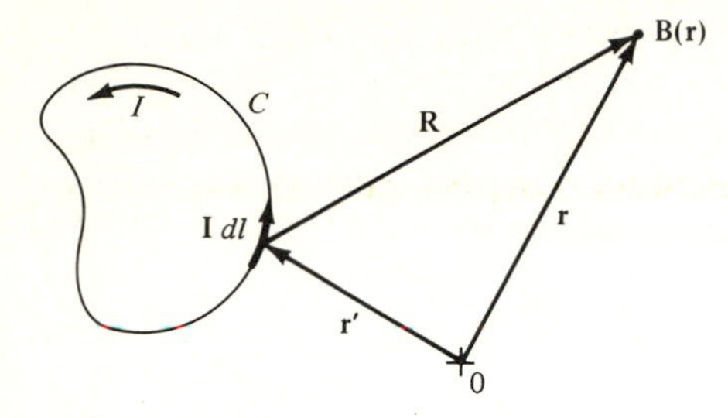

Figure 6.6 The magnetic field B due to loop C is obtained by integrating the effects of the current elements **I** dl.

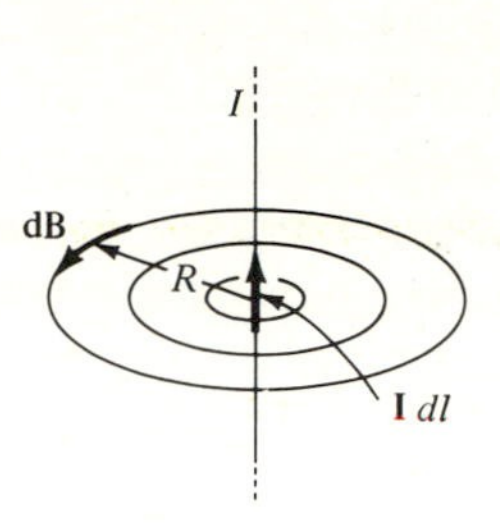

Figure 6.7 Magnetic field lines around a wire carrying current I.

produce a magnetic field $\mathbf{B}(\mathbf{r})$ at point r which is given by

$$\mathbf{B}(\mathbf{r}) = \frac{\mu_0 I}{4\pi} \oint_c \frac{\mathbf{dl}(\mathbf{r}') \times \hat{\mathbf{R}}}{R^2} \tag{6.8}$$

where $R = |\mathbf{r} - \mathbf{r}'|$, $\hat{\mathbf{R}} = \mathbf{R}/R$, and I is constant around the circuit. The above expression was first deduced by Biot and Savart and is usually referred to as the *Biot-Savart law*.

A more general expression for postulate 2 than (6.6) can be given by re-expressing (6.6) in terms of current density **J**. Since **I** dl can be written as $\mathbf{I}\,dl = \mathbf{J}\,dv_{ol}$, where **J** is a current density (amperes per meter squared) in the elemental volume dv_{ol}, the magnetic field created by the total current contained in a volume v_{ol} is

$$\mathbf{B}(\mathbf{r}) = \frac{\mu_0}{4\pi} \iiint_{v_{ol}} \frac{\mathbf{J}(\mathbf{r}') \times \hat{\mathbf{R}}}{R^2}\, dv_{ol} \tag{6.9}$$

For a moving charge, since the equivalence between a current element and a moving point charge is $\mathbf{I}\,dl = q\mathbf{v}$, postulate 2 can be written as

$$\mathbf{B} = \frac{\mu_0}{4\pi} \frac{q\mathbf{v} \times \hat{\mathbf{R}}}{R^2} \tag{6.10}$$

which gives the magnetic field $\mathbf{B}(\mathbf{r})$ produced at the observation point **r** when the charge is at the source point $\mathbf{r}'$ and is moving with velocity v (see coordinates of Fig. 6.6). The vector product implies again [see Eq. (6.7)] that the magnetic field in the direction of charge motion is zero.

To summarize we can say that unlike electric field lines that end on charges, magnetic field lines, because they have no scalar sources, are continuous and close on themselves. The closure of B lines is implied by Eq. (6.6) and is shown in Fig. 6.7 where the addition of all dB's at a fixed R gives a magnetic field whose field lines are circles about $I\,dl$. The right-hand rule, in which the thumb points in the direction of current and the curled fingers in the direction of the magnetic field, is an expression of the cross product of (6.6).

6.3 THE MAGNETIC FORCE BETWEEN TWO CURRENT ELEMENTS AND BETWEEN TWO MOVING CHARGES

Combining the two postulates, we can express the force $\mathbf{F}_{12}$ between two current elements as

$$\boxed{\mathbf{F}_{12} = \frac{\mu_0}{4\pi} \frac{I_2\, \mathbf{dl}_2 \times (I_1\, \mathbf{dl}_1 \times \hat{\mathbf{R}}_{12})}{R_{12}^2}} \tag{6.11}$$

since each current element finds itself in the magnetic field of the other. For parallel currents, (6.11) becomes

$$F_{12} = \frac{\mu_0}{4\pi R^2} I_1\, dl_1\, I_2\, dl_2 \tag{6.12}$$

Note that the above expressions (like Coulomb's law) relate the force between two points: one point defined by differential element dl_1, and the other point by dl_2. For (6.11) to be of practical use, it must be integrated over the entire path of currents I_1 and I_2. Although Ampère's initial experiments were measurements of forces between two long parallel conducting wires which were connected to batteries, we have expressed the results of these experiments for differential length as postulates 1 and 2.

In Prob. 6.5 it is shown that, for two complete circuits, $\mathbf{F}_{12} = -\mathbf{F}_{21}$, which implies that when two complete loops are considered, the net force on them is zero. Newton's third law is thus not violated. On the other hand two isolated current elements oriented in the same plane but not parallel to each other each have a component of force in the same direction, as shown in Fig. 6.8, inducing the system of two current elements to move rectilinearly. This difficulty with (6.11) is avoided when two complete circuits are considered.

Now let us consider two moving charges. Just like for the current elements, each moving charge will find itself in the magnetic field created by the other moving charge and thus will be subjected to a magnetic force. Combining postulates 1 and 2 for moving charges, (6.4) and (6.10), we have for the magnetic force between the charges:

$$\boxed{\mathbf{F}_{12} = \frac{\mu_0}{4\pi} \frac{q_1 q_2}{R_{12}^2} \mathbf{v}_2 \times (\mathbf{v}_1 \times \hat{\mathbf{R}}_{12})} \tag{6.13}$$

where $\mathbf{R}_{12}$ is the vector distance between the charges, as shown in Fig. 6.9.

Let us compare the maximum magnetic force (charges moving parallel to each other) which is

$$F_m = \frac{\mu_0}{4\pi R^2} q_1 q_2 v^2 \tag{6.14}$$

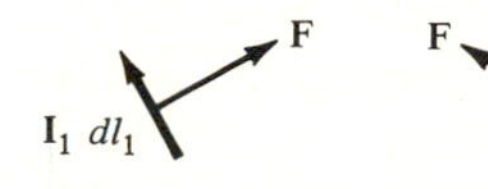

Figure 6.8 The force between two current elements.

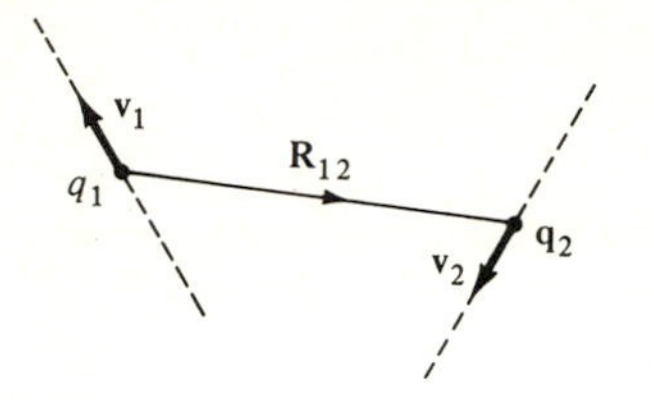

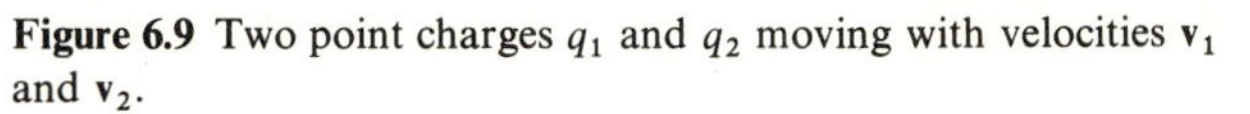
Figure 6.9 Two point charges q_1 and q_2 moving with velocities $\mathbf{v}_1$ and $\mathbf{v}_2$.

with the electric Coulomb force $F_e = q_1 q_2/4\pi\varepsilon_0 R^2$. Referring to Fig. 6.10, we obtain

$$\frac{F_m}{F_e} = \varepsilon_0 \mu_0 v^2 = \left(\frac{v}{c}\right)^2 \tag{6.15}$$

because the velocity of light is given by $c = 1/\sqrt{\mu_0 \varepsilon_0} = 3 \times 10^8$ m/s.† In practice, the velocities of the electric charges are much smaller than the velocity of light. Thus ordinarily the magnetic force is very much weaker than the electric Coulomb force.

It is interesting to note that we can write the magnetic force F_m between two moving charges as the Coulomb force F_e multiplied by the factor $(v/c)^2$; that is, $F_m = F_e(v/c)^2 = (q_1 q_2/4\pi\varepsilon_0 R^2)(v/c)^2$. This indicates that the magnetic force is the resultant of charges in relative motion. One can speculate—as was already done in the Introduction to this chapter—that Ampère's result for the magnetic force between moving charges could also be obtained from Coulomb's law and the special theory of relativity. That such a generalization of Coulomb's law is indeed possible will be shown in Chap. 12, "Relativity and Maxwell's Equations."

Magnetic Force on Current-Carrying Conductor

The question that now arises is the following: If the magnetic force between two moving charges is such a tiny force, why is it that the magnetic force

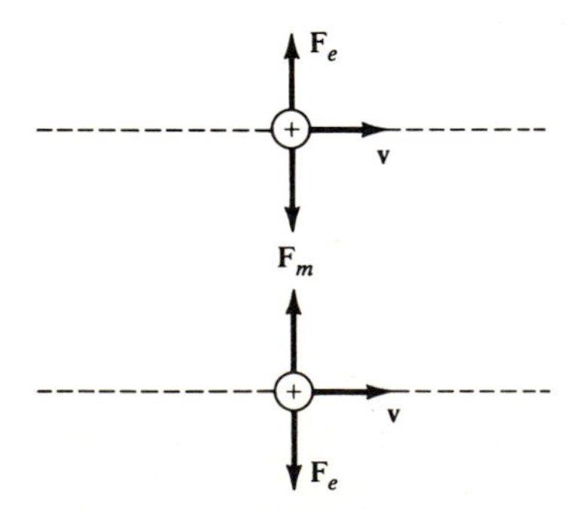

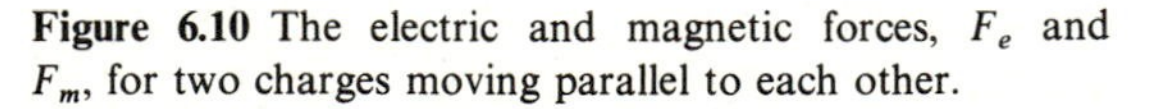
Figure 6.10 The electric and magnetic forces, F_e and F_m, for two charges moving parallel to each other.

† The derivation of $c^2 = 1/\mu_0 \varepsilon_0$ will have to wait until we derive the wave equation for electromagnetic fields from Maxwell's equations.

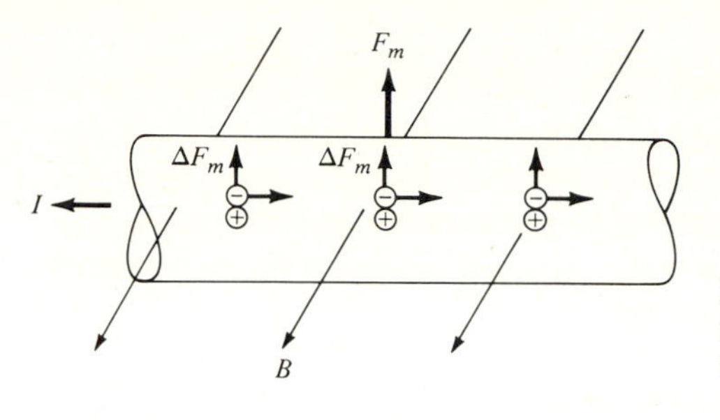

Figure 6.11 Magnetic force ΔF on the conduction electrons in a wire carrying current I. The wire is at right angles to the magnetic field **B**.

between current-carrying conductors can be quite large? After all, the principles of large electric generators and motors are based on postulates 1 and 2, and the currents in the conductors of such devices are electrons that are moving along the wire, as shown in Fig. 6.11. The magnetic force on each electron is shown as the upward force ΔF_m and is transferred to the lattice ions of the conductor, which are indicated by $\oplus$. The conducting wire as a whole is therefore subject to an upward force F_m. To answer the above question, we note that the conducting wire is neutral—the electric forces due to electrons are balanced by those due to the positive lattice ions. Hence the tiny magnetic effects can manifest themselves, and since there are about 5×10^{22} free electrons/cm^3 in a conductor such as copper, the resultant magnetic force can be very sizable.

To give some quantitative data on the magnetic force, let us consider a No. 12 copper wire (0.2 cm in diameter). Since there are 5×10^{22} free conduction electrons, or 8000 C, per cubic centimeter of copper, a 1-cm long No. 12 wire will contain 250 C of negative and a similar amount of positive charge. For purposes of illustration we can say that two such wires 1-m apart would experience an attractive electric Coulomb force of 1.1×10^{15} N (1.2×10^{11} tons) which is balanced by a similar repulsive force. What remains is the magnetic force. If the wire carries a current of 1 A, the drift velocity of the free electrons is approximately 2.2×10^{-5} m/s. Using Eq. (6.12), (6.14), or (6.15), we obtain for the magnetic force $F_m \cong 10^{-11}$ N.

6.4 THE TOTAL FORCE ON A MOVING CHARGE: LORENTZ FORCE

The two postulates of magnetism are expressed in terms of current I which is related to its current density **J** by $I = \iint \mathbf{J} \cdot \mathbf{dA}$. No distinction was made between conduction current ($J = \sigma E$) and convection current ($J = \rho v$) in these postulates. We did however consider each of these currents separately. This was relatively easy to do since the equivalent current element for a moving point charge q is given by $I\, dl = qv$. This relationship can also be derived simply from the definition of current as $I = dQ/dt$. If we have a small discrete charge ΔQ moving in the l direction with velocity v, I is given as $I = \Delta Q/\Delta t = \Delta Q/(\Delta l/v) = v\, \Delta Q/\Delta l$. Therefore $I\, \Delta l = \Delta Q\, v$. Similarly, if an amount of charge $\Delta Q = \rho\, \Delta v_{ol} = \rho\, \Delta A\, \Delta l$ is contained in the volume element $\Delta v_{ol} = \Delta A\, \Delta l$ which is moving in the direction l

with velocity v, current is given as $I = \Delta Q/\Delta t = \rho\ \Delta A\ \Delta l/\Delta t = \rho\ \Delta A\ v$. Therefore the current density J for moving charges can be expressed as $J = \rho v$.

The difference that remains between current in a conductor and current in free space is that the conduction current produces no electric field outside the conductor since the electric forces are neutralized between electrons and ions. On the other hand, moving charges in free space are always subject to electric fields, and the total force on a point charge q is given by

$$\boxed{\mathbf{F} = q(\mathbf{E} + \mathbf{v} \times \mathbf{B})} \tag{6.16}$$

Thus knowing **E** and **B** at all points in space in addition to the initial velocity of q, the Lorentz force relation can be used to predict the future motion of a charged particle. Similarly if we have a charge cloud with charge density ρ at each point, the force on each elemental volume is given by

$$\boxed{\frac{\Delta \mathbf{F}}{\Delta v_{ol}} = \rho(\mathbf{E} + \mathbf{v} \times \mathbf{B})} \tag{6.17}$$

It should be noted that the $\mathbf{v} \times \mathbf{B}$ term is equivalent to an electric field; that is, $\mathbf{E}' = \mathbf{v} \times \mathbf{B}$ since an E field is defined as $E = F/q$.

Point Charge q Moving in Constant E Field

In Sec. 1.7, under the example of electrostatic deflection in an oscilloscope, we showed that a moving charge in a uniform E field follows a parabolic path; i.e., from Eq. (1.41) we have $y = (qE/2mv^2)x^2$. The axis of the parabola is parallel to the direction of E as depicted in Fig. 6.12. If the motion **v** is initially parallel to **E**, the parabola reduces to a straight line path.

Figure 6.4 and the discussion preceding it shows that when a charge q moves in a constant B field, it follows a helical path. The axis of the helix is parallel to the direction of **B**. Again, when the initial velocity **v** of the charged particle is parallel to **B**, the magnetic field has no effect on the motion of the charge since the magnetic force term $\mathbf{v} \times \mathbf{B} = 0$.

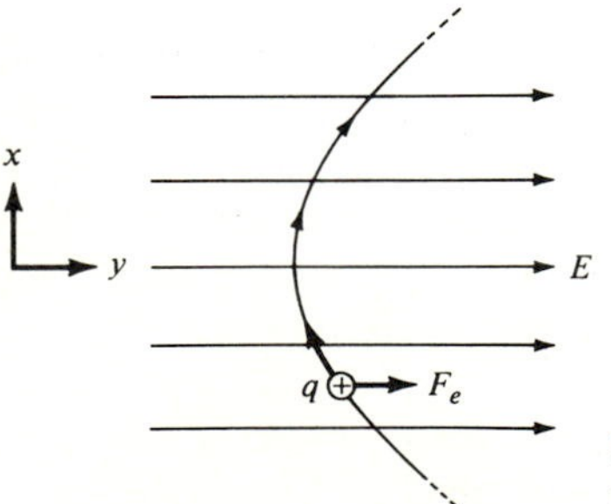

Figure 6.12 A moving charge in a uniform E field follows a parabolic path.

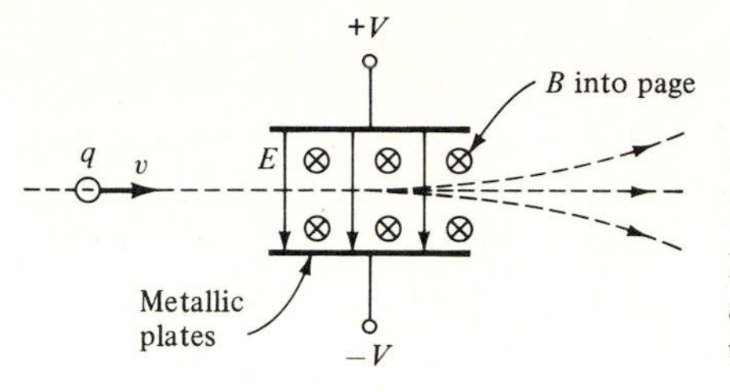

Figure 6.13 Particle q moving at right angles to **E** and **B.** This arrangement can be used to measure the velocity of q.

Point Charge q Moving in Crossed E and B Fields

When the magnetic and electric fields are perpendicular to each other, as shown in Fig. 6.13, we see that the electric field E due to the potential V on the plates deflects a negative point charge $-q$ upward and the magnetic field B deflects it downward. The initial motion of the charge is at right angles to both **E** and **B**; that is, the directions of **v**, **B**, and **E** are mutually perpendicular. If the magnitudes of the uniform E and the uniform B fields are so adjusted that the forces they exert on the moving particles are equal, that is, $F_e = F_m$, the particles pass through the region without deflecting from their original path. The condition for the electric force F_e to equal the magnetic force F_m is obtained from Lorentz' force as

$$0 = q(\mathbf{E} + \mathbf{v} \times \mathbf{B}) \tag{6.18}$$

or

$$qE = qvB \tag{6.19}$$

since the fields are orthogonal. Hence

$$v = \frac{E}{B} \tag{6.20}$$

Therefore, for a given particle speed v, the condition for zero deflection can be satisfied by adjusting E or B according to (6.20). All charged particles of the same speed, irrespective of their mass or charge, will pass undeflected. Such a device can be used as a velocity selector. If we pass a beam of particles of different energies through it, only those particles with speeds given by $v = E/B$ will pass undeflected. Thomson, in 1897, used such a device to measure the q_e/m ratio of the electron as 1.7×10^{11} C/kg. After Millikan's oil-drop experiment was used to determine the mass m of the electron as 9.1×10^{-31} kg, the charge of the electron was found to be $q_e = -1.6 \times 10^{-19}$ C.

If the initial velocity of the particle is not perpendicular to E and B, the trajectory of the particle is more complex, usually a cycloid trajectory. Such motion in the crossed electric and magnetic fields is important in the operation of devices such as the magnetron, a high-power source of microwaves.

Point Charge q Moving in Parallel E and B Fields

Figure 6.14 shows a field where **E** and **B** are parallel. A charged particle with zero initial velocity placed in such a field will begin to accelerate in the direction of the

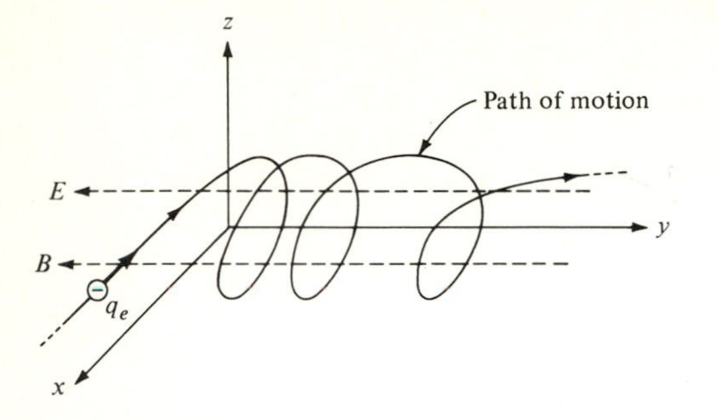

Figure 6.14 An electron entering a parallel electric and magnetic field will be accelerated in the y direction, resulting in a helix with an increasing pitch.

E field. For uniform E the acceleration will be constant, with motion specified by

$$v_y = v_{0y} - at \qquad y = v_{0y}t - \tfrac{1}{2}at^2 \tag{6.21}$$

where $a = qE/m$ is the magnitude of the acceleration and v_{0y} is the initial velocity. The negative sign results from the fact that the direction of acceleration of a negative charge is opposite to the direction of the electric field **E**.

If $-q_e$ moves along the y axis, the magnetic field will exert no force on it. If it enters with a velocity v_x (parallel to the xz plane and at right angles to the E and B fields), it will begin to spiral about the B field. The radius of the circular path is given from Fig. 6.4 as

$$\boxed{r = \frac{vm}{q_e B}} \tag{6.22}$$

However, as the electron is accelerated in the $+y$ direction because of the E field, the path becomes helical with an ever-increasing pitch, as shown in Fig. 6.14. Even though the pitch changes with time, the diameter of the helix remains constant, because neither of the fields can contribute energy to the component of velocity perpendicular to them.

Point Charge q Moving in a Converging B Field

We have shown that charged particles injected in a uniform magnetic field **B** travel at constant speed in a helix wrapped around the **B** field. In the special case when the velocity is perpendicular to **B**, the particles travel in a circular motion about **B** with a radius given by (6.22). When injected parallel to the **B** field, they continue traveling in a straight line unaffected by the magnetic field. It is interesting to observe that charged particles of one kind (like electrons), when injected into a **B** field, complete one loop about **B** in precisely the same time as all other electrons, despite differences in their energies, velocities, or initial direction of their path. In other words, the cycling frequency of all electrons is the same. This can be seen from Fig. 6.4 and Eq. (6.22). Replacing v by $v = \omega r$ in (6.22), where v is actually $v_\perp$ and is the perpendicular component of the particle velocity, we obtain for the

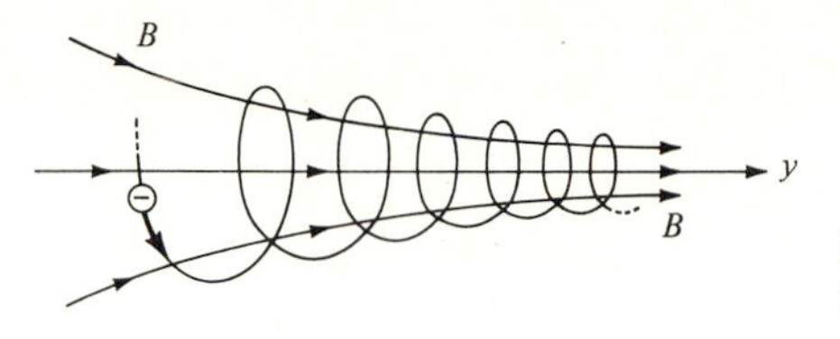

Figure 6.15 Particle motion in a converging magnetic field.

angular velocity

$$\boxed{\omega = \frac{q_e B}{m}} \tag{6.23}$$

Since this expression is independent of particle speed v and orbit r, we conclude that the angular frequency ω for all electrons is the same. Dividing ω by 2π, we obtain $f = \omega/2\pi$, which is called the cyclotron frequency.

Suppose charged particles are injected in a converging **B** field, a **B** field that increases in strength as shown in Fig. 6.15. Since the **B** field increases in the direction of the axis of helical motion of the particles, the radius of orbit r must decrease as given by (6.22) and the cycling frequency (6.23) must increase. As a result, the particles spiral faster and faster in ever-tightening loops. Now, we have observed that a magnetic field alone cannot change the kinetic energy of a charged particle, because a magnetic field causes a force which is always normal to the particle velocity. This was shown in the discussion in the paragraph preceding Eq. (6.6). Therefore, as the spiraling velocity $v_\perp$, which is normal to the increasing B lines increases, the velocity component parallel to B, $v_\parallel$, must decrease because the energy of the particles must remain constant. That is,

$$\mathbf{v} = \mathbf{v}_\perp + \mathbf{v}_\parallel = \text{constant} \tag{6.24}$$

The particle velocity $v_\parallel$ in the direction of the increasing magnetic field will be decreased, and may actually be reduced to zero or reversed. The point in the converging **B** field at which a particle is actually reflected is known as the *magnetic mirror region*. The kinetic energy of the particle remains constant during this reflection process, which means that the backward particle velocity will be equal to the incoming velocity for the same points in space.

The recently discovered Van Allen belts which surround the earth, as shown in Fig. 6.16, are captured particles, trapped by the earth's magnetic field. As these particles (protons and electrons) which are of cosmic origin enter the inhomogeneous earth's magnetic field, they are trapped and spiral back and forth between the two magnetic mirrors formed by the converging B lines near the north and south poles. Note that the magnetic field of the earth is roughly a dipole field. Thus the lines converge in the polar regions of the earth.

The magnetic mirror is also an important principle in the confinement of extremely hot plasmas. These plasmas are so hot that contact with the walls of an ordinary container would melt the material of the container. To confine a plasma,

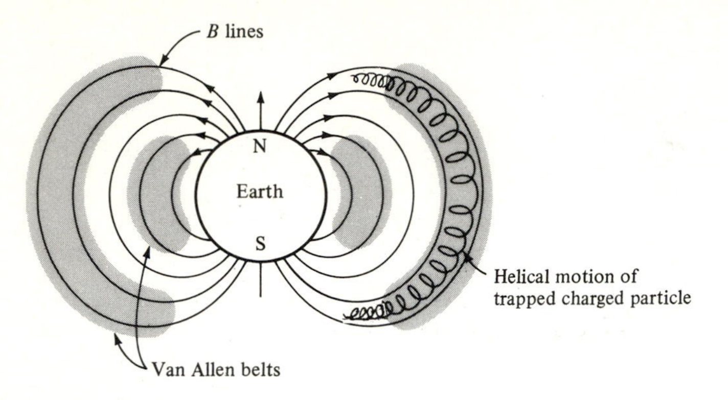

Figure 6.16 The Van Allen belt of charged particles which are captured by the earth magnetic field B.

an inhomogeneous field is used to trap the charged particles in what is known as a *magnetic bottle*. More on this in the next section.

The Pinch Effect in the Confinement of a Plasma

Let us consider a column of moving charged particles, as shown in Fig. 6.17. The streaming particles constitute a current of current density $J = \rho v$ which produces a B field that will surround the column according to the right-hand rule. As the distributed current finds itself in its own magnetic field, the result is, according to postulate I, that a $\mathbf{J} \times \mathbf{B}$ term of force exists on each volume element of the column. This force is always toward the center of the column. From Eq. (6.5) we have then for the pinching force

$$\frac{\Delta \mathbf{F}_{\text{pinch}}}{\Delta v_{ol}} = \mathbf{J} \times \mathbf{B}_{\text{self}} \tag{6.25}$$

where $\mathbf{B}_{\text{self}}$ is the self-magnetic field. The column of streaming particles is in effect "squeezed" or "pinched," hence the name *pinch effect*. This is of considerable interest in thermonuclear work. For example, as shown in Fig. 6.18, the current in the plasma produces a magnetic field which encircles the current and, in turn,

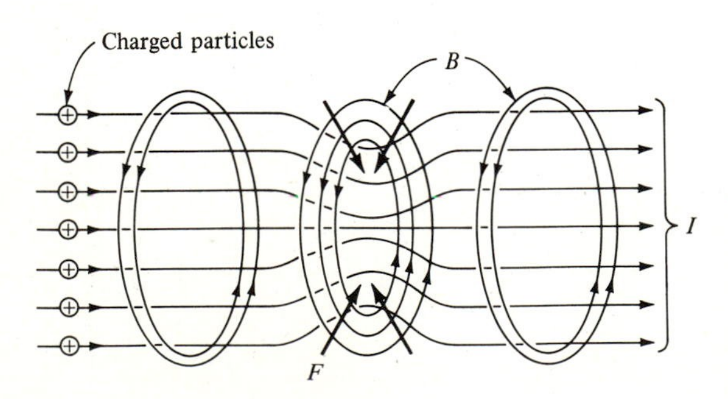

Figure 6.17 The pinching force F on a column of streaming particles. The motion of the charges results in current I which, in the presence of the self-magnetic field B, results in an inward force on the column.

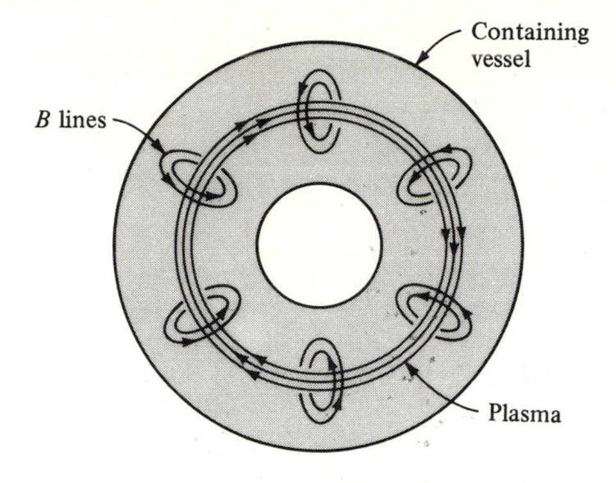

Figure 6.18 A ring-shaped plasma confined to the interior of a doughnut-shaped containing vessel. The plasma can be a rapidly moving ionized gas.

exerts an inward force on the particles which pulls the plasma from the material walls of the vessel. If such a pinched configuration were stable for sufficiently long times, it would be possible to confine very high-temperature plasmas without burning up the walls of the container. Unfortunately such a simple setup as that in Fig. 6.18 has many instabilities and breaks up rapidly since small instabilities are amplified quickly. At every point in the fluid the magnetic pressure is balanced by the fluid hydrostatic pressure in the most delicate way. For example, a *sausage or neck instability*, shown in Fig. 6.17, can result when the magnetic pressure exceeds the hydrostatic pressure. Since at this point the column is squeezed more, resulting in a larger magnetic field (B is inversely proportional to distance from axis of current) which in turn results in stronger pinching, the temperature will rise to such an extent that this section will virtually explode when the hydrodynamic limit is reached. Another instability known as a *kink instability* results when the column kinks slightly. Again, on the inside of the kink, the magnetic field lines are more concentrated, hence, rapidly increasing the kink.

One question that arises immediately, especially in view of our discussion following (6.15), is the effectiveness of the repulsive electric forces between charges in the column. Such repulsive forces would be opposite to the pinching magnetic forces. So as not to introduce too many things at once, we neglected to mention that most plasmas are electrically neutral.† *Plasmas* are ionized gases which are electrically conducting by virtue of their charged particles, but which are electrically neutral because the number of positive charges and the number of negative charges are equal. The lighter charges, usually electrons, are the major contributors to the current; the heavier ions primarily serve to neutralize the plasma. Because of charge neutrality, electrostatic forces can be ignored.

Another medium that is subject to magnetic forces is a conducting liquid, such as mercury. If a voltage is applied across the ends of a long column of mercury, a current will flow in the column. Since the conducting medium is a fluid, the magnetic body or volume force $\mathbf{J} \times \mathbf{B}$ will influence the motion of the fluid. The study of the coupling or interaction of a moving conducting fluid with a magnetic

† Note also the discussion of plasmas centered around Fig. 2.17.

field is known as *magnetohydrodynamics.* It is important in the study of the stars, sunspots, ionized gases, supersonic ionized shock waves, ion motors, and the generation of electricity by passing hot plasmas through a dc magnetic field. Such power generation will be considered further in Sec. 6.13 of this chapter.

6.5 MAGNETIC FIELD OF LONG STRAIGHT WIRE CARRYING A CURRENT

Let us now consider a few simple geometrics of current-carrying conductors and the magnetic field that is associated with them. Perhaps the simplest geometry is that of a long straight wire. The magnetic field surrounds the wire as shown in Fig. 6.19*b*. Of course, an infinite straight wire is not realistic. Also, there would have to be return leads. However, since the results for such infinite structures usually turn out to have a simple form, they are very useful in obtaining approximate answers for wires of finite length. For example, in the region near the middle of a finite wire, the infinite-wire formula gives accurate results† for **B.**

Let us first consider the magnetic field **dB** produced by the current element $I\,dl$ in Fig. 6.19*a*. For the upward current direction shown, **B** to the right of the wire is into the page, and **B** to the left of the wire is out of the page, according to the right-hand rule. The angle between the current **I** and distance **r** is θ. The contribution of the element **dl** to the magnetic field **B** at **r** is, from Eq. (6.6), given by

$$dB = \frac{\mu_0}{4\pi}\frac{I\,dl\sin\theta}{r^2} \tag{6.26}$$

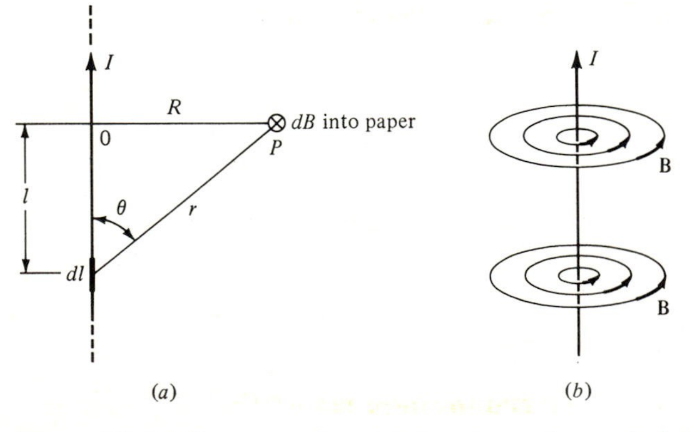

Figure 6.19 (*a*) Geometry of an infinite wire used to calculate the B field; (*b*) the magnetic field surrounds the infinite wire in concentric circles about the wire.

† For an expression of magnetic field from a finite wire, see Prob. 6.7.

Since every current element of the wire produces a field into the paper at point P, an integration of Eq. (6.26) over the entire length of the wire will give the total field at P as

$$B = \frac{\mu_0 I}{4\pi} \int_{-\infty}^{\infty} \frac{\sin\theta \, dl}{r^2} \tag{6.27}$$

To carry out this integration, let us find the relationship between the variables r, θ, l and express them in terms of the single variable θ. Since $l \tan\theta = R$, where R is a constant perpendicular distance from the wire to point P, we have by differentiation that $\cos\theta \sin\theta \, dl = -l \, d\theta$. Substituting this and using $r \sin\theta = R$ and $r \cos\theta = l$ to simplify the integrand, we obtain

$$\boxed{B = \frac{\mu_0 I}{4\pi R} \int_0^{\pi} \sin\theta \, d\theta = \frac{\mu_0 I}{2\pi R} \quad \text{Wb/m}^2} \tag{6.28}$$

which is a magnetic field in the form of concentric circles about the straight conductor. The conductor diameter was assumed to be sufficiently small when compared to R, and so it could be ignored in the above calculation. To get an idea of the magnitude of the magnetic field from an infinitely long wire, let us assume that a current of 10 A flows in the wire. At a distance $R = 2$ cm from the wire, we have

$$B = \frac{(4\pi \times 10^{-7} \text{ Wb/A} \cdot \text{m})(10 \text{ A})}{2\pi(2 \times 10^{-2} \text{ m})} = 10^{-4} \text{ Wb/m}^2 = 1 \text{ G}$$

which is on the order of the earth's magnetic field.

Because the wire is infinite, B drops off as $1/R$, which is slower than the $1/R^2$ from a small current element. The same phenomenon was observed for electric fields. For a point charge, Coulomb's law predicts a fall off inversely as R^2, whereas for an infinite-line charge, (1.25) gives a fall off as $1/R$.

6.6 FORCE BETWEEN TWO PARALLEL WIRES

If we consider two thin infinite wires which are parallel and separated by a distance R, as in Fig. 6.20, we see that each wire will find itself in a magnetic field

$$B = \frac{\mu_0 I}{2\pi R}$$

which is produced by the other wire. To find the force per unit length acting on wire 2, we apply (6.1) and obtain

$$\boxed{\frac{F}{l} = I_2 \frac{\mu_0 I_1}{2\pi R}} \tag{6.29}$$

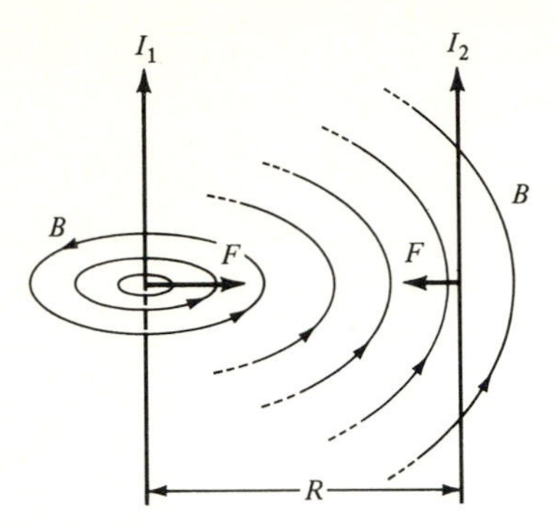

Figure 6.20 The force between two wires. The B field of wire 1 and the current I_2 interact to give a force between the wires according to the FBI rule (see Fig. 6.2).

which is a simple expression because I, B, and F are at right angles to each other for parallel wires. It gives $F = IBl$ for the force. Since the above expression is symmetric in I_1 and I_2, the same force per unit length acts on wire 1. The two conductors experience an attractive (repulsive) force when I_1 and I_2 are in the same (opposite) direction. For example, two conductors, each carrying a current of 100 A and separated by 1 cm, will have a magnetic force per meter on either conductor of

$$\frac{F}{l} = \frac{(4\pi \times 10^{-7}\ \text{Wb/A}\cdot\text{m})(100\ \text{A})^2}{2\pi(10^{-2}\ \text{m})} = 0.2\ \text{N/m}$$

It is instructive to compare the attraction between like currents to the pinch effect on a plasma column which was discussed in Sec. 6.4. Viewing the cross section of two wires, each carrying a current, as shown in Fig. 6.21, we see that the magnetic field at some distance from the two wires surrounds both conductors. Hence one can think of the two wires as two filaments of current in a plasma column which are pinched or forced together by the magnetic force.

6.7 MAGNETIC FIELD FROM A WIRE LOOP

As another example of the use of postulates 1 and 2, let us derive the magnetic field that a thin wire bent into a circular loop will produce. For the sake of simplicity let us find the field on the axis of the loop only. As shown in Fig. 6.22, the plane of

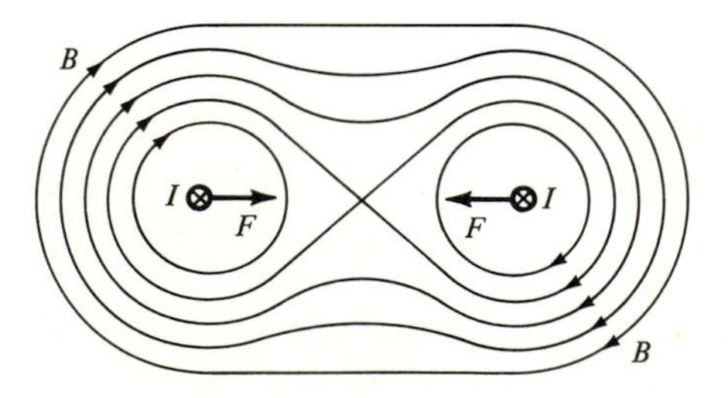

Figure 6.21 Cross-sectional view of two currents I flowing into the paper, and their magnetic field resulting in an attractive force between the wires.

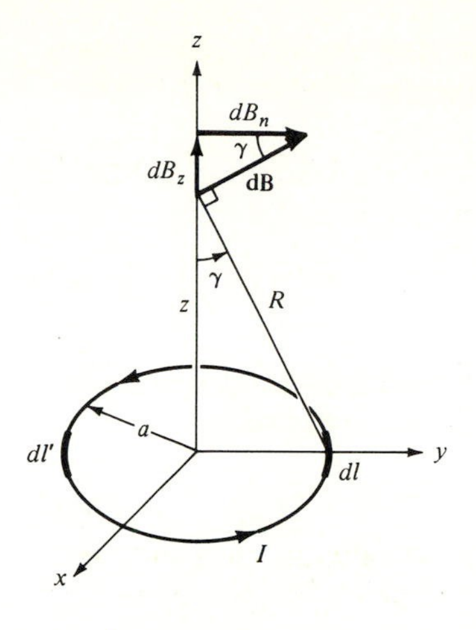

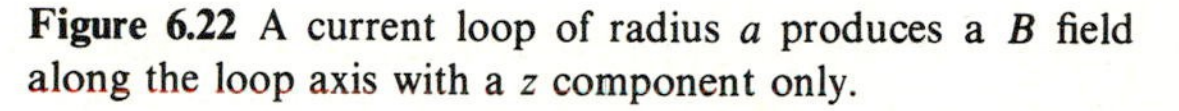

Figure 6.22 A current loop of radius a produces a B field along the loop axis with a z component only.

the loop lies in the xy plane with center at the origin and loop axis coincident with the z axis. The contribution of a small segment dl to the magnetic field on the axis is given from (6.6) as

$$dB = \frac{\mu_0}{4\pi} \frac{|\mathbf{I} \times \hat{\mathbf{R}}|}{R^2} dl = \frac{\mu_0 I \, dl}{4\pi R^2} \tag{6.30}$$

The last term is obtained because $\hat{\mathbf{R}}$ is orthogonal to any $\mathbf{I}\, dl$ in this geometry. The $\mathbf{dB}$ field of current element $I\,\mathbf{dl}$ is at right angles to $\mathbf{R}$, but when we consider the contribution $\mathbf{dB}'$ from a diametrically opposite element dl', we see that the dB_n components which are perpendicular to the axis cancel. As a matter of fact, since this cancellation occurs for all diametrically opposite pairs of dl elements around the loop, only the component $dB_z = dB \sin\gamma$ parallel to the loop axis remains. The total field will be given by summing all dB_z:

$$B_z = \int dB_z = \frac{\mu_0 I \sin\gamma}{4\pi R^2} \oint dl \tag{6.31}$$

where R, I, and γ are taken outside the integral because they are constant. The dl integration around the loop is just $2\pi a$. Expressing the magnetic field in terms of the loop radius a and distance z from the origin, we obtain for the field along the z axis

$$B = B_z = \frac{\mu_0 I a^2}{2(a^2 + z^2)^{3/2}} \qquad \text{Wb/m}^2 \tag{6.32}$$

The magnetic field at the center of the loop can now be obtained by setting $z = 0$ in the above expression. This gives

$$B = \frac{\mu_0 I}{2a} \tag{6.33}$$

For example, Eq. (6.33) can be used to find the field at the center of a short solenoid of N closely packed turns simply by multiplying it by N.

Comparing the expression for the B field at the center of a loop to that from an infinite wire, $B = (\mu_0 I/2\pi a)$, where a is now the distance from the infinite wire, we see that the loop expression is larger than the infinite-wire expression. This is as expected, because when a long wire is formed into a loop, the remote elements which contribute negligibly to the long-wire formula are now at the same distance from the center of the loop as all other elements.

The magnetic field at points on the axis far from the current loop can be obtained by approximating $a^2 + z^2 \cong z^2$ in (6.32). Then for $z \gg a$,

$$B_z = \frac{\mu_0 I a^2}{2z^3} \tag{6.34}$$

To calculate the fields not on the axis of the loop is very laborious and results in complicated expressions. For distances far from the current loop, however, these expressions simplify because the current loop can then be treated like a magnetic dipole. In spherical coordinates, for $r \gg a$, the magnetic field is

$$\mathbf{B} = B_r\hat{\mathbf{r}} + B_\theta\hat{\boldsymbol{\theta}} = \frac{\mu_0 I a^2}{4r^3}(2\cos\theta\,\hat{\mathbf{r}} + \sin\theta\,\hat{\boldsymbol{\theta}}) \tag{6.35}$$

where r is the distance from the origin to a field point and θ is the angle that r makes with the z axis. This expression is derived rigorously in Sec. 8.7. In the above expression for **B**, the ϕ component is absent. This is because the current in the loop flows in the ϕ direction, and the B field must always be at right angles to the current. Also, when $\theta \to 0$, $B \to B_z$, because in spherical coordinates $z = r\cos\theta$. We can now construct the magnetic field lines about a current loop. Figure 6.23 shows a cross section of the loop and the B field around it. Notice the similarity of the field to that of a bar magnet and to the electric field of an electric dipole which is shown in Fig. 4.8.

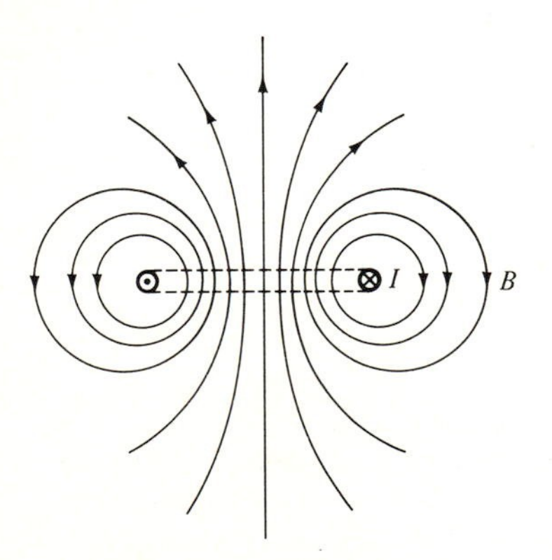

Figure 6.23 Lines of magnetic field surrounding a circular loop of current. The cross represents current flowing into the paper; the dot, current out of the paper.

6.8 FORCE BETWEEN TWO WIRE LOOPS

A current loop immersed in a uniform B field will not experience a net translational force. For example, if the constant B field is at right angles to the plane of the loop, the *FBI* rule gives a force on the loop which at every point on the periphery of the loop is radially outward or inward, depending on the direction of the B field. Similarly, the self-magnetic field of a loop results in an outward force which is trying to stretch the loop into a larger radius loop.

Let us consider now the force between two current loops, each of radius a and separated by a distance d as shown in Fig. 6.24. We shall first find the force dF on the current element $I'\,dl'$ due to the magnetic field B of the other loop. From Eq. (6.1) this is given by $dF = I'\,dl'\,B$. If we make the simplifying assumption that d is small compared to radius a, we can neglect the curvature of the loop and say that the current element $I'\,dl'$ is in a magnetic field of a long wire which, from (6.28), is given by $B = \mu_0 I/2\pi d$. The force on the current element is therefore

$$dF \cong \frac{I'\,dl'\,\mu_0 I}{2\pi d} \tag{6.36}$$

Integrating this around the circumference of the loop, the net force on the loop is given by

$$F \cong \frac{\mu_0 I I'}{2\pi d} \oint dl' = \frac{\mu_0 I I' a}{d} \qquad \text{N} \tag{6.37}$$

This expression [or the equivalent one for two parallel wires, Eq. (6.29)] can be used to define the *ampere* as the unit of electric current in the SI system of units. If we measure length in meters and force in newtons, I will be given in amperes.

The contribution of the current in the wires which are attached to the loops can be neglected if they are kept close together and are twisted. Since the magnetic

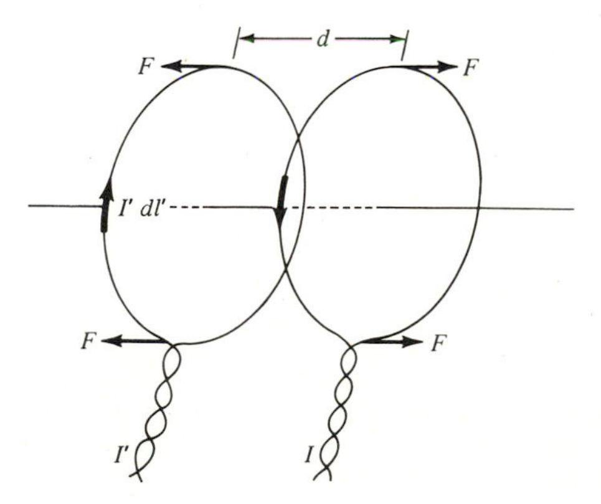

Figure 6.24 The force between two current loops of radius a will be attractive (repulsive) depending whether the currents in the two loops are in the same (opposite) direction.

field lines around each wire are concentric circles, two wires close together with currents in opposite directions will produce a vanishingly small B field due to cancellation.

6.9 MAGNETIC FIELD IN A SOLENOID AND TOROID

A configuration used to produce strong magnetic fields is a helical coil called a *solenoid*. It is shown in Fig. 6.25*a*. To find the field along the axis of the solenoid, we will use the results for a wire loop which were obtained in Sec. 6.7. The same symmetry arguments as were used for the single loop tell us that along the axis of the solenoid only a z component of B exists. Once we know the field on the axis of

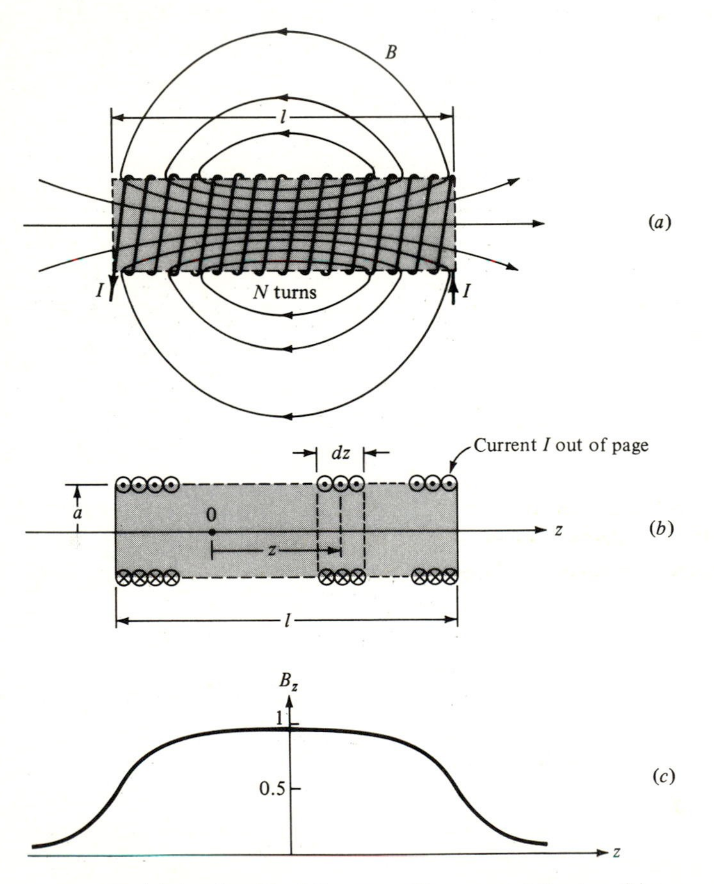

Figure 6.25 (*a*) A solenoid of N turns, with current I flowing; (*b*) a cross section of the solenoid. The total current that flows into or out of the page is NI, and the amount in section dz is $NI(dz/l)$; (*c*) magnetic field on the axis of a four to one length-to-diameter ratio ($l = 8a$) solenoid, relative to the field of an infinitely long solenoid.

a single loop, which from (6.32) is given by

$$B_z = \frac{\mu_0 a^2}{2(a^2 + z^2)^{3/2}} I$$

we can then regard the solenoid as a stack of current loops and calculate the B field using this formula. This approximation becomes more accurate if the turns N are many and are closely spaced. The contribution of a small section of the solenoid of width dz to a point 0 on the axis of the solenoid is, therefore,

$$dB_z = \frac{\mu_0 a^2}{2(a^2 + z^2)^{3/2}} \frac{NI}{l} dz \tag{6.38}$$

where the total current in the small section dz is the proportionate amount of current loops, that is, $NI(dz/l)$. This is shown in Fig. 6.25*b*. The total B field at point 0 can now be obtained by integrating (6.38) over the length of the coil. Choosing 0 at the center of the coil, we have

$$B_z = \frac{\mu_0 NIa^2}{2l} \int_{-l/2}^{l/2} \frac{dz}{(a^2 + z^2)^{3/2}}$$

$$= \frac{\mu_0 NI}{(4a^2 + l^2)^{1/2}} \tag{6.39}$$

For a solenoid that is much longer than its radius ($l \gg a$), we can neglect a with respect to l and obtain for the axial field

$$\boxed{B = \frac{\mu_0 NI}{l}} \tag{6.40}$$

This is a rather simple expression which gives the magnetic field, strictly speaking, at the center of an infinite solenoid. However, at the center of a coil with a 4 to 1 length-to-diameter ratio, the field is nearly as large as that for an infinite coil. Expression (6.40) is, therefore, very useful for engineering purposes.

To obtain the B field at different points, we let 0, which is the origin of the z axis, coincide with the point on the axis of the solenoid at which the B field is desired.† For example, the B field a quarter length inside the coil is obtained by letting the limits in the integral of Eq. (6.39) be $-l/4$ and $3l/4$. For the B field at the ends of the coil, the limits of integration are changed to 0 and l, which gives

$$B = \frac{\mu_0 NI}{2(a^2 + l^2)^{1/2}} \tag{6.41}$$

† For an expression valid anywhere on the axis of the solenoid, see Eq. (9.67) and replace M in that formula by NI/l.

In the case of a long solenoid ($l \gg a$), this can be approximated by

$$B = \frac{\mu_0 NI}{2l} \tag{6.42}$$

Thus at the ends of the coil the B field drops to one-half its value at the center. The picture that we now have for the axial magnetic field of a long coil is that B is fairly constant throughout the coil but drops to one-half its center value at the ends. The axial variation is shown in Fig. 6.25*c*.

The quantity NI/l, which is the total number of ampere-turns divided by the length of the coil, can be treated as a linear current density $K = NI/l$, A/m. As a matter of fact, if the turns consist of fine wire closely spaced, we can consider the solenoid as a single current sheet of width l, carrying a current NI which is bent into a cylindrical form of radius a. Figure 6.25*b* would be a cross section of such a tube if the discrete wires would be fused into a single sheet of current. The B field at the center of such a cylindrical tube would be $B \cong \mu_0 K$.

The Short Coil

We can use (6.39) to obtain the B field of a many-turn solenoid that is much shorter than its radius ($l \ll a$), as shown in Fig. 6.26. Neglecting l with respect to a, we obtain

$$B = \frac{\mu_0 NI}{2a} \qquad l \ll a \tag{6.43}$$

as the magnetic field at the center of such a coil. Note that this is the B field at the center of a wire loop, multiplied by the number of turns N. From (6.33) we have that $B = \mu_0 I/2a$; multiplying by N, (6.43) follows. This serves as an independent check on the solenoid formula (6.39).

The Toroid

A *toroidal coil* is a long solenoid bent into a circular shape, as shown in Fig. 6.27. It is a useful geometric shape and, like the solenoid, finds wide application. Because of its symmetry, all the magnetic field is within the coil (not completely true unless the winding is assumed to be a current sheet).

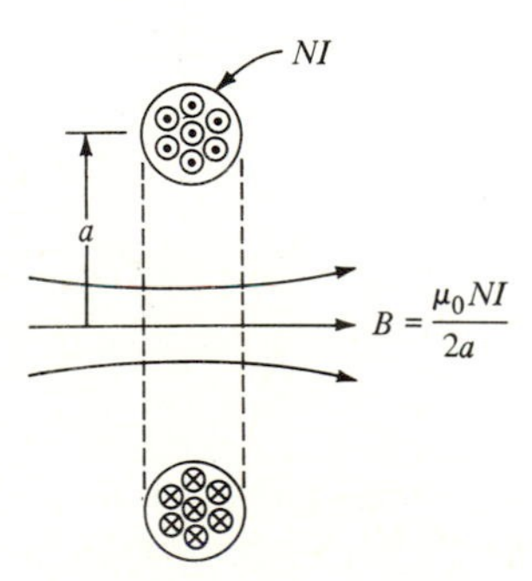

Figure 6.26 The cross section of a short coil.

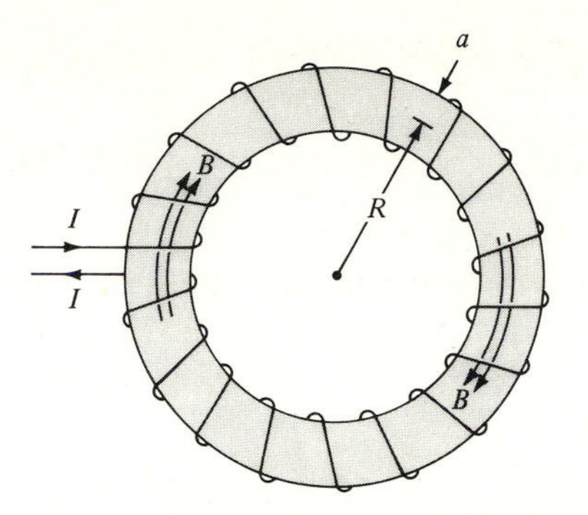

Figure 6.27 A toroid of N turns carrying current I.

To find the B field, we simply take the long-solenoid formula (6.40) and substitute the mean circumference of the toroid, which is $2\pi R$, for the length l of the solenoid. Thus

$$B = \frac{\mu_0 NI}{2\pi R} \tag{6.44}$$

This completes the study of fundamental shapes that produce magnetic fields by currents. To recap, we have derived the B field from a long wire, a wire loop, a solenoid, and a toroid.

6.10 TORQUE ON A WIRE LOOP: THE MAGNETIC DIPOLE

Let us now return to the study of forces that act on current-carrying conductors which are immersed in constant magnetic fields. As observed at the beginning of Sec. 6.8, a wire loop in a uniform B field experiences no net translational force but will experience a rotational force. The understanding of this torque is of utmost importance, as it not only explains the basis of electric motors and generators, but also explains some interaction of fields with matter (we will show later that spinning electrons in material bodies are equivalent to circulating currents).

To begin with, consider a rectangular loop with its axis perpendicular to a uniform B field, as shown in Fig. 6.28*a*. The current I that flows in the wires of length l produces a torque on the loop which tends to rotate it clockwise for the direction of B shown. Since *torque* is defined as force times moment arm, where moment arm is $d/2$ and the force is $F_n = F \sin\theta$, we obtain for the torque

$$T = 2F_n \frac{d}{2} = IBld \sin\theta \tag{6.45}$$

where according to Eq. (6.1) the force F on the wire of length l is $F = |\mathbf{I} \times \mathbf{B}l| = IBl$. Observing that the area of the loop $A = ld$, we can write for torque

$$T = IAB \sin\theta \tag{6.46}$$

The product of current times area is defined as the *magnetic moment* $m = IA$. It is a vector and has the direction of the normal to the area A. Thus the magnetic

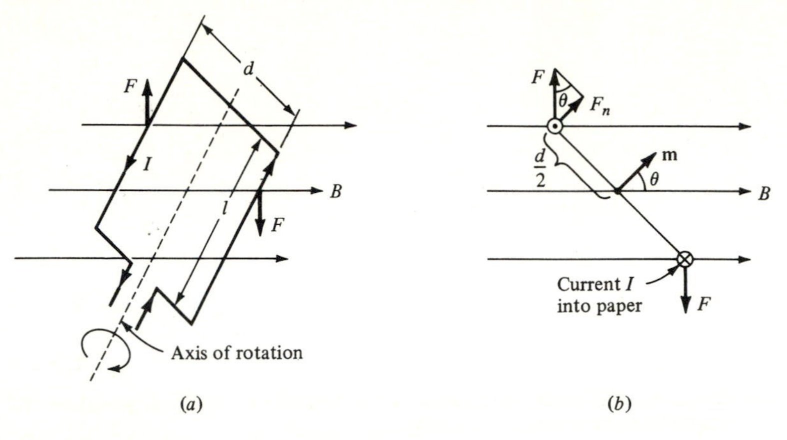

Figure 6.28 (*a*) A rectangular loop of area $A = ld$ carrying a current I; (*b*) cross section of the loop showing direction of magnetic moment **m**.

moment is

$$\boxed{\mathbf{m} = I\mathbf{A} = IA\hat{\mathbf{n}}} \tag{6.47}$$

where $\hat{\mathbf{n}}$ is the unit vector of A and is in the direction of the right-hand thumb if the fingers of the right hand are curled in the direction of the current I circulating in the loop (right-hand rule). The relation between the direction of **m** and the current I for a circular loop is shown in Fig. 6.29.

Torque is a vector quantity and is in the direction of the axis of rotation. We can express torque in a general form using the vector product as

$$\boxed{\mathbf{T} = \mathbf{m} \times \mathbf{B} \qquad \text{N} \cdot \text{m}} \tag{6.48}$$

For the rectangular loop shown in Fig. 6.28*b*, the direction of torque is into the paper. We see now that the magnetic torque on the loop turns it in the clockwise sense until $\hat{\mathbf{m}}$ aligns with B, at which point $\theta = 0$ and the torque becomes zero. The magnitude of the torque is expressed by (6.46).

Although (6.47) and (6.48) were derived for a rectangular loop, they correctly give the magnetic moment and torque for a current loop of any shape. The magnetic moment depends only on IA and not on the shape. To see this, we can take an arbitrarily shaped plane loop and approximate it by a sum of rectangular loops as shown in Fig. 6.30. Since the currents in the adjacent long legs of the

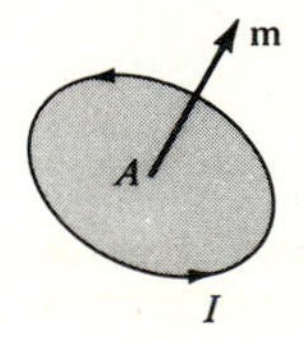

Figure 6.29 A current loop of area A. Magnetic moment **m** is normal to the surface and is related to the current direction through the right-hand rule.

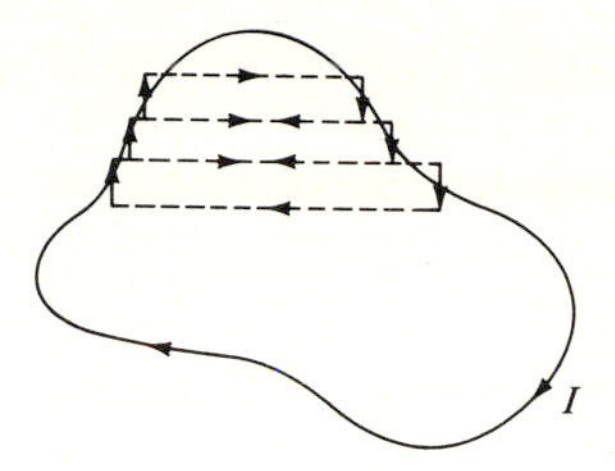

Figure 6.30 The conducting wire of an arbitrarily shaped loop is approximated by small straight segments.

small rectangles cancel each other, only the peripheral currents remain, which approximate the original loop as closely as desired. If we add the torques on the individual rectangles, the sum will be the total torque on the loop and therefore depends only on the total area and current of the loop.

Magnetic Dipole

A small current loop is called a *magnetic dipole*. Small, as used here, means that the loop appears small at observation distances which are large in terms of loop diameters. We observe that the behavior of the magnetic dipole m in a magnetic field B is the same as that of the electric dipole p in an electric field E. For example, the torque on the electric dipole is given by (4.3) as $\mathbf{T} = \mathbf{p} \times \mathbf{E}$, which is the same as (6.48) with $\mathbf{p}$ replaced by $\mathbf{m}$ and $\mathbf{E}$ replaced by $\mathbf{B}$. A comparison of Figs. 4.5 and 6.28*b* shows this readily. As a matter of fact the analogy goes much deeper than this: The shape of the field lines for both dipoles, as shown in Fig. 6.31, is the same; both dipoles when immersed in uniform fields experience no resultant force, and both electric and magnetic dipoles are subject to resultant forces in nonuniform electric and magnetic fields, respectively. Using Eq. (4.7), the net force on a magnetic dipole is

$$\mathbf{F} = (\mathbf{m} \cdot \nabla)\mathbf{B} \tag{6.49}$$

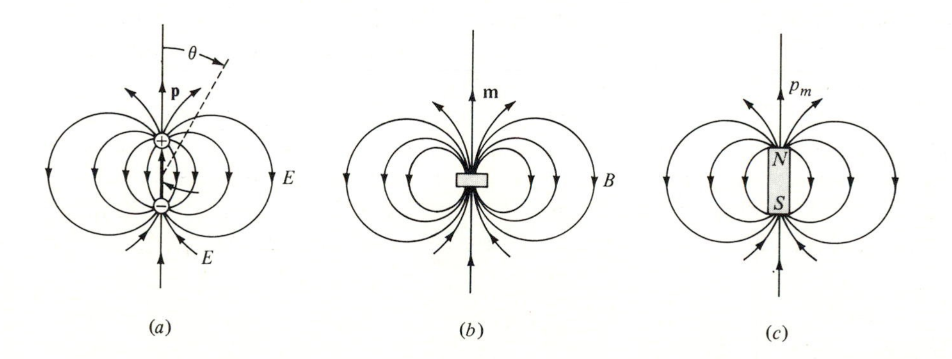

Figure 6.31 (*a*) The electric field lines around an electric dipole $\mathbf{p} = q\mathbf{d}$; (*b*) the magnetic field lines around a magnetic dipole $\mathbf{m} = I\mathbf{A}$; (*c*) a bar magnet is equivalent to a magnetic dipole of moment $p_m = q_m d$.

As in the electrical case, magnetic dipoles are attracted toward regions of stronger fields. The potential energy of an electric dipole in an electric field was derived in Eq. (4.5) as $W = -\mathbf{p} \cdot \mathbf{E}$. Similarly the potential energy of a magnetic dipole in a magnetic field is given by

$$W = -\mathbf{m} \cdot \mathbf{B} \tag{6.50}$$

The maximum potential energy mB occurs when the dipole is aligned against the magnetic field, whereas the minimum energy $-mB$ occurs when the dipole moment and the B lines are aligned.

To show that the shape of the field lines around each dipole is the same, we first observe that the electric field on the axis of an electric dipole p is given by $E = E_r = E_z = p/2\pi\varepsilon_0 r^3$. This is obtained from Eq. (4.14) by letting $\theta = 0$ and noting hat $r = z$ on the z axis. The equivalent expression for the magnetic dipole can be obtained from (6.32) by letting $z \gg a$ which gives

$$B_z = \frac{\mu_0 I a^2}{2r^3} = \frac{\mu_0 m}{2\pi r^3} \tag{6.51}$$

for the far field on the axis of a circular loop, where the magnetic moment for a circular current loop is $m = I\pi a^2$. Note the similarities between E_z and B_z. Without proving it (it will be shown rigorously in Sec. 8.7), we can now use Eq. (4.14) to state the magnetic field about a magnetic dipole as

$$\boxed{\mathbf{B}(\mathbf{r}) = \frac{\mu_0 m \cos\theta}{2\pi r^3}\hat{\mathbf{r}} + \frac{\mu_0 m \sin\theta}{4\pi r^3}\hat{\boldsymbol{\theta}}} \tag{6.52}$$

For comparison we have drawn the E lines for an electric dipole and the B lines for the magnetic dipole in Fig. 6.31*a* and *b*, respectively. The only difference between the fields is that the electric lines leave and terminate on charges, while the magnetic lines are continuous loops. This difference is not revealed in (6.52) and (4.14) because these two expressions are valid only for observation points r which are far from the dipoles, that is, $r \gg a$ and $r \gg d$ for the magnetic and electric dipoles, respectively. The distance restriction was applied because it allowed the use of mathematical approximations which resulted in the relatively simple expressions (4.14) and (6.52) for the dipole fields.

Traditional View of Magnetism

In the introduction we stated that magnetic fields are the results of charges in motion. One might call this the modern viewpoint. The early theory of magnetism was based on concepts similar to those used in electrostatics and involved the notion of isolated magnetic charge, or, in other words, a magnetic monopole. Even though such monopoles do not exist in nature, a formulation using monopoles can simplify problems involving permanent magnets. Many students introduced to magnetism seem to prefer to think in terms of magnetic charges residing at the ends of bar magnets.

There is a similarity in the behavior of a permanent magnet, such as a compass needle, to that of a dipole. We have drawn the magnetic field of a small bar magnet in Fig. 6.31*c* and observe that it is a dipole field. The magnetic effects seem to be concentrated at the ends of the bar magnet, just as if it had two magnetic charges there which would give the bar magnet a dipole moment of $p_m = q_m d$, where d is the length of the bar and q_m is an isolated magnetic charge. If we immerse such a magnet in an external B field, it will experience a magnetic torque

$$\mathbf{T} = q_m \mathbf{d} \times \mathbf{B} \tag{6.53}$$

tending to align it with the B field. Just as an electric charge can be acted upon by a force $\mathbf{F} = q\mathbf{E}$, the magnetic charge on each end of the bar magnet can be considered to be acted upon by a force

$$\mathbf{F} = q_m \mathbf{B} \qquad \text{N} \tag{6.54}$$

The above expression allows us to define magnetic field as $B = F/q_m$, similar to that in electrostatics. Comparing (6.54) to (6.3), we find that q_m must be equal to $q_m = I\,dl$, which also determines the dimensions of magnetic charge. The torque acting on a bar magnet is depicted in Fig. 6.32.

A bar magnet and current loop are equivalent if a given B field produces the same torque on each. Thus equating the torques, we have

$$mB = p_m B \tag{6.55}$$

which gives

$$IA = q_m d \tag{6.56}$$

for equivalence. Let us state again that magnetic charges do not exist in nature; at least to date the most subtle experiments were not able to detect the presence of magnetic charge. All magnetic effects are attributable to electric charges in motion. That is also why magnetic lines never end but always close upon themselves. If they were to end, they would have to terminate on magnetic charge. In conclusion, we can say that even though the magnetic effects of a permanent magnet are caused by the equivalent circulating currents of the spinning electrons, the concept of magnetic charge is a fruitful approach in the understanding of magnetism.

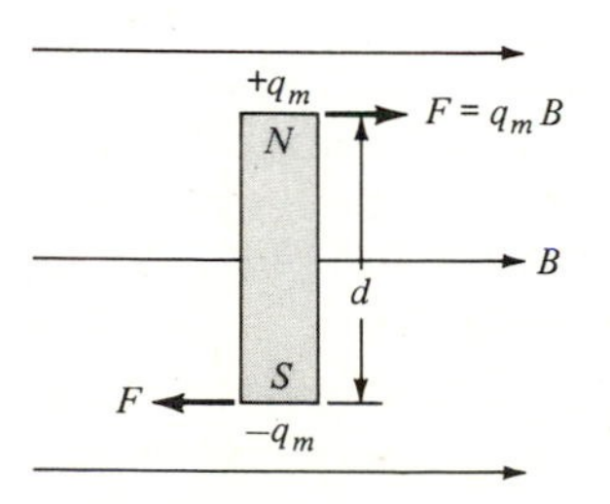

Figure 6.32 The torque on a bar magnet with magnetic moment $q_m d$, immersed in an external B field.

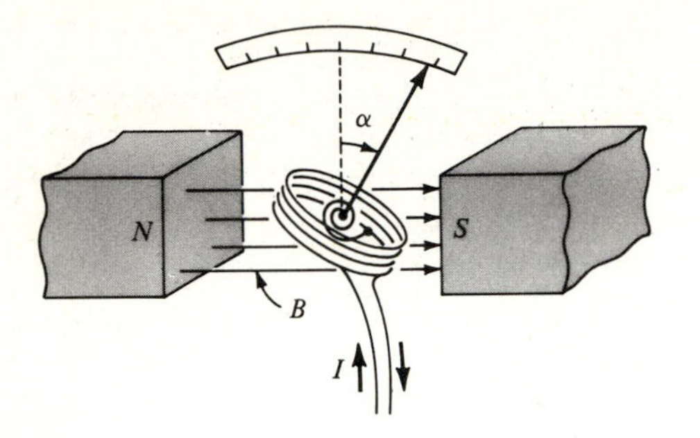

Figure 6.33 A galvanometer moving-coil is placed between magnet poles. The indicator needle will come to rest when the magnetic torque equals that of the restoring hair spring.

Example: Galvanometer Most voltmeters and ammeters use a *galvanometer*, also known as a d'Arsonval movement, which is a moving-coil instrument for measuring current. As shown in Fig. 6.33, it consists of a magnetic field produced by a permanent magnet and a coil of N turns, which is free to rotate against the restoring torque of a hair spring. The rotating torque is provided by the current I that flows through the coil and is to be measured. The torque produced by the movement is $Bm \cos \alpha$, where the magnetic dipole moment is $m = NIA$ and α is the angle of rotation from the zero-current equilibrium position. The restoring torque $k\alpha$ of the hair spring, where k is the spring constant, will stop the rotation when $k\alpha$ becomes equal to the magnetic torque. To obtain the equilibrium position α with current I flowing, we equate the two torques,

$$k\alpha = BNIA \cos \alpha \tag{6.57}$$

If the deflection angle α is small, $\cos \alpha \cong 1$, and the deflection α is directly proportional to the current I. Thus when no current flows, $\alpha = 0$, and the hair spring positions the plane of the coil parallel to the magnetic field. A current in the loops causes the plane of the coil to rotate toward the perpendicular to the B field against the restoring torque of the hair spring.

6.11 MOTORS AND GENERATORS: MOTIONAL EMF

The ideas developed in the last section can now be applied to obtain motor action. Let us first consider a simple dc motor as shown in Fig. 6.34. It consists of a loop, usually of many turns of wire, called an *armature* which is immersed in the uniform field of a magnet. In practice, small motors use a permanent magnet, whereas larger ones use an electromagnet to establish the constant B field. The armature is connected to a commutator which is a divided slip ring. The purpose of the commutator is to reverse the current at the appropriate phase of rotation so that

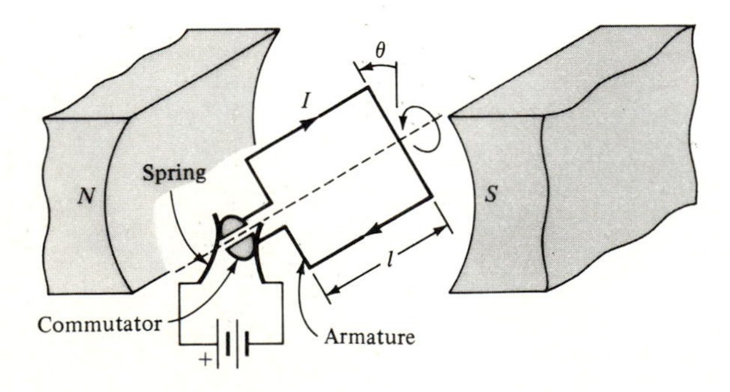

Figure 6.34 The elements of a simple dc motor. The commutator reverses the current in the armature so that the torque on the armature remains in the same direction.

the torque on the armature always acts in the same direction. The current is supplied from a battery through a pair of springs or brushes which rest against the commutator. The torque of the motor, using (6.45) or (6.48), can be given as

$$T = NBIA \sin \theta \tag{6.58}$$

where N and A are turns and area of the armature, respectively. When the angle $\theta = 0$, no current flows through the armature, since the commutator is just about to reverse the current and, in the process, momentarily shorts the battery. Once the motor has started, the rotational inertia carries the armature through the zero-torque region.

DC Generator

It is easy to make a dc generator of the motor shown above simply by removing the battery and substituting a load resistance R in place and providing an external rotary force on the armature. In the discussion following (6.4), we have shown that when a wire of length l moves through a uniform B field, a voltage V called a *motional emf* will appear across its ends. This voltage, from (6.4), is

$$V = \int \mathbf{E} \cdot \mathbf{dl} = \int \frac{\mathbf{F}}{q} \cdot \mathbf{dl} = \int \mathbf{v} \times \mathbf{B} \cdot \mathbf{dl} \tag{6.59}$$

For a straight wire of length l moving with uniform velocity at right angles to the magnetic field, we have $V = vBl$ for the induced voltage.

Another way to derive this equation is to consider the force expression on a current-carrying wire in a magnetic field, which is $F = IBl$ (see geometry shown in Fig. 6.3*a*). Let us use this expression in a different sense (see Lenz' law) by saying that if a force F is applied to the wire, a current I will flow. Now, if the wire is forced to move a small distance ds at right angles to B and l, the incremental work done will be $dW = F\,ds$. But voltage is defined as work per unit charge, that is, $V = dW/dq$, which gives $V = IBl\,ds/dq$. Since current is the time rate of charge flow, $I = dq/dt$, we have $V = (dq/dt)Bl(ds/dq)$. Canceling the dq's and identifying ds/dt as the velocity of the piece of wire, we obtain the desired result, $V = Blv$.

In our simple generator the parts of the armature that produce the voltage are the two wires of length l. The armature rotates in the B field which gives for the voltage across the spring connectors:

$$V = |2vBl \sin \theta| = |\omega BA \sin \omega t| \tag{6.60}$$

where $\theta = \omega t$ and the angular velocity ω in radians per second is related to linear velocity v by $v = \omega d/2$. See Fig. 6.28 for dimensional notation and note that the area of a loop is $A = ld$. The absolute sign is the result of the commutator reversing the voltage every half cycle. Therefore, as the armature coil is rotated, it supplies a pulsating but unidirectional voltage across the load resistance R. The current that flows through R is given by $I = V/R$. The generated voltage in our simple generator is not very smooth, but is nevertheless dc voltage. In practical generators one has many, equally spaced loops in the armature, as shown in

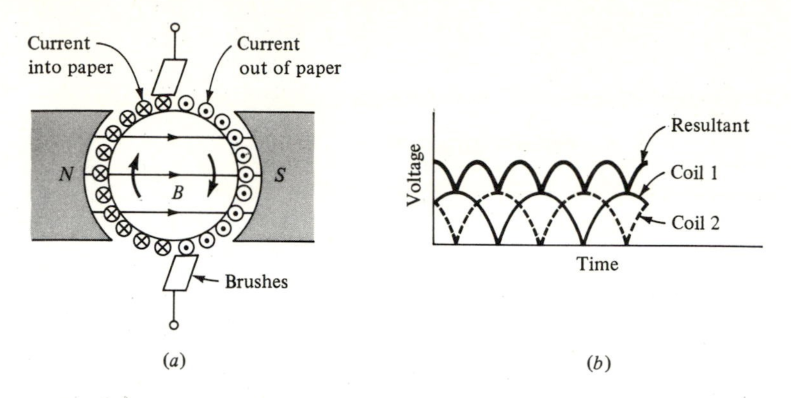

Figure 6.35 (*a*) A multi-coil armature in a bipolar generator. More coils or loops in the armature, equally spaced, produce a smoother dc voltage; (*b*) example of reduction of voltage fluctuations with two coils, spaced at right angles to each other and connected in series.

Fig. 6.35. Combining this with a commutator (not shown) which consists of many segments will give a steady dc voltage with a small amount of ripple. The commutator keeps the current in the left-hand belt of conductors directed away from the observer (indicated by the cross) and that in the right-hand belt directed toward the observer.

Example A bipolar generator driven at 1200 r/min has 60 armature coils arranged in four parallel paths. The magnetic field from one pole to the other multiplied by the area of each loop is given as $BA = 0.05$ W. Since

$$1200\frac{\text{r}}{\text{min}} = 1200\frac{\not{\text{r}}}{\not{\text{min}}}\ \frac{1\ \not{\text{min}}}{60\ \text{s}}\ \frac{2\pi\ \text{rad}}{1\ \not{\text{r}}} = 126\frac{\text{rad}}{\text{s}} = \omega$$

we have from (6.60) for the voltage induced in each bank of 15 series-connected coils, $V = (15) \times (126)(0.05) = 95$ V. This voltage is somewhat high since the generated voltage per loop which we have used in this calculation is not the maximum as given by Eq. (6.60). The reason for paralleling is to obtain a larger current capacity for the generator.

AC Generator

The expression for induced voltage when a conductor moves or cuts through a B field, also referred to as motional emf, was derived in Eq. (6.59). It forms the basis of operation of all types of electric generators. The alternating-current generator is one in which current in a circuit that is connected to it traverses the circuit first in one direction and then the other. An elementary ac generator is shown in Fig. 6.36. The induced emf appears at the terminals which are connected to slip rings. The loop rotates with uniform angular velocity ω in radians per second. Since in place of the commutator we now have two slip rings, the voltage that will be produced is, from (6.60),

$$V = \omega BA \sin \omega t \qquad (6.61)$$

and is seen to be sinusoidally varying as shown in Fig. 6.36. In place of a permanent magnet we have two coils connected to a battery. The current that flows in

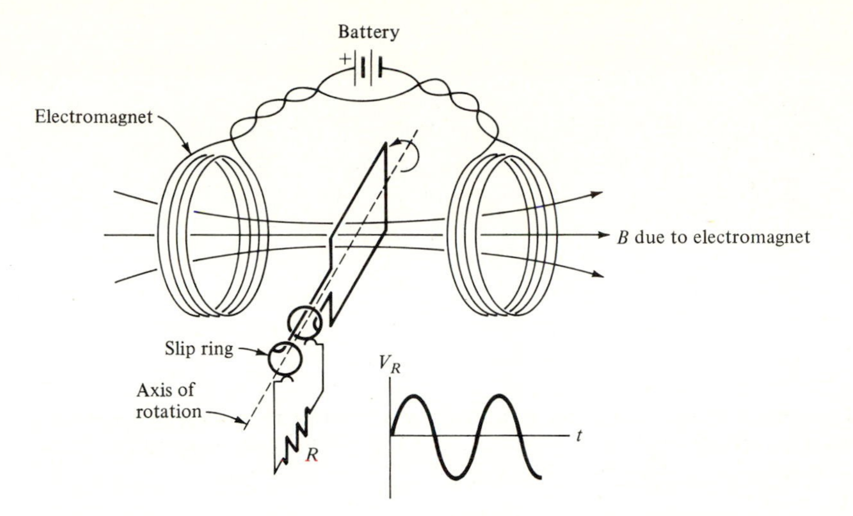

Figure 6.36 A simple ac generator. The two coils produce a uniform magnetic field B in the vicinity of the armature loop. In the loop, rotating with angular velocity ω, a sinusoidally varying motional emf is induced which appears across the load resistor R as V_R.

the multiturn, short solenoidal coils produces a faily uniform B field in the vicinity of the armature.

Counter EMF. Lenz' Law

We have considered the principles of simple generators and motors. But have we not forgotten something? Since these devices are energy converters that convert mechanical to electric energy in a generator and vice versa in a motor, a look at the energy balance will lead to the discovery of back emf, or counter emf. First, the machine shown in Fig. 6.34 can act either as a motor or as a generator. This reciprocity can be demonstrated simply by observing that turning the shaft of such a machine will generate a voltage V across a load resistor R connected between the commutator springs. The power consumed in the resistor is given by $P = V^2/R = I^2R$, and at least as much mechanical power must be supplied to the shaft. But what is the mechanism in the generator that makes it accept the mechanical power and convert it to I^2R? Similarly, applying a voltage V to the commutator springs of the machine in Fig. 6.34 turns the machine into a motor with a torque at the shaft given by (6.58). Since power is the time rate of doing work, and work is torque times angle, that is, $P = dW/dt = dT\theta/dt$, we have from (6.58)

$$P = \frac{d}{dt} T\omega t = NBIA\omega(\omega t \cos \omega t + \sin \omega t) \tag{6.62}$$

where $\theta = \omega t$ with ω assumed constant. This is a surprising result which says that the power of the motor will continue to increase with time t. Equation (6.62) therefore cannot possibly represent a realistic situation.

To clarify some of these points, we observe now that when an armature is rotating, motor and generator action is present in the armature simultaneously. For example, in the case of the motor, as the armature rotates because of motor action, the conductors cut magnetic field and an emf is induced in the conductors, exactly as in the armature of a generator. To determine the direction of this emf, suppose the additional currents in the armature conductors flow as shown in Fig. 6.34. The additional force on the loop is acting in a direction to assist the motion of the armature loop. The armature would therefore speed up without any additional power input from the battery, or, better yet, one could disconnect the battery entirely, and the armature would continue to rotate by virtue of this induced current. Clearly this cannot be. As a matter of fact, the induced emf is such that this current is opposite to that due to the battery. As the motor armature rotates, a counter, or back, emf is induced in its armature with a magnitude proportional to its shaft speed. Its value can be calculated in the same way as for a generator, that is, from Eq. (6.60). Thus, when a source of voltage V is connected to a motor, the counter emf V_c will reduce the applied voltage to the value $V - V_c$ which is now effective in producing the current in the armature. If the armature resistance is R, the current in the armature will be

$$I = \frac{V - V_c}{R} \tag{6.63}$$

Increasing the mechanical load on a motor will cause it to slow down, thereby reducing the counter emf. As a result, the armature current I will increase, developing a larger torque. The motor, therefore, accepts more power from the battery, and the increased armature current drives the greater load at a slightly reduced speed.

In the case of the generator, we find that, just as a motor when turning develops a counter emf, a generator which is delivering current develops an opposing torque. The armature current of a generator under load, being in a magnetic field, sets up a torque which is opposite in direction to that of the applied torque. This breaking torque must be overcome by the driving source. The phenomenon that induced current always counters applied current is an essential physical fact and is just one more manifestation of the tendency of systems to resist change. In this particular case it is traditionally called *Lenz' law*, which says that the direction of the induced emf is such as to oppose any change in current; induced current is opposite to applied current when the applied current is increasing, and is in the direction of the applied current when applied current is decreasing.

Linear Motor and Generator†

As a further example of the two basic concepts (force on a current-carrying conductor and motion-induced voltage in a conductor) which are the basis of

† E. R. Laithwaite and S. A. Nasar, Linear-Motion Electrical Machines, *Proc. IEEE*, April 1970.

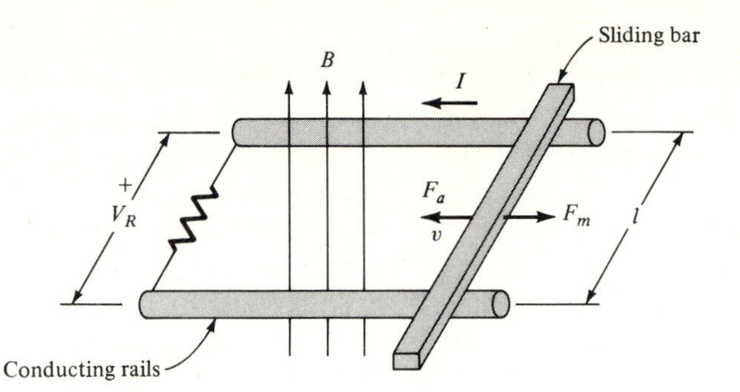

Figure 6.37 A linear generator consisting of a "loop" at right angles to a B field. If force F_a is applied, the bar moves to the left; as a result an induced emf will appear across the load resistor R.

electromechanical systems, let us consider a linear generator as shown in Fig. 6.37. A sliding bar, two conducting rails, and load resistor R form a closed loop which is threaded by a uniform B field at right angles to the loop. If a force is applied to the sliding bar, a voltage V_R will be generated across R. A linear motor which produces a force on the sliding bar is obtained with the same configuration, simply by replacing R with a battery. This example also demonstrates Lenz' law in a clear way, since in the generator mode the induced emf produces a force which opposes the applied force on the bar. Similarly, in the motor mode the induced emf opposes the applied emf and thus reduces the current in the loop which in turn reduces the force on the sliding bar.

Generator. If a force F_a is applied which moves the bar to the left, as shown in Fig. 6.37, the induced voltage from (6.59) is

$$V_R = vBl \tag{6.64}$$

which appears across the resistor R because the resistance of the sliding bar and conducting rails is assumed to be negligible. The polarity of V_R and the direction of I is as shown. Because of the induced current I that exists in the circuit, the sliding bar experiences a magnetic force F_m, which from (6.1) is given by

$$F_m = IBl \tag{6.65}$$

and is directed opposite to the applied force F_a. To keep the bar in motion, a force F_a, equal and opposite to F_m, must be applied. Since the induced current is $I = V_R/R = vBl/R$, the applied force must be equal to

$$F_a = F_m = IBl = vB^2l^2/R \tag{6.66}$$

Noting that power is the time rate of doing work, $P = dW/dt$, we find that the mechanical work done per unit time by this force in pushing the sliding bar is

$$P = F_a v = \frac{v^2B^2l^2}{R} \tag{6.67}$$

But this is equal to the power dissipated in the load resistance which is $P = I^2R$, as the induced current is related by $IR = vBl$. Thus, when the mechanical force F_a is larger than the opposing magnetic force F_m, energy is converted from mechanical to electric.

Motor. Let us now consider the linear motor. If we replace the load R with a battery of voltage V having polarity opposite to that shown on V_R, the sense of current I will remain as shown in Fig. 6.37. We now have a linear motor in which the bar will move in the direction of F_m. Using Kirchhoffs voltage law to write the loop equation for the closed circuit of Fig. 6.37, we have $V - V_c - IR = 0$, where V_c is the induced counter emf just as in Eq. (6.63). The resistance R, in this case, is assumed to be concentrated primarily in the sliding bar l. The magnetic force F_m that accelerates the bar is given by

$$F_m = IBl = \frac{V - V_c}{R} Bl = \frac{V - vBl}{R} Bl \tag{6.68}$$

where the counter emf is given by $V_c = vBl$. Note that v is now in the direction of F_m. At the start when $v = 0$, the accelerating force F_m on the bar, which is assumed to have a mass m, is the greatest. As soon as the bar begins to move, a voltage will be induced in the circuit that reduces the current through the bar. In addition to this we have the frictional force between the rails and the sliding bar. Both of these will reduce the accelerating force to a value at which the bar achieves an equilibrium velocity. If we treat the sliding bar as a free body, summing the forces on it will give the equation of motion for it. Thus from Newton's second law we have

$$m \frac{dv}{dt} = F_m - F_{\text{fric}} \tag{6.69}$$

where F_m is given by (6.68). After the proper friction force for the problem under consideration is chosen, we can solve the simple differential equation and thus characterize the motion of the bar.

The energy-balance equation can be obtained from Kirchhoff's loop equation for voltage, $V - V_c - IR = 0$, by multiplying the loop equation by current I. This gives

$$VI = V_c I + I^2 R \tag{6.70}$$

where the left side is the energy supplied by the battery at the rate of VI; the motor produces mechanical work at the rate $V_c I$; and energy is dissipated as heat in the resistance of the circuit at the rate I^2R. If we replace I by using $F_m = IBl$ and the counter emf by $V_c = vBl$ in the $V_c I$ term of (6.70), we obtain another expression for the energy balance per unit time which is

$$VI = F_m v + I^2 R \tag{6.71}$$

The mechanical power is now given by the $F_m v$ term, which is similar in nature to the equivalent term for the linear generator as given by (6.67). The frictional energy dissipation can be lumped with the I^2R term.

6.12 EXAMPLE: THE MAGNETIC BRAKE

A useful and interesting application of the magnetic force and the electromagnetic reaction principle (Lenz' law) occurs in the magnetic brake. Let us consider a pendulum apparatus shown in Fig. 6.38. It consists of a permanent magnet with a uniform B field through which a copper-plate pendulum swings. If the field of the magnet were not present, the pendulum would swing freely back and forth. In the presence of the B field, the motion of the copper plate as it approaches the magnet gap is stopped. Let us see what happens as the plate begins to enter the magnetic field. The leading edge of the plate can be thought of as a wire of length l which will have a voltage induced along it of magnitude $V = El = vBl$, as given by (6.59). The induced current density J which is related to the induced electric field E by Ohm's law as

$$\mathbf{J} = \sigma\mathbf{E} = \sigma\mathbf{v} \times \mathbf{B} \tag{6.72}$$

is downward (in the direction of $\mathbf{v} \times \mathbf{B}$), where σ is the conductivity of the plate material. This induced current, being in a magnetic field, will transfer to the plate a magnetic force per unit volume which from Eq. (6.5) is

$$\frac{\Delta\mathbf{F}_m}{\Delta v_{ol}} = \mathbf{J} \times \mathbf{B} = \sigma(\mathbf{v} \times \mathbf{B}) \times \mathbf{B} \tag{6.73}$$

If all the vectors are at right angles as in Fig. 6.38, the magnetic force on each unit volume of plate material in the leading-edge region is given by

$$\Delta F_m = \sigma v B^2 \, \Delta v_{ol} \tag{6.74}$$

and is in a direction opposite to the motion of the plate. If the conductivity is high but not infinite, as in the case of copper material, the plate will first be slowed down and then stopped. As the induced currents are dissipated in the I^2R losses (or $JE = \sigma E^2$ losses in watts per Δv_{ol}), the plate slowly moves further inside the

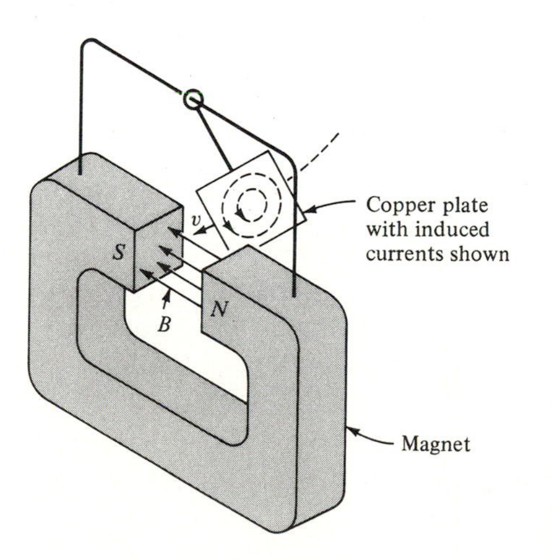

Figure 6.38 A pendulum consisting of a square copper plate of dimensions l by l. Braking results when the induced current produces a magnetic force opposite to the direction of motion. Since the plate is finite, the current will loop around as shown and is usually referred to as an eddy current.

gap and eventually settles down inside the magnetic field at the same position that it would come to rest in the absence of a B field. We observe that the braking magnetic force is proportional to σ. If the plate were of perfect conducting material in which $\sigma \to \infty$, the induced currents would be so great that the plate would bounce back and come to rest outside the magnetic field.† We have shown the currents forming closed loops. This is a consequence of the finite dimensions of the plate. Since the currents cannot leave the material of the plate, they "deflect" off the edges to form closed loops. Such circulating currents are known as *eddy currents.* Clearly if the plate were of infinite dimensions, the current direction would be downward only. Eddy currents in electric machinery and transformers can produce considerable losses, and one goes to great lengths to reduce them. This is done by assembling transformer or rotor cores from insulated sheets of iron material, called *laminates.* The laminates are in such directions as to minimize the eddy currents. Similarly in our pendulum, if we were to cut vertical slots in the copper plate, the induced currents would be greatly decreased as would the braking action.

Another way to look at this problem is strictly from the point of view of Lenz' law. Before the pendulum enters the B field of the magnet, the magnetic field and current in the copper plate are zero. As the plate enters the gap, the magnetic field starts to penetrate the plate. The result is that a current which tries to oppose the establishment of such a field is generated in the plate. The loop currents are in a direction such that the magnetic field which they produce is opposite to that produced from the permanent magnet. Since the conductivity of copper is finite, the eddy currents induced can never be strong enough to establish a field that would cancel the magnetic field of the magnet (part of the eddy current goes into heating the plate, i.e., the I^2R losses). However, for a perfect conductor, as the plate approaches and enters the B field of the magnet, the field through the plate would remain zero. The eddy currents create an equal and opposite field, with the result that the plate bounces back.

6.13 EXAMPLE: MAGNETOHYDRODYNAMIC POWER GENERATION

In recent years a large number of different methods for producing electric power from heat have been undergoing experimental development. In conventional steam power plants the energy stored in the fossil fuel is converted into thermal energy, then into mechanical energy by use of steam turbines, and finally into electric energy by use of rotating generators. Among methods that do not employ rotating machinery, magnetohydrodynamic (MHD) generation is considered to offer greatest promise.

† To make the example using a perfectly conducting material tractable, we are assuming a fictitious magnetic field, one that decays away from the magnet and becomes identically zero after a finite distance.

The fundamental principle of electrodynamic power generation remains unchanged: An electric conductor is moved through a magnetic field, thereby generating an electric current. In conventional generators, the conductor or armature is a solid moving part. In the MHD generator, the armature is replaced by a moving ionized gas—a plasma. In its simplest form, the MHD generator is little more than a pipe through which the plasma flows, driven by a pressure gradient. Electrodes are placed in the pipe, and magnetic coils alongside it. As the plasma flows through the magnetic field, a voltage is induced between anode and cathode. Power is extracted by connecting the electrodes to a load. Except for the replacement of an armature by the plasma, the basic principle of operation is the same as that of conventional rotating machinery. Because it is a homopolar machine, it is best suited for the production of dc power.

The MHD generator must operate at high temperatures. For the working fluid to be a good conductor, it must be at least partially ionized. Thermal ionization in gases usually takes place at temperatures of 4000 K or higher. However, the addition of small amounts of easily ionized "seed" materials, such as potassium salts, significantly lowers the temperature at which adequate conductivity is attained. Seeded air becomes sufficiently conducting at 2500 K; seeded argon, at 2000 K. In both cases, the potassium seed material need be no more than 1 percent of the total plasma volume. Thus, adequate conductivities can be obtained only at temperatures that are quite high in comparison to those in steam turbogenerators (which operate at about 850 K) but such temperatures are nonetheless within the limitations of available materials and heat sources. MHD generators do not become efficient until a critical size is reached. Because the best gas conductivity that we can hope to attain is roughly five orders of magnitude less than that of copper, the plasma of an MHD generator must contain a much bigger volume of moving conductor material than does the armature of a conventional machine. This also implies that a much bigger volume must be filled with a magnetic field, usually a difficult problem.

As shown in Fig. 6.39, the MHD generator consists of a combustion chamber in which fuel is burned at a high temperature and the gaseous products of combustion are made electrically conducting by injection of a seed material, such as potassium carbonate. The electrically conducting gas is expelled and travels at high velocity through a magnetic field and in the process produces a flow of direct current in the load. The interaction of the high-velocity stream with a magnetic field thus converts the kinetic energy of the stream into electric energy. The stream of gas is slowed down because of the induced force F'_{ind} shown in Fig. 6.39*b* which acts in a direction opposite to the velocity v of the stream. This force is the reaction force which "extracts" energy from the stream of gas particles.

Let us describe in detail this process using the coordinate system shown in Fig. 6.39*b*. For convenience the applied magnetic field B_{appl} is chosen to point upward. The conducting fluid is moving with velocity v through the B field which, according to Lorentz' force equation or Eq. (6.59), induces an electric field

$$\mathbf{E}_{\text{ind}} = \mathbf{v} \times \mathbf{B}_{\text{appl}} \tag{6.75}$$

Since the conducting fluid is isotropic, we can use Ohm's law to relate the induced

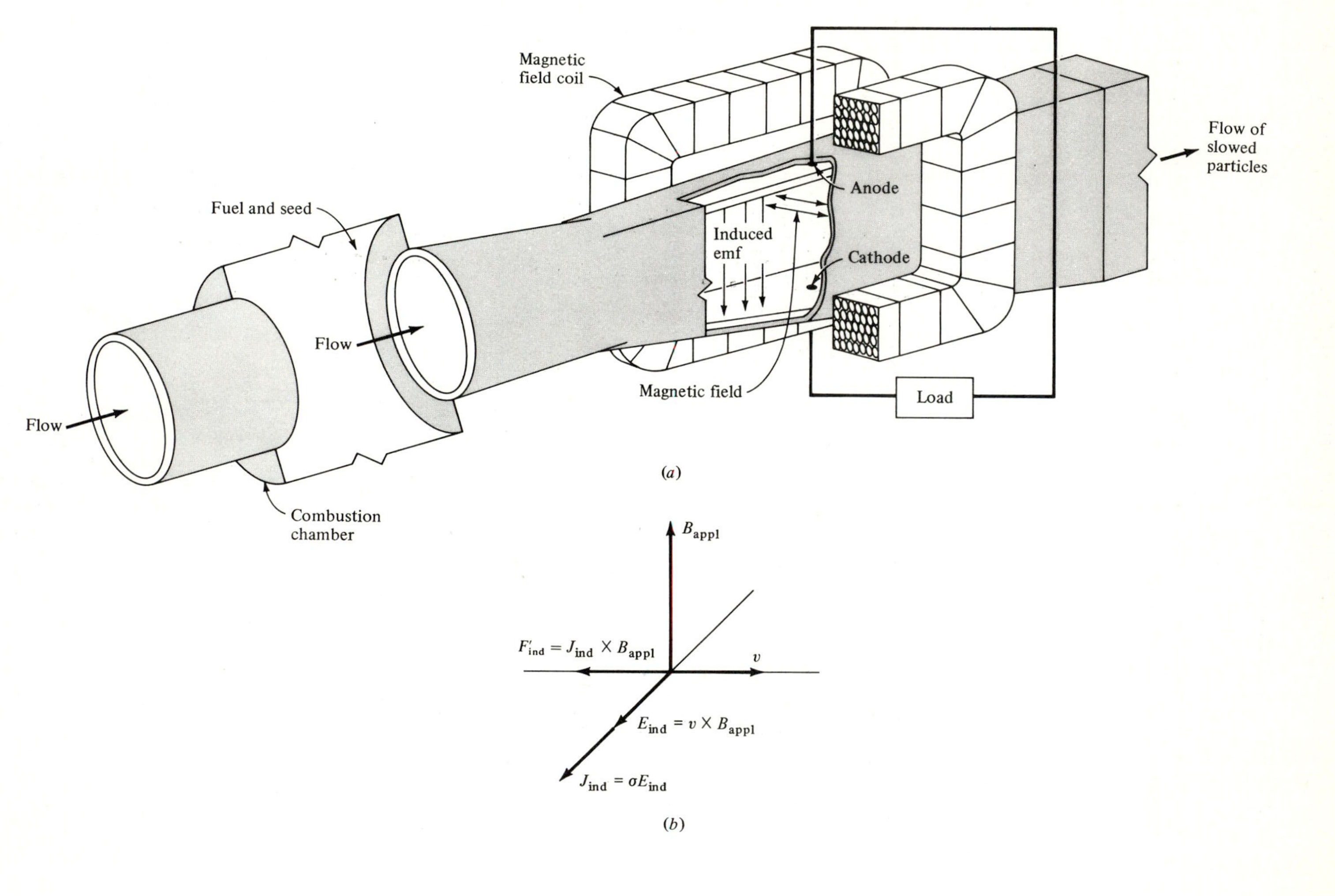
Magnetic field coil
Fuel and seed
Flow of slowed particles
Anode
Induced emf
Cathode
Flow
Flow
Magnetic field
Load
Combustion chamber
(a)
B_{appl}
$F'_{ind} = J_{ind} \times B_{appl}$
v
$E_{ind} = v \times B_{appl}$
$J_{ind} = \sigma E_{ind}$
(b)

electric field to an induced current density $\mathbf{J}_{\text{ind}}$ by

$$\mathbf{J}_{\text{ind}} = \sigma \mathbf{E}_{\text{ind}} = \sigma \mathbf{v} \times \mathbf{B}_{\text{appl}} \tag{6.76}$$

where σ is the scalar conductivity of the fluid. The induced current cuts the lines of magnetic field and experiences a force per unit volume [see Eq. (6.5)]:

$$\mathbf{F}'_{\text{ind}} = \mathbf{J}_{\text{ind}} \times \mathbf{B}_{\text{appl}} \tag{6.77}$$

which is transmitted to the fluid. Since this force is opposite to $\mathbf{v}$, it slows the moving fluid. It is the same magnetic force F_m given by (6.65) which we found for the linear generator. If the directions of the respective vectors are at right angles to each other, we can write, for the induced magnetic force per unit volume,

$$F'_{\text{ind}} = \sigma v B^2_{\text{appl}} \tag{6.78}$$

This equation is again similar to the corresponding one, (6.66), for the linear generator. The power P, force $\mathbf{F}$, and velocity $\mathbf{v}$ of a stream are related by $P = \mathbf{F} \cdot \mathbf{v}$. The power obtained from this generator is therefore given by $P' = F'v = \sigma v^2 B^2_{\text{appl}}$ in watts per unit volume.

The expression for power tells us that high electric conductivity σ is a requirement in the choice of working fluids. Liquid-metals, such as mercury, potassium, and sodium have high σ's, and even though liquid-metals are difficult to handle, they show some promise. Hot gases which use a seed material (usually potassium carbonate) to make them conductive have also been successfully used.

Power Generation by Nuclear Fusion†

Nuclear fusion, the basic energy process of the stars, is well known. In 1952, the first thermonuclear test explosion resulted in an uncontrolled release of massive amounts of fusion energy. This test showed that fusion energy could be released on a large scale by raising the temperature of a high-density gas of charged particles, a plasma, to about 50 million kelvin, thereby igniting a fusion reaction within the ionized gas. Soon after the successful explosion of the hydrogen bomb, the search for a more controlled release of fusion energy was begun.

Here is how the fusion process works. If two light atomic nuclei are brought together with enough force to overcome the repelling Coulomb force, they fuse, yielding a heavier nucleus and at least one other particle. This particle can be a

† T. Fowler and R. Post, Progress toward Fusion Power, *Sci. Am.*, pp. 21–31, December 1966. D. Steiner, The Technological Requirements for Power by Fusion, *Proc. IEEE*, pp. 1568–1608, November 1975. L. A. Booth, et al., Prospects of Generating Power with Laser-Driven Fusion, *Proc. IEEE*, pp. 1460–1482, October 1976.

Figure 6.39 (*a*) The magnetohydrodynamic power generator; (*b*) vector diagram of velocity $\mathbf{v}$ of conducting fluid, induced emf, and reaction force $\mathbf{F}_{\text{ind}}$.

proton or a neutron, depending on the reaction. The reaction products are characterized by extremely high energy. For example, one possible fusion reaction involves deuterium and tritium, which are two isotopes of hydrogen. When they fuse, they yield an alpha particle and a neutron with an energy gain of 17.6 million electron volts.

A fusion reactor could be a container holding a mixture of fully ionized deuterium and tritium nuclei at a very high temperature. Fusion reactions would occur in such a hot plasma whenever the ignition temperature was reached, which is around 50 million kelvin. Since no solid material can exist at these temperatures, the principal emphasis from the beginning has been on the use of magnetic fields to confine the plasma. Most of the magnetic bottles are of the toroidal form, with a combination of radial and axial pinches for confinement.† The high-velocity charged particles produced by the thermonuclear reaction are usually trapped in such a way as to generate electricity directly.

6.14 EXAMPLE: THE PLASMA ENGINE

The $\mathbf{J} \times \mathbf{B}$ force can be used to accelerate a plasma, as shown in Fig. 6.40. Such an accelerator when used to generate thrust becomes a simple rocket motor. The accelerating force is produced when an ionized gas containing a neutral mixture of ions and electrons is passed through a crossed E and B field. Since the plasma is electrically neutral, its density (proportional to thrust) is not limited by electrostatic forces as in a pure ion beam. The principle of the engine is shown in the vector diagram of Fig. 6.40*b*. A dc magnetic field B_{appl} and a dc electric field E_{appl} are applied as shown. The total current density which is transverse to the flow $\mathbf{v}$ of the plasma is now given by

$$\mathbf{J} = \sigma(\mathbf{E}_{\text{appl}} + \mathbf{v} \times \mathbf{B}_{\text{appl}}) \tag{6.79}$$

The force per unit volume acting on this current is the $\mathbf{J} \times \mathbf{B}$ force, which is, from (6.5),

$$\mathbf{F}' = \sigma(\mathbf{E}_{\text{appl}} + \mathbf{v} \times \mathbf{B}_{\text{appl}}) \times \mathbf{B}_{\text{appl}} \tag{6.80}$$

Now if $\mathbf{E}_{\text{appl}} > \mathbf{v} \times \mathbf{B}_{\text{appl}}$, we have an accelerator. If the respective vectors are at right angles as shown in Fig. 6.40*b*, a force given by

$$F'_{\text{mot}} = \sigma(E_{\text{appl}} - vB_{\text{appl}})B_{\text{appl}} \tag{6.81}$$

will act in the direction v on the plasma particles and will thus accelerate them. On the other hand, if $E_{\text{appl}} < vB_{\text{appl}}$, we have generator action, and the force which is retarding the flow of particles is

$$F'_{\text{gen}} = \sigma(vB_{\text{appl}} - E_{\text{appl}})B_{\text{appl}} \tag{6.82}$$

† B. Coppi and J. Rem, The Tokamak Approach in Fusion Research, *Sci. Am.*, pp. 65–75, July 1972.

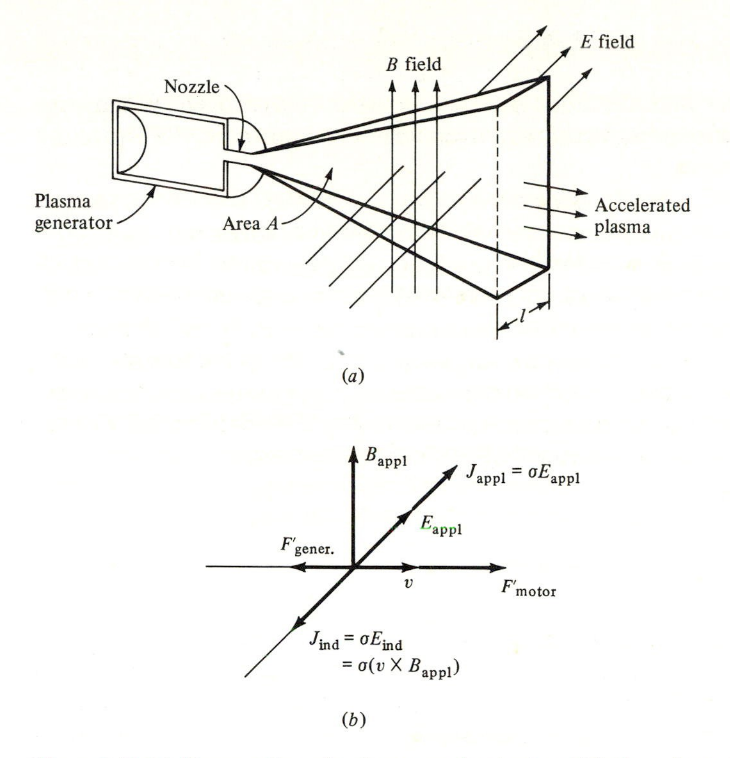

Figure 6.40 (*a*) Cross-section of a plasma accelerator in which the volume of the plasma acted upon by the applied E and B fields is approximately Al; (*b*) diagram of vectors involved.

The energy given up by the slowed particles goes now into "charging" the source of the E_{appl} field. In the motor configuration the power transferred to each incremental volume Δv_{ol} at the velocity **v** is given by

$$\Delta P = \mathbf{F}' \cdot v \, \Delta v_{ol}$$

$$= (\sigma v E_{appl} B_{appl} - \sigma v^2 B_{appl}^2) \, \Delta v_{ol} \tag{6.83}$$

where the first term in the parentheses is the motor-action term and the second is the generator-action term. Rewriting the power expression in terms of the total current density **J**, Eq. (6.79), across the moving plasma gives $\Delta P = JBv \, \Delta v_{ol}$. If the region of the plasma which is acted upon by the applied E and B fields is approximately equal to the volume Al, we have for the power

$$P = IBvl \tag{6.84}$$

where the current I is given by $I = JA$. This is an expression for the kinetic power delivered to the plasma by the electrical source which generates the current I. The amount of power can be significant for practical values of the parameters involved. For example, for values of $I = 2$ A, $B = 0.1$ Wb/m^2, $v = 10^5$ m/s, and $l = 0.2$ m, the power, using Eq. (6.84), is 4000 W.

6.15 EXAMPLE: THE HALL-EFFECT VOLTAGE

When a conductor that carries a current is placed in a uniform magnetic field, an electrostatic field appears whose direction is perpendicular both to the magnetic field and to the current. Figure 6.41 shows this effect for a metallic bar through which a current I flows. Since the current carriers in a metal are electrons, they will experience a downward force, which is the Lorentz force and is given by (6.4) or (6.16) as $\mathbf{F}_m = -q_e\mathbf{v} \times \mathbf{B}$. The electrons will be deflected and pile up on the bottom leaving the top positively charged. The electron concentration produces a negative space charge along the lower surface, which discourages further deposition of electrons. The negative space charge gives rise to an electric field $\mathbf{E}_H$ across the bar. The direction of $\mathbf{E}_H$ is such that it repels electrons from the bottom. The build-up of space charge and consequently of $\mathbf{E}_H$ continues until the force on the electrons $-q_e\mathbf{E}_H$ just balances the force $\mathbf{F}_m$ due to the magnetic field $\mathbf{B}$, after which the electron stream flows straight through the bar. The electric field known as the *Hall field* reaches equilibrium in the order of 10^{-14} s and is given by

$$\mathbf{E}_H = \mathbf{v} \times \mathbf{B} \tag{6.85}$$

If the current density as given by Eq. (2.6) is $\mathbf{J} = nq_e\mathbf{v}$ (n and q_e being the density and charge of the carriers, respectively) and is characterized by a single electron velocity, the Hall field becomes

$$E_H = \frac{JB}{nq_e}\sin\theta \tag{6.86}$$

where θ is the angle between the direction of current density $\mathbf{J}$ and magnetic field $\mathbf{B}$. Since E_H is inversely proportional to the carrier density, the Hall effect is much more pronounced in semiconductors than it is in metals.† The Hall voltage between the top and bottom faces is

$$V_H = E_H d \tag{6.87}$$

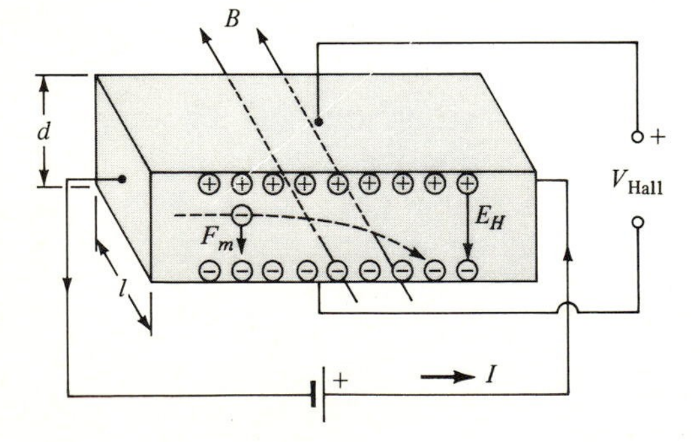

Figure 6.41 The electrons experience a downward deflection due to the magnetic force F_m. This leads to a Hall voltage as shown.

† In copper, as in other typical metallic conductors, there is approximately one free electron per atom, which gives $n = 8.4 \times 10^{28}\ \text{m}^{-3}$. The electron density for n-type germanium is typically $n = 10^{22}\ \text{m}^{-3}$.

and can be related to the total current $I = Jld$ through the bar by

$$V_H = \frac{IB}{nq_e l} \tag{6.88}$$

where it is assumed that the current is perpendicular to the B field.

If the bar were a sample of a p-type semiconductor, rather than an n-type, the holes or positive-charge carriers would be deflected downward again, but E_H and V_H would be reversed in direction. Thus, the sign of V_H indicates whether the sample contains predominantly holes or electrons as the current carriers. For example, an intrinsic semiconductor for which $nq_e v$ is equal to $nq_h v$, V_H would be equal to zero because equal numbers of holes and electrons would be deflected in the same direction, and the resultant equal space charges would cancel. In general, the measurement of V_H allows one to determine whether the sample is p-type, n-type, or near intrinsic. Another application is the gaussmeter, a device used to measure the strength of magnetic fields by measuring the induced Hall voltage. Other applications of the Hall effect have been proposed.†

PROBLEMS

$1\ \text{T} = 1\ \text{Wb/m}^2 = 10^4\ \text{G} \qquad \mu_0 = 4\pi \times 10^{-7}\ \text{H/m} \qquad \varepsilon_0 = 8.85 \times 10^{-12}\ \text{F/m} \cong \frac{1}{36\pi} \times 10^{-9}\ \text{F/m}$

6.1 A wire lying in the x axis carries a current of 20 A in the positive x direction. A uniform magnetic field $\mathbf{B} = 10\hat{\mathbf{x}} + 5\hat{\mathbf{y}}$ exists. Find the force $\mathbf{F}$ per meter length of wire.

6.2 Calculate the force due to the earth's magnetic field on a 10-m-long wire lying in a horizontal plane, pointing due north, and carrying a current of 100 A. The earth's magnetic field at this location is 0.5 G at a 60° dip angle (angle between the field lines and a north-pointing horizontal).

6.3 An electron is to be fired at the equator (east or west?) so it circles the earth at a constant speed. If the magnetic field of the earth at the equator is $\frac{3}{4}$ G, is horizontal, and points north, what should the electron's speed be?

6.4 The accompanying figure shows an electron injected into a magnetic field B with velocity v_0 of 10^7 m/s in a direction lying in the plane of the paper and making an angle of 30° with B. If the length l is 0.1 m, what must the value of B be in order that the electron will pass through the point P?

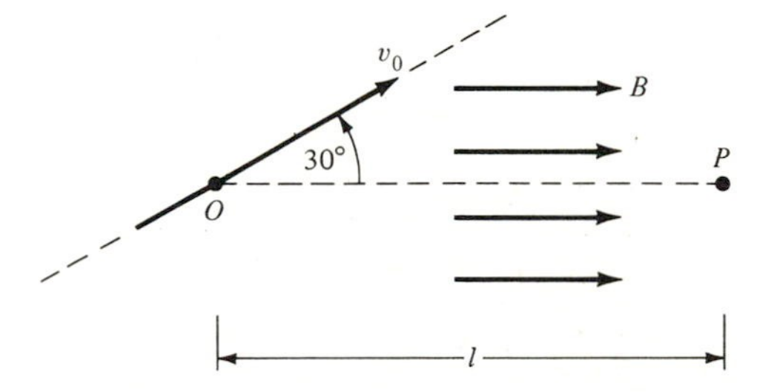

† M. Epstein, Hall-Effect Devices, *IEEE Trans. Magnetics*, pp. 352–359, September 1967.

6.5 (*a*) Referring to Fig. 6.9 and using the Lorentz force (6.16), we can state that the total force $\mathbf{F}_{12}$ exerted by the first particle on the second is given by

$$\mathbf{F}_{12} = q_2\mathbf{E}_1 + q_2\mathbf{v}_2 \times \mathbf{B}_1$$

Show that this can be written explicitly in the symmetric form as

$$\mathbf{F}_{12} = \frac{q_1 q_2 \mathbf{R}_{12}}{4\pi\varepsilon_0 R_{12}^3} + \frac{\mu_0 q_1 q_2 \mathbf{v}_2 \times (\mathbf{v}_1 \times \mathbf{R}_{12})}{4\pi R_{12}^3}$$

(*b*) Expression (6.11) for the two closed loops shown in the figure can be written as

$$\mathbf{F}_{12} = \frac{\mu_0}{4\pi} I_1 I_2 \oint_1 \oint_2 \frac{\mathbf{dl}_1 \times (\mathbf{dl}_2 \times \hat{\mathbf{R}})}{R^2}$$

where $\mathbf{F}_{12}$ is the total force on loop 1 exerted by current I_2 in the second loop. Because $\mathbf{dl}_1$ and $\mathbf{dl}_2$ are not symmetric in the integrand, it is not apparent that $\mathbf{F}_{12}$ is equal to $\mathbf{F}_{21}$. This could be quite puzzling if $\mathbf{F}_{12} \neq \mathbf{F}_{21}$, since then a net force would exist on the system of the two loops inducing the system to move, a contradiction of Newton's third law. Show by expanding the vector triple product, using identity (A1.56), $\mathbf{dl}_1 \times (\mathbf{dl}_2 \times \hat{\mathbf{R}}) = \mathbf{dl}_2(\mathbf{dl}_1 \cdot \hat{\mathbf{R}}) - \hat{\mathbf{R}}(\mathbf{dl}_1 \cdot \mathbf{dl}_2)$ that $\mathbf{F}_{12} = -\mathbf{F}_{21}$; hence the net force on the system is zero.

Hint: The double integral of the first term on the right is zero, leaving a symmetric expression in dl_1 and dl_2.

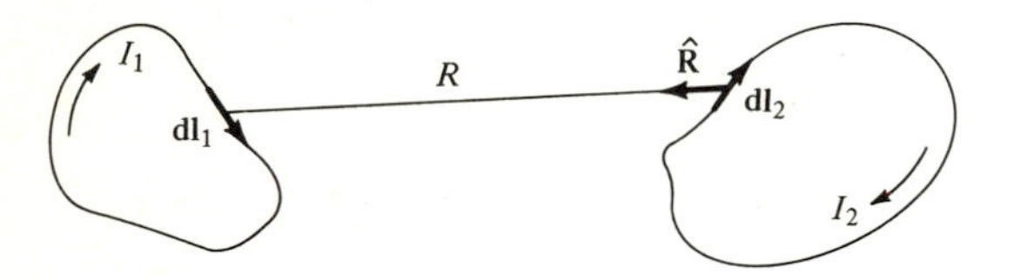

6.6 Two charges separated by distance R move along parallel paths as shown in Fig. 6.10. Is there any circumstance under which the magnetic force of attraction is equal to or larger than the electric force of repulsion? Discuss.

6.7 Show that the magnetic field of a straight wire of length l and carrying a current I is given by $B = \mu_0 I/4\pi R(\cos\theta_1 - \cos\theta_2)$. Note that a current element as well as a finite length of wire which has an abruptly beginning and abruptly ending current has no physical meaning unless such an element is part of a closed circuit.

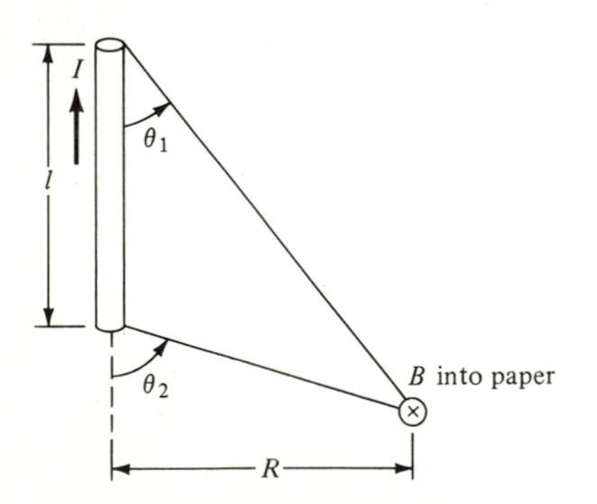

6.8 Show that the magnetic field for an infinitely long straight wire, given in Eq. (6.28), can be obtained from the magnetic field of a finite wire, derived in the previous problem.

6.9 Find the magnetic field at the center of a slender, rectangular loop carrying current I. The sides of the rectangle are such that $l \gg h$.

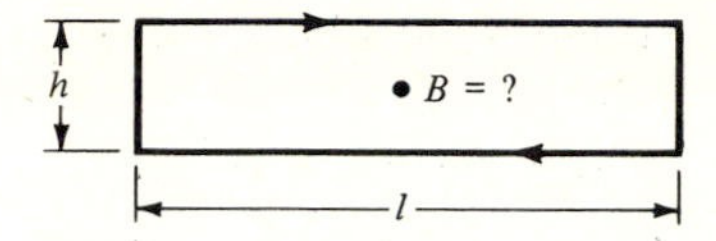

6.10 A long piece of wire carrying current I has a half loop of radius a bent into it as shown. Find the magnetic field at the center of the half loop.

6.11 A long piece of wire carries a current of $I = 10$ A and is bent in the shape shown. Find the magnetic field at the center of the half loop which has a radius of $a = 1$ cm.

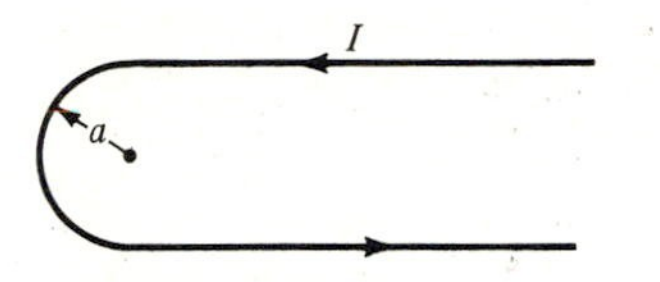

6.12 A square loop of wire having sides a has a dc current I flowing through it. If the plane of the loop is the xy plane, show that the magnetic field at the center of the loop is given by $\mathbf{B} = (2\sqrt{2}\mu_0 I/\pi a)\hat{\mathbf{z}}$, where $\hat{\mathbf{z}}$ is a unit vector in the z direction.

6.13 A rectangular loop of wire having sides a and b, respectively, has a dc current I flowing through it. Calculate the magnetic field B on the axis of the loop at a point z from the plane of the loop.

6.14 A long wire carrying current I_1 is parallel to the plane of a rectangular loop of wire which carries current I_2 as shown.

(*a*) Find the magnitude and direction of the force acting on the loop and on the long wire.

(*b*) Find the torque on the loop and the long wire.

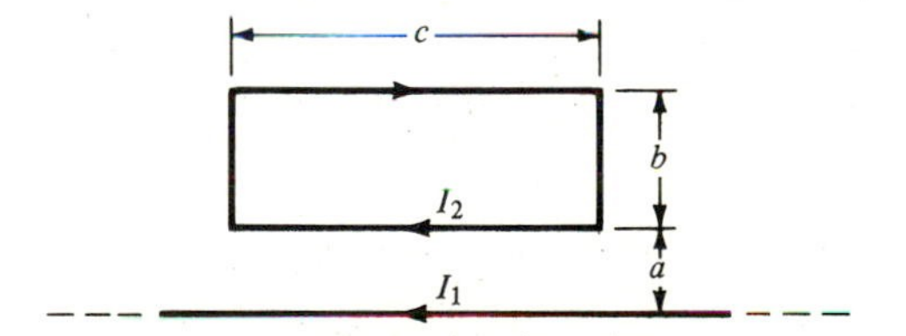

6.15 Determine the magnetic field B at the center of a flat coil with inner radius a and outer radius b, with N turns wound as shown. Assume N is large, the wire is very fine, and the coil closely wound. Check your answer with Eq. (6.43) by letting $b \to a$ in your answer.

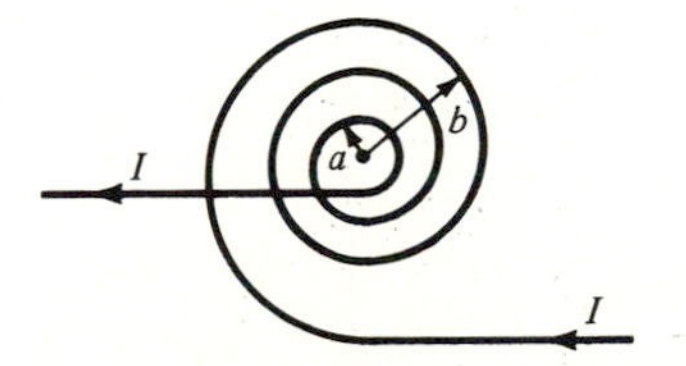

6.16 Two parallel conducting rods are tipped at an angle θ to the horizontal. They form a rigid inclined plane on which a conducting rod of length l and mass m can slide without friction. If a constant magnetic field B acts upward, what must the current I through the rod be in order for the weight mg of the movable rod to be in equilibrium on the inclined plane?

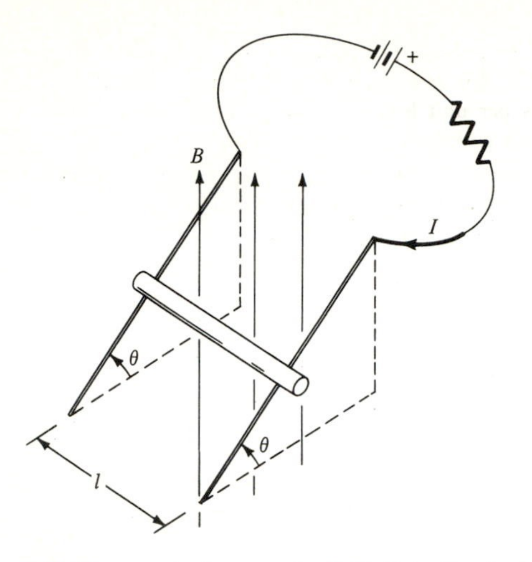

6.17 The constant magnetic field B in the figure of the previous problem is replaced by a long current-carrying conductor parallel to the movable rod and at the base of the inclined plane. If the current in the long wire is I_B and is opposite to the current I in the movable rod, find the equilibrium distance between the two current-carrying conductors.

6.18 Determine the equivalent current of the rotating charge for the following cases:

(*a*) Show that a point charge q rotating with angular frequency ω is equivalent to a current $I = q\omega/2\pi$.

(*b*) A ring of line charge ρ_L rotates about its center with an angular velocity ω. If the radius of the ring is R, show that it is equivalent to a current $I = \rho_L \omega R$, where ρ_L is measured in coulombs per meter.

(*c*) A washer-shaped surface charge ρ_s of inside radius a and outside radius b rotates about its axis with an angular velocity ω. Show that this is equivalent to a current sheet $K = \rho_s \omega r$, where $a < r < b$ and K is measured in amperes per meter or is equivalent to a total current $I = \rho_s \omega(b^2 - a^2)/2$, where the dimensions of ρ_s are in coulombs per square meter.

(*d*) A sphere of radius R has a constant charge density ρ throughout its interior. If the sphere rotates about its center with an angular velocity ω, show that this is equivalent to a current $I = \rho\omega 2R^3/3$.

6.19 An insulating, thin, circular washer of inner radius a and outer radius b has a uniformly distributed static charge of ρ_s coulombs per square meter. If the washer rotates about its axis with an angular velocity ω, find the magnetic field at the center of the washer.

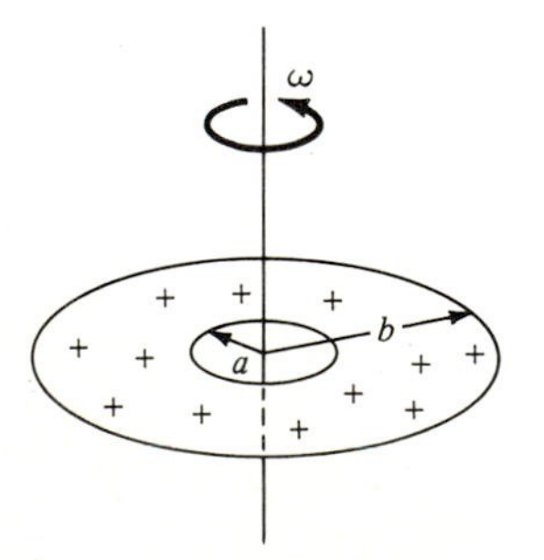

6.20 Show that the magnetic field at the center of a charged circular disc of radius b which rotates about its axis with an angular velocity ω is given by $B = \mu_0 \rho_s \omega b/2$, where ρ_s is the surface charge density in coulombs per square meter.

6.21 The current under ordinary conditions in two parallel bus bars which are used to supply energy to a low-voltage high-current process plant is 200 A. Under short-circuit conditions the current may reach 4000 A. Calculate the force between the bars per unit length under both conditions if the separation between the bars is 20 cm. Give the answer in pounds per foot as well as newtons per meter.

6.22 Calculate the magnetic field B at the center of a solenoid which has 6000 turns, is 30 cm long, and has a diameter of 1 cm. A current of 4 mA flows through the coil.

6.23 Calculate the maimum torque on the solenoid of the above problem if the solenoid is placed in a constant magnetic field of 2 T.

6.24 Calculate the magnetic moment of a loop of 20 turns which has an area of 0.005 m^2. A current of 1 A flows through the loop.

6.25 Find an expression for the force tending to open a circular wire carrying current I.

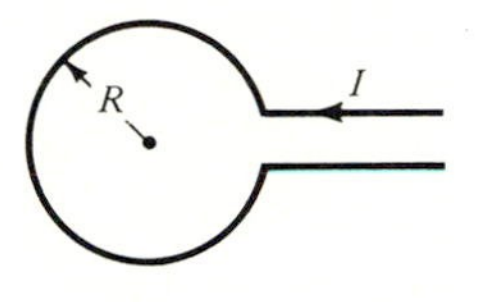

6.26 Find an expression for the force acting on one rectangular loop due to the second loop. The planes of the loops are parallel to each other and separated by d.

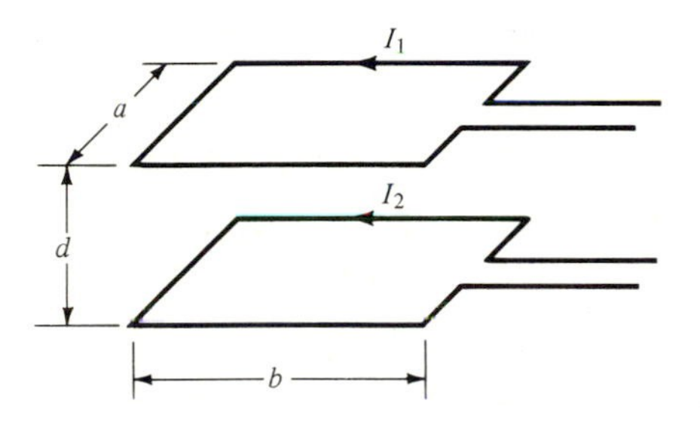

6.27 A conducting rod of length l rotates with an angular frequency ω cycles per second about one of its ends in a steady magnetic field, as shown in the figure. Find an expression for the voltage that will appear across the ends of the rod. The magnetic field B is out of the paper and the rod rotates in the plane of the paper.

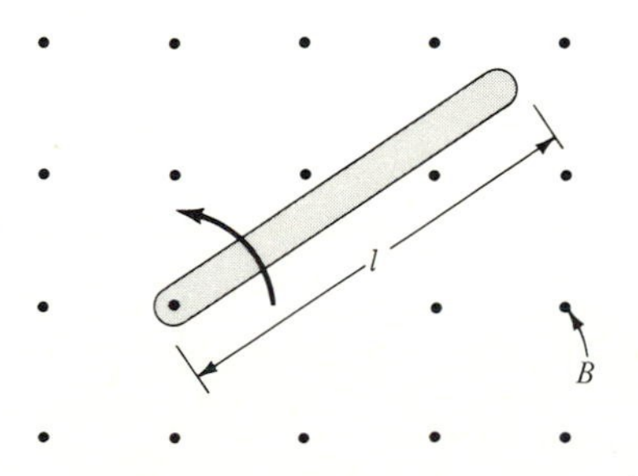

6.28 A Faraday disc generator is a thin metallic disc of radius l situated with its plane normal to a uniform magnetic field B. The disc rotates with an angular velocity ω. Show that a voltage $V = \omega B l^2/2$ will be generated between the axis and the periphery of the disc, i.e., between points 0 and P.

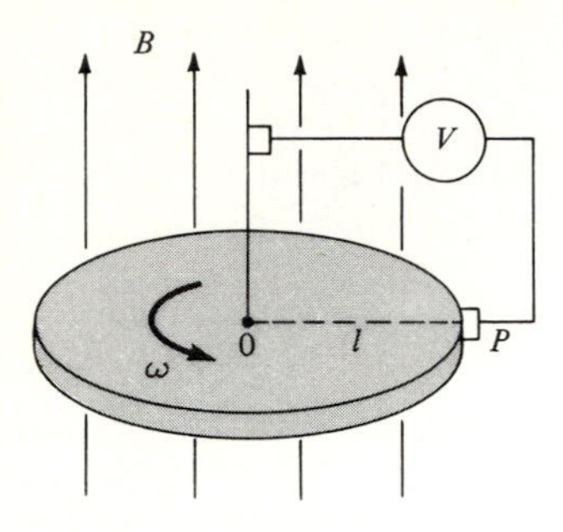

6.29 An ac generator produces a voltage having an amplitude of 160 V and a frequency of 60 Hz. The area of the generator coil is 10^{-2} m^2, and a constant magnetic field of 1 T is applied. Find the number of turns in the generator coil.

6.30 Show that the induced voltage that will appear across the ends of a wire bent into a semicircle of radius a is given by

$$V = \pi^2 f B a^2 \sin(2\pi f t)$$

if the wire rotates at a frequency f in a uniform magnetic field B as shown in the accompanying figure.

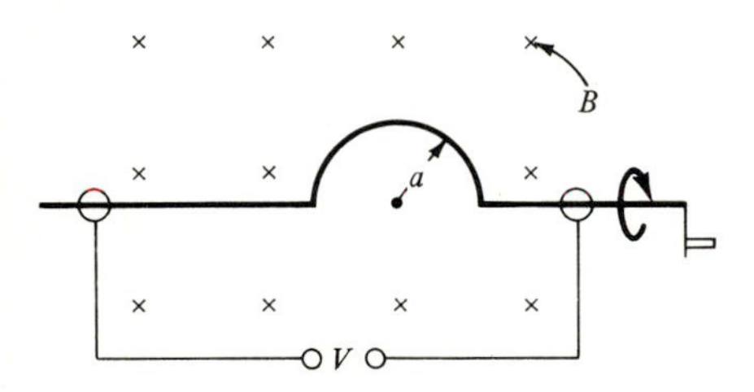

6.31 A conducting rod is 1 m long and slides over two parallel straight conductors at a constant speed of 2 m/s. The structure is immersed in a constant B field as shown. Because of induced current in the moving rod, it is subject to a retarding force of 1 N. What force must be applied to the rod to move it at a speed of 5 m/s?

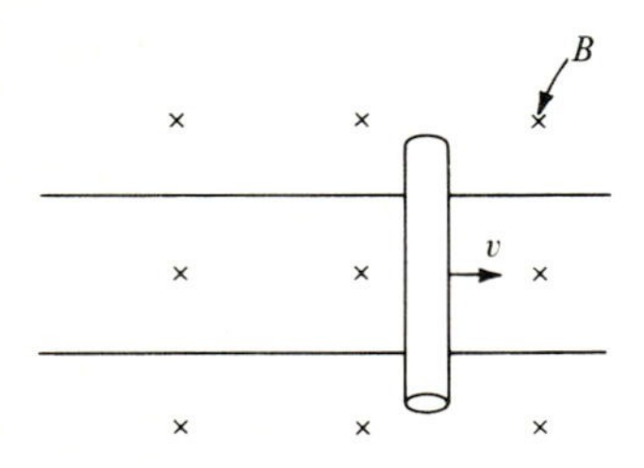

6.32 A linear motor is shown in the figure. Assume R is the concentrated resistance of the battery, conducting rails, and movable rod. For a constant frictional force (F_{fric} = constant) show that the velocity of the rod as a function of time is given by

$$v(t) = v_{\text{eq}}(1 - e^{-t/\tau})$$

where $\tau = Rm/(Bl)^2$, m is the mass of the rod, and the equilibrium velocity $v_{\text{eq}} = \lim_{t\to\infty} v(t) = V/lB - F_{\text{fric}}R/(lB)^2$.

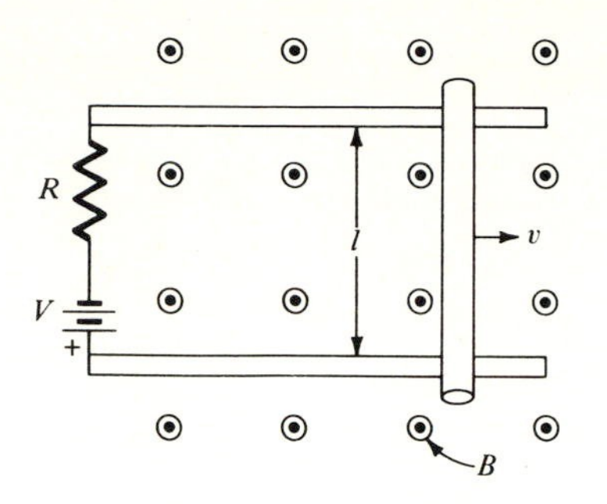

6.33 A movable rod of mass m and length l rests on two conducting rails which are connected to a resistance R as shown. A constant magnetic field B exists normal to the plane of the two rails.

(a) If the initial velocity of the rod is v_0, show that the velocity decreases as a function of time according to $v(t) = v_0 e^{-t/\tau}$.

(b) Find τ.

(c) Find the total distance the rod travels before it comes to rest.

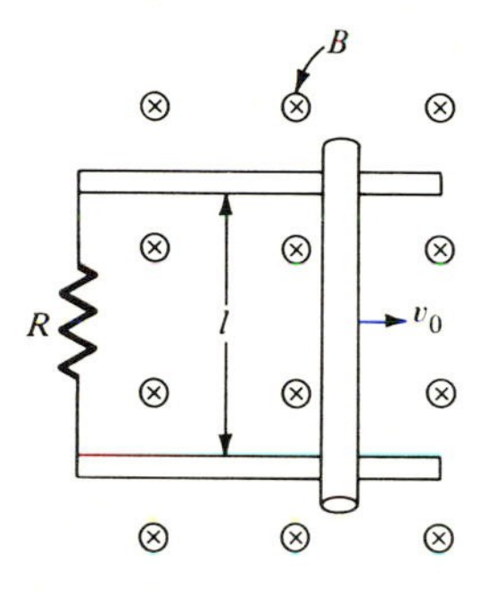

6.34 A large conducting copper sheet of thickness t falls with a velocity v through a uniform magnetic field B. Velocity v is at right angles to B, and a small section of the copper sheet is illustrated. Show that a force, $F = \sigma v t B^2$, per unit area of sheet exists which resists the motion of the conductor.

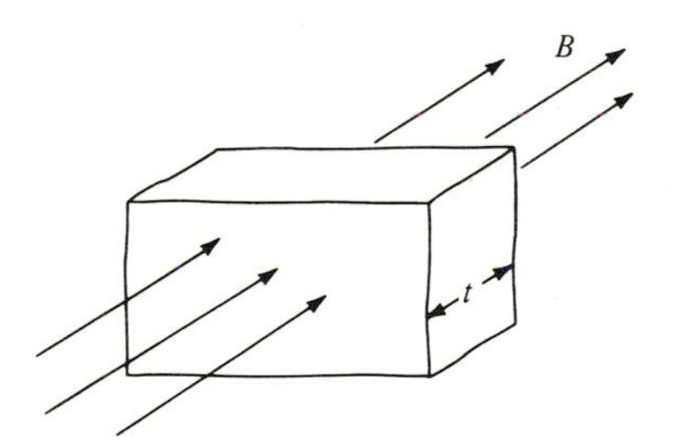

6.35 A 5-mm by 5-mm square metallic conductor carries a current of 10 A. If the Hall voltage induced by a magnetic field of 2 T is 2.5×10^{-6} V, find the density n of carriers in the conductor.

CHAPTER

SEVEN

AMPÈRE'S LAW, INDUCTANCE, AND ENERGY IN THE MAGNETIC FIELD

In the previous chapter the magnetic **B** field was introduced via two postulates. The remainder of that chapter concerned itself with the many applications of the magnetic field and force. There are other properties, however, that the magnetic field possesses which either are not contained in these two postulates or at least are not readily apparent from them. In this chapter the magnetic field will be characterized more completely.

7.1 MAGNETIC FLUX AND FLUX DENSITY

The magnetic **B** field defined by Eqs. (6.6) and (6.7) varies inversely as the square of the distance from the current element that produces it. Because of this inverse square dependence, the B field is also referred to as the magnetic flux density. In the SI system of units, B is given in teslas (T) or in webers per meter squared ($\mathrm{Wb/m^2}$). We can now define the *magnetic flux* through an elemental area dA as

$$d\phi = \mathbf{B} \cdot \mathbf{dA} = \mathbf{B} \cdot \hat{\mathbf{n}}\, dA \qquad \mathrm{Wb} \tag{7.1}$$

as shown in Fig. 7.1. The *total magnetic flux* through a finite area is then given by

$$\boxed{\phi = \iint \mathbf{B} \cdot \mathbf{dA}} \tag{7.2}$$

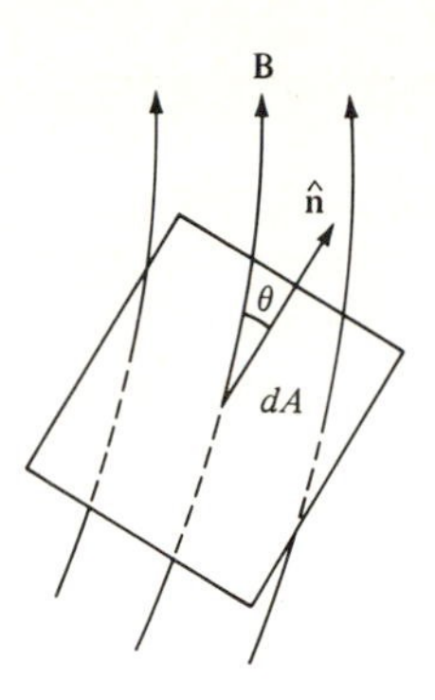

Figure 7.1 Flux through a small area dA. The $\hat{\mathbf{n}}$ is the normal to the area, and θ is the angle between the **B** field and the normal; that is, $\cos\theta = \hat{\mathbf{n}} \cdot \hat{\mathbf{B}}$.

7.2 GAUSS' LAW FOR MAGNETIC FIELDS

A law similar to Gauss' law for electric flux

$$\Psi = \oiint \mathbf{D} \cdot \mathbf{dA} = Q$$

can be written for magnetic flux ϕ:

$$\phi = \oiint \mathbf{B} \cdot \mathbf{dA} = Q_M$$

where Q_M is the total magnetic charge inside a closed surface. The difference between electric and magnetic fields is that no magnetic charge exists; that is, $Q_M = 0$. Therefore, for magnetic fields, we have

$$\boxed{\oiint \mathbf{B} \cdot \mathbf{dA} = 0} \tag{7.3}$$

The implication of (7.3) is that one cannot find magnetic field lines terminating on anything but themselves; in other words, magnetic lines form closed loops. For any closed surface, we have as many magnetic lines entering the surface as are leaving it. If this were not the case, then a volume exists out of which a net number of magnetic lines stick out or stick in. The net number of lines sticking out of a volume would imply that a positive magnetic charge Q_M resides in the volume. Now it just so happens that no magnetic charge or magnetic monopoles exist. Even in lunar rock magnetic charge has not been found to be present.

By applying Gauss' divergence theorem (1.94) to Eq. (7.3), we can convert (7.3) to a differential expression

$$\boxed{\nabla \cdot \mathbf{B} = 0} \tag{7.4}$$

Equation (7.4), together with the integral expression (7.3), forms one of Maxwell's equations.†

Before concluding this section, let us examine the field that arises from a permanent magnet vis-à-vis (7.3) or (7.4). An examination of the distribution of metal filings around a permanent magnet appears to indicate that field lines emanate from one end of the magnet and terminate on the other end. However, a B line that appears to start on the north pole and terminate on the south pole in fact continues inside the magnetic material to form a closed loop. Hence, (7.3) is not violated. Since permanent magnets always have a north and south pole, cutting one in half simply forms two smaller magnets, each one with a north and south pole.

7.3 AMPÈRE'S LAW

An alternative to postulate 2 of the previous chapter which relates magnetic fields and electric currents is given by Ampère's law. This law will be particularly useful in finding the magnetic field from currents which have a degree of symmetry. An extended form of Ampère's law is also one of Maxwell's equations.

To arrive at Ampère's law, let us first consider the magnetic field produced by current I flowing in an infinitely long wire. At a distance R from the wire, the B field is given from (6.28) as

$$B = \frac{\mu_0 I}{2\pi R} \tag{7.5}$$

As shown in Fig. 6.19, the B field forms concentric circles about the current-carrying wire. An integration of $\mathbf{B}$ around a circular path of radius R which encloses the wire gives

$$\oint \mathbf{B} \cdot \mathbf{dl} = \oint B\, dl = \frac{\mu_0 I}{2\pi R} \oint dl = \frac{\mu_0 I}{2\pi R} 2\pi R = \mu_0 I \tag{7.6}$$

We now observe that the result of the above integration is valid not only for a simple circular path which coincides with the circular B field of the infinite wire but also for arbitrarily shaped closed loops about the current I. To demonstrate this, let us consider Fig. 7.2a which shows a current I inside a noncircular closed path. Integrating the B field around the path, we obtain

$$\begin{aligned} \oint \mathbf{B} \cdot \mathbf{dl} &= \int_0^{\pi/2} \frac{\mu_0 I}{2\pi b} (b\, d\theta) + \int_{\pi/2}^{2\pi} \frac{\mu_0 I}{2\pi a} a\, d\theta \\ &= \frac{\mu_0 I}{2\pi} \frac{\pi}{2} + \frac{\mu_0 I}{2\pi} \frac{3\pi}{2} \\ &= \mu_0 I \end{aligned} \tag{7.7}$$

† Note that, as shown in Prob. 7.1, the statement $\nabla \cdot \mathbf{B} = 0$ is contained in the Biot-Savart law (6.9).

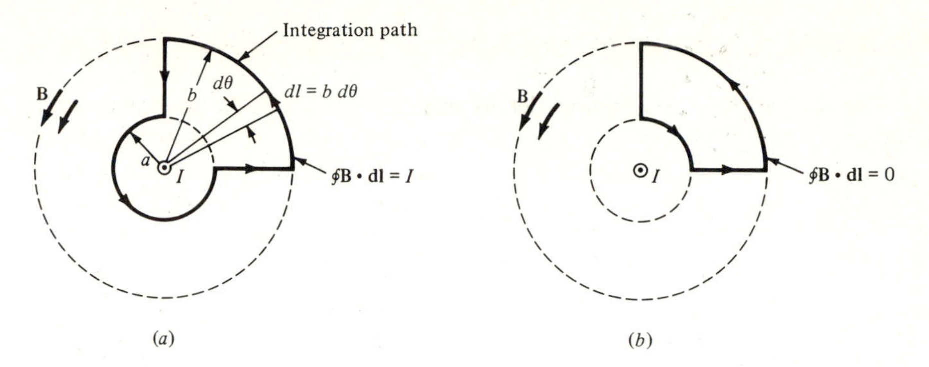

Figure 7.2 Two paths of integration for Ampère's law. (*a*) Path encloses current I (out of paper); (*b*) path does not enclose current I.

Note that the radial segments of the path do not contribute to the integral because there the B field is normal to the path. Hence the line integral of **B** around any closed path is equal to $\mu_0 I$, because a closed path of arbitrary shape can always be replaced by many small circular arcs and radial segments. The above statement expresses Ampère's law and is valid for any magnetic field and current, not just for fields produced by a current in an infinitely long wire.

What if the integration path does not enclose current, as, for example, the path shown in Fig. 7.2*b*? The right side of Ampère's law is then zero, because the contribution to the integral along the outer arc cancels that of the inner arc. That is, the outer arc gives a value of $\mu_0 I/4$, whereas the inner one yields $-\mu_0 I/4$. The signs are different because along the outer arc the B field is along the direction of integration, whereas along the inner arc **B** and **dl** are opposite in direction. To conclude, we can now state *Ampère's law* as follows:

$$\oint_l \mathbf{B} \cdot d\mathbf{l} = \begin{cases} \mu_0 I & \text{path } l \text{ encloses } I \\ 0 & \text{path } l \text{ does not enclose } I \end{cases} \tag{7.8}$$

The current I is the net current enclosed by path l.

Suppose the current is distributed through space with a current density **J**, then Ampère's law can be written as

$$\oint_{l(A)} \mathbf{B} \cdot d\mathbf{l} = \mu_0 \iint_{A(l)} \mathbf{J} \cdot d\mathbf{A} \tag{7.9}$$

where $l(A)$ is path l that encloses area A and $A(l)$ is area A bounded by path l. Thus only current that flows through area A contributes to the line integral of B. Currents outside the contour l make no contribution. Conversely, knowing the value of the line integral of the magnetic field around a closed loop, we know at once the net current flowing through the loop.

As a final remark it should be pointed out that Ampère's law is valid for steady currents (and approximately valid for slowly varying currents) in free space.† Ampère's law will require some modification (Chap. 9) in the presence of magnetic material such as iron. However, inside nonmagnetic materials such as copper (note that μ for copper is approximately that of free space, μ_0) it can be used as presented here.

7.4 AMPÈRE'S LAW IN RELATION TO A CURRENT ELEMENT AND MAGNETIC CHARGE

In the previous section we pointed out that Ampère's law is an alternative formulation of the magnetic field to that given by postulate 2 of Chap. 6. This law is valid in general for any magnetic field produced by any type of current. Let us take the expression (6.6) of postulate 2 and integrate it about a closed path and see if it reduces to Ampère's law. Referring to Fig. 6.5 or 6.7, we integrate the B field from a current element in a plane perpendicular to the current element $I\ dl'$. Thus

$$\begin{aligned}\oint \mathbf{B} \cdot \mathbf{dl} &= \frac{\mu_0\, dl'}{4\pi R^2} \oint (\mathbf{I} \times \hat{\mathbf{R}}) \cdot \mathbf{dl} \\ &= \frac{\mu_0\, dl'}{4\pi R^2} \oint I\, dl = \frac{\mu_0\, dl'}{4\pi R^2} \cdot I 2\pi R \\ &= \mu_0 I \frac{dl'}{2R} \end{aligned} \tag{7.10}$$

Note that the vector $\mathbf{I} \times \hat{\mathbf{R}}$ is parallel to $\mathbf{dl}$. Hence the dot product in the above integrand is simply $I\ dl'$. Since I is constant, it was taken outside the integral leaving an integration around the circumference of a circle of radius R, which is $2\pi R$. The result that we have obtained in (7.10) is different from $\mu_0 I$ which Ampère's law predicts. As a matter of fact $\mu_0 I$ is multiplied by the dimensionless quantity $dl'/2R$ which is the angle subtended by the dl' element at the observation point R (located on the integration circle of circumference $2\pi R$). In a sense the different result obtained does not violate Ampère's law since a current element $I\ dl'$ cannot exist by itself in nature, since currents are continuous. For example, if we integrate all the infinite current elements that make up an infinite wire, we obtain Ampère's law; that is, the right side of (7.10) will then give $\mu_0 I$ (see Prob. 7.2).

Ampère's law, which relates the B field and steady currents, describes a situation which is quite different from that in electrostatics. Recall Eq. (1.12), which states that the line integral of $\mathbf{E}$ around any closed path is zero:

$$\oint \mathbf{E} \cdot \mathbf{dl} = 0 \tag{1.12}$$

This had two implications: First, the work done in carrying a test charge around a closed path in an electric field is always zero; i.e., the electrostatic field is conservative. Second, (1.12) implied that the electrostatic field $\mathbf{E}$ can always be derived from a scalar potential V; that is, $\mathbf{E} = \nabla V$. In the magnetostatic case we therefore conclude that in general we cannot derive the B field from some scalar potential ϕ, such that $\mathbf{B} = \nabla\phi$. Only in regions that are void of currents and permanent magnets (which are in some sense currents) can we derive $\mathbf{B}$ from a scalar potential, since in such regions Ampère's law is

$$\oint \mathbf{B} \cdot \mathbf{dl} = 0 \tag{7.11}$$

† In the more general time-dependent case in which there is coupling between electric and magnetic fields, we shall see (Chap. 11) that a changing electric flux can create a magnetic field and hence, can contribute to the line integral of the magnetic field.

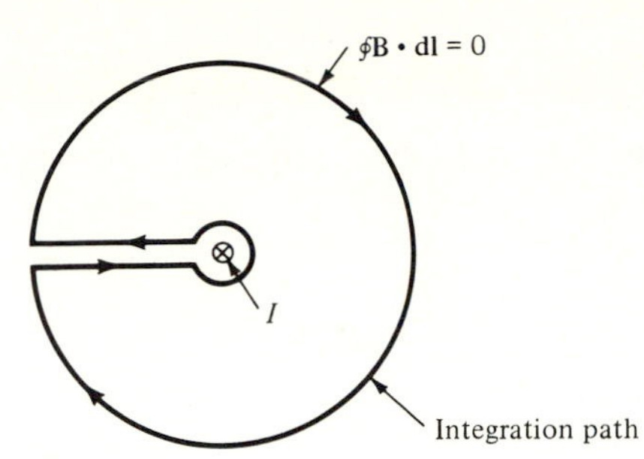

Figure 7.3 Since the path shown does not enclose any current, Ampère's law is equal to $\oint \mathbf{B} \cdot \mathbf{dl} = 0$.

In electrostatics, **E** is a force field and the potential V is a work field. A statement such as (1.12) implies that no work is done by a test charge that is moved about a closed path. Let us borrow this interpretation and apply it to magnetostatics. Again assume that we have an isolated magnetic test charge Q_m which has the property that, when immersed in a magnetic B field, it will be acted on by a force given by

$$\mathbf{F} = \mathbf{B}Q_m \tag{7.12}$$

as discussed in the text following (6.7). Ampère's law can now be written as

$$\oint \mathbf{B} \cdot \mathbf{dl} = \oint \frac{\mathbf{F}}{Q_m} \cdot \mathbf{dl} = \mu_0 I \tag{7.13}$$

Since the line integral of force is the work W done in carrying Q_m about a closed path, we can interpret the above statement as

$$\frac{W}{Q_m} = \mu_0 I \tag{7.14}$$

that is, Ampère's law says that the work in carrying a magnetic charge Q_m once around a current I is $W = Q_m \mu_0 I$. Perhaps it is now clearer that Ampère's law applied to a path, as shown in Fig. 7.3, gives $\oint \mathbf{B} \cdot \mathbf{dl} = 0$ since no current is enclosed. Interpreting this in terms of (7.14), we find that the work in carrying Q_m about the small circle is equal and opposite to the work about the large circle. Hence no net work is done.

7.5 MAGNETIC FIELD STRENGTH H AND MAGNETOMOTIVE FORCE $\mathscr{F}$

The relationships developed for the magnetic field thus far are only valid in free space. No provisions have been made to include the presence of materials of any kind. This explains the appearance of the free-space permeability μ_0 in the expressions for the magnetic field. The question that now arises is the following: If a material were to occupy all of free space, would the equations be applicable simply by replacing μ_0 by some value appropriate for the material? The answer to this question is yes if the magnetic properties of the materials themselves can be described simply by a scalar permeability μ.

Permeability

Before proceeding, we must reexamine permeability. Postulate 2, Eq. (6.6), which is based upon many experiments, gives the magnetic field from a current element in free space. This expression shows that permeability μ_0 is principally a ratio of the B field to the current I that produces the B field. For free space, B depends linearly on I; hence the ratio μ_0 is a constant. There is a large class of materials for which a linear relationship between B and I also exists and with which a constant permeability may be associated. The materials in this class (which includes all dielectrics and metals) are nonmagnetic as their permeability, for all practical purposes, is equal to μ_0. All equations developed previously are applicable in the presence of these materials.

There exists a small class of materials that show strong magnetic effects. These belong to the iron group and are known as the *ferromagnetic materials*, but are usually referred to simply as the magnetic materials.† They are strongly magnetic in the sense that a given current produces a much larger B field in these materials than in free space. Permeability μ for the ferromagnetic materials is, therefore, much higher than μ_0 but is also nonlinear since μ varies over a wide range with variations in current I.

In electric machinery and in most other applications of these materials, a linear relationship between B and I is desired. A linearization of the magnetic materials is effected by restricting the excursions of I with the result that μ can be approximated by a constant. Materials that are suitable for linearization are known as the *soft ferromagnetic materials.*

Magnetic materials which are suitable for permanent magnets are known as *hard ferromagnetic materials.* For these it is difficult to give a meaning in terms of μ because the relationship between B and I is too complicated. For example, we know that a permanent magnet produces a magnetic field in the absence of current. This implies that μ cannot simply be a scalar, because (6.6) states that B, within some factor, is equal to the product of μ and I. The free-space equations that involve μ_0 cannot be used, therefore, for hard ferromagnetic materials without some modification. This modification will come in Chap. 9 where it will be shown that the magnetic B field has *two* sources—one, the current I, and the other, the magnetization M. The new source M represents the contribution to B that originates entirely in the medium. Clearly, in the case of the permanent magnet, the source of the B field must be magnetization M.

Magnetic Field Strength H

Let us try to separate the two sources for the magnetic B field. If we introduce a new magnetic field vector **H**, called the *magnetic field strength*, by

$$\boxed{\mathbf{B} = \mu \mathbf{H}} \tag{7.15}$$

† The usefulness of the ferromagnetic materials in electric machinery is expressed by postulates 1 and 2, given by (6.1) and (6.6), respectively. Postulate 2 states that a current of a given value will produce a stronger magnetic field in a medium of permeability μ if that medium has a higher value of permeability than μ_0. This fact explains why cores of transformers contain magnetic materials. Postulate 1 states that the force on a current-carrying wire is stronger if the magnetic field is stronger, which makes the use of magnetic materials in motors obvious.

we will find that **H** is independent of the medium.† That is, **H** remains unchanged when a magnetic material is substituted for free space and vice versa. If we divide expression (6.6) for **B** by μ_0, the medium-independent quantity **H** is obtained as

$$d\mathbf{H} = \frac{\mathbf{I} \times \hat{\mathbf{R}}}{4\pi R^2}\, dl \tag{7.16}$$

The above expression shows explicitly that the source for **H** is solely current I.

The introduction of the auxiliary‡ field vector **H** allows us once more to conveniently divide magnetic effects associated with material media into three groups. In the first group which includes free space and the nonmagnetic materials, current is the only source for the B field; hence $\mathbf{B} = \mu_0 \mathbf{H}$.

The second group is that of the soft ferromagnetic materials. The fact that the magnetic B field can be increased without increasing the current I, by filling the free space surrounding the current with magnetic material for which $\mu > \mu_0$, implies that there is a second source for the magnetic field. This source must lie in the material, and its effect is expressed by a μ in (7.15) which is larger than μ_0. Let us define a relative permeability§ μ_r as

$$\mu_r = \frac{\mu}{\mu_0} \tag{7.17}$$

where μ_0 is the permeability of vacuum ($\mu_0 = 4\pi \times 10^{-7}$ H/m). If we replace μ_0 by μ in equations that relate magnetic B field to current, such as Eq. (6.6), we see that μ_r acts as an amplification factor for current I in these equations. It is as if the stronger magnetic B field that is obtained in a magnetic material for which $\mu_r > 1$ is now the result of a larger current $\mu_r I$. Materials in this group are said to be linear if μ_r remains constant with variations in H. All previous equations are then applicable simply by changing μ_0 to $\mu = \mu_r \mu_0$. However, soft ferromagnetic materials are nonlinear. Before such a change can be made, they must be

† For isotropic media, which are of most interest to us because most practical materials are isotropic, μ is a scalar and **H** and **B** are therefore in the same direction. Expression (7.15) serves as a convenient definition of permeability; μ is simply the ratio of B and H. Each material has a BH curve associated with it. For example, the BH curve for free space and nonmagnetic materials is a straight line. The magnetic materials are nonlinear, because $\mu = \mu(H)$. See Figs. 9.28 and 9.31 for examples.

‡ The use of the field vector **H** is analogous to the use of the displacement flux density **D** in electrostatics. There **D** was introduced in order to avoid direct reference to polarization charges. **D** was related to true charges as **H** is related to true currents. In electrostatics we worked with **E** and **D** instead of **E** and polarization **P**. Similarly, for some magnetic materials, it is easier to work with **B** and **H** than with **B** and magnetization **M**. We should emphasize that even though **H** is directly related to I, the fundamental field is **B**, as **B** enters in the force relations. Similarly, in electrostatics, the fundamental quantity is **E** and not **D**.

§ Most materials are nonmagnetic; their relative permeability has a value which is very nearly unity. For paramagnetic materials, μ_r is slightly greater than unity (for example, air is $1 + 3.8 \times 10^{-7}$, and aluminum is $1 + 2.3 \times 10^{-5}$), whereas for diamagnetic materials, μ_r has a value slightly less than unity (water: $1 - 9 \times 10^{-6}$; copper: $1 - 8.8 \times 10^{-6}$). On the other hand ferromagnetic materials such as iron have values of relative permeability in the tens of thousands.

linearized. This is done by operating over a limited portion of the BH curve such that μ remains approximately constant.

The third group is that of the hard ferromagnetic materials which are used in permanent magnets. Since a permanent magnet material produces a magnetic B field when H is zero, (7.15) implies that μ acts as the source term. As it is practically impossible to give a simple expression of μ for the hard ferromagnetic materials, the relationship between B and H is given by a graph (the BH curve) which is experimentally obtained for each material. An example of such a graph is given in Fig. 9.28; it shows that μ is a multivalued function of H. Hence, for these materials, the representation in terms of μ as given by (7.15) is not very useful. From the previous discussion we know that a permeability which is different from μ_0 indicates that magnetization sources are present. These are sources which have their origin in the material medium. Therefore, in the case of hard ferromagnetic materials, B should be related to magnetization M directly instead of through μ. For the linearized soft ferromagnetic materials, there was no need to bring in magnetization because μ effectively expresses magnetization in a simple and useful way.

Ampère's Law and Magnetomotive Force $\mathscr{F}$

Ampère's law for **H** can now be written as

$$\boxed{\oint \mathbf{H} \cdot d\mathbf{l} = I} \tag{7.18}$$

In a sense this is a more general form of Ampère's law. Unlike (7.8), it is independent of the medium that exists within the closed contour of integration. The direct relationship of **H** to I is borne out by (7.18), and gives the units of H as amperes per meter.

If a path encloses a current N times, the right side of (7.18) becomes simply NI. Let us explain why H is sometimes referred to as magnetizing force. If, for example, a current I flows in a coil of N turns and length l, we can say that an H of $H = NI/l$ exists inside the coil.† The magnetic flux density B in the coil is obtained by multiplying H by μ_0, which gives the total flux as $\phi = \mu_0 HA$, where A is the coil cross section. Now, if in place of the air core, a magnetic core is substituted, the core becomes magnetized because as before an $H = NI/l$ acts on the core. Multiplying this H by μ of the core material and by A, the total magnetic flux $\phi = \mu HA$ through the core is determined. As $\mu \gg \mu_0$, the flux is much larger than before. H can, therefore, be pictured as driving a flux through a medium.

We can now define a magnetomotive force (mmf) $\mathscr{F}$, similar to electromotive force, by

$$\boxed{\mathscr{F} = NI = \oint \mathbf{H} \cdot d\mathbf{l}} \tag{7.18a}$$

† For most coil shapes the approximation of the line integral (7.18) by Hl is a very good one.

Returning again to the example of a core with N turns of wire wrapped around it, we see that by passing a current I through the wire, a magnetomotive force of NI ampere-turns is applied to the core. This results in a magnetizing force $H = NI/l$ which pushes the magnetic flux through the core.

The operation of devices such as relays, transformers, lifting magnets, electric machines, etc., is based on the total magnetic flux ϕ. Hence it is important to understand how such a flux is established in the cores of these devices. It should also be understood that even though μ is widely used in engineering design, it is not practical to treat hard ferromagnetic materials in terms of permeability μ. Such materials are discussed in detail in Chap. 9.

7.6 APPLICATIONS OF AMPÈRE'S LAW

In situations where symmetry exists, the calculation of magnetic fields is greatly simplified by the use of Ampère's law. Let us illustrate with a few examples.

Thick Conducting Cylinder and Coaxial Transmission Line

A solid cylindrical conductor of radius a has a steady current I distributed uniformly over the cross section, as shown in Fig. 7.4. The current density J is equal to $J = I/\pi a^2$. The objective is to find the B field inside and outside the cylinder. Outside, the magnetic field is given by Ampère's law as

$$B = \frac{\mu_0 I}{2\pi r} \qquad r \geq a \tag{7.19}$$

and forms concentric circles about the cylinder just as in the case of the thin, infinitely long wire. Inside the cylinder, Ampère's law tells us that the B field is determined solely by the current flowing inside a circle of radius r, where $r \leq a$.

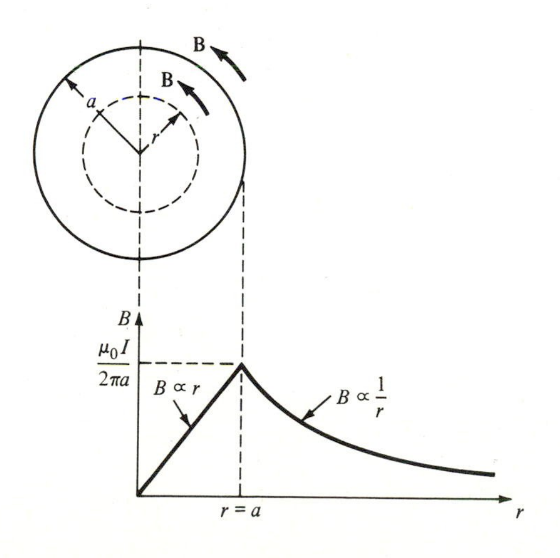

Figure 7.4 The magnetic field inside and outside a conductor carrying a uniformly distributed current.

Thus from (7.9) and the symmetry of the problem

$$\oint \mathbf{B} \cdot \mathbf{dl} = \mu_0 \iint \mathbf{J} \cdot d\mathbf{A}$$

$$B2\pi r = \mu_0 J \pi r^2$$

$$B = \frac{\mu_0 J r}{2}$$

$$= \frac{\mu_0 I}{2\pi a^2} r \qquad r \leq a \tag{7.20}$$

Hence inside the cylinder, the magnetic field increases linearly from zero on the axis to a maximum at the surface of the conductor and then decreases inversely with distance outside the cylinder. Note that the permeability μ of the conducting material is that of free space; that is, $\mu = \mu_0$.

If we have two coaxial conductors, as in the coaxial transmission line shown in Fig. 5.5, the inner and outer conductors carry a current I in opposite directions. Hence the field in the region between the two cylinders is solely determined by current I flowing in the inner conductor and is given by (7.19). Also from Ampère's law, we know that the B field outside the coaxial line is equal to zero because the net current enclosed $I - I$ is zero.

Solenoid

The B field from a solenoid was already considered in Sec. 6.9 and sketched in Fig. 6.25. Rederiving it using Ampère's law will show the power of Ampère's law and, at the same time, permit us to learn more about the B field.

A cross section of a solenoid which is a tightly wound helix of conducting wire is shown in Fig. 7.5. Let us designate the number of turns per unit length as n; that is, $n = N/l$. For an infinite ideal solenoid† the field everywhere inside must be

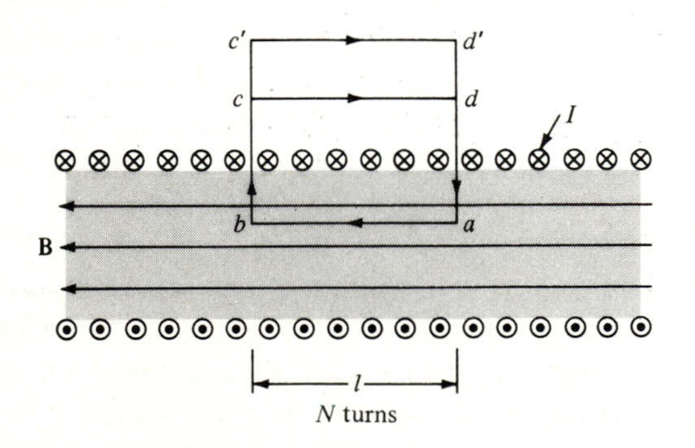

Figure 7.5 A cross section of a solenoid showing the B field and a rectangular contour for use with Ampère's law. The crosses represent currents going into the page, whereas the dots are currents coming out of the page.

† In an ideal solenoid the winding is assumed to be a current sheet as discussed in the paragraph following Eq. (6.42). A closely spaced winding of fine wire is an approximation to a cylindrical current sheet.

parallel to the axis of the solenoid. If it is not, as for example in Fig. 6.25, then the solenoid is not infinite in length. We also observe that the field outside is very small compared to the field inside if the solenoid is long compared to its diameter. For an infinitely long ideal solenoid, the field outside is zero; therefore along path *cd*, the line integral of $\mathbf{B} \cdot \mathbf{dl}$ is zero. If it is not, and a finite value along *cd* is obtained, then this would imply that the same value along *c'd'* must be obtained. This in turn implies that a constant field everywhere outside the coil exists which is contrary to experimental observation. Ampère's law gives us

$$\oint \mathbf{B} \cdot \mathbf{dl} = \int_a^b \mathbf{B} \cdot \mathbf{dl} + \int_b^c \mathbf{B} \cdot \mathbf{dl} + \int_c^d \mathbf{B} \cdot \mathbf{dl} + \int_d^a \mathbf{B} \cdot \mathbf{dl}$$

$$= \mu_0 NI \tag{7.21}$$

Since the field along *cd* is zero, and along *bc* and *da* the *B* field is at right angles to the path, we are left with

$$\int_a^b \mathbf{B} \cdot \mathbf{dl} = \mu_0 NI$$

$$B = \frac{\mu_0 NI}{l} = \mu_0 nI \tag{7.22}$$

which is the same result obtained earlier in (6.40). This infinite solenoid result is an excellent approximation for the field of long solenoids. It shows that *B* is independent of the solenoid diameter, and that *B* is constant over the cross section of the solenoid.

In terms of the magnetomotive force $\mathscr{F}$ we can say that the *H* inside the solenoid is given by

$$\oint \mathbf{H} \cdot \mathbf{dl} = NI = \mathscr{F}$$

$$H = \frac{\mathscr{F}}{l} \tag{7.23}$$

Thus the mmf $\mathscr{F}$ divided by the length of a solenoid (or per unit length of an infinite solenoid) gives us the magnetic field strength or magnetizing force *H* inside the solenoid. Multiplying by μ_0 and cross section *A*, the total magnetic flux $\phi = \mu_0 ANI/l$ is obtained.

It is easy to show that the flux at either end of a long solenoid must be half that at its center. To see this, let us cut an infinite solenoid in two. If the current in the two parts is maintained at the same value, the *B* field at the newly created ends must drop to one-half its original value, otherwise the *B* field would not have its original value when the two ends are reconnected. This implies that half the lines that exist at the center of a long solenoid leak out through the solenoid turns somewhere between the center and one end, as shown in Fig. 6.25*a*.

Toroid

Using Ampère's law to calculate the B field in a toroid is a more rigorous approach than that used in Sec. 6.9. This approach will tell us not only that the magnetic field outside the doughnut-shape configuration is zero but also that the field inside the toroidal winding varies inversely with distance as shown in Fig. 7.6. In this figure we are showing the cross section of a toroid with current I in the N turns of the winding. Ampère's law for a circular path $r \leq a$ gives $\oint \mathbf{B} \cdot \mathbf{dl} = 0$ because no current is enclosed. For a path $a < r < b$, we have

$$\oint \mathbf{B} \cdot \mathbf{dl} = \mu_0 NI \tag{7.24}$$

Because of symmetry, the B field is constant at a fixed radius r. This gives $B2\pi r$ for the integral, and the magnetic field is thus

$$B = \frac{\mu_0 NI}{2\pi r} \qquad a < r < b \tag{7.25}$$

For a circular path $r \geq b$ Ampère's law tells us that

$$\oint \mathbf{B} \cdot \mathbf{dl} = \mu_0 N(I - I) = 0$$

$$2\pi r B = 0$$

$$B = 0 \qquad r \geq b \tag{7.26}$$

that is, the field outside a toroid is zero (not completely true unless current sheets are assumed) because no net current is enclosed. This is also expected from the symmetry of the toroid, because if a field were assumed to come out, where would it come out and what shape would the field lines have? Therefore the magnetic field is confined to the interior of the winding.

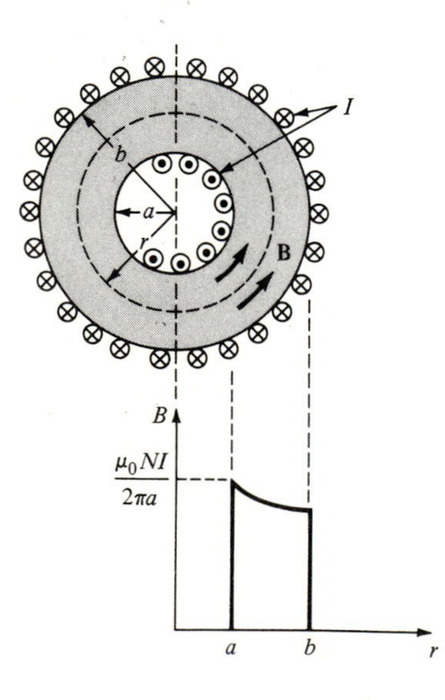

Figure 7.6 The magnetic field from a toroid. Ampère's law tells us that the B field is confined to the interior of the toroid where it varies inversely with distance r from the center.

7.7 INDUCTANCE

Like capacitance, inductance L is a property of a physical arrangement of conductors. It is a measure of magnetic flux which links the circuit when a current I flows in the circuit. It is also a measure of how much energy is stored in the magnetic field of an inductor, such as a coil, solenoid, etc.

The definition of inductance rests on the concept of flux linkage. It is not a very precise concept unless one is willing to introduce a complicated topological description. For our purposes it will be sufficient to define *flux linkage* Λ as the flux that links all the circuit, multiplied by the number of turns N. For example, in the case of the solenoid shown in Fig. 7.7, flux linkage will be given by

$$\Lambda = N\phi = N \iint \mathbf{B} \cdot \mathbf{dA} \cong NBA \qquad \text{Wb} \tag{7.27}$$

that is, only the flux that goes through the inside of the solenoid and therefore links all turns is used. The small flux loops about each turn are ignored in a first-order analysis because they link only one or two turns and flow through a small area. The area A is that area through which the flux that links all turns flows. For the solenoid of Fig. 7.7, a good approximation to A is the cross section of the solenoid.

Because the medium is assumed to be linear, the magnetic flux is proportional to the current causing the flux. We can now define *inductance* L of an inductor as the ratio of flux linkages to the current through the inductor:

$$L = \frac{\Lambda}{I} = \frac{N \iint \mathbf{B} \cdot \mathbf{dA}}{I} \qquad \text{H} \tag{7.28}$$

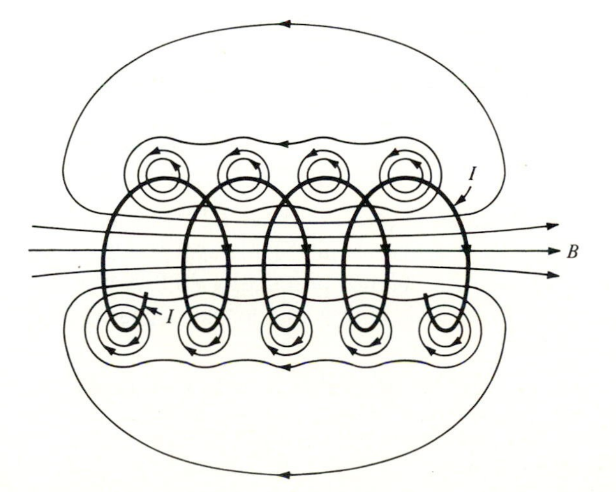

Figure 7.7 A solenoid and its magnetic field.

The unit of inductance is the henry (H). Inductors for filter applications in power supplies are usually wire-wound solenoids on an iron core with inductances in the range from 1 to 10 H. Inductors found in high-frequency circuits are air-core solenoids with values in the millihenry (mH) range. The definition for inductance, (7.28), even though it is derived for steady currents, is valid up to very high frequencies.

Let us calculate L for some useful geometries.

Solenoid

A good approximation of the B field in a solenoid that links all turns is the B field at the center of the solenoid; that is, $B = \mu_0 NI/l$ from (7.22) or (6.40). There is some leakage at the ends of the solenoid (recall that the value of the B field drops to one-half at the ends), which we will ignore because it occurs mainly at the ends. The inductance L of a solenoid is therefore

$$L = \frac{\Lambda}{I} = \frac{NBA}{I} = \frac{\mu_0 N^2 A}{l} \tag{7.29}$$

where l is the length and A is the cross section of the solenoid.

If we have a short solenoid of N turns, that is, one where the length l is smaller than radius a, we can use Eq. (6.43), which gives the magnetic field as $B = \mu_0 NI/2a$. Using this expression for the interior of the solenoid, we obtain for inductance:

$$L = \mu_0 N^2 \frac{A}{2a} = \mu_0 N^2 \frac{\pi a}{2} \tag{7.30}$$

As a special case of this we find that the inductance of a one-turn loop is $L = \mu_0 \pi a/2$.

As pointed out before, the expressions for inductance, except for μ, depend only on geometrical factors. In these expressions one always finds μ since flux is proportional to it, and one always finds the number of turns as N^2 (once as the N in $N\phi$ and once again since ϕ is proportional to N).

For example, the inductance of a long solenoid of $N = 1000$, $l = 50$ cm, $a = 2$ cm, with an air core ($\mu_0 = 4\pi \times 10^{-7}$ H/m) gives the value 3.1 mH for L.

Toroid

For a toroid, shown in Fig. 7.6, flux leakage is small because a toroid, unlike a solenoid, has no ends. Flux leakage between turns is negligible and is zero if we assume a current sheet in place of the winding. Hence practically all flux links all turns, and calculations for inductance using (7.28) are very accurate. Using (7.25) to express the B field in a toroid, we obtain for L:

$$L = \frac{\Lambda}{I} = \frac{\mu_0 N^2 A}{2\pi r} \tag{7.31}$$

where A is the cross section and r is the mean radius of the toroid.

For example, the inductance of a 2000-turn toroid having a cross-sectional area of 1 cm^2 and mean radius of 5 cm is

$$L = (4\pi \times 10^{-7}\ \text{H/m})(2000)^2(10^{-4}\ \text{m}^2)/2\pi(0.05\ \text{m}) = 1.6\ \text{mH}$$

If the toroid were filled with iron instead of air, the inductance could be increased many thousand fold.

Note that we have neglected the variation of B across the cross section of the toroid. By using an average r, as, for example, $r = (a + b)/2$, we have in effect used an average value of B in the calculation of inductance. If this is not sufficiently accurate, the variation of B should be considered by integrating (7.25) between a and b.

Coaxial Transmission Line

The student usually does not have any difficulty in grasping the concept of inductance as long as the geometries involve windings (such as in coils and toroids). In the following examples flux linkage is used in a broader sense and should clarify that concept further. Figure 7.8 shows a longitudinal and transverse cross section of a coaxial line (already considered in Sec. 5.5 when the capacitance per unit length was calculated). The current I flows in the center conductor and returns in the outer conductor. The magnetic flux in the space $a < r < b$ is the flux that links the inner and outer conductors. As discussed in Sec. 7.6, B in the region between the two conductors is established by the current flowing in the inner conductor and is therefore given by

$$B = \frac{\mu_0 I}{2\pi r} \qquad a < r < b \tag{7.32}$$

Since the B field is not constant but varies inversely with r, we must integrate the B field. The total flux linkage in a transmission line of length l is therefore

$$\Lambda = \iint \mathbf{B} \cdot \mathbf{dA} = l \int_a^b \frac{\mu_0 I}{2\pi r}\,dr = \frac{l\mu_0 I}{2\pi} \ln \frac{b}{a} \tag{7.33}$$

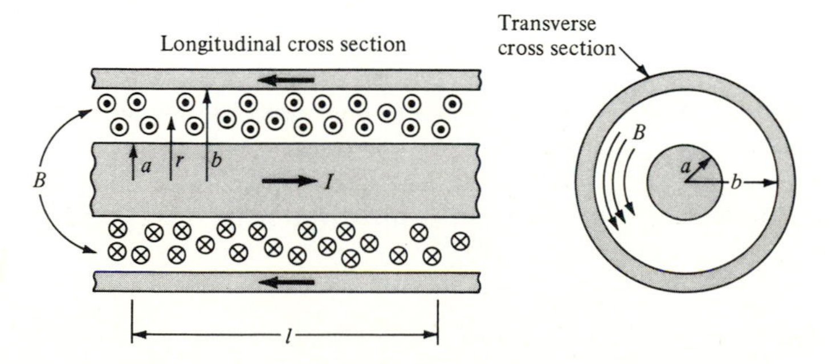

Figure 7.8 A longitudinal and transverse cross section of a coaxial cable.

The inductance per unit length L/l is

$$\frac{L}{l} = \frac{\Lambda}{Il} = \frac{\mu_0}{2\pi} \ln \frac{b}{a} \qquad \text{H/m} \tag{7.34}$$

For an air-filled coaxial line the above expression can be written as $L/l = 0.2 \ln (b/a)$ microhenrys per meter (μH/m).

We have ignored the contribution of the magnetic field inside the inner conductor for several reasons.† First, as shown in Fig. 7.4, the magnetic flux within the inner conductor (assuming the current I is distributed uniformly throughout the cross section of the inner conductor, which is a valid assumption for direct current and for current at low frequencies) links only a fraction of that conductor; that fraction is proportional to $(r/a)^2$ because $I_{\text{at}\, r} = (r/a)^2 I$. Second, at the higher frequencies the current is effectively confined to a thin layer (skin depth) at $r = a$ for the inner conductor and at $r = b$ for the outer. Third, most practical transmission lines use a small inner conductor and a thin-walled outer conductor. Hence the flux linkages within the conductors can be neglected, and (7.34) is an accurate expression for inductance per unit length.

Note that we have now characterized the coaxial transmission line. Knowing capacitance and inductance per unit length, C/l [see Eq. (5.14)] and L/l, respectively, permits us to obtain the characteristic impedance $Z_0 = \sqrt{L/C}$ and phase velocity of an electromagnetic wave:

$$v_p = \frac{1}{\sqrt{LC}} = 1 \bigg/ \sqrt{\left[\frac{\mu_0}{2\pi} \ln \left(\frac{b}{a}\right)\right] \left[\frac{2\pi\varepsilon_0}{\ln (b/a)}\right]} = \frac{1}{\sqrt{\mu_0 \varepsilon_0}} = 3 \times 10^8 \text{ m/s}$$

for an air-filled transmission line.

Parallel-Wire Transmission Line

The 300-Ω TV twin lead is perhaps the best known among the two-wire transmission lines. In Sec. 5.6 the capacitance per unit length of that line was calculated. Let us now find the inductance per unit length for such a cable. Fortunately this calculation will be much simpler than the corresponding one for capacitance.

Figure 7.9 shows the longitudinal and transverse cross sections of the parallel line. Here we may divide the magnetic field B into two parts: That produced by the current in the lower conductor and linking the upper conductor, and that produced by the current in the upper conductor and linking the lower one. The flux linkages of each conductor may be determined separately because the medium is assumed to be linear. Hence, the total flux linkage is just twice the integral of (7.32) from a to $d - a$, or

$$\Lambda = 2l \int_a^{d-a} \frac{\mu_0 I}{2\pi r} \, dr \cong \frac{l\mu_0 I}{\pi} \ln \frac{d}{a} \tag{7.35}$$

† In Prob. 7.15 it is shown that the contribution to inductance due to the flux inside the inner conductor is $L/l = \mu_0/8\pi$. A more accurate expression of (7.34) is therefore $L/l = \mu_0/8\pi + (\mu_0/2\pi) \ln (b/a)$. An excellent discussion of partial flux linkages is given in J. B. Walsh, "Electromagnetic Theory and Engineering Applications," sec. 4-9. The Ronald Press Company, New York, 1960.

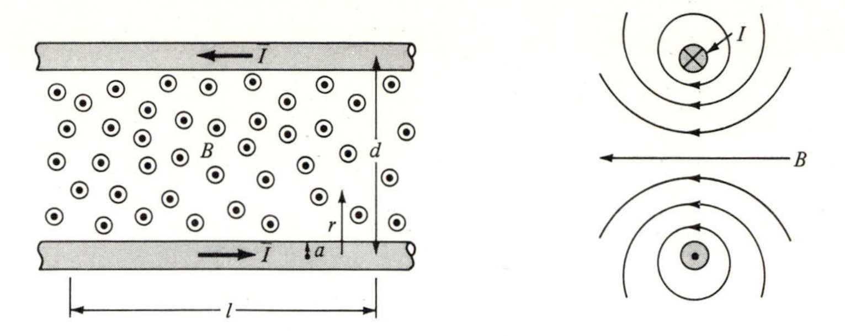

Figure 7.9 Cross sections of the parallel-wire transmission line.

We have approximated the upper limit $d - a$ by d because for practical transmission lines $d \gg a$. This approximation also accounts for the flux from the lower conductor which partly links the current inside the upper wire. As a matter of fact it can be shown that the replacement of $d - a$ by a gives an exact result for the flux linkages.† The inductance per unit length L/l is then

$$\frac{L}{l} = \frac{\mu_0}{\pi} \ln \frac{d}{a} \tag{7.36}$$

For an air medium we can substitute $\mu_0 = 4\pi \times 10^{-7}$ H/m and obtain $L/l = 0.4 \ln (d/a)$, μH/m.

7.8 ENERGY STORED IN AN INDUCTOR

Just as a capacitor stores energy in its electric field, so does an inductor in its magnetic field. A measure of the effectiveness of energy storage in the electric field is the capacitance ($W = \frac{1}{2}CV^2$); in the magnetic field, it is the inductance ($W = \frac{1}{2}LI^2$).

To derive the expression for the storage of energy in the magnetic field of an inductor, let us begin with Kirchhoff's law (3.8) for a general RLC circuit shown in Fig. 7.10, which is

$$V = RI + L\frac{dI}{dt} + \frac{1}{C}\int I\,dt \tag{7.37}$$

Since the instantaneous power is $P = VI = dW/dt$, we can obtain the energy in the inductor L by integrating power $P_L = V_L I = LI\,dI/dt$. We obtain

$$W = \int_0^t P_L\,dt = L\int_0^t I\frac{dI}{dt}\,dt = L\int_0^I I\,dI = \tfrac{1}{2}LI^2 \tag{7.38}$$

† E. W. Kimbark, "Electrical Transmission of Power and Signals," sec. 2-11, John Wiley and Sons, Inc., New York, 1949. L. F. Woodruff, "Electric Power Transmission," pp. 17 and 22, McGraw-Hill Book Company, New York, 1938.

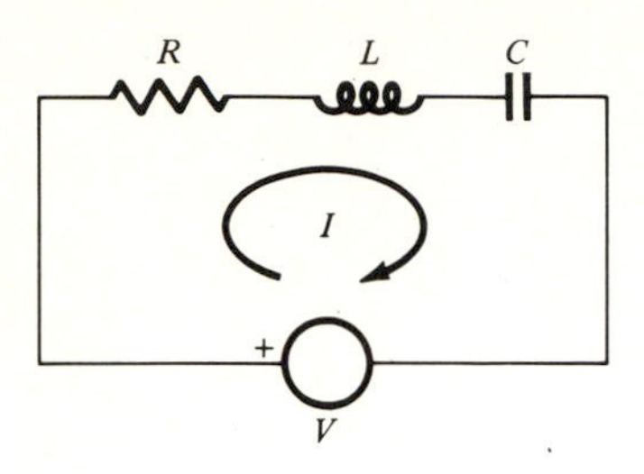

Figure 7.10 An RLC series circuit. Over a large range of current values, the parameters RLC are constant.

which is the desired result and gives the total stored magnetic energy in an inductance L carrying current I. For example, a solenoid with an inductance of 8 H and a current of $\frac{1}{2}$ A has an energy stored of $W = \frac{1}{2}LI^2 = 1$ J.

LR Circuits

We can now make the important observation that an inductor has inertia for current and will oppose any changes of current that flow in it. Because an inductor L stores energy and because energy cannot change instantaneously [for this would require infinite power $(P = dW/dt)$], it follows that current I cannot change instantaneously. This effect is made use of in filters where an inductance is used to smooth current variations, just as capacitors are used to smooth voltage variations.

Because at any instant an inductor stores energy of $\frac{1}{2}LI^2$, energy is required to build up a current in a coil. An alternative point of view, in terms of Lenz' law, is therefore as follows: A counter-emf voltage is induced in a coil to oppose the change of current through it, which from (7.37) is $V_L = L\,dI/dt$. The current through an inductor can therefore only change "slowly." To obtain the behavior of the current versus time during "charge" and "discharge" of an inductor,† let us consider the LR circuit of Fig. 7.11. If the switch is in position b for a long time, the current I through R and L can be assumed to be zero. Throwing the switch to position a will connect a battery of voltage V to the series RL circuit. Even though moments after the switch is thrown the current in the circuit remains zero, it will start to build up as described by the solution to the following circuit equation:

$$V = RI(t) + L\frac{dI(t)}{dt} \tag{7.39}$$

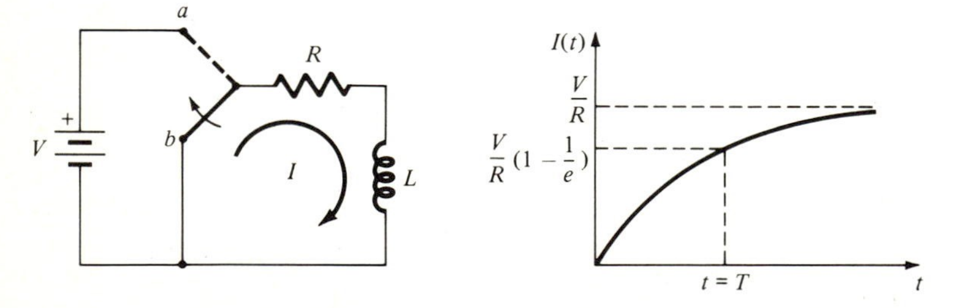

Figure 7.11 An RL circuit used to show the build-up of current when the switch is thrown from b to a.

† We are using the terms "charge" and "discharge" figuratively to denote the build-up and decay of current in an inductor.

Rearranging Eq. (7.39) as

$$\frac{dI}{dt} + \frac{R}{L} I = \frac{V}{L} \tag{7.40}$$

we can write the solution by inspection as

$$I(t) = Ae^{-(R/L)t} + B \tag{7.41}$$

where A and B are constants to be determined by boundary conditions (in this case, initial and final conditions for the current). Initially the current is zero, which gives $I(0) = 0 = A + B$. After a long time ($t \to \infty$) has elapsed, the current will settle down to $I(\infty) = B = V/R$. The solution for the current is therefore

$$I(t) = \frac{V}{R}(1 - e^{-t/(L/R)}) \tag{7.42}$$

and is plotted in Fig. 7.11*b*. Note that just as in the case of a capacitor which was discussed in Secs. 5.2 and 5.9, we can associate a time constant $t = T = L/R$ with the build-up of current in an LR circuit. After the time equal to one time constant L/R has elapsed, the current I differs from its final value V/R by $1/e$, or 37 percent.

We now have the following picture of the physical process involved: During the time that the current I rises from zero, the battery continuously supplies energy to the resistor R at the rate of I^2R and to the inductor L at the rate of $LI\, dI/dt$. In practical situations we can say that after four time constants, the current in the circuit has reached its final value of $I = V/R$. After this time the battery supplies energy to R at the rate of $I^2R = V^2/R$ but none to the inductor. The inductor merely stores an amount of energy equal to $W = \frac{1}{2}LI^2 = \frac{1}{2}LV^2/R^2$. This stored energy in L can be released (again "slowly") by throwing the switch to position b. The equation describing the new current is

$$L\frac{dI}{dt} + RI = 0 \tag{7.43}$$

The initial current after the switch has been thrown to b is $I = V/R$. The final current, after a long time has elapsed, is of course zero. This gives for a solution to the above equation:

$$I = \frac{V}{R}e^{-t/(L/R)} \tag{7.44}$$

The current is thus seen to decrease exponentially from an initial value of V/R to zero. As the current decreases, the stored energy in L also decreases. The energy released by L is dissipated as heat in the resistor (the I^2R loss).

7.9 ENERGY STORED IN A MAGNETIC FIELD

In Chap. 5, we showed that the site of stored energy in a capacitor can either be the charge on the capacitor plates or the electric field between the plates. We determined that the energy per unit volume in the electric E field is $w = \frac{1}{2}\varepsilon E^2$. Similarly, energy is stored in a magnetic field; the energy density will turn out to be $\frac{1}{2}\mu H^2$. To show this in the simplest way, we look for an inductor with a uniform field which is well confined. A long solenoid is suitable as is a toroid which has the magnetic field confined entirely to the region within the windings.

A toroid with a diameter that is large compared with that of its cross section has an almost uniform field, as shown in Fig. 7.6. The energy density at each point in the magnetic field is then

$$w = \frac{W}{v} = \frac{\frac{1}{2}LI^2}{A2\pi r} = \frac{1}{2}\mu_0\left(\frac{NI}{2\pi r}\right)^2 = \frac{B^2}{2\mu_0} = \frac{1}{2}\mu_0 H^2 \tag{7.45}$$

where A = cross section
$2\pi r$ = mean circumference
v = volume, $v = A2\pi r$
L = inductance, $L = \mu_0 N^2 A/2\pi r$
B = magnetic field of the toroid, $B = \mu_0 H = \mu_0 NI/2\pi r$

We state now without proof that (7.45) expresses the energy density in any magnetic field. The total magnetic energy stored in the field of an inductor can therefore be obtained by integrating the magnetic energy density:

$$W = \iiint w\, dv \tag{7.46}$$

The integration is over the entire volume in which the field exists and must, of course, be equal to $\frac{1}{2}LI^2$.

Alternative Definition of Inductance

An interesting application of the magnetic energy density is in an alternative definition of inductance. From (7.46) we can express inductance L as follows:

$$L = \frac{2W}{I^2} = \frac{\iiint \mu H^2\, dv}{I^2} = \frac{\iiint BH\, dv}{I^2} \tag{7.47}$$

For example, the inductance of a coaxial transmission line of length l, which was derived in (7.34), and for which the magnetic field between the inner and outer conductors is given by $B = \mu I/2\pi r$, becomes

$$L = \frac{1}{I^2}\int_a^b \frac{\mu I^2}{4\pi^2 r^2} l 2\pi r\, dr = \frac{\mu l}{2\pi} \ln \frac{b}{a} \tag{7.48}$$

Expression (7.48) is the same as obtained previously. The medium between conductors is usually air or some polystyrene foam; the permeability is then $\mu \cong \mu_0$.

Field Cells and the Meaning of Permeability μ_0

In Sec. 5.11 we gave an interpretation of permittivity ε_0. A similar picture can be presented for μ_0. In order to do this, we first look for a structure with a uniform magnetic field. Two infinite conducting parallel planes carrying equal and opposite currents, as shown in Fig. 7.12, produce a uniform field inside and zero field outside.†

Imagine a cubical section, as shown in the three-dimensional view of Fig. 7.12*b*, is cut from the two parallel sheets. A structure such as this is called a

† We can use the same proof as that for an infinitely long solenoid, shown in Fig. 7.5, to show that $B = 0$ outside the parallel planes.

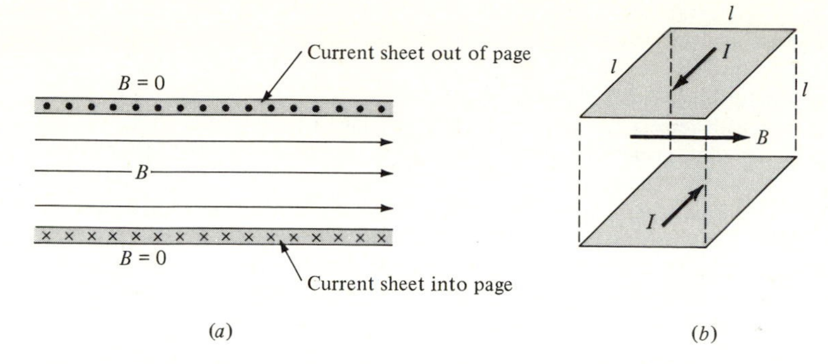

Figure 7.12 (*a*) The cross section of two infinite parallel current sheets; (*b*) a field cell is a cubical cut from the infinite sheets.

field cell if its dimensions are small. Current I in the field cell can be related to the H field by Ampère's law as

$$Hl = I \tag{7.49}$$

The inductance of the cell is

$$L = \frac{\Lambda}{I} = \frac{Bl^2}{I} = \frac{\mu_0 H l^2}{Hl} = \mu_0 l \tag{7.50}$$

The permeability μ_0 of free space can then be interpreted as the inductance per unit length of a field cell; that is,

$$\mu_0 = \frac{L}{l} \qquad \text{H/m} \tag{7.51}$$

Hence each point in space has associated with it such an inductance irrespective of whether a magnetic field exists at the point or not.

To summarize, we have shown that each point in free space has inductance and capacitance associated with it. In that sense, free space is similar to a transmission line which also has inductance and capacitance. And as in the case of the transmission line along which waves can travel, free space can similarly support traveling waves. The velocity of propagation of such waves is given in terms of the inductance and capacitance of free space. In free space or vacuum this velocity is the velocity of light $c = 1/\sqrt{\mu_0 \varepsilon_0} = 3 \times 10^8$ m/s.

PROBLEMS

$$\varepsilon_0 = 8.85 \times 10^{-12} \text{ F/m} \qquad \mu_0 = 4\pi \times 10^{-7} \text{ H/m}$$

7.1 The Biot-Savart law (6.9):

$$\mathbf{B}_1 = \frac{\mu_0}{4\pi} \iiint_v \frac{\mathbf{J}_2 \times \hat{\mathbf{R}}}{R^2} \, dv_2$$

is a generalization of experimental data. An experimental fact is that magnetic fields generated by currents form closed lines about such currents; hence $\nabla \cdot \mathbf{B} = 0$ should be reflected in the Biot-Savart law. Show this by applying the divergence with respect to field point 1 to the expression (6.9). Note that integration is with respect to source point 2; hence we can take ∇_1 inside the integral (in this case when the differentiation is with respect to x_1, y_1, z_1 and integration is with respect to x_2, y_2, z_2, the differentiation and integration operators can be interchanged). Apply identity (A1.21) and note that $\hat{\mathbf{R}}/R^2 = -\nabla_1(1/R)$ [see Eq. (8.44)].

7.2 When the magnetic field from a current element $I\,dl'$ was integrated about a closed path, a result given by Eq. (7.10) was obtained which did not agree with Ampère's law; that is, $\oint \mathbf{B} \cdot \mathbf{dl} \neq \mu_0 I$. This contradiction is explainable because a current element is nonphysical; that is, it can exist only as part of a complete circuit. Hence for a closed-path integration of a **B** field produced by a continuous current (i.e., a continuum of current elements), we obtain

$$\oint \mathbf{B} \cdot \mathbf{dl} = \oint_c \frac{\mu_0}{4\pi} I \oint_{c'} \frac{\mathbf{dl}' \times \hat{\mathbf{R}}}{R^2} \cdot \mathbf{dl}$$

For this statement to agree with Ampère's law, the right side must be equal to†

$$\oint_c \oint_{c'} \frac{\mathbf{dl}' \times \hat{\mathbf{R}}}{R^2} \cdot \mathbf{dl} = 4\pi$$

Show this. For example, in the special case when path c' is an infinite straight wire-carrying current I, and path c is circular, we obtain

$$\begin{aligned}\oint \mathbf{B} \cdot \mathbf{dl} &= \oint_c \frac{\mu_0}{4\pi} I \int_{-\infty}^{\infty} \frac{\sin\theta\, dl'}{R^2}\, dl \\ &= \frac{\mu_0 I}{4\pi} \int_{-\infty}^{\infty} \frac{\sin\theta\, dl'}{R^2} (2\pi R_0) \\ &= \frac{\mu_0 I}{2} \int_0^{\pi} \sin\theta\, d\theta = \mu_0 I\end{aligned}$$

where the dl integration about path c yields $2\pi R_0$, $\sin\theta = |\hat{\mathbf{dl}}' \times \hat{\mathbf{R}}|$, $R_0 = R \sin\theta$, and $dl' \sin\theta = R\, d\theta$.

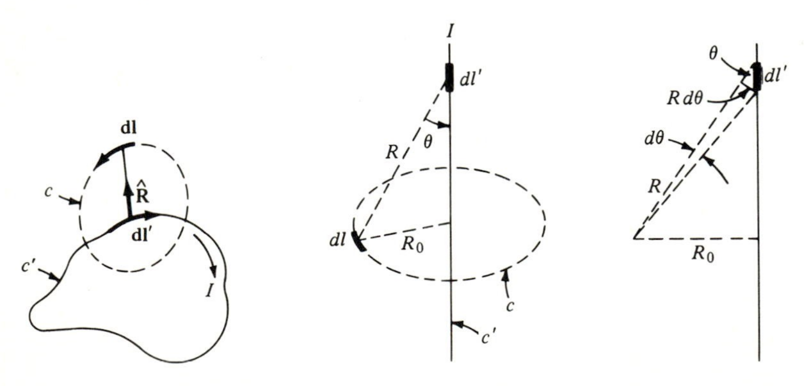

† A proof for an arbitrary path is given in J. C. Slater and N. H. Frank, "Electromagnetism," p. 60, McGraw-Hill Book Company, New York, 1947.

7.3 Show that the magnetic field inside a long straight wire of radius a carrying a current with uniform density J is given by $B = \mu_0 I'/2\pi r$, where $r < a$ and I' is the total current enclosed in the area πr^2; that is, $I' = J\pi r^2$. Note that this looks like the expression for the B field outside a current-carrying wire. Compare this expression to Eq. (7.20).

7.4 Calculate the magnetic field strength **H** and magnetic flux density **B** at a distance of 1 m from a wire which carries a current of 1 A if

(*a*) The wire is immersed in free space.

(*b*) The wire is immersed in a magnetic material with relative permeability $\mu_r = \mu/\mu_0 = 100$.

7.5 Determine the magnetic field due to a current I flowing in the wall of a cylindrical tube, assuming the wall to have negligible thickness. Give the field both inside and outside the tube.

7.6 Calculate the **B** field for all r created by a hollow cylindrical conductor which carries a current I. The inner and outer radii of the conducting cylindrical shell are a and b, respectively.

7.7 Show that the magnetic field in a coaxial cable is given by

$$B = \frac{\mu_0 I}{2\pi a^2} r \qquad \text{for } r \le a$$

$$B = \frac{\mu_0 I}{2\pi r} \qquad \text{for } a \le r \le b$$

$$B = \frac{\mu_0 I(c^2 - r^2)}{2\pi r(c^2 - b^2)} \qquad \text{for } b \le r \le c$$

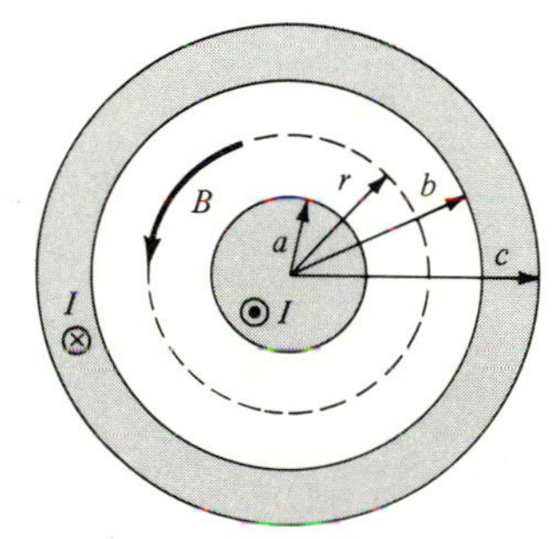

7.8 Find the magnetic field B at the center of a cylindrical hole of radius b which is displaced from the center of a cylindrical conductor of radius a. The conductor carries a current density J amperes per square meter. The geometry is shown in the accompanying sketch.

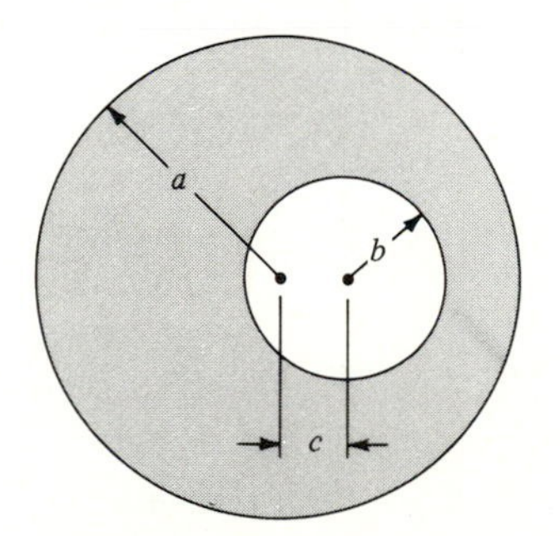

7.9 Calculate the magnetic flux through a closed surface A which encloses a small loop which carries current I.

7.10 Find the magnetic field B outside an infinite very thin planar current sheet carrying a sheet current **K** amperes per meter. A section of the current sheet is shown.

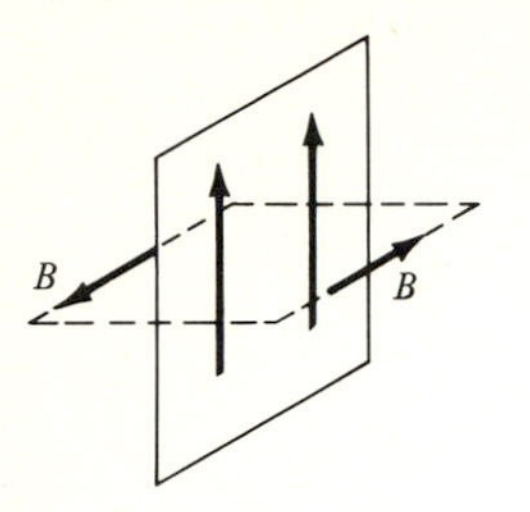

7.11 Show that the magnetic fields between and outside two infinite planar current sheets, each carrying a sheet current **K** (which is given in amperes per unit width), is as shown in the diagram. An edge view of the current sheets is shown, with B into the paper.

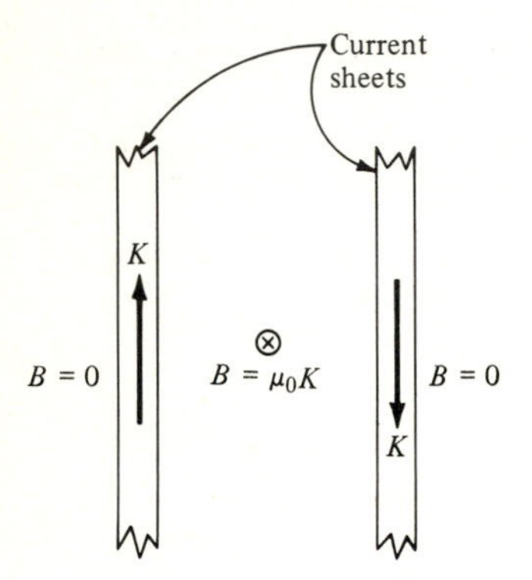

7.12 If a circular loop of conducting wire has inductance L, what is the inductance of a coil of the same radius with N turns?

7.13 State the inductance L for a solenoid of N turns using (6.39), which gives the B field at the center of the solenoid as the approximate expression anywhere in the cross section of the coil.

(*a*) Give a radius-independent expression for L valid for a long and slender solenoid, i.e., one for which $l \gg a$.

(*b*) Give a length-independent expression for L valid for a short and fat coil, i.e., one for which $a \gg l$.

7.14 Show that the expression for the inductance of a washer-shaped toroid of rectangular cross section is given by

$$L = \frac{\mu_0 N^2 h}{2\pi} \ln \frac{b}{a}$$

where the height of the washer-shaped toroid is h and the inner and outer radii are a and b, respectively. The area of the cross section then is $h(b - a)$.

7.15 Show that the inductance for a length l of a long wire associated with the flux inside the wire is $\mu_0 l/8\pi$, and is thus independent of the wire radius.

Hint: The expression for inductance, (7.47), is applicable to this problem.

7.16 Determine the inductance L and the internal resistance R of a coil, given that the inductor's current decays to $1/e$ of its initial value in 0.6 ms when a 4-Ω resistor is placed in series with the inductor. If the additional series resistor is removed, the initial current falls to $1/e$ of its initial value in 0.8 ms.

7.17 Show that, if the time constant L/R is long compared with the period of the input signal, the circuit shown can be considered to be an integrating circuit; that is,

$$V_0 \propto \int V_i \, dt.$$

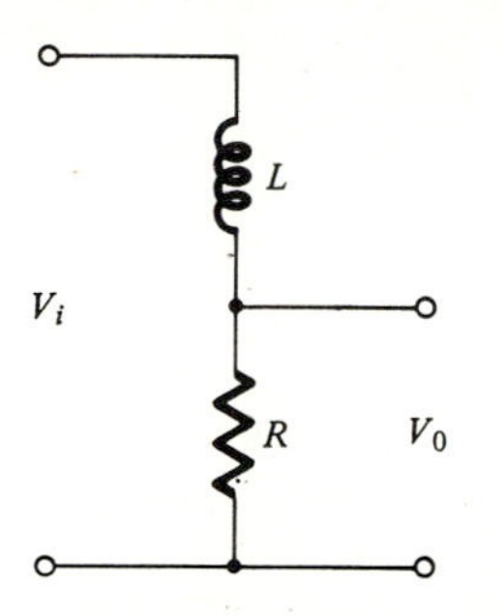

7.18 Find the total thermal energy that is dissipated in a resistor R when a coil of $L = 0.1$ H carrying a current of 2 A is suddenly disconnected and connected across R.

7.19 A coil has an inductance of 4 H and a resistance of 10 Ω. What is the energy stored in the magnetic field when an emf of 100 V is applied and after the current has built up to its maximum value?

CHAPTER

EIGHT

DESCRIPTION OF THE MAGNETIC FIELD IN DIFFERENTIAL FORM

Up to this point the description of the static magnetic field was given primarily in terms of integral expressions such as Ampère's law, $\oint \mathbf{H} \cdot \mathbf{dl} = I$, and Gauss' law for magnetic fields, $\oiint \mathbf{B} \cdot \mathbf{dA} = 0$. One could argue that postulates 1 and 2 introduced in Chap. 6 are differential expressions which are valid at a point in space. This is certainly true, but the postulates are given in terms of a current element $\mathbf{I}\, dl$ which is somewhat restrictive since a current element cannot exist in nature by itself. What is desired is a more general relationship between the magnetic field and current density $\mathbf{J}$ at a point in space. For example, in electrostatics the integral expression of Gauss' law $\oiint \mathbf{D} \cdot \mathbf{dA} = Q$ was converted into the differential form $\nabla \cdot \mathbf{D} = \rho$, which is valid at any point in space.

To express Gauss' law of electrostatics in differential form required the introduction of Gauss' theorem and the concept of divergence. Similarly, to convert Ampère's law to the differential form will require the introduction of Stokes' theorem and the concept of curl. Since curl is a more difficult abstraction than divergence, its derivation will be presented in two places. In this chapter we will concentrate on concepts and leave the detailed derivation for the appendix (Sec. A1.1).

8.1 CIRCULATION AND CURL OF A VECTOR FIELD

One of the important properties in characterizing a vector field $\mathbf{F}$ (such as a turbulent fluid-water, wind, etc.) is its rotation. The strength of rotation can be measured by the *circulation* C, defined by

$$\boxed{C = \oint_{l(A)} \mathbf{F} \cdot \mathbf{dl}} \qquad (8.1)$$

Thus the higher the number C, the stronger the rotation of the fluid in the area A which is bounded by the closed contour l. A sketch of a vector field $\mathbf{F}$ with circulation is given in Fig. A1.1. The vector field can be a rotating fluid, such as a vortex in water, or a rotating wind, such as is found in a tornado. To measure the circulation C over any area A of a turbulent flow, a paddle wheel, shown in Fig. A1.5, can be used. If the paddle wheel turns, one can say that the speed of rotation is proportional to circulation C.†

Figure 8.1 shows two examples of fluid flow. The first is a linear flow with no variation at right angles to the flow; hence circulation is zero. A paddle wheel stuck inside the area bounded by contour l will not turn since the pressure on opposite paddles is the same. On the other hand, the flow in Fig. 8.1*b* has a transverse variation which results in a net circulation about contour l.

Circulation by itself is limited as a measure of turbulence or rotation for two reasons. The first is that a rotating fluid has a direction which is the axis of rotation. This direction coincides with the direction of the axis of the spinning paddle wheel if and only if the paddle wheel is oriented in such a way that its spinning rate is a maximum. If the paddle wheel axis is at an angle to the maximum rotation direction, only a component of circulation is measured. For example, if it is desired to measure the rotation of a vortex on the water surface of a river, a paddle wheel must be oriented so that its axis is vertical. As a matter of fact, orienting the paddle wheel axis horizontally in such a vortex would give zero circulation because the paddle wheel would not rotate. Second, it is desirable to measure the rotation of a fluid at an infinitesimally small area, that is,

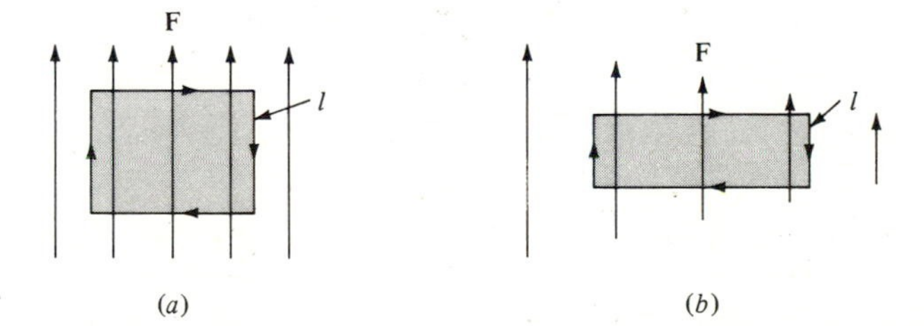

Figure 8.1 A vector field **F**. The strength and direction is represented by the arrows. (*a*) A conservative field; (*b*) a nonconservative field because the integral about the closed path l is not zero.

† Recall that in Chap. 1 the electrostatic field **E** was found to be conservative, which means that the circulation C is zero because $\oint \mathbf{E} \cdot \mathbf{dl} = 0$, as given by (1.12). It also means that the work done on a test charge which is moved about a closed path in the field **E** is zero. Ampère's law (7.8) or (7.13), on the other hand, states that the circulation of **B** about a current I is not zero, which implies that work must be done in moving a magnetic charge in a closed path about I.

at a point in the field. Circulation C as defined by (8.1) is not only related to **F** but also to the area enclosed by l. The dependence on l limits circulation (8.1) as a measure of rotation. What is clearly needed, especially in the case of a three-dimensional flow, is a vector measure of rotation. This is provided by the curl of **F** which can be defined as

$$\text{curl } \mathbf{F} = \max \lim_{A \to 0} \hat{\mathbf{n}} \frac{\oint_{l(A)} \mathbf{F} \cdot d\mathbf{l}}{A} \tag{8.2}$$

Curl of **F** is therefore a vector whose magnitude is the maximum circulation of **F** per unit area. The area is infinitesimal and oriented in a way which gives maximum circulation, and the direction $\hat{\mathbf{n}}$ of the area is the direction of curl **F**. If $\hat{\mathbf{x}}$ is a unit vector in the direction of the x axis, the component of curl in that direction can be written as

$$(\text{curl } \mathbf{F})_x = \text{curl } \mathbf{F} \cdot \hat{\mathbf{x}} = \lim_{A_x \to 0} \frac{\oint \mathbf{F} \cdot d\mathbf{l}}{A_x} \tag{8.3}$$

where $A_x = \Delta y \, \Delta z$. Of course, curl **F** can be written in terms of its rectangular components as

$$\text{curl } \mathbf{F} = (\text{curl } \mathbf{F} \cdot \hat{\mathbf{x}})\hat{\mathbf{x}} + (\text{curl } \mathbf{F} \cdot \hat{\mathbf{y}})\hat{\mathbf{y}} + (\text{curl } \mathbf{F} \cdot \hat{\mathbf{z}})\hat{\mathbf{z}} \tag{8.4}$$

Since it is not always convenient to calculate curl **F** from the definition (8.2), a form in terms of derivatives using the gradient operator ∇ can be given. This form, called *del cross* **F**, is derived in the appendix (see A1.8) and is

$$\text{curl } \mathbf{F} = \nabla \times \mathbf{F} = \left(\frac{\partial F_z}{\partial y} - \frac{\partial F_y}{\partial z}\right)\hat{\mathbf{x}} + \left(\frac{\partial F_z}{\partial x} - \frac{\partial F_x}{\partial z}\right)\hat{\mathbf{y}} + \left(\frac{\partial F_y}{\partial x} - \frac{\partial F_x}{\partial y}\right)\hat{\mathbf{z}} \tag{8.5}$$

This is the form that is used when an explicit calculation of curl in rectangular coordinates has to be made. The form of curl in spherical and cylindrical coordinates is given on the inside of the back cover.

The physical significance at a point at which curl is finite is that the line integral in the neighborhood of the point has a finite value; that is, $C \neq 0$. This does not imply, however, that the field lines must have curvature. As shown in Fig. 8.1*b*, the field lines can be straight and the field have a nonzero curl. At the same time, a field that has rotation, such as the circular flow shown in Fig. A1.4*b*, can have a zero curl. In general we can say that a field which has a nonzero curl must have a variation in a direction transverse to the field lines. Note that the curl operator of (8.5) shows this explicitly, since, for example, the F_z term is differentiated with respect to y and x but not with respect to z. This is in contrast to the divergence operation $\nabla \cdot \mathbf{F}$. There, the field must vary in the direction of the field lines in order for the field to have a nonzero divergence (note that the terms in divergence are of the type $\partial F_z / \partial z$).

Let us now give two examples. The first will be from electrostatics and will show a field with zero curl but nonzero divergence. The second will be from magnetostatics and will show a field with nonzero curl and zero divergence.

Example 1 Figure 1.22 shows a uniformly charged sphere, the variation of the fields with distance r from the sphere, and the variation of charge density ρ. We know that $\nabla \cdot \mathbf{E}$ must be proportional to ρ, since Gauss' law (1.92) states that $\nabla \cdot \mathbf{E} = \rho/\varepsilon$. We can show this explicitly by differentiating the electric fields inside and outside the sphere, (1.84d) and (1.84a), respectively. Thus, from the inside of the back cover, the divergence in spherical coordinates gives us

$$\nabla \cdot \mathbf{E} = \nabla \cdot E_r \hat{\mathbf{r}} = \frac{1}{r^2} \frac{\partial}{\partial r} (r^2 E_r)$$

$$= \frac{1}{r^2} \frac{\partial}{\partial r} \left(r^2 \frac{Q_T}{4\pi\varepsilon r^2} \right) = 0 \qquad r \geq a \tag{8.6}$$

$$= \frac{1}{r^2} \frac{\partial}{\partial r} \left(r^2 \frac{\rho r}{3\varepsilon} \right) = \frac{\rho}{\varepsilon} \qquad r \leq a \tag{8.7}$$

The divergence of the electric field is sketched in Fig. 8.2.

Since an electrostatic field is conservative, that is, from (1.12) circulation $\oint \mathbf{E} \cdot d\mathbf{l} = 0$, we can immediately conclude that the curl of $\mathbf{E}$ must be zero everywhere. To show this explicitly, we go to the back cover where curl in spherical coordinates is given and obtain

$$\nabla \times \mathbf{E} = \nabla \times E_r \hat{\mathbf{r}} = \frac{1}{r \sin\theta} \frac{\partial E_r}{\partial \phi} \hat{\boldsymbol{\theta}} - \frac{1}{r} \frac{\partial E_r}{\partial \theta} \hat{\boldsymbol{\phi}}$$

$$= 0 \qquad r \geq 0 \tag{8.8}$$

since E_r does not depend on either ϕ or θ.

Example 2 To show a field with zero divergence and nonzero curl, we resort to Fig. 7.4. This figure shows the magnetic field inside and outside a conducting cylinder which carries a uniformly distributed current. Gauss' law for magnetic fields, (7.4), states that $\nabla \cdot \mathbf{B}$ must be zero

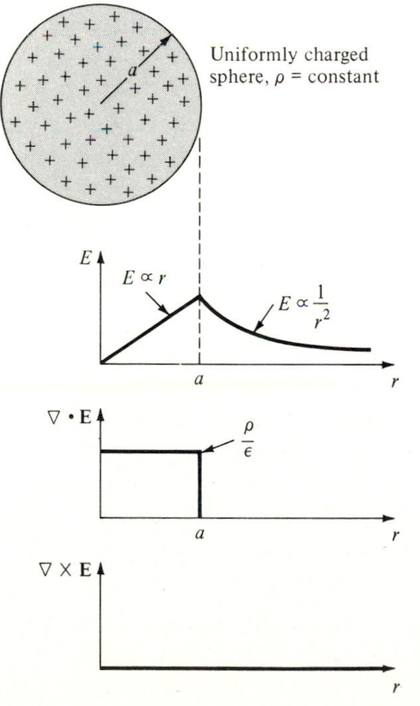

Figure 8.2 Variation of the electric field, divergence, and curl from a uniformly charged sphere. See Fig. 1.22 for other details.

everywhere. To show this explicitly, we can differentiate the magnetic field inside and outside the conducting cylinder. From the inside of the back cover, divergence of **B** in cylindrical coordinates gives

$$\nabla \cdot \mathbf{B} = \nabla \cdot B_\phi \hat{\boldsymbol{\phi}} = \frac{1}{r}\frac{\partial B_\phi}{\partial \phi}$$

$$= 0 \qquad r \geq 0 \tag{8.9}$$

because the magnetic field of a cylinder, (7.19) and (7.20), only has a ϕ component (the axis of the cylinder is aligned along the z axis), and that ϕ component is independent of ϕ; therefore $\partial/\partial\phi \equiv 0$.

We expect that the curl for this field is not zero because the circulation, which from Ampère's law is given as $\oint \mathbf{B} \cdot \mathbf{dl} = \mu_0 I$, is nonzero. To show this explicitly, we use the curl of **B** in cylindrical coordinates which gives

$$\nabla \times \mathbf{B} = \nabla \times B_\phi \hat{\boldsymbol{\phi}} = -\frac{\partial B_\phi}{\partial z}\hat{\mathbf{r}} + \frac{1}{r}\frac{\partial}{\partial r}(rB_\phi)\hat{\mathbf{z}}$$

$$= \frac{\mu_0 I}{\pi a^2}\hat{\mathbf{z}} \qquad r \leq a \tag{8.10}$$

$$= 0 \qquad r \geq a \tag{8.11}$$

The variation of **B**, $\nabla \cdot$ **B**, and $\nabla \times$ **B** is shown in Fig. 8.3.

Finally, the reason for using a fluid flow as an example is that for this type of a vector field, curl is easily visualized, especially with the aid of the paddle wheel.

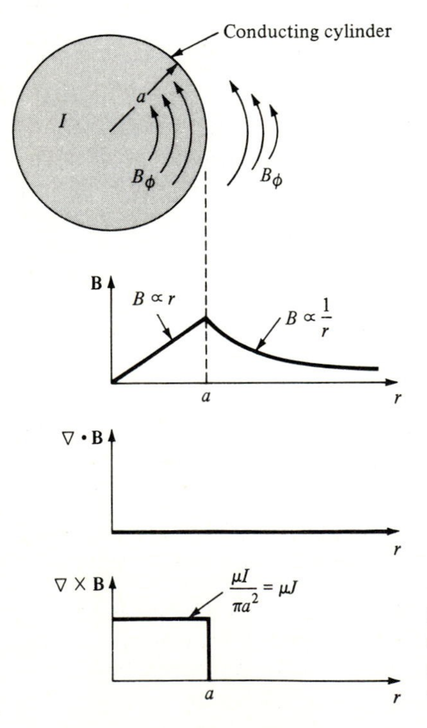

Figure 8.3 A conducting cylinder which carries a total current I uniformly distributed; $I = J\pi a^2$. Variation of B, $\nabla \cdot B$, and $\nabla \times$ **B** is shown. See Fig. 7.4 for additional details.

For other vector fields a simple physical interpretation is not always possible.† However, the interpretation that curl is a vector whose direction is the axis of circulation and whose magnitude is the circulation per unit area is always valid.

8.2 DIFFERENTIAL FORM OF AMPÈRE'S LAW

Using the concept of curl, we can now convert Ampère's law in integral form:

$$\oint_{l(A)} \mathbf{B} \cdot \mathbf{dl} = \mu_0 I = \mu_0 \iint_{l(A)} \mathbf{J} \cdot \mathbf{dA} \tag{7.8}$$

to the differential form. The differential form gives the relationship between magnetic field $\mathbf{B}$ and current density $\mathbf{J}$ at a point.

Let us apply Ampère's law to a small area ΔA, which is bounded by the differential path $l(\Delta A)$. Equation (7.8) becomes then

$$\oint_{l(\Delta A)} \mathbf{B} \cdot \mathbf{dl} = \mu_0 \mathbf{J} \cdot \hat{\mathbf{n}}\, \Delta A \tag{8.12}$$

where $\hat{\mathbf{n}}$ is the normal to the area ΔA; that is, $\mathbf{\Delta A} = \hat{\mathbf{n}}\, \Delta A$. In the limit as $\Delta A \to 0$, we obtain a point relationship. An inspection reveals that both sides of (8.12) vanish as $\Delta A \to 0$. However, dividing (8.12) by ΔA and going to the limit gives

$$\lim_{\Delta A \to 0} \frac{\oint \mathbf{B} \cdot \mathbf{dl}}{\Delta A} = \mu_0 \mathbf{J} \cdot \hat{\mathbf{n}} \tag{8.13}$$

We observe that the limit of this ratio must be nonzero at all points for which the n component of current density $\mathbf{J}$ is nonzero. By use of (8.3) the left side of (8.13) can be identified as the projection of the vector curl $\mathbf{B}$ onto the normal $\hat{\mathbf{n}}$ of area ΔA. Thus we can write

$$\text{curl}\, \mathbf{B} \cdot \hat{\mathbf{n}} = \mu_0 \mathbf{J} \cdot \hat{\mathbf{n}} \tag{8.14}$$

and observe that the direction of curl $\mathbf{B}$ coincides with that of $\mathbf{J}$. This allows us to write Ampère's law in differential form as‡

$$\boxed{\text{curl}\, \mathbf{B} = \mu_0 \mathbf{J}} \tag{8.15}$$

† We could construct a "curl meter" for an electric field by attaching positive charges at the ends of the paddles of a paddle wheel. If such a device was inserted in an electric field, it would tend to rotate wherever curl $\mathbf{E}$ was not zero.

‡ Ampère's law, as stated in the free-space form of (7.8) and (8.15), is limited because $\mathbf{B}$ is not necessarily equal to $\mu_0 \mathbf{H}$ in the presence of material media. On the other hand, starting with (7.18) instead of (7.8) gives the general Ampère's law curl $\mathbf{H} = \mathbf{J}$, which is valid in free space and in material media.

which is the desired point relationship between **B** and the current density **J** in free space. For media other than free space, (8.15) is valid if μ_0 can be replaced by μ, as discussed in Sec. 7.5.

Example What is the current distribution required to produce a magnetic field $\mathbf{B} = \hat{\mathbf{x}}y^3$? Apply (8.15) in rectangular coordinates to obtain

$$\mathbf{J} = \frac{1}{\mu_0}\nabla \times \mathbf{B} = \frac{1}{\mu_0}\nabla \times B_x\hat{\mathbf{x}} = \frac{1}{\mu_0}\left(\frac{\partial B_x}{\partial z}\hat{\mathbf{y}} - \frac{\partial B_x}{\partial y}\hat{\mathbf{z}}\right)$$

$$= -\frac{1}{\mu_0}\frac{\partial}{\partial y}y^3\hat{\mathbf{z}} = -\frac{3}{\mu_0}y^2\hat{\mathbf{z}}$$

Another good example is the cylindrical conductor carrying a uniformly distributed current, which was considered in Figs. 7.4 and 8.3. Referring to Fig. 8.3, we see that curl $\mathbf{B} \neq 0$ inside the cylinder where a finite current density $J = I/\pi a^2$ exists. Outside the cylinder, $\mathbf{J} = 0$, and therefore $\nabla \times \mathbf{B} = 0$.

At this point it might be worthwhile for the student to try to obtain (8.15) by directly taking the curl of the expression for **B** in Eq. (6.9). For a hint see Prob. 8.8.

8.3 STOKES' THEOREM

We have seen that Gauss' divergence theorem, (1.94), was useful in transforming integral statements (such as Gauss' law) to their equivalent differential form. Stokes' theorem, in that sense, is analogous to the divergence theorem; it relates the curl of a vector field **F** inside a contour to the circulation of **F** along the contour. The detailed derivation is given in Sec. A1.2, where it is shown that Stokes' theorem is†

$$\oint_{l(A)} \mathbf{F} \cdot \mathbf{dl} = \iint_{A(l)} \nabla \times \mathbf{F} \cdot \mathbf{dA} \tag{8.16}$$

We can see that this theorem is an extension of the definition of $\nabla \times \mathbf{F}$ as given in (8.3). For example, if we write (8.16) for an infinitesimally small area ΔA, we obtain

$$\oint_{l(\Delta A)} \mathbf{F} \cdot \mathbf{dl} = \nabla \times \mathbf{F} \cdot \hat{\mathbf{n}}\,\Delta A \tag{8.17}$$

which is the same as (8.3). Note that for an infinitesimal area, the surface integral in (8.16) can be approximated by its integrand, but the line integral on the left-hand side cannot.

The principal use of Stokes' theorem here is twofold: It allows us to interpret curls as vector sources of a field (Sec. 8.4) and permits the conversion of integral

† As shown in Fig. A1.2, the direction of l and that of $\mathbf{dA}$ are related by the right-hand rule.

statements involving circulation to their equivalent differential form. For example, take Ampère's law,

$$\oint_{l(A)} \mathbf{B} \cdot \mathbf{dl} = \mu_0 \iint_{A(l)} \mathbf{J} \cdot \mathbf{dA} \tag{8.18}$$

Applying Stokes' theorem to the left side, we have

$$\iint_{A(l)} \nabla \times \mathbf{B} \cdot \mathbf{dA} = \mu_0 \iint_{A(l)} \mathbf{J} \cdot \mathbf{dA} \tag{8.19}$$

Since this statement must be true for any area A (or equivalently for any set of limits), we can equate the integrands of (8.19). This gives

$$\nabla \times \mathbf{B} = \mu_0 \mathbf{J}$$

which is the desired differential statement of Ampère's law derived previously in (8.15).

8.4 DIVERGENCE AND CURL AS MEASURES OF SCALAR AND VECTOR SOURCES

We have seen that derivative operations on a field involve the operator ∇. The gradient operation on a scalar field ϕ resulted in the vector $\nabla\phi$. Two kinds of derivatives were involved when operating on a vector field, a curl and a divergence. Summarizing, we can state that

$$\text{grad } \phi = \nabla\phi = \text{a vector}$$

$$\text{div } \mathbf{F} = \nabla \cdot \mathbf{F} = \text{a scalar}$$

$$\text{curl } \mathbf{F} = \nabla \times \mathbf{F} = \text{a vector} \tag{8.20}$$

The integral theorems connected with these functions are the gradient theorem

$$\phi(b) - \phi(a) = \int_a^b \nabla\phi \cdot \mathbf{dl} \tag{1.64}$$

which gives the relationship between the scalar potential and the line integral of its gradient over any curve from a to b, Gauss' divergence theorem

$$\oint\!\!\oint_{A(v)} \mathbf{F} \cdot \mathbf{dA} = \iiint_{v(A)} \nabla \cdot \mathbf{F} \, dv \tag{1.94}$$

which gives the relationship of the total flux flowing through the surface A to the divergences inside v, and Stokes' theorem

$$\oint_{l(A)} \mathbf{F} \cdot \mathbf{dl} = \iint_{A(l)} \nabla \times \mathbf{F} \cdot \mathbf{dA} \tag{8.16}$$

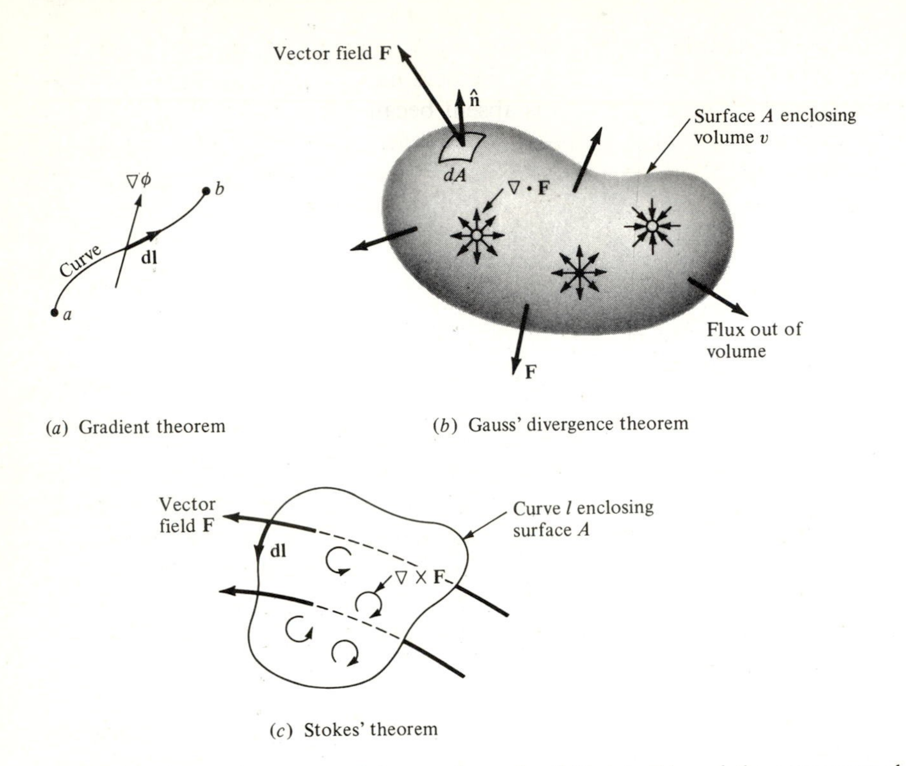

Figure 8.4 The maximum rate of change of a scalar field ϕ is $\nabla\phi$, and the component along the curve element $\mathbf{dl}$ is $\nabla\phi \cdot \mathbf{dl}$. (*a*) The difference of ϕ at the points a and b is given by the integral of the gradient variation along the curve between a and b. (*b*) The total flux out of volume v equals the sum of divergences inside. (*c*) Net circulation about l equals the sum of curls inside.

which gives the relationship between the circulation over a curve l and the curls inside A. These relationships are summarized in Fig. 8.4.

Let us first consider divergence as a measure of scalar sources. The electrostatic field is described in integral form by

$$\oiint \varepsilon\mathbf{E} \cdot \mathbf{dA} = Q \qquad \oint \mathbf{E} \cdot \mathbf{dl} = 0 \tag{8.21}$$

or in differential form by

$$\nabla \cdot \varepsilon\mathbf{E} = \rho \qquad \nabla \times \mathbf{E} = 0 \tag{8.22}$$

Using Gauss' law, (1.73) or (8.21), we can reason as follows: If we surround a point in space with a closed surface and if any net flux sticks out of that volume, charge Q must be contained within the volume. Charge must, therefore, be the cause for the net flux. The scalar sources can be obtained directly from the field by determining the divergence of the field. Wherever $\nabla \cdot \mathbf{E}$ is finite, we have a scalar source present. Summing all these sources gives the net flux out of the volume as stated by Gauss' divergence theorem (1.94).

A field that has only scalar sources is called an irrotational field because circulation $\oint \mathbf{F} \cdot \mathbf{dl} = 0$. Since $\nabla \cdot \mathbf{F} \neq 0$, the field has a variation in the direction of the field lines. Sideways variation is absent because $\nabla \times \mathbf{F} = 0$.

Let us now turn to Stokes' theorem and the interpretation of curl as a vector source. Consider first a field that has no scalar sources, such as the magnetic field **B** for which $\nabla \cdot \mathbf{B} = 0$. Stating the behavior of **B** in integral form, we have

$$\oint \mathbf{B} \cdot \mathbf{dl} = \mu_0 I \qquad \oiint \mathbf{B} \cdot \mathbf{dA} = 0 \tag{8.23}$$

In differential form, we obtain

$$\nabla \times \mathbf{B} = \mu_0 \mathbf{J} \qquad \nabla \cdot \mathbf{B} = 0 \tag{8.24}$$

Now a field for which $\nabla \cdot \mathbf{B} = 0$ everywhere is called a *solenoidal* field. Such a field has no net flux sticking out of any volume. A solenoidal field, unable to terminate or emanate from "charges" must have field lines which close upon themselves. Such a field has circulation, and the mechanism that induces this circulation must be the source of this type of a field. For the B field a source is the current, and its presence can be verified by calculating the curl of **B.** Since curl **B,** which is equal to **J,** is a vector, we conclude that a solenoidal field has vector sources. Stokes' theorem simply states that the net circulation about l is equal to the curls inside curve l.

The presence of vector sources is thus detected by the presence of sideways derivatives ($\partial F_x/\partial y$, $\partial F_x/\partial z$, etc.). Thus, a field created by vector sources will show a dependence transverse to the direction of the field lines.

From this discussion we see that the existence of a field is related to the existence of divergence and curl. If it is specified that curl and divergence are zero everywhere, then there can be no field, since a field without sources cannot exist.

8.5 TWO VECTOR IDENTITIES AND THEIR RELATIONSHIP TO POTENTIALS

Among the many vector identities listed in the appendix, there are two which are particularly significant in understanding and deriving the electromagnetic potentials. If $\phi(x, y, z)$ and $\mathbf{A}(x, y, z)$ are scalar and vector functions, respectively, the two identities are, from (A1.26),

$$\text{curl grad } \phi = \nabla \times \nabla\phi = 0 \tag{8.25}$$

and from (A1.27),

$$\text{div curl } \mathbf{A} = \nabla \cdot \nabla \times \mathbf{A} = 0 \tag{8.26}$$

The importance of these identities becomes apparent in their relationship to irrotational and solenoidal fields. For irrotational fields, we can state a corollary to (8.25) as follows:

$$\text{If } \nabla \times \mathbf{F} = 0, \text{ there exists a scalar } \phi \text{ such that } \mathbf{F} = \nabla\phi \tag{8.27}$$

Identity (8.25) guarantees that an irrotational field can always be derived from a scalar field, usually called a *scalar potential.*

For solenoidal fields, we can state a corollary to (8.26) as follows:

$$\text{If } \nabla \cdot \mathbf{F} = 0, \text{ there exists a vector } \mathbf{A} \text{ such that } \mathbf{F} = \nabla \times \mathbf{A} \tag{8.28}$$

Therefore, identity (8.26) implies that a solenoidal field can be derived from a vector field, usually called a *vector potential.*

In conclusion we can say that a vector field can always be derived from a scalar and vector potential. Putting it another way: Any vector field can be resolved into an irrotational and a solenoidal part. Combining this with the results found in the last section (a vector field has scalar and vector sources), we see that the potentials are a convenient way to relate a field to its scalar and vector sources. For example, consider the irrotational electrostatic field **E** and the statement of (8.27). The scalar sources for this field are given by taking the divergence of **E**:

$$\text{Scalar source} = \nabla \cdot \mathbf{E} = -\nabla^2 \phi = \frac{\rho}{\varepsilon} \tag{8.29}$$

A solution of (8.29) expresses the potential ϕ in terms of the source density ρ. Mathematically, it is a simpler expression than that for **E**. The advantages and use of the potential were discussed in detail in Chap. 1. There, the symbol V was used for electrostatic potential [see Eq. (1.49)], and the relationship between potential and charge density ρ was given by Eq. (1.96*c*).

A similar procedure can be followed for the solenoidal magnetostatic field **B** and statement (8.28). The vector sources for **B** are obtained by taking the curl of **B**:

$$\text{Vector source} = \nabla \times \mathbf{B} = \nabla \times \nabla \times \mathbf{A} = \mu_0 \mathbf{J} \tag{8.30}$$

where the relationship between **B** and current density **J** is given by Ampère's law. The vector potential **A** is obtained from the solution of $\nabla \times \nabla \times \mathbf{A} = \mu_0 \mathbf{J}$. Unlike the case of the scalar potential, it is not apparent at this time how the use of **A** will simplify the calculation of **B**. The next section will show how.

8.6 THE MAGNETIC VECTOR POTENTIAL

In Chap. 1 we saw that not only did the scalar potential have physical significance (it was the energy per unit charge), but also it simplified the calculation for the electric field **E**. The reason for this was twofold: The scalar potential V was simply related to the sources ρ and, hence, was easy to calculate; once V was obtained, the **E** field was then found by a derivative operation on V, usually a straightforward procedure. Can we expect that **A** will do the same thing for **B**, as V did for **E**? The answer is yes. It will simplify the calculation of **B**, but the physical meaning of **A** is not as obvious as that of V.

To find **A**, we examine the characteristic equations of the magnetostatic field in differential form which are $\nabla \cdot \mathbf{B} = 0$ and $\nabla \times \mathbf{B} = \mu_0 \mathbf{J}$. The first of these implies that $\mathbf{B}$ is solenoidal; hence a vector field $\mathbf{A}$ must exist such that

$$\mathbf{B} = \nabla \times \mathbf{A} \tag{8.31}$$

Substituting this in the second equation, we obtain the partial differential equation for $\mathbf{A}$:

$$\nabla \times \nabla \times \mathbf{A} = \mu_0 \mathbf{J} \tag{8.32}$$

which when solved will determine $\mathbf{A}$.

Before attempting to solve (8.32), let us see if we can find some important characteristics of $\mathbf{A}$ by examining the magnetic $\mathbf{B}$ field. The magnetic field generated by a current whose current density is $\mathbf{J}$ is given by (6.9) as

$$\mathbf{B}(\mathbf{r}) = \frac{\mu_0}{4\pi} \iiint \frac{\mathbf{J}(\mathbf{r}') \times \hat{\mathbf{R}}}{R^2}\, dv' \qquad \text{Wb/m}^2 \tag{8.33}$$

where $\mathbf{r}$ = observation coordinates
$\mathbf{r}'$ = source coordinates
$\hat{\mathbf{R}} = \mathbf{R}/R$
$R(\mathbf{r}, \mathbf{r}')$ = distance between $\mathbf{r}$ and $\mathbf{r}'$

that is,

$$R = |\mathbf{r} - \mathbf{r}'| = [(x - x')^2 + (y - y')^2 + (z - z')^2]^{1/2}$$

as shown in Figs. 6.6 or 8.5. We observe that the integrand of (8.33) varies inversely as the square of the distance from the current element $J\, dv'$. Since $\mathbf{B}$ is obtained from $\mathbf{A}$ by differentiation with respect to space coordinates, we conclude that $\mathbf{A}$ must vary inversely with respect to R.

We observe further that $\mathbf{B}$ does not specify $\mathbf{A}$ uniquely, because $\mathbf{B}$ is obtained from $\mathbf{A}$ by differentiation. Hence, adding a constant to $\mathbf{A}$ does not change anything. In fact, we can add a gradient of a scalar field to $\mathbf{A}$ and still obtain the same $\mathbf{B}$ because of identity (8.25). To see this, let us choose $\mathbf{A}$ and $\mathbf{A}'$, where $\mathbf{A}' = \mathbf{A} + \nabla\Psi$, and Ψ is some scalar field. Both of these give the same $\mathbf{B}$ field:

$$\mathbf{B} = \nabla \times \mathbf{A} = \nabla \times \mathbf{A}' \tag{8.34}$$

because $\nabla \times \nabla\Psi \equiv 0$. That this should be so is not surprising since we are trying to determine $\mathbf{B}$ from the "transverse" properties of $\mathbf{A}$, leaving the "longitudinal" properties (which would be given by $\nabla \cdot \mathbf{A}$) unspecified. Longitudinal refers to the direction parallel to $\mathbf{A}$. Recall that to specify a vector completely, all its components must be given. Since we are trying to specify $\mathbf{B}$ in terms of differential operators, this means that not only the curl of $\mathbf{A}$ but also the divergence of $\mathbf{A}$ must be specified. Since the equations of the magnetostatic field, $\nabla \cdot \mathbf{B} = 0$ and $\nabla \times \mathbf{B} = \mu_0 \mathbf{J}$, do not require that $\nabla \cdot \mathbf{A}$ be specified in any particular way, we will choose for simplicity

$$\nabla \cdot \mathbf{A} = 0 \tag{8.35}$$

This is usually called a *gauge condition*, and in magnetostatics it is known as the *Coulomb gauge*. We have now removed the ambiguity in $\mathbf{A}$, and our new definition for A will be $\nabla \times \mathbf{A} = \mathbf{B}$ and $\nabla \cdot \mathbf{A} = 0$. We will shortly see that a Coulomb gauge for $\mathbf{A}$ will yield a Poisson's equation for $\mathbf{A}$, similar to that for the scalar potential V in electrostatics.

Method 1

To find an explicit expression for $\mathbf{A}$, let us substitute the identity $\nabla \times \nabla \times \mathbf{A} = -\nabla^2 \mathbf{A} + \nabla(\nabla \cdot \mathbf{A})$ in (8.32). This gives

$$\nabla^2 \mathbf{A} - \nabla(\nabla \cdot \mathbf{A}) = -\mu_0 \mathbf{J} \tag{8.36}$$

Since we have chosen to set $\nabla \cdot \mathbf{A} = 0$, we obtain

$$\nabla^2 \mathbf{A} = -\mu_0 \mathbf{J} \tag{8.37}$$

This vector equation is really three scalar equations†

$$\nabla^2 A_x = -\mu_0 J_x \qquad \nabla^2 A_y = -\mu_0 J_y \qquad \nabla^2 A_z = -\mu_0 J_z$$

Each of these equations is similar to Poisson's equation (1.96c) for the electrostatic potential V; that is,

$$\nabla^2 V = -\frac{\rho}{\varepsilon} \tag{8.38}$$

which had the general solution (1.101), or

$$V(\mathbf{r}) = \frac{1}{4\pi\varepsilon} \iiint \frac{\rho(\mathbf{r}')\, dv'}{R} \tag{8.39}$$

By comparison to (8.38) and (8.39), the solutions for the component equations of (8.37) are

$$A_x = \frac{\mu_0}{4\pi} \iiint \frac{J_x\, dv'}{R} \qquad A_y = \frac{\mu_0}{4\pi} \iiint \frac{J_y\, dv'}{R} \qquad A_z = \frac{\mu_0}{4\pi} \iiint \frac{J_z\, dv'}{R} \tag{8.40}$$

Combining the three solutions, we obtain the vector potential $\mathbf{A}$:

$$\mathbf{A}(\mathbf{r}) = A_x \hat{\mathbf{x}} + A_y \hat{\mathbf{y}} + A_z \hat{\mathbf{z}} = \frac{\mu_0}{4\pi} \iiint \frac{\mathbf{J}(\mathbf{r}')\, dv'}{R} \qquad \text{Wb/m} \tag{8.41}$$

If the current is confined to a fine wire of cross section dA, then $J\, dv' = (J\, dA)\, dl' = I\, dl'$, and the vector potential becomes

$$\mathbf{A}(\mathbf{r}) = \frac{\mu_0}{4\pi} \oint \frac{\mathbf{I}(\mathbf{r}')\, dl'}{R} \tag{8.42}$$

The geometry is shown in Fig. 8.5.

† Notice that after examining the ∇^2 operator in various coordinate systems, the separation of the vector equation (8.37) into three scalar equations as shown above works only in cartesian coordinates. That is, in cylindrical coordinates $\nabla^2 \mathbf{A} \neq \nabla^2 A_\rho \hat{\boldsymbol{\rho}} + \nabla^2 A_\phi \hat{\boldsymbol{\phi}} + \nabla^2 A_z \hat{\mathbf{z}}$.

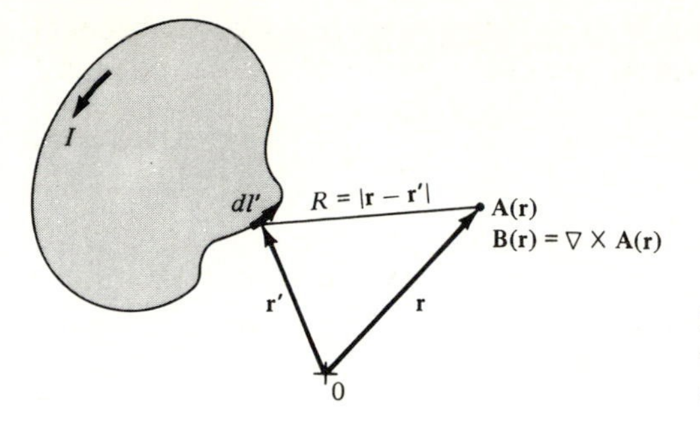

Figure 8.5 The vector potential at the observation point **r** is obtained by integrating the current I around the circuit.

We have shown that **A** is obtained by solving three fictitious problems in electrostatics which have the charge distributions $\rho_1 = \varepsilon\mu_0 J_x$, $\rho_2 = \varepsilon\mu_0 J_y$, and $\rho_3 = \varepsilon\mu_0 J_z$, respectively. Knowing the current distribution **J**, **A** can be found. Once **A** is known at **r**, we can obtain **B**(**r**) as $\mathbf{B}(\mathbf{r}) = \nabla \times \mathbf{A}(\mathbf{r})$.

Method 2

We can also obtain the vector potential directly from the expression for the magnetic field, (8.32), which is

$$\mathbf{B}(\mathbf{r}) = \frac{\mu_0}{4\pi} \iiint \frac{\mathbf{J}(\mathbf{r}') \times \hat{\mathbf{R}}}{R^2} dv' \tag{8.43}$$

by rearrangement of terms. First, let us observe that

$$\nabla\left(\frac{1}{R}\right) = -\nabla'\left(\frac{1}{R}\right) = -\frac{\hat{\mathbf{R}}}{R^2}$$
$$= -\frac{1}{R^2}\left[\frac{(x-x')}{R}\hat{\mathbf{x}} + \frac{(y-y')}{R}\hat{\mathbf{y}} + \frac{(z-z')}{R}\hat{\mathbf{z}}\right] \tag{8.44}$$

where $\mathbf{R} = (x - x')\hat{\mathbf{x}} + (y - y')\hat{\mathbf{y}} + (z - z')\hat{\mathbf{z}}$ and $R = [(x - x')^2 + (y - y')^2 + (z - z')^2]^{1/2}$. Some details of (8.44) are given in Prob. 8.5. The integrand of (8.43) can be written with the aid of (8.44) as

$$J' \times \frac{\hat{\mathbf{R}}}{R^2} = \nabla\left(\frac{1}{R}\right) \times J'$$
$$= \nabla \times \frac{\mathbf{J}'}{R} - \frac{1}{R}\nabla \times J' \tag{8.45}$$

where the last line was obtained by use of the vector identity (A1.19). Remembering that ∇ operates only on the unprimed coordinates $\mathbf{r} = (x, y, z)$, we see that $\nabla \times J' = 0$. The **B** field expression of (8.43) can now be written as

$$\mathbf{B}(\mathbf{r}) = \frac{\mu_0}{4\pi} \iiint \nabla \times \frac{\mathbf{J}'}{R} dv'$$
$$= \nabla \times \frac{\mu_0}{4\pi} \iiint \frac{\mathbf{J}'}{R} dv' \tag{8.46}$$

The curl operator can be taken outside the integral because it does not operate on the primed integration coordinates. Since **B** can be expressed in terms of a vector potential **A** as $\mathbf{B} = \nabla \times \mathbf{A}$,

Eq. (8.46) gives the expression for **A** as

$$\mathbf{A}(\mathbf{r}) = \frac{\mu_0}{4\pi} \iiint \frac{\mathbf{J}(\mathbf{r}')}{R} dv' \tag{8.47}$$

which is the desired result.

Divergence of A

It remains for us to show that our solution for **A**, Eq. (8.41), satisfies $\nabla \cdot \mathbf{A} = 0$. To do this, we will have to make use of the ∇ operator with respect to observation coordinates $\mathbf{r} = (x, y, z)$ and source coordinates $\mathbf{r}' = (x', y', z')$. With respect to these two coordinates, we will denote the del operator as ∇ and ∇', respectively. Taking the divergence of **A**, we obtain

$$\nabla \cdot \mathbf{A}(\mathbf{r}) = \nabla \cdot \frac{\mu_0}{4\pi} \iiint \frac{\mathbf{J}(\mathbf{r}')\, dv'}{R(\mathbf{r}, \mathbf{r}')} = \frac{\mu_0}{4\pi} \iiint \left(\nabla \cdot \frac{\mathbf{J}'}{R}\right) dv' \tag{8.48}$$

where $\mathbf{J}(\mathbf{r}') \equiv \mathbf{J}'$ and $R(\mathbf{r}, \mathbf{r}')$ is given by 8.44). The ∇ operator can be moved inside the integral because the integration is with respect to the primed variables. Using vector identity (A1.18), which expresses the divergence of the product of a scalar and vector function, we can write the above as

$$\nabla \cdot \mathbf{A}(\mathbf{r}) = \frac{\mu_0}{4\pi} \iiint \left[\frac{\nabla \cdot \mathbf{J}'}{R} + \mathbf{J}' \cdot \nabla\left(\frac{1}{R}\right)\right] dv' \tag{8.49}$$

The first term in the brackets is zero because J' is a function of the primed source coordinates only (which makes $\nabla \cdot \mathbf{J}' = 0$). Using the fact that $\nabla(1/R) = -\nabla'(1/R)$, which is shown in (8.44), we can state the divergence of **A** as

$$\nabla \cdot \mathbf{A} = -\frac{\mu_0}{4\pi} \iiint \mathbf{J}' \cdot \nabla'\left(\frac{1}{R}\right) dv' \tag{8.50}$$

Applying identity (A1.18) again, we obtain

$$\nabla \cdot \mathbf{A} = \frac{\mu_0}{4\pi} \iiint \left[\frac{1}{R}\nabla' \cdot J' - \nabla' \cdot \left(\frac{\mathbf{J}'}{R}\right)\right] dv' \tag{8.51}$$

The divergence of $\mathbf{J}'$ is again zero. However, the reason now is that the continuity equation (2.11), which is given by $\nabla' \cdot \mathbf{J}' + (\partial \rho'/\partial t) = 0$, reduces to $\nabla' \cdot \mathbf{J}' = 0$ for steady currents. The remaining term in the volume integral can be converted to a simpler surface integral by the application of Gauss' divergence theorem (1.94); that is,

$$\nabla \cdot \mathbf{A} = -\frac{\mu_0}{4\pi} \oiint \frac{1}{R} \mathbf{J}' \cdot dA' \tag{8.52}$$

If the closed surface of integration encloses all currents, there is no current flow through the bounding surface; hence the integral vanishes, leaving $\nabla \cdot \mathbf{A} = 0$. Thus we see that our choice of $\nabla \cdot \mathbf{A} = 0$ was not as arbitrary as it seemed. Besides giving us a simplified differential equation for **A**, it was a choice that was based on the continuity equation.

8.7 APPLICATIONS OF THE VECTOR POTENTIAL

Let us give a few examples of the use of the vector potential. As we will see, it is not always clear that the use of the vector potential simplifies the calculations for the magnetic field. In some cases it is simpler to start with the expression for the magnetic field, (8.33), directly. However, in more complicated problems, especially those including time-varying fields (radiation and propagation), the absence of a cross product in the integral for **A** makes it easier to solve for the vector potential first, and then determine **B** from it by differentiation (usually a trivial operation).

Vector Potential and B Field of a Current Element

Figure 8.6 shows a current element $I\,dl$ positioned at the origin along the z axis in a spherical coordinate system. This is one example where the magnetic field can be given directly by postulate 2 [Eq. (6.6)] as

$$\mathbf{dB} = \hat{\boldsymbol{\phi}}\,\frac{\mu_0}{4\pi}\,\frac{I\,dl}{r^2}\sin\theta \tag{8.53}$$

The magnetic field **B** has only a $\hat{\boldsymbol{\phi}}$ component.

To use the method of the vector potential, we first observe that for an infinitesimal current element, $R \gg dl$. This means that the distance R from the observation point to various points on the current element can be considered constant (approximately equal to the distance r to the origin) and therefore taken outside the integral of (8.41). That is,

$$\mathbf{A}(\mathbf{r}) = \frac{\mu_0}{4\pi}\iiint \frac{\mathbf{J}(\mathbf{r}')\,dv'}{R} = \frac{\mu_0}{4\pi r}\iiint \mathbf{J}'\,dv' \tag{8.54}$$

Since the current element is a thin, short piece of wire carrying current I, the above expression further simplifies to

$$\mathbf{A}(\mathbf{r}) = \frac{\mu_0}{4\pi r}\iiint \mathbf{J}'\,dA'\,dl' = \frac{\mu_0}{4\pi r}\int_{-dl/2}^{dl/2} \mathbf{I}'\,dl' \tag{8.55}$$

Noting that the current is constant over the length of the element, we can complete the integration and obtain

$$\mathbf{A}(\mathbf{r}) = \hat{\mathbf{z}}\,\frac{\mu_0}{4\pi r}\,I\,dl \tag{8.56}$$

To find the magnetic **B** field, we perform the curl operation on **A**. We can do this either in the rectangular coordinate system or the spherical coordinate system. As the spherical coordinates are

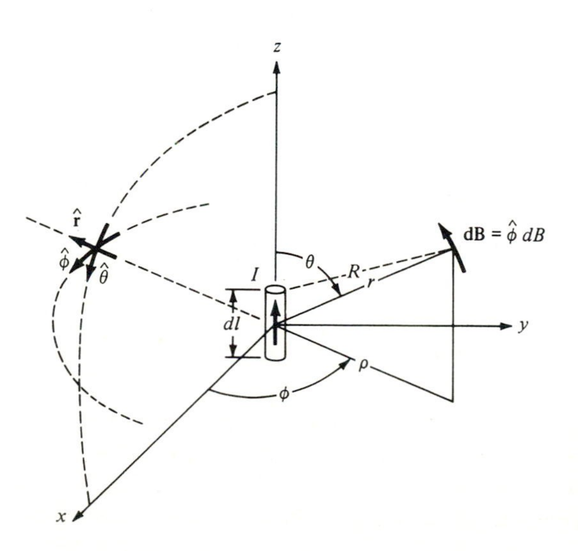

Figure 8.6 A current element $I\,dl$ located at the origin and parallel to the z axis produces a magnetic field in the ϕ direction only.

more convenient, we begin by finding the components of $\hat{\mathbf{z}}$ in spherical coordinates. The unit vectors $\hat{\mathbf{r}}$, $\hat{\boldsymbol{\theta}}$, $\hat{\boldsymbol{\phi}}$ are shown in Fig. 8.6. The vector potential is therefore

$$\mathbf{A} = \hat{\mathbf{z}}A_z = \hat{\mathbf{r}} \cos \theta A_z - \hat{\boldsymbol{\theta}} \sin \theta A_z$$
$$= \hat{\mathbf{r}}A_r - \hat{\boldsymbol{\theta}}A_\theta \tag{8.57}$$

Taking the curl, we obtain

$$\mathbf{B} = \nabla \times \mathbf{A} = \hat{\boldsymbol{\phi}} \frac{1}{r} \left[\frac{\partial}{\partial r}(rA_\theta) - \frac{\partial A_r}{\partial \theta} \right]$$
$$= \hat{\boldsymbol{\phi}} \frac{1}{r} \left[\frac{\partial}{\partial r} \left(-r \sin \theta \frac{\mu_0}{4\pi r} I\, dl \right) - \frac{\partial}{\partial \theta} \cos \theta \frac{\mu_0}{4\pi r} I\, dl \right]$$
$$= \hat{\boldsymbol{\phi}} \frac{\mu_0}{4\pi r^2} I\, dl \sin \theta \tag{8.58}$$

which is the same expression as (8.53). The B field is everywhere in the ϕ direction, forming closed circles concentric with the z axis. Since a current element is a small localized source, the vector potential decreases to zero as $1/r$ and the magnetic field as $1/r^2$.

Vector Potential and B Field of a Long, Straight Wire

The magnetic field for an infinitely long, straight wire carrying current I was obtained in Sec. 6.5 as $B = \mu_0 I/2\pi R$. The field lines are circles about the wire, as sketched in Fig. 6.19.

To use the vector potential method, let us obtain **A** from the solution of an equivalent problem in electrostatics. Orienting the infinitely long wire along the z axis, we can say that the current density in the wire of radius a is $J_z = I/\pi a^2$, because the current is uniformly distributed throughout the cross section of the wire. The vector potential has therefore only a z component, $\mathbf{A} = \hat{\mathbf{z}}A_z$. The other components are zero; that is, $A_x = A_y = 0$ because $J_x = J_y = 0$. Recall that the expression (8.41) for **A** was derived from the solution of three related electrostatic problems, with the equivalence of charge density ρ to current density as $\rho = \mu_0 \varepsilon J_z$. The equivalent electrostatic problem is thus an infinite line charge oriented along the z axis. The solution of such a problem was carried out in Sec. 1.7, with the result for the electrostatic potential V given by Eq. (1.26) as†

$$V = -\frac{\rho \pi a^2}{2\pi\varepsilon} \ln \frac{R}{R_0} \tag{8.59}$$

where the connection between line charge density ρ_L and volume charge density ρ is $\rho_L = \rho \pi a^2$. Using $\rho = \mu_0 \varepsilon J_z$, which relates the electrostatic to the magnetostatic problem, we have for the vector potential

$$\mathbf{A} = \hat{\mathbf{z}}A_z = -\hat{\mathbf{z}} \frac{\mu_0 I}{2\pi} \ln \frac{R}{R_0} \tag{8.60}$$

The magnetic **B** field can now be obtained by performing the curl operation in the cylindrical or rectangular coordinate system. It is most convenient to use the cylindrical coordinates, which from the inside of the back cover gives us for curl

$$\mathbf{B} = \nabla \times \hat{\mathbf{z}}A_z = -\hat{\boldsymbol{\phi}} \frac{\partial}{\partial R} \left[-\frac{\mu_0 I}{2\pi} \ln \frac{R}{R_0} \right]$$
$$= \hat{\boldsymbol{\phi}} \frac{\mu_0 I}{2\pi R} \tag{8.61}$$

which agrees with the result (6.28) obtained by direct integration.

† Notice that the perpendicular distance from the line charge in Fig. 1.9 is r, whereas in Fig. 6.19, which we are using now, it is R.

It should be pointed out at this time that although it is tempting to obtain **A** for the infinite wire directly from (8.41) or, what is equivalent, by considering a long wire as composed of many small current elements whose contributions [given by Eq. (8.56)] can be integrated to obtain A for a long wire—it cannot be done this way. The reason is that the resulting integral for **A** diverges because of the infinite extent of the wire. The same difficulty was encountered in obtaining the absolute potential for an infinite line charge, as discussed in the paragraph following Eq. (1.26).

The principal use of the results obtained for infinite structures is that they serve as approximations for practical structures. For example, if the observation distance R from a long wire l is such that $R \ll l$ and if the observation point is near the center of the wire, the results for an infinite wire and those for a finite wire are practically indistinguishable.

Vector Potential and B Field of a Magnetic Dipole

The last two examples did not require the use of the vector potential, since it was easier to calculate **B** directly from (8.33). However, in the case of the magnetic dipole, calculating **B** by first obtaining the vector potential **A** is an easier approach than calculating **B** directly. In Sec. 6.7 we tried to calculate the **B** field from a wire loop, but because the mathematics got too involved, we limited ourselves to obtaining the magnetic field on the axis of the loop only. We did obtain the **B** fields for a magnetic dipole in a later section [see Eq. (6.52)] by noting an analogy to the fields of an electric dipole. Now we will calculate the **B** fields rigorously, except for the restriction that the observation distance from the loop be large compared with the diameter of the loop.

Figure 8.7 shows a current-carrying wire loop of radius a lying in the xy plane. In place of the rectangular coordinate system we will use spherical coordinates because the symmetry of the loop is then utilized. A loop positioned as shown has ϕ symmetry, which implies that $\partial/\partial\phi = 0$. The **A** and **B** fields therefore have no ϕ variation. Recalling that the current element I **dl**′ produces a **dA** which is parallel to **dl**′, we see with the help of Fig. 8.7 that the contributions to **A** at point P' of two symmetrically located elements, **dl** and **dl**′, will cancel in the direction $0'P'$ and add in the direction normal to the plane $0PP'0'$. In spherical coordinates the direction normal to plane $0PP'0'$ is the ϕ direction. Hence two symmetrically located current elements give rise to an **A** which at P' has only a ϕ component, or

$$dA_\phi = 2\frac{\mu_0 Ia\, d\phi' \cos(\phi' - \phi)}{4\pi R} \tag{8.62}$$

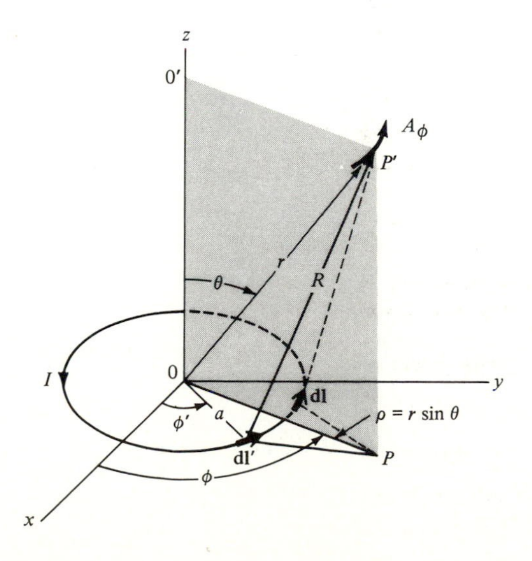

Figure 8.7 The geometry of a wire loop. The observation point P' is located at (r, θ, ϕ). The integration point is at $(a, \pi/2, \phi')$.

where $dl' = a\,d\phi'$, the symmetry plane is chosen at the angle ϕ, and $\cos(\phi' - \phi)$ accounts for the component normal to the plane. The total vector potential at P' is therefore

$$\mathbf{A} = \hat{\boldsymbol{\phi}}\frac{\mu_0 Ia}{2\pi}\int_{\phi}^{\phi+\pi}\frac{\cos(\phi' - \phi)\,d\phi'}{R} \tag{8.63}$$

The distance from the observation point P' to the source point dl' is R, which varies with ϕ' as the integration around the loop is performed. We note that†

$$\mathbf{R} = \mathbf{r} - \mathbf{a} \tag{8.64}$$

and

$$R^2 = \mathbf{R}\cdot\mathbf{R} = r^2 - 2\mathbf{r}\cdot\mathbf{a} + a^2$$

where the projection of $\mathbf{a}$ on $\mathbf{r}$ is equal to $\mathbf{r}\cdot\mathbf{a} = ra\sin\theta\cos(\phi - \phi')$. Substituting the expression for R into (8.63) and changing variables from $\phi' - \phi$ to ϕ', we obtain

$$A_\phi = \frac{\mu_0 Ia}{2\pi}\int_0^{\pi}\frac{\cos\phi'\,d\phi'}{(r^2 + a^2 - 2ar\sin\theta\cos\phi')^{1/2}} \tag{8.65}$$

This is an *elliptical integral* and cannot be evaluated in terms of elementary functions. However, in most practical cases the observation distance is much larger than the size of the loop; that is, $r \gg a$. This allows us to approximate R as

$$\begin{aligned} R &= r\left[1 - 2\frac{a}{r}\sin\theta\cos\phi' + \left(\frac{a}{r}\right)^2\right]^{1/2} \\ &\cong r\left(1 - 2\frac{a}{r}\sin\theta\cos\phi'\right)^{1/2} \end{aligned}$$

Using the binomial expansion for $(1 + \Delta)^{\pm n} \cong 1 \pm n\Delta + \cdots$ when $\Delta \ll 1$, we can write

$$\begin{aligned} A_\phi &\cong \frac{\mu_0 Ia}{2\pi r}\int_0^{\pi}\left(1 + \frac{a}{r}\sin\theta\cos\phi'\right)\cos\phi'\,d\phi \\ &= \frac{\mu_0 Ia^2\sin\theta}{4r^2} \end{aligned} \tag{8.66}$$

which is the vector potential for a small loop. Small, as used here, means that the loop appears small at observation distances which are large in terms of loop dimensions; that is, $r \gg a$.

Expression (8.66) for a small, z-directed loop whose moment is $\mathbf{m} = m\hat{\mathbf{z}} = I\pi a^2\hat{\mathbf{z}}$ can be written in vector form as

$$\begin{aligned} \mathbf{A}(\mathbf{r}) &= \mu_0\frac{m\sin\theta}{4\pi r^2}\hat{\boldsymbol{\phi}} \\ &= \mu_0\frac{m}{4\pi r^2}\hat{\mathbf{z}}\times\hat{\mathbf{r}} \\ &= \frac{\mu_0}{4\pi r^2}\mathbf{m}\times\hat{\mathbf{r}} \\ &= -\frac{\mu_0}{4\pi}\mathbf{m}\times\nabla\left(\frac{1}{r}\right) \end{aligned} \tag{8.67}$$

† The cosine of the angle Ψ between two radial vectors $\mathbf{r}(r, \theta, \phi)$ and $\mathbf{r}'(r', \theta', \phi')$ is given by

$$\cos\Psi = \cos\theta\cos\theta' + \sin\theta\sin\theta'\cos(\phi - \phi')$$

where for a dipole centered at the origin $r' = 0$, (8.44) gives $\nabla(1/r) = -\hat{\mathbf{r}}/r^2$. The last expression in (8.67) is valid for any orientation of the magnetic dipole **m**. To make (8.67) valid for a dipole at any distance r' from the origin, we can replace r by R. By use of (8.44), the vector potential **A** of an arbitrarily located dipole is thus

$$\mathbf{A}(\mathbf{r}) = -\frac{\mu_0}{4\pi}\mathbf{m} \times \nabla\left(\frac{1}{R}\right) \tag{8.68}$$

where $R = |\mathbf{r} - \mathbf{r}'|$, as shown in Fig. 8.5.

The **B** field for a z-directed dipole located at the origin can now be obtained by performing the curl operation. In spherical coordinates we have

$$\mathbf{B} = \nabla \times \mathbf{A} = \nabla \times A_\phi \hat{\boldsymbol{\phi}} = \frac{\hat{\mathbf{r}}}{r \sin\theta}\left[\frac{\partial}{\partial\theta}(\sin\theta A_\phi)\right] + \frac{\hat{\boldsymbol{\theta}}}{r}\left[-\frac{\partial}{\partial r}(rA_\phi)\right]$$

$$= \frac{\mu_0 m}{4\pi r^3}(2\cos\theta\hat{\mathbf{r}} + \sin\theta\hat{\boldsymbol{\theta}}) \tag{8.69}$$

The shape of the **B** lines is shown in Fig. 6.23. In Chap. 6 the magnetic field B_z on the axis of a wire loop was derived as (6.34). Expression (8.69) reduces to the field on the axis by letting $\theta \to 0$.

It should be noted that (8.69) was derived previously in (6.52) by comparing the magnetic dipole to the electric dipole. It is interesting that such a similarity between the **E** fields of an electric dipole $p = qd$ and the **B** fields of a magnetic dipole $m = IA$ exists. After all, electrostatic fields are derived from $\nabla \cdot \mathbf{E} = \rho/\varepsilon$ and $\nabla \times \mathbf{E} = 0$ and magnetostatic fields from $\nabla \times \mathbf{B} = \mu_0 \mathbf{J}$ and $\nabla \cdot \mathbf{B} = 0$, which are different laws. To resolve this, we note that dipole fields as given by (4.14) and (8.69) are valid only at distances far from the sources, i.e., for $r \gg d$ and $r \gg a$, respectively. Since no sources exist in the far-field region, both types of fields are divergence- and curl-free. The fields far from the sources, therefore, obey the same laws; that is, $\nabla \cdot \mathbf{E} = \nabla \times \mathbf{E} = 0$ and $\nabla \times \mathbf{B} = \nabla \cdot \mathbf{B} = 0$, which explains the similarity between the two types of fields. However, near the electric and magnetic dipoles the shape of the field lines is different, as can be seen from Fig. 6.31.

8.8 BOUNDARY CONDITIONS FOR THE MAGNETIC FIELD

In Sec. 1.13 the boundary conditions for the electric field were derived. We shall now determine the behavior of the magnetic field as it crosses the boundary between two different materials.

Normal Components of B Field at Boundary

We can obtain the boundary conditions for the normal components from the result (1.87*b*) obtained for the electric field. Noting that Gauss' law for electrostatic fields is $\oiint \mathbf{D} \cdot d\mathbf{A} = Q$ and Gauss' law for magnetostatic fields is $\oiint \mathbf{B} \cdot d\mathbf{A} = 0$, we can use Fig. 1.24 and Eq. (1.87*b*) to conclude that

$$\hat{\mathbf{n}} \cdot (\mathbf{B}_1 - \mathbf{B}_2) = 0$$

or

$$B_{n1} = B_{n2} \tag{8.70}$$

Hence the normal components of **B** are continuous across an interface between two different materials. This relation holds in the static and time-varying case.

Parallel Components of H Field at Boundary

We can again try to use the results obtained for the tangential components of the electrostatic field, derived in Sec. 1.13. However, we should note that for the electrostatic field the condition $\oint \mathbf{E} \cdot d\mathbf{l} = 0$ holds; i.e., the electrostatic field is conservative. The magnetostatic field, on the other hand, can be nonconservative, which is expressed by Ampère's law (8.23) as† $\oint \mathbf{H} \cdot d\mathbf{l} = I$. Therefore [by comparing with Eq. (1.86)], the tangential components of the **H** field across an interface between two media are continuous:

$$H_{t1} = H_{t2} \tag{8.71}$$

if and only if no current is allowed to flow in a thin sheet along the interface.

If a current in a layer of vanishing thickness flows at the boundary, then (8.71) does not hold. For example, consider two media, one metal, the other air. Let a current flow in a thin sheet just inside the metal, as shown in Fig. 8.8. Apply Ampère's law

$$\oint_{abcd} \mathbf{H} \cdot d\mathbf{l} = I = \iint J\, dA \tag{8.72}$$

to the rectangle *abcd* which lies with its longest sides (of length Δl) parallel to the interface separating the two media. If we let *ab* and *cd* which are of length Δn become vanishingly small, but in such a way that the boundary or interface is always between the two longer sides Δl, we have

$$\int_a^d \mathbf{H}_1 \cdot d\mathbf{l} - \int_c^b \mathbf{H}_2 \cdot d\mathbf{l} = \int_0^{\Delta l} \int_0^{\Delta n} J\, dl\, dn$$

$$(\mathbf{H}_1 - \mathbf{H}_2) \cdot \Delta\mathbf{l} = J\, \Delta l\, \Delta n$$

$$H_{\tan 1} - H_{\tan 2} = J\, \Delta n = K \qquad \text{A/m} \tag{8.73}$$

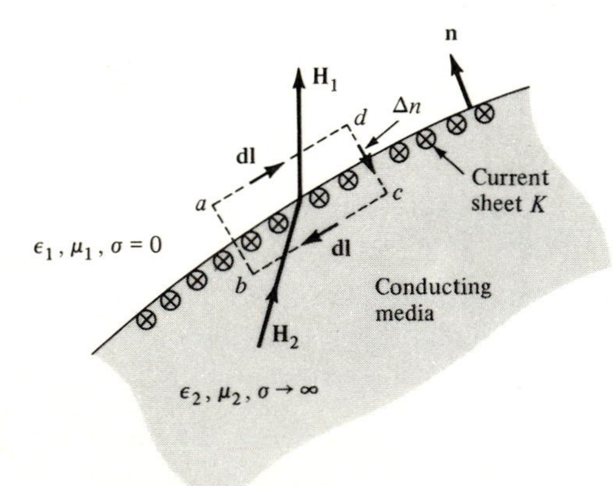

Figure 8.8 The boundary between a conducting and a nonconducting media. The surface current K (amperes per meter) flows into the paper.

† Note that for the derivation of boundary conditions, we are using the media-independent statement of Ampère's law in place of $\oint \mathbf{B} \cdot d\mathbf{l} = \mu I$ which depends on the permeabilities μ on both sides of the interface.

as $\Delta n \to 0$. The minus sign enters because tangential $\mathbf{H}_1$ is along $\mathbf{dl}$ and tangential $\mathbf{H}_2$ is opposite to $\mathbf{dl}$. The line integral of H over the path Δn does not contribute because if H remains finite as $\Delta n \to 0$, the integral over Δn vanishes.† Therefore the tangential components of the $\mathbf{H}$ field are discontinuous by the current sheet density K that flows at the interface. If no such current sheet exists, then the tangential H components are continuous. Also, if current flow along the boundary is distributed over a thick sheet, then $K \to 0$ as $\Delta n \to 0$ because J remains finite inside the rectangle *abcd*, and we have again that $H_{\tan 1} = H_{\tan 2}$. Clearly current flow must be confined to an infinitesimal layer for $H_{\tan}$ to be discontinuous. Therefore, if J is the current density in the sheet, and K is defined by

$$K = J\,\Delta n \qquad \text{A/m} \tag{8.75}$$

then K is finite as $\Delta n \to 0$ only if current density $J \to \infty$ within the sheet. The reason is that the product of two factors can remain finite only if, as one of the factors goes to zero, the other factor goes to infinity.

Difficulty with the Tangential Boundary Condition

The concept of a surface current sheet is, at times, confusing to a student. One can compare a current sheet K on a conducting surface to a charge layer (as shown in Figs. 1.25 and 2.6) which is obtained when a charge Q is deposited on a conducting body. Q will distribute itself in an infinitesimally thin layer on the surface of density ρ_s, given by (1.87*a*).

The problem is more subtle for a current sheet, though. First, the depth Δn of the current sheet depends on the conductivity σ of the material and the time rate (frequency f) of change of the field. To clarify the variation in depth of a current sheet, we have to borrow the concept of skin depth δ, which will be developed in Sec. 13.6. It says that fields and currents‡ decay exponentially with distance into the conducting medium, falling to $1/e$ of their value at the surface in a distance called the skin depth δ, where

$$\delta = \frac{1}{\sqrt{\pi f \mu \sigma}} \tag{8.76}$$

and f and σ are the frequency and conductivity of the medium, respectively. Thus as either f or σ tends toward infinity, the fields and currents tend to concentrate in a progressively thinner layer near the surface of the conducting medium. However, for static fields (which is the dc case corresponding to $f = 0$) we see that even the

† Since the relationship between the directions of $\mathbf{H}$ and $\mathbf{J}$ can be confusing, it is best to write (8.73) in vector notation as

$$\hat{\mathbf{n}} \times (\mathbf{H}_1 - \mathbf{H}_2) = \mathbf{K} \tag{8.74}$$

which relates the directions in terms of the normal $\hat{\mathbf{n}}$ to the interface. Note that the $\hat{\mathbf{n}} \times$ operator picks out the tangential component of $\mathbf{H}$, since $|\hat{\mathbf{n}} \times \mathbf{H}| = H \sin\theta = H_{\tan}$.

‡ Note that (8.24) implies that current (of current density $\mathbf{J}$) cannot exist unless it is accompanied by a spatially varying magnetic field.

best conductors such as silver or copper ($\sigma \approx 6 \times 10^7$ S/m; $\mu = \mu_0 = 4\pi \times 10^{-7}$ H/m) have a skin depth which approaches infinity; that is, $\delta \cong 0.06/\sqrt{f} \to \infty$. This means that dc current does not flow in a thin surface layer but is uniformly distributed throughout the entire conducting body. Hence J in (8.73) remains finite as $\Delta n \to 0$, which implies that $K = 0$. The appropriate boundary condition for the tangential components of a magnetostatic field on the interface between dielectric and metal is therefore $H_{\tan 1} = H_{\tan 2}$. In other words, the static magnetic field penetrates the metallic body completely. On the other hand, at the higher frequencies the penetration depth goes to zero rapidly, and the appropriate boundary condition for a metallic surface is $H_{\tan} = K$. It thus appears logical to separate the boundary conditions for static fields and time-varying fields. This will be done in the next section.

Let us give an example which demonstrates the relationship between the static magnetic fields at the boundary between air and a magnetic material. This relationship is important in the design of dc machinery.

Example A magnetic **B** field that crosses from one medium into another will in general have a different direction after passing. This is shown in Fig. 8.9 where the boundary separates two media of permeabilities μ_1 and μ_2. The boundary conditions that apply for a static magnetic field are (8.70) and (8.71); that is, $B_{n1} = B_{n2}$ and $H_{t1} = H_{t2}$ for two media of any conductivities, permeabilities, and permittivities. The change in direction of **H** or **B** is given by

$$\frac{\tan \theta_1}{\tan \theta_2} = \frac{\mu_1}{\mu_2} \tag{8.77}$$

which is derived by noting that $B_{n1} = B_1 \cos \theta_1 = \mu_1 H_1 \cos \theta_1$, $B_{n2} = B_2 \cos \theta = \mu_2 H_2 \cos \theta_2$, $H_{t1} = H_1 \sin \theta_1$, and $H_{t2} = H_2 \sin \theta_2$. These relations apply for isotropic media, where **B** and **H** have the same direction (μ is a scalar).

The bending depends only on differences in permeabilities μ. For example, if medium 2 is soft iron of the type used in dc machinery and in electromagnets, the relative permeability might be 6000; that is, $\mu_r = \mu/\mu_0 = 6000$. If medium 1 is air, with $\mu_r = 1$, this implies that $\tan \theta_1 = (\mu_1/\mu_2) \tan \theta_2 \ll 1$ for $0 \le \theta_2 < \pi/2$. In medium 1, $\tan \theta_1 \ll 1$ means that $\tan \theta_1 \approx \theta_1 \approx 0°$. Therefore, a B field in iron at angles of θ_2, say between 86 and 0°, will for practical purposes emerge normally in air. The property that the magnetic field in air is essentially normal to the boundary of a high-μ medium is important for engineering purposes and in mapping magnetic fields.

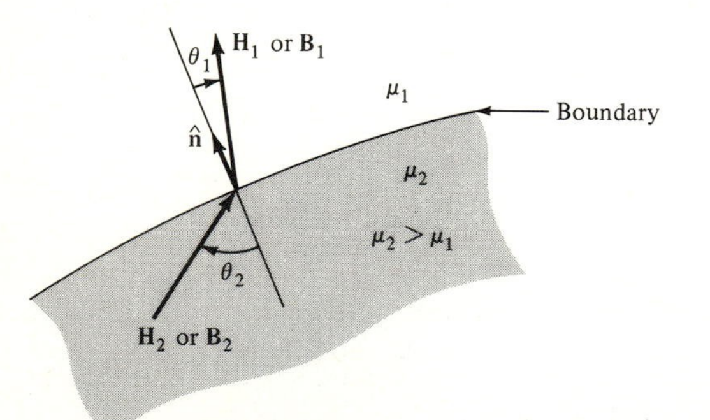

Figure 8.9 The magnetic field changes direction upon crossing the boundary between two media of different permeabilities μ. The angles θ are measured from the normal $\hat{\mathbf{n}}$ to the boundary.

8.9 SUMMARY OF BOUNDARY CONDITIONS FOR ELECTRIC AND MAGNETIC FIELDS

Boundary conditions were considered in Secs. 1.13 and 8.8. A summary of boundary conditions follows. They are valid for static and for time-varying fields.

General Case

$$\hat{\mathbf{n}} \cdot (\mathbf{D}_1 - \mathbf{D}_2) = \rho_s \tag{8.78}$$

Normal components of **D** are discontinuous by surface charge layer ρ_s, where ρ_s is a layer of free charge (bound charge, such as polarization charge, does not count). For development see Sec. 1.13.

$$\hat{\mathbf{n}} \cdot (\mathbf{B}_1 - \mathbf{B}_2) = 0 \tag{8.79}$$

Normal components of **B** are continuous because no magnetic charge exists. For development see Eq. (8.70).

$$\hat{\mathbf{n}} \times (\mathbf{E}_1 - \mathbf{E}_2) = 0 \tag{8.80}$$

Tangential components of **E** are continuous. For development see Sec. 1.13.

$$\hat{\mathbf{n}} \times (\mathbf{H}_1 - \mathbf{H}_2) = \mathbf{K} \tag{8.81}$$

Tangential components of **H** are discontinuous by surface current sheet **K**. For development see Eq. (8.73).

As shown in Fig. 8.10, the unit normal $\hat{\mathbf{n}}$ points from medium 2 into medium 1. The relationships $\mathbf{D} = \varepsilon\mathbf{E}$ and $\mathbf{B} = \mu\mathbf{H}$ can be used to express variations of the above boundary conditions.

The general boundary conditions can be specialized to two dominant classes of materials: dielectrics and conductors.

Dielectrics

Since in dielectric materials charge cannot move about freely (the rearrangement time for free charges in a good dielectric is on the order of days; see Sec. 4.1 or 2.8), an electric field in a dielectric forces the positive and negative charges of all molecules to separate slightly and form dipoles throughout the interior of

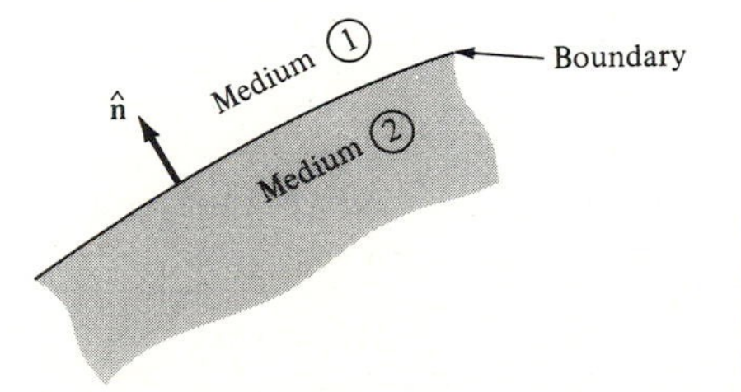

Figure 8.10 The unit normal points from medium 2 into 1. Note that operation $\hat{\mathbf{n}}\,\cdot$ selects the normal component to the boundary, whereas $\hat{\mathbf{n}} \times$ selects the tangential component.

the material. The charges in a dielectric are "bound" charges. It is important to realize that the free charge layer ρ_s and sheet current density K are zero for a dielectric. Hence we have

$$\hat{\mathbf{n}} \cdot (\mathbf{D}_1 - \mathbf{D}_2) = 0 \tag{8.82}$$

$$\hat{\mathbf{n}} \cdot (\mathbf{B}_1 - \mathbf{B}_2) = 0 \tag{8.83}$$

$$\hat{\mathbf{n}} \times (\mathbf{E}_1 - \mathbf{E}_2) = 0 \tag{8.84}$$

$$\hat{\mathbf{n}} \times (\mathbf{H}_1 - \mathbf{H}_2) = 0 \tag{8.85}$$

The normal components of **D** and **B**, as well as the tangential components of **E** and **H**, are continuous. However, other components ($\mathbf{D} = \varepsilon\mathbf{E}$, and $\mathbf{B} = \mu\mathbf{H}$) can be discontinuous. For example, at the interface of air-dielectric, the normal components of **E** are discontinuous and are given from (8.82) as $\varepsilon_0 E_n = \varepsilon_2 E_{n2}$.

Good Conductors

In practical problems we often treat good conductors as though they were perfect conductors. This is a good approximation since the metallic conductors, such as copper, do have high conductivities ($\sigma \approx 6 \times 10^7$ S/m). However, only superconductors have infinite conductivity and are truly perfect conductors.

As pointed out in the previous section, it is best to separate the boundary conditions for good conductors into time-independent and time-dependent cases. Thus for *static fields*

$$\hat{\mathbf{n}} \cdot \mathbf{D}_1 = \rho_s \tag{8.86}$$

$$\hat{\mathbf{n}} \cdot (\mathbf{B}_1 - \mathbf{B}_2) = 0 \tag{8.87}$$

$$\hat{\mathbf{n}} \times \mathbf{E}_1 = 0 \tag{8.88}$$

$$\hat{\mathbf{n}} \times (\mathbf{H}_1 - \mathbf{H}_2) = 0 \tag{8.89}$$

where subscript 2 denotes the conducting medium. As shown in Sec. 2.9, the electrostatic fields inside the conducting medium are zero, which is expressed in (8.86) and (8.88). Free charge can exist on the surface of a conductor, which makes the normal component of **D** discontinuous. Tangential components of **E** just inside the conductor must be zero even if the surface is charged (for if they were not zero, the tangential E field would force the charges to move until they were distributed along the surface in such a manner that $\hat{\mathbf{n}} \times \mathbf{E}_2 = 0$).

The electric and magnetic fields in the static case are independent. A static magnetic field can thus exist inside a metallic body [see discussion following Eq. (8.76)] even though an **E** field cannot. The normal component of **B** and the tangential components of **H** are thus continuous across the interface.

In the case of *time-varying fields*, the boundary conditions for good conductors are

$$\hat{\mathbf{n}} \cdot \mathbf{D} = \rho_s \tag{8.90}$$

$$\hat{\mathbf{n}} \cdot \mathbf{B} = 0 \tag{8.91}$$

$$\hat{\mathbf{n}} \times \mathbf{E} = 0 \tag{8.92}$$

$$\hat{\mathbf{n}} \times \mathbf{H} = \mathbf{K} \tag{8.93}$$

where the subscripts have been deleted, because in this case the only nonvanishing fields are fields outside the conducting body. That is, from the definition (8.76) of skin depth δ, we see that as $\sigma \to \infty$ for finite frequency, $\delta \to 0$. Thus, all fields are excluded from the interior of a good conductor, and all currents flow in a thin layer at the surface. For example, δ for copper at 1 MHz is $\delta = 6 \times 10^{-5}$ m.

Summarizing, expression (8.90) is the same as in the static case; i.e., the normal component of **D** just outside a conducting body is equal to the surface charge layer ρ_s in the conducting body. Equations (8.91) and (8.92) imply that the normal component of **B** and the tangential component of **E** just outside the surface must vanish, whereas the tangential component of **H** just outside the surface is equal to the surface current sheet, $K = H_{\tan}$.

Another way of looking at the boundary conditions is as follows: If a time-varying magnetic field is directed against a metallic surface, the tangential component of the total magnetic field will induce a current density K in the conductor. Conversely, if a current sheet exists in the surface of a conductor, it will give rise to a magnetic field on the free-space side of the conducting surface.

For a perfect conductor, such as a superconductor, for which $\sigma = \infty$, the skin depth $\delta = 1/(\pi f \mu \sigma)^{1/2}$ is equal to zero even for zero frequencies. Hence for a superconductor, the boundary conditions for static and time-varying fields is given by Eqs. (8.90) to (8.93).

8.10 SUMMARY OF STATIC ELECTRIC AND MAGNETIC FIELDS

Table 8.1 summarizes the relations developed in the previous chapters. It shows us at a glance the similarities and differences between the electric and magnetic fields. The emphasis is on free-space relations because we have not yet covered magnetic fields and magnetization in material bodies (although polarization was discussed in Chap. 4).

We have learned that electrostatic and magnetostatic effects can exist independently. Only when motion is considered or time-variation allowed, do the **E** and **B** field couple. The expressions which relate this coupling are Maxwell's equations, which will be covered in Chap. 11. The equations presented here are therefore not true in general because time-dependent terms are missing.

Table 8.1 Comparison of electric and magnetic equations

Description of equation	Electric	Magnetic
Experimental force law	Coulomb's law $\mathbf{F} = \frac{Q_1 Q_2}{4\pi\varepsilon_0 R^2}\hat{\mathbf{R}}$ 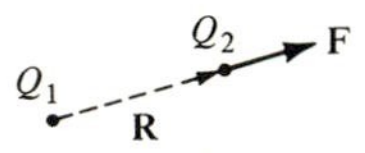	Force law between current elements $\mathbf{dF} = \frac{\mu_0}{4\pi}\frac{I_2\,\mathbf{dl}_2 \times (I_1\,\mathbf{dl}_1 \times \hat{\mathbf{R}})}{R^2}$ 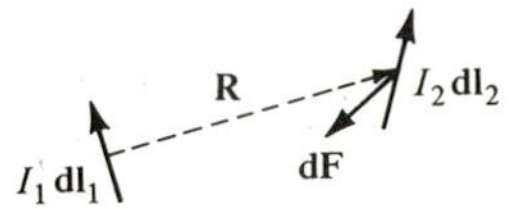
Definition of field from force law[1]	$\mathbf{F} = Q\mathbf{E}$	$\mathbf{dF} = \mathbf{I} \times \mathbf{B}\,dl$ current element $= \mathbf{J} \times \mathbf{B}\,dv_{ol}$ distributed current element $= q\mathbf{v} \times \mathbf{B}$ moving charge
General force law (Lorentz' force)	$\mathbf{F} = q(\mathbf{E} + \mathbf{v} \times \mathbf{B})$ $\mathbf{dF} = (\rho\mathbf{E} + \mathbf{J} \times \mathbf{B})\,dv_{ol}$	$dQ = \rho\,dv_{ol}$
Fields from an elementary source	$\mathbf{E} = \frac{Q}{4\pi\varepsilon R^2}\hat{\mathbf{R}}$	$\mathbf{dB} = \frac{\mu_0}{4\pi}\frac{\mathbf{I} \times \hat{\mathbf{R}}}{R^2}\,dl$ $= \frac{\mu_0}{4\pi}\frac{\mathbf{J} \times \hat{\mathbf{R}}}{R^2}\,dv_{ol}$
Fields from a distributed source	$\mathbf{E}(\mathbf{r}) = \frac{1}{4\pi\varepsilon_0}\iiint \frac{\rho(\mathbf{r}')}{R^2}\hat{\mathbf{R}}\,dv'_{ol}$ 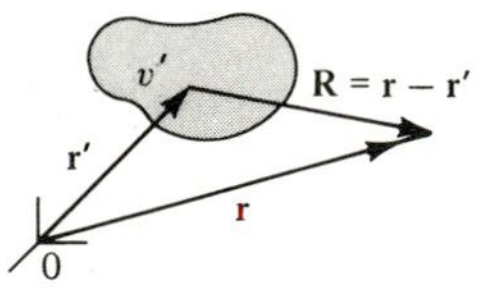	$\mathbf{B}(\mathbf{r}) = \frac{\mu_0}{4\pi}\iiint \frac{\mathbf{J} \times \hat{\mathbf{R}}}{R^2}\,dv'_{ol}$

Gauss' law in integral form	$\oiint \mathbf{D} \cdot \mathbf{dA} = Q$	$\oiint \mathbf{B} \cdot \mathbf{dA} = 0$
Gauss' law in differential form	$\nabla \cdot \mathbf{D} = \rho$	$\nabla \cdot \mathbf{B} = 0$
Definition of irrotational and solenoidal field[2]	$\nabla \times \mathbf{E}_c = 0$	$\nabla \cdot \mathbf{B} = 0$
Definition of scalar and vector potential[2]	$E_c = -\nabla V$	$\mathbf{B} = \nabla \times \mathbf{A}$
Poisson's equation for potential function	$\nabla^2 V = -\frac{\rho}{\varepsilon}$	$\nabla^2 \mathbf{A} = -\mu_0 \mathbf{J}$
Potential function (solution to above equation)	$V = \frac{1}{4\pi\varepsilon_0} \iiint \frac{\rho'}{R} dv'_{ol}$	$\mathbf{A} = \frac{\mu_0}{4\pi} \iiint \frac{\mathbf{J}'}{R} dv'_{ol}$
Gauss' law (enclosing charge) and Ampère's law (enclosing current)	$\oiint \mathbf{D} \cdot \mathbf{dA} = \iiint \rho \, dV$ $\nabla \cdot D = \rho$	$\oint \mathbf{H} \cdot \mathbf{dl} = I$ $\nabla \times \mathbf{H} = \mathbf{J}$
Constitutive relations	$\mathbf{D} = \varepsilon \mathbf{E}$ $\mathbf{D} = \varepsilon_0 \mathbf{E} + \mathbf{P}$	$\mathbf{B} = \mu \mathbf{H}$ $\mathbf{B} = \mu_0 \mathbf{H} + \mu_0 \mathbf{M}$ (9.11)
Definition of relative permittivity and permeability	$\varepsilon_r = \frac{\varepsilon}{\varepsilon_0}$ $\varepsilon_0 = 8.854 \times 10^{-12}$ F/m	$\mu_r = \frac{\mu}{\mu_0}$ $\mu_0 = 4\pi \times 10^{-7}$ H/m
Energy density of a field	$w_E = \frac{1}{2}\varepsilon E^2$	$w_M = \frac{1}{2}\mu H^2$
Capacitance and inductance	$C = \frac{Q}{V}$	$L = \frac{\Lambda}{I}$

(continued on next page)

Table 8.1 Comparison of electric and magnetic equations *(continued)*

Description of equation	Electric	Magnetic
Capacitance and inductance of a field cell	$\varepsilon_0 = \frac{C}{l}$	$\mu_0 = \frac{L}{l}$
Energy stored in capacitance and inductance	$W = \frac{1}{2}CV^2$	$W = \frac{1}{2}LI^2$
Closed-path integration, with and without sources[2]	$\oint \mathbf{E} \cdot \mathbf{dl} = \mathscr{V}$ emf $\oint \mathbf{E}_c \cdot \mathbf{dl} = 0$	$\oint \mathbf{H} \cdot \mathbf{dl} = NI = \mathscr{F}$ mmf $\oint \mathbf{H} \cdot \mathbf{dl} = 0$ no enclosed current
Electric and magnetic dipole moments	$\mathbf{p} = q\mathbf{d}$	$\mathbf{m} = I\mathbf{A}$
Torque on dipole immersed in a field	$\mathbf{T} = \mathbf{p} \times \mathbf{E}$	$\mathbf{T} = \mathbf{m} \times \mathbf{B}$
Force on dipole immersed in a field	$\mathbf{F} = (\mathbf{p} \cdot \nabla)\mathbf{E}$	$\mathbf{F} = (\mathbf{m} \cdot \nabla)\mathbf{B}$
Potential energy of dipole	$W = -\mathbf{p} \cdot \mathbf{E}$	$W = -\mathbf{m} \cdot \mathbf{B}$
Far-field from z-oriented dipole at origin	$\mathbf{E} = \frac{p}{4\pi\varepsilon_0 r^3}(2\cos\theta\,\hat{\mathbf{r}} + \sin\theta\,\hat{\boldsymbol{\theta}})$	$\mathbf{B} = \frac{\mu_0 m}{4\pi r^3}(2\cos\theta\,\hat{\mathbf{r}} + \sin\theta\,\hat{\boldsymbol{\theta}})$

[1] dv_{ol} stands for an element of volume and should not be confused with velocity v. Symbols for charge are Q and q, with q usually denoting an infinitesimal point charge.

[2] E_c is the electrostatic field created by charges only. On the other hand, $\mathbf{E}$ can be a general electric field, which can include emf-producing fields (due to batteries, etc.).

Since one of our goals is to derive Maxwell's equations, let us summarize to the extent that we have already derived these from a purely static point of view. Terms due to time-varying fields will be missing. These will be denoted by "time-varying" and will be derived in the following chapters. Maxwell's equations in integral and differential form are then

$$\oint\!\!\oint \mathbf{D} \cdot \mathbf{dA} = Q \qquad \nabla \cdot \mathbf{D} = \rho$$

$$\oint\!\!\oint \mathbf{B} \cdot \mathbf{dA} = 0 \qquad \nabla \cdot \mathbf{B} = 0$$

$$\oint \mathbf{E} \cdot \mathbf{dl} = 0 + (\text{time-varying}) \qquad \nabla \times \mathbf{E} = 0 + (\text{time-varying})$$

$$\oint \mathbf{H} \cdot \mathbf{dl} = I + (\text{time-varying}) \qquad \nabla \times \mathbf{H} = \mathbf{J} + (\text{time-varying})$$

To this set we might add the continuity equations

$$\oint\!\!\oint \mathbf{J} \cdot \mathbf{dA} + \frac{dQ}{dt} = 0 \qquad \nabla \cdot \mathbf{J} + \frac{\partial \rho}{\partial t} = 0$$

Even though the differential form of Maxwell's equations is better known, it is the integral form which represents the experimental laws and is thus more fundamental. The differential form can be derived from the integral form by applying Stokes' theorem and the divergence theorem. Furthermore, the differential form is valid only for regions without discontinuities in ε or μ. The boundary conditions can therefore be derived only from the integral forms.

PROBLEMS

$$\mu_0 = 4\pi \times 10^{-7}\ \text{H/m} \qquad \varepsilon_0 = 8.85 \times 10^{-12}\ \text{F/m}$$

8.1 Show that $\nabla \times \nabla\phi \equiv 0$.

(*a*) Show this for a specific coordinate system by performing the mathematical operations in, say, the rectangular coordinate system.

(*b*) Show this in general (irrespective of a coordinate system) by integrating over an arbitrary surface and applying Stokes' theorem.

8.2 An infinitely long solenoid has n turns per meter and carries current I. If the solenoid is oriented parallel to the z axis and if the magnetic field is zero outside the solenoid and does not vary with distance along the axis, show that the magnetic field for any point inside the solenoid has the value

$$H_z = nI$$

8.3 A cylindrical conductor of radius a oriented coaxial with the z axis carries a nonuniformly distributed current $\mathbf{J} = \rho\hat{\mathbf{z}}$.

(*a*) Using $\nabla \times \mathbf{H} = \mathbf{J}$, calculate the magnetic field strength $\mathbf{H}$ inside and outside the conductor.

(*b*) Repeat, using $\oint \mathbf{H} \cdot \mathbf{dl} = I$.

8.4 Prove that the circulation of the vector potential **A** about a path is equal to the flux enclosed by the path; that is,

$$\oint_{l(S)} \mathbf{A} \cdot d\mathbf{l} = \iint_{S(l)} \mathbf{B} \cdot d\mathbf{S}$$

8.5 Prove Eq. (8.44); that is, show that $\nabla(1/R) = -\nabla'(1/R) = -\hat{\mathbf{R}}/R^2$. Let the vector **R** be directed from $P'(x', y', z')$ to $P(x, y, z)$.

(*a*) If point P is fixed and point P' is allowed to move, show that the gradient of $1/R$ is given by $\nabla'(1/R) = \hat{\mathbf{R}}/R^2$, where $\hat{\mathbf{R}}$ is the unit vector directed along **R** and the prime on the gradient operator denotes differentiation with respect to the primed quantities x', y', z'. Show that the above expression is the maximum rate of change of $1/R$.

(*b*) Similarly, show that if P' is fixed and P allowed to move, $\nabla(1/R) = -\hat{\mathbf{R}}/R^2$.

8.6 Show that $\nabla \times \mathbf{H} = \mathbf{J}$ is satisfied inside the conductor of Fig. 8.3 and inside each conductor of the coaxial transmission line of Prob. 7.7.

8.7 A cylindrical conductor of radius a, whose axis coincides with the z axis of a cylindrical coordinate system, carries a current I of current density $\mathbf{J} = J_0 e^{-r/a}\hat{\mathbf{z}}$.

(*a*) Find the total current I flowing in the conductor.

(*b*) Find **H** at $r < a$ and $r > a$.

(*c*) Show that $\nabla \times \mathbf{H} = \mathbf{J}$.

8.8 Starting with the integral expression for **B** as given by (8.33) or (6.9), obtain the differential form of Ampère's law

$$\nabla \times \mathbf{B} = \mu_0 \mathbf{J}$$

by taking the curl of the integral expression.

Hint: See Prob. 7.1.

8.9 The axis of a cylindrical conductor of radius a is coincident with the z axis. A uniformly distributed current I flows in the conductor; that is, $I = \pi a^2 J \hat{\mathbf{z}}$. If we assume that $\mathbf{A} = 0$ at $r = a$, determine the vector potential **A** within the conductor, using $\nabla^2 \mathbf{A} = -\mu_0 \mathbf{J}$.

8.10 The magnetic field from an infinitely long, solid conducting cylinder of radius a, carrying a total current I, is given by Eqs. (7.19) and (7.20), and the **B** field is sketched in Fig. 7.4. Show that the magnetic vector potential **A** for such a conductor, carrying a uniformly distributed current $I = \pi a^2 J$, is given by

$$A_z = -\frac{\mu_0 I}{2\pi} \ln r + C_1 \qquad \text{for } r > a$$

$$A_z = -\frac{\mu_0 J r^2}{4} + C_2 \qquad \text{for } r < a$$

where C_1 and C_2 are constants. Assume the cylinder axis is coincident with the z axis of a cylindrical coordinate system. Check the above answers with Eqs. (7.19) and (7.20) after performing the curl operation; that is, $\mathbf{B} = \nabla \times \mathbf{A}$.

8.11 Show that the vector potential inside and outside an infinitely long solenoid of radius a, with n turns per unit length, is given by

$$A = \frac{\mu_0 n I r}{2} \qquad \text{for } r < a$$

$$A = \frac{\mu_0 a^2 n I}{2r} \qquad \text{for } r > a$$

where I is the current through the windings.

Hint: Use the result of Eq. (7.22).

8.12 Determine the vector potential **A** of a piece of thin wire of length l, carrying a constant current I. Assume the wire is located along the z axis and centered at the origin. Calculate the **B** field using this result and compare it to the **B** field from a straight wire given in Prob. 6.7.

8.13 Calculate the vector potential **A** for a piece of thin wire of length l using the geometry of Prob. 6.7. Obtain the **B** field expression.

8.14 Using the result of Prob. 8.13, show that the vector potential **A** of two parallel, infinite straight wires with currents I of equal strength but in opposite directions is given by

$$\mathbf{A} = A_z \hat{\mathbf{z}} = \frac{\mu_0 I}{2\pi} \ln \frac{r_2}{r_1} \hat{\mathbf{z}}$$

where $r_1(r_2)$ is the distance from wire 1(2) to the observation point, as shown in Fig. 5.6 (assume the wires coincide with the position of the line charges). The wires are parallel to the z axis.

8.15 Derive the boundary condition on the normal components of **B** by starting with $\nabla \cdot \mathbf{B} = 0$.

8.16 Derive the boundary condition on the tangential components of **E** by starting with $\nabla \times \mathbf{E} = 0$.

8.17 In Secs. 6.9 and 7.6 it was shown that all the magnetic field associated with a toroid is within the toroid. However, from boundary condition (8.73) or (8.74) one could assume that an outside magnetic field exists as shown. The discontinuity in H, treating the current in the wires as a sheet of current, is given by $H_1 - H_2 = K$.

(*a*) Is it appropriate to apply the above boundary condition to this problem?

(*b*) Assuming it is appropriate, give reasons why H_2 is zero.

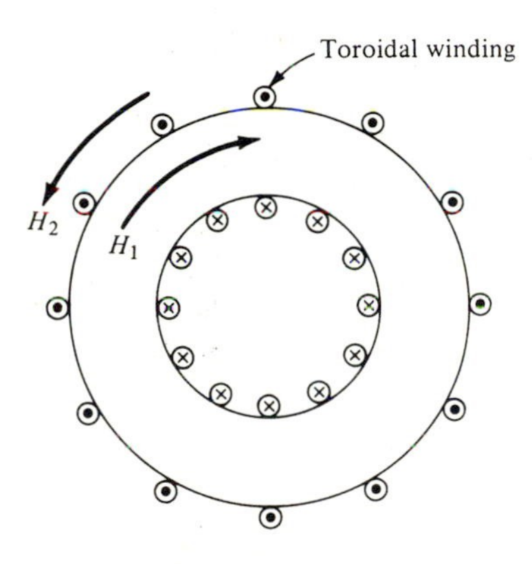

8.18 The boundary condition at the interface between two dielectrics is given by (8.82) as $\hat{\mathbf{n}} \cdot (\mathbf{D}_1 - \mathbf{D}_2) = 0$. If region 1 is free space with $D_1 = \varepsilon_0 E_1$, show that $\hat{\mathbf{n}} \cdot (\mathbf{E}_1 - \mathbf{E}_2) = \hat{\mathbf{n}} \cdot \mathbf{P}_2/\varepsilon_0$ at the boundary. This shows that the discontinuity in **E** is caused by the polarization **P**. Refer to Fig. 8.10 or 1.24 for the direction of the normal $\hat{\mathbf{n}}$.

8.19 The boundary condition on the normal components of **B** is given by $\hat{\mathbf{n}} \cdot (\mathbf{B}_1 - \mathbf{B}_2) = 0$. If region 1 is free space with $\mathbf{B}_1 = \mu_0 \mathbf{H}_1$, show that $\hat{\mathbf{n}} \cdot (\mathbf{H}_1 - \mathbf{H}_2) = \hat{\mathbf{n}} \cdot \mathbf{M}_2$ at the boundary. That is, the discontinuity in **H** is caused by the magnetization **M** in the second medium. Magnetization **M** is defined by Eq. (9.11) as $\mathbf{B} = \mu_0(\mathbf{H} + \mathbf{M})$.

8.20 The **E** field in air above a block of wax is at an angle of 45° with respect to the plane surface of the block. If the dielectric constant ε_r of wax is 2, find the angle between **E** and the surface in the wax.

8.21 Find the electric field inside a parallel-plate capacitor with plate separation d if a dielectric slab of relative permittivity ε_r and thickness a occupies part of the interior of the capacitor. Neglect fringing. This problem can be treated either as a circuits problem with two capacitors in series or as a field problem, i.e., a boundary-value problem. Obtain the answer using both methods.

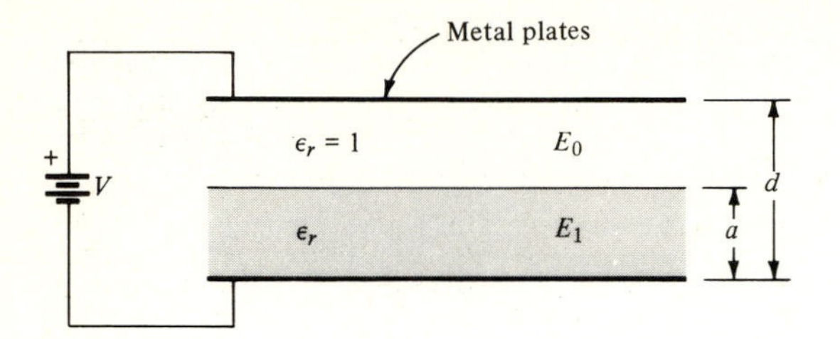

8.22 Free space occupies region 1 ($z > 0$), and a dielectric medium with $\varepsilon_2 = 3\varepsilon_0$ occupies region 2 ($z < 0$). At the boundary, $z = 0$, and a surface charge density $\rho_s = 0.3$ C/m^2 exists. Given that $\mathbf{D}_1 = 3x\hat{\mathbf{x}} + 4y^2\hat{\mathbf{y}} + 3\hat{\mathbf{z}}$ and $\mathbf{H}_2 = 2\hat{\mathbf{x}} + 5y^2\hat{\mathbf{y}} + 5\hat{\mathbf{z}}$, find $\mathbf{E}_1$, $\mathbf{H}_1$, $\mathbf{B}_1$, $\mathbf{B}_2$, $\mathbf{E}_2$, and $\mathbf{D}_2$ at the boundary.

8.23 At the boundary between air and a ferromagnetic material ($\mu_r = 8000$), the **B** field makes an angle of 0, 45, 60, and 87° with the normal to the surface in the ferromagnetic material. Determine the corresponding directions of the **B** field in air.

8.24 A superconductor is a perfect diamagnetic material for which $\mu = 0$. Assume that in Fig. 8.9, medium 1 is superconducting and medium 2 is free space. If a magnetic field could go across the boundary, find the allowable direction that the magnetic field would have in the superconductor for arbitrary directions of the magnetic field in free space.

CHAPTER

NINE

MAGNETIC MATERIALS, MAGNETS, AND SUPERCONDUCTORS

Guide to the Chapter

The material in this chapter can be divided into three parts.† The first part covers traditional engineering magnetics. It should provide the student with an understanding of magnetization and the behavior of magnetic objects. The first four sections of this chapter, plus "Ferromagnetism," "Hysteresis in Ferromagnetism," and "Soft and Hard Ferromagnetic Materials" of Sec. 5 should be covered in a basic course on electromagnetics.

The second part covers material in Secs. 5 to 8 in which the microscopic nature of magnetism is examined. It is unfortunate that a thorough study of magnetism cannot be confined to the macroscopic properties only. Eventually questions arise which can be answered only by looking into the microscopic characteristics. These sections may be omitted without loss of continuity. However, if course time permits or if the student is interested in some of the "whys" of hysteresis and soft and hard ferromagnetism, this part should provide a working knowledge.

The subject matter of the last part is that of superconductivity. It follows the first two parts naturally, as a superconductor, besides being a perfect conductor, is a perfect diamagnetic material.

† A good reference book to the material of this chapter is B. D. Cullity, "Introduction to Magnetic Materials," Addison-Wesley Publishing Company, Inc., Reading, Mass., 1972.

9.1 SOURCES OF THE MAGNETIC FIELD

As pointed out in the introduction to Chap. 6, the sources of the magnetic field are electric charges in motion. Since moving charges constitute a current, it follows that current is the source of the magnetic field.

Current

In the previous chapters we studied some practical arrangements for producing B fields. They were current-carrying wires twisted into various shapes, such as shown in Fig. 9.1.

Permanent Magnets

We know that a permanent magnet produces a magnetic field in the space surrounding the magnet. Since ordinary currents cannot be involved, what type of currents do in fact exist inside a magnet that produce the external B field? Modern physics has shown that, in addition to the orbital motion of the electrons around the nucleus, the electrons spin around their own axis.† It is the spin of the electrons which causes the large magnetic fields of permanent magnets.

The charge of the electron spinning about its own axis can be considered to be equivalent to a tiny current loop‡ with a magnetic moment $m = IA$, where I and A are the equivalent current and area, respectively. As shown in Fig. 9.2, each micro-

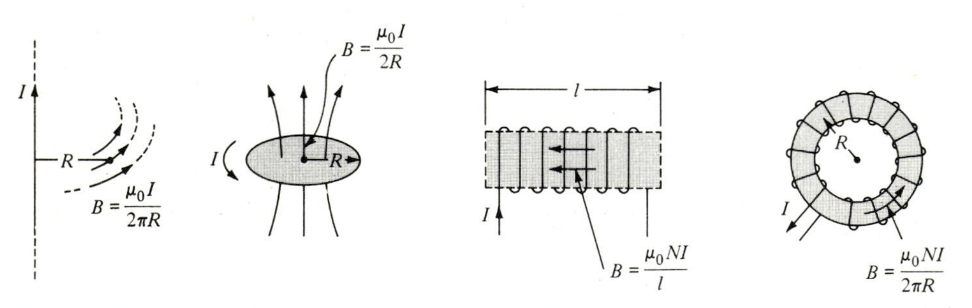

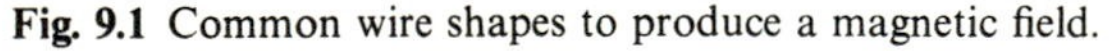

Fig. 9.1 Common wire shapes to produce a magnetic field.

† Electron spin, strictly speaking, is a quantum-mechanical effect. On an atomic scale a spinning electron possesses angular momentum, which in turn can be related to the magnetic moment.

The orbital motion of the electrons is a form of a current loop which also produces a magnetic field. This effect is called *diamagnetism* and is of a very weak nature; i.e., the magnetic field produced by the orbital motion is so weak that it is usually ignored. Although all materials are diamagnetic, only those materials which produce magnetic fields due to spin are of practical interest in motors, generators, transformers, permanent magnets, etc.

‡ Using a macroscopic analog of a rotating charged sphere, we see that the charge around the equator moves faster than the charge near the poles, which is equivalent to a current loop around the equator.

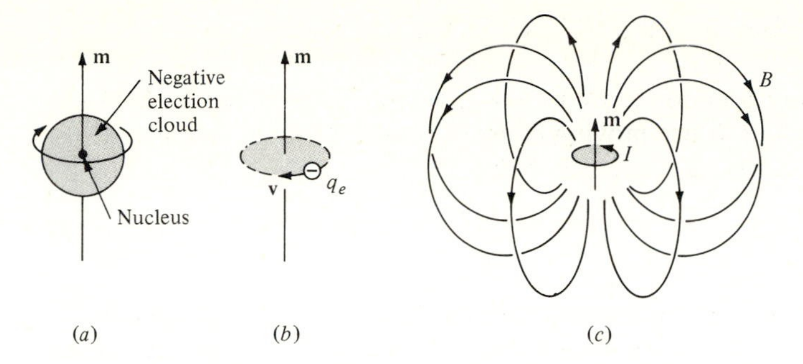

Figure 9.2 (*a*) The electron can be thought of as a ball of negative charge spinning around its own axis with a magnetic moment m as shown; (*b*) the spinning spherical charge (macroscopically speaking) is equivalent to a circulating charge q_e; (*c*) the circulating charge acts like a dipole **m** with the magnetic field as shown.

scopic current loop produces a small magnetic field in the same manner as the wire loop of Fig. 9.1. Ordinarily, because of the randomizing thermal agitation (see Sec. 2.1) the molecular magnets in a slab of material are oriented randomly, producing zero net magnetic field. However, the molecular moments in a permanent magnet are aligned, and the combined effect of billions of them results in a significant field about the permanent magnet. This effect is called *ferromagnetism.* The field outside a permanent bar magnet and the aligned molecular magnets inside the bar are shown in Fig. 9.3. Notice the similarity in the exterior field of the bar magnet and that of the spinning electron. The exterior field is the superposition of the fields from many spinning electrons and hence looks like that from a large dipole.

Because the field of a bar magnet resembles that of an electric dipole, with positive charge at one end and negative charge at the other, historically one thought of a bar magnet with north and south poles situated at the ends of the bar as shown in Fig. 9.4. The north and south poles were considered as the seats of positive and negative magnetic charges, respectively. The quest to isolate a mag-

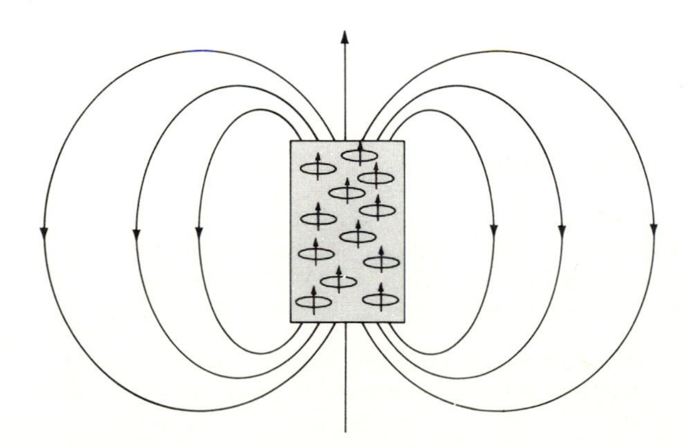

Figure 9.3 A cross section of a bar magnet showing the interior distribution of molecular magnetic moments and the exterior magnetic field that these produce.

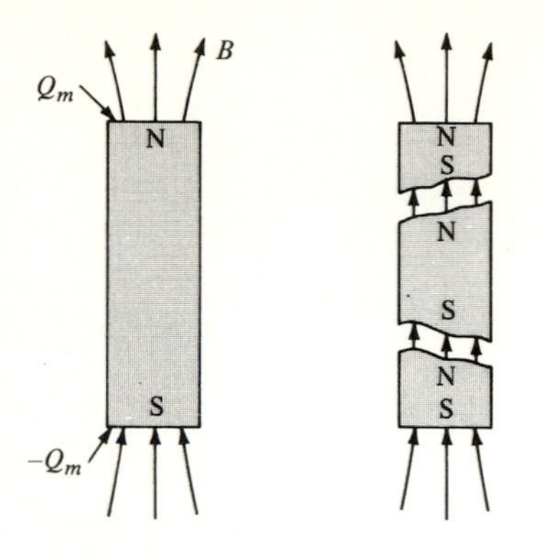

Figure 9.4 A magnetic pole cannot be isolated by breaking a bar magnet. As shown, a new set of poles will always appear at the broken surfaces.

netic charge by breaking a long magnetized needle into smaller and smaller pieces failed, however, since it only resulted in producing progressively smaller pieces of magnets, each with a north and south pole. It is clear that this should happen if one thinks that billions of spinning electrons, each a microscopic magnet with a north and south pole, fill the interior of a bar magnet. This is also the reason why Gauss' law for magnetism reads $\oiint \mathbf{B} \cdot \mathbf{dA} = 0$ and not $\oiint \mathbf{B} \cdot \mathbf{dA} = \mu_0 Q_m$; that is, any finite volume, no matter how small, contains as many negative as positive charges.

Although a magnetic pole or charge cannot be isolated, a good approximation to the field of a magnetic monopole is given by the field near one pole of a long magnetized needle, as shown in Fig. 9.5. Iron filings, for example, can be used to construct such a field.

The equivalence between a small bar magnet and a small current loop is apparent from Fig. 9.6, where it is seen that the B fields at a distance from either structure are identical. This equivalence was discussed at length in Sec. 6.10. The strength of the fields will be the same if the dipole moment of the small magnet, $p_m = q_m l$, and the dipole moment of the loop, $m = IA$, are the same; that is,

$$q_m l = IA \tag{9.1}$$

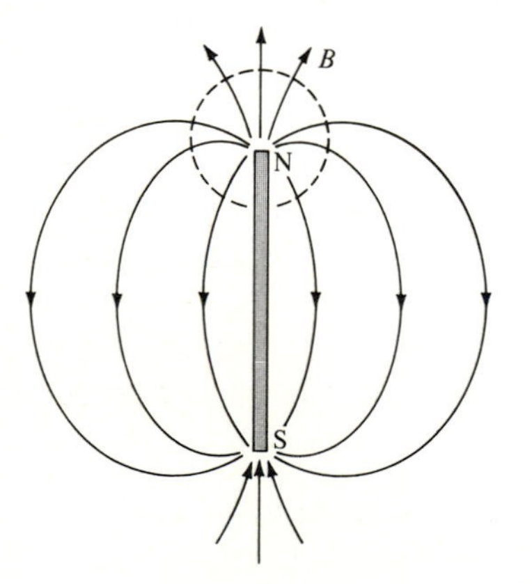

Figure 9.5 The field inside the dashed circle due to a long magnetized needle approximates that of a positive magnetic charge.

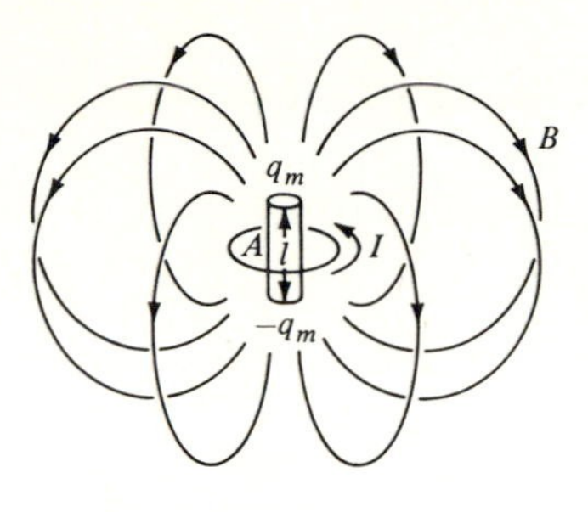

Figure 9.6 The magnetic far fields of a bar magnet and an electromagnet which consists of a single loop of current are identical.

Therefore we are presented with a dual way of looking at both a large bar magnet and an infinitesimal magnet, which is the spinning electron.

Magnetization M from Two Points of View

We have shown that the tiny magnets created by the circulating atomic currents are the real sources of the B field of permanent magnets and magnetizable materials. This concept can be formalized by introducing the *magnetization* **M** which is defined as the average dipole moment per unit volume, or

$$\boxed{\mathbf{M} = \frac{\mathbf{m}}{\Delta v} \qquad \text{A/m}} \tag{9.2}$$

where it is understood that Δv is a macroscopically small volume that contains many atomic dipoles (Δv is microscopically large). Hence **m** is an average dipole moment, localized in Δv, caused by billions of tiny circulating atomic currents in Δv. The magnetization **M** is thus a macroscopic quantity like the magnetic flux density **B** or the magnetic field strength **H**. Knowing **M** for a material, we do not have to consider the individual atomic magnets, as **M** will provide the link between the atomic world and the macroscopic one. To understand Eq. (9.2) for magnetization, Sec. 4.2 should be reread at this time. The concepts outlined there are directly applicable here merely by substituting magnetization **M** for polarization **P**.

A magnetic material that is placed in a magnetic field will become magnetized; that is, the material itself becomes a magnet and in turn contributes to the external magnetic field. A measure of the strength of induced magnetism is the magnetization **M** which exists continuously throughout the material. In the hard ferromagnetic materials a large part of the induced **M** remains after the external magnetic field is withdrawn, leaving us with a permanent magnet. In soft ferromagnetic materials the remaining **M** is small and is usually approximated by zero.† In the paramagnetic or diamagnetic materials the induced **M** vanishes as

† The tendency of a magnetized substance to persist in its state of magnetization is known as *hysteresis*. An attempt to classify magnetic materials was made in Sec. 7.5.

the external field goes to zero. If the material of an object becomes uniformly magnetized, we can obtain the total magnetic moment of the object simply as $\mathbf{m} = \mathbf{M}v$, where v is the volume of the object. If it becomes nonuniformly magnetized with $\mathbf{M}$, a known function of position, the magnetic moment at each point, is $\mathbf{m} = \mathbf{M}\,\Delta v$ and that of the object is then given by

$$\mathbf{m} = \iiint \mathbf{M}\,dv \qquad \mathrm{A \cdot m^2} \tag{9.3}$$

View 1. Equivalent-charge formulism A cylindrical bar or rod magnet is shown in Fig. 9.7. Let us relate its magnetization $\mathbf{M}$ to the equivalent pole strength Q_m. Assume the bar magnet is uniformly magnetized with $\mathbf{M}$ parallel to the axis of the cylinder. In view 1 the bar magnet is assumed to have a positive magnetic charge Q_m (north pole) and a negative magnetic charge $-Q_m$ (south pole), located as shown in Fig. 9.7*b*. This comes about when we visualize the interior of the magnet as an array of aligned tiny magnets. The adjacent pole pieces of the interior magnets cancel each other, leaving a large uncanceled charge on top and bottom, $Q_m = \sum q_m$. Because $\mathbf{M}$ is assumed uniform throughout the material, we have from Eq. (9.2)

$$\boxed{M = \frac{q_m\,\Delta l}{\Delta A\,\Delta l} = \frac{q_m}{\Delta A} = \frac{Q_m}{A} = \rho_{sm} \qquad \mathrm{A/m}} \tag{9.4}$$

where $\Delta v = \Delta A\,\Delta l$ is a small macroscopic volume element that contains a dipole of net moment $q_m\,\Delta l$. Thus we may regard the pole surfaces as having a pole

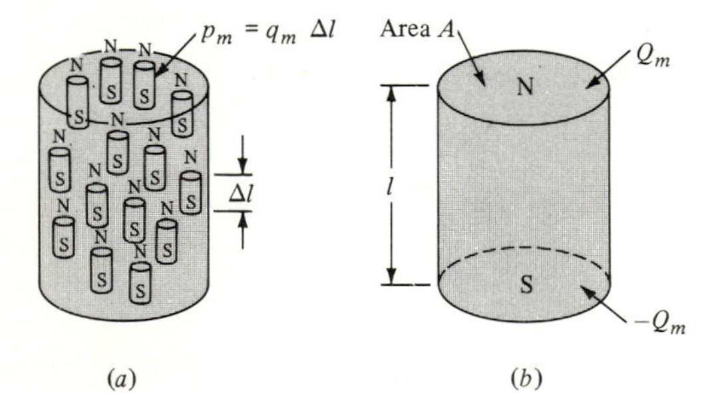

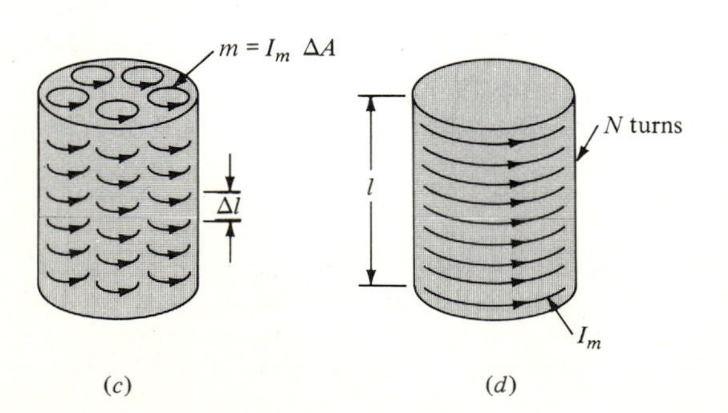

Figure 9.7 Two views of a bar magnet. In (*a*) and (*b*) the magnet is imagined to have a distribution of aligned small bar magnets throughout the interior. Because of cancellation of adjacent poles, we are left with one large top and bottom surface charge. In (*c*) and (*d*) the magnet is viewed as having a distribution of small current loops which are aligned. Because of cancellation of the interior currents, we are left with a current sheet on the cylindrical surface only.

surface density ρ_{sm} which is equal to the magnetization **M**. In this point of view the seat of the source of the magnetic field is a charge $Q_m = \rho_{sm} A$ located on the top and bottom of the bar magnet.

View 2. Equivalent-current formulism In the second point of view, originally due to Ampère, we treat the permanent bar magnet as an equivalent electromagnet, that is, as a solenoid with current I_m flowing through turns of a fictitious winding, as shown in Fig. 9.7*d*. Here the magnetization of the bar magnet material arises because of the alignment of the spinning electrons, which are now viewed as tiny current loops.† This approach comes about as follows:

1. The external magnetic field of a bar magnet (Fig. 9.3) and that of a solenoid (Fig. 6.25 or 7.7) are similar.
2. The adjacent currents of the aligned current-loop dipoles cancel throughout the interior of the magnet, leaving, in effect, only a net current circling around the cylindrical surface of the bar magnet. This is shown in Fig. 9.7*c* and *d*.

If we assume that current I_m flows through a solenoidal winding of N/l turns per unit length, we have completed our analogy. The external field of the bar magnet is created by I_m in this fictitious winding. The field at the center of a long bar magnet is then given by Eq. (6.40) as

$$B = \mu_0 \frac{NI_m}{l} \tag{9.5}$$

If we imagine a winding with many turns of fine wire, we can visualize the magnetization current as flowing in a current sheet $K_m = NI_m/l$ around the periphery of the magnet [as discussed for a real solenoid in the paragraph following Eq. (6.42)]. The magnetization **M** can be related to this current sheet as follows:

$$\boxed{M = \frac{m}{\Delta v} = \frac{I_m \, \Delta A}{\Delta A \, \Delta l} = \frac{I_m}{\Delta l} = \frac{NI_m}{l} = K_m \qquad \text{A/m}} \tag{9.6}$$

where the length l of the bar is related to Δl through N; that is, $l = N\,\Delta l$. The magnetic field at the center of the bar magnet is then given by $B = \mu_0 M$.

In concluding this section we observe that the magnetization **M** is analogous to the electric polarization **P** discussed in Chap. 4, where polarization was defined as the net dipole moment per unit volume, or $\mathbf{P} = \mathbf{p}/\Delta v$. As a source of electric field, **P** was overshadowed by electric charges Q. Here, however, **M** is the primary source for steady magnetic fields from permanent magnets since isolated magnetic charges do not exist.

† The same interpretation can be applied to explain the increase in the magnetic field when a magnetic core is inserted in an inductor that carries a current I, such as a solenoid. We could achieve the same increase of B without a core material by either increasing the current I in the solenoid or by having an additional current I_m circulating in the same direction. The magnetization **M** of the core provides such an additional current I_m.

Example For a uniformly magnetized cylindrical rod magnet of dimensions $l = 10$ cm and $A = 1\ \text{cm}^2$ and a pole strength of 1 A · m, an equivalent 100-turn solenoidal electromagnet must have a current of $I = 10$ A, obtained as follows:

$$M = \frac{m}{v} = \frac{Q_m l}{Al} \equiv \frac{NI_m}{l}$$

Equating I_m to I:

$$I = I_m = \frac{Q_m l}{AN} = \frac{(1)(0.1)}{(10^{-4})(100)} = 10 \text{ A}$$

9.2 RELATIONSHIP BETWEEN B, H, AND M

To find the relationship between these three vectors, we make use of the toroidal solenoid, because the toroid is the only finite structure in which the magnetic field is confined to the interior. Fringing and leakage is minimal if the winding is of fine wire with the turns closely spaced. Furthermore, B is uniform along the toroid, which makes this a unique structure, particularly suitable when relations of a general nature are to be derived. From (6.44), (7.25), or directly from Ampère's law $\oint \mathbf{B} \cdot d\mathbf{l} = \mu_0 I$, which can be easily integrated for the case of a toroid, the magnetic flux density for a large air-wound toroid is given by

$$B_0 = \mu_0 \frac{NI}{l} = \mu_0 \frac{NI}{2\pi r} \tag{9.7}$$

where the path length of the toroid is $l = 2\pi r$, N is the number of turns, I is the real current flowing, and subscript zero denotes a magnetic field in free space. The geometry is shown in Fig. 9.8*a*.

The strength of the magnetic field inside the toroid can be measured by a secondary winding which is connected to a voltmeter (galvanometer). The deflection of the voltmeter when current I in the primary is switched on (or interrupted) is proportional to changes in the magnetic flux $\phi = B_0 A$ that will flow in the core. This is so because Faraday's law states that the induced voltage in the secondary coil is given by $V = N\, d\phi/dt$. The physical process involved in the deflection of the voltmeter is as follows: Switching on current I in the primary causes an H field in the toroid. This can be seen from Ampère's law, which states that $\oint \mathbf{H} \cdot d\mathbf{l} = NI = \mathscr{F}$. In other words, we have turned on a source of magnetomotive force equal to $\mathscr{F} = NI$ which is also equal to $\mathscr{F} = Hl$. In a cause-effect relationship, $\mathscr{F}$ is the cause, Hl the effect. Multiplying H by μ_0 gives the magnetic flux density B_0. Multiplying B_0 by cross-sectional area A gives us total magnetic flux for the toroid; that is, $\phi = B_0 A$, which links the secondary winding. Any change with time in the flux will induce a voltage $V = N\, d\phi/dt$ in the secondary which will activate the voltmeter.

If we now insert a core of magnetic material (Fig. 9.8*b*), we find that the deflection of the voltmeter is different for identical conditions of current interruption. We would find that for diamagnetic material the deflection is slightly less, for paramagnetic material slightly more, and for ferromagnetic material much greater

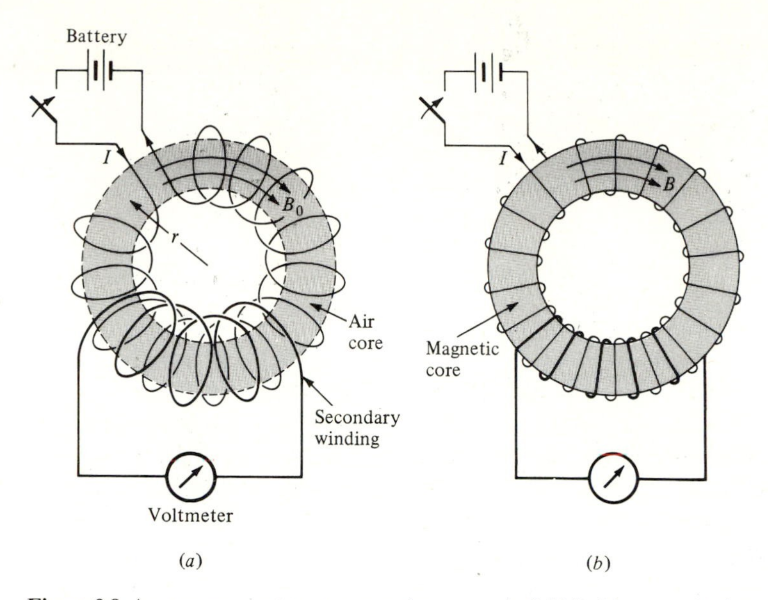

Figure 9.8 An arrangement to measure the magnetic field inside a toroid (often referred to as the Rowland ring). In (*a*), the subscripted B_0 denotes that the interior of the toroid is void of magnetic materials. In (*b*), we have a magnetic core present which changes the magnetic flux density to B.

than that for the air core. We conclude from this experiment that the total magnetic flux density B inside the core has changed from B_0. In the presence of magnetic materials Ampère's law in the form $\oint \mathbf{B} \cdot d\mathbf{l} = \mu_0 I$ is, therefore, not valid anymore. As the right side $\mu_0 I$ remains the same for both cases (air and magnetic core), B should also remain the same, a conclusion contrary to fact. The change in B with the magnetic core present can be accounted for correctly by the equivalent magnetization current I_m. With the material core in place, Ampère's law can be correctly written as $\oint \mathbf{B} \cdot d\mathbf{l} = \mu_0(I + I_m)$, which gives for B in the toroid

$$B = \mu_0 \frac{N}{l}(I + I_m)$$

$$= \mu_0 \left(\frac{NI}{l} + M \right) \tag{9.8}$$

where (9.6) was used to relate I_m to magnetization M. The above result can be rewritten as

$$\frac{B - \mu_0 M}{\mu_0} = \frac{NI}{l} \tag{9.8a}$$

The right side of this equation is a function of only I, which is the real current that flows in the winding. This being the case, the medium-independent form of Ampère's law (7.18), which is also a relationship only in terms of I,

$$\oint \mathbf{H} \cdot d\mathbf{l} = NI \tag{9.9}$$

can be used to express (9.8a) in terms of the magnetic field strength H as follows: Eq. (9.9) for a toroid with N turns gives $Hl = NI$ or $H = NI/l$. Expression (9.8a) can therefore be generalized to

$$\boxed{\frac{\mathbf{B} - \mu_0 \mathbf{M}}{\mu_0} = \mathbf{H}} \tag{9.10}$$

or

$$\boxed{\mathbf{B} = \mu_0 \mathbf{H} + \mu_0 \mathbf{M}} \tag{9.11}$$

This is the desired relationship between the three quantities **B**, **H**, and **M**. It is general and is valid for linear or nonlinear magnetic materials. The $\mu_0 \mathbf{H}$ term gives the contribution to the flux density **B** due to the real current I in the windings of the toroid, whereas $\mu_0 \mathbf{M}$ gives the additional magnetic flux density due to the induced magnetization **M** in the core material. Assuming the magnetic core is not permanently magnetized, what causes the magnetization **M** to be induced? It is the current I when switched on that creates an H field, $H = NI/l$; $\mu_0 H$ in turn aligns the random magnetic dipole moments of the electrons. The field of the aligned dipoles then adds to the $\mu_0 H$ field. H is thus the magnetizing force, or cause, with B being the result, or effect.

In conclusion we observe that the new generalized expression (9.10) for H makes Ampère's law in the form (9.9) the most general statement of Ampère's law. The reason for this is that H and I are quantities which are independent of the presence or absence of magnetic materials. This is in contrast to Ampère's law expressed in terms of B which requires that magnetization currents must be taken into consideration explicitly. In the presence of magnetic materials, this is an important distinction. Until this chapter, magnetic materials were not considered and both forms of Ampère's law, $\oint \mathbf{B} \cdot d\mathbf{l} = \mu_0 I$ and $\oint \mathbf{H} \cdot d\mathbf{l} = I$, could be used without leading to difficulties.

Relative Permeability†

For an air core, when magnetization $\mathbf{M} = 0$, we have $\mathbf{B}_0 = \mu_0 \mathbf{H}$. We can write a similar linear relationship when **M** is not zero; that is,

$$\mathbf{B} = \mu \mathbf{H} \qquad \text{for linear magnetic materials} \tag{9.12}$$

Using (9.11),

$$\begin{aligned} \mathbf{B} &= \mu_0(\mathbf{H} + \mathbf{M}) \\ &= \mu_0 \left(1 + \frac{M}{H}\right) \mathbf{H} \end{aligned} \tag{9.13}$$

† Section 7.5 should be reread at this time as relative permeability was introduced there. It is in the present section, however, that we relate relative permeability to magnetization.

The general permeability μ is therefore

$$\mu = \mu_0 \left(1 + \frac{M}{H}\right) \tag{9.14}$$

and is equal to μ_0 when $M = 0$. The relative permeability μ_r is given by $\mu_r = \mu/\mu_0$, which was already defined in (7.16). We have assumed that **H**, **B**, and **M** are parallel, as they are for isotropic media. The dimensionless ratio M/H is known as the *magnetic susceptibility* χ_m and gives a measure of the extent to which a material can be magnetized by **H**. That is,

$$\mathbf{M} = \chi_m \mathbf{H} \qquad \text{for linear magnetic materials} \tag{9.15}$$

The relative permeability can be written in terms of χ_m as

$$\mu_r = \frac{\mu}{\mu_0} = 1 + \chi_m \tag{9.16}$$

For diamagnetic and paramagnetic materials, χ_m is very small compared to unity. It is negative for diamagnetic materials, as they produce magnetic fields which oppose the applied **H** field. For paramagnetic materials, the induced dipoles align parallel to the applied field, and χ_m is positive. It is the ferromagnetic materials which have large relative permeabilities and for which (9.15) does not hold because these materials are nonlinear and have hysteresis. [Note that Eq. (9.11) always holds.] Permeability is then a function of H, which is usually obtained from the BH curve of a ferromagnetic material. The expression (9.12) can still be written as $\mathbf{B} = \mu(H)\mathbf{H}$, showing the explicit dependence of μ on H. A detailed discussion of magnetic materials will be given in Sec. 9.5, which will explain the nonlinear μ for ferromagnetic materials.

9.3 H AND B FIELDS FOR PERMANENT MAGNETS—EQUIVALENT–MAGNETIC CHARGE FORMULATION

A description of the magnetic field in terms of equivalent magnetic charge as outlined in view 1 of Sec. 9.1 will now be developed. Using **H** as given by (9.10), we can state Ampère's law in terms of real currents as

$$\oint \mathbf{H} \cdot dl = I \tag{9.17}$$

which is valid in the presence or absence of magnetic matter. For example, in the absence of magnetic matter, $\mathbf{M} = 0$, and (9.17) with the use of (9.10) gives us $\oint (\mathbf{B}/\mu_0) \cdot \mathbf{dl} = I$; comparing this result to (9.7) shows that $\mathbf{B} = \mathbf{B}_0$, as expected.

On the other hand, if a permanent magnet is the only source of the magnetic field, then, because of the absence of real currents,

$$\oint \mathbf{H} \cdot \mathbf{dl} = 0 \tag{9.18}$$

even if the contour pierces the magnetized body. This implies that the magnetic field strength **H** in this case is a conservative field and may be derived from a scalar potential Φ as†

$$\boxed{\mathbf{H} = -\nabla\Phi} \tag{9.19}$$

As discussed in Sec. 8.4, the sources for potential fields are charges, and the relationship between potential and charge density is given by Poisson's equation (1.96*c*) as

$$\boxed{\nabla^2\Phi = -\rho_m} \tag{9.20}$$

The **H** field can therefore be thought of as beginning and ending on magnetic charges Q_m, similar to the conservative electrostatic **E** field. Magnetic charge density ρ_m and the divergence of **H** can now be related as follows:

$$\nabla \cdot \mathbf{H} = -\nabla \cdot \nabla\Phi = -\nabla^2\Phi = \rho_m \tag{9.21}$$

This equation is analogous to Poisson's equations for the electric scalar potential. The charge density ρ_m can be related to magnetization **M** by making use of $\nabla \cdot \mathbf{B} = 0$. Taking the divergence of (9.11), we obtain

$$\nabla \cdot \mathbf{B} = \mu_0 \nabla \cdot \mathbf{H} + \mu_0 \nabla \cdot \mathbf{M} = 0$$

or

$$\boxed{\nabla \cdot \mathbf{H} = -\nabla \cdot \mathbf{M} \equiv \rho_m} \tag{9.22}$$

The magnetic volume charge density ρ_m is therefore equivalent to the space rate of change of magnetization, or $\rho_m = -\nabla \cdot \mathbf{M}$. To obtain the **H** field inside and outside a permanent magnet, we can solve Poisson's equation (9.20) for the scalar potential Φ and then obtain **H** from $\mathbf{H} = -\nabla\Phi$. The solution of Poisson's equation is given by (1.16) or (1.101) as

$$\boxed{\Phi(\mathbf{r}) = -\frac{1}{4\pi} \iiint \frac{\nabla' \cdot \mathbf{M}'(\mathbf{r}')}{|\mathbf{r} - \mathbf{r}'|} dv} \tag{9.23}$$

where the integration is over the volume in which ρ_m exists, **r** is the observation point which can be inside or outside the permanent magnet, and **r**′ is the source point inside the magnetized material, as shown in Fig. 9.9. The prime on the operator ∇ denotes differentiation with respect to the source variable **r**′.

For smooth variations of *M*, the divergence derivatives in (9.23) are well behaved. However, in practical problems *M* is discontinuous; that is, magnetization *M* drops abruptly to zero outside a magnetized volume. The $\nabla' \cdot \mathbf{M}'$ term is then infinite on part of the boundary of a magnet, and the integral in (9.23) does not exist. To get around this difficulty, we evaluate the singularity on the bound-

† Note that in (1.61) it was shown that $\oint \nabla\Phi \cdot dl \equiv 0$.

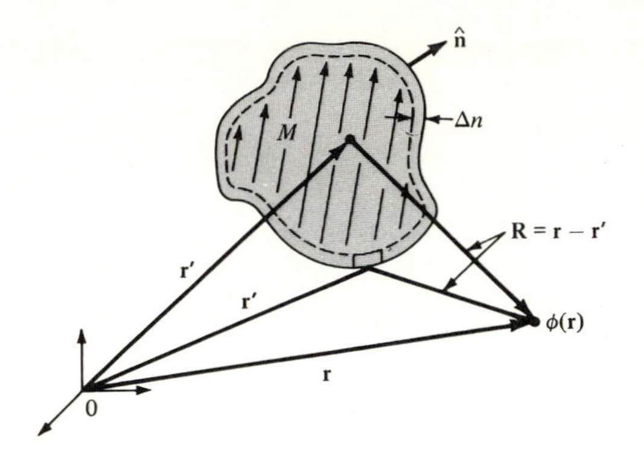

Figure 9.9 A magnetized object showing an interior source point and one on the surface. A volume element on the surface is $\Delta v = \Delta n\, \Delta A$, where Δn is the thickness of the outer layer.

ary separately as follows: If we divide the volume v of a magnetized object into an interior volume and a shell-like volume ∂v over the surface of the object, as shown in Fig. 9.9, we have for (9.23)

$$\phi(\mathbf{r}) = -\frac{1}{4\pi} \iiint_{\partial v} \frac{\nabla' \cdot \mathbf{M}'}{R}\, dn\, dA - \frac{1}{4\pi} \iiint \frac{\nabla' \cdot \mathbf{M}'}{R}\, dv \tag{9.24}$$

where $dv = dn\, dA =$ volume element of shell
$dn =$ thickness of shell
$\mathbf{dA} = \hat{\mathbf{n}}\, dA$
$\hat{\mathbf{n}} =$ outward normal of the object
$R = |\mathbf{r} - \mathbf{r}'|$

The volume integral of the shell can be reduced to a surface integral by noting that the integrand can be written as

$$\nabla \cdot \mathbf{M}\, dn\, dA = \frac{\partial M_n}{\partial n}\, dn\, dA = \Delta M_n\, dA = -M_n\, dA \tag{9.25}$$

where the normal component of $\mathbf{M}$ is $M_n = \mathbf{M} \cdot \hat{\mathbf{n}}$ and the differential ΔM_n is the difference in M_n between the inside and outside of the object. That is, $\Delta M_n = M_n^{\text{out}} - M_n^{\text{in}} = -M_n^{\text{in}} = -\mathbf{M} \cdot \hat{\mathbf{n}}$, because magnetization $\mathbf{M}$ in the free space outside the magnetized object is zero. Hence we can write (9.24) as

$$\boxed{\phi(\mathbf{r}) = \frac{1}{4\pi} \oiint_{A(v)} \frac{\mathbf{M}' \cdot \hat{\mathbf{n}}}{R}\, dA - \frac{1}{4\pi} \iiint_{v(A)} \frac{\nabla' \cdot \mathbf{M}'}{R}\, dv} \tag{9.26}$$

where $v(A)$ is the interior of the volume of the magnetized object and $A(v)$ is the surface area that bounds volume v. The magnetic field of a permanent magnet is thus seen to be produced by a surface layer of magnetization charge ρ_{sm}, which was already developed in (9.4), and is given by

$$\boxed{\rho_{sm} = \mathbf{M} \cdot \hat{\mathbf{n}}} \tag{9.26a}$$

and a volume distribution ρ_m which is

$$\boxed{\rho_m = -\nabla \cdot \mathbf{M}} \tag{9.26b}$$

Of course in objects in which M goes to zero smoothly as the boundary of the object is approached, $\Delta M = 0$, and we are left with the volume integral only.

The magnetic field strength is obtained from the potential by

$$\mathbf{H}(\mathbf{r}) = -\nabla\Phi(\mathbf{r}) = \frac{1}{4\pi}\iiint \frac{\rho'_m}{R^2}\hat{\mathbf{R}}\,dv + \frac{1}{4\pi}\oiint \frac{\rho'_{sm}}{R^2}\hat{\mathbf{R}}\,dA \tag{9.27}$$

noting that $\nabla(1/R) = -\hat{\mathbf{R}}/R^2 = -(\mathbf{r}-\mathbf{r}')/|\mathbf{r}-\mathbf{r}'|^3$ from (8.44).

Summarizing, we can say that the magnetic scalar potential method gives the correct **H** field inside and outside a magnetized body. The magnetized body is replaced by magnetic charge densities which are imagined to exist in vacuum. For example, the permanent magnet of Fig. 9.10, for purposes of calculating the **H** field, is equivalent to the top and bottom layers of charge, all other space assumed to be vacuum. Once **H** is known inside and out, the **B** fields can be obtained as follows: Outside the magnet, **B** is simply **H** of (9.27) multiplied by the free-space permeability; that is, $\mathbf{B} = \mu_0\mathbf{H}$. Inside, the **B** field can be obtained by adding magnetization **M** to **H** of (9.27); that is, $\mathbf{B} = \mu_0(\mathbf{H} + \mathbf{M})$. We should note that **H**

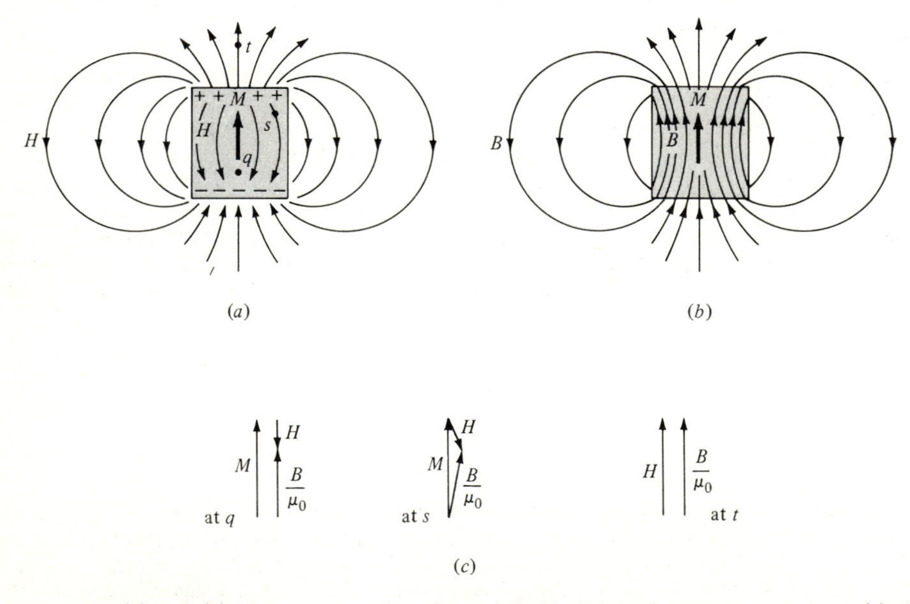

Figure 9.10 (*a*) and (*b*) The **H** and **B** lines for a cylindrical-shaped permanent magnet; (*c*) the relationship between **H, B,** and **M** for two interior points q and s, and for an exterior point t is shown. Note that the relationship between the three vectors is given by Eq. (9.11) as $\mathbf{B} = \mu_0(\mathbf{H} + \mathbf{M})$.

and **B** outside have the same direction, but inside **H** and **B** have, in general, different directions.†

An Alternative Derivation

We could have used an alternative approach in deriving the **H** field, (9.27), due to a permanent magnet. A magnetized object has a distribution of dipoles $\mathbf{p}_m = q_m\mathbf{l}$ throughout the interior, where $\mathbf{p}_m$ is the magnetic dipole moment (we are using $\mathbf{p}_m$ to denote a magnetic charge dipole, whereas **m** is used to denote a current-loop dipole). Since the potential at **r** from a dipole at **r**′ has been derived before, for example, in Eq. (4.12*a*), we have for the potential

$$\Phi(\mathbf{r}) = \frac{\mathbf{p}_m(\mathbf{r}') \cdot \hat{\mathbf{R}}}{4\pi R^2} \tag{9.28}$$

where $R = |\mathbf{r} - \mathbf{r}'|$. From Eq. (9.2), magnetization **M** is expressed as dipole moment per unit volume, or $\mathbf{p}_m = \mathbf{M}\,\Delta v$. Substituting this in the above equation and integrating over the volume of the magnetized object,

$$\Phi = \frac{1}{4\pi}\iiint \frac{\mathbf{M}' \cdot \hat{\mathbf{R}}}{R^2}\,dv \tag{9.29}$$

gives the potential for the magnetized object. This expression can be reduced to (9.26) by the use of vector identities. The **H** field can then be obtained from $\mathbf{H} = -\nabla\Phi$. Note that this development parallels that for polarization **P**, discussed in Eqs. (4.25) to (4.30).

Example: Uniformly magnetized rod magnet The magnetization **M** for a permanent magnet is usually specified by the manufacturer of the magnetic material. Most magnets have a uniform **M** throughout the interior, which makes the divergence of **M** zero. That is, in Eq. (9.27), $\nabla \cdot \mathbf{M} = 0$ because the derivative of a constant **M** is zero. The magnetic field is thus produced by the surface charges, where **M** changes abruptly from a constant value inside to zero outside the magnet. Let us consider an example of a cylindrical bar magnet with constant **M**, as shown in Fig. 9.10. (The surface layer of magnetic charge is shown in Fig. 9.10*a*.) Note that a completely uniform **M** is not realizable in practice, as **M** near the top and bottom edges of the bar tends to be nonuniform. However, the assumption of uniform **M** throughout the volume of the magnet is a very good approximation.

The magnetic scalar potential for the cylindrical bar magnet is thus, from (9.26),

$$\Phi(\mathbf{r}) = \frac{1}{4\pi}\oiint \frac{\mathbf{M} \cdot \hat{\mathbf{n}}}{R}\,dA$$

$$= \frac{1}{4\pi}\iint_{\text{top}} \frac{M}{R}\,dA + \frac{1}{4\pi}\iint_{\text{bottom}} \frac{-M}{R}\,dA \tag{9.30}$$

The evaluation of these integrals is left as a problem (see Probs. 9.31 and 9.32). The simpler problem of the field on the axis of the cylinder can be obtained by using the results for the potential field of a charged disc, Eq. (1.34). Since the magnet is composed of two charged discs, a superposition of result (1.34) will give Φ everywhere on the axis. Magnetic field strength is then obtained from $\mathbf{H} = -\nabla\Phi$.

† The most general situation is one in which real currents flow in the magnetic material. For example, iron is a good conductor and may carry a real current via its free electrons. The total magnetic field is then obtained by adding (6.9) to the magnetic field due to magnetization, (9.27).

This superposition was already carried out for the analogous problem of a permanently polarized ferroelectric rod. The potential Φ on the axis is given by (4.58*a*). The only modification needed is the substitution of ρ_{sp} by $\rho_{sm} = M$ and the deletion of ε_0. This modification can be seen by comparing the electric and magnetic dipole expressions (4.12*a*) and (9.28).

We can obtain the fields off the axis if we restrict the observation point to be far from the cylinder. In the far field, the cylinder looks like a dipole. The far-field potential Φ is then similarly given by (4.58), with p given by (4.58*c*) as

$$p_m = Mv\frac{1 + l^2}{4a^2} \tag{9.31}$$

where a = radius of cylindrical rod
l = height
v = volume

The H far field is then given by a dipole expression such as (4.14), in which **E** is replaced by **H** and ε is deleted.

As is shown in Fig. 9.10*a*, the sources of the **H** field are the negative and positive charge layers on the top and bottom of the cylinder. The H lines originate there, or saying it a different way, the H lines begin where the M lines end, and end where the M lines begin. We observe that **H** inside is opposite to **M**, and that **H** changes direction in traversing the top and bottom surfaces. The change in direction of **H** allows the closed-line integral of **H** to be zero, as required by (9.18). Note that the H lines bulge outward and cross the cylindrical wall of the material as if the wall was not there. The bulging can be explained from the fact that the charged top and bottom surfaces constitute a dipole with the charges smeared out in the shape of discs. The shape of the H lines is therefore similar to the shape of the bulging E lines near an equivalent electric dipole, as shown in Fig. 6.30*a*.

Example: Uniformly magnetized sphere The geometry of a uniformly magnetized sphere is shown in Fig. 4.17, except for the substitution of P by M. All the results of Sec. 4.9 are directly applicable to the problem of the magnetized sphere. Any modifications are obtained by comparing the electric and magnetic dipole expressions (4.12*a*) and (9.28). Thus from (4.63) the potential exterior to the sphere is given by

$$\Phi(\mathbf{r}) = \frac{Ma^3}{3}\frac{\cos\theta}{r^2} \qquad r \geq a \tag{9.32}$$

where M is the magnetization and a is the radius of the sphere. The $\cos\theta/r^2$ term identifies it as a dipole field. The relationship between the dipole moment p_m and magnetization M is given by (4.64) and is

$$p_m = \frac{4\pi}{3}a^3M \tag{9.33}$$

where $4\pi a^3/3$ is the volume of the sphere.

The **H** field outside the sphere is that of a dipole and is given by (4.14) or (6.52) as

$$\mathbf{H} = -\nabla\Phi = \frac{p_m\cos\theta}{2\pi r^3}\hat{\mathbf{r}} + \frac{p_m\sin\theta}{4\pi r^3}\hat{\boldsymbol{\theta}} \tag{9.34}$$

Note that when dealing with magnetic charge, we use p_m to denote the magnetic dipole moment, whereas the dipole moment of a current loop is customarily denoted by m. From (9.1) we see that $p_m = m$.

The **H** field inside a uniformly magnetized sphere is constant and by use of (4.66) is given by

$$\mathbf{H}_i = -\tfrac{1}{3}\mathbf{M} \tag{9.35}$$

The interior **B** field is thus

$$\mathbf{B}_i = \mu_0(\mathbf{H} + \mathbf{M}) = \mu_0\tfrac{2}{3}\mathbf{M} \tag{9.36}$$

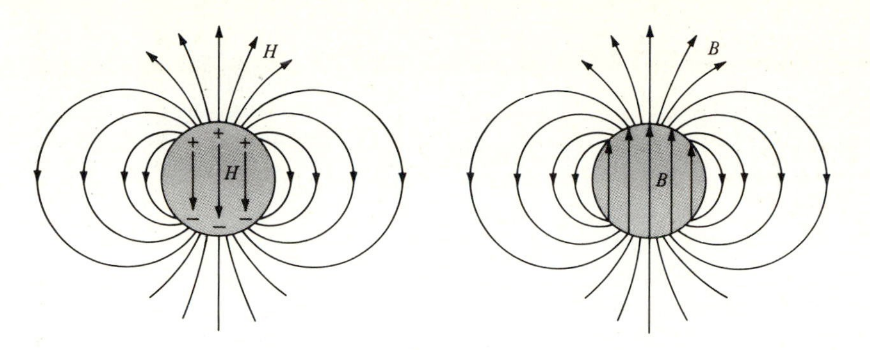

Figure 9.11 The **H** and **B** fields for a uniformly magnetized sphere.

The **H** and **B** fields are sketched in Fig. 9.11. The **B**, **H**, and **M** fields are parallel and uniform inside the sphere, which is a property of spherical shapes. The internal fields, for example, are not parallel for the cylindrical rod geometry shown in Fig. 9.10.

Example: Sphere in a uniform magnetic field If the sphere is not permanently magnetized but the magnetization **M** is induced by an external field $B_0 = \mu_0 H_0$, the results of the analogous problem discussed in Sec. 4.11 apply again.† As pointed out in that section, the sphere when immersed in a uniform field is one of the ellipsoidal shapes in which the induced fields are uniform. The internal fields in a sphere placed in an external field $\mathbf{H}_0 = H_0\hat{\mathbf{z}} = H_0(\hat{\mathbf{r}} \cos\theta - \hat{\boldsymbol{\theta}} \sin\theta)$ are, from (4.79) and (4.80),

$$\Phi_i = -\frac{3H_0 z}{\mu_r + 2} \qquad r \leqslant a \tag{9.37}$$

$$\mathbf{H}_i = \frac{3}{\mu_r + 2}\mathbf{H}_0 \qquad r \leqslant a \tag{9.38}$$

where $\mu_r = \mu/\mu_0$. The magnetization induced in the sphere by the external $\mathbf{H}_0$ field is, from (4.84),

$$\mathbf{M} = 3\mathbf{H}_0 \frac{\mu_r - 1}{\mu_r + 2} \tag{9.39}$$

The external fields, in spherical coordinates, are given [from Eqs. (4.78) and (4.87)] by

$$\Phi_e = -H_0 z + a^3 H_0 \frac{\mu_r - 1}{\mu_r + 2}\frac{\cos\theta}{r^2} \qquad r \geqslant a \tag{9.40}$$

$$\mathbf{H}_e = \mathbf{H}_0 + a^3 H_0 \frac{\mu_r - 1}{\mu_r + 2}\left(\frac{2\cos\theta}{r^3}\hat{\mathbf{r}} + \frac{\sin\theta}{r^3}\hat{\boldsymbol{\theta}}\right) \qquad r \geqslant a \tag{9.41}$$

and are seen to be a superposition of the original H_0 field and a dipole field with magnetic dipole moment

$$p_m = 4\pi a^3 H_0 \frac{\mu_r - 1}{\mu_r + 2} = \frac{4\pi a^3}{3} M \tag{9.42}$$

† Notice that in a charge-free region, the electrostatic field is described by $\nabla^2 V = 0$, $\mathbf{E} = -\nabla V$, $\mathbf{D} = \varepsilon_0\mathbf{E} + \mathbf{P}$, and $\mathbf{D} = \varepsilon\mathbf{E}$. In a current-free region, the magnetostatic field is given by $\nabla^2\Phi = 0$, $\mathbf{H} = -\nabla\Phi$, $\mathbf{B} = \mu_0(\mathbf{H} + \mathbf{M})$, and $\mathbf{B} = \mu\mathbf{H}$. The boundary conditions are the same because the normal and tangential components of E, D, H, and B are continuous across a boundary. Solution to a magnetostatic problem can therefore be obtained from the analogous electrostatic problem by changing V to Φ, $\mathbf{E}$ to $\mathbf{H}$, $\mathbf{D}$ to $\mathbf{B}$, and ε to μ.

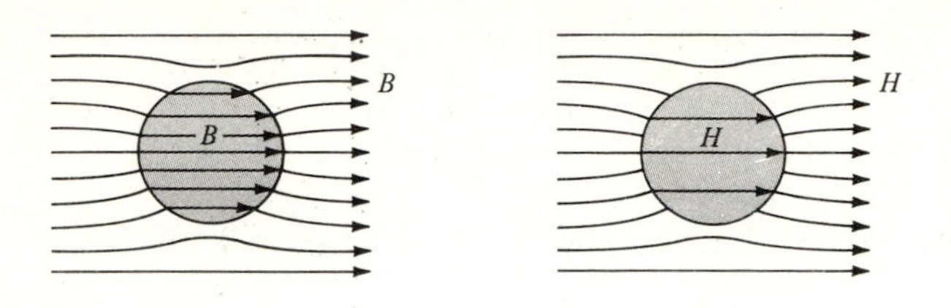

Figure 9.12 **B** and **H** field lines of a magnetic sphere in an external uniform field. Before placement of the sphere, the field lines are straight lines.

The dipole is assumed to be concentrated near the origin. The fields (9.40) and (9.41) are therefore valid right down to the surface of the sphere.

The fields can be sketched, using Figs. 4.21 and 4.22 as a reference. Figure 9.12 shows the lines of **B** and **H**. The internal and external flux density is obtained by multiplying by the appropriate μ; that is, $\mathbf{B}_i = \mu \mathbf{H}_i$ and $\mathbf{B}_e = \mu_0 \mathbf{H}_e$. Notice that in Fig. 9.12 the **B** lines appear to concentrate inside the sphere. This is characteristic of paramagnetic or ferromagnetic materials and comes about because $\mu_r > 1$. If $\mu_r < 1$, as in the case of diamagnetic materials, some of the **B** lines would be expelled (a superconductor is a perfect diamagnet; hence all **B** lines are expelled).

The strong concentration of flux in ferromagnetic materials for which $\mu_r \gg 1$ can be utilized in magnetic shielding. For example, a spherical shell immersed in an external field will have a significantly decreased **B** field inside because of the concentration of lines in the shell material.† This is shown in Fig. 9.13. As a matter of fact, since the field inside a solid sphere is uniform, we can use the above results to find the field inside a small spherical cavity which is cut into the large sphere. The interior field of a large sphere immersed in an external H_0 field is given by (9.38). The H field inside the small spherical cavity is given by the same formula but with μ and μ_0 interchanged and H_i and H_0 interchanged. Thus

$$H_{\text{cav}} = \frac{3\mu_r}{1 + 2\mu_r} H_i = \frac{9\mu_r}{(\mu_r + 2)(1 + 2\mu_r)} H_0 \cong \frac{9}{2\mu_r} H_0 \Big|_{\mu_r \gg 1} \tag{9.43}$$

The cavity field is thus much smaller than the external field. For $\mu_r = 10^3$, $H_{\text{cav}} \cong 4.5 \times 10^{-3} H_0$. Thus, the cavity effectively shields the outside field. Using boundary conditions $B_{n1} = B_{n2}$ and $H_{t1} = H_{t2}$, we can find the field in other cavities, such as the disc-shaped and needle-shaped cavities (see Sec. 4.12).

The above results can be applied to a ferromagnetic sphere with permanent magnetization. From (9.15) we have $\mathbf{M} = \chi_m \mathbf{H}_i = (\mu_r - 1)\mathbf{H}_i = 3\mathbf{H}_0(\mu_r - 1)/(\mu_r + 2)$. For a linear medium, $\mathbf{B}_i = \mu \mathbf{H}_i$, and we obtain for the interior fields

$$\mathbf{H}_i = \frac{3}{\mu_r + 2} \mathbf{H}_0 = \mathbf{H}_0 - \frac{\mathbf{M}}{3} \tag{9.44}$$

$$\mathbf{B}_i = \mu_0(\mathbf{H}_i + \mathbf{M}) = \mu_0(\mathbf{H}_0 + \tfrac{2}{3}\mathbf{M}) \tag{9.45}$$

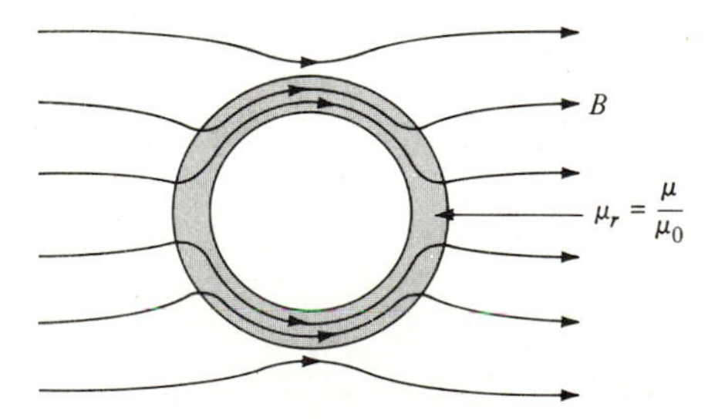

Figure 9.13 Shielding effect of a spherical shell of magnetic material for which $\mu_r \gg 1$. The inside of the shell is relatively free of B.

† For a discussion of magnetic shielding, see Sec. 10.2.

When the exterior field $\mathbf{H}_0$ vanishes, we are left with the fields inside a permanently magnetized sphere, which were derived previously in (9.35) and (9.36). We can now apply this to a sphere of nonlinear ferromagnetic material by making use of the hysteresis or BH curves for the particular ferromagnetic material. The magnetization is given by $\mathbf{M} = \mathbf{B}_i/\mu_0 - \mathbf{H}_i$.

Example: Torque on a magnetized sphere in a uniform field Another observation that can now be made is that the torque on a permanently magnetized sphere which is placed in a uniform external field $\mathbf{B}_0$ is given by [see Eqs. (6.48) and (6.53)]

$$T = p_m B_0 \sin\alpha = \frac{4\pi a^3}{3} M B_0 \sin\alpha \tag{9.46}$$

where the magnetic dipole moment p_m of the sphere is given by (9.33) and α is the angle between the magnetization M and the external field B_0.

What if we were to immerse a magnetized sphere in a medium other than free space, for example, in a medium whose permeability is μ. When calculating the torque on a magnetized sphere, one cannot simply take (9.46) and substitute $B_0 = \mu H_0$, where B_0 is the uniform magnetic flux density assumed to exist in the medium of permeability μ before the magnetized sphere is placed in the medium. The reason is that a hole the size of the sphere must first be cut out of the medium before the sphere can be placed into the medium. Such a free-space cavity will distort the initially uniform B_0 field. Therefore the field acting on the magnetized sphere will be the distorted field which is uniform inside the cavity and from (9.38) is given by (interchange H_i and H_0 and interchange μ and μ_0)

$$B_{\text{cav}} = \mu_0 H_{\text{cav}} = \frac{3\mu_0\mu_r}{1 + 2\mu_r} H_0 \tag{9.47}$$

The torque on a permanently magnetized sphere of dipole moment $p_m = 4\pi a^3 M/3$ placed in this cavity is then $T = p_m B_{\text{cav}} \sin\alpha$ or

$$T = \frac{4\pi a^3 M B_0}{1 + 2\mu_r} \sin\alpha \tag{9.48}$$

When the exterior medium is free space, $\mu_r \to 1$, and (9.48) reduces to (9.46).

No net force exists on the sphere if the external field is uniform. However, in a nonuniform field, magnetized objects which tend to concentrate magnetic fields (paramagnetic and ferromagnetic) will be pulled toward higher B-field regions, whereas diamagnetic objects are repelled by stronger B fields and tend to move toward regions of weaker flux.

9.4 H AND B FIELDS FOR PERMANENT MAGNETS—EQUIVALENT-CURRENT FORMULATION

In the previous section we showed that the magnetic charge formulation can be obtained by starting with the generalized expression for magnetic field as given by (9.11):

$$\mathbf{B} = \mu_0(\mathbf{H} + \mathbf{M})$$

and taking the divergence of it. The equivalent-current formulation can be initiated by considering the curl of $\mathbf{B}$:

$$\nabla \times \mathbf{B} = \mu_0 \nabla \times \mathbf{H} + \mu_0 \nabla \times \mathbf{M} \tag{9.49}$$

According to Ampère's law (9.9), the curl of **H** is related to real or free current, $\nabla \times \mathbf{H} = \mathbf{J}$. The above equation can then be written as

$$\boxed{\nabla \times \mathbf{B} = \mu_0 \mathbf{J} + \mu_0 \nabla \times \mathbf{M}} \tag{9.50}$$

At points where no magnetization **M** is present, $\mathbf{B} = \mu_0 H$ and (9.50) reduces to $\nabla \times \mathbf{B} = \mu_0 \mathbf{J}$. The $\nabla \times \mathbf{M}$ term can now be identified as an equivalent-current density $\mathbf{J}_m$ in amperes per square meter due to magnetization. Because $\mathbf{J}_m$ is due to atomic magnetic moments, it is also called a bound, as opposed to free, current density. Recall that free current density **J** is an ordinary conduction current due to free charges (electrons) that flows over a large-scale path and can be measured by an ammeter. $\mathbf{J}_m$ is the macroscopic average of the atomic bound currents. Therefore Ampère's law relates **H** to the free current ($\oint \mathbf{H} \cdot \mathbf{dl} = I$), whereas it relates **B** to the total current, free and bound ($\oint \mathbf{B} \cdot \mathbf{dl} = \mu_0 I + \mu_0 I_m$).

We are interested in the case when real currents are absent and the source of the magnetic field is a permanent magnet. Equation (9.50) reduces in this case to

$$\nabla \times \mathbf{B} = \mu_0 \mathbf{J}_m \tag{9.51}$$

The sources of the **B** field are thus distributed over points for which **M** shows a variation, i.e., where the derivative as expressed by $\nabla \times \mathbf{M}$ is nonzero. For example, a uniformly magnetized rod magnet, as shown in Fig. 9.10, has no sources in the interior; the curl vanishes because **M** is constant inside. The sources are at the surface because **M** changes abruptly at the surface giving a value for curl on the cylindrical surfaces.

In order to find the field of a magnetized object, we can follow the same procedure that was developed for finding the **B** field of any current distribution. For example, we can use (6.9) and substitute for **J** in the integrand, $\mathbf{J}_m = \nabla \times \mathbf{M}$. Or we can start with the vector potential **A** as developed in (8.41) or (8.47). Following the latter approach, **A** at the observation point **r** due to magnetization sources located at **r**′ is given by

$$\mathbf{A}(\mathbf{r}) = \frac{\mu_0}{4\pi} \iiint \frac{\nabla' \times \mathbf{M}'(\mathbf{r}')}{R}\, dv \tag{9.52}$$

where ∇' denotes differentiation with respect to r', the geometry is shown in Fig. 9.9, and the integration is over the volume in which J_m exists. If **M** for a magnet is known, we can substitute it in (9.52), calculate **A**, and then obtain the magnetic field from $\mathbf{B} = \nabla \times \mathbf{A}$.

The only difficulty arises when M is discontinuous, for example, when M changes abruptly from a finite value inside a magnet to a zero value outside the magnet. Then the derivative in $\nabla' \times M'$ is infinite, and the integral (9.52) is undefined on such surfaces. To get out of this difficulty, we can again split off the surface integral as was done in Eqs. (9.24) to (9.26) and evaluate it separately, or we can use a different approach.

Choosing the latter, we start with the vector potential for a magnetic dipole $\mathbf{m}'$ located at the source point $\mathbf{r}'$, which from (8.68) is given by

$$\mathbf{A}(\mathbf{r}) = \frac{\mu_0}{4\pi}\mathbf{m}' \times \nabla'\left(\frac{1}{R}\right) \tag{9.53}$$

Since magnetization $\mathbf{M}$ is defined by (9.2) as $\mathbf{M} = \mathbf{m}/dv$, the contribution to $\mathbf{A}$ from an element of volume dv which contains the dipole moment $\mathbf{M}\,dv$ is

$$\mathbf{dA}(\mathbf{r}) = \frac{\mu_0}{4\pi}\mathbf{M}' \times \nabla'\left(\frac{1}{R}\right) dv \tag{9.54}$$

Integrating over the volume of the magnetized object,

$$\mathbf{A}(\mathbf{r}) = \frac{\mu_0}{4\pi}\iiint \mathbf{M}' \times \nabla'\left(\frac{1}{R}\right) dv \tag{9.55}$$

This expression is easily put into the desired form by the use of the vector identity (A1.19) which is $\mathbf{F} \times \nabla\phi = \phi\nabla \times \mathbf{F} - \nabla \times (\phi\mathbf{F})$. With $\phi = 1/R$ and $\mathbf{F} = \mathbf{M}$, we obtain

$$\mathbf{A} = \frac{\mu_0}{4\pi}\iiint \frac{\nabla' \times \mathbf{M}'}{R} dv - \frac{\mu_0}{4\pi}\iiint \nabla' \times \frac{\mathbf{M}'}{R} dv \tag{9.56}$$

The last term can be reduced to the desired surface integral by applying the vector transformation (A1.32):

$$\mathbf{A} = \frac{\mu_0}{4\pi}\iiint \frac{\nabla' \times \mathbf{M}'}{R} dv + \frac{\mu_0}{4\pi}\oint\!\!\oint \frac{\mathbf{M}' \times \hat{\mathbf{n}}}{R} dA \tag{9.57}$$

where $\mathbf{dA} = \hat{\mathbf{n}}\,dA$ is an area element and should not be confused with the vector potential $\mathbf{A}$. The vector potential is now given as a volume and surface contribution. Note that the volume integral in (9.57) is only over the interior volume of the magnetized object where M is assumed to be a continuous function. The contribution of the outer layer of volume where M jumps from a finite value to zero is given by the surface integral. For magnetized objects in which M varies smoothly and goes to zero gradually at the boundaries of the object, the contribution of the surface integral vanishes and only the volume integral remains. However, for most practical magnets which have a constant M throughout their interior, the magnetic field is expressed entirely by the surface integral.

The fictitious current densities which enable us to calculate the vector potential are

$$\boxed{\mathbf{J}_m = \nabla \times \mathbf{M} \qquad \text{A/m}^2} \tag{9.58}$$

and

$$\boxed{\mathbf{K}_m = \mathbf{M} \times \hat{\mathbf{n}} \qquad \text{A/m}} \tag{9.59}$$

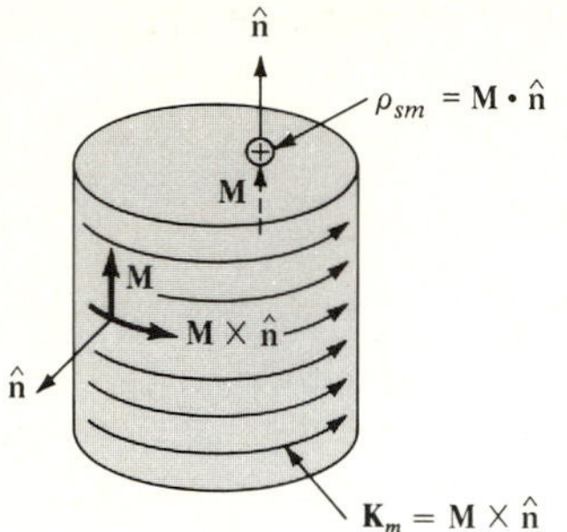

Figure 9.14 A cylindrical bar magnet with uniform magnetization **M.** A fictitious solenoidal current sheet $\mathbf{K}_m = \mathbf{M} \times \hat{\mathbf{n}}$ exists on the cylindrical surface. On the top and bottom surface, a fictitious current density $\rho_{sm} = \mathbf{M} \cdot \hat{\mathbf{n}}$ exists.

If (9.59) is compared to (9.6), we see that K_m is an equivalent solenoidal current sheet flowing on the cylindrical surfaces of a bar magnet, as shown in Fig. 9.7*d* or 9.14. As a matter of fact, if we were to wind a single-layer solenoid of fine wire on a cardboard cylinder and pass a current of $I = K_m l/N$ through it, where l is the length and N the number of turns in the solenoid, we could duplicate the exterior and interior **B** fields of the bar magnet. Note the similarity of K_m to that of a real current sheet K, (8.74). A discontinuity in M plays the same role in K_m that a discontinuity in H plays in K.

The magnetic field can be obtained from $\mathbf{B}(\mathbf{r}) = \nabla \times \mathbf{A}(\mathbf{r})$, which gives

$$\mathbf{B}(\mathbf{r}) = \frac{\mu_0}{4\pi} \iiint \nabla \times \left(\frac{\mathbf{J}'_m}{R}\right) dv + \frac{\mu_0}{4\pi} \oiint \nabla \times \left(\frac{\mathbf{K}'_m}{R}\right) dA \tag{9.60}$$

where $R = |\mathbf{r} - \mathbf{r}'|$ and ∇ is moved inside the integral. Note that the differentiation ∇ is with respect to the observation variable $\mathbf{r}$, but the integration is over the source variable $\mathbf{r}'$. Using identity (A1.19), the integrand can be rearranged as

$$\begin{aligned} \nabla \times \left(\frac{J'}{R}\right) &= \nabla\left(\frac{1}{R}\right) \times \mathbf{J}' + \frac{1}{R} \nabla \times J' \\ &= \nabla\left(\frac{1}{R}\right) \times \mathbf{J}' \end{aligned} \tag{9.61}$$

because the differentiation is with respect to $\mathbf{r}$, and J' is a function of $\mathbf{r}'$; hence $\nabla \times \mathbf{J}' = 0$. Noting from (8.44) that $\nabla(1/R) = -\hat{\mathbf{R}}/R^2$, we obtain

$$\boxed{\mathbf{B} = \frac{\mu_0}{4\pi} \iiint \frac{\mathbf{J}'_m \times \hat{\mathbf{R}}}{R^2} dv + \frac{\mu_0}{4\pi} \oiint \frac{\mathbf{K}'_m \times \hat{\mathbf{R}}}{R^2} dA} \tag{9.62}$$

This expression is similar to Eq. (6.9) developed in Chap. 6, except for the explicit statement of the surface current contribution. The equivalent current formulation, gives the correct **B** field inside and outside a magnet, whereas the equivalent charge formulation gave the correct **H** field inside and outside a magnet. We can now think of a permanent magnet as being equivalent to a distribution of current K_m on the surface and a distribution of current J_m throughout the volume of the magnet. The equivalent currents are assumed to exist in free space, hence the appearance of μ_0 in the above equations.

The expression (9.62) can be generalized to include real currents by adding B of (6.9) to it.

Example: Uniformly magnetized rod magnet We have already considered a rod magnet by using the equivalent-charge formulation. Let us now repeat it by using the equivalent-current point of view. For a uniformly magnetized cylindrical magnet shown in Figs. 9.10 and 9.14, the volume current $\mathbf{J}'_m = 0$ because the derivative of a constant M is zero. This leaves a current on the cylindrical surface only, which for a rod oriented so its axis and its magnetization M are parallel to the z axis, is

$$\mathbf{K}'_m = \mathbf{M} \times \hat{\boldsymbol{\rho}} = M\hat{\mathbf{z}} \times \hat{\boldsymbol{\rho}} = M\hat{\boldsymbol{\phi}} \tag{9.63}$$

where the normal to the cylindrical surfaces is $\hat{\mathbf{n}} = \hat{\boldsymbol{\rho}}$ and the unit vector parallel to the z axis is $\hat{\mathbf{z}}$. The magnetic field, using (9.62), is therefore

$$\mathbf{B}(\mathbf{r}) = \frac{\mu_0 M}{4\pi} \int_{-l/2}^{l/2} \int_0^{2\pi} \frac{\hat{\boldsymbol{\phi}}' \times \hat{\mathbf{R}}}{R^2} \, dA \tag{9.64}$$

where $\mathbf{R} = \mathbf{r} - \mathbf{r}'$ and $dA = a\,d\phi'\,dz'$ for a cylindrical surface of radius a. For arbitrary observation points $\mathbf{r}$, this integral is rather formidable. Let us only calculate the $\mathbf{B}$ field on the axis of the rod magnet, which is the z axis shown in Fig. 9.15. The observation point is then $r = z$. On the axis the magnetic field has two components:

$$d\mathbf{B} = dB(\hat{\boldsymbol{\phi}} \times \hat{\mathbf{R}}) = dB\,(\sin\alpha\,\hat{\mathbf{z}} + \cos\alpha\,\hat{\boldsymbol{\rho}}) = dB_z\hat{\mathbf{z}} + dB_\rho\hat{\boldsymbol{\rho}} \tag{9.65}$$

which are produced by the current element $K_m\,dA = M\,dA = (M\,dz')a\,d\phi'$, where $M\,dz' = \Delta I_m$ is the magnetization current that flows in each imaginary loop of width dz', and $a\,d\phi'$ is a small element of the loop. If we integrate ϕ' around a loop, the B_ρ components will cancel (this can be easily seen by considering two diametrically opposite current elements which produce opposing B_ρ components),

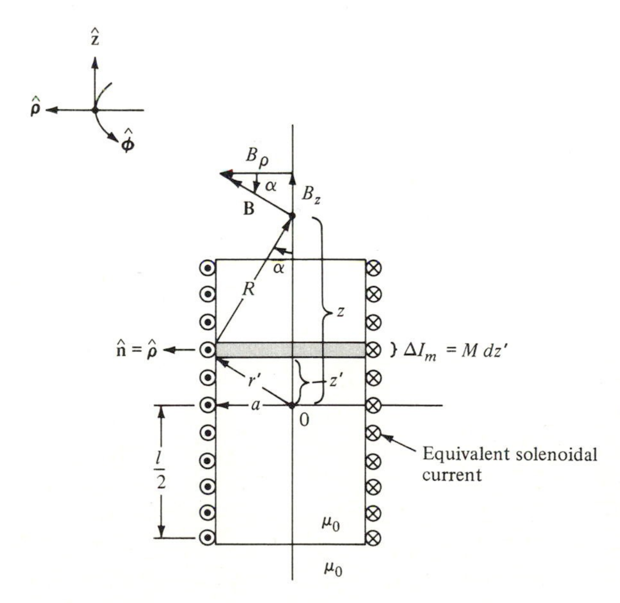

Figure 9.15 Geometry of a permanent magnet of length l and radius a which is used to calculate the magnetic field on the axis of the rod. The equivalent solenoidal sheet of current is shown by the dots and crosses, which denote current flow out and into the paper, respectively.

leaving the axial component

$$B_z = \frac{\mu_0 M}{4\pi}(2\pi a)\int_{-l/2}^{l/2} \frac{dz'}{R^2}\sin\alpha \tag{9.66}$$

Using Fig. 9.15, we see that $\sin\alpha = a/R$ and that $R^2 = a^2 + (z - z')^2$. This gives

$$\begin{aligned} B_z &= \frac{\mu_0 M a^2}{2}\int_{-l/2}^{l/2} \frac{dz'}{[a^2 + (z - z')^2]^{3/2}} \\ &= \frac{\mu_0 M a^2}{2} \left.\frac{z' - z}{a^2[a^2 + (z - z')^2]^{1/2}}\right|_{-l/2}^{l/2} \\ &= \frac{\mu_0 M}{2}\left\{\frac{1 - 2z/l}{[(2a/l)^2 + (2z/l - 1)^2]^{1/2}} + \frac{1 + 2z/l}{[(2a/l)^2 + (2z/l + 1)^2]^{1/2}}\right\} \end{aligned} \tag{9.67}$$

which is the axial field at any distance z from the center of the rod magnet. The term $2a/l$ is the shape factor of the magnet or the equivalent solenoid. Notice that we can use Eq. (9.67) for a real solenoid by replacing the linear current density M with $M = K = NI/l$, where N is the number of turns and I the current in a solenoidal coil, such as shown in Fig. 6.25. The graph of the axial field given in Fig. 6.25*c* also applies, which is for an $l = 8a$ solenoid. The field at the center of the magnet is given by

$$B_z(z = 0) = \frac{\mu_0 M}{[(2a/l)^2 + 1]^{1/2}} \cong \mu_0 M\Big|_{a \ll l} \tag{9.68}$$

and at the ends by

$$B_z(z = \pm l/2) = \frac{\mu_0 M}{[(2a/l)^2 + 4]^{1/2}} \cong \frac{\mu_0 M}{2}\Big|_{a \ll l} \tag{9.69}$$

Thus for a long, slender magnet, for which $a \ll l$, the field at the ends is one-half that of the field at the center.

As pointed out before, the equivalent-current formulation gives the **B** field inside and outside a magnet (whereas the equivalent-charge formulation gives the **H** field inside and outside a magnet). In the current formulation the magnet is replaced by an equivalent air-core solenoid whose windings carry the current $I_m = K_m l/N = Ml/N$. The **H** field outside the magnet is simply given by $H_z = B_z/\mu_0$. However, to find the **H** field inside the magnet, the magnetization M must be considered. Thus from (9.10) the magnetic field strength H is given by

$$H_z = \frac{B_z}{\mu_0} - M \tag{9.70}$$

where B_z is given by (9.67) and $M = M_z$. For example, at the center of the magnet, the H field is

$$H_z(z = 0) = M\left\{\left[\left(\frac{2a}{l}\right)^2 + 1\right]^{-1/2} - 1\right\} = -M\left|1 - \left[1 + \left(\frac{2a}{l}\right)^2\right]^{-1/2}\right| \tag{9.71}$$

which shows that the H field inside a permanent magnet is opposite to that of magnetization **M**. For an infinitely long magnet, the inside H field vanishes. This can be seen from (9.71) by taking the limit as $l \to \infty$ or from the fact that $B = \mu_0 M$ inside an infinitely long magnet,† as given by (9.68). The latter implies that H inside must be zero because, in general, $B = \mu_0(H + M)$.

The variation and some values of the axial fields for a 4 to 1 rod magnet ($l = 8a$) are given in Fig. 9.17.

† An infinitely long magnet has no poles, hence no magnetic charge. Since the sources for the **H** field in a permanent magnet are magnetic charges, it follows that **H** is zero.

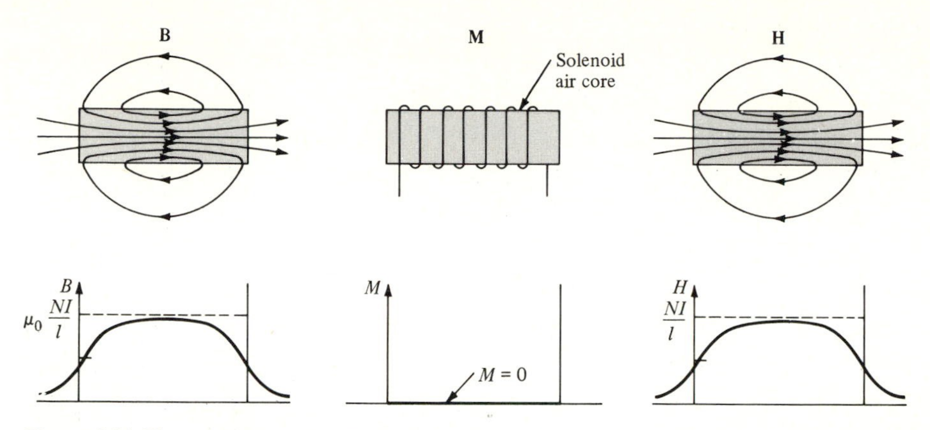

Figure 9.16 The top figures give the shape of the field lines for a solenoid. The figures on the bottom show the magnitude of the axial field.

Summary of Methods Used for a Rod Magnet and Solenoid

We have now calculated the field from a solenoid or cylindrical magnet by four different methods. In Sec. 6.9 we integrated the fields from successive loops, in Sec. 7.6 we used Ampère's law, in Sec. 9.3 the equivalent-charge formulation, and here we have applied the equivalent-current method. The axial fields from a solenoid and rod magnet are summarized in Figs. 9.16 and 9.17, respectively.

The equivalent-charge formulation gives the **H** field inside and outside a permanent magnet. It replaces the magnet by free space and a surface layer of charge ($\rho_{sm} = M$). If M is not constant throughout the body, an additional volume

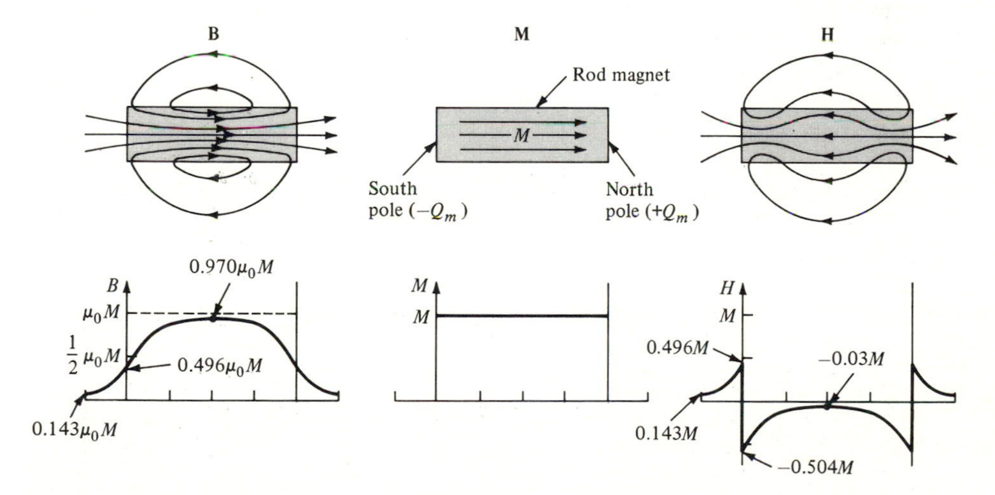

Figure 9.17 The top figures give the shape of the field lines for a rod magnet. The figures on the bottom show the magnitude of the axial field. (Numerical values are for a 4 to 1 shape.)

charge density $\rho_M = \nabla \cdot \mathbf{M}$ is required. The advantage of this method, also known as the *magnetic scalar potential method*, is that it reduces the problem to an analogous one in electrostatics, with H produced by scalar magnetic charges.

The equivalent-current method gives the **B** field inside and outside a permanent magnet. It replaces the magnet by free space and a solenoidal current ($K_m = M$). If M is not constant throughout the body, an additional volume current density $\mathbf{J}_m = \nabla \times \mathbf{M}$ is required. It can be argued that **B** is a more fundamental field than **H** because it arises from currents and not from magnetic charges. Recall that sources of magnetic field are moving charges, which makes the interpretation of the field from a permanent magnet in terms of circulating atomic currents more fundamental. A further supporting argument of this premise is that a charge moving in a magnetized medium is acted upon by B and not H, as stated by Lorentz' force law $\mathbf{F} = q(\mathbf{v} \times \mathbf{B})$.

Example: Torque on a bar magnet in a uniform field The torque on a magnetic dipole when immersed in an external B_0 field is given by $T = mB_0 \sin \alpha$, where α is the angle between the external magnetic field and the direction of the dipole. Dipole moment m for a current loop is $m = I\pi a^2$. Hence, the torque acting on each elemental current loop of a bar magnet, as shown in Fig. 9.15, is given by $\Delta T = (\Delta I_m \pi a^2)B_0 \sin \alpha = (M\, dz' \pi a^2)B_0 \sin \alpha$. Integrating over the length l of the magnet, we obtain for the total torque

$$T = (Ml\pi a^2)B_0 \sin \alpha \tag{9.72}$$

where M is the uniform permanent magnetization and $l\pi a^2$ is the volume of the bar magnet. Hence, the torque T will tend to align the magnet until its axis is parallel to B. The above expression is similar to that for a magnetized sphere, (9.46), which permits us to generalize and give the torque for any uniformly magnetized object of volume v as $T = vMB_0 \sin \alpha$.

For small angles α, the restoring torque is proportional to angular displacement ($\sin \alpha \cong \alpha$). A magnet in an external field B_0 will therefore oscillate about the equilibrium condition with a frequency f given by

$$f = \frac{1}{2\pi}\sqrt{\frac{vMB_0}{I}} \tag{9.73}$$

where I is the moment of inertia of the magnet about its center of oscillation.

9.5 DIAMAGNETIC, PARAMAGNETIC, AND FERROMAGNETIC MATERIALS

In the previous two sections we have calculated the magnetic fields from permanent magnets, i.e., from objects with a permanent magnetization **M** throughout. We assumed that **M** existed and were not concerned with the origin of **M**. In this section let us examine the nature of magnetization **M**.

We have mentioned briefly that several kinds of magnetic materials exist. Some that exhibit magnetization when exposed to an external field retain a part of the induced M after the external field is withdrawn. Such materials are said to be *ferromagnetic*. The other materials lose all their induced magnetization when the external field is withdrawn. These are known as the *paramagnetic* and *diamagnetic* materials. Let us characterize magnetic materials in terms of the effect they have

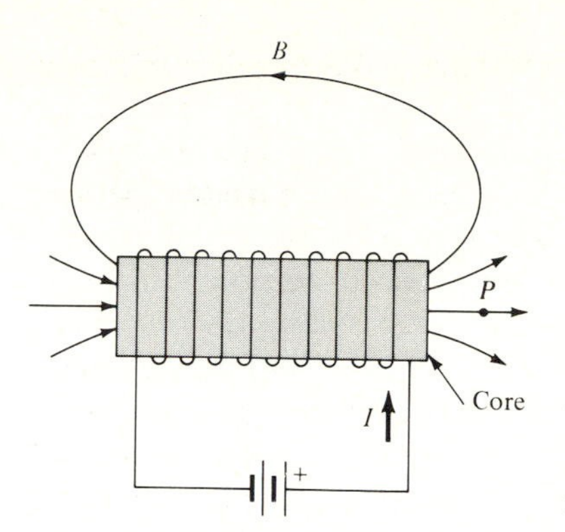

Figure 9.18 The **B** field at the point P of a solenoid depends on the core material. The field at P can be measured by a pick-up loop whose induced voltage (when I is switched on) is proportional to B. It can also be measured by the deflection of a compass needle.

on the magnetic field in a solenoid, shown in Fig. 9.18. When no magnetic material is present inside the solenoid, i.e., the solenoid has an air core, the magnetic flux density at point P is B_0. Introducing various core materials, we will find that B_0 changes to B. In terms of this change we can classify magnetic materials as†

Diamagnetic, if $\quad \dfrac{B}{B_0} \lesssim 1$

Paramagnetic, if $\quad \dfrac{B}{B_0} \gtrsim 1$

Ferromagnetic, if $\quad \dfrac{B}{B_0} \gg 1 \qquad (9.74)$

Using (9.11), we can relate B to B_0 as‡

$$\frac{B}{B_0} = \frac{\mu_0(H + M)}{\mu_0 H} = \frac{\mu}{\mu_0} = \mu_r = 1 + \chi_m \tag{9.75}$$

where $\mu_r = \mu/\mu_0$ is the relative permeability and $\chi_m = M/H$ is the magnetic susceptibility as given by (9.15). Table 9.1 illustrates the range of μ_r for various materials.

The induced magnetization M in the core material is given from (9.75) as

$$M = \chi_m H = (\mu_r - 1)H \tag{9.76}$$

where H is the core magnetic field strength due to current flowing in the solenoid. Using Fig. 9.19, we observe that the substances labeled diamagnetic and paramagnetic show almost no response to a magnetizing force H; that is, $M \cong 0$ because $\chi_m \cong 0$; therefore $B \cong B_0$. Thus, the assumption that the permeability for these materials (which includes metals) is that of free space, $\mu = \mu_0$, is a very good

† Since inductance L of a solenoid depends on the **B** field that exists inside the coil, we could have also used the change in inductance L/L_0 to classify magnetic materials, similarly to (9.74).

‡ Note that (9.75) is strictly valid when point P is inside the core material, where $M \neq 0$. However, (9.67) can be used for points P outside the core with the result that B/B_0 is as given by (9.75).

Table 9.1 Relative permeability μ_r and susceptibility χ_m of some substances

Substance	Relative permeability, $\mu_r = 1 + \chi_m$	Susceptibility, χ_m	
Bismuth	0.99983	-1.66×10^{-4}	Diamagnetic
Mercury	0.999968	-3.20×10^{-5}	Diamagnetic
Gold	0.999964	-3.60×10^{-5}	Diamagnetic
Silver	0.99998	-2.60×10^{-5}	Diamagnetic
Lead	0.999983	-1.7×10^{-5}	Diamagnetic
Copper	0.999991	-0.98×10^{-5}	Diamagnetic
Water	0.999991	-0.88×10^{-5}	Diamagnetic
Vacuum	1	0	
Air	1.00000036	3.6×10^{-7}	Paramagnetic
Aluminum	1.000021	2.5×10^{-5}	Paramagnetic
Palladium	1.00082	8.2×10^{-4}	Paramagnetic
Cobalt	250		Ferromagnetic; nonlinear μ_r
Nickel	600		Ferromagnetic; nonlinear μ_r
Commercial iron (0.2 impurity)	6000		Ferromagnetic; nonlinear μ_r
High-purity iron (0.05 impurity)	2×10^5		Ferromagnetic; nonlinear μ_r
Supermalloy (79% Ni, 5% Mo)	1×10^6		Ferromagnetic; nonlinear μ_r

approximation. Also, the susceptibility χ_m for these materials is independent of the magnetizing force H; that is, χ_m is constant, as shown in Fig. 9.19. On the other hand, a ferromagnetic material has large μ_r which can assume many values depending on H. Large μ_r implies $M \cong \mu_r H$, which means that a small H is very effective in causing a large M to be induced within the material; therefore $B \gg B_0$. Figure 9.19 shows the linear dependence of magnetization M on H for diamagnetic and paramagnetic materials and the highly nonlinear dependence for ferromagnetic materials. The saturation magnetization is indicated by M_s and occurs when all atomic magnetic moments are aligned along the direction of the magnetizing force H.

In view of the results presented in Table 9.1 and Fig. 9.19, we can say that three basic magnetic properties can be noticed which require explanation. The *first* property is exhibited by *all* materials and is a magnetic response in a direction opposite to that of an externally applied H field. This negative magnetization, called *diamagnetism*, is too weak to be of any practical value. Diamagnetic materials display no permanent magnetization; that is, when the applied H field is withdrawn, M vanishes. The *second* property, *paramagnetism*, is exhibited by materials that show a response in the direction of an applied field. The induced magnetization M is parallel and proportional to the applied H. Like diamagnetism, it is very weak in nature and cannot be used to make a permanent magnet.

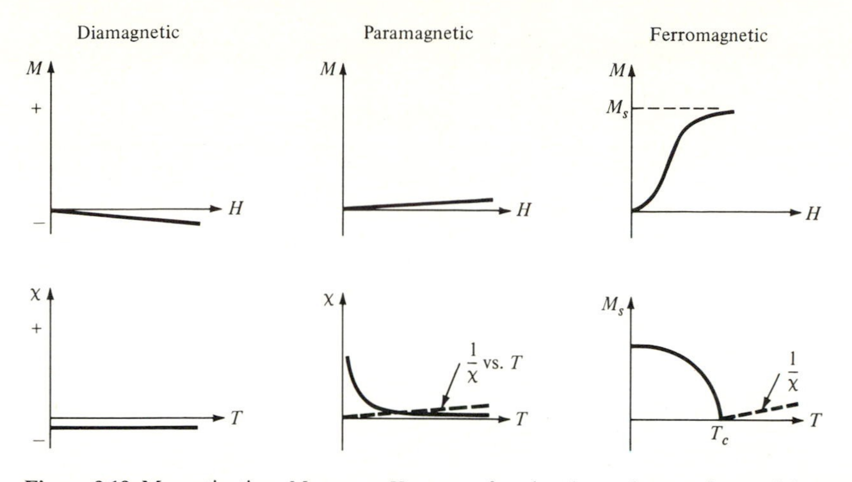

Figure 9.19 Magnetization M versus H curves for the three classes of materials are shown in the top three figures (M_s = saturation magnetization). The lower figures show temperature dependence. The negative diamagnetic susceptibility χ is approximately independent of temperature. T_c is the Curie temperature at which ferromagnetic materials lose their magnetism. Above T_c these substances exhibit paramagnetism.

Unlike diamagnetism, it has a temperature dependence (induced magnetization decreases with an increase in temperature). The *third*, and most important, property is the strong magnetic field created by some materials to which a comparatively weak magnetic field is applied. This property is called *ferromagnetism* and disappears when the material is heated above the Curie temperature, as shown in Fig. 9.19. Above T_c, ferromagnetic materials become paramagnetic.

In general, diamagnetism is the weakest effect, followed by paramagnetism which is somewhat stronger, which in turn is followed by ferromagnetism, a very strong effect. Since ferromagnetism turns into paramagnetism above T_c, and since paramagnetic susceptibilities decrease with temperature, while the diamagnetic χ stays essentially constant, it follows that all substances become diamagnetic at sufficiently high temperatures.

Another classification that is convenient to use is that paramagnetic and ferromagnetic materials are attracted toward magnetic fields, whereas diamagnetic materials are repelled. Let us bring a piece of bismuth which is diamagnetic near the north pole of a magnet, as shown in Fig. 9.20. Since the induced magnetization M is opposite to the B field of the magnet, the bismuth object has an induced north pole which is nearest the north pole of the magnet; hence the bismuth object is repelled (the attractive force on the south pole is weaker because it is in a weaker field). Diamagnetic objects, therefore, can be made to float in a magnetic field. Just the opposite occurs for a piece of iron material, as shown in Fig. 9.20.

We have already mentioned in the introduction to this chapter that there are two possible atomic origins of magnetism. They are the orbital motion of the electron around the nucleus (which causes diamagnetism) and the spin motion of

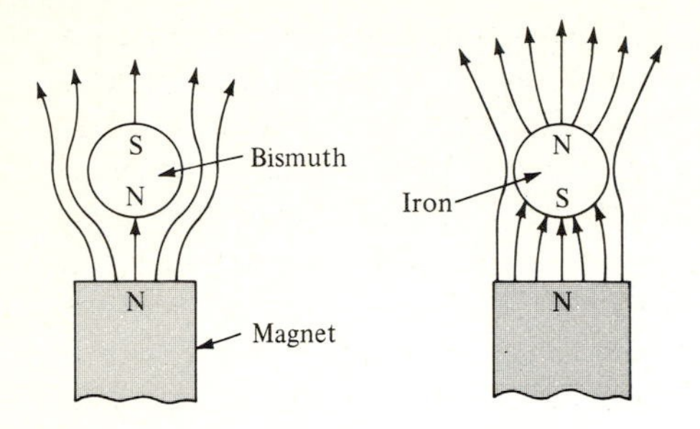

Figure 9.20 Diamagnetic bismuth partially expels a B field from its interior and is repelled by a magnetic field. Iron, on the other hand, concentrates flux and is attracted by magnetic fields. Note that the magnetic field of the magnet is nonuniform—the part of the object nearer the magnet is in a stronger field.

the electron about its own axis (which causes paramagnetism and ferromagnetism). Both of these motions circulate a charge which produces a magnetic field equivalent to that of a magnetic dipole. Strictly speaking, these are quantum-mechanical concepts. However, the angular momentum that quantum mechanics predicts for the electron motion can be viewed classically as an orbital and spin motion, which can be related to magnetic moment as follows: An electron $-e$ moves in a circular orbit of radius r at an angular velocity ω, making $\omega/2\pi$ turns per second. Since current is defined as charge per unit time, this motion constitutes a current of $I = -e\omega/2\pi$ amperes, where $-e$ is the electric charge of a single electron. The magnetic moment m of a loop of current I is given by (6.47) as $m = IA$, where A is the cross section of the loop. The magnetic moment of the circulating electron is thus

$$m = -\frac{e\omega}{2\pi}(\pi r^2) = -\tfrac{1}{2}e\omega r^2 \tag{9.77}$$

Since angular momentum is given by $L = m_e\omega r^2$, where m_e is the electron mass, the relation between magnetic moment and orbital angular momentum is

$$m = -\frac{e}{2m_e}L \tag{9.78}$$

The relation between spin angular momentum L_s and spin magnetic moment is given by $m = -eL_s/m_e$, a relation very similar to that for orbital motion, but differing by a factor of 2.

Diamagnetism, a Property of All Materials

Diamagnetism is a result of the orbital motion of the electrons. Each circulating electron acts as a current loop and produces a magnetic field. Why, then, don't we notice the magnetic fields of all electrons which are orbiting in the atoms of every substance? The answer is that actually two electrons travel in each orbit of diamagnetic atoms; since they circulate in opposite directions, no net magnetic moment is produced. The orbital pairing of two oppositely traveling electrons explains the zero magnetic effect of atoms in the absence of an applied field. Now what happens when a magnetic field is applied to a substance? The orbital pairing will be unbalanced. Currents in the equivalent loops of the circulating electrons will be induced in such a way as to oppose any change in the existing magnetic field of the atom.† That is, the induced currents will produce a magnetic field that will oppose the applied field. If the magnetic field of an electron orbit happens to be in the direction of the applied field, the electron will resist the applied field by slowing down immediately. The magnetic field produced by the orbiting electron will now decrease, because the slower moving electron is equivalent to a smaller current ($I = -e\omega/2\pi$). The net change of magnetic field across the atom has thus been reduced. If, on the other hand, the magnetic field of an electron orbit is initially opposite to that of the applied field, the electron will immediately speed up in its orbit so as to increase its own opposing field. This is Lenz' law on an atomic scale.

Let us formalize some of these ideas. Before the magnetic field is applied, the electron is in equilibrium in its orbit. An electric centripetal force $F_e = m_e a = m_e \omega_0^2 r$ holds the electron to its atom, where ω_0 is the angular frequency of the electron in its orbit and the centripetal acceleration is $a = \omega_0^2 r$, as shown in Fig. 9.21. Application of a magnetic field B exerts an additional magnetic force $\mathbf{F}_m = -e\mathbf{v} \times \mathbf{B}$ on the moving electron which can be centripetal or centrifugal depending on the direction of the magnetic field.‡ Summing forces that act on the electron, Newton's second law gives for the new angular velocity ω

$$F_e \pm F_m = m_e a \tag{9.79}$$

$$m_e \omega_0^2 r \pm e\omega r B = m_e \omega^2 r \tag{9.80}$$

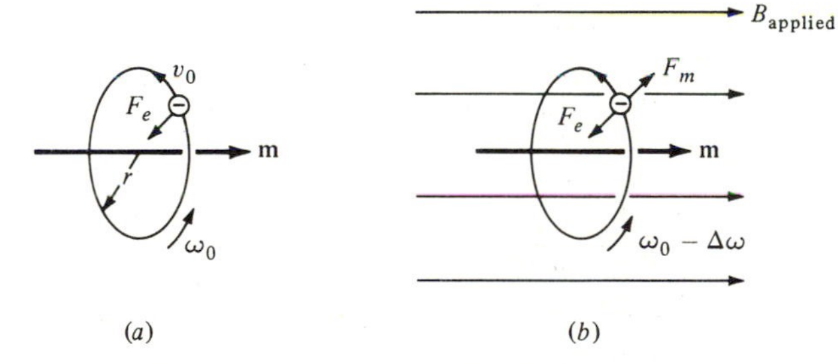

Figure 9.21 (*a*) An electron orbiting with velocity $v_0 = \omega_0 r$ has a magnetic moment **m** as shown. (*b*) The electron will react to an applied field by slowing down to $\omega_0 - \Delta\omega$. Magnetic moment **m** decreases, so that $\Delta\mathbf{m}$ is opposite B. If the applied B were reversed, the electron would speed up to $\omega_0 + \Delta\omega$, **m** would increase, F_m would be in the same direction as F_e, and $\Delta\mathbf{m}$ would again be opposite B.

† This electromagnetic reaction, better known as Lenz' law, is well known as counter emf in motors and generators and was discussed in Secs. 6.11 and 6.12. It states that when the flux through an electric circuit is changed, an induced current is set up in such a direction as to oppose the flux change. Since the "resistance" in an electron orbit is zero, the diamagnetic moment will persist as long as the external field is applied.

‡ Since the electron orbits are quantized, and since the additional magnetic force is very small, quantum-mechanical considerations tell us that the electron will stay in the same orbit but will change its velocity slightly.

Rearranging, we obtain

$$\pm e\omega B = m_e(\omega - \omega_0)(\omega + \omega_0) \tag{9.81}$$

Since the change in velocity $\Delta\omega$ will be very small, we let $\omega - \omega_0 = \Delta\omega$ and approximate $\omega + \omega_0 \cong 2\omega_0$, which gives

$$\Delta\omega \cong \pm\frac{e}{2m_e}B \tag{9.82}$$

as the increase or decrease in angular velocity when the B field is applied. The minus sign corresponds to the case when B and the magnetic moment m of the orbiting electron are in the same direction, whereas the plus sign is for opposite B and m. In either case, the component of induced magnetic moment corresponding to the change in angular velocity predicted by (9.82) is always opposite to B and is given by

$$\Delta\mathbf{m} = -\frac{e^2r^2}{4m_e}\mathbf{B} \tag{9.83}$$

This can be seen by studying Fig. 9.21 and using Eq. (9.77) which shows that m and ω are proportional; that is, $|\Delta m| = \frac{1}{2}er^2|\Delta\omega|$. The induced magnetization $\mathbf{M}$, defined as dipole moment per unit volume, is therefore opposite to the applied B field, or

$$\boxed{\mathbf{M} = -N\frac{e^2r^2}{4m_e}\mu_0\mathbf{H}} \tag{9.84}$$

where $B = \mu_0 H$
N = number of molecules per unit volume
r = average radius of atomic orbit†

The diamagnetic susceptibility is then given by

$$\boxed{\chi_m = -N\frac{e^2r^2}{4m_e}\mu_0} \tag{9.85}$$

Of course, not all orbiting electrons will be parallel or antiparallel to the applied field. In general, because of the random arrangement, we will have dipole moments in all directions, as shown in Fig. 9.22. Thus the above expression should be multiplied by $\cos\theta$, where θ is some average angle between the orbit axis and the applied field. Figure 9.22*b* shows the directions of the induced dipoles in an applied B field. This figure also shows that the magnetization of all materials (except for the ferromagnetic ones) in the absence of an external magnetic field is zero. As an additional effect, we should note that an applied B field will produce a

† Note that for diamagnetic material the internal field that acts on the molecules differs very little from the applied field [see Eq. (9.92)].

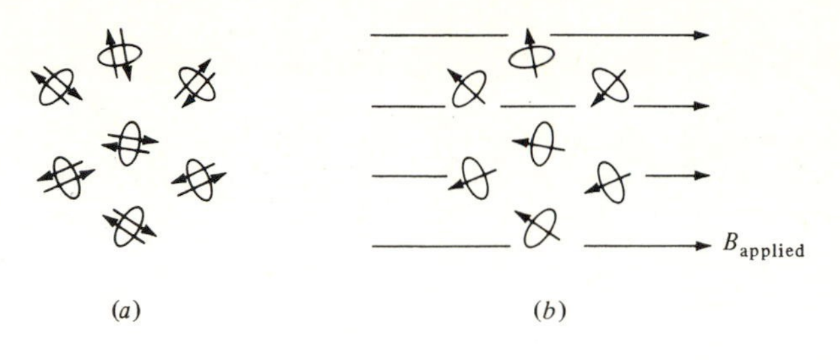

Figure 9.22 (*a*) The equivalent dipoles of the paired orbiting electrons in a material have a random arrangement; (*b*) applying a field to the dipoles of (*a*) will induce magnetic moments in each dipole opposite to the applied field.

torque ($\mathbf{T} = \mathbf{m} \times \mathbf{B}$) on each dipole which is not completely aligned with the field. The effect of this torque will be that the axis of each dipole will precess about the direction of the applied field, similar to a spinning top which is placed in a gravitational field. The precession frequency is the same as that given by (9.82), namely, $(e/2m_e)B$ and is known as the *Larmor frequency*. The precession of a magnetic dipole about the magnetic B field is shown in Fig. 9.23.

To reiterate, diamagnetism in which magnetization **M** is opposite to **B** is present in all materials. This effect is temperature insensitive because diamagnetism involves induced magnetic moments that are independent of the orientation of the atoms. Thus, thermal agitation does not affect the diamagnetic susceptibilities. This is shown by the horizontal χT curve in Fig. 9.19. In paramagnetic or ferromagnetic substances, the weak diamagnetic effect is completely masked. The permeability μ in strictly diamagnetic materials is typically less than μ_0 by about 1 part in 10^6. The negative magnetization explains the tendency of diamagnetic materials to be expelled from a magnetic field. Negative magnetization also reduces the interior fields. A superconductor is an example of a perfect diamagnetic material; it expels all magnetic fields.

Example Let us calculate the change in magnetic moment for an orbiting electron when a strong magnetic field of 2 Wb/m^2 (2×10^4 G) is applied at right angles to the plane of the orbit.

Using $r = 5.1 \times 10^{-11}$ m, which corresponds to the radius of a hydrogen atom in its normal state, we obtain from (9.83)

$$\Delta m = \frac{e^2 r^2 B}{4m_e} = \frac{(1.6 \times 10^{-19}\ \text{C})^2 (5.1 \times 10^{-11}\ \text{m})^2 (2\ \text{Wb/m}^2)}{4(9.1 \times 10^{-31}\ \text{kg})} = 3.7 \times 10^{-29}\ \text{A} \cdot \text{m}^2$$

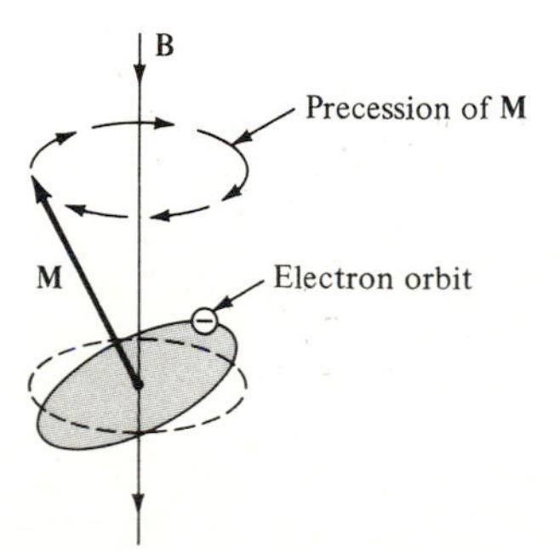

Figure 9.23 A magnetic dipole M will precess about the magnetic field with angular velocity $eB/2m_e$.

To see the change that this represents from the unperturbed orbital magnetic dipole moment m, we must first calculate m, which is given by (9.77). In order to use that expression, we need to obtain the angular velocity ω of the orbiting electron. Since the electron is held to its atom by electric forces, Coulomb's law can be used to give the force term in Newton's second law, $F = ma$. The centripetal acceleration is $a = v^2/r = \omega^2 r$, which gives for Newton's law $e^2/4\pi\varepsilon_0 r^2 = m_e a = m_e \omega^2 r$. Solving for ω and substituting in (9.77), we have

$$m = \tfrac{1}{2}e\omega r^2 = \frac{e^2}{4}\sqrt{\frac{r}{\pi\varepsilon_0 m_e}} = \frac{(1.6 \times 10^{-19})^2}{4}\sqrt{\frac{5.1 \times 10^{-11}}{\pi(8.9 \times 10^{-12})(9.1 \times 10^{-31})}} = 9.1 \times 10^{-24}\ \mathrm{A \cdot m^2}$$

The change in magnetic moment induced by the external B field is therefore

$$\frac{\Delta m}{m} = \frac{3.7 \times 10^{-29}}{9.1 \times 10^{-24}} = 4 \times 10^{-6}$$

which is rather small but is on the order of the induced magnetization in diamagnetic materials.

Paramagnetism

An electron possesses a second kind of angular momentum which is related to spin of the electron about its own axis. The spin angular momentum L_s gives rise to a magnetic moment (called *Bohr magneton*) which is given by $m = -eL_s/m_e$. Quantum mechanics tells us that the spin angular momentum can assume only two values: $L_s = \pm h/4\pi$, corresponding to left-handed and right-handed spins, where h is Planck's constant. The magnetic moment of electron spin is thus

$$\boxed{m = \pm\frac{eh}{4\pi m_e} = \pm 9.3 \times 10^{-24}\ \mathrm{A \cdot m^2}} \qquad (9.86)$$

as shown schematically in Fig. 9.24. We can think of each spinning electron as a tiny permanent magnet with north and south poles, each possessing moment m. This is different from diamagnetism where magnetization was induced by an applied field. On an atomic level, the response of diamagnetic materials was an induced dipole moment that would oppose the applied field. Paramagnetic materials, on the other hand, have permanent dipole moments which are independent of any applied field. Because of thermal vibrations the axes of the spins are distributed over all possible orientations. Thus a piece of paramagnetic material shows no net magnetization. Applying a B field to such a material, the axes of the

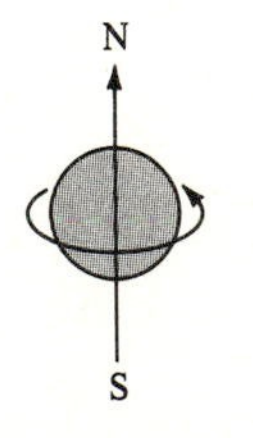

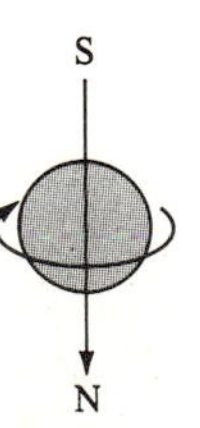

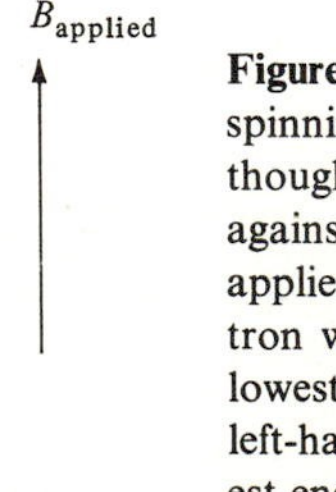

Figure 9.24 The two orientations of a spinning electron, $\pm m$, which can be thought of as being aligned with or against an applied magnetic field. If the applied B field is as shown, an electron with right-handed spin is in the lowest energy state, whereas one with left-handed spin would be in the highest energy state.

spins will tend to align parallel to the direction of the field. This effect is very similar to the rotation of a compass needle toward the minimum energy position. The alignment will only be partial since the thermal agitation both at room temperature and higher temperatures is much stronger than the energy of the magnetic field which rotates a dipole. For example, the mean thermal energy of agitation is given by Eq. (2.2), and at room temperature, $T = 300$ K, the thermal energy is equal to

$$W_T = \tfrac{3}{2}kT = (\tfrac{3}{2})(1.38 \times 10^{-23})(300) = 6 \times 10^{-21}\ \text{J} \tag{9.87}$$

The potential energy of interaction of a dipole with an applied magnetic field is given by (4.5) or (6.50) as $W = -\mathbf{m} \cdot \mathbf{B}$. The work that it takes to turn a dipole from the minimum energy direction in which it is aligned with the field to a direction antiparallel to the field is given by† $W = 2mB$. Even for a strong external field of 2 Wb/m^2 (20,000 G), we obtain

$$W = 2mB = 2(9 \times 10^{-24})(2) = 3.6 \times 10^{-23}\ \text{J} \tag{9.88}$$

which is weaker by a factor of 170 when compared to the kinetic vibration energy calculated in (9.87). Thus even if a dipole should align parallel to the external field, the thermal vibrations are so strong that it will soon be knocked from that position. A partial alignment of the dipoles is shown in Fig. 9.25. If there are N atoms per unit volume, the magnetization M would be $M = Nm$, for perfect alignment of all atomic dipoles. In solids a representative number for the density is $N = 10^{29}$ atoms/m^3. Assuming one spin per atom, we obtain $M_{\max} = Nm = (10^{29}\ \text{m}^{-3})(9 \times 10^{-24}\ \text{A} \cdot \text{m}^2) = 9 \times 10^5$ A/m. However, because of thermal randomizing, a field applied to a paramagnetic object is able to induce a magnetization M in the direction of the applied field which is only a small fraction of Nm. Applying a larger field will increase that fraction linearly, as shown in Fig. 9.19. The linearity of the curve makes χ_m and μ_r constant for a given parametric material.

We now have two forces acting on each dipole. One is the force of the applied B field tending to align the dipole, and the other is due to thermal energy which acts in such a way as to produce a completely random orientation of the dipole. How successful an applied B field is in aligning the dipoles under such conditions is given by the Langevin formula for magnetization (see Prob. 9.34):

$$\boxed{M = Nm\left(\coth\frac{mB}{kT} - \frac{kT}{mB}\right)} \tag{9.89}$$

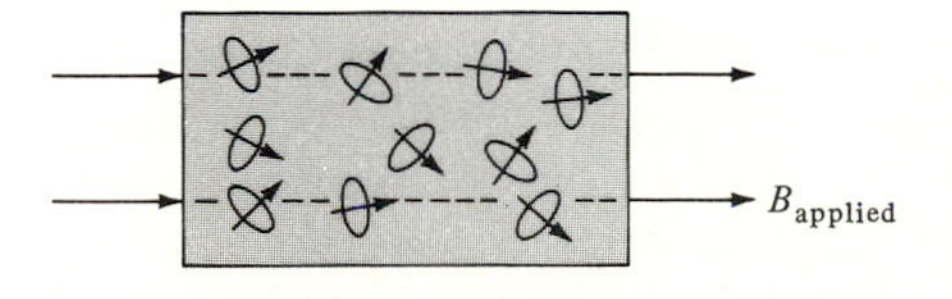

Figure 9.25 The permanent dipole moments of the spins in a paramagnetic material will tend to align parallel to the applied B field. The alignment is greatly exaggerated.

† This is similar to the work required to rotate a compass needle by 180° from its rest position.

which is derived under the assumption that the number of dipoles with a given orientation relative to B is given by Boltzmann statistics. Except for temperatures near absolute zero, the ratio $mB/kT \ll 1$, as can be seen from (9.87) and (9.88). We can then approximate (9.89) by the following equation which is known as Curie's law:

$$\boxed{M \cong \frac{Nm^2}{3kT} B = \frac{Nm^2}{3kT} \mu_0 H} \tag{9.90}$$

where we have assumed that the field acting on each molecule is the applied B field; that is, $B_{\text{molecular}} = B = \mu_0 H$. The reason for this can be seen from the expression of the total macroscopic B field, $B = \mu_0(H + M)$, inside a material. Since the magnetization M for paramagnetic materials is much smaller than H, we have $B \cong \mu_0 H$. The physics of the problem is nicely displayed if we write (9.90) in the form

$$M = M_{\text{max}} \frac{mB}{3kT} \tag{9.90a}$$

where $M_{\text{max}} = Nm$. This equation shows that the maximum alignment is reduced by the ratio of alignment energy mB and the thermal randomizing energy kT. Since the randomizing energy (9.87) is much larger, M is only a small fraction of M_{max}. The above expression (9.90) leads directly to the paramagnetic susceptibility $\chi_m = M/H$ as

$$\boxed{\chi_m = \frac{Nm^2 \mu_0}{3kT}} \tag{9.91}$$

A representative value for χ_m is $\chi_m \cong 10^{-3}$ if $N \cong 10^{29}$, $m \cong 10^{-23}$, $\mu_0 = 4\pi \times 10^{-7}$, and $3kT \cong 10^{-20}$.

It should be pointed out at this time that the field acting on each molecule, $B_{\text{molecular}}$, sometimes also called the local field, need not be the same as $B = \mu_0(H + M)$, which is the total macroscopic interior field that exists inside a magnetic material. The reason is that the macroscopic field was obtained by averaging over a small macroscopic volume which includes billions of molecules, whereas B_{mol} is some average field acting on an individual molecule. It can be shown† that

$$B_{\text{mol}} = \mu_0(H + \tfrac{1}{3}M) \tag{9.92}$$

For small values of M, which is characteristic of paramagnetic and diamagnetic materials, we can say that $B_{\text{mol}} \cong \mu_0 H = B$, which is the same approximation used in (9.90). However, as will be shown in the next section, for ferromagnetic materials the M field will be the dominant field, being much larger than H.

† C. Kittel, "Introduction to Solid State Physics," John Wiley & Sons, Inc., New York, 1956.

Curie's law, Eq. (9.90), shows that $M \propto B/T$. That is, if B is increased, M increases linearly because of an increased alignment of the dipoles, whereas if T is increased, the alignment is decreased. Figure 9.26 shows the Langevin function and Curie's law. Equation (9.89) can be expressed in terms of the Langevin function as $M = NmL(a)$. We can make this graph readily usable by noting that the magnetic moment m of most atoms is on the order of a few Bohr magnetons [which is given by Eq. (9.86) as 9×10^{-24}]. Since Boltzmann's constant k ($= 1.38 \times 10^{-23}$) has almost the same value, we can approximate the abscissa in Fig. 9.26 as $a \cong B/T$. Our strongest magnets usually do not exceed 10 Wb/m^2. Thus for any temperature above 10 K, the ratio $a < 1$. Curie's law as given by (9.90), and the susceptibility χ_m derived from it, accounts for the values of χ_m given in Table 9.1. Of course for temperatures near zero with simultaneous large fields, Langevin's function must be used.

How do we account for the variations of μ_r or χ_m for the paramagnetic materials that are listed in Table 9.1? We have shown that molecules have a magnetic moment which is the resultant of orbital and spin moments of the various electrons in the molecule. In most kinds of atoms there is a tendency for the spin (and orbital) angular momentums of adjacent electrons to cancel each other by the formation of antiparallel pairs. In such a pair, the spin of one electron, for example, will be right-handed, and the spin of the other left-handed. Their total momentum is therefore zero, and so is their total magnetism. It is this tendency of oppositely directed spins to pair off near each other—or rather slight deviations from it—that accounts for the very weak magnetism of "nonmagnetic" materials. When the application of a magnetic field slightly unbalances the orbital pairing of electrons, the result is diamagnetism. When it slightly unbalances the spin pairing, or when atoms or ions are present having an odd number of electrons which will cause incomplete cancellation thus leaving unpaired spins, the result is paramagnetism. As pointed out before, the magnetic moment of paramagnetic materials ranges over a few Bohr magnetons which accounts for the variations in the values given in Table 9.1.

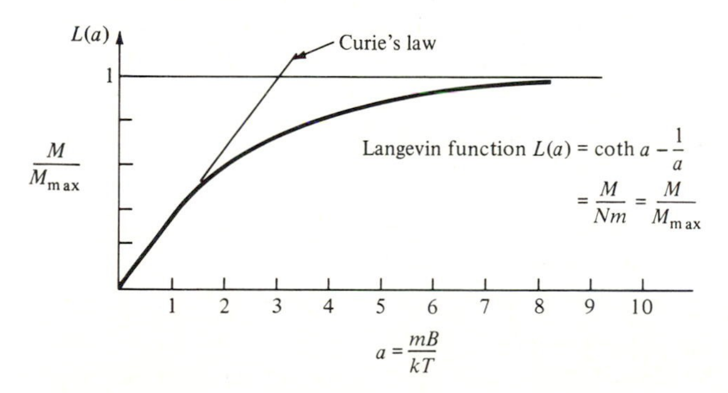

Figure 9.26 The Langevin function $L(a)$ for magnetization gives the fraction of dipoles that are aligned with the applied B field. Note that the ordinate is the ratio of the magnetization M actually induced in a paramagnetic material to that if all dipoles were aligned, which is $M_{max} = Nm$.

Ferromagnetism

The term "magnetic materials" is generally applied to substances exhibiting ferromagnetism. Their strong magnetism as compared with that of nonmagnetic materials (a term that we use to designate diamagnetic and paramagnetic materials) is what makes them useful in magnetic technology (motors, transformers, relays, etc.). Ferromagnetic materials are the elements Fe, Ni, Co, and their alloys, some M_n compounds, and some rare earth elements. Why do the ferromagnetic materials exhibit such a large magnetization M, which in turn yields large internal and external magnetic fields that are so useful to us? The answer was already given by Weiss in 1906 who postulated that a powerful internal molecular field must exist which acts on the individual electron spins of the molecules and aligns them parallel to each other over small volumes called *domains*.† The domains are small, but macroscopic, with dimensions on the order of 10^{-3} to 10^{-6} m (10^{-6} m is about 10^4 atomic diameters), or volumes ranging from 10^{-9} to 10^{-18} m^3. Since there are 8.5×10^{28} atoms of iron (Fe) per cubic meter, on the average, a domain contains 10^{16} atoms. As a result of the strong molecular field, there is complete alignment of the spins within each domain. The alignment within the domains is spontaneous; that is, no external fields need to be applied.

Why then are not all ferromagnetic materials strong permanent magnets? The answer is that although the molecular magnets align themselves spontaneously within each domain, the domains are randomly oriented with respect to each other, as shown in Fig. 9.27. The reason for this is that all systems tend toward a minimum energy configuration. A system of strongly magnetized domains, with all domains aligned in the same direction, would produce a powerful external magnetic field. The energy density at each point in such a field is given by $w = \mu_0 H^2/2 = B^2/2\mu_0$, where $B = \mu_0 H$ is the external field produced by the aligned domains. The energy stored in such an external field could be used to do work, for example, to attract an iron object. An arrangement that maximizes the external energy is contrary to nature. To obtain such a high-energy configuration,

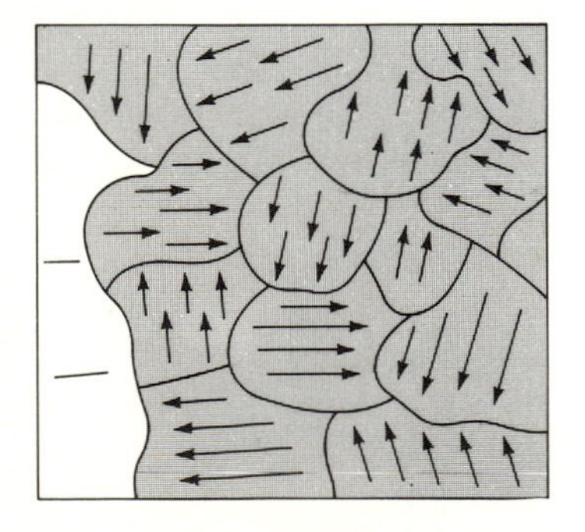

Figure 9.27 Domain structure in a polycrystalline ferromagnetic material. The spins in each domain line up spontaneously. However, the domains are randomly oriented with respect to each other which makes the specimen as a whole appear to be unmagnetized. The external magnetic field will therefore be zero. Note that for simplicity we have assumed very small grains, so that one domain occupies each crystal grain. In the usual polycrystalline materials, such as iron, many domains are contained in each grain.

† Note the similarity between paramagnetism and ferromagnetism. Both depend on the magnetism of the permanent spin moments of unpaired electrons. The difference is that ferromagnetic materials show domain formation.

we must first put work into the specimen, usually by subjecting it to a very strong magnetic field which magnetizes or aligns all domains in the specimen. Without an external work input, the domains will arrange themselves in the minimum-energy configuration.

Another peculiar, but extremely useful, property of ferromagnetic materials is that the application of a very weak magnetic field can cause complete alignment of the domains, which is indicated by M_s, the saturation magnetization, on the MH plot in Fig. 9.19. As an illustration, Fig. 9.28 gives the magnetization curve of fairly pure iron. It shows that an application of a field H of only 80 A/m (≈ 1 Oe) will change the magnetization from an initial value of zero to almost the saturation value of M_s which corresponds to $B_s = \mu_0(H + M_s) \cong 1.3$ Wb/m^2 on the graph. Note, this is an approximate result since it takes larger values than 80 A/m to obtain saturation. However, the values of H will never be so large that the result of (9.93) is affected; that is,

$$M_s = \frac{B_s}{\mu_0} - H \cong \frac{1.3}{4\pi \times 10^{-7}} - 80 \cong 10^6 - 80 \cong 10^6 \cong \frac{B_s}{\mu_0} \tag{9.93}$$

Thus, in a ferromagnetic material the applied magnetizing force H contributes very little to the magnetization but is very effective in aligning the domains, which results in a large magnetization. The domains once aligned yield a large magnetic field.

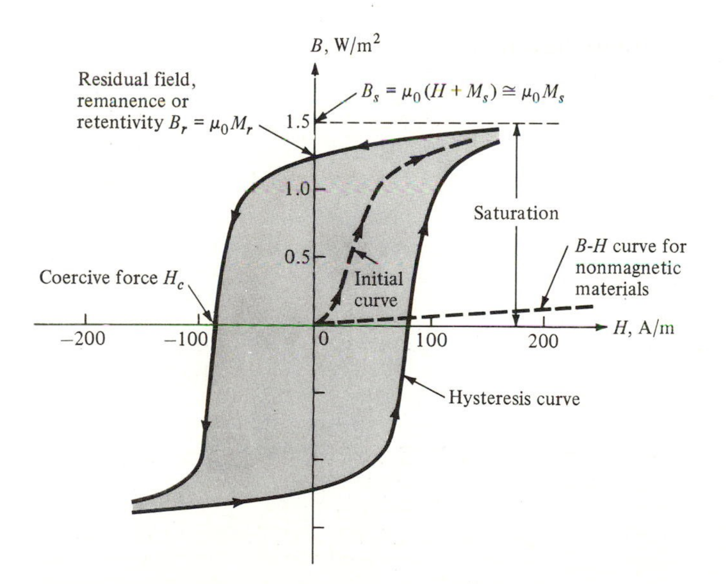

Figure 9.28 Initial magnetization curve and hysteresis curve for an ordinary polycrystalline metal which is ferromagnetic, such as fairly pure iron. For comparison the BH curve for non-magnetic material (wood, copper, air, etc.) is also plotted (the slope which is $B/H = \mu_0 = 4\pi \times 10^{-7}$ is highly exaggerated; for the given B scale it should almost coincide with the H axis).

To get an idea of how small $H = 80$ A/m actually is, let us consider the solenoid of Fig. 9.18 or the toroid of Fig. 9.8 with an air core and with a ferromagnetic core. The applied field is given by $H = NI/l = 80$ A/m and does not depend on the core material. Assuming a winding of 100 turns/cm and coil length of 10 cm, we obtain for current $I = Hl/N = 0.8$ A, which is needed to produce an H of 80 A/m in the coil. If the core of the coil is air, wood, or some other nonmagnetic material, for which $M = 0$, the magnetic flux density inside the coil is $B = \mu_0 H = (4\pi \times 10^{-7}) \times (80) \cong 10^{-4}$ Wb/m^2 = 1 G. This is a very small quantity which is on the order of the earth's magnetic field (about $\frac{1}{2}$ G).

On the other hand, if a ferromagnetic material such as iron is used as a core material, the induced flux density for the same value of H, using Fig. 9.28, is $B \cong 1.3$ Wb/m^2. To produce a B of this magnitude in an air-core solenoid would require a value for $H = B/\mu_0 = 10^6$ A/m, a very large value. Relative permeability μ_r can be obtained by calculating the ratio of flux density with and without a magnetic core material. For iron $\mu_r = B/B_0 = \mu H/\mu_0 H \cong 1.3/10^{-4} = 1.3 \times 10^4$, and using (9.93), $\chi_m \cong M_s/H = 10^6/80 = 1.3 \times 10^4$. Note how much higher these values are than the μ_r's and χ_m's for the nonmagnetic materials given in Table 9.1.

It is surprising that a magnetizing force of only $H = 80$ A/m applied to a ferromagnetic material will result in a net magnetization of $M = 10^6$ A/m in the material. Even this value of $H = 80$ A/m is high considering that it was obtained from the hysteresis curve of ordinary polycrystalline† iron. Had we used a carefully prepared large single crystal of silicon-iron, the hysteresis curve would have been similar to that of Fig. 9.28, except it would have been much narrower, with the H intercept occurring at about $H = 0.8$ A/m. In other words, a magnetizing force smaller by a factor of 100 orients all domains. The susceptibility or relative permeability of such a material is $\chi_m = M_s/H \cong 10^6$, a very high value.

We can now view a ferromagnetic material as paramagnetic with the addition of a powerful local or molecular field that spontaneously orients all molecular spins over a small macroscopic region called a domain. It takes a very small external field (as compared with the molecular field) to align all domains so the material has a large net magnetization as a whole and gives rise to a powerful magnetic field. Weiss pictured the domains as a group of weather vanes which, when oriented randomly, will produce a magnetic field which is very small, even zero. The application of an external magnetic field aligns the magnetism of the

† A crystalline state implies a well-ordered arrangement of the atoms or molecules in a substance. In a crystal one can view the arrangement of atoms as a three-dimensional lattice with constant spacing between the lattice centers. The spacing between the atoms is about 10^{-10} to 5×10^{-10} m (note that the wavelength of light is about 5×10^{-7} m). The physical size of crystals can be very small, as, for example, when the lattice has no more than 10 atoms on a side. On the other hand, crystals can be as large as several inches or more. We can obtain large crystals by growing them in a furnace, starting with a small seed crystal (large crystals for germanium or silicon transistors are obtained this way), or they can be found in nature, such as chunks of quartz. A crystal of ferromagnetic material has "easy" and "hard" axes of magnetization. Hence, orienting the magnetizing force which is applied to a large single crystal along one of the easy axes of magnetization will require a small H for the magnetization to reach the saturation value M_s. On the other hand, commonly used ferromagnetic materials are polycrystalline; that is, a piece of the material consists of a tremendous number of small single crystals of random orientation. In each single crystal there are many domains. Hence only a small fraction of the microcrystals will be lined up with their easy axis of magnetization along the applied H field. To reach saturation will therefore require a much stronger applied H field.

domains in the same way that the wind will align a group of weather vanes, thus greatly increasing the material's total magnetic field. The applied field need not be particularly strong: a light breeze turns weather vanes as effectively as a gale does. Thus the introduction by Weiss of the powerful molecular field and the domain structure, which to many at that time appeared a fanciful speculation, seemed to explain ferromagnetism. Weiss had not the slightest notion of the origin of the mysterious field. Twenty years later, Heisenberg showed that very subtle quantum-mechanical effects produce forces, usually referred to as *exchange forces*, which give rise to the powerful molecular field that orients the spins parallel to each other within a domain.

There are two more effects to be explained: the hysteresis loop and the occurrence of T_c, the Curie temperature, above which a ferromagnetic material becomes an ordinary paramagnetic substance.

Hysteresis in Ferromagnetism

Let us start with an initially unmagnetized ferromagnetic core material and apply an increasing H field to it, as, for example, in the toroidal core of Fig. 9.29*a*. We will find that the magnetization M and the induced magnetic flux density B also increase. The increase will follow the initial (also known as the normal, or virgin) magnetization curve shown in Fig. 9.28 until we reach saturation given by

$$B_s = \mu_0(H + M_s) \cong \mu_0 M_s \tag{9.94}$$

If we start to decrease the current I in the magnetizing coil, which in turn decreases the magnetizing force H, the initial curve will not be retraced, but instead B will decrease along a curve, which implies there is a lag or delay in the reversal of the domains. This effect is known as *hysteresis* after the Greek word meaning "to lag." As a matter of fact when H is reduced to zero (corresponding to $I = 0$ in the magnetizing coil of Fig. 9.29), a net magnetization indicated by B_r (the remanence) remains in the core. The induced magnetization becomes a permanent magnetization which is the state that exists in permanent magnets. Thus, by increasing the current in the magnetizing coil and then reducing it to zero, we have in effect created a permanent magnet.

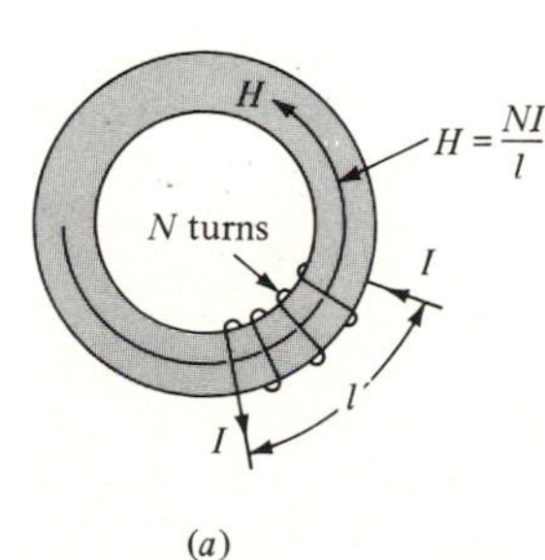

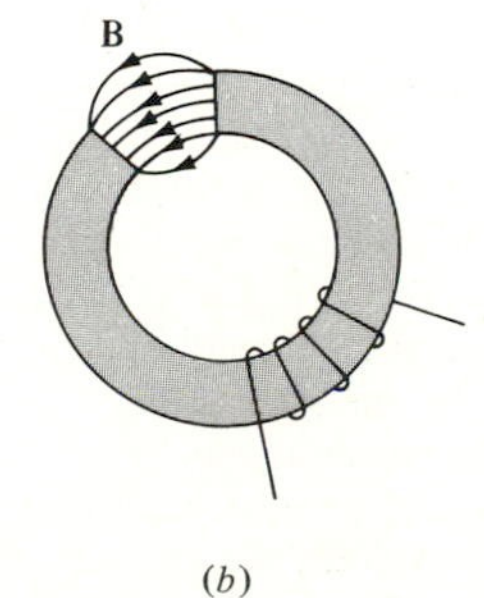

Figure 9.29 (*a*) A ferromagnetic toroid with a magnetizing coil which produces a magnetomotive force $\mathscr{F} = NI = Hl$ in the toroid; (*b*) the large internal magnetic field that exists in the toroid becomes accessible when a gap is cut into the toroid as shown.

Returning to Fig. 9.28, we observe that the internal magnetization will be reduced to zero if the H field has increased in the negative direction to the value H_c, called the *coercive force*. Apparently what happens in the material is that a sufficient number of domains have been reversed which cancel the effect of those still pointing in the original direction. Increasing the current in the negative direction still further, more domains (the more "sluggish" one) will be reversed until we reach the saturation level at which point all domains can be considered aligned in the opposite direction. If the applied current is periodically reversed in this manner, the hysteresis loop shown in Fig. 9.28 will be traced out. It shows the multivalued and nonlinear nature of magnetization in ferromagnetic materials.

What causes hysteresis, and what determines the shape of the hysteresis loop? Crystal imperfections, inclusions, cavities, and crystalline anisotropy all cause the domain walls to stick during domain growth† or reversal. The number of imperfections gives an indication of how much friction can be expected. The area of the hysteresis loop (as will be shown later) is a measure of the energy lost to heat per cycle; it is therefore related to the energy that must be dissipated by a magnetic material in an alternating field. Any generator, motor, relay, or transformer would operate with maximum efficiency if no magnetization remained after the applied external field dropped to zero, a fact of great practical importance.

Soft and Hard Ferromagnetic Materials

From the user's point of view, magnetic materials divide into two broad classes. "*Soft*" materials are used in generators, motors, and transformers to increase the flux density B when a current flows in their circuits. Under ac operation these materials can change their magnetization rapidly without much friction. They are characterized by a narrow but tall hysteresis loop of small area, as shown in Fig. 9.30. For such materials it is desirable to have H_c as small as possible; Table 9.2 shows that it is usually less than 100 A/m. To make a magnetic material soft, the motion of domain boundaries should be made as easy as possible. Materials which are soft have a uniform structure, free from inclusion and local strains with well-aligned crystal grains and low crystalline anisotropy.

On the other hand, it seems logical that materials for permanent magnets should have directly opposite properties. Good permanent magnets should show great resistance to demagnetization. Hence the coercive force H_c should be as large as possible. Such materials are referred to as "*hard*." For the hardest material known, this force can be 10 million times as high as the intrinsic coercive force for the softest materials. As can be seen from Table 9.2, a representative value for permanent magnetic materials is $H_c = 8 \times 10^4$ A/m ($\cong 1000$ Oe).

† We will show in the next sections that magnetization in a ferromagnetic specimen increases first by domain growth. When small currents I are applied, those domains whose magnetic moments are most nearly parallel to H increase in size at the expense of those in other directions. For larger applied currents, the remaining domains rotate in a direction parallel to those that have been increasing. For still larger currents, all domains rotate until they are exactly parallel to the applied H field.

Table 9.2 Properties of magnetic materials and magnetic alloys

Material (composition[1])	Initial relative permeability, μ_i/μ_0	Maximum relative permeability, μ_{max}/μ_0	Coercive force H_c, A/m (Oe)	Residual field B_r, Wb/m^2 (G)	Saturation field B_s, Wb/m^2 (G)	Electrical conductivity ρ, $\times 10^{-8}\ \Omega \cdot m$	Uses
Soft							
Commercial iron (0.2 imp.[2])	250	9,000	≈ 80 (1)	0.77 (7,700)	2.15 (21,500)	10	Relays
Purified iron (0.05 imp.)	10,000	200,000	4 (0.05)	—	2.15 (21,500)	10	
Silicon-iron (4 Si)	1,500	7,000	20 (0.25)	0.5 (5,000)	1.95 (19,500)	60	Transformers
Silicon-iron (3 Si)	7,500	55,000	8 (0.1)	0.95 (9,500)	2 (20,000)	50	Transformers
Silicon-iron (3 Si)	—	116,000	4.8 (0.06)	1.22 (12,200)	2 (20,100)	50	Transformers
Mu metal (5 Cu, 2 Cr, 77 Ni)	20,000	100,000	4 (0.05)	0.23 (2,300)	0.65 (6,500)	62	Transformers
78 Permalloy (78.5 Ni)	8,000	100,000	4 (0.05)	0.6 (6,000)	1.08 (10,800)	16	Sensitive relays
Supermalloy (79 Ni, 5 Mo)	100,000	1,000,000	0.16 (0.002)	0.5 (5,000)	0.79 (7,900)	60	Transformers
Permendur (50 Cs)	800	5,000	160 (2)	1.4 (14,000)	2.45 (24,500)	7	Electromagnets
Mn-Zn ferrite	1,500	2,500	16 (0.2)	—	0.34 (3,400)	20×10^6	Core material for coils
Ni-Zn ferrite	2,500	5,000	8 (0.1)	—	0.32 (3,200)	10^{11}	

Material (composition)	Coercive force H_c, A/m $\times 10^3$ (Oe)	Residual field B_r, Wb/m^2 (G)	Energy product $(BH)_{max}$, J/m^3 $\times 10^3$ (G $\times$ Oe $\times 10^6$)	Uses
Hard				
Carbon steel (0.9 C, 1 Mn)	4 (50)	1 (10,000)	1.6 (0.2)	Permanent magnets
Chromium steel (3.5 Cr, 1 C, 0.5 Mu)	5.2 (66)	0.95 (9,500)	2.2 (0.27)	
Alnico V (14 Ni, 24 Co, 8 Al, 3 Cu)	44 (550)	1.2 (12,000)	40 (5)	
Alnico VIII (15 Ni, 35 Co, 7 Al, 4 Cu, 5 Ti)	126 (1,600)	1.04 (10,400)	44 (5.5)	
Alnico IX	126 (1,600)	—	88 (11)	
Iron powder (100% Fe)	61 (770)	0.57 (5,700)	12.8 (1.6)	
Iron powder (elongated)	63 (790)	1.02 (10,200)	36 (4.5)	
Ba ferrite (Ferroxdure)	120 (1,500)	0.2 (2,000)	8 (1)	
Cobalt-samarium	560 (7,000)	0.84 (8,400)	128 (16)	

[1] Percent by weight; remainder is Fe.
[2] imp. = impurities.

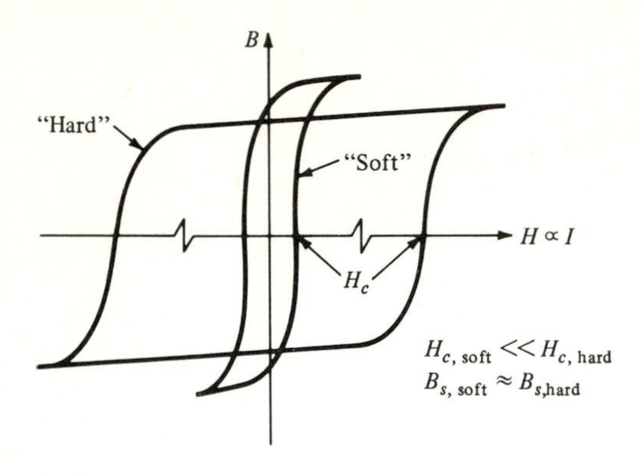

Figure 9.30 Hysteresis loops of soft magnetic materials, which are easy to magnetize and demagnetize, and those of hard magnetic materials. The former are useful in transformers and machinery, whereas the latter find application in permanent magnets.

The table for soft magnetic materials suggests that there is a large variation in relative permeability $\mu_r = \mu/\mu_0$. Values of μ_r for the hard materials are not given since such materials are used primarily in permanent magnets where knowledge of H_c and the energy product are more important. The values of μ_r are obtained from the initial magnetization curve in Fig. 9.28. Note that μ is not the slope of the curve, which is given by dB/dH, but is equal to the ratio B/H. The highly nonlinear variation of $\mu = B/H$ for a ferromagnetic material such as commercial iron is shown in Fig. 9.31. The permeability at low flux densities, called the *initial permeability*, is much less than the permeability at higher flux densities. This fact is of particular importance in communication equipment, where the current is commonly very weak. The maximum permeability occurs at the "knee" of the curve. The reason for the knee is that as H increases, the magnetization M reaches a maximum value M_s, the saturation magnetization in the material. The flux density **B**, given by

$$\mathbf{B} = \mu_0(\mathbf{H} + \mathbf{M}) \tag{9.95}$$

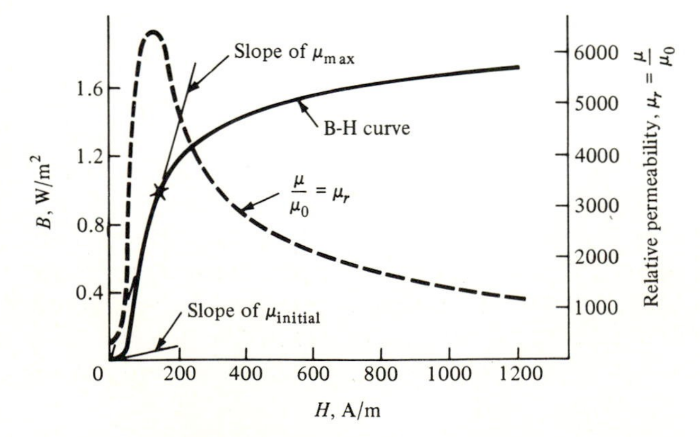

Figure 9.31 Magnetization curve of commercial iron. Permeability is given by the ratio B/H.

continues to increase for large values **H** only because of the presence of the $\mu_0 H$ term. The slope above the knee approaches

$$\frac{dB}{dH} = \mu_0 \left(\frac{dH}{dH} + \frac{dM}{dH}\right) \cong \mu_0 = 4\pi \times 10^{-7} \tag{9.96}$$

which for the scales of the above figure is practically a straight horizontal line.

In many practical applications, a small alternating H field is superposed on a large constant biasing field (loudspeakers, chokes in power supplies, etc.). The effective incremental permeability is then given by

$$\mu_{\text{inc}} = \frac{\Delta B}{\Delta H} \tag{9.97}$$

and is equal to the slope of the small hysteresis loop traced out by the ac field at the point of the constant biasing field. Since the inductance of a coil, as shown in Sec. 7.7, is proportional to the permeability, for many practical devices, (9.97) should be used for μ. A further discussion of this is to be found in Sec. 10.10.

To recap, we can say that soft materials are associated with low values of coercive force H_c (1 to 100 A/m), high saturation flux densities B_s, high values of permeability μ, and no permanent magnetic qualities; i.e., no appreciable magnetization remains after the magnetizing force H has been removed. Such materials are employed when high flux densities for low magnetizing currents are needed and when low losses are required. All transformers, relays, motors, magnetic recording and playback heads fall within this category.

Hard ferromagnetic materials are characterized by high coercive values (H_c from 10^3 to 10^5 A/m), lower flux densities at saturation, low permeability values (μ_r from 1 to 10), and high remanence values B_r. Materials with such properties are used in the manufacture of permanent magnets and magnetic recording tape, since a length of recorded tape constitutes a series of permanent magnets (see Sec. 10.11). Figure 9.30 gives an idea of the relative shapes of the loops for a soft and hard material.

Temperature Dependence of Ferromagnetism

We have shown that ferromagnetism is like paramagnetism with the addition of a strong molecular field H_m. This molecular field spontaneously orients (against the randomizing thermal $\frac{3}{2}kT$ field) the magnetic spins m over small regions called domains. Even though $B_m = \mu_0 H_m$ is very powerful, it gradually loses its effect as the temperature rises, as shown in the $M_s T$ curve of Fig. 9.19. The Curie temperature T_c is the temperature at which a ferromagnetic substance becomes paramagnetic. We can reason, therefore, that at T_c, the molecular field must be on the order of the thermal field. We can use this to find the strength of the molecular field by equating the aligning energy [see Eq. (9.88)] of the molecular field mB_m to the randomizing thermal energy; that is,

$$kT_c \cong mB_m \tag{9.98}$$

Table 9.3 Saturation magnetization M_s and Curie temperature T_c for some ferromagnetic materials

Substance	$\mu_0 M_s$ at 293 K, Wb/m^2	$\mu_0 M_s$ at 0 K, Wb/m^2	T_c, K
Fe	2.14	2.2	1043
Co	1.7	1.82	1394
Ni	0.61	0.64	631
Gd	0	2.49	293
MnBi	0.75	0.85	670
$Y_3Fe_5O_{12}$	0.17	0.25	560

Using iron for which the Curie temperature $T_c \cong 1000$ K (as given in Table 9.3), we obtain

$$B_m \cong \frac{kT_c}{m} = \frac{(1.38 \times 10^{-23})(10^3)}{9 \times 10^{-24}} = 1.5 \times 10^3 \text{ Wb/m}^2 \tag{9.99}$$

or $H_m = B_m/\mu_0 = 1.5 \times 10^9$ A/m. This is an extremely powerful field, considering that our strongest laboratory magnets have difficulty achieving a field of $B = 10$ Wb/m^2. Such a field is too powerful to come from the adjacent spins in the lattice, since the magnetic field at one lattice point arising from the magnetic moment of an electron at a neighboring lattice point is of the order of

$$B \cong \frac{\mu_0 m}{4\pi r^3} \cong 10^{-1} \text{ Wb/m}^2 \tag{9.100}$$

which is smaller than the molecular field by a factor of 10^4 (a typical lattice constant is 2×10^{-10} m). We now know from quantum mechanics that the molecular field which holds the magnetic moments parallel within a domain has its origin in the quantum-mechanical exchange force.

Above the Curie temperature, thermal agitation is strong enough to overcome the aligning power of the intrinsic molecular field B_m. Table 9.3 gives the saturation magnetization M_s and Curie temperature T_c for several ferromagnetic materials. The saturation magnetization is equal to the spontaneous magnetization of a single domain.

Let us now derive the temperature dependence of ferromagnetism. We have shown that both paramagnetism and ferromagnetism derive their magnetization from the spin of the electrons. Langevin's formula, (9.89),

$$M = Nm\left(\coth a - \frac{1}{a}\right)$$

where $a = mB/kT$ and $M_{\max} = Nm$, gives the degree of alignment of the magnetic moments m of a system of spinning electrons. For a paramagnetic substance the local field acting on each electron was given by (9.92), and could be approximated by the externally applied field $B_{\text{mol}} \cong \mu_0 H = B$. Ferromagnetic substances, on the other hand, are characterized by a powerful local field (the exchange force field) that acts on the system of spins and is much larger than any applied H field could be. Expression (9.92), if generalized, can still be used to represent this field. Let us replace the one-third factor by γ which gives

$$B_{mol} = \mu_0(H + \gamma M) \cong \mu_0 \gamma M \tag{9.101}$$

where γM represents the exchange force field ($\gamma M \gg H$). We can find what γ is by noting that B_{mol} is given by (9.99) as 1.5×10^3 and the spontaneous magnetization within a domain is given by $M = Nm \cong (10^{29}\ \text{atoms/m}^3)(9 \times 10^{-24}\ \text{A} \cdot \text{m}^2) \cong 10^6$ A/m. This gives $\gamma \cong 10^3$, very much higher than the one-third value for paramagnetic materials.

If we use the above B_{mol} and substitute it for B into Langevin's formula (9.89) and also for B into $a = mB/kT$, which can then be rewritten as

$$M = \frac{akT}{\gamma \mu_0 m} \tag{9.102}$$

we will have two expressions for magnetization M versus temperature T. The solution of these will give us the temperature-dependent spontaneous magnetization $M(T)$—spontaneous because the external field H in B_{mol} of (9.101) is taken as zero. If we plot the two curves for M as shown in Fig. 9.32, their intersection gives the solution $M(T)$ to both equations. As the temperature is increased, the straight line curve becomes steeper, whereas the other curve representing Langevin's formula is unchanged. The result is that the intersection of the curves moves to the left, giving a lower value for the spontaneous magnetization M. If the temperature continues to increase, the magnetization will be reduced to zero. This happens when the straight line is tangent to the Langevin curve (9.89) at the origin. This condition corresponds to the Curie temperature. At this and higher temperatures the spontaneous magnetization is zero, and the material exhibits ordinary paramagnetic properties.

Figure 9.33*a* shows how $M(T)$ decreases with temperature T for a ferromagnetic material. It is the same curve shown in Fig. 9.19. Notice that $M(T)$ is the spontaneous magnetization in a domain, which is equal to the saturation magnetization $M_s(T)$ of a large piece of ferromagnetic material (assuming all domains are perfectly aligned). Also $M(0) = M_{max} = Nm \cong 10^6$.

To demonstrate the temperature dependence of a ferromagnetic material such as iron, a simple experiment can be made. An iron nail is heated until red-hot, which makes it paramagnetic. A magnet is held above the nail at a height at which it would ordinarily pick up a nail. As the nail cools and its temperature goes below T_c, it becomes ferromagnetic and is suddenly picked up by the magnet.

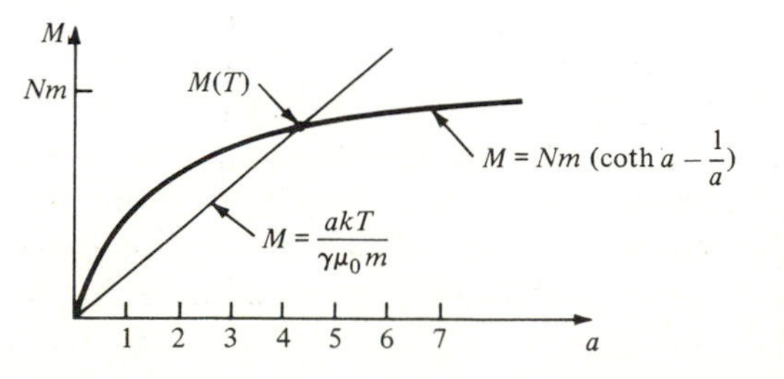

Figure 9.32 The intersection of the Langevin function graph and the straight line gives the spontaneous magnetization $M(T)$ at temperature T for a ferromagnetic material.

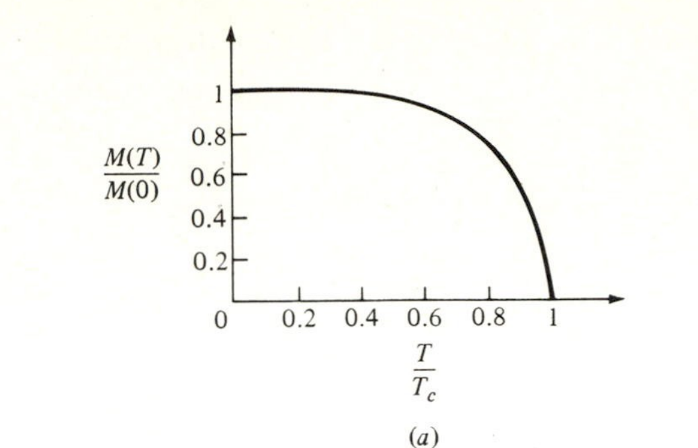

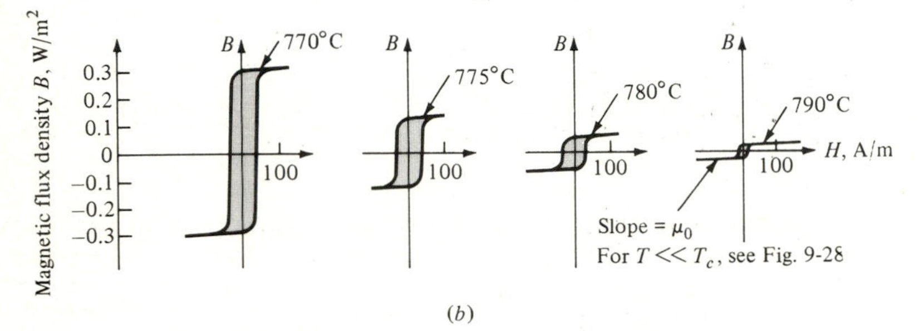

Figure 9.33 (*a*) Curve showing the decrease of spontaneous magnetization of a ferromagnetic material as the Curie temperature T_c is approached; (*b*) change in shape of the hysteresis loops of iron near the Curie temperature ($T_c = 770°C$).

9.6 FORMATION OF DOMAINS IN FERROMAGNETIC MATERIALS

Why does the volume of a ferromagnetic material break spontaneously into small subvolumes, called domains? A block of iron, for example, which has been heated above its Curie temperature $T_c = 770°C$ (1043 K) is paramagnetic. As it cools to below T_c, small regions of the material become spontaneously magnetized to the saturation magnetization M_s for that temperature. The spontaneously magnetized domains are randomly oriented with respect to each other (as shown in Fig. 9.27), and so the block of ferromagnetic iron shows no net external field. A second question therefore arises: In the absence of an external magnetic field, why do the domains align themselves randomly as a ferromagnetic material is allowed to cool to below its paramagnetic threshold? We also know that for soft materials it takes little external field H_c to saturate the specimen (align all domains so the specimen looks like one large domain), whereas for hard materials it takes a large coercive force H_c to saturate the specimen (about 10^5 A/m). At first sight it seems contradictory, in view of the 10^9 A/m molecular field, to suppose that an applied field of a few amperes per meter, or even a field as large as $H = 10^5$ A/m, could alter the magnetization of the specimen by an appreciable amount.

To answer these questions, we must first point out that four basic energies are involved in ferromagnetism. They are

$$W = W_{\text{exchange}} + W_{\text{magnetostatic}} + W_{\text{anisotropic}} + W_{\text{magnetostrictive}} \tag{9.103}$$

Coupling this with the principle that all systems tend toward a state possessing minimum energy, we can show that domain formation is the outcome of an energy-minimization contest between these four energies (see also discussion on page 350). The four energies are

1. *Exchange energy.* This energy is minimized when all atomic dipole moments are parallel.
2. *Magnetostatic energy.* This energy is related to the external H field produced by the magnetic material. It is a minimum when the integral of $\frac{1}{2}\mu H^2$ over all space is a minimum.
3. *Anisotropic energy.* The magnetocrystalline anisotropic energy of a ferromagnetic crystal is a minimum when the magnetization is along certain preferred directions called axes of "easy" magnetization.
4. *Magnetostrictive energy.* This energy is a minimum when the material is oriented so changes in its dimensions are along the magnetization axis.

Exchange Energy

The large molecular field $B_m = \mu_0 H_m$ which holds the moments of neighboring atoms in ferromagnetic materials parallel to each other has its origin in the quantum-mechanical exchange forces. This exchange interaction, which aligns adjacent spins and is also known as the *spin-spin interaction*, depends critically on the distance between atoms. Let us briefly review the characteristics of spacing between atomic particles.

The *Pauli exclusion principle* requires that no two electrons of the same spin and orbital momentum occupy the same region of space. The only way that two electrons can occupy the same region is by having opposite spin, as shown schematically in Fig. 9.34. This is the reason for the formation of antiparallel pairs of adjacent electrons in solid materials where the atoms are close to each other. The spins and angular momentums of the paired electrons cancel each other, with

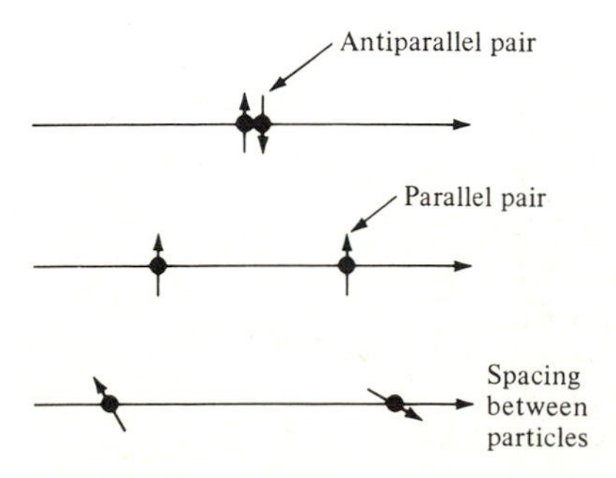

Figure 9.34 Allowable orientations of two electrons versus spacing between them. Starting from the bottom: Two particles, widely separated, can have any orientation with respect to each other. For certain critical distances, they can have parallel orientation. Electrons of opposite spin can occupy the same region of space. However, the magnetism of such a pair is zero.

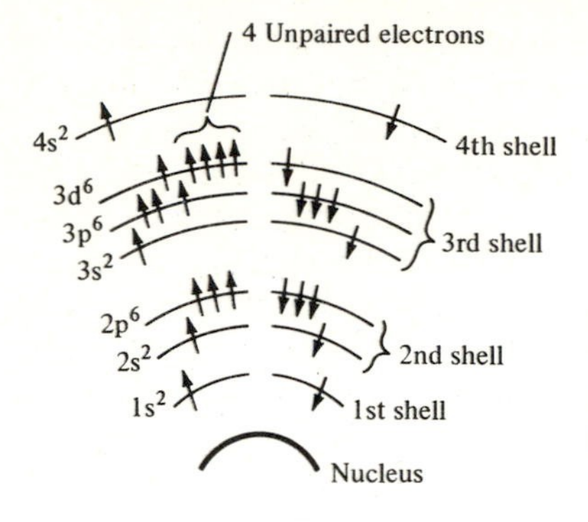

Figure 9.35 Diagram of an iron atom which has 26 electrons arranged as shown. The third subshell, labeled $3d$, has 4 unpaired electrons which give iron its magnetic properties. The up spins are denoted by ↑, and down spins by ↓.

the result that their magnetism is also zero. This tendency of the outer electrons of adjacent atoms to form antiparallel pairs explains why all solids are nonmagnetic, except for a very few, such as iron, cobalt, and nickel. When an application of an external magnetic field to such nonmagnetic materials slightly unbalances the antiparallel spin coupling, the result is a weak magnetism (paramagnetism); when it unbalances the orbital pairing, the result is diamagnetism.

How, then, do the ferromagnetic materials avoid the tendency of the electrons to form antiparallel pairs? Ferromagnetic materials have their unpaired electrons not in the outermost electron cloud or shell but in the inner ones where they are not able to form pairs with electrons in other atoms. For example, Fig. 9.35 shows the shell structure and the number of electrons with "up" and "down" spins in each shell of the ferromagnetic material iron.† It is these unpaired electron spins which are aligned by the molecular field of the exchange force. For a material to be ferromagnetic, not only must the material atoms have "shielded" unpaired electrons, but the electrons of neighboring atoms must be spaced at specific distances. Figure 9.34 suggests that at only certain critical distances can two particles have their spins point in the same direction. The Bethe curve, sketched in Fig. 9.36, shows that only the ferromagnetic materials have an interatomic separation which coincides with a peak in the exchange force. The quantum-mechanical

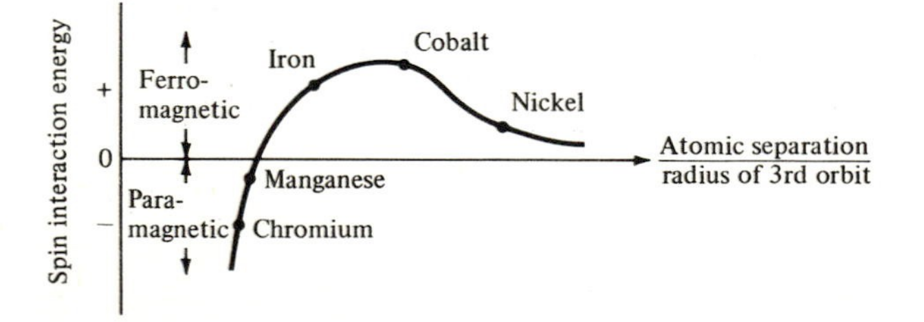

Figure 9.36 The Bethe curve shows that ferromagnetic substances have a favorable ratio of distance between atoms to radius of unfilled inner electron shell.

† Note that in iron, which is a good conductor, the conduction electrons come from the 4s shell, whereas the ferromagnetic electrons are in the 3d shell.

exchange forces are negligible when the atoms are several times as far apart as they usually are in crystals. As two atoms are brought nearer to each other from a distance, these forces at first cause the electron spins in the two atoms to align parallel (positive interaction). As the atoms are brought nearer, this force increases to a peak when the spin moments are held parallel most firmly. It then decreases until it becomes zero, and with still closer approach, the spins set themselves antiparallel with relatively strong forces (negative interaction). The Bethe curve also explains why manganese when alloyed becomes ferromagnetic; apparently mixing it with other substances (for instance, bismuth) gives it, on the average, a larger interatomic separation, bringing the alloy in the positive interaction region of the Bethe curve.

The exchange energy which can be expressed by

$$W_{\text{spin-spin}} = -C \cos \Psi \tag{9.104}$$

where C is a positive constant and Ψ is the angle between moments of adjacent atoms, is minimized when all spins in a block of ferromagnetic material are aligned, as shown in Fig. 9.37*a* (that is, $\Psi = 0$ gives the minimum energy

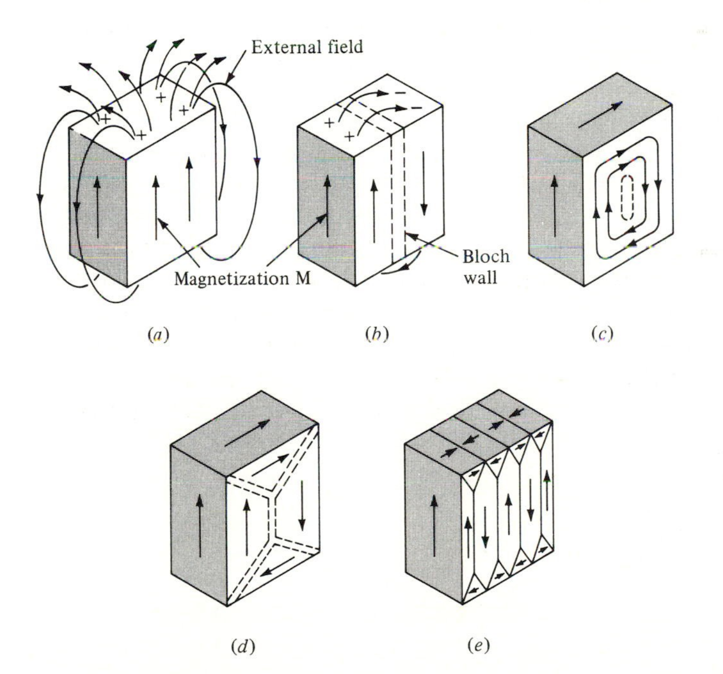

Figure 9.37 (*a*) Evolution into domain structure. A block of uniformly magnetized material with all the spins pointing in the same direction minimizes the exchange energy but maximizes the external magnetic field energy; (*b*) division into two domains lowers the external field energy, as is evident from the weaker external magnetic lines; (*c*) without anisotropy and magnetostriction present, the closed-loop arrangement of magnetic moments minimizes the total energy; (*d*) with anisotropy but no magnetostriction energy present, the spin arrangement of (*c*) is forced into discrete domains; (*e*) anisotropy and magnetostriction energies present, the energy-minimization contest results in many distinct domains.

$W = -C$, whereas for an antiparallel alignment of adjacent dipoles, $\Psi = \pi$, and the energy would rise to $W = +C$). Equivalently one can say that in a ferromagnetic material the exchange energy manifests itself in the form of a torque that aligns the spins. However, such alignment would create a powerful magnet with a large external magnetic field; in other words, parallel alignment would maximize the magnetostatic energy. Since it is the sum of these two energies that tends toward a minimum, we can see that if we allow the exchange energy to rise (not all spins point in the same direction), the magnetostatic energy can then decrease to give an overall energy which might be less than if all spins were perfectly aligned. A possible configuration that this might take is shown in Fig. 9.37*b*. The external field is now reduced; the exchange energy is increased somewhat because the spins in the two domains are opposite.

Since the exchange force is a short-range force (it acts over one or two atomic distances), the exchange energy is highest at the boundary between the two domains. The boundary wall (denoted by the dashed lines in Fig. 9.37) separates regions of opposite spin. The transition could conceivably be only one atomic diameter thick with the magnetic moments pointing up to the left and down to the right of the dashed line. The exchange energy would be very high in such a sharp transition. If, on the other hand, the transition region is wider, the change to opposite spin involves a lower energy, because the dipoles can arrange themselves so the transition is smooth, as shown in Fig. 9.38. This is called a 180° Bloch wall. The angle between adjacent spins is very small; hence, according to (9.104), the exchange energy between adjacent spins is almost at the minimum. The size of the Bloch wall can be 300 atomic diameters long to allow a smooth transition. The energy of these walls is so low that the walls move quite freely within the material, when exposed to a small external field. One can visualize the motion of a wall by imagining the dipoles within a wall to rotate through 180° as the wall moves to the left or right. Wall motion between domains is the primary mechanism in the growth of magnetization in ferromagnetic substances. When a ferromagnetic material is subjected to a magnetic field, the domains with magnetization nearest to the direction of the applied field will grow at the expense of the other ones. Only when this mechanism is exhausted, and only if the strength

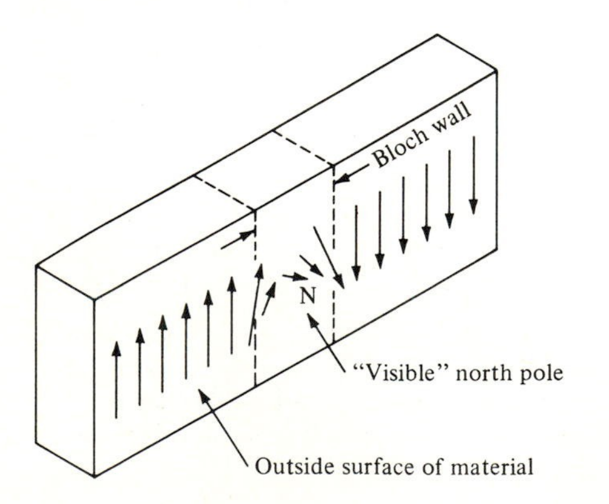

Figure 9.38 The zone between domains is known as a Bloch wall. The rotation of spin dipoles with respect to each other in a 180° wall is shown.

of the applied field can be sufficiently increased, will the domains begin to rotate until they are all parallel with the applied field.

The domain walls can be made visible using a technique developed by F. Bitter. One notices from examining Fig. 9.37*e* that magnetization is primarily parallel to the *exterior* surfaces of a ferromagnetic substance. Such an arrangement minimizes the external magnetic field energy, because any magnetization ending normally on a surface continues into free space as an H field; this H field in turn would give rise to a larger value for the external magnetic field energy, (9.105). However, in the Bloch zone, some north poles of the atomic spins, denoted by N in Fig. 9.38, which are turning through 180° come to the surface of the ferromagnetic material. The spins parallel to the surface that are shown as finite length vectors in Fig. 9.38 are actually components of the continuous magnetization vector **M** which exists throughout the interior. Such spins, therefore, do not contribute to any external magnetic field because an H field is created only at places where M is discontinuous [see discussion following Eqs. (9.20) and (9.31)]. Bitter's technique consists of spreading magnetic iron-oxide particles over the surface. These particles collect along the lines where the north poles of the atomic magnets in each Bloch wall point toward the surface, making the domain boundaries plainly visible through a microscope.†

The total energy can be further reduced by dividing the block of Fig. 9.37*b* into many rectangular bar-shaped domains. This would reduce the external magnetic field to almost zero but would significantly raise the exchange energy by the creation of many walls. A more likely distribution of the magnetic moments is shown in Fig. 9.37*c* where the closed-loop magnetization creates zero external field but still allows the spins over large regions to point in the same direction (the magnetic field is now entirely internal and points in the direction of the magnetization). The exchange energy is higher than that in Fig. 9.37*a*, but since the external field is zero, the sum of these two energies is minimized.

Magnetostatic Energy

This energy resides in the magnetic field of the magnet and is given by (7.46) as

$$W = \frac{1}{2} \iiint \mu H^2 \, dv \tag{9.105}$$

It decreases with the external field of the magnet, and is the energy that we usually associate with a permanent magnet. This energy is minimized whenever M assumes a continuous configuration. Any discontinuous M, as, for example, when M ends normally on a surface of a material, must continue into free space as an H field, thereby increasing the magnitude of the above integral.

At this time the minimum and maximum energy orientations of a spin in a magnetic field should be reviewed. Using Fig. 4.6, the minimum energy position of a spin dipole minimizes the energy integral (9.105) because it decreases by partial

† There are two excellent films available on free loan from Bell Telephone Laboratories which show domains and domain wall motion. They are *Ferromagnetic Domains*, No. 682, and *Domains and Hysteresis in Ferromagnetic Materials*, No. 850.

cancellation the magnetic field in which it is immersed. This is also the reason why a compass needle aligns with the field, because this action lowers the energy of the system. Similarly, the maximum energy orientation of a dipole maximizes the energy integral. Therefore, in a system in which the dipoles are not free to rotate, applying a field can raise the energy of the system.

Anisotropic Energy

Anisotropy is a term which means having different properties in different directions. Ferromagnetic materials, such as iron, crystallize in the cubic system. Magnetic anisotropy is the preference of the atomic magnets to align themselves with the axis of the crystal. The reason for this is not very well understood at present. There are three axes of easy magnetization in an iron crystal. Figure 9.39 shows a body-centered cubic crystal of iron with "easy," "medium," and "hard" directions of magnetization. From this we see that the spins in a ferromagnetic material not only like to point in the same direction but also prefer to lie along a cube edge of the crystal. It requires considerably more energy to magnetize a crystal to saturation in a hard direction. The excess energy required in the hard as compared with the easy direction is the anisotropic energy. If we were to take a large crystal of iron and place it between the poles of a magnet, it would align itself quickly along one of the three 001 axes of the crystal. To swivel it to another 001 axis would require energy since all the spins would have to swing through less favorable orientation.

Such a hindrance to spin rotation (crystalline anisotropy) is what makes a permanent magnet possible as it causes retention of magnetization (hysteresis). Anisotropy is also the reason that domains in a ferromagnetic material become magnetized in certain directions as the material cools down to below its Curie temperature T_c. Another puzzling question which is answered by anisotropy relates to the Bloch wall. It was implied before that the exchange energy of a Bloch

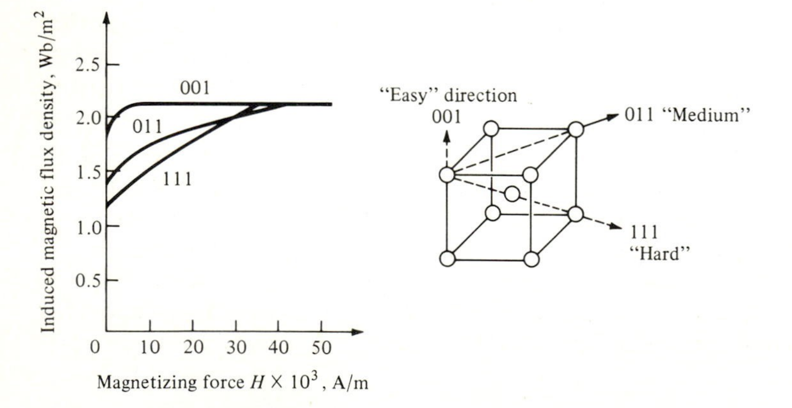

Figure 9.39 Magnetization curves along three axes for a single crystal of iron. The axes are denoted by the three Miller indices, 001, 011, and 111.

wall between domains is inversely proportional to its thickness. Hence, the wall might spread out until it filled most of the crystal were it not for the restraining effect of the anisotropic energy, which acts to limit the width of the transition layer. The actual thickness and energy of the transition layer is the result of a balance between the competing forces of exchange energy and anisotropic energy; the former tends to increase, whereas the latter tends to decrease the thickness.

How does the anisotropy of the crystalline structure in ferromagnetic materials enter into the formation of domains? Figure 9.37*c* would be the arrangement that spins would take were it not for the crystalline anisotropic torque tending to turn the moments into the preferred directions. This new torque effectively squares up the corners of the loop arrangement of Fig. 9.37*c* until the magnetization breaks into linear domains shown in Fig. 9.37*d*. Wherever there was a sudden change in magnetization direction, a domain wall now appears. As the figure shows, we have now two types of domain walls. At the corners we have 90° domain walls, and at the center a 180° domain wall. The magnetic field is still confined to the interior and is along the closed magnetization lines. Recall that an external B field will exist only if there is a discontinuity in the normal component of magnetization M across a surface between the interior and the exterior. The top and bottom domains of triangular shape are called *domains of closure* because they complete the magnetization within the crystal and thus allow the magnetic field to be also completed within. Figure 9.37*d* is thus the shape of the domain arrangement which minimizes the sum of the exchange, external magnetic, and anisotropic energies.

Magnetostrictive Energy

Some crystalline substances when exposed to a magnetic field experience a stress in their lattice structure, and the substance as a whole will change dimensions (strain) to relieve the stress. Conversely, when a ferromagnetic material strains under stress (tension or pressure), its magnetization is affected; i.e., its permeability will increase or decrease depending on the nature of the material. This is called *magnetostriction*. The deformation $\Delta l/l$ due to magnetostriction (denoted by the symbol λ) is on the order of 10^{-5} to 10^{-6} and is shown in Fig. 9.40 for positive ($\lambda > 0$) and negative ($\lambda < 0$) magnetostriction.† Figure 9.41 shows the effect of tensile stress on the induced magnetic field of two ferromagnetic materials. A constant tension is applied while the field is increased from zero. Permeability μ, defined as the ratio of B to H, increases for 68 Permalloy (a 68 percent nickel 32 percent iron alloy) under tension and decreases for nickel under tension.

Nickel is a material with a negative magnetostriction; it contracts in the direction of an applied magnetic field. Conversely, under pressure (tension), the magnetization increases (decreases) if the stress is applied parallel to the direction

† The resistivity of ferromagnetic materials also changes with magnetization. There is a close relation between changes in length and changes in resistivity.

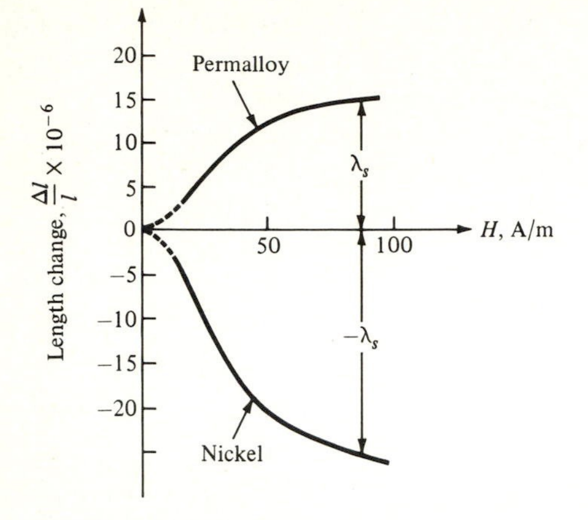

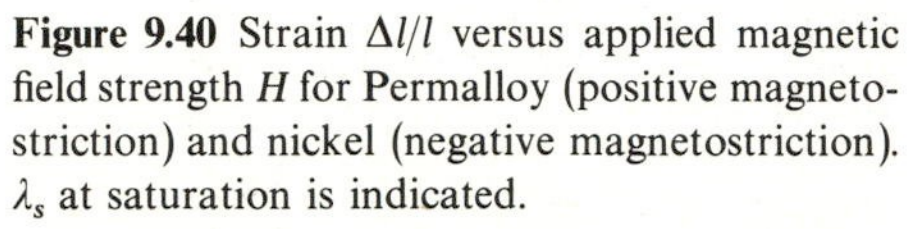
Figure 9.40 Strain $\Delta l/l$ versus applied magnetic field strength H for Permalloy (positive magnetostriction) and nickel (negative magnetostriction). λ_s at saturation is indicated.

of the external magnetic field. On the other hand, when pressure (tension) is applied at right angles to the magnetic field of a sample of nickel, magnetization decreases (increases). We therefore conclude that in a material with negative magnetostriction, the domains tend to align or rotate in the direction of the applied compression, whereas a tensile stress will tend to rotate the domains to a position perpendicular to the direction of the applied tension. These effects are schematically shown in Fig. 9.42.

For a material like iron which has positive magnetostriction,† the effects are opposite to those shown in Fig. 9.42. Let us use a different schematic representa-

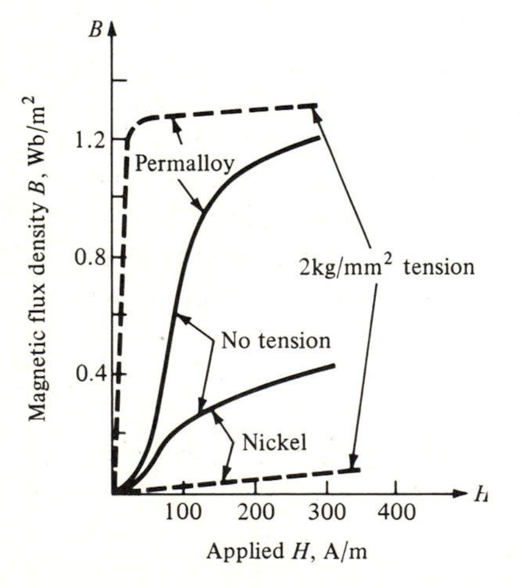

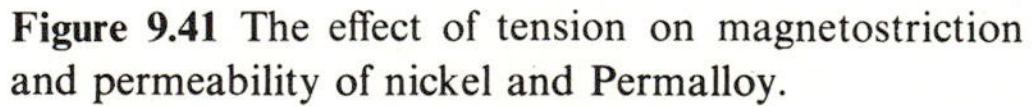
Figure 9.41 The effect of tension on magnetostriction and permeability of nickel and Permalloy.

† The magnetostrictive changes in length of an iron crystal depend on the direction of the applied field. For an applied field parallel to the 001 crystal axis, iron expands; for a field parallel to the cube diagonal 111, it contracts; and for a field applied parallel to a face diagonal 011, it first expands and then contracts as the field is increased. Nickel is less complicated; it contracts in all field strengths in all directions.

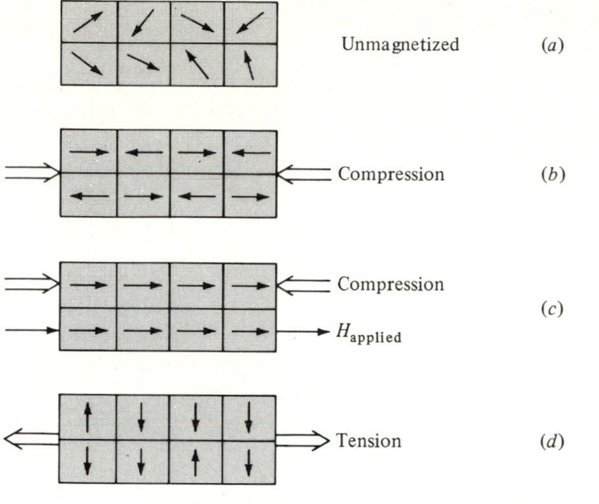

Figure 9.42 The effect on the domains of stress and magnetic field strength H applied to a material with negative ($\lambda < 0$) magnetostriction.

tion, Fig. 9.43, to show this. Referring to Fig. 9.43, we can say that the effect of stress on the magnetization **M** of a material with positive ($\lambda > 0$) magnetostriction will be

Tension $T_{\parallel}$ increases $M_{\parallel}$, decreases $M_{\perp}$.
Pressure $P_{\parallel}$ decreases $M_{\parallel}$, increases $M_{\perp}$.
Tension $T_{\perp}$ decreases $M_{\parallel}$, increases $M_{\perp}$.
Pressure $P_{\perp}$ increases $M_{\parallel}$, decreases $M_{\perp}$.

Magnetostrictive and anisotropic energies are interrelated because the crystalline anisotropy depends on the state of deformation of the lattice. A crystal will deform spontaneously if doing so will lower the anisotropic energy. The magnetostrictive energy is defined as zero for an unstrained lattice. The direction of the internal contraction or elongation in a ferromagnetic material prefers to point along the direction of the internal magnetization. Thus, if the direction of the internal strain in a small particle of material is different from that of the magnetic moment, a torque T exists on the particle; this torque is given by

$$T = -\tfrac{3}{2}\lambda S v \sin 2\theta \tag{9.106}$$

and tends to align the magnetic moment. In the above formula, S is the stress (positive for tension, negative for compression), v is the particle volume, λ is the

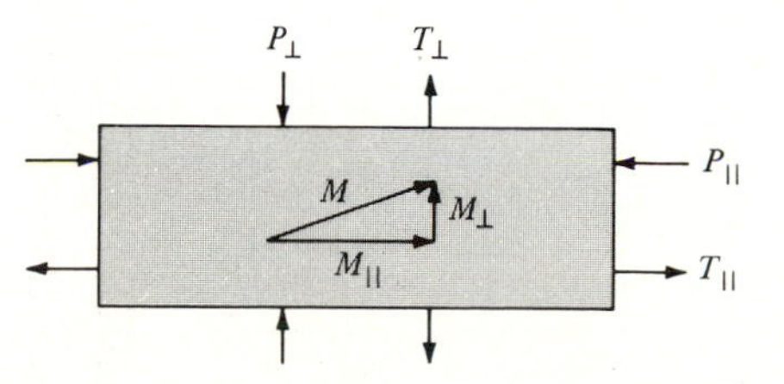

Figure 9.43 The effect of stress on magnetization **M** for a material with positive magnetostriction.

magnetostriction constant, and θ is the angle between the strain axis and the magnetic moment.

Thus, we have shown that there are four competing torques acting on each particle of a ferromagnetic material. They are: Torque exerted by the magnetic dipoles on each other (exchange), torque exerted by an externally applied magnetic field, torque exerted by the preference of magnetic moments to lie along particular crystallographic directions, and torque exerted by strain in the ferromagnetic material.

Let us now return to the discussion of domain formation and how magnetostriction enters into it. If the magnetic moments themselves cause lattice distortion in the moment direction, it follows that a specimen of ferromagnetic material will deform appreciably when all domains reach parallel alignment. Referring to Fig. 9.37*d*, we see that the crystal will be severely distorted as the domains elongate along the direction of magnetization. The elongation of the triangular end domains and the side domains, which are perpendicular to each other, will result in an elastic strain. To relieve, in part, the internal stresses and thereby lower the magnetostrictive energy, more domains will form, as shown in Fig. 9.37*e*. The total strained volume is now decreased because strains in adjacent domains are opposite. As long as the increase in exchange energy which accompanies the creation of more domain walls is less than the decrease of the magnetostriction energy, the domain arrangement of Fig. 9.37*e* is a stable one. The division of a large crystal into many domains of the type shown in Fig. 9.37*e* is an outcome of the energy-minimization contest between the four energies. For a crystal which is weakly anisotropic and weakly magnetostrictive, the energy-minimization contest might result in the stable formation of many domains of the type shown in Fig. 9.37*b*.

Ordinary samples of iron are polycrystalline. As shown in Fig. 9.27, the crystals are oriented randomly, with each grain having its own easy direction of magnetization. Usually the grain size is such that it contains many domains. Even though it appears that iron is not isotropic, for large enough specimens the average direction of **M** is the same as that of **H**. Hence, for practical purposes, we can consider it isotropic. Iron can also be manufactured so the crystals are oriented. For example, the Si-Fe steels become grain oriented in the rolling direction, which in use (such as in high-quality transformers) becomes the principal direction for the applied magnetic field.

9.7 RELATIONSHIP OF DOMAIN ROTATION AND DOMAIN WALL MOTION TO THE HYSTERESIS CURVE

The concepts introduced in the preceding two sections can now be used to explain the characteristics of the hysteresis curve in terms of the motion of domain walls and rotation of domains. When a magnetic field is applied to a specimen containing domains, the minimizing balance of the four fundamental energies is upset. Because the magnetization vector of many domains is not initially parallel to the applied field, the magnetostatic energy of many domains is raised. The tendency of an applied field to raise the magnetostatic energy was explained in the discussion following (9.105). To

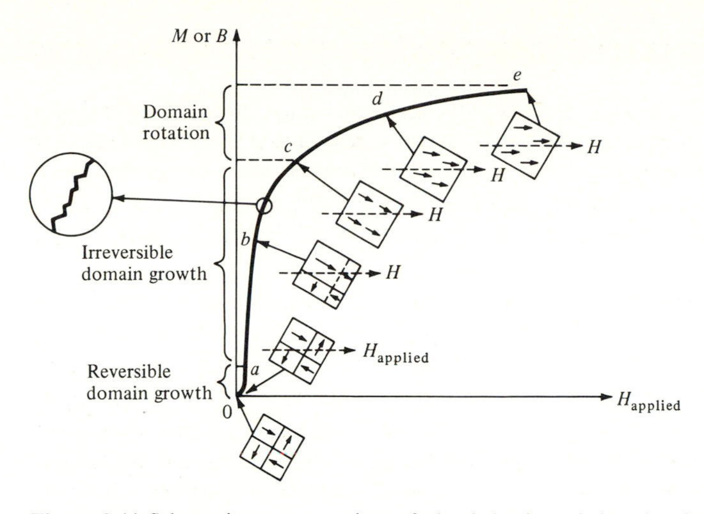

Figure 9.44 Schematic representation of the behavior of domains in a ferromagnetic material during the magnetization process. The enlarged section of path shows the Barkhausen effect.

lower the total energy, the magnetization vectors of the domains tend to align with the applied field. This in turn raises the external magnetostatic energy, lowers the exchange energy, etc. Clearly, in the magnetization of the ferromagnetic specimen by the external field, all energies readjust to minimize the total energy. With the aid of Fig. 9.44, let us follow through how the domains become aligned and the specimen as a whole functions as one big domain.

Let us begin with a specimen of ordinary iron which has a polycrystalline structure. As shown in Fig. 9.27, each crystal grain has a crystal axis along which the magnetization points.† Assume the specimen is initially unmagnetized and apply current I to a magnetizing coil (Fig. 9.29, for example) in which the specimen is inserted. As the magnetic field strength $H = NI/l$ rises, the magnetization M in the specimen increases, and with it the magnetic flux density B, since B is given by

$$B = \mu_0(H + M) \tag{9.107}$$

Of course, M increases so much faster than the applied H, that for practical purposes the above expression can be written as $B \cong \mu_0 M$. For H, corresponding to the lowest part of the magnetization curve, the domains do not rotate into alignment at all, but the more favorably oriented domains increase in volume as the Bloch wall expands into domains which are not favorably oriented. Recall that the energy of a Bloch wall is low; therefore, even a small applied field can readily move it. To rotate a domain, on the other hand, the magnetic moments would have to swing through unfavorable orientations which exist between the easy axes of magnetization. Rotation, therefore, must overcome the anisotropic energy which is usually much higher than wall energy.‡ The growth of domains with increasing H, corresponding to a on the magnetization curve (Fig. 9.44), is reversible. This is because the domain walls have not moved over appreciable distances to encounter many crystal imperfections or inhomogeneities, especially if we are talking about a soft material. If we remove H, the domain walls will move back to their original position, as shown by the domain picture of 0, which corresponds to the demagnetized condition in the specimen.

† For ordinary iron this figure should be modified since each crystal grain is large in the sense that it contains many domains of the type shown in Fig. 9.37*d*.

‡ Note that a hard material, which makes a good permanent magnet, should have large anisotropy and be full of obstacles to give a fat hysteresis loop, whereas a good soft material should have few imperfections and vanishing anisotropy which results in a narrow hysteresis loop.

Increasing the H field further, the moving Bloch walls begin to encounter various microscopic obstructions, such as inhomogeneities, imperfections, impurities, nicks, and holes. The walls tend to stick to these and, with increasing fields, break loose and shoot forward until other impedances are met. Such motion of the walls is irreversible and is accompanied by sudden discontinuous changes in magnetization called *Barkhausen jumps*, which are shown in the enlarged section of the magnetization path.† In this region some domains will also begin to swing into more favorable directions since the energy corresponding to the increased H is now sufficient to perform such rotations. These also contribute to Barkhausen jumps. The domain picture corresponding to this region of the curve is that shown at point b in Fig. 9.44. This process will be completed in a relatively low H field, roughly at point c of Fig. 9.44, which is the knee of the magnetization curve. At point c, most of the domains have rotated into a preferred direction closest to that of the applied H field. They are all aligned; the specimen acts like one large domain, except that the magnetization $\mathbf{M}$ is not in the direction of the applied H. If the specimen had been a large crystal with the applied field along the 001 crystal axis (see Fig. 9.39), the magnetization process would now be complete; i.e., the magnetization curve (Fig. 9.44) would be flat after point c because the crystal had reached saturation. However, in a polycrystalline material, the magnetization curve is a combination of the separate curves of Fig. 9.39 for the many single crystals of which it is composed. In Fig. 9.44 the composite curve between points c and e shows in what manner magnetization for a polycrystalline material reaches saturation. Above c, most boundary movement, smooth or jerky, has been completed, and the domains must now, by sheer application of magnetizing force H, be gradually rotated into the direction H. This process is reversible and is completed at point e, when the specimen is saturated; i.e., it has the same value of magnetization as the individual domains.

Should we now reduce the applied field, the magnetization curve will be retraced up to point c or d, but then the curve will follow the hysteresis loop for that material. At this point, reverse domains will begin to form in the single large domain. The nucleation mechanisms for the reverse domains are probably the imperfections, dislocations, impurities, etc., which exist in abundance throughout the microscopic structure of any material.

In view of the above discussion, we conclude that a complete hysteresis loop for a soft material, such as polycrystalline iron, has two characteristic parts. As shown in Fig. 9.45a, a rectangular area is primarily attributable to irreversible domain wall motion which brings the magnetization to about two-thirds saturation, and a triangular area represents reversible domain rotation and brings it to saturation. The distinctness of these two parts becomes clearer when we consider the magnetization curve, Fig. 9.45b, of a single crystal of iron when the direction of the applied H field is not parallel to an axis of easy magnetization in the crystal. In this figure, the finite area of the rectangular part represents loss due to irreversibility of wall motion; the curve which takes us to saturation has no area, hence, no loss, and therefore represents the reversible process of domain rotation. The relationship between Fig. 9.45a and b is that a polycrystalline material, such as ordinary iron, can be considered to be an assembly of many crystal grains which, in general, do not have their axes along the direction of the applied field.

Why is there an energy loss in changes of magnetization well below saturation? It is simply because energy is taken from the applied field in nucleating (creating) domain walls and moving them past obstructions such as inhomogeneities, impurities, etc. The area of the rectangle represents the energy lost in creating and moving walls past obstructions, which shows up as heat in the material and is not recoverable.

Why is there no energy loss in bringing the specimen from a point on top of the rectangle to saturation, and why does it require such large applied fields to do it? The energy supplied by the field in rotating the domains is used mostly in overcoming the crystal's anisotropy. This energy is returned to the field as the field is decreased, just as a compressed spring returns its energy when released (in our polycrystalline material, some wall motion is still possible in the high-field region which accounts for the broadening of the tips so that a smooth loop results). The anisotropic energy is much higher than

† Barkhausen jumps can be madc audible by connecting a pick-up coil, amplifier, and speaker to a specimen undergoing magnetization. A series of clicks will be heard as the applied H is varied.

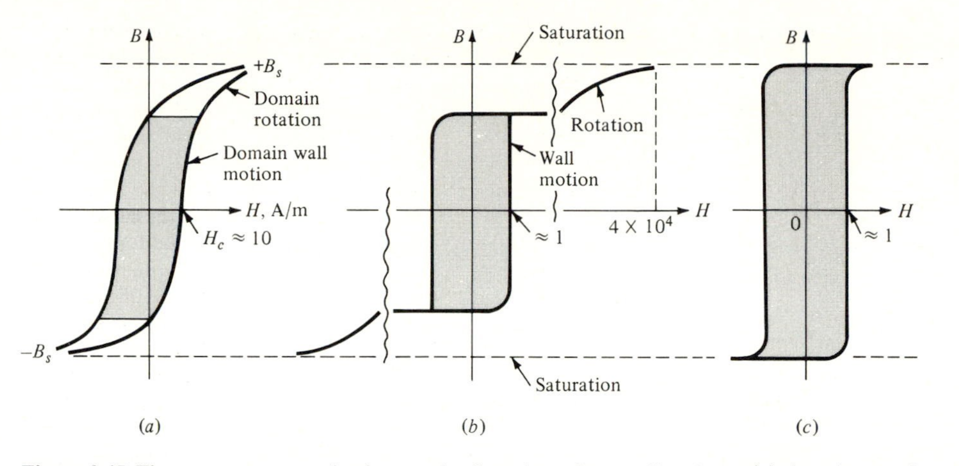

Figure 9.45 The separate magnetization mechanisms in polycrystalline iron: (*a*) domain rotation in the triangular areas and domain wall motion in the rectangle; (*b*) a pure crystal of iron with its axis of easy magnetization not parallel to the applied field, and the resulting hysteresis loop; (*c*) a pure crystal of iron with its axis parallel to the applied field can be magnetized to saturation by domain wall motion alone.

the energy to move a Bloch wall. It takes a field which is up to 50,000 times stronger than that required for Bloch-wall motion to overcome the iron crystal's anisotropy, i.e., to rotate the domain into final alignment with the applied field.

For comparison, the magnetization curve for a crystal of iron in which the crystal axis is oriented to be parallel to the applied field is shown in Fig. 9.45*c*. Notice that this loop, which results from wall motion alone, is nearly rectangular right up to saturation. A similar behavior and a similar loop to that shown in Fig. 9.45*c* is found for Permalloy, which is a soft material containing about 70 percent nickel and 30 percent iron. The opposing magnetostrictive constants for nickel and iron yield an alloy with hardly any anisotropy or magnetostriction. In Permalloy, the domains rotate easily, and it can attain saturation in relatively weak fields. Additional research produced Supermalloy, a material with zero magnetostriction and anisotropy. This material becomes magnetized in almost negligible fields.

9.8 FERRITES†

The magnetism of ferrites is called *ferrimagnetism*. It differs from ferromagnetism in that the spins of adjacent atoms are opposite. If the adjacent spins are equal and opposite, as, for example, in chromium, the material as a whole will not show a net magnetization nor an external magnetic field. From Fig. 9.36 we can see that the exchange force for manganese and chromium orients adjacent spins into an antiparallel alignment. We call such materials *antiferromagnetic*. Figure 9.46 shows the orientation of electron spins in such a material. There is a third type of material, the ferrites, which shows a net magnetization with an antiparallel arrangement of adjacent spins, as suggested in Fig. 9.46. The net magnetization comes about because the magnetization of one spin direction is weaker than the opposite spin. As a direct consequence of this, the maximum magnetization of ferrites is substantially below that of the ferromagnetic materials, typically about 0.3 Wb/m^2 (3000 G) versus 2 Wb/m^2

† G. F. Dionne, A Review of Ferrites for Microwave Application, *Proc. IEEE*, pp. 777–789, May 1975.

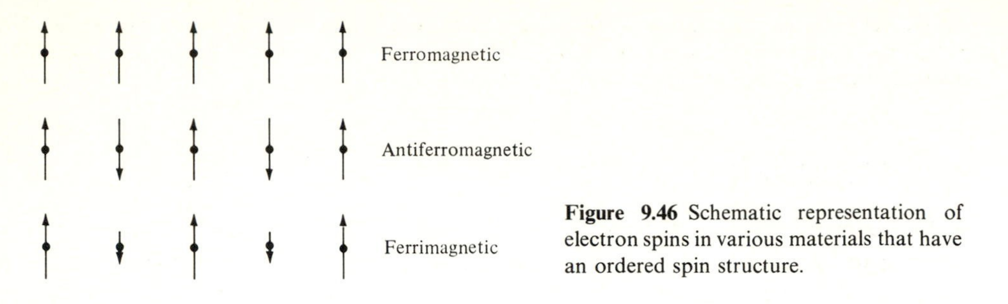

Figure 9.46 Schematic representation of electron spins in various materials that have an ordered spin structure.

(2×10^4 G) for ferromagnets. The low magnetization rules ferrites out for most generator, motor, and transformer applications. In these devices, the generated voltage, torque, etc., is proportional to magnetic flux density B; hence, a material with a large saturation magnetization which can be easily induced is needed.

Ferrites belong to the chemical group having the formula

$$(MO)(Fe_2O_3) \tag{9.108}$$

where M is a divalent metal such as iron (Fe), manganese (Mn), cobalt (Co), nickel (Ni), magnesium (Mg), zinc (Zn), cadmium (Cd), etc. Ferrites are prepared by mixing powdered iron oxide Fe_2O_3 and the metallic oxide MO, firing the mixture, which sinters it together into a spinel crystal structure. The spinel crystal consists of two crystal lattices interlaced. The magnetic atoms in one crystal lattice point opposite to those in the other. For example, if there are N_a atoms having moments m_a all oriented in the same direction in group A (called A sites), and N_b atoms of moment $-m_b$ in group B (the B sites), the resultant volume magnetization of the ferrimagnetic material is

$$M = N_a m_a - N_b m_b \tag{9.109}$$

Magnetite, which is the ancient lodestone, is a ferrite Fe_3O_4 whose composition can be rewritten in the form (9.108) as $(Fe^{++}O)(Fe_2^{++}{}^{+}O_3)$. The two triply ionized Fe atoms have oppositely oriented magnetic moments, which leaves Fe^{++} to account for the magnetization of the material. This is indicated schematically in Fig. 9.47. Thus, the oldest known magnet is not a ferromagnet at all, but a ferrite magnet. Other ferrites are formed by the substitution of other magnetic ions for the Fe^{++} in (9.108).

The unique characteristic that distinguishes ferrites from iron and other ferromagnetic materials is that ferrites with their bonded-alloy powder structure are insulators. Typical resistivities of ferrites range from 1 to $10^4\ \Omega \cdot$ m as compared with $10^{-7}\ \Omega \cdot$ m for iron. Such high resistivities make them immune to eddy currents, with a result that ferrites can be used at high frequencies as core material in audio and rf coils, in television "flyback" transformers, in memory cores of computers, etc. At microwave frequencies the devices that utilize ferrites do so because the microwave fields are able to get inside and propagate through the insulating material without undergoing large attenuation or reflection, whereas such fields would be kept out by the induced eddy currents in highly conducting magnetic materials such as iron. Why then are ferrites not used in power equipment that operates at lower frequencies and where painstaking efforts are made to avoid eddy-current losses. First, since only part of a ferrite is magnetic (the remaining ceramic that bonds the structure is nonmagnetic), the saturation field B_s is much lower than that of entirely ferromagnetic materials (see Table 9.2). Second, they are relatively expensive to manufacture. Third, their low mechanical strength and brittleness make them unsuitable in power equipment where mechanical strength is often a consideration. To avoid eddy-current losses in power equipment, one first considers the silicon steels which have increased resistivity.

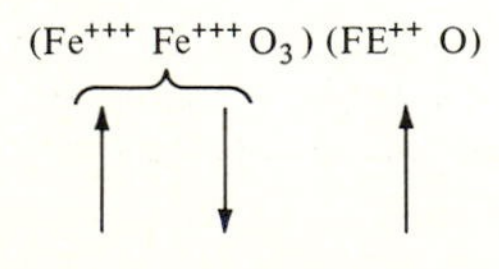

Figure 9.47 Schematic representation of the spins that contribute to magnetization in magnetite.

If this is not sufficient, then the core material must be laminated. Laminating effectively breaks up the path of the eddy currents that are induced by the time-varying magnetic flux in the cores of power equipment. Eddy currents are discussed in detail in Sec. 10.2.

9.9 SUPERCONDUCTIVITY†

A discussion of superconductivity fits particularly well at the end of a chapter on magnetism because (1) superconductors are an example of perfect diamagnetic materials that expel all magnetic fields from their interior, and (2) their zero resistivity permits a current I to circulate practically forever in a loop of superconducting material, such as the one shown in Fig. 9.48. The latter property provides us with a permanent magnet in the form of a macroscopic loop which is like a greatly enlarged version of one of the microscopic circulating currents in an ordinary bar magnet. Such a superconducting magnet sheds further light on the two views of magnetism represented in Fig. 9.7 and gives the equivalent-current formulism even more appeal.

The two independent properties mentioned above characterize the superconducting state of matter. One is the abrupt disappearance of resistance in certain metals at temperatures near absolute zero, which was discovered by H. K. Onnes in 1911. The other is the tendency of a superconductor to expel a magnetic field from its interior, which was discovered by W. Meissner in 1933 and is referred to

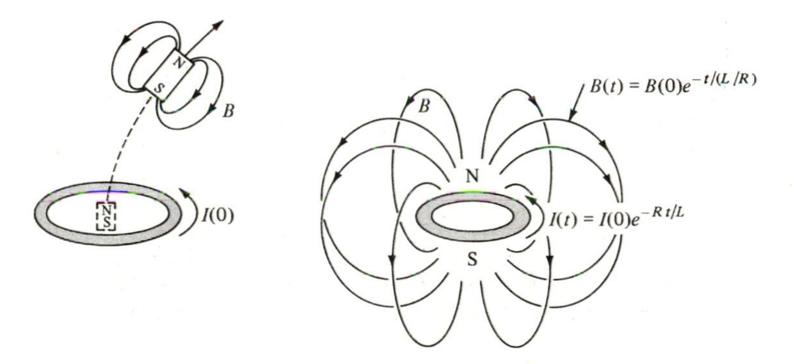

Figure 9.48 A current induced in a wire loop by withdrawal of a permanent magnet from within the loop will decay in fractions of a second. The time constant of decay is $\tau = L/R$ as given by (7.44), where L is the inductance of the single-turn coil (7.30) and R is the resistance of the metal wire. If the loop is superconducting, the current will persist almost indefinitely because $\tau \to \infty$ as $R \to 0$. No measurable decay has been detected after several years of current flow in such coils, which permits us to put an upper limit of 10^{-25} $\Omega \cdot$ m on the resistivity of superconductors; that is, $\rho < 10^{-25}$ $\Omega \cdot$ m.

† D. Fishlock, "A Guide to Superconductivity," MacDonald & Co., Publishers, Ltd., London, 1969. E. A. Lynton, "Superconductivity," Methuen & Co., Ltd., London, 1969. D. Greig, "Electrons in Metals & Semiconductors," McGraw-Hill Book Company, New York, 1969. Special Issue on Applications of Superconductivity, *Proc. IEEE*, vol. 61, January 1973.

as the Meissner effect. These are independent in the sense that knowledge of one property does not lead one to anticipate the other basic property.

The superconducting state will be destroyed by a rise in temperature, a rise in the applied magnetic field, or a rise in the current being conducted. Superconductivity is, therefore, confined to the interior of an onion-shaped region of temperature-field-current space, as shown in Fig. 9.49. Early superconductors, now known as type I, would generate only about 0.01 Wb/m^2 (100 G) to perhaps 0.2 Wb/m^2 (2000 G). These values are typical of fields produced by permanent magnets, and they would carry no more current per unit area than normal conductors. Beginning in the 1950s, the situation was dramatically changed with the discovery of a new class of superconductors, known as type II, that remained superconducting while carrying substantial amounts of current (up to a million amperes per square centimeter) and remained superconducting to 18 K and to 20 Wb/m^2 (200 kG). Let us now discuss the theory behind superconductivity.

In Secs. 2.1 and 2.5, we introduced the free-electron gas model, first proposed by K. L. Drude in 1900, which was used with good results to explain conduction in

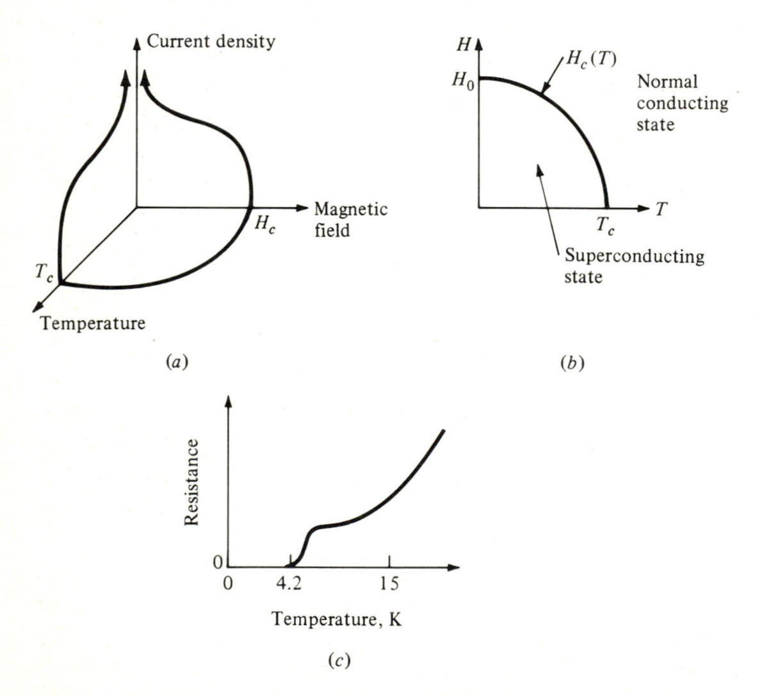

Figure 9.49 (*a*) A substance remains superconducting if temperature, magnetic field, and current density are within the onion-shaped region. Outside this region, superconductivity ceases. Temperature and magnetic field have critical values, but current density can increase if temperature and magnetic field decrease to zero asymptotically. (*b*) Phase diagram for a type I superconductor. When H and T fall in the region labeled "normal," all electrons are unpaired. In the superconducting state, below the $H_c(T)$ curve, some are paired and some are unpaired. (*c*) Resistance of a sample of mercury as a function of temperature, showing that resistance vanishes entirely at about 4.2 K.

metals. We observed that current is transmitted by the motion of electrons which are driven through the metal's crystal lattice by the applied voltage. The cloud of free or unattached electrons, which exists in all good conductors, fills the open latticework of the crystal and collides with the atoms in the lattice as the electron gas slowly drifts in the direction of the applied field. This impedance to the motion constitutes the conductor's electric resistance. The resistance increases as the temperature increases because, as shown in Fig. 2.1, the atoms vibrate quicker and over wider distances from their lattice positions and therefore interfere with the electron's motion more strongly. On the other hand, as pointed out in Sec. 2.7, a reduction of temperature to 0 K would still leave the "zero-point vibration" in the lattice atoms, which prevents the resistance of a good conductor, such as copper, from becoming zero (a residual resistivity of $0.02 \times 10^{-8}\ \Omega \cdot m$ remains).

It is ironic that the lattice vibrations which are the cause of resistance in ordinary conductors also account for the zero resistance of superconductors. At low temperatures in superconductors an ordering effect between the lattice atoms and electrons takes place which permits the electrons to move through the lattice without resistance. The ordering effect is a synchronization between the vibrations of the lattice atoms and the motion of paired electrons (of opposite spin and momentum), first described by Bardeen, Cooper, and Schrieffer in 1957, and now simply known as the BCS theory of superconductivity. This synchronization is very delicate and can exist in the relative "quietness" near absolute-zero temperatures. As the temperature rises, the thermal energy quickly breaks the bonds between the paired electrons, the coordinated lattice vibration and electron motion stops, and the superconductor reverts to being an ordinary conductor. Because of this, there is some reason to believe that superconductivity cannot exist at temperatures much higher than 20 K. In view of the fact that it is the lattice vibrations which make a metal superconducting, it is not surprising to find that poor conductors at normal temperatures are most apt to be superconductors at low temperatures. The strong scattering effect due to large lattice vibrations of these metals which make them poor conductors at room temperature causes a strong ordering between the lattice and electron motions at the low temperatures. Let us discuss how the vibrating atoms in a crystal lattice no longer obstruct the flow of electrons but, on the contrary, begin abruptly to conduct this flow at low temperatures.

In an ordinary metal, when a net current is absent, the electron velocities are randomly distributed. For any direction there are as many electrons moving to the left as to the right, giving a symmetric distribution of electrons in which the average velocity of the electrons in any one direction is zero, and the net current is zero. When a voltage is applied to the metal, the resultant electric field forces the electrons to move in one direction; consequently a current starts to flow. To the random velocity of each electron must be added the component of this drift velocity, resulting in an asymmetric distribution of electrons. This asymmetric distribution has a higher energy than the original symmetric distribution. Hence, when the applied voltage is removed, the asymmetric distribution decays to the symmetric one as the higher-energy electrons lose their additional drift-velocity energy by collision with the lattice, and current ceases to flow. Clearly, in a superconductor, once a current is established in the material, the asymmetric electron distribution must persist. It does so by an attractive interaction of electrons that binds them together in pairs, commonly called *Cooper pairs*. This effect is represented in Fig. 9.50. Two negatively charged electrons are connected by a spring which allows them to oscillate with respect to each other. The spring represents the lattice and permits an exchange of momentum between the lattice and the pair.

The pair coupling comes about as follows: As an electron passes near a positive lattice ion, the ion is attracted momentarily to the negatively charged electron as it passes. The distortion of the lattice is shown in Fig. 9.51. It takes the form of a "puckering up" in the vicinity of the electron. A second electron is attracted to the excess positive charge created by the higher density of the ions in the puckered region of the lattice. It is thereby indirectly attracted to the first electron and can be considered to form a pair with the first electron. The ions, being heavier, move more slowly than the electrons, and by the time they respond to the sharp pull of a passing electron and move slightly together, the electron will have traveled a considerable distance. Therefore, for a second electron, it is energetically most favorable to take advantage of the accumulation of the positive charge at a large distance behind the first electron. This explains the reason for the large coherence length in the Cooper-paired electrons. Some quantitative values are: The electrons move about 100 times faster than the ions, the coherence length is about 10^{-7} m [1000 Angstroms (Å)], and the separation between ions in the lattice (lattice constant) is about 10^{-10} m (1 Å). The large separation between Cooper-paired electrons remove, in a sense, the objection to pairing of charges of the same sign.

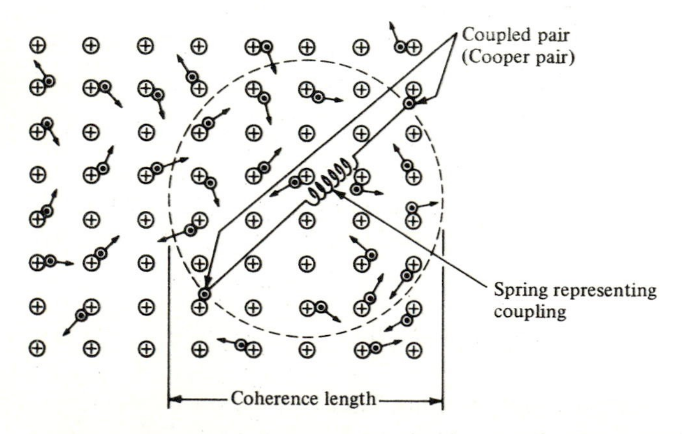

Figure 9.50 The formation of a Cooper pair at a temperature below the critical T_c. The distance between the paired electrons, also known as the *coherence length*, is indicated by the diameter of the dashed circle. The pluses represent the ions of the lattice, whereas the electrons and their velocities are represented by the dots with arrows.

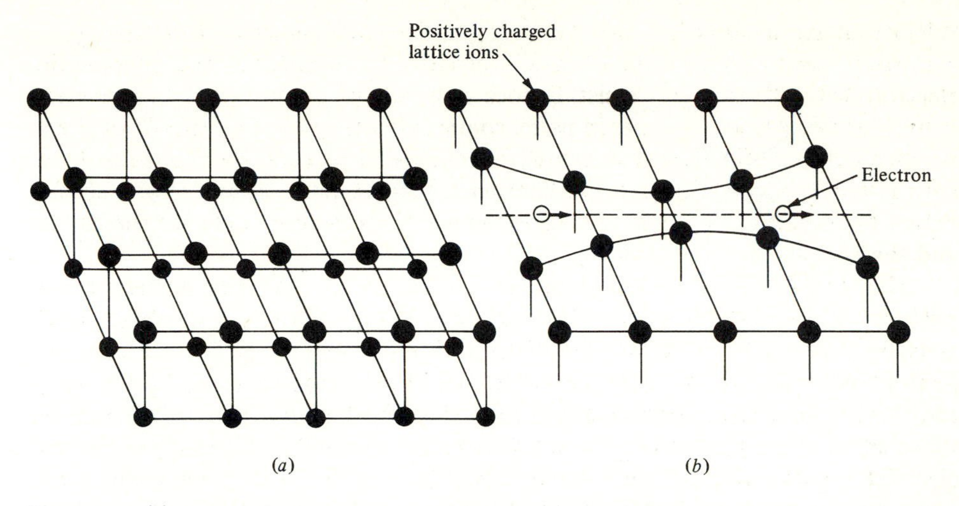

Figure 9.51 (*a*) A section of an undistorted lattice; (*b*) the elasticity of the lattice allows the positively charged ions some freedom of motion about their equilibrium positions. A moving electron causes the ions to move closer to each other, presenting a region of increased positive charge which will attract a second electron. This effect is shown for a single lattice plane.

The involvement of the ionic lattice was suspected when it was discovered that the transition temperature T_c is inversely proportional to the square root of the mass of the atoms in superconducting materials; that is, it was found that

$$m^{1/2} T_c = \text{constant} \tag{9.110}$$

Since a lattice of ions has vibrational frequencies that also depend on the square root of the mass of the ions [see Eq. (9.73)], it suggested involvement of the lattice in superconductivity. The lighter isotopes of an element would become superconducting sooner than the heavier ones, because the lighter ones, being livelier, interact more strongly with the electrons. This proved to be true, and Eq. (9.110) played a crucial role in the development of the BCS theory.

We can regard the Cooper-paired electrons as a new particle with twice the charge and mass of an electron. When a material is sufficiently cooled and goes into the ordered superconducting state, normal electrons condense into the paired quasi-particles. The concentration of the normal and paired electrons, given by N_n and N_s ($N_s/2$ pairs), respectively, is given by

$$N_n = N_0 \left(\frac{T}{T_c}\right)^4 \qquad N_s = N_0 \left(1 - \frac{T}{T_c}\right)^4 \qquad T \leq T_c \tag{9.111}$$

where N_0 is the concentration of valence electrons. Therefore, as the temperature drops, the density of normal electrons N_n drops and the density of the paired electrons N_s increases, as shown in Fig. 9.52, until at zero temperature, all the electrons have formed Cooper pairs. The motion of all Cooper pairs is the same. Either they are at rest or, if the superconductor carries a current, they move with

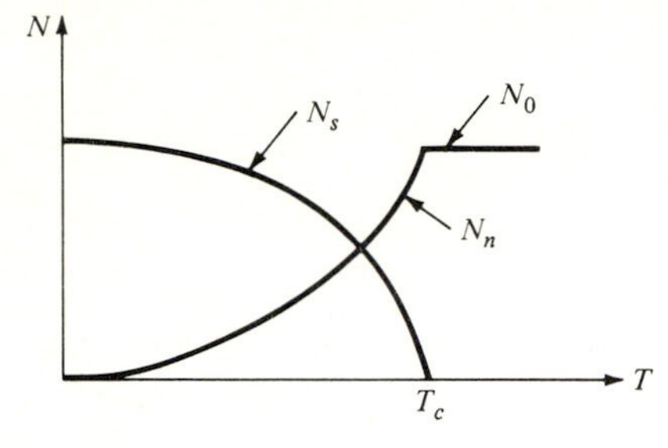

Figure 9.52 Concentration of paired electrons N_s and unpaired electrons N_n as a function of temperature T. T_c is the transition temperature above which no electrons exist in the paired state.

identical velocity in the direction of the current. Since the density of Cooper pairs is quite high, even large currents require only a small velocity. The small velocity of the Cooper pairs, combined with the precise ordering in which an individual pair has the same velocity as the entire pair assembly, minimizes collision processes. The extremely rare collisions of the Cooper pairs with the lattice allow the electric resistance of the superconductor to vanish. A process that leads to infinite conductivity when collisions are ignored was already discussed in Eqs. (2.19), (2.20), and (2.22). It involves letting the mean free time between collisions, τ, to go to infinity. To repeat, the equation of motion for an electron of charge e and mass m moving with velocity v in an electric force field E is given by

$$m\frac{dv}{dt} = eE \tag{9.112}$$

Since current density J is the product of charge density eN and average velocity v, that is, $J = eNv$, the supercurrent will obey the equation

$$\frac{dJ_s}{dt} = \frac{e^2 N_s}{m} E \tag{9.113}$$

This predicts an infinite current if the electric field E is maintained, or more to the point, the current can persist in the absence of an electric field. A high concentration of pairs needed to maintain a large current is possible since the large space between a pair is filled with electrons of other pairs.

Conduction in a superconductor can now be pictured as a long row of closely spaced electrons, each correlated with the lattice and another electron far down the row, moving down a corridor of positive lattice ions. This is possible if the centers of mass of the pairs all have the same momentum. Each electron sets up a vibration in the ions since it is elastically bound to the lattice. Vibrations of this nature are very weak and are completely swamped by the uncorrelated thermal lattice vibrations at higher temperatures. But, it is these vibrations when correlated with the moving electron "train" that account for superconductivity. Also, such correlation lowers the energy, permitting the asymmetric electron distribution to exist without an applied voltage, as was discussed on page 377.

The fact that just below T_c only a fraction of the electrons condense into the superconducting state, leaves the remaining normal electrons to participate in the conduction process in the usual way, which is by collision with the lattice and

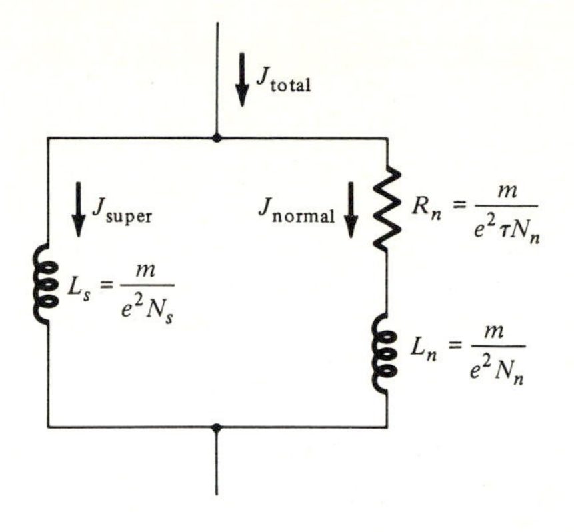

Figure 9.53 Equivalent circuit for the impedance of a superconductor. The two branches represent the flow of current due to normal and paired electrons. Mass m, electron charge e, scattering time for normal electrons τ, and particle concentration enter as shown in the equivalent circuit.

the resulting generation of heat. If sufficient heat is generated, the superconducting state will cease. The equivalent circuit, representing the two contributions to conduction current, is shown in Fig. 9.53. Since the impedance of an inductance L is given by ωL, where ω is related to frequency f by $\omega = 2\pi f$, we see that only direct current ($\omega = 0$) will experience no resistance, because for direct current the inductances L_s and L_n act like short circuits. For direct current the parallel branch carrying J_{normal} plays no part, and all current is transported by the paired electrons in J_{super} without dissipation and zero potential drop across the superconductor. When sinusoidal fields ($\omega \neq 0$) are applied, a voltage across L_s will be developed. The sinusoidal electric field associated with this voltage will accelerate the normal electrons which, in turn, will give rise to dissipation. Hence, for ac operation some loss will occur as a potential drop develops across the superconductor. The loss is appreciable only at very high frequencies since it is possible to make cavities with superconducting walls that have extremely high Q even at microwave frequencies. However, at infrared frequencies the losses approach that of a normal conductor. This suggests that it is advantageous to operate well below T_c, because N_n increases rapidly as T_c is approached (as shown in Fig. 9.52).

Type I Superconductors

As pointed out at the beginning of this section, we can divide superconducting materials into two categories according to their magnetic behavior. All type I superconductors exclude applied magnetic fields up to a certain critical value H_c. Figure 9.54 shows this effect. For fields greater than H_c, the material makes an abrupt transition to the normal state, as shown in Fig. 9.49*b*.

In order for a material to exclude an applied magnetic field from its interior, it is necessary that a current in the material be induced which will produce in the interior an equal and opposite field to the applied field. The only place such a current can flow is in a thin layer on the surface of the material, because from Ampère's law (8.15)

$$\nabla \times \mathbf{B} = \mu_0 \mathbf{J} \tag{9.114}$$

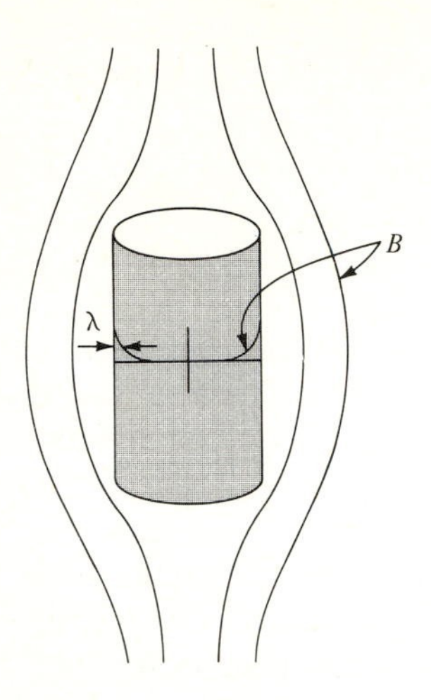

Figure 9.54 A superconducting rod excludes magnetic fields from the interior except for a thin layer at the surface. The penetration depth λ is the distance at which the field has fallen to $1/e$ of its value on the surface.

if a current is present inside the material, it must be accompanied by a spatially varying magnetic field. The existence of a surface current, however, implies [again using Eq. (9.114)] that the applied magnetic field can penetrate a small distance, called the *London penetration depth* λ, into the material. It can be shown that if the field on the surface is $B(0)$, the field will decay with distance into the material as given by

$$B(r) = B(0)e^{-r/\lambda} \tag{9.115}$$

and as shown in Fig. 9.54. A typical value for the magnetic penetration depth is $\lambda = 10^{-5}$ cm, which is on the order of the coherence length of the paired electrons. Surface current which accompanies the penetration of the magnetic field into the superconductor varies in a manner similar to that of the B field. The reason that type I superconductors are unable to carry much current is precisely that they do not allow a magnetic field to penetrate their interior.

The skin depth (at which the field is $1/e$ of its value on the surface) for ordinary conductors is given by (8.76) as

$$\delta = (\pi f \mu_0 \sigma)^{-1/2} \tag{9.116}$$

and is seen to depend on frequency. Hence, for direct current ($f = 0$) a static magnetic field completely penetrates the interior of a conductor. However, for a superconductor, the behavior is given entirely by the paired electrons, and the penetration depth λ is related to e^2N/m of (9.113) by $m/e^2N_s = \mu_0 \lambda^2$; that is,†

$$\delta = \lambda = \left(\frac{m}{\mu_0 N_s e^2}\right)^{1/2} \tag{9.117}$$

which is seen to be independent of frequency. Therefore, λ gives the penetration depth of dc or ac fields into a superconductor. The penetration depth is strongly

† E. A. Lynton, "Superconductivity," Methuen & Co., Ltd., London, 1969.

dependent on temperature and becomes very large as T_c is approached. Explicitly,

$$\left(\frac{\lambda_T}{\lambda_0}\right)^2 = \frac{1}{1-(T/T_c)^4} \tag{9.118}$$

where λ_T is the penetration depth at temperature T and λ_0 is the penetration depth at temperature $T = 0$.

To recap, we can view type I superconductors in terms of a two-fluid model. In this model, a fraction of the conduction electrons "condense" in an ordered state and form a superfluid which does not interact with the crystal lattice in the usual way, exhibits no resistance to flow (incapable of energy loss or transport of heat), and is responsible for the characteristic superconducting properties. The remaining fraction of conduction electrons which are in the normal state (ordinary conduction electrons in a metal) form the other fluid which is responsible for ac resistance and heat conduction. These two fluids are interpenetrating and noninteracting. Type I superconductors exclude applied magnetic fields up to a critical value H_c. When this value is exceeded, the superconducting state ceases and a transition to the normal state is made. The temperature dependence of H_c is shown in Fig. 9.49*b* and is given by

$$H_c = H_0\left[1-\left(\frac{T}{T_c}\right)^2\right] \tag{9.119}$$

where H_0 is the critical field at $T = 0$. The destruction of superconductivity by a magnetic field means that a superconducting wire of radius r cannot carry a current exceeding the critical current I_c, which from (7.19) can be given as

$$I_c = 2\pi r H_c \tag{9.120}$$

where I_c is the current which will just generate the critical field at the edge of the wire.

Examples of type I materials are the elements aluminum ($T_c = 1.2$ K), indium (3.4 K), tin (3.7 K), and lead (7.2 K). For reference note that the boiling point of liquid helium is 4.2 K.

Type II Superconductors

The disadvantage of type I materials is that H_c is relatively low (less than 1000 G) which limits their usefulness for carrying large currents. Type II superconductors, also known as high-field or hard superconductors, differ substantially from the ideal superconductors (type I). First, they have two critical magnetic field intensities, H_{c_1} and H_{c_2}. Below H_{c_1}, these materials act like ideal superconductors; that is, they expel all magnetic fields. Second, between H_{c_1} and H_{c_2}, they allow the magnetic field to penetrate while still remaining superconducting. Third, magnetic fields exceeding H_{c_2} destroy the superconductivity, and the material becomes an ordinary metal.

Clearly, the region of interest is between H_{c_1} and H_{c_2}, especially since H_{c_2} can be as high as 20 Wb/m^2 (200,000 G). The retention of superconductivity in such

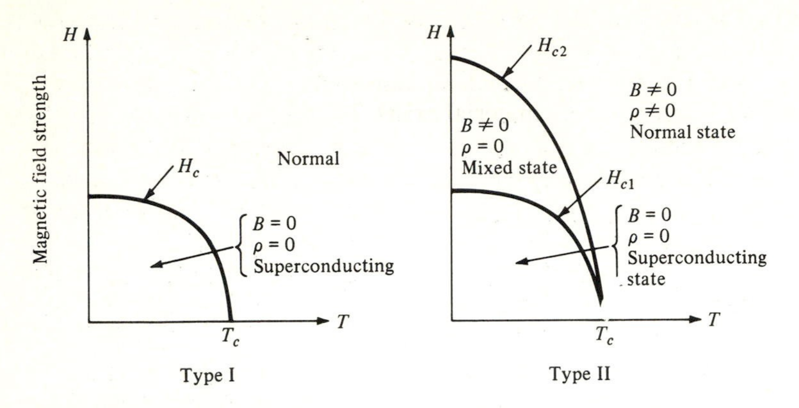

Figure 9.55 The HT plane for type I and II superconductors, showing regions where resistivity ρ and magnetic B field are zero.

high magnetic fields makes type II materials very attractive for use in magnetic coils. Figure 9.55 shows the difference between type I and type II materials. The dependence of H_{c_2} on temperature has a form similar to that for H_c. The region labeled "mixed state" puzzled many scientists for a long time, because it could not be understood how a material could remain superconducting ($\rho = 0$) while allowing penetration by a magnetic field ($B \neq 0$). The following theory was finally worked out: If a field H is applied to a sample of type II material (again use the shape in Fig. 9.54), it will stay in the perfectly diamagnetic state until the value H_{c_1} is reached. If the applied field is increased further, it becomes energetically more favorable to admit a single flux quantum, equal to $\phi_0 = 2.07 \times 10^{-15}$ Wb, rather than have the superconductor exclude the field H_{c_1}. Associated with the quantum of flux, which is in the direction of the applied field, is a circulating current which shields the flux bundle from the remaining superconducting material. The combination is called a *vortex* or a *fluxoid* and is shown in Fig. 9.56. In the center of the fluxoid, the superconducting material is in the ordinary, or normal, state of conduction. However, outside the fluxoid, which has a diameter of about two penetration depths, the material remains in the superconducting state. As the applied field is further increased, more individual quanta enter, thus increasing the internal flux density. When the flux density reaches the value corresponding to H_{c_2}, the entire material goes into the normal state because the flux lines of the

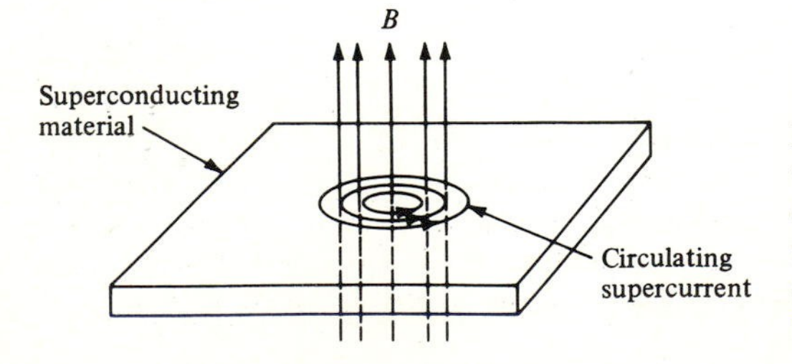

Figure 9.56 A single quantum of flux, ϕ_0, which has penetrated a type II superconducting material. Accompanying the B field is a circulating current which extends to about the distance of a penetration depth from the center of the fluxoid.

individual vortices begin to overlap and occupy the whole volume. We can now picture a type II material as threaded by many nonsuperconducting cylindrical regions, leaving a perforated material which remains superconducting so long as a continuous path of superconducting material exists. Thus, penetration of a B field in quantized steps allows a type II material to carry current throughout its interior, unlike type I materials for which the current is confined to the surface.

Incidentally, the phenomenon that trapped flux inside a type II superconductor can only be changed in small macroscopic steps which are whole-number multiples of a single *fluxoid quantum* ϕ_0, is an example of macroscopic quantum mechanics. If it were not for the extremely low temperatures, quantum states might have been discovered much before the 1920s.

Figure 9.57 shows a block of type II material with an applied H field and a current I which is normal to H passing through the block. The applied field penetrates the block, creating an array of vortices or fluxoids. To clarify this picture further, we can compare the induced-versus-applied magnetic field and the induced internal magnetization M versus the applied magnetic field for type I and II materials. This is shown in Fig. 9.58. Since the internally induced magnetic field B can be related to the applied H and induced magnetization M by

$$B = \mu_0(H + M) = \mu H \tag{9.121}$$

we can see from the figure that type I and type II materials remain perfectly diamagnetic ($\mu = 0$) until H_c or H_{c_1} is reached; that is,

$$H = -M \qquad \text{and} \qquad B = 0, \qquad \text{when } H < H_c \tag{9.122}$$

For type I, when H_c is exceeded, the magnetic field enters the material abruptly; that is,

$$-M = 0 \qquad \text{and} \qquad B = \mu_0 H \qquad \text{when } H > H_c \tag{9.123}$$

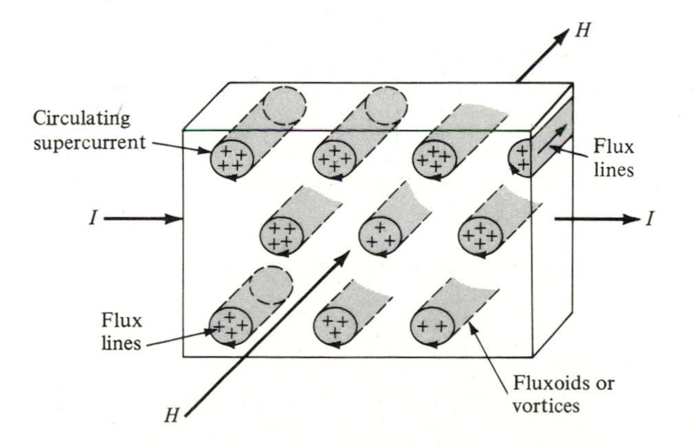

Figure 9.57 A block of type II superconducting material showing an array of fluxoids caused by the normally applied H field ($H > H_{c_1}$). A current I can pass freely, but it will give rise to a magnetic force, $\mathbf{I} \times \mathbf{H}$, which will tend to move the fluxoids in the vertical direction.

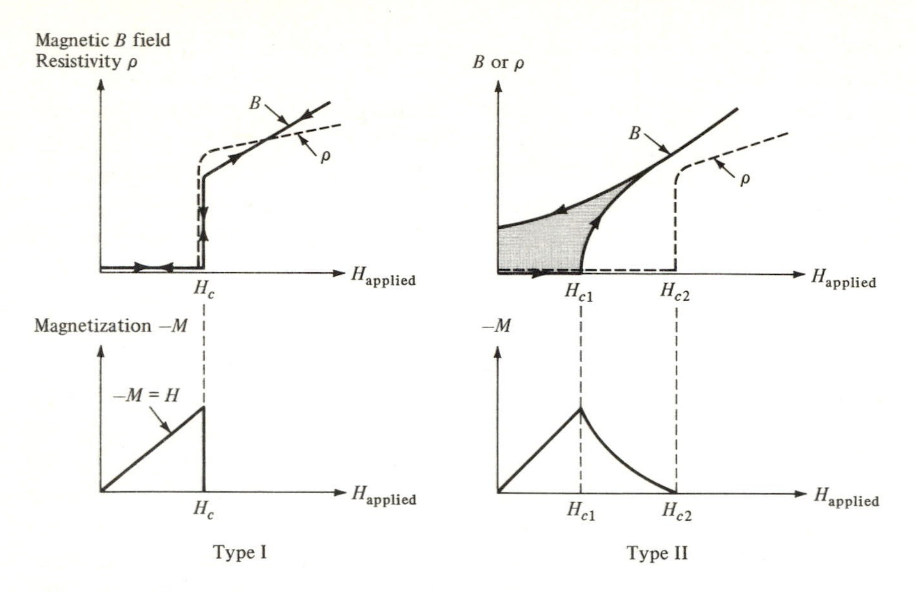

Figure 9.58 Comparison of induced B, induced M, and resistivity ρ for type I and II superconductors. The shaded area represents a quadrant of a hysteresis loop which is caused by trapped fluxoids in the material when H is reduced to zero.

and the resistivity changes to the normally conducting values. For type II materials, a gradual intrusion into the material is permitted for fields above H_{c_1}. Measurements with sufficiently high resolution would show that the magnetization curve or the induced magnetic field curve is not gradual but consists of small quantum steps, each equivalent to a multiple of the fluxoid quantum ϕ_0. The resistance graph shows that the material remains superconducting up to H_{c_2}.

Another difference between type I and II materials is apparent from the figures. On reducing the applied field, a hysteresis effect in type II materials occurs. We see that much flux remains frozen in when the external field strength is reduced to zero. The amount of frozen-in flux depends largely on imperfections and structural defects in the crystals of the superconducting materials.

Type II superconductors are not entirely without dc losses. First, there are two mechanisms that force the fluxoids to move. There is a repulsion between fluxoids because of the circulating supercurrent about each flux bundle. The reason for the repulsion is that two coplanar loops carrying current in the same sense repel each other (equivalent to repulsion between two bar magnets which lie side by side). From Fig. 9.57, we also see that the transport current I running through the slab will produce a magnetic force $\mathbf{J} \times \mathbf{B}$ on the flux bundles, tending to make them move in the vertical direction. According to Lorentz' law (6.17), a motion of flux lines in the vertical direction gives rise to an electric field in the horizontal direction (parallel to current I), which corresponds to ordinary resistance. Note that a motion of the flux lines parallel to current I would give rise to a Hall electric field in the vertical direction [see Eq. (6.85)]. Second, the motion of fluxoids is impeded by imperfections present in the superconducting material, which also gives rise to the aforementioned hysteresis effect. A motion of fluxoids past imperfections requires a transfer of energy from current I to the stationary metal. This leads to dissipation of power and results in heating of the metal which usually would raise the temperature past T_c, destroying the superconducting state. To avoid this, the fluxoids should remain stationary. This can be done by introducing imperfections which act to pin the fluxoids and prevent or restrict their movement. The dc resistance of the mixed state can be maintained virtually at zero; i.e., the superconductor can carry current without dissipating power, because of the pinning effect. However, for alternating current, the situation is different. The pinning effect restricts the motion of the fluxoids, which results in a hysteresis magnetization curve, shown in Fig. 9.58. The alternating field drives the material constantly around its cycle of magnetization with the result that defect-containing type II materials exhibit large hysteresis losses. If the fields produced by the applied current exceed H_{c_1}, hysteresis losses will appear which, at low frequencies (say, 0 to 5000 Hz), increase linearly with frequency. To avoid such losses, penetration of flux into the material must be avoided. This means that current I or magnetic field strength H must be less than H_{c_1} if a type II material is used, or less than H_c for a type I superconductor. It can therefore be said that type II materials which find dc application, such as in high-field magnets, will likely not be acceptable for ac applications because of their high ac losses when H_{c_1} is exceeded.

Examples of type II materials are the element niobium ($T_c = 9.3$ K) and the compounds niobium-tin (18.1 K), vanadium-gallium (16.8 K), niobium-zirconium (10.8 K), and the latest metal alloy consisting of niobium, aluminum, and germanium (20.6 K). All are well above the boiling point of liquid helium (4.2 K). Therefore ordinary liquid-helium cooling techniques can be used to lower the temperatures of these materials to T_c.

To summarize this section, we show in Fig. 9.59 the distribution of fluxoids in a strip of type II superconductor carrying a current I. The fluxoids are created by

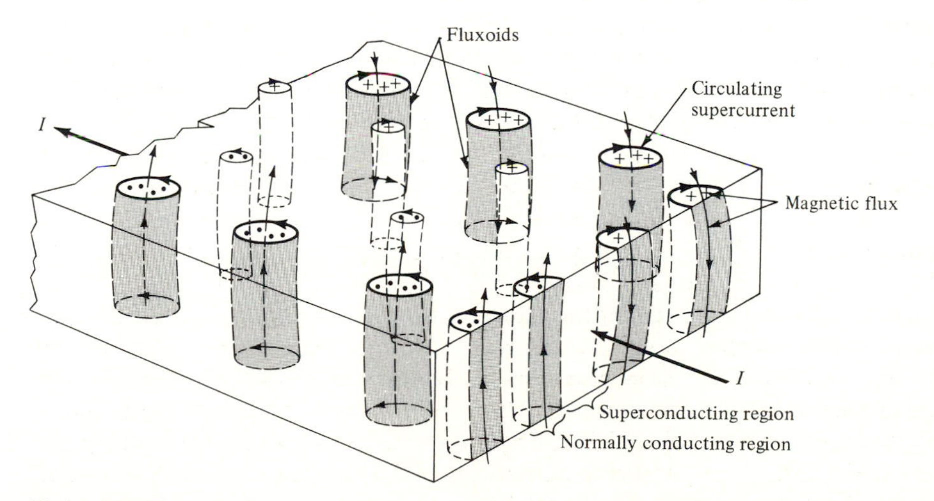

Figure 9.59 The creation of fluxoids in a strip of type II superconducting material by the self-magnetic field of current I.

the self-magnetic field of the current and are driven from the surface to the interior of the conductor. They begin at the surface because initially the current, and hence the magnetic field, is confined to a thin layer on the surface. With the penetration by fluxoids of the interior, current is allowed to flow there. The concentration of fluxoids should always be highest at the surface because, as shown in Fig. 7.4, the magnetic field inside a conductor carrying a uniformly distributed current is highest at the surface. If the current increases to a value such that the upper critical field H_{c_2} is exceeded, the number of fluxoids increases to such an extent that they begin to overlap. Hence the flux which exists in the normally conducting cores of the fluxoids also overlaps, and the whole interior becomes a normal conductor.

PROBLEMS

$$\varepsilon_0 = 8.85 \times 10^{-12}\ \text{F/m} \qquad \mu_0 = 4\pi \times 10^{-7}\ \text{H/m}$$

9.1 A solenoidal electromagnet (air filled) is 5 cm long and has a cross section of 1 cm^2. A current of 2 A flows through the winding which has 1000 turns. Find the magnetic pole strength in amperemeters of an equivalent uniformly magnetized cylindrical rod magnet of the same dimensions.

9.2 An iron bar, shown in Fig. 9.3, has a dipole moment of $1.8 \times 10^{-23}\ \text{A} \cdot \text{m}^2$ associated with each iron atom. If it is assumed that all the atomic dipole moments are aligned, and if the bar is 6 cm long and has a cross section of 1 cm^2, find

(*a*) The dipole moment of the iron bar.

(*b*) The torque which must be exerted to hold this magnet at right angles to an external field of 2 T. Note that $1\ \text{T} = 1\ \text{Wb/m}^2 = 10^4\ \text{G}$.

9.3 The volume of a uniformly magnetized bar magnet is 0.02 m^3. If the magnetic moment of the magnet is 700 $\text{A} \cdot \text{m}^2$ and the flux density **B** inside the bar is $3 \times 10^{-2}\ \text{Wb/m}^2$, find the magnetization **M** and magnetic intensity **H** inside the bar.

9.4 A cylindrical rod magnet has a length of 5 cm and a 1-cm diameter. If it has a uniform magnetization of 10^3 A/m,

(*a*) Find the magnetic moment of the rod.

(*b*) Find the equivalent current sheet at the surface of the rod.

9.5 If the magnetic dipole moment of the earth is $6.4 \times 10^{21}\ \text{A} \cdot \text{m}^2$, find the current that would have to flow in a single turn of wire going around the earth at the equator that would produce the same magnetic field at distances far from the earth surface. Earth radius is 6.4×10^3 km.

9.6 Magnetization of iron can contribute as much as 2 Wb/m^2 to the magnetic field in iron. If the magnetic moment of each electron is $0.9 \times 10^{-23}\ \text{A} \cdot \text{m}^2$, how many electrons per atom contribute to the magnetization?

9.7 If a *BH* curve is given by the analytical expression $B = \mu_0 H^2$, find **M** and μ.

9.8 A copper wire and an iron wire of length l carry a current I and are perpendicular to a uniform magnetic field H_0. Is the force on the iron wire equal, greater, or smaller than the force on the copper wire? Give reasons for your answer.

9.9 A sphere (radius a) is of "approximately linear" magnetic material for which $\mathbf{B} = \mu\mathbf{H}$ is immersed in a uniform magnetic field $\mathbf{B}_0 = \mu_0\mathbf{H}_0$. Referring to Fig. 9.12,

(*a*) Find the internal magnetic field $\mathbf{B}_i$ and the external magnetic field $\mathbf{B}_e$ of the sphere.

(*b*) For magnetic material with high permeability ($\mu_r \gg 1$), show that $\mathbf{B}_i \cong 3\mathbf{B}_0$ and $\mathbf{M} \cong 3\mathbf{H}_0$.

Soft ferromagnetic materials which have negligible permanent magnetization (that is, $B \cong 0$ when $H = 0$) can usually be considered to be "approximately linear" magnetic materials.

9.10 It was shown in Sec. 9.3 that a uniformly magnetized sphere has **B** and **H** fields which are also uniform inside. Calculate the equivalent current sheet corresponding to the uniform **M**. Is it possible to design a current winding which will produce a uniform magnetic field in a spherical region of space?

9.11 Consider a long, uniformly magnetized rod magnet of length l and cross-sectional area A. Show that the permeability μ at the center of the magnet is given by $-\mu_0 \pi l^2/2A$.

9.12 A long cylindrical iron rod has a permanent and uniform magnetization **M** which is parallel to the axis of the cylinder.

(*a*) Find **B** and **H** in the iron, neglecting end effects.

(*b*) If a long needle-shaped cavity which is parallel to the axis is cut into the interior of the cylinder, find **B** in the center of the cavity.

9.13 An iron sphere is immersed in a uniform magnetic field.

(*a*) Show that the internal field $\mathbf{B}_i$ is a uniform magnetic field and is given by

$$\mathbf{B}_i = \frac{3\mu_r}{2 + \mu_r} \mu_0 H_0$$

(*b*) If a small disc-shaped cavity which is normal to the internal field is cut into the sphere, find the magnetic field in the center of the cavity.

9.14 Verify that a small cavity inside a magnetic object screens out external magnetic fields. That is, verify that (9.43) gives the **H** field inside a small spherical cavity which is cut into a large magnetic sphere that in turn is immersed in a uniform external field H_0. What is the shielding ratio S, which is given by $S = H_0/H_{\text{cav}}$?

9.15 A small piece of equipment is to be shielded from a magnetic field of $B = 2$ T. If it can be enclosed by a thick spherical shell whose inside radius is much smaller than its outside radius, find the relative permeability of the shell material that is required to reduce the field to $B = 2$ mT.

9.16 A magnetic shield is formed by a spherical shell, shown in Fig. 9.13, which has internal and external radii of a and b, respectively. The magnetic material of the shell has a relative permeability μ_r. If the shell is immersed in an external uniform field B_0, show that the internal field is reduced to the value given by

$$B_3 = \frac{9\mu_r B_0}{(2\mu_r + 1)(\mu_r + 2) - 2(\mu_r - 1)^2(a/b)^3}$$

where B_3 is the magnetic field for $r < a$.

Hint: Use the scalar potential method of Sec. 4.11 or 9.3.

9.17 Compare the magnetic field expression for a spherical shell given in the previous problem to the more limited expression (9.43). Under what conditions are they the same?

9.18 The shielding ratio S is defined by $S = H_0/H_3$. Show that for a spherical ferromagnetic shield, where the permeability is large, S is given by

$$S \cong \frac{2\mu_r}{9}\left(1 - 2\frac{a^3}{b^3}\right)$$

Hint: Use the magnetic field given in Prob. 9.16.

9.19 An infinitely long cylinder of magnetic material (constant relative permeability μ_r) is immersed in an initially uniform magnetic field H_0 which is normal to the axis of the cylinder. Show that the internal $\mathbf{H}_i$ and external $\mathbf{H}_e$ fields are given by

$$H_i = H_0 \frac{2}{1 + \mu_r}(\cos\theta\, \hat{\mathbf{r}} - \sin\theta\, \hat{\boldsymbol{\theta}})$$

$$H_e = H_0 \left[\left(1 - \frac{a^2}{r^2}\frac{1 - \mu_r}{1 + \mu_r}\right)\cos\theta\, \hat{\mathbf{r}} - \left(1 + \frac{a^2}{r^2}\frac{1 - \mu_r}{1 + \mu_r}\right)\sin\theta\, \hat{\boldsymbol{\theta}}\right]$$

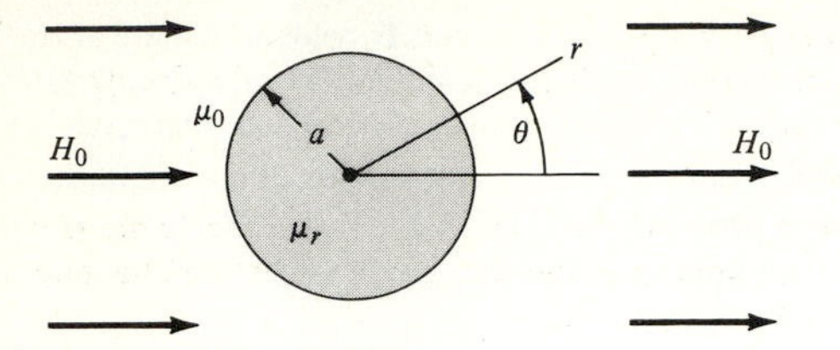

9.20 Using the results of the above problem, calculate the induced magnetization **M** in the cylinder.

9.21 The field of a magnetic shield in the form of a hollow cylinder of permeability μ_r which is placed initially in a uniform field $\mathbf{H}_0$ (at right angles to the axis of the cylinder) is shown in the figure. Show that the field inside the cylindrical cavity is uniform and is given by

$$\mathbf{H}_3 = \frac{4\mu_r H_0}{(\mu_r + 1)^2 - (a/b)^2(\mu_r - 1)^2}\,\hat{\mathbf{z}}$$

where $\hat{\mathbf{z}}$ is a unit vector in the z direction.

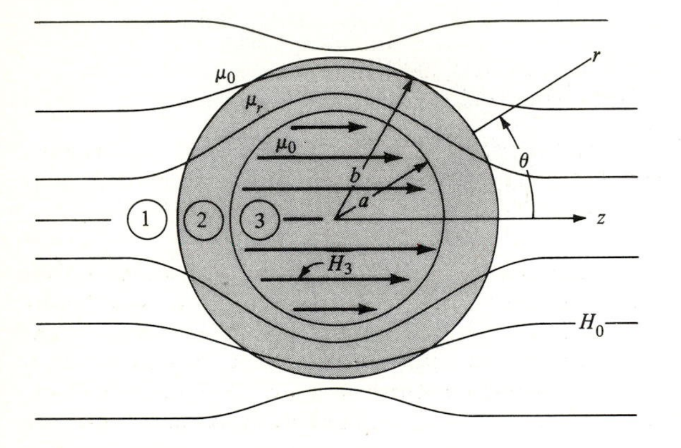

9.22 The shielding ratio S is defined by $S = H_0/H_3$. Show that for a cylindrical ferromagnetic shield of large permeability, S becomes

$$S = \frac{\mu_r}{4}\left(1 - \frac{a^2}{b^2}\right)$$

Use the expressions developed in the previous problem.

9.23 Compare the effectiveness of shielding by a cylindrical and a spherical ferromagnetic shield.

9.24 The expressions (9.68) and (9.69) give the **B** field at the center and ends of a cylindrical bar magnet. Show that these also give the **B** field at the center and ends of a solenoid which has the same outer dimensions as the bar magnet. See Sec. 6.9 for solenoid expressions.

9.25 Show that Eq. (9.53) gives the vector potential **A** for a magnetic dipole which is located at the point $\mathbf{r}'$ in a coordinate system.

9.26 Prove that the scalar potential method, outlined in Sec. 9.3, cannot be used to determine the **B** field inside a magnetized body. Can it be used outside a magnetized body to determine B?

9.27 Starting with the vector potential for magnetized objects, (9.52), show that it leads to (9.57), which is an expression that separates the contributions of the surface and the interior of the magnetized object.

9.28 Starting with the scalar potential for magnetized objects, (9.29), show that it reduces to the form given by (9.26).

9.29 The magnetic field from a magnetized object is given by (9.62) or by (9.27).

(*a*) For exterior fields, show that the **B** field obtained from (9.62) is identical to that from (9.27) after the H of that expression is first multiplied by μ_0.

(*b*) What is the relationship for interior fields?

9.30 A thin iron disc of radius a and thickness t has a permanent magnetization **M** parallel to the axis of the disc. Calculate the **H** and **B** fields everywhere on the axis of the disc. Make a sketch of the axial field like those of Figs. 6.25 or 9.17 which are for a 4 to 1 cylinder.

9.31 Determine the magnetic field on the axis of a rod magnet shown in Fig. 9.10, using the equivalent charge formulation. Assume it is permanently magnetized with M parallel to the axis of the rod. The length of the magnet is l, and its radius is a. As the answer to this problem is already obtained in (9.67) by a different method, state your answer to this problem in the form of (9.67). Use your result to check the plots of H and B shown in Fig. 9.17 for a 4 to 1 magnet.

9.32 Derive the expression (9.67) for the axial magnetic field from a rod magnet with a uniform magnetization M, as shown in Fig. 9.10, starting with Eq. (4.58a), which is an expression that gives the potential from two charged discs, each of radius a and separated by distance d. This problem demonstrates the attractiveness of the magnetic scalar potential method, since it reduces problems involving magnetized objects to problems in electrostatics.

9.33 The magnetic far field from a rod magnet shown in Fig. 9.10 reduces to a dipole field

$$\mathbf{H} = \frac{p_m \cos\theta}{2\pi r^3}\hat{\mathbf{r}} + \frac{p_m \sin\theta}{4\pi r^3}\hat{\boldsymbol{\theta}}$$

where the dipole moment p_m of the rod magnet is given by (9.31). By letting $\theta = 0°$, we obtain the magnetic far field along the axis (assumed to be coincident with the z axis) of the rod magnet as $H_z = p_m/2\pi z^3$. Starting with the exact expression (9.67) for a rod magnet, show that it reduces to $H_z = p_m/2\pi z^3$, where p_m is given by (9.31).

9.34 Derive Langevin's formula (9.89) for magnetization, $M = NmL(a)$, where $L(a) = \coth a - 1/a$ and $a = mB/kT$. When a field is applied, there is a torque $\mathbf{m} \times \mathbf{B}$ tending to align each molecule with the field. Thermal agitation prevents that alignment. For such a situation Maxwell-Boltzmann statistics apply, which give the probability P of finding the dipole axis within the angles θ and $\theta + d\theta$ as $P = e^{-W/kT}$, where W is the energy of the dipole in the external field ($W = -\mathbf{m} \cdot \mathbf{B} = -mB\cos\theta$), k is Boltzmann's constant (1.38×10^{-23} J/K), and T is the temperature in kelvins. If we take the direction of the applied B field to be parallel to the z axis (θ measured from z axis), then the average dipole moment $\langle m \rangle$ in the direction of the applied B field, in terms of which magnetization M will be expressed, is given by $\langle m \rangle = m\langle \cos\theta \rangle$. The average cosine $\langle \cos\theta \rangle$ can now be expressed in terms of the probability $P(\theta) = e^{mB\cos\theta/kT}$ as

$$\langle \cos\theta \rangle = \frac{\int_0^\pi \int_0^{2\pi} (P(\theta)\cos\theta)\sin\theta\, d\phi\, d\theta}{\int_0^\pi \int_0^{2\pi} P(\theta)\sin\theta\, d\phi\, d\theta} = L(a)$$

Calculate the fraction M/M_{max} of magnetization induced for a paramagnetic material whose atomic dipole moment is 2 Bohr magnetons, $T = 300$ K, and $B = 1$ Wb/m² (10^4 G).

9.35 For a superconducting metal, both **B** and **E** are zero inside.

(*a*) Find μ and χ_m for a superconductor.

(*b*) The boundary condition for the **E** field on a superconductor is that at the surface **E** must be normal, which means that the tangential component must be zero. Find the boundary condition for the magnetic field.

CHAPTER

TEN

APPLICATIONS OF MAGNETISM

10.1 ENERGY LOST IN A HYSTERESIS CYCLE

In Secs. 9.5 and 9.7, we stated that the area of the hysteresis loop represents energy dissipated as heat. Thus, if H is increased and decreased, for example, when an alternating field of frequency f is applied to a ferromagnetic core, the hysteresis power loss P_h in the core is given by

$$P_h = W_h f \qquad \text{W} \tag{10.1}$$

where W_h is the energy dissipated during each hysteresis cycle. Let us derive the expression for W_h.

In Sec. 7.9 it was shown that magnetic energy is distributed over all points in space at which a magnetic field exists. The total energy is therefore a summation of incremental energies dW which are contained in small volume cells dv surrounding these points; that is,

$$W = \iiint w \, dv \tag{10.2}$$

where w is the energy density. When the medium is linear (permeability μ a constant), the energy density at a point in a magnetic field was shown to be $w_M = \frac{1}{2}\mu H^2$, as given by (7.45). However, for a ferromagnetic material, μ is obtained from the hysteresis loop as the ratio $\mu = B/H$ and is not constant (see Fig. 9.31 which shows the BH curve for iron). We must, therefore, derive a more general expression for w_M.

Let us refer again to the toroidal coil shown in Fig. 9.8 or 9.29. If we increase current I by an amount dI, the H field will increase by an amount dH and the B

field by an amount dB. If this increase takes place in a time interval dt, then by Lenz' law [Sec. 6.11 or Eq. (11.1)] an electromotive force

$$\mathscr{V} = -N\frac{d\phi}{dt} \qquad \text{V} \tag{10.3}$$

will be induced in the winding, tending to oppose the increase in current. The induced emf is opposite to the emf of the generator causing the increase of current. To increase I, the generator must furnish energy of amount

$$\Delta W = -\mathscr{V}I\,dt = NI\,d\phi \tag{10.4}$$

If the cross section of the ring is small, we may say that the flux density B is uniform over the cross section A, and hence, $\phi = BA$ and $d\phi = A\,dB$. Furthermore, Ampère's law gives the H field for the toroid, (7.25), as $H = NI/2\pi r$. Substituting, we obtain

$$\Delta W = 2\pi r A H\,dB \tag{10.5}$$

where the volume of the toroid is $v = 2\pi r A$. Integrating from zero to B_{max} on the initial magnetization curve of Fig. 10.1, we obtain the work done by the generator. in increasing the flux density from zero to B_{max}; that is,

$$W = v\int_0^{B_{max}} H\,dB \tag{10.6}$$

The toroidal coil is now uniformly magnetized to B_{max}. The energy density at each point of the core can be given, similarly to (7.45), as

$$\boxed{w = \frac{W}{v} = \int_0^B H\,dB \qquad \text{J/m}^3} \tag{10.7}$$

This is a more general expression than (7.45), but it reduces to (7.45) for linear media; that is, when $B = \mu H$ and $dB = \mu\,dH$, w of (10.7) gives $w = \frac{1}{2}\mu H^2$. We can now state that the work done by external sources in establishing a magnetic field

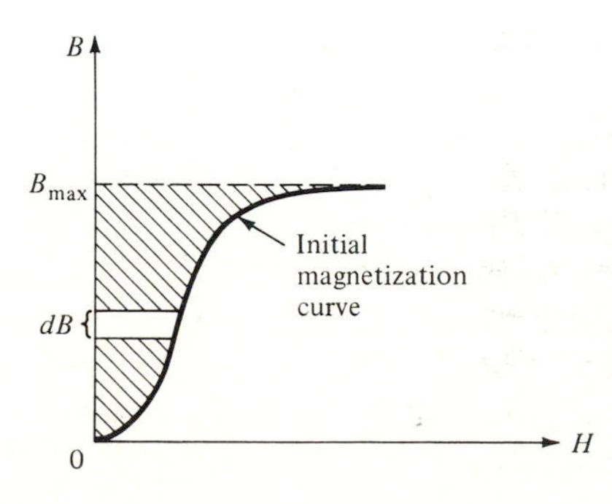

Figure 10.1 The striped area to the left of the initial magnetization curve represents the work done by an external source in establishing a magnetic field of B_{max}.

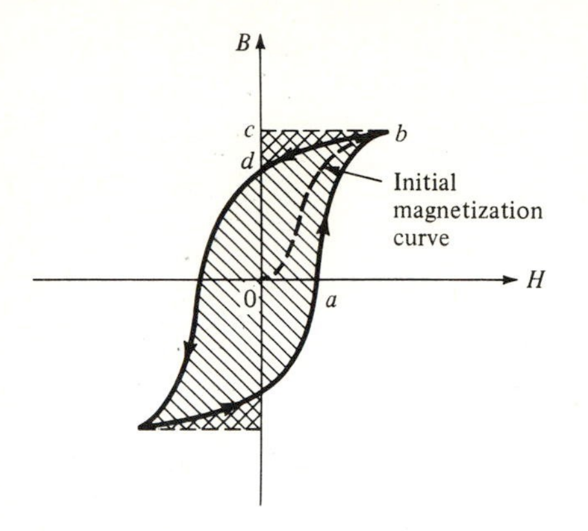

Figure 10.2 The hysteresis loop for a ferromagnetic specimen (the dashed curve is the initial magnetization curve). The striped area is the net work done by an external source. The cross-hatched area is the work returned by the specimen to the source.

can be given by

$$W = \iiint_v \left(\int_0^B H\, dB \right) dv \qquad \text{J} \tag{10.8}$$

where v is the volume throughout which the magnetic field exists. We may note that the integrand is $H\,dB$ and not $B\,dH$. The area in Fig. 10.1 representing the work done is therefore the striped area to the left of the magnetization curve.

Returning to hysteresis loss W_h, we see from Fig. 10.2 that the striped area, which is the area of the hysteresis loop, is equal to W_h. This comes about as follows: In order for the external source to increase the H field from a to a value corresponding to b, it must do work on the ferromagnetic specimen equivalent to area $0abc$. When the H field is reduced to zero, the ferromagnetic material will do work on the source equivalent to the cross-hatched area bcd, because H remains positive while dB changes sign. If we follow this through a complete cycle, the area of the hysteresis loop represents the net work done on the specimen per cycle by the external source.

Example As an illustration of the above, let us calculate the hysteresis loss for a transformer whose "soft" core material has the hysteresis loop shown in Fig. 9.28. The volume of the core is 100 cm^3, and the transformer is to be operated at 400 Hz.

Various ways can be used to obtain the integral (10.7). We can calculate the area by dividing the hysteresis loop into many small squares and counting them, or we can use a planimeter, etc. Since we are interested only in obtaining a figure for a typical transformer, let us approximate (10.7) by

$$w = (1.5 \text{ Wb/m}^2)(100 \text{ A/m}) = 150 \text{ Wb} \cdot \text{A/m}^3 = 150 \text{ J/m}^3$$
$$= 1.5 \times 10^{-4} \text{ J/cm}^3$$

Multiplying this figure by the volume and by the frequency, as in Eq. (10.1), we obtain for the hysteresis loss

$$P_h = (1.5 \times 10^{-4})(100)(400) \text{ J/s} = 6 \text{ W}$$

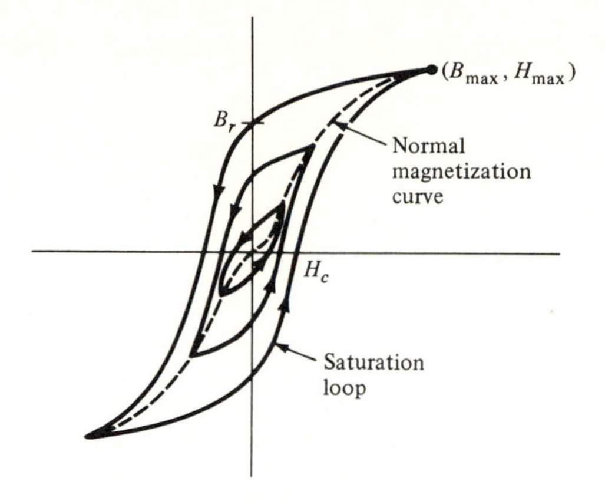

Figure 10.3 Hysteresis loops for low, intermediate, and high B_{max}. Values of retentivity B_r and coercive force H_c are stated in terms of the saturation loop. Note the similarity between the normal magnetization curve and the initial magnetization curve of Fig. 10.1.

The Steinmetz Equation

We have shown that hysteresis loss is proportional to the area of the hysteresis loop. In electric machinery, flux densities commonly reach the maximum flux density B_{max}; it is then useful to know the hysteresis loss in terms of B_{max}. The normal magnetization curve, which is the locus of the tips of a series of hysteresis loops, each of smaller size, is shown in Fig. 10.3. Steinmetz, using the results of many tests, derived an empirical formula which, in the practical range of $B_{max} = 1000$ to 15,000 G, varies approximately as the 1.6 power of B_{max}; that is,

$$p_h = \eta f B_{max}^{1.6} \qquad \text{W/m}^3 \tag{10.9}$$

where η is a constant and is equal to 0.001 for good silicon steel, 0.002 to 0.004 for soft iron, and as high as 0.03 for hard cast steel. To obtain the power loss P_h in watts, one can integrate the power density p_h over the volume of the core. To repeat, Steinmetz's formula is only an approximation. For very small values of B_{max}, P_h varies as B_{max}^3. For some of the newer alloys, the exponent in (10.9) might not be accurate either.

10.2 EDDY-CURRENT LOSSES

We have seen in Sec. 6.11 that a counter emf is induced in a circuit whenever the flux through the circuit changes. This is usually known as Lenz' law. As long as the resistance R of the circuit is not infinite, there will be a current I associated with the counter emf which will give rise to I^2R losses in the circuit known as *eddy-current losses.*

Let us consider a magnetic circuit which consists of a core and a winding that is connected to an alternating current, as shown in Fig. 10.4*a*. The flux in the core is continually rising, falling, and reversing because the core is subjected to an alternating magnetizing force. Accordingly, emf's are induced, as indicated throughout the entire circuit, their direction reversing as the rate of change of flux

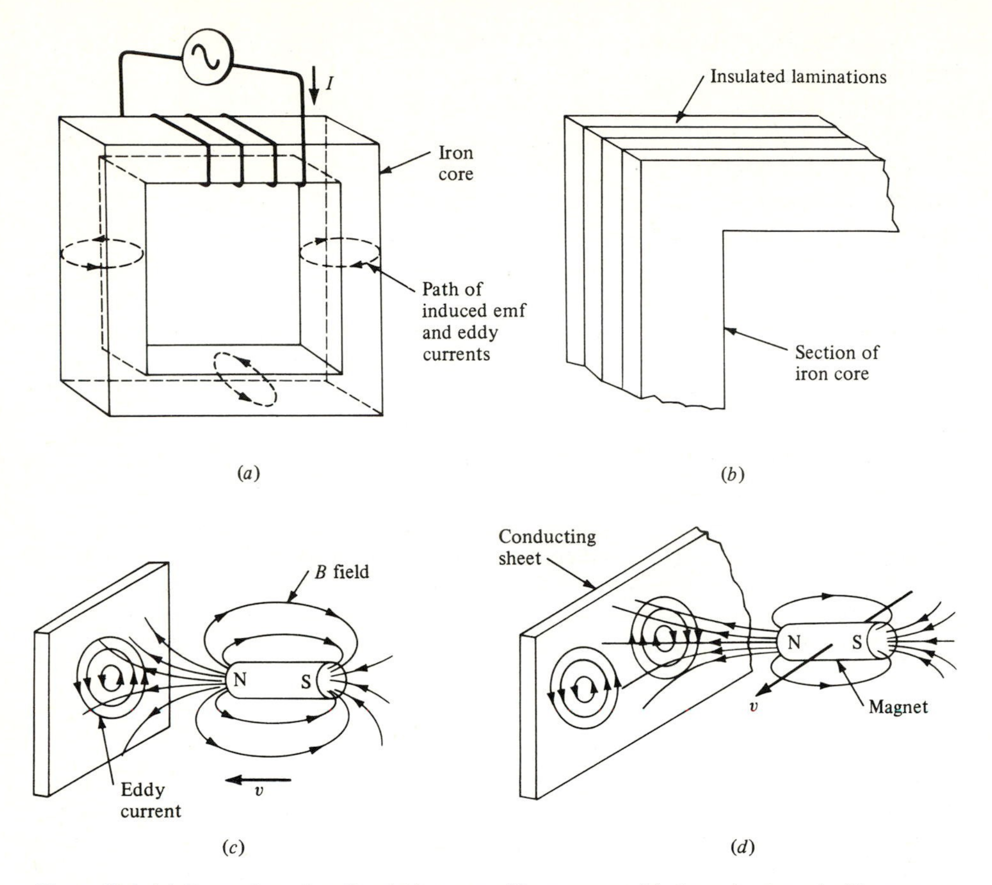

Figure 10.4 (*a*) Formation of emf's which cause eddy currents; (*b*) the reduction of eddy currents by laminating the core; (*c*) a magnet moving toward a conducting sheet will induce eddy currents as shown; (*d*) eddy currents induced by a magnet moving parallel to the sheet.

reverses. Most magnetic materials (iron, steel, nickel, etc.) have comparatively low resistivity, and if the core were solid, these emf's would give rise to large currents (eddy currents). With such a core a relatively small magnetizing force H, even for slow time variation (low frequency), would generate eddy currents sufficiently high to heat the core to melting temperatures in a short time. Eddy currents are avoided† by substituting thin laminations, insulated one from another, in place of the solid core, as shown in Fig. 10.4*b*. The insulation, usually a coating of nonconducting paint on one side of each lamination, breaks the continuity of the eddy-current-paths and reduces the losses. In general, the thinner the laminations, the lower the losses, since the eddy-current losses are proportional to

† For a discussion on the use of ferrites to reduce eddy currents, see the last paragraph in Sec. 9.8.

t^2, where t is the thickness of the laminations.† Since, as shown in Eq. (10.11), the eddy-current losses increase with frequency squared, they play a definite part in degrading the high-frequency response of electric systems.

If the circuit is a large conducting body, the induced eddy currents will flow in closed paths, as shown in Fig. 10.4. In part (*c*) of this figure we are showing a magnet approaching a flat conducting plate, and in part (*d*) the magnet is moving parallel to the conducting plate. We notice that the eddy currents always tend to flow perpendicular to the flux and in such a direction as to oppose any changes in flux. In part (*c*), the direction of the eddy current is such as to oppose the flux from the magnet. Hence circular magnetic-dipole currents will be induced in the conducting sheet and will produce as nearly as possible a magnetic field that is equal and opposite to that of the approaching magnet. In part (*d*), the direction of the eddy-current loop in front of the magnet will oppose the flux of the magnet, while in back of the magnet the eddy-current loop has a direction which will aid the flux. Just as in the magnetic brake (see Sec. 6.12), the motion of the magnet will be impeded by the induced eddy currents. The reason is that the eddy currents dissipate energy as heat, and the energy lost must come from the kinetic energy of the magnet. The opposing force is given by Eq. (6.74).

The I^2R heating losses resulting from eddy currents,‡ or more specifically the $JE = \sigma E^2$ losses in W/m^3, can be obtained from (6.72) as

$$\boxed{JE = \sigma E^2 = \sigma v^2 B^2} \tag{10.10}$$

where J, E, σ, B, and v are the current density, electric field, conductivity, flux density, and velocity, respectively. When the applied magnetic field is alternating sinusoidally with frequency f, the induced counter field is $E \propto dB/dt \propto fB$. The eddy-current losses are then

$$\boxed{\sigma E^2 \propto \sigma f^2 B^2} \tag{10.11}$$

Since the concept of eddy currents is so fundamental, the following examples of induced currents which normally are not identified as eddy currents but really are special cases of them, should be reviewed at this time. They are:

The opposing magnetic force F_m due to induced currents in the linear motor, Fig. 6.37.
The braking action due to eddy currents in the pendulum, Fig. 6.38.
The opposing force F_{ind} due to induced currents in the MHD generator, Fig. 6.39.

† EE Staff MIT, "Magnetic Circuits and Transformers," p. 137, John Wiley & Sons, Inc., New York, 1943.

‡ The existence of eddy currents can be shown by the physical development of Sec. 6.11 in terms of motional emf, restated as Lenz' law (10.3), or simply by stating Faraday's law (11.1). In any case, eddy-current loss is proportional to the square of the induced voltage, and induced voltage, in turn, is proportional to the rate of change of the flux density.

The induced current J_{ind} in the plasma engine, Fig. 6.40.
The currents due to back emf in a motor.
The currents produced by a generator.

An application of eddy currents that we are all familiar with is the automobile speedometer. A magnet rotates in proportion to the speed v of the car and induces eddy currents in a cup which is otherwise free to rotate about the magnet. The induced current in turn interacts with the field B of the magnet and produces a magnetic force per unit volume f_m:

$$f_m = \sigma v B^2 \tag{6.74}$$

which drags the cup along. A linear spring retards the cup from continuous rotation. The electromagnetic torque which the cup experiences is proportional to the speed v. Thus, a pointer connected to the cup undergoes an angular displacement which is proportional to the speed of the vehicle.

Total Core Loss

We can now state the total power losses in the conducting ferromagnetic cores of electric equipment such as transformers, generators, motors, etc., as a combination of hysteresis losses, (10.1), and eddy-current losses. That is,

$$P_{h+e} = W_h f + k\frac{B^2 f^2}{\rho} \tag{10.12}$$

where ρ is the resistivity ($\rho = 1/\sigma$) and k is a constant that depends on the geometry. Hysteresis losses may be kept low by using a soft material with a narrow hysteresis loop (small H_c). Eddy-current losses may be reduced by increasing the electric resistance of the core materials (silicon steels and ferrites†) or by breaking up eddy-current paths by using laminations rather than solid cores. Note that by using Eq. (10.9), we can write the total core loss in terms of $B_{\max}$ as $P_{h+e} = k_1 f B_{\max}^{1.6} + k_2 f^2 B_{\max}^2/\rho$, where k_1 and k_2 are constants.

Magnetic Shielding

The shielding of static magnetic fields is principally different from that of high-frequency magnetic fields. This stems from the fact that electric and magnetic fields in the static case are independent. Hence a static magnetic field can exist inside a metallic body even though an electric field cannot. In practice, to keep a magnetostatic field out of a region, one surrounds that region with a shell of high-permeability material, such as Mu metal for which $\mu_r \cong 2 \times 10^4$ (see Table 9.2). The function of such a material is to concentrate the magnetic flux in it. This concentration of flux is shown in Figs. 9.12 and 9.13 and is similar to the concentration of electric field in a dielectric material, as discussed in Sec. 4.11.

† Ferrites have such a high resistance that eddy currents are negligible. However, their high cost, low saturation field B_s, and low mechanical strength make them unsuitable in power equipment (see Sec. 9.8).

Referring to Fig. 9.13, we see that as the magnetic field $B_0 = \mu_0 H_0$ passes into the medium of high permeability, it leaves a relatively field-free space inside the hollow sphere. The field in a small spherical cavity inside the sphere is given as

$$\boxed{B_{\text{cavity}} \cong \frac{4.5B_0}{\mu_r}} \tag{9.43}$$

and thus, can be easily made smaller by a factor of at least 1000 over that of the exterior field B_0. As the shell is made thinner, the shielding effect decreases; the field inside such a shell with inside and outside radii of a and b, respectively, is given by†

$$B_{\text{cavity}} = \frac{9\mu_r}{(2\mu_r + 1)(\mu_r + 2) - 2(\mu_r - 1)^2(a/b)^3} B_0$$

$$\cong \frac{9}{2\mu_r(1 - a^3/b^3)} B_0$$

where the last expression is an approximation valid when $\mu_r \gg 1$. For a thick shell, i.e., when $a/b \ll 1$, we can drop this term and obtain the simple result as given by (9.43).

Since soft magnetic materials with very high μ_r's are not available, magnetic shielding which is $1/\mu_r$-dependent can only be partially effective. For example, from Table 9.2, we see that an exceptionally good magnetic material such as Mu metal has a working permeability μ_r of about 2×10^4. Thus a magnetic material with $\mu_r \to \infty$ would be clearly the ideal shielding material. The reason that electric fields can be completely shielded by metal enclosures (see Fig. 2.10*c*) is that a metal acts, in effect, as a dielectric of infinite permittivity; that is, $\varepsilon_r \to \infty$, as discussed in Sec. 4.7.

Shielding of Time-Varying Magnetic Fields

For static fields and for fields with slow time changes, i.e., quasi-static fields, the only method of shielding a space is by surrounding it with a high-permeability material. However, for higher frequencies a sheet of metal will act very effectively as a shield for magnetic fields because of the induced eddy currents in it. As shown in Sec. 8.8, an alternating magnetic field incident on a metal will induce eddy currents on the metal surface which decrease exponentially to $1/e$ in a distance δ called the *skin depth*. The skin depth is given by (8.76) or by the more rigorous development of Sec. 13.6 as

$$\delta = \frac{1}{(\pi f \mu \sigma)^{1/2}} \tag{8.76}$$

† J. D. Jackson, "Classical Electrodynamics," p. 200, John Wiley & Sons, Inc., New York, 1975.

The time-varying magnetic field inside the metal is limited by the skin depth. If the skin depth δ is sufficiently small, the magnetic field is effectively kept out of the metal; that is, the magnetic field cannot penetrate the metal and emerge on the other side of it if the thickness of the metal sheet is at least several skin depths. Thus, a nonmagnetic but highly conducting material makes an excellent shield; the thickness of the shield can be less if the frequency is higher. Thin aluminum shields are therefore widely used at radio frequencies. Of course if the frequency is low, the skin depth δ will be rather large. Then a combination of magnetostatic and eddy-current shielding must be used. A practical shield takes the form of alternating layers of Mu metal and copper, for example.† Such practice must be resorted to in the audio-frequency range, where, for example, an eddy-current shield of copper ($\sigma = 5.8 \times 10^7$ S/m) has a skin depth of

$$\delta = \frac{0.066}{\sqrt{f}} = \begin{cases} 1 \text{ cm at } f = 60 \text{ Hz} \\ 2 \text{ mm at } f = 10^3 \text{ Hz} \\ 0.5 \text{ mm at } f = 1.5 \times 10^4 \text{ Hz} \end{cases}$$

Therefore, at the lower audio frequencies a skin-effect shield must be usually supplemented by a layer of high-permeability material. At high frequencies, on the other hand, the skin depth is so small that any metal sheet, except the very thinnest, is an effective shield. In general we can say that for a given thickness of the shield, iron is a better material than copper for low frequencies, and the reverse is true for high frequencies. In certain tests iron and copper proved to be equally good at 1300 Hz; higher than this, copper was better, and vice versa.

Frequency Dependence of Permeability μ

We have seen that a changing magnetic field induces eddy currents in a conducting medium. Since the field produced by induced eddy currents is opposite to the applied field, the effect of eddy currents is to prevent the applied field from penetrating *immediately* to the interior of the medium. Therefore, it takes time for the applied field to overcome the back field created by eddy currents and to penetrate a conducting material. Penetration comes only after energy as I^2R losses is dissipated. A perfect conductor ($\sigma = \infty$) is never penetrated by an applied magnetic field because the skin depth δ is zero. If, on the other hand, the material is a perfect insulator ($\sigma = 0$), $\delta = \infty$, the field penetrates with approximately the velocity of light. Now, if the magnetic field alternates with frequency f, then the magnetic field strength in the interior of a conducting material is never more than a fraction of the field strength at the surface. The magnetic field, therefore, decreases from the surface exponentially (see Sec. 13.6) toward the interior, and the skin depth δ denotes the distance at which the field has decreased to $1/e$ of its value at the surface.

The exclusion of time-varying magnetic fields from the interior of a magnetic material markedly reduces the effective permeability of the material. Most of the magnetic material is thus removed from participating in magnetization and to this end acts as if it were not present. We can express the effective μ by

$$\mu_{\text{eff}} = \mu \frac{H_{\text{eff}}}{H_a}$$

† W. G. Gustafson, "Magnetic Shielding of Transformers at Audio Frequencies," *B.S.T.J.*, vol. 17, p. 416, July 1938.

where H_a is the applied field. The effective magnetic intensity can be written as

$$H_{\text{eff}} = H_a \frac{\delta}{t} \qquad \text{for } \delta < t$$

where t is the sample thickness and δ is the skin depth, $\delta = (\pi f \mu \sigma)^{-1/2}$. The effective permeability when the skin depth δ is smaller than the thickness t of the magnetic material is then

$$\mu_{\text{eff}} = \frac{1}{t}\sqrt{\frac{\mu}{\pi f \sigma}} \qquad \text{for } \delta < t$$

Thus μ_{eff} decreases as $f^{1/2}$, and the flux-carrying ability of a material of given conductivity σ and thickness t is reduced to being proportional to $\mu^{1/2}$. When the penetration depth δ is larger than the thickness t, we have

$$\mu_{\text{eff}} \approx \mu \qquad \text{for } \delta > t$$

We have shown that for ac applications, eddy currents reduce the permeability of soft magnetic materials. For high frequencies, the ac permeability can be substantially below the dc permeability μ. A low conductivity σ in a magnetic material is desirable in suppressing eddy-current losses for ac applications. Eddy currents can also be suppressed by subdividing (laminating) the material or by using it in powdered form. Both methods, in effect, decrease the conductivity of the magnetic material.

10.3 ENERGY STORED IN A MAGNET

Permanent magnets can store energy because of the irreversible property of hysteresis, which has its origin in crystal imperfections. Imperfections cause the domain walls to stick instead of moving smoothly. For example, when the current of an external source is increased and then reduced to zero (thereby also reducing the H field to zero), Fig. 10.2 shows that energy proportional to area $0bd$ remains stored in the ferromagnetic specimen. If the ferromagnetic material is hard instead of soft (see Fig. 9.30), then this area represents an appreciable amount of stored work and is, of course, the reason why hard materials make good permanent magnets.

Hysteresis was discussed in detail in the last chapter. Nevertheless, a simple model to show that work must be done on a ferromagnetic material in order to make it into a permanent magnet is given in Fig. 10.5. Let us represent the domains by small bar magnets connected by a string or rod. If left alone, the bar magnets will arrange themselves in a configuration approximating that shown in Fig. 10.5a, as this minimizes the total energy of the configuration. The external

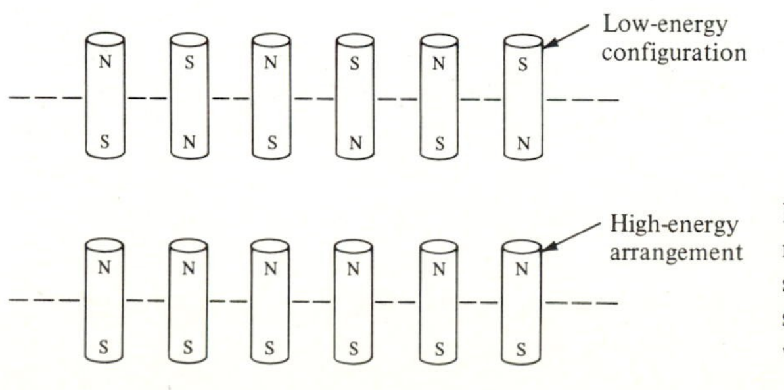

Figure 10.5 (a) A series of bar magnets, free to rotate, will arrange themselves a shown. (b) The configuration shown is obtained by performing work on alternate magnets.

field is also minimized. Figure 10.5*a* is thus representative of the unmagnetized state in a ferromagnetic material. To obtain the arrangement of Fig. 10.5*b*, we must rotate alternate bar magnets while holding the others fixed. This requires work and creates a large magnetic field external to the configuration. Figure 10.5*b* is thus representative of the domain arrangement in a magnetized permanent magnet.

To show explicitly that the energy stored resides in the magnetization M, we can write the integrand of the energy density (10.7) as

$$dw = H\,dB = \mu_0 H\,dH + \mu_0 H\,dM \tag{10.13}$$

where the definition $B = \mu_0(H + M)$ was used. The first term represents the work done on free space, whereas the second term $\mu_0 H\,dM$ represents the work done on the ferromagnetic material. If we note that the integral of $H\,dH$ vanishes for a complete cycle, as well as for a half cycle, we can write for the stored energy density in a magnet

$$\boxed{w = \mu_0 \int_{\text{half cycle}} H\,dM} \tag{10.14}$$

which is equal to area $0bd$ in Fig. 10.2 if we start with an initially unmagnetized material. Similarly, we can write for the hysteresis loss

$$w_h = \mu_0 \oint H\,dM \tag{10.15}$$

which is the area of the hysteresis loop shown in Fig. 10.2.

10.4 MAGNETIC CIRCUIT

A simple magnetic circuit and its equivalent electric circuit are shown in Fig. 10.6. For the electric circuit, we can write

$$\text{Resistance } R = \frac{\text{emf}}{\text{current}} = \frac{\mathscr{V}}{I} = \frac{\oint \mathbf{E} \cdot \mathbf{dl}}{\iint \mathbf{J} \cdot \mathbf{dA}} \tag{10.16}$$

The electric circuit was discussed at length in Secs. 3.1 and 3.8. A magnetic circuit also involves a source which sets a flux ϕ flowing around the magnetic circuit. This flow is resisted by the reluctance of the magnetic path defined as

$$\boxed{\text{Reluctance } \mathscr{R} = \frac{\text{mmf}}{\text{flux}} = \frac{\mathscr{F}(=NI)}{\phi} = \frac{\oint \mathbf{H} \cdot \mathbf{dl}}{\iint \mathbf{B} \cdot \mathbf{dA}}} \tag{10.17}$$

Magnetomotive force $\mathscr{F}$, often written in abbreviated form as mmf, was discussed in Secs. 7.5 and 9.2. The reason why such a simple approach, which parallels that of dc circuits, is possible is that we are assuming the flux to be confined to the

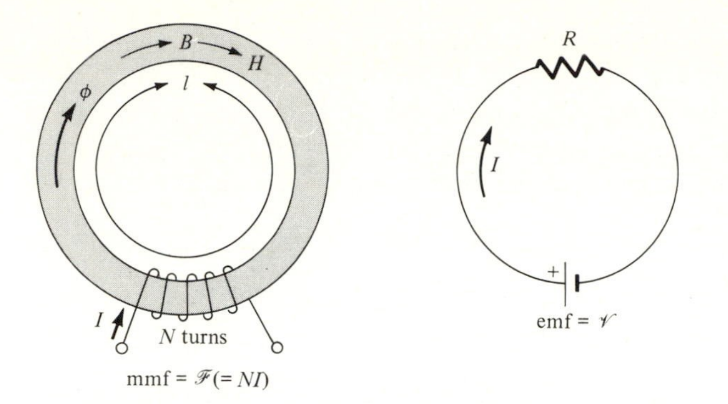

Figure 10.6 A magnetic circuit consisting of a ferromagnetic ring with a concentrated winding. Note that since the flux is confined to the interior of the toroid, there is no difference between a winding uniformly distributed around the ring or one concentrated in a small sector. The equivalent electric circuit is also shown.

ferromagnetic core (no flux leakage). Hence, the lines of B are parallel to the physical circuit. This is a good approximation for the ferromagnetic materials which have a large permeability μ.

The analogous quantities for the magnetic and electric circuit are thus $\oint \mathbf{H} \cdot \mathbf{dl} = \mathscr{F}(=NI)$, $\Delta\phi = B\,\Delta A$, $\mathbf{B} = \mu\mathbf{H}$, $\oiint \mathbf{B} \cdot \mathbf{dA} = 0$ for the magnetic circuit and $\oint \mathbf{E} \cdot \mathbf{dl} = \mathscr{V}$, $\Delta I = J\,\Delta A$, $\mathbf{J} = \sigma\mathbf{E}$, $\oiint \mathbf{J} \cdot \mathbf{dA} = 0$ for the electric circuit. Since Kirchhoff's law (3.8) for any closed electric circuit states that the summation of voltage rises equals the summation of voltage drops, $\sum \mathscr{V} - \sum RI = 0$, similarly for any closed magnetic circuit we can write

$$\sum NI - \sum \mathscr{R}\phi = 0 \tag{10.18}$$

At any node, we also have $\sum I = 0$ for the electric circuit and

$$\sum \phi = 0 \tag{10.19}$$

for the magnetic circuit.

For the simple magnetic circuit of Fig. 10.6, the reluctance can be expressed as

$$\boxed{\mathscr{R} = \frac{\oint H\,dl}{\iint B\,dA} = \frac{Hl}{BA} = \frac{l}{\mu A}} \tag{10.20}$$

because the field B and the cross-sectional area A are uniform. The path length l is a mean length, usually that of a path along the middle of the iron. The dimensions of μ are henry per meter which gives reluctance $\mathscr{R}$ the dimension of reciprocal henry.

Example For the toroid of Fig. 10.6, Ampère's law gives H inside the iron as $H = NI/l$. The magnetic flux density B can be obtained by multiplying H by μ of the iron, or

$$B = \mu H = \frac{\mu NI}{l} \tag{10.20a}$$

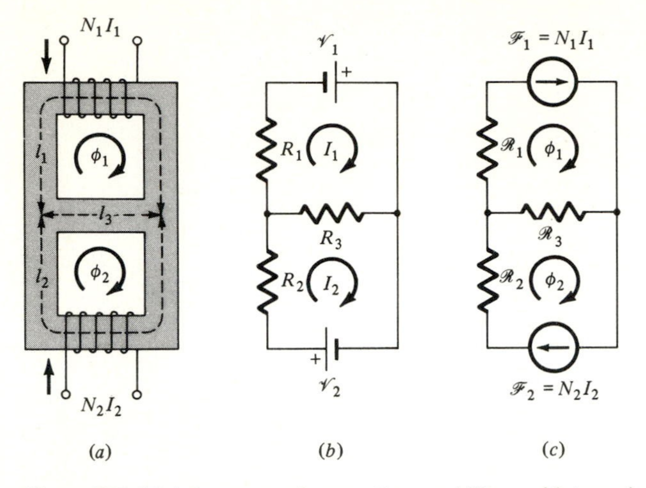

Figure 10.7 (*a*) A ferromagnetic core of permeability μ with two windows. The center arm has length l_3 and cross section A_3. (*b*) The electrical equivalent of the magnetic circuit. (*c*) The equivalent magnetic circuit.

This can be reexpressed by the use of (9.13) or (9.14) as

$$B = \mu_0(H + M) = \frac{\mu_0 NI}{l} + \mu_0 M \tag{10.20b}$$

which shows explicitly that B inside the iron has two contributions, one from the winding, which is the $\mu_0 NI/l$ term, and one from the magnetization M.

The magnetic field in a toroid with mean circumference of $l = 10$ cm, 1000 turns, and carrying a current of 10 mA results in an H field which is $H = NI/l = 100$ A/m. If the core is of iron, the BH curve of Fig. 9.31 gives a $\mu_r = \mu/\mu_0 = 4500$ for this value of H. This results in a $B = (4500)(4\pi \times 10^{-7})(100) = 0.57$ T in the core.

Figure 10.7 shows a more complicated circuit and its electrical equivalent. If we are able to determine the currents in the electrical case, we will then know the magnetic fluxes through the branches of the magnetic circuit. For example, Kirchhoff's law (10.18) for the top window of the magnetic circuit gives

$$N_1 I_1 = \oint \mathbf{H} \cdot \mathbf{dl} = H_1 l_1 + H_1 l_3 - H_2 l_3 = \frac{l_1}{\mu A_1}\phi_1 + \frac{l_3}{\mu A_3}\phi_1 - \frac{l_3}{\mu A_3}\phi_2 \tag{10.21}$$

Note that any part of a magnetic circuit which has cross section A and permeability μ can be expressed as a lumped reluctance $\mathscr{R} = l/\mu A$.

The analogies between electric and magnetic circuit quantities are summarized in Table 10.1.† Units of the magnetic field were discussed in Sec. 6.1.

† See also Secs. 7.5 and 9.2.

Table 10.1 Analogous electric and magnetic circuit quantities

Electric	Magnetic
Electric field strength $\mathbf{E}$	Magnetic field strength $\mathbf{H}$
Current density $\mathbf{J} = \sigma\mathbf{E}$	Flux density $\mathbf{B} = \mu\mathbf{H}$
Conductivity σ	Permeability μ
Current $I = \iint J\,dA \cong JA$	Flux $\phi = \iint B\,dA \cong BA$
emf $\mathscr{V} = \oint \mathbf{E}\cdot\mathbf{dl} \cong El$	mmf $\mathscr{F} = \oint \mathbf{H}\cdot\mathbf{dl} \cong Hl$
Ohm's law $I = V/R = GV$	$\phi = \mathscr{F}/\mathscr{R} = \mathscr{P}\mathscr{F}$
Resistance $R = l/\sigma A$	Reluctance $\mathscr{R} = l/\mu A$
Conductance $G = \sigma A/l$	Permeance $\mathscr{P} = \mu A/l$
Electromotive force (emf) $= \mathscr{V}$	Magnetomotive force (mmf) $= \mathscr{F}$

Example If the second source in the magnetic circuit of Fig. 10.7 is removed ($I_2 = 0$), find the flux ϕ_1 through branch l_1.

The reluctance that the mmf source $N_1 I_1$ now sees is composed of reluctance $\mathscr{R}_1 = l_1/\mu A_1$ in series with the parallel combination of reluctances $\mathscr{R}_2 = l_2/\mu A_2$ and $\mathscr{R}_3 = l_3/\mu A_3$; that is,

$$\mathscr{R} = \mathscr{R}_1 + \frac{\mathscr{R}_2\mathscr{R}_3}{\mathscr{R}_2 + \mathscr{R}_3} = \frac{l_1}{\mu A_1} + \frac{(l_2/\mu A_2)(l_3/\mu A_3)}{l_2/\mu A_2 + l_3/\mu A_3} = \frac{l_1}{\mu A_1}\left(1 + \frac{1}{l_1 A_3/l_3 A_1 + 1}\right)$$

The last expression is obtained because in the structure of Fig. 10.7 we have assumed that $l_1 = l_2$ and $A_1 = A_2$. The flux ϕ_1 that flows is now given by

$$\phi_1 = \frac{\text{mmf}}{\mathscr{R}} = \frac{N_1 I_1}{\mathscr{R}}$$

If the dimensions of the physical structure are $l_1 = 15$ cm, $l_3 = 5$ cm, $A_1 = 2\text{ cm}^2$, $A_3 = 1\text{ cm}^3$, and $\mu = 500\mu_0$, where $\mu_0 = 4\pi \times 10^{-7}$, we obtain for the reluctance

$$\mathscr{R} = \frac{0.15}{(500 \times 4\pi \times 10^{-7})(2 \times 10^{-4})}\left[1 + \frac{1}{(0.15)(10^{-4})/(0.05)(2 \times 10^{-4}) + 1}\right]$$
$$= 1.7 \times 10^6\ \text{H}^{-1}$$

10.5 MAGNETIC CIRCUIT WITH AIR GAPS ELECTROMAGNETS

As we will shortly see, a magnetic circuit with an air gap has most of its mmf dropped across the air gap. This fact is of practical importance since Kirchhoff's law for such a magnetic circuit can then be approximated as

$$\text{mmf} = \phi\mathscr{R} \cong \phi\mathscr{R}_{\text{gap}} \tag{10.22}$$

Figure 10.8 shows a toroidal core with an air gap of length l_g cut into it. The concentrated winding with a current flowing through it serves as the source of mmf. A configuration such as this is known as an *electromagnet*, since by cutting a gap in the toroid, the magnetic field, which before was confined to the interior of

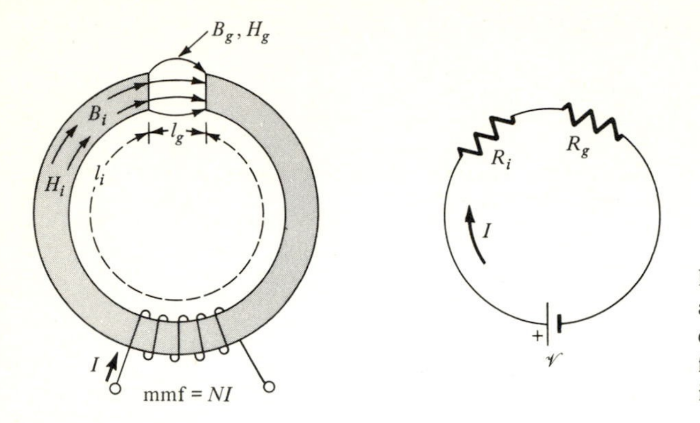

Figure 10.8 The magnetic circuit and its electrical equivalent of an electromagnet. The core is a soft ferromagnetic material—usually iron.

the toroid, is now accessible. Because the toroid with a gap is a series circuit, the same flux flows through the iron as flows through the gap; that is, $\phi_i = \phi_g$, where $\phi = BA$ and A is the cross section. If fringing in the air gap is negligible, the flux in the gap is confined to the same cross section as in iron; that is, $A_i = A_g$ and $B_i = B_g$. Kirchhoffs law for the series circuit can now be written as

$$\begin{aligned} \text{mmf} &= \phi\mathscr{R} = \phi(\mathscr{R}_{\text{iron}} + \mathscr{R}_{\text{gap}}) \\ &= \phi\left(\frac{l_i}{\mu A_i} + \frac{l_g}{\mu_0 A_g}\right) \\ &= \frac{\phi}{\mu_0 A}\left(\frac{l_i}{\mu_r} + l_g\right)\bigg|_{A_i = A_g} = \frac{B}{\mu_0}\left(\frac{l_i}{\mu_r} + l_g\right) \end{aligned} \tag{10.23}$$

where the relative permeability of the ferromagnetic core is given by $\mu_r = \mu/\mu_0$ and $A_i = A_g = A$.

Example† The flux density B in the iron as well as in the gap of the electromagnet of Fig. 10.8 can be explicitly written by substituting mmf $= NI$ and $\phi = BA$ in (10.23). This gives

$$B = \frac{\mu_0 NI}{l_i/\mu_r + l_g} \tag{10.23a}$$

By substituting $\mu_r = 1 + M/H_i$ from (9.14) in the above equation, we can again show explicitly the two contributions to the B field. Thus, (10.23a) can be written as

$$B = \frac{\mu_0 NI}{l} + \mu_0 M\left(1 - \frac{l_g}{l}\right) \tag{10.23b}$$

where $l = l_i + l_g$ is the total average length of the magnetic path. The first term on the right is the contribution to $\mathbf{H}$ of the current I which flows in the winding; that is, $H = NI/l$; the other is the contribution of the magnetization M in the magnetic material, which is $M(1 - l_g/l)$.

The flux density in the gap, when the winding is the same as that of the first example

† As this is an important example, the student should check that Eqs. (10.23a) and (10.23b) give the same results for the limiting cases: $l_g = 0$, $l_g = l$ (air core), and $\mu_r = 1$ (air core).

($N = 1000$, $I = 10$ mA) in Sec. 10.4, is given by (10.23a). For a gap length of $l = 1$ cm, we have

$$B \cong \frac{\mu_0 NI}{l_g} = \frac{(4\pi \times 10^{-7})(10^3)(10^{-2})}{10^{-2}} = 0.0013 \text{ T}$$

where l_i/μ_r was ignored with respect to l_g because $\mu_r \gg 1$. Because of the presence of the high-reluctance air gap, the flux density is now substantially less.

For iron or other ferromagnetic materials, (10.23) almost always simplifies to

$$\text{mmf} \cong \frac{\phi}{\mu_0 A} l_g = \phi \mathcal{R}_g \tag{10.24}$$

because $l_i/\mu_r \ll l_g$ for most practical gap geometries. It is true that usually the path in the iron is longer than that in the gap, $l_i/l_g \gg 1$, but for most ferromagnetic materials μ_r is sufficiently large so that $\mu_r \gg l_i/l_g$. The above formula permits us immediately to calculate the ampere-turns in an electromagnet once the desired magnetic field B is specified.† For example, solving for B in (10.24), we obtain

$$B = \frac{\mu_0 NI}{l_g} \tag{10.25}$$

Example It is desired to produce a magnetic field of 1 Wb/m² in the air gap of an electromagnet shown in Fig. 10.8. The cross section of the iron ring is 1 cm², the mean length $l_i + l_g = 10$ cm, gap width $l_g = 5$ mm, and the BH curve of soft iron which is to be used is shown in Fig. 9.31.

1. Find the ampere-turns required to produce the desired flux density using (10.25).
2. Find the ampere-turns required to produce the desired flux density without ignoring the reluctance $\mathcal{R}_i$ of the iron path.

SOLUTION 1

$$NI = \frac{Bl_g}{\mu_0} = \frac{(1)(5 \times 10^{-3})}{4\pi \times 10^{-7}} = 4000 \text{ ampere-turns}$$

SOLUTION 2 We can use (10.23) which gives

$$NI = \frac{B}{\mu_0}\left(\frac{l_i}{\mu_r} + l_g\right) = \frac{1}{4\pi \times 10^{-7}}\left(\frac{9.5 \times 10^{-2}}{5000} + 5 \times 10^{-3}\right) \cong 4000 \text{ ampere-turns}$$

where the BH curve for iron was used to find that $\mu_r = 5000$ at $B = 1$ Wb/m². We can also use an alternative method which involves Ampère's law. Thus from (10.21) we have

$$NI = \oint \mathbf{H} \cdot \mathbf{dl} = H_i l_i + H_g l_g$$

† Note, that the analysis developed in this section is for soft ferromagnetic materials for which the flux for all intents and purposes falls to zero when the applied magnetizing force is removed.

The boundary condition of the normal components of B relates H_i and H_g; that is, $\mu H_i = \mu_0 H_g$ or $\mu_r = H_g/H_i$ which gives us

$$NI = H_i(l_i + \mu_r l_g) \cong 4000 \text{ ampere-turns}$$

where again the BH curve for iron was used to find that at a flux density of $B = 1$ Wb/m^2, the relative permeability $\mu_r = 5000$, and the magnetic field strength in the iron is $H_i = 160$ A/m.

From the above example we see that if the reluctance of the air gap is sufficiently large, the reluctance of the iron path can usually be ignored. Some fringing of the magnetic field will occur around the air gap, but for many practical purposes we can ignore this especially if l_g is small.

Flux Leakage

If a second winding is wound on the toroidal core, as, for example, the secondary in the transformer of Fig. 10.18, we usually find that not all the flux that exists in the first coil links the second. Some flux is lost as it passes around the circuit. The reason for the flux leakage is that the relative permeabilities μ_r of ferromagnetic materials (whose purpose is to confine the flux) range between 10^2 and 10^5, whereas that of the surrounding medium, usually air, is 1. Therefore, air is not a particularly good insulator for the magnetic circuit, and some magnetic flux will always find a path outside the core. It would be nice if a magnetic insulator existed, one whose reluctance $\mathscr{R}$ would be very high. Then we could simply wrap it around the core and thereby keep the flux from leaking off.

To fully appreciate the difficulty that leakage flux can present in some magnetic circuits, let us take a look at an electric circuit. Practically none of the electric current leaks off the electric circuit, even when such circuits are miles long. The reason for this is that the conductivity of a good electric conductor is over 10^{21} times that of a good insulator such as mica. The leakage path for electric current has resistances which for practical purposes are infinitely large. We can, therefore, talk about electric insulators. However, as long as the largest ratio of reluctance for available materials is only about 10^5, we cannot confine flux completely to the magnetic circuit and conclude that a magnetic insulator for flux does not exist in the same sense as one does for electric currents.

Flux Fringing

Fringing occurs when the magnetic circuit is interrupted by a gap. Due to *fringing*, which is a spreading out of the flux about the gap as shown in Fig. 10.9, the flux density B is smaller in the gap than it is in the iron. For gaps which are narrow and whose gap faces are relatively large (a low-reluctance gap), we can ignore fringing and obtain good accuracy in our calculations. On the other hand, if the length l_g of the gap is on the same order of magnitude as the cross-sectional dimensions of the core, then the error becomes sufficiently large that the whole circuit approach might have to be abandoned or at best a "cut-and-try" process substituted for finding the reluctance of the gap. Between these two extremes we can make a

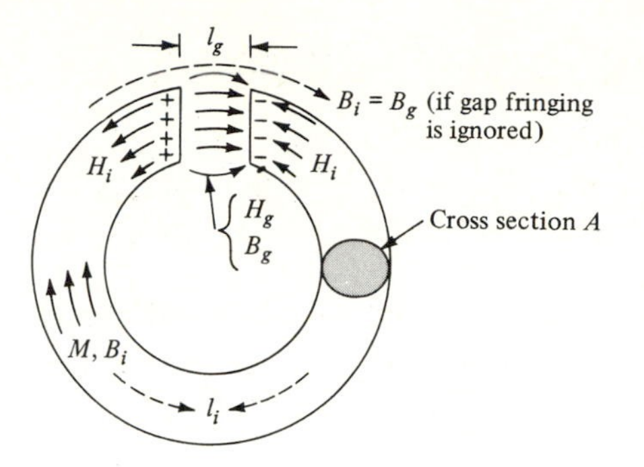

Figure 10.9 A permanent magnet in the form of a toroid with an air gap. The internal fields in the iron are denoted by the subscript i. The magnetic surface charges which are the source for the H field are shown on the pole faces.

compensating correction for the error caused by fringing. This is done by increasing the cross-sectional dimension by the length of the gap. Thus, if the core has a rectangular cross-section with dimensions a and b, the effective area is

$$A = (a + l_g)(b + l_g)$$

To show how the flux lines spread around the edges of the gap, we can make use of the fact (derived in the example of Sec. 8.8) that for an air-iron interface the flux on the air side of the boundary must always be normal to the iron surface. By following this simple rule, surprisingly accurate pictures for the flux lines can be drawn.

It is tempting to say that fringing and leakage is always undesirable. Certainly, the ideal magnetic circuit would not have any leakage or fringing of the magnetic flux. However, like friction which can be useful or not depending on the application, there are many situations in which an increase in fringing is desirable. For example, tape recording (Sec. 10.11) could hardly be possible were it not for the fringing fields of the recorder-head gap which magnetizes the tape in proportion to an electric current in the recorder-head coil.

10.6 MAGNETIC CIRCUIT OF A PERMANENT MAGNET

Thus far we have considered two toroidal configurations, one without an air gap (Fig. 10.6), and one with an air gap (Fig. 10.8). Both had a winding which acted as the source of magnetomotive force (mmf = NI) and which produced the core flux. But perhaps the most interesting case is the permanently magnetized toroid with an air gap, as shown in Fig. 10.9. Such a configuration often leads to confusion when the BH curve for the ferromagnetic material is used. One usually concludes, using Ampère's law $\oint \mathbf{H} \cdot \mathbf{dl} = NI$, that the H field in the material must be zero because the absence of a winding implies that NI is zero. However, as we will shortly see, an absence of true currents in no way implies that H is zero.

Let us assume that the ferromagnetic core is of Alnico V, a hard material used in many magnets. The hysteresis loop is shown in Fig. 10.10. To magnetize the toroid, let us begin with the original toroid without an air gap, wrap a coil around

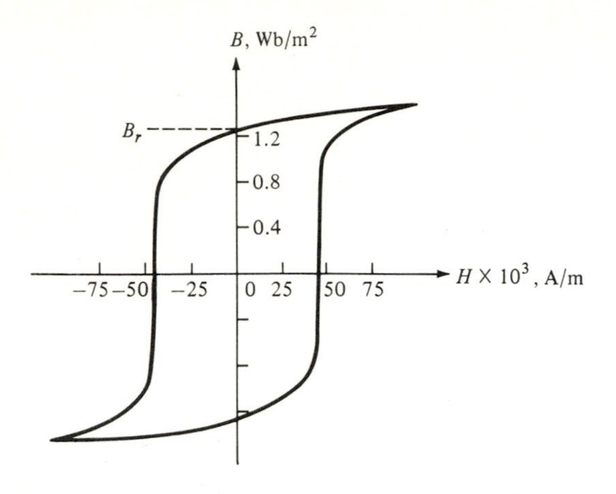

Figure 10.10 The *BH* curve for a typical permanent-magnet material such as Alnico V.

it, and pass a large steady current through it. Upon removal of the coil, the value of H goes to zero, but B remains at about $B_r = 1.2$ Wb/m^2. In other words, the remaining magnetization M is now the source for the B field in the toroid. This can be seen from the general expression for magnetic flux density which is $B = \mu_0(H + M)$ and which in the absence of NI (or H) gives $B = \mu_0 M$.

To make the magnetic field trapped inside the toroid available, we cut an air gap of length l_g into the iron ring. This greatly increases the reluctance of the magnetic path through which the flux ϕ flows. Hence it is expected that the flux ϕ and flux density B will decrease from $B_r A$ and B_r, respectively. To locate the point on the hysteresis loop of Fig. 10.10 which will correspond to the new flux density in the iron (and therefore also in the gap, since $B_i = B_g$), we need another equation in terms of the coordinates of Fig. 10.10, which are H_i and B_i. We can obtain this from Ampère's law $\oint \mathbf{H} \cdot \mathbf{dl} = NI$, with NI equal to zero, which gives us for the circuit of Fig. 10.9

$$H_g l_g + H_i l_i = 0 \tag{10.26}$$

where H_i is the magnetic field strength inside the iron material. Again from the continuity of flux in the series circuit: $\phi_i = \phi_g$, which gives for the normal components of flux density (ignoring fringing):

$$B_i = B_g \tag{10.27}$$

which can be written as $H_g = B_i/\mu_0$. Substituting this in (10.26), we obtain an equation of a straight line

$$\boxed{B_i = \frac{-l_i}{l_g} \mu_0 H_i} \tag{10.28}$$

which is the additional relationship between B and H in the iron and which involves the geometry of the circuit. This line is known as the shearing line and is shown intersecting the *BH* curve in Fig. 10.11. The point of intersection is the

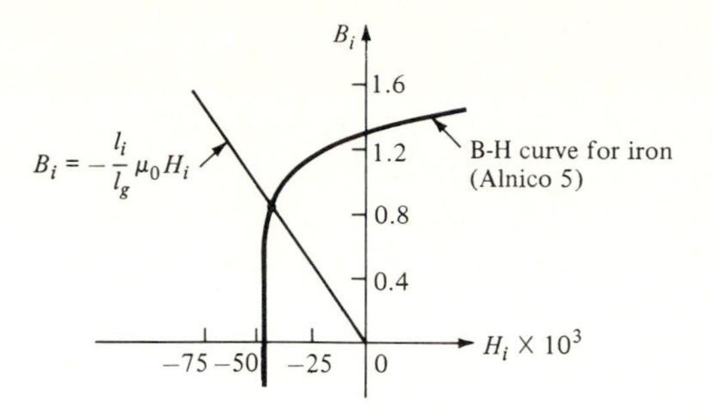

Figure 10.11 The intersection between the shearing line and the demagnetization curve gives the magnetic field in the permanent magnet of Fig. 10.9.

solution of two simultaneous equations [Eq. (10.28)] and the equation for the hysteresis loop. The problem is now solved because knowing B_i, we know B_g and the total flux $\phi = B_g A$ in the gap. Notice that increasing the reluctance of the gap by making the gap l_g larger, shifts the point of intersection farther down the *BH* curve. The result is a smaller flux in the gap.†

Several points merit additional discussion. The equation of the shearing line tells us that an *H* field now exists in the iron and furthermore that the *H* field is opposite in direction to B_i, as shown in Fig. 10.9. The *H* field inside the iron and in the gap is created by the interruption of the magnetization *M* which results when the gap is cut into the toroid. As shown in Sec. 9.3, this leads to a magnetic surface charge density $\rho_{sm} = \mathbf{M} \cdot \hat{\mathbf{n}}$ on the gap faces which acts as the source for the *H* field. The surface charge density is shown in Fig. 10.9 and can be determined by first finding **M** as follows: The point of intersection in Fig. 10.11 determines B_i and H_i. Using the relation $\mathbf{B}_i = \mu_0(\mathbf{H}_i + M)$, we can determine *M* as‡

$$\mathbf{M} = \frac{\mathbf{B}_i}{\mu_0} - \mathbf{H}_i \tag{10.29}$$

The presence of the magnetic charges on the pole faces leads to the negative *H* field inside the magnet which has a demagnetizing effect on the material. We can now say, in general, that whenever a magnetic flux is created in free space by a permanent magnet, the magnet itself is subject to a demagnetizing field.

For practical purposes we have assumed here that the field in the gap is uniform and the fringing field outside the gap can be ignored as being small. Should we need a more accurate solution to this problem, however, we can use the rigorous analysis developed in Secs. 9.3 and 9.4. Recall that in these sections the fields were derived in terms of magnetization **M** without specifying how **M** is to be obtained. The primary contribution of this section, besides an approximate solution to the magnetic field in the gap of a permanent magnet, is the determination

† To ensure permanence in a magnet, an aging process is employed during manufacture. See P. P. Cioffi, Stabilized Permanent Magnets, *Trans. AIEE*, vol. 67, pp. 1540–1543, 1948.

‡ The relation $\mathbf{B}_i = \mu\mathbf{H}_i$, when used for a permanent magnet, gives a negative permeability μ. Such a result should not be surprising because the operating point on the *BH* curve falls in the second quadrant, where H_i is negative.

of M from the hysteresis curve of the permanent magnet materials and the geometry of the magnet as incorporated in the equation of the shearing line. It should be also pointed out that the rod magnet shown in Figs. 9.10 and 9.15 can be considered as a special case of the toroid of Fig. 10.9 with a large gap. However, in both cases the flux leakage is so great that the magnetic circuit concept (in which the flux is confined to well-defined paths) is difficult to apply.

Example A permanent magnet is composed of three parts, as shown in Fig. 10.12. The source of the magnetic field is a bar of permanently magnetized hard material at the bottom of the magnet labeled **PM**. The cross section of the **PM** part is A_i. The horn-shaped sides are of soft ferromagnetic material such as soft iron, and their function is to concentrate the flux and guide it to the gap. The cross section is A_{si}. The third part is the air gap which makes the magnetic field available. The cross section of the gap is A_g. If the **PM** part of the permanent magnet is composed of Alnico V, and if the dimensions of the structure are $l_i = 5$ cm, $l_g = 2$ mm, $A_i = 2$ cm^2, $A_g = 1$ cm^2, determine the flux in the gap neglecting leakage.

It should be first pointed out that soft iron is not permanent-magnet material. The hysteresis is negligible as compared with that of a hard material. Using Ampère's law, we can write for the circuit of Fig. 10.12

$$H_i l_i + H_{si} l_{si} + H_g l_g = 0$$

Since the flux is continuous, $\phi = B_{si} A_{si} = B_g A_g = B_i A_i$, and we can express the above equation as

$$H_i l_i + B_{si} A_{si} \mathscr{R}_{si} + B_g A_g \mathscr{R}_g = 0$$

where the reluctances of the soft-iron path and of the air-gap path are $\mathscr{R}_{si} = l_{si}/\mu_{si} A_{si}$ and $\mathscr{R}_g = l_g/\mu_0 A_g$, respectively. We observe that $\mathscr{R}_g \gg \mathscr{R}_{si}$ because $\mu_{si} \gg \mu_0$. Neglecting $\mathscr{R}_{si}$, the above equation becomes

$$H_i l_i + B_g A_g \frac{l_g}{\mu_0 A_g} \cong 0$$

which leads to the final expression for the shearing line

$$B_i = -\mu_0 H_i \frac{l_i A_g}{l_g A_i} = -1.57 \times 10^{-5} H_i \tag{10.30}$$

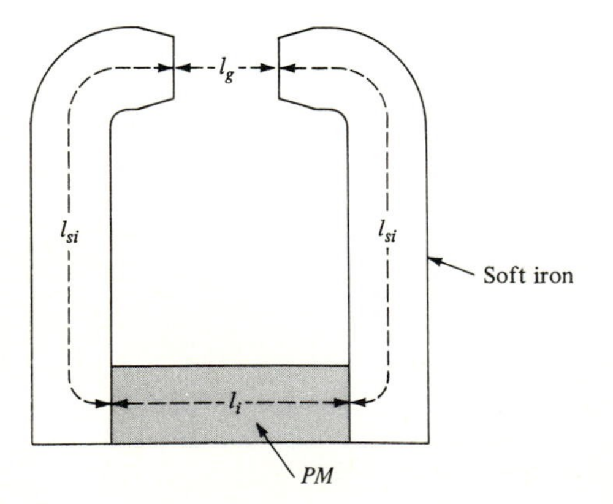

Figure 10.12 A typical permanent magnet consisting of a hard ferromagnetic material, a soft-iron material which guides the flux to the gap, and the gap itself.

When this line is superimposed on the demagnetization curve for Alnico V (Fig. 10.10), we obtain a point of intersection at (H_i, B_i = 43,000 ampere-turns/m, 0.66 Wb/m^2). The flux in the gap is, therefore,

$$\phi = B_i A_i = B_g A_g = (0.66 \text{ Wb/m}^2)(2 \times 10^{-4} \text{ m}^2) = 1.3 \times 10^{-4} \text{ Wb}$$

Quality Criterion for Permanent Magnets

In Table 9.2 the energy product $(BH)_{\max}$ is listed for various permanent-magnet materials. This figure is probably the best single figure for judging the quality of hard magnetic materials. A magnet that operates at $(BH)_{\max}$ provides a specified amount of flux with a minimum of magnetic material. This criterion is derived as follows: The volume v_i of the magnetic material can be obtained from $H_i l_i + H_g l_g = 0$ and $B_i A_i = B_g A_g$ as

$$v_i = l_i A_i = \frac{B_g^2 l_g A_g}{\mu_0(-B_i H_i)} \tag{10.31}$$

If the flux density in the gap B_g and the volume of the gap $l_g A_g$ is specified, the smallest volume of permanent-magnet material occurs when $B_i H_i$ is a maximum. The energy density product $(BH)_{\max}$ thus plays an important role in the practical design of permanent magnets. For a given material one should therefore choose to operate the magnet at a point on the demagnetization curve at which the product $B_i H_i$ is a maximum. Demagnetization curves for some hard materials and the points of maximum $B_i H_i$ are shown in Fig. 10.13.

As mentioned before, during the magnetization process of a permanent magnet, a *keeper*, which is a piece of soft iron, is inserted in the slot. This decreases

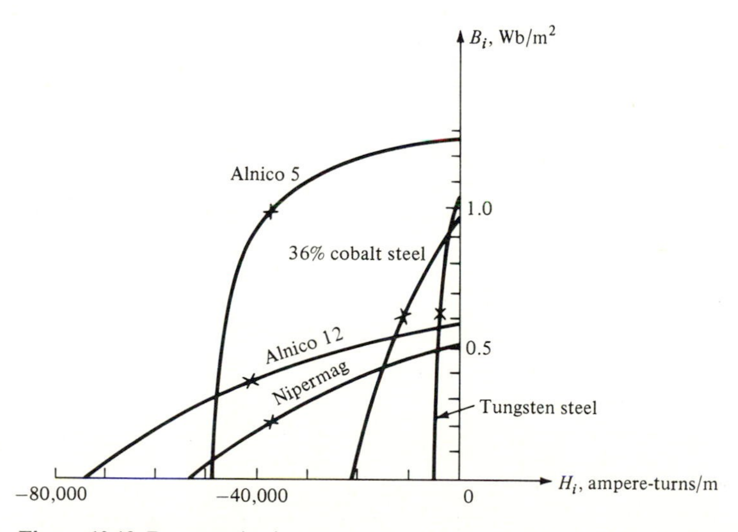

Figure 10.13 Demagnetization curves for Alnico V, Alnico XII, Nipermag, tungsten steel, and 36 percent cobalt steel. The points of maximum BH are marked with a cross.

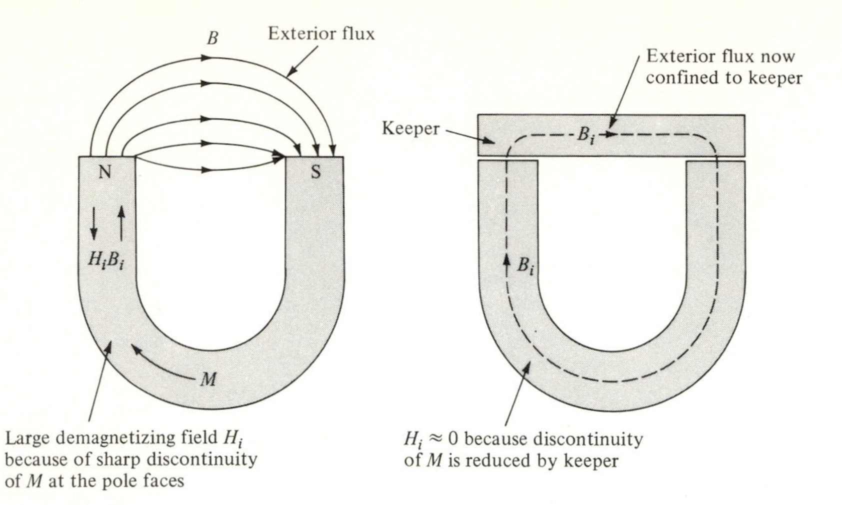

Figure 10.14 A soft-iron keeper provides a low-reluctance path for the field in the air gap. It also decreases the demagnetizing field H_i in the magnet.

the reluctance of the magnetizing path and permits the B field in the permanent-magnet material to reach the maximum value for the material. Similarly to reduce the high demagnetizing field that is present in a permanent magnet, a soft-iron keeper, as shown in Fig. 10.14, should be placed in the air gap of a permanent magnet that is not in use. Otherwise the permanent magnet has a tendency to demagnetize spontaneously.

Demagnetization

If it is desired to demagnetize a permanent magnet, it can be accomplished by heating or severely jarring the magnet. A more suitable method, however, might be demagnetizing or deperming by application of an alternating B field which decreases in amplitude. For example, many of us are familiar with the demagnetization of tape recorder heads which produce noise and hiss when acquiring a permanent magnetization. A device which produces an alternating B field is held next to the recorder head and then slowly withdrawn. The demagnetization process is shown in Fig. 10.15. If the rate of withdrawal is slow, the device to be demagnetized will be left in a demagnetized state with $B = 0$ because magnetization M was reduced to zero.

10.7 LIFTING FORCE OF MAGNETS

In Secs. 5.14 and 5.16 the forces on charged conductors and polarized bodies were derived, using the law of conservation of energy. For example, (5.72) gave the force when the charges in a system were held constant (external source did not supply

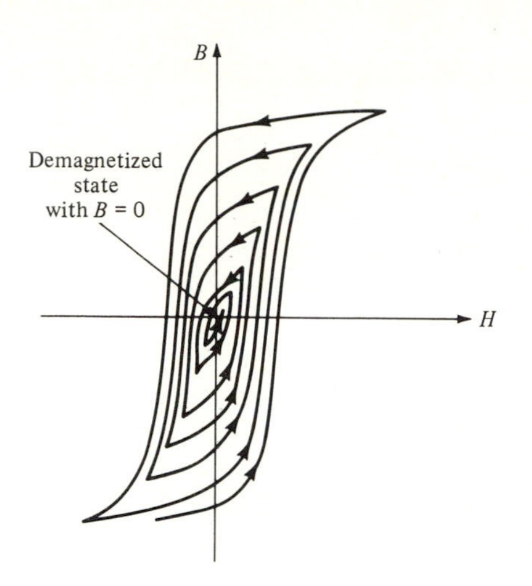

Figure 10.15 An object that has acquired a permanent magnetization can be demagnetized by application of a reversing and decreasing magnetic field.

energy to the system), whereas (5.79) gave the force when the potential was kept constant (external source now supplies or removes energy in order to hold V constant).

Using the law of conservation of energy, we can derive the forces or torques acting on current-carrying conductors and on magnetized bodies. For the case where the flux in a circuit is kept constant (the sources supply no energy to the system), the x component of the magnetic force acting on a rigid body is given by†

$$\boxed{F_x = -\left.\frac{\partial W}{\partial x}\right|_{\phi \text{ constant}}} \tag{10.32}$$

where ∂W is the change of energy stored in the magnetic field. The mechanical work that is done in this case is accompanied by a corresponding decrease of magnetic energy. The magnetic force when the currents are kept constant is given by

$$\boxed{F_x = +\left.\frac{\partial W}{\partial x}\right|_{I \text{ constant}}} \tag{10.33}$$

In this case, the mechanical work done is accompanied by an increase of magnetic energy, both being supplied by the external sources.

Before we consider the lifting force of a magnet, let us observe that magnetic materials, such as iron, will always move or be pulled into regions of stronger magnetic field. The reasons for this phenomenon are the same as those for the analogous case of dielectric objects which also tend to move toward regions of stronger electric field (see Sec. 5.16). What happens is as follows: A ferromagnetic

† In general we can say that force $\mathbf{F}$ (a vector) and energy W (a scalar) are related by $\mathbf{F} = -\nabla W$ and $W = -\int \mathbf{F} \cdot d\mathbf{l}$.

object which is brought into the field of a magnet will have poles induced in such a way that the induced north pole is closer to the south pole of the magnet or the induced south pole is closer to the north pole of the magnet. The result is a net attraction of the object by the magnet.† If the object is diamagnetic, however, it will experience a net repulsion, as shown in Fig. 9.20. The diagrams shown in Fig. 5.19 apply directly to the case here if we simply replace the dielectric object by a ferromagnetic object and replace all plus and minus charges by plus magnetic charges (north poles) and minus magnetic charges (south poles).

An alternative approach to understanding why magnets attract iron objects is to view the space between an object and magnet as a gap. Since magnetic poles of opposite polarity exist on the ends of the air gap, they will attract each other, and the effect of the magnetic field will be to exert a force which tends to close the gap. This is the reason why a relay shown in Fig. 10.16*a* closes and why an electromagnet shown in Fig. 10.16*b* can support weight.

Let us determine the lifting force of the electromagnet shown in Fig. 10.16*b*. The cross-sectional area is A, and the small air gap x between the upper and lower sections allows us to neglect fringing. The energy stored in each air gap is given, using (7.45), as $W = \frac{1}{2}\mu_0 H^2 v$, where $v = Ax$ is the volume of the air gap. If we assume that the current I in the coil is allowed to change so as to keep the flux $\phi = BA$ through the magnetic circuit constant, a change in the magnetic energy ΔW of the system can occur only as change of energy stored in the air gap.‡ Therefore, if a mechanical force pulls the two half sections apart an additional distance Δx, the work it must do is given by

$$\begin{aligned}\Delta W_{\text{magn}} &= 2[\tfrac{1}{2}\mu_0 H^2 A(x + \Delta x) - \tfrac{1}{2}\mu_0 H^2 Ax] \\ &= \mu_0 H^2 A\,\Delta x \qquad (10.34)\end{aligned}$$

where the factor 2 accounts for the presence of two air gaps. The two energies involved must be equal, $\Delta W_{\text{mech}} = \Delta W_{\text{magn}}$, which gives us for the magnetic force

$$\boxed{F = \mu_0 H^2 A = \frac{B^2 A}{\mu_0} = \frac{\phi^2}{\mu_0 A} \qquad \text{N}} \qquad (10.35)$$

If only one air gap is present (as in Fig. 10.16*a*), the magnetic force is half that given by (10.35). The magnetic pressure, which is defined as force per unit area, on each pole face can now be expressed as $P = F/A = \frac{1}{2}\mu_0 H^2$ newtons per square meter and is seen to be equal to the energy density of the magnetic field in the air gap. The magnetic pull could have also been calculated directly from (10.32).

An interesting observation about (10.35) is that the magnetic pull is inversely

† Of course, if the field that induces the north–south dipole in the object is uniform, the object experiences no net force. As shown by (6.49) or (4.7), a dipole does not move in a uniform field.

‡ As long as the flux density in the air gap is the same, the lifting force produced by the magnetic field will be the same whether it is an electromagnet or a permanent magnet. Hence, the equations apply also to permanent magnets.

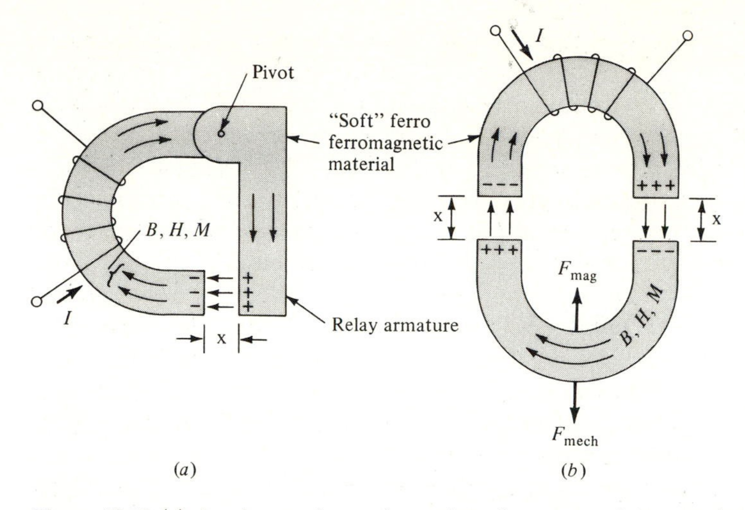

Figure 10.16 (*a*) A relay tends to close when the current I is turned on. (*b*) The horseshoe electromagnet exerts an upward force F_x on the horseshoe-shaped iron bar. Magnetic charges due to interruption of the induced magnetization M are shown.

proportional to the area A of the pole faces. This means that for a constant flux ϕ around the circuit, greater magnetic pull will be obtained when the flux is forced through a smaller pole-face area. This effect is put to use in lifting magnets by rounding the pole faces. However, there is a limit to how far the pole faces can be decreased before excessive fringing takes place, the effect of which is to increase the pole-face area A again. Also, the reluctance of the magnetic circuit increases with excessive rounding, and the magnetic material may saturate which in turn decreases the total flux ϕ through the circuit. The following example will clarify these concepts further.

Example An electromagnet shown in Fig. 10.17 has two unequal pole faces. Find the lifting force of each pole if pole P_1 has a cross-sectional area of 6 cm^2 and P_2 an area of 3 cm^2. Assume the coil produces a flux of $\phi = 10^{-3}$ Wb around the circuit. Using (10.35), we obtain for the magnetic pull of P_1

$$F_1 = \frac{(10^{-3})^2}{(2)(4\pi \times 10^{-7})(6 \times 10^{-4})} = 660 \text{ N}$$

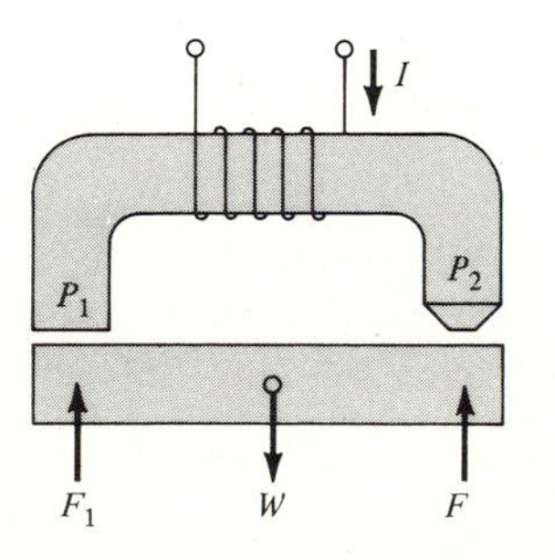

Figure 10.17 An electromagnet with unequal pole faces.

which is equal to 660 × 0.225 = 148 lb. The magnetic pull of the rounded pole P_2 is

$$F_2 = \frac{(10^{-3})^2}{(2)(4\pi \times 10^{-7})(3 \times 10^{-4})} = 1320 \text{ N}$$

which is equal to 1320 × 0.225 = 296 lb. The pull of the smaller pole is therefore twice as much even though the magnetic flux is the same for both poles, provided there is no excessive fringing at P_2. This effect can be easily demonstrated by attaching a weight W at the center of the lower bar. Increasing W, we will always find that pole P_1 opens before P_2 lets go.

10.8 THE TRANSFORMER

A transformer and its schematic representation are shown in Fig. 10.18. A transformer is a device that allows a change in ac voltage and in impedance to be made without appreciable loss of power. A magnetic flux ϕ induced in the ferromagnetic core (a soft material, usually iron) by the current I_1 flowing in the primary winding N_1 is guided to and intercepted by the secondary winding N_2. According to Faraday's law [Eq. (10.3) or (11.1)], a changing magnetic flux through a coil will induce a voltage in it, tending to oppose the change in flux through the coil; that is,

$$\mathscr{V}_1 = -N_1 \frac{d\phi}{dt} \qquad \mathscr{V}_2 = -N_2 \frac{d\phi}{dt} \tag{10.36}$$

The induced emf in the primary is thus equal and opposite to the applied ac voltage (we are neglecting the resistance of the coils). Note that we have assumed that all the flux flows through the coil and none exists outside the core. This assumption is valid because the high permeability of the iron results in a concentration of the flux ϕ in the core. The above expressions give us the basic transformer equation which expresses the voltage ratio in terms of the turns ratio as

$$\boxed{\frac{\mathscr{V}_1}{\mathscr{V}_2} = \frac{N_1}{N_2}} \tag{10.37}$$

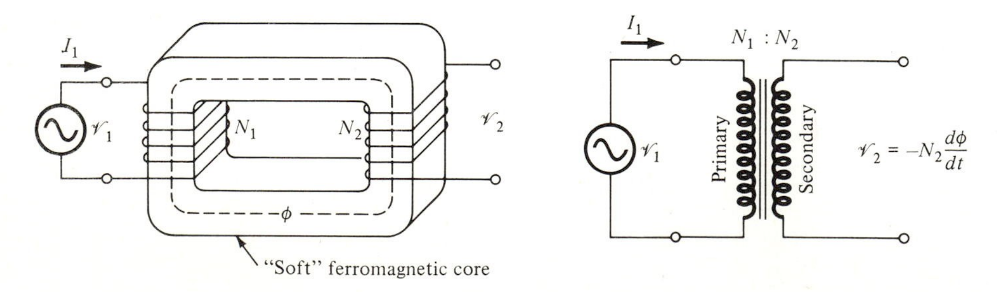

Figure 10.18 A transformer and its schematic representation. In the setup shown, a source of voltage $\mathscr{V}_1$ is connected to the primary with the result that a current I_1 flows in the primary and a voltage $\mathscr{V}_2$ is induced in the open-circuited secondary.

Thus a changing voltage $\mathscr{V}_1$ can be stepped up or stepped down by changing the turns ratio appropriately.

Let us now calculate the relationship between current and voltage in the primary, again with no load connected to the secondary. Since the flux is confined to the transformer core, it does not matter whether the primary winding is concentrated or spread uniformly along the entire core. Therefore, we can use the results for a toroid [Eqs. (6.44) or (7.25)], which give the relationship between the flux and the current as $\phi = BA = \mu HA = \mu(N_1 I_1/l)A$. Furthermore, if the current varies sinusoidally as $I_1 = I_0 \sin \omega t$, the induced voltage in the primary is (for simplicity we are approximating μ of the ferromagnetic core by a constant μ)

$$\mathscr{V}_1 = -N_1^2 \frac{\mu A}{l} I_0 \omega \cos \omega t = \frac{-N_1^2 I_0 \omega \cos \omega t}{\mathscr{R}} \tag{10.38}$$

where $\mathscr{R}$ is the reluctance of the magnetic circuit. This equation shows that flux and current both vary sinusoidally and are in phase but that current I_1 and induced voltage $\mathscr{V}_1$ are out of phase by 90°, a result to be expected since we have assumed the coil to be a pure inductance. If the reluctance $\mathscr{R}$ for the magnetic circuit is small, as it is for good transformer material, the current I_1 that flows in the primary (with the secondary open-circuited) is usually very small. This small current is needed to periodically magnetize the core in accordance with the alternations of current (hysteresis losses). Thus, in practical situations there is some power loss because of hysteresis. Additional losses occur because of eddy currents and the finite resistance of the primary winding. The induced emf in the transformer is then just different enough from the applied voltage to allow this small magnetizing current to flow. We should keep in mind that for transformers of good quality the magnetizing loss is a small fraction of the energy consumed by a load which is placed across the secondary. With a loaded secondary the magnetizing current is a small fraction of the total current that flows in the primary circuit.

To find the relationship between the current I_1 in the primary and current I_2 in the secondary, we must connect a load R_L to the secondary winding. This allows current I_2 to flow in the secondary loop, as shown in Fig. 10.19. The new current will contribute an additional flux $\phi_2 = B_2 A = \mu(N_2 I_2/l)A = N_2 I_2/\mathscr{R}$ in the core which will oppose the flux produced by I_1. The reason for the opposing flux is given by Lenz' law which states that in case of a change in a magnetic

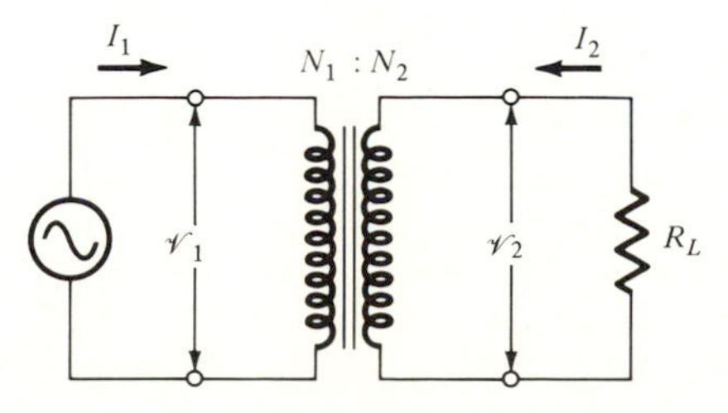

Figure 10.19 A transformer with a load R_L connected to the secondary. R_L allows a current I_2 to flow in the secondary.

system, a flux will be induced which opposes the change. Hence, the new, decreased flux in the core is given by

$$\phi = \phi_1 - \phi_2 = \frac{1}{\mathscr{R}}(N_1 I_1 - N_2 I_2) \tag{10.39}$$

However, if the applied voltage to the primary, $\mathscr{V}_1$, remains constant, the flux ϕ through the core must also remain constant. This can be seen from (10.36), which for a sinusoidally varying flux $\phi = \phi_0 \sin \omega t$ gives $\mathscr{V}_1 = -N_1 \omega \phi_0 \cos \omega t$. If $\mathscr{V}_1$ remains the same, so must ϕ_0. Therefore, (10.39) is incorrect. For the flux to remain constant, the current in the primary must increase to compensate for the effect of the secondary current; that is,

$$\phi = \frac{1}{\mathscr{R}}[N_1(I_1 + I_1') - N_2 I_2] \tag{10.40}$$

where I_1' is the additional current that flows in the primary. Equating (10.40) to $\phi = N_1 I_1/\mathscr{R}$, we obtain $N_1 I_1' = N_2 I_2$. Since in a loaded transformer the alternating magnetizing current is usually quite small, I_1' is a good approximation to the total primary current, that is, $I_1 + I_1' \approx I_1'$. Dropping the prime on I_1', we obtain

$$\boxed{\frac{I_1}{I_2} = \frac{N_2}{N_1}} \tag{10.41}$$

as the relationship between the primary and secondary current.† Unlike the voltage ratio of (10.37), the current ratio is inversely proportional to the turns ratio N_1/N_2.

To recap: The effect of the current in the secondary will be to decrease the flux and hence to decrease the induced emf in the primary. Since the applied voltage to the primary remains constant, more primary current can flow, and just enough flows to balance the mmf ($= N_2 I_2$) of the current in the secondary. This brings the emf's and the flux in the core back to the same value they had before a load was connected to the secondary.

Combining (10.37) and (10.41), we can write

$$V_1 I_1 = V_2 I_2 \tag{10.42}$$

which shows that the power input is equal to the power output. This is only true if losses are neglected. For good transformers, losses amount to a few percent. Hysteresis losses are reduced by using a soft magnetic material for the core. Eddy-current losses (see Sec. 10.2) which increase with frequency are minimized at the low frequencies (such as 60 Hz) by breaking up the eddy-current path. This is done by slicing the core into many thin strips. The resulting laminations are then covered with insulating material and reassembled into a form of the original core.

† Combining (10.37) and (10.41), we see that impedances are transformed as $Z_1/Z_2 = (N_1/N_2)^2$.

The flow of the induced eddy current is now greatly reduced since the eddy current cannot flow across the insulated laminates. For higher frequencies, the cores consist of finely divided iron particles held together by an insulating material. For still higher frequencies, the resistance of the core must be even higher. Ferrites, which are insulating magnetic materials, are then used (see Sec. 9.8 and Table 9.2).

Example A power transformer is designed to operate at 110 V and a frequency of 440 Hz. If this is a well-designed transformer, can it operate at 110 V and 60 Hz?

Before we can answer this question, we must derive the relationship between voltage, frequency, and flux. If the applied voltage is sinusoidal, we can assume that the flux is also sinusoidal. From (10.36) and (10.38) it can be seen that if $\phi = \phi_{max} \sin \omega t$, the counter emf is given by

$$\mathscr{V}_1 = -N_1 \phi_{max} \omega \cos \omega t \tag{10.43}$$

Again, if we can ignore the resistance of the primary, the induced voltage (10.43) must be very nearly equal to the applied voltage. Expressing the above voltage in terms of rms values, we obtain

$$\boxed{\mathscr{V}_{rms} = 4.44 N f \phi_{max} = 4.44 N f A B_{max}} \tag{10.44}$$

where f is the frequency ($\omega = 2\pi f$), A is the cross section of the core, and $\mathscr{V}$ is in volts if flux ϕ is expressed in webers. Equation (10.44) is one of the most useful expressions for transformer design. It shows that for the voltage to remain constant, the flux ϕ must increase as the frequency is decreased. If the transformer is well designed for operation at 440 Hz, the current in the primary swings up to the "knee" on the hysteresis loop, as shown in Fig. 10.20. Thus in a well-designed transformer the flux is limited in how much it can increase; flux increases beyond the knee can only be small because B becomes saturated. As the frequency is decreased, the flux must increase so that the counter emf generated by the changing flux nearly equals the applied voltage, and the current in turn must readjust to produce this flux. In our example flux must increase by a factor of 440/60 because frequency was reduced by this factor. The current I will therefore quickly exceed tolerable values. The I^2R losses will increase to the point that the transformer will overheat and soon become inoperable. This is known in the trade as not having enough "iron" to operate at 60 Hz. Equation (10.44) shows that by increasing the cross section A of the core sufficiently, B_{max} can be held to values below the knee of the hysteresis curve. This also explains why high-quality audio transformers which must operate at low frequencies are bulky.

The converse of the above case, a transformer designed to be used at 60 Hz will operate with too much "iron" at 440 Hz. Only a small portion of the linear part of the BH curve is used as the current swings back and forth, clearly very inefficient. To increase the efficiency, the cross section A of the core should be decreased until B_{max} reaches values near the knee of the BH curve.

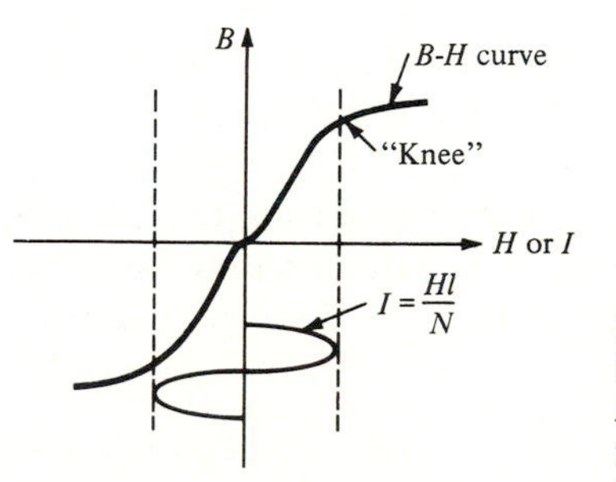

Figure 10.20 A typical BH curve for soft iron which is used in the cores of power transformers is shown. From Ampère's law the H axis is proportional to the current I that flows in the primary.

An alternative way of looking at this problem is to treat the primary winding as an RL circuit. From Eq. (7.37) the relation between current and voltage for such a circuit is given by

$$\mathscr{V} = RI + L\frac{dI}{dt} \tag{10.45}$$

If the current varies sinusoidally as $I = I_0 \sin \omega t$ and if the resistive voltage drop is quite small, we can write

$$\mathscr{V} = RI + L\omega I_0 \cos \omega t \cong L\omega I_0 \cos \omega t \tag{10.46}$$

where L is the inductance of the primary which, by comparison to (10.38), is given by

$$\boxed{L = \mu \frac{N^2 A}{l} = \frac{N^2}{\mathscr{R}}} \tag{10.47}$$

The permeability μ, which appears in the expression for inductance, is proportional to the slope of the BH curve (see Fig. 10.23). Thus as the current increases and drives H beyond the region of the knee into the flat portion of the BH curve, the inductance decreases. If this happens, the current I must increase even more to compensate for the decreasing L. The result will be that the inductive term decreases and the resistive term increases to the point that the approximation in (10.46) is not valid anymore. If this is a power transformer, lowering the frequency will overheat the transformer (due to the large I^2R losses) unless the applied voltage is correspondingly lowered. In an audio transformer this loss of inductance will result in a decrease of response at the lower frequencies.

10.9 SELF-INDUCTANCE AND MUTUAL INDUCTANCE

Inductance was already discussed in Eqs. (7.28) and (7.47). In this section we will introduce mutual inductance and reexamine inductance. We will show in the next section that the introduction of an iron core in a coil greatly increases the inductance, but because ferromagnetic materials have a nonlinear BH curve, the inductance will also be nonlinear.

Self-Inductance

If we use the definition given in Sec. 7.7, we can state *self-inductance* as

$$\boxed{\begin{aligned} L = L_{11} &= \frac{\text{flux linking circuit } C_1 \text{ due to current in } C_1}{\text{current in } C_1} \\ &= \frac{\Lambda_{11}}{I_1} = \frac{N\phi_{11}}{I_1} \end{aligned}} \tag{10.48}$$

Using again the toroidal configuration (it concentrates flux to the inside of the coil and therefore minimizes stray flux), we obtain from (7.31) or Ampère's law

$$L = \frac{\Lambda}{I_1} = \frac{N\phi_{11}}{I_1} = \frac{N_1 BA}{I_1} = \frac{N_1(\mu N_1 I_1/l)A}{I_1} = N_1^2 \frac{\mu A}{l} \tag{10.49}$$

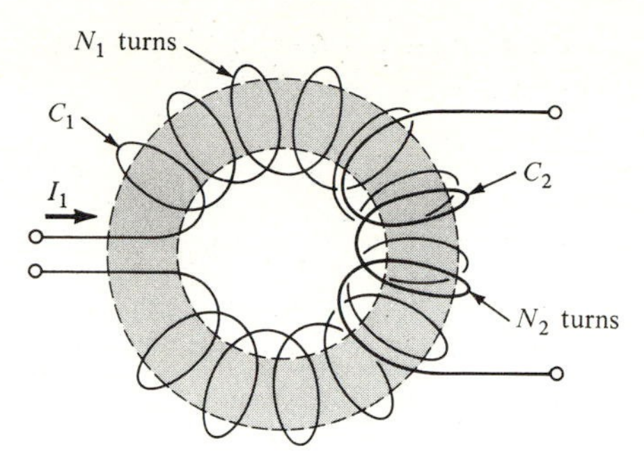

Figure 10.21 Circuit C_1 is a toroidal coil. Circuit C_2 is wound on top of C_1 so that maximum flux linkage exists between C_1 and C_2.

where A and l are the cross section and length of the toroidal path shown in Fig. 10.21. The definition of reluctance which is $\mathcal{R} = l/\mu A$ can now be used to express self-inductance as

$$L = \frac{N^2}{\mathcal{R}} \tag{10.50}$$

Note that for purposes of expressing the self-inductance of C_1, the effect of the secondary winding C_2 is ignored; i.e., we assume that winding C_2 in Fig. 10.21 is open-circuited or does not exist. The self-inductance is thus proportional to the number of turns squared and inversely proportional to the reluctance of the magnetic path. Although the above derivation was applied to a toroidal coil, we can apply (10.50) to any coil provided we can express reluctance readily. This should include all cases for which leakage flux is negligible [iron-cored coils, solenoids, but not the short solenoid of (7.30)].

Mutual Inductance

If two circuits are linked by a magnetic field, we can talk about *mutual inductance* and define it as

$$M = L_{12} = \frac{\text{flux linking circuit } C_2 \text{ due to current in } C_1}{\text{current in } C_1} = \frac{\Lambda_{12}}{I_1} \tag{10.51}$$

In circuit C_2 of Fig. 10.21, the flux created in C_1 will link C_2 and will induce a voltage in C_2 if the flux changes with time. The magnitude of this induced voltage is given by Faraday's law (10.36) as $\mathcal{V}_2 = -N_2\, d\phi/dt$. Since the flux is given by $\phi = BA = (\mu N_1 I_1/l)A$, we can write the induced voltage as

$$\mathcal{V}_2 = N_2 \frac{d\phi}{dt} = N_1 N_2 \frac{\mu A}{l} \frac{dI_1}{dt} \tag{10.52}$$

Using Kirchhoff's law for a series RLC circuit, Eq. (7.37), the relationship between voltage and current in an inductor is given as $\mathcal{V} = L\, dI/dt$. By comparison, the term in (10.52) which multiplies dI_1/dt is an inductance and therefore, must be the mutual inductance M:

$$M = L_{12} = N_1 N_2 \frac{\mu A}{l} \tag{10.53}$$

For magnetic circuits with little flux leakage the reluctance of the magnetic path is $\mathscr{R} = l/\mu A$. Mutual inductance can then be written as

$$M = \frac{N_1 N_2}{\mathscr{R}} \tag{10.54}$$

which is similar to the expression (10.50) for inductance. Both are proportional to the number of turns squared and inversely proportional to the reluctance of the magnetic path. Therefore, decreasing the reluctance (such as by the introduction of a ferromagnetic core) will increase the mutual and self-inductances.

Referring to Fig. 10.21, we see that the open-circuited voltage $\mathscr{V}_2$ can be written as

$$\mathscr{V}_2 = M \frac{dI_1}{dt} \tag{10.55}$$

Similarly, if the source is connected to coil 2, the induced voltage in coil 1 is $\mathscr{V}_1 = M\, dI_2/dt$.

For the case of maximum flux linkage between two coils, we can express mutual inductance M in terms of the self-inductances of the coils, which from (10.50) can be given as $L_1 = N_1^2/\mathscr{R}$ and $L_2 = N_2^2/\mathscr{R}$. Using the definition of M given in (10.54), we can express M as

$$M^2 = \frac{N_1^2 N_2^2}{\mathscr{R}^2} = L_1 L_2 \tag{10.56}$$

Since it was assumed that all the flux through one coil links the other coil, (10.56) gives the maximum possible value for M; that is,

$$M_{\text{max}} = \sqrt{L_1 L_2} \tag{10.57}$$

In general, the relation between M and the self-inductances is written as

$$M = k\sqrt{L_1 L_2} \tag{10.58}$$

where k is a coefficient of coupling ranging between 0 and 1. The coefficient is equal to 1 if all the flux from one coil links all the turns of the other (such as in iron-core transformers), a condition referred to as *tight coupling*. If much flux leakage exists, as, for example, when two coils in free space are far apart, the coefficient is small or might be near zero depending on the remaining mutual flux linkage.

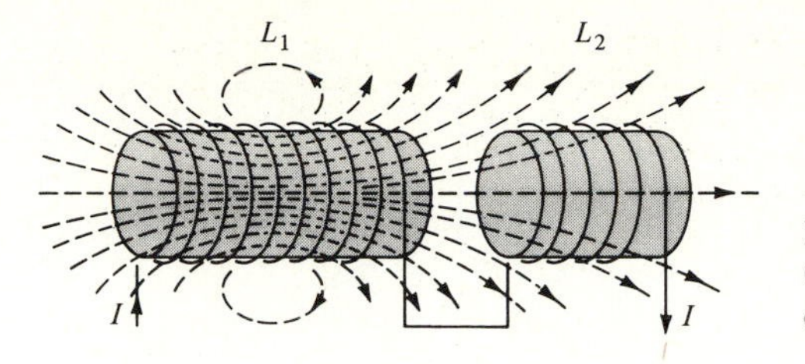

Figure 10.22 Two coils connected in series. Only the flux lines due to current flowing in coil 1 are shown.

Figure 10.22 shows two coils which are linked by mutual inductance. If a current I flows as shown, the inductance of the combination will not be simply $L_1 + L_2$ but will be modified by the mutual inductance M. The voltage across the pair of coils is given by

$$\mathscr{V} = L_1 \frac{dI}{dt} + L_2 \frac{dI}{dt} \pm 2M \frac{dI}{dt} \tag{10.59}$$

The sign depends on the phasing of the two coils (+ if both coils produce flux in the same direction, which is the case shown in the figure). The mutual inductance must be counted twice, once for the voltage induced in L_2 by dI_1/dt, and once for the voltage induced in coil 1 by current changes in coil 2. The self-inductance of the series combination of the two coils is therefore given by

$$L = L_1 + L_2 \pm 2M \tag{10.60}$$

10.10 IRON-CORE INDUCTORS AND TRANSFORMERS

Introducing a core of ferromagnetic material such as iron greatly increases the inductance of a coil. At alternating currents the nonlinear magnetization characteristics of the core cause the inductance to be variable. For nonconstant μ we define inductance as the ratio of infinitesimal change in flux linkage to infinitesimal change in current producing it, or

$$\boxed{L = \frac{d\Lambda}{dI}} \tag{10.61}$$

For linear media (μ constant) this expression reduces to (7.28), which is $L = \Lambda/I$. Using a toroidal coil configuration of length l, cross section A, and N turns, for which $\Lambda = N\phi = NAB$, $NI = Hl$, and $dI = (l/N)\, dH$, we obtain the result (valid approximately for any coil with small flux leakage)

$$\boxed{L = \frac{N^2 A}{l} \frac{dB}{dH}} \tag{10.62}$$

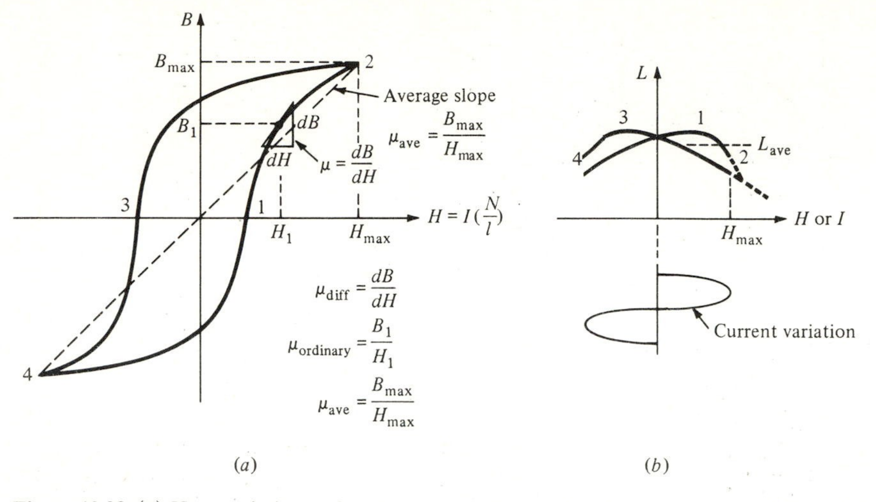

Figure 10.23 (*a*) Hysteresis loop of an iron-core inductor showing average and differential permeability; (*b*) inductance variations in a typical iron-core coil. Note that dB/dH is different from the ordinary permeability which is $\mu = B_1/H_1$.

Comparing this with (10.47) or (10.49), we observe that the inductance depends on the differential permeability $\mu = dB/dH$, which is the slope of the hysteresis curve as shown in Fig. 10.23. As the alternating current I (or magnetizing force H) sweeps through one cycle, the flux moves once around the hysteresis loop indicated by 12341 and differential inductance varies as shown in Fig. 10.23*b*. In practical situations one assumes an average value for inductance by using an average value for permeability; that is, $L_{ave} = (N^2A/l)\mu_{ave}$, where $\mu_{ave} = (dB/dH)_{ave} = B_{max}/H_{max}$. We observe now that as the current drives the flux into saturation, the inductance drops to its lowest value. The variations in inductance can be minimized by limiting the current swings since the smaller BH loops are more linear. As shown in Fig. 10.3, they are more oval-shaped, but become S-shaped as B is driven into saturation. For linear media, $\mu_{diff} = \mu_{ord} = \mu_{ave} =$ constant.

When transformers were considered (Sec. 10.8), the approximation that μ is a constant was used, which implied that the magnetization curve of iron was linear. This in turn implied that flux ϕ and current I vary sinusoidally when the applied voltage $\mathscr{V}$ varies sinusoidally. Taking the nonlinear magnetization characteristics of iron into consideration, we can show that the waveform of current I cannot be sinusoidal even when flux ϕ varies sinusoidally. If we apply a sinusoidal voltage source across an inductor (with negligible resistance), we have from Faraday's law that

$$\mathscr{V} = N\frac{d\phi}{dt} \qquad \text{or} \qquad \phi = \frac{1}{N}\int \mathscr{V}\, dt \tag{10.63}$$

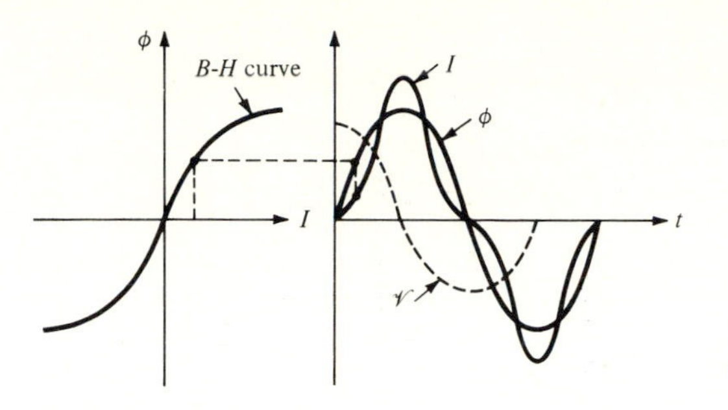

Figure 10.24 Distortion in current I by the nonlinearity of the BH curve.

which shows that flux varies sinusoidally when the voltage does. However, as shown in Fig. 10.23, flux is a nonlinear function of current, so that current variations associated with a sinusoidal flux variation must be obtained graphically by utilizing the hysteresis loop. For a very thin hysteresis loop the procedure to be utilized and the result are indicated in Fig. 10.24. Current is seen to be nonsinusoidal and peaks when ϕ is a maximum. The distortion in the current is due to the nonlinearity of the BH curve. The current peak can reach abnormally high values if the applied voltage is too large and drives the core deeply into saturation [see discussion following Eqs. (10.44) and (10.47)]. Additional distortion in I will be introduced when the finite width of the hysteresis loop is considered. In some circumstances, the harmonics introduced by the distorted current can be troublesome.

Iron-Core Inductors with a DC Current

In many applications an inductor must carry direct current as well as alternating current. For example, dc "choke" coils are used in power supplies to smooth ripple currents that exist on the large dc current delivered by the power supply. We will show that the presence of a dc biasing current decreases the inductance.

Figure 10.25a shows a large magnetizing force H_0 present, corresponding to the large dc current

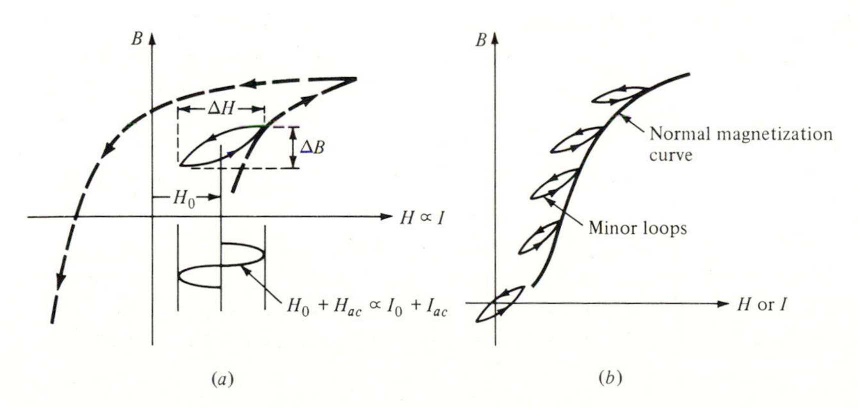

Figure 10.25 (a) Minor hysteresis loop due to superposed direct I_0 and alternating current I_{ac}; (b) minor loops for different values of direct current.

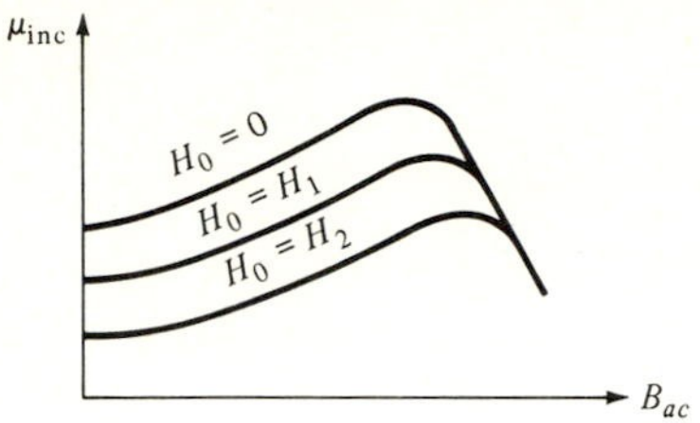

Figure 10.26 Typical incremental permeability versus amplitude of alternating flux density for iron. $H_2 > H_1$.

I_0 $(H_0 l = NI_0)$. The alternating current about I_0 sweeps out a minor hysteresis loop centered about H_0. The incremental permeability is defined as the average slope of the minor loop:

$$\mu_{inc} = \frac{\Delta B}{\Delta H} \tag{10.64}$$

where ΔB and ΔH are as shown in Fig. 10.25. The ac component thus experiences an inductance, given by (10.62) as $L = N^2 A\mu_{inc}/l$. Figure 10.25*b* shows the variation in minor hysteresis loops when the alternating component of the current is kept the same but the direct component is increased. Since the slope of the minor loops decreases, we conclude that μ_{inc} decreases for increasing dc biasing current. This is shown in Fig. 10.26 as a family of curves for different values of direct current and alternating current.

We can now make several observations: (1) The slope of the minor loops is much less than the slope of the normal magnetization curve. (2) The larger the direct component of the magnetizing force, the smaller the incremental permeability becomes. (3) Up to a certain value, the larger the alternating component of the magnetizing force, the larger μ_{inc}. (4) The minor hysteresis loops are oval-shaped.

10.11 EXAMPLE: MAGNETIC TAPE RECORDING†

As an example of the application of magnetism, we will study the principles of magnetic recording. Besides being useful, it will help us gain a better understanding of magnetism.

Magnetic recording can be divided into two parts. Transferring information from an electric signal onto the magnetic tape as a pattern of permanent magnetization is the first part. It is known as the *recording process*. Recovering the information which is in the distribution of permanent magnets that has been recorded on the magnetic tape and converting it back to an electric signal is the second part. It is known as the *playback process*.

Storing Electric Signals on Magnetic Tape

Magnetic tape is a composition of a plastic base and a coating of iron-oxide particles. The iron-oxide particles remain bonded in their random position, and it is only the magnetization **M** within the particles that changes in response to a

† H. F. Olson, "Acoustical Engineering," Van Nostrand Reinhold Company, New York, 1957. J. C. Mallinson, "Tutorial Review of Magnetic Recording," *Proc. IEEE*, February 1976.

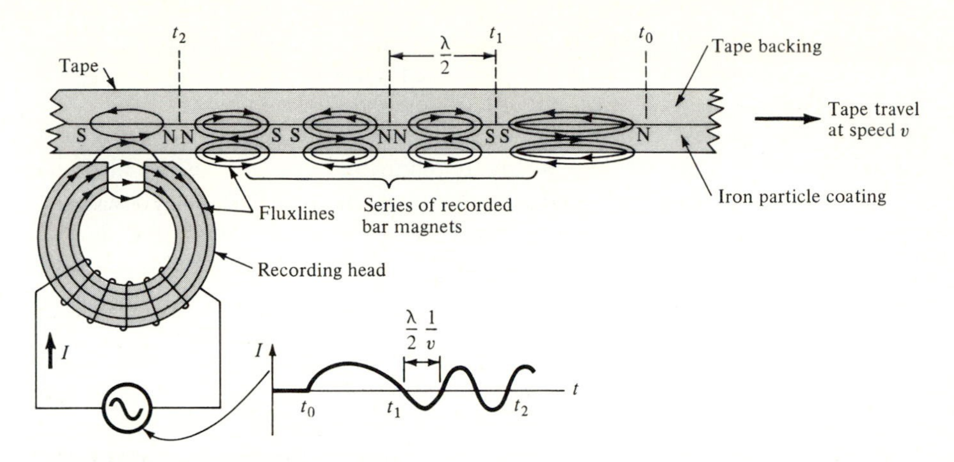

Figure 10.27 The leakage flux from the recorder head magnetizes the tape and leaves it with a pattern of permanent magnets which corresponds to the signal current I flowing in the recorder coil.

magnetizing force from the recording head. The magnetizing force H is derived from the current I flowing in the recording coil, as shown in Fig. 10.27. In this figure the tape is shown in direct contact with the gap of an electromagnet (the recording head). While the tape moves past the electromagnet, the fringing magnetic field from the gap penetrates the magnetic tape and magnetizes the iron-oxide coating. The magnetization of the tape is in a series of magnets with north and south poles in response to an alternating current in the coil. Two adjacent magnets, over which the polarity changes north to south and back again to north, correspond to one cycle of the recording signal; in other words, two adjacent but oppositely oriented magnets constitute one wavelength of the signal as recorded on the tape. This wavelength λ is equal to the tape speed v divided by the frequency f of the signal, or

$$\lambda = \frac{v}{f} \tag{10.65}$$

Thus, at a tape speed of 7.5 in/s, the wavelength of a 1500-Hz signal is 7.5/1500 = 0.005 in. The process of recording produces a series of bar magnets along the tape, each made up of many iron-oxide particles.† The strength of the bar magnets depends on the amplitude of the signal, i.e., the amplitude of the current I flowing in the coil. The length of the individual bar magnets depends on the tape speed and the frequency of the signal being recorded. For low audio frequencies the length of the magnets is about an inch, whereas for high audio frequencies the length is on the order of one-thousandth of an inch. Since 0.001 in (1 mil) is about

† The iron-oxide particles are of a hard ferromagnetic material which gives the tape coating a good permanent-magnet quality.

the upper limit for the length of the magnets, we can readily see from (10.65) that a higher tape speed will improve the high-frequency response.

Unlike magnetic tapes which are of hard ferromagnetic material with a retentivity B_r on the order of 0.05 to 0.1 Wb/m^2 (500 to 1000 G) and a coercive force H_c of about 2.4×10^4 A/m (300 Oe), magnetic recording heads are of soft magnetic material, high permeability μ, and very low coercivity, which leave recording heads essentially with no permanent-magnet qualities. This means that the recording coil can be driven with small currents and that no appreciable magnetization remains after the current or the magnetizing force H has been removed. The hysteresis loss, because of the small area of the hysteresis loop of soft magnetic material, has only a slight effect on the recording-head characteristics. Eddy-current losses in recorder heads are more important, especially at the higher frequencies. These are avoided by making the core material of thin lamination, insulated one from the other, which breaks up the eddy-current paths (for a discussion of eddy currents, see Sec. 10.2).

The gap in the recording heads, besides making the magnetic field available for recording on tape, has another important effect. It makes the magnetization curve of the recorder-head material more linear. Let us illustrate this with the aid of Fig. 10.28, which shows a recorder head with typical dimensions of $l_{\text{gap}} = 0.003$ cm and $l_{\text{iron}} = 1$ cm. For assembly convenience in manufacture, recorder heads have two gaps and two coils, as shown in Fig. 10.28*a*. It was shown in Eq. (10.24) that for materials with large μ_r most of the applied magnetomotive force $\mathscr{F}$, where $\mathscr{F} = NI = Hl$, is dropped across the air gap. However, since the air gap of recorder heads and playback heads is very small, an appreciable portion of the mmf is dropped in the iron path also. Curve $0AB$ shows the magnetization curve of a typical core material for recorder heads. For convenience we have

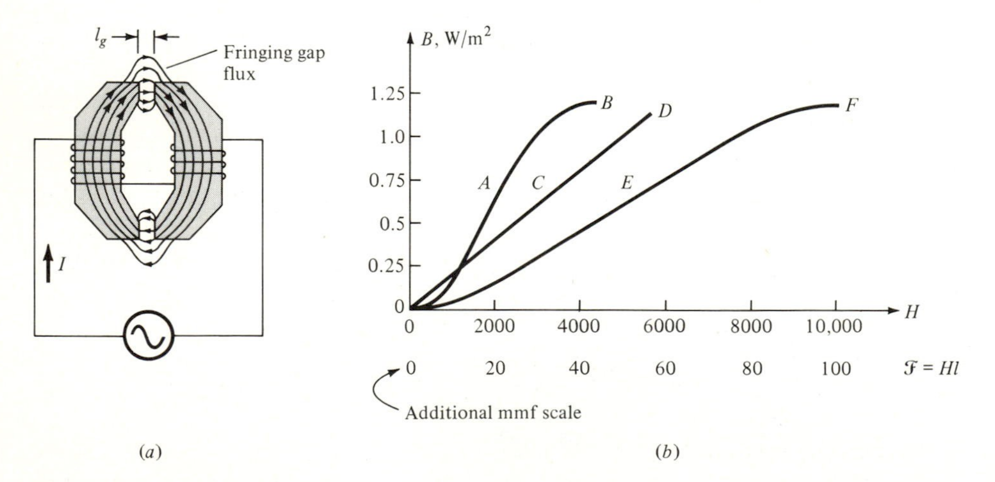

Figure 10.28 (*a*) Magnetic recording head, showing path of flux; (*b*) the effect of an air gap on the magnetization curve is shown by $0EF$. Note that magnetomotive force $\mathscr{F}$ is often abbreviated as mmf; that is, $\mathscr{F}$ = mmf.

drawn a second mmf scale, in addition to the H scale, which was obtained by multiplying H by $l_i + l_g \cong l_i = 1$ cm. The curve $0AB$ shows the mmf required for a given B in a core without an air gap. Cutting an air gap in the core will require more mmf to maintain the same flux density B, since the additional reluctance of the air gap must be overcome. Curve $0CD$ shows the linear relationship between B and $\mathscr{F}$ for the air gap and was obtained as follows: From (10.25) the mmf required to maintain a given B across the reluctance of an air gap is

$$\mathscr{F} = NI = H_g l_g = \frac{B}{\mu_0} l_g \qquad (10.66)$$

For $l_g = (2)(0.003) = 0.006$ cm and $\mu_0 = 4\pi \times 10^{-7}$ H/m, the slope of the linear graph $0CD$ is given by $B/\mathscr{F} = 1/50$. By adding together the corresponding values of mmf of the two curves $0AB$ and $0CD$, we obtain curve $0EF$ which gives the mmf required to maintain a given B in the magnetic circuit with an air gap. When this curve is compared to $0AB$, we see immediately that the gapped circuit increased the required mmf. Hence, a recording head with a larger gap is not as sensitive as one with a smaller gap. However, the introduction of an air gap makes the magnetization curve markedly more linear, which is essential for low distortion in the recording process.

Linearization of Magnetic Tape

The hard ferromagnetic materials used in tapes are of necessity nonlinear because of their wide hysteresis loops. If the signal to be recorded were applied directly to the recorder head, the magnetization produced on the tape would not be linear, and severe distortion would result upon playback. The nonlinearity can best be seen from a plot of residual magnetic flux density B_r versus magnetizing force H, which is shown in Fig. 10.29*a*. This curve is obtained by applying a small magnetizing force H and measuring the residual flux density B_r (also referred to as residual or remanent induction) after the magnetizing force H has been removed. The material is then demagnetized and the measurement repeated for a higher value of H. The resulting $B_r H$ curve is mostly nonlinear near low and high values of H, with a fairly linear region in between. Clearly, applying a recording signal current centered about zero will produce a severely distorted flux density in the tape. We can make use of the linear region by applying a fixed magnetizing force, a so-called dc bias, which brings the operating point halfway up the linear part, as shown in Fig. 10.29*b*. Such a biased operation produces superior results to unbiased operation, although the presence of dc magnetization increases the noise level of the tape. The fixed magnetic field can be produced by a permanent magnet or by a fixed direct current in the recorder-head winding in addition to the signal current to be recorded.

While the introduction of dc bias was a considerable improvement, a high-frequency bias technique developed later gives much better results with respect to signal-to-noise ratio and dynamic range of the recording. Consequently, it is used almost universally in audio recording. In this technique the recording signals are

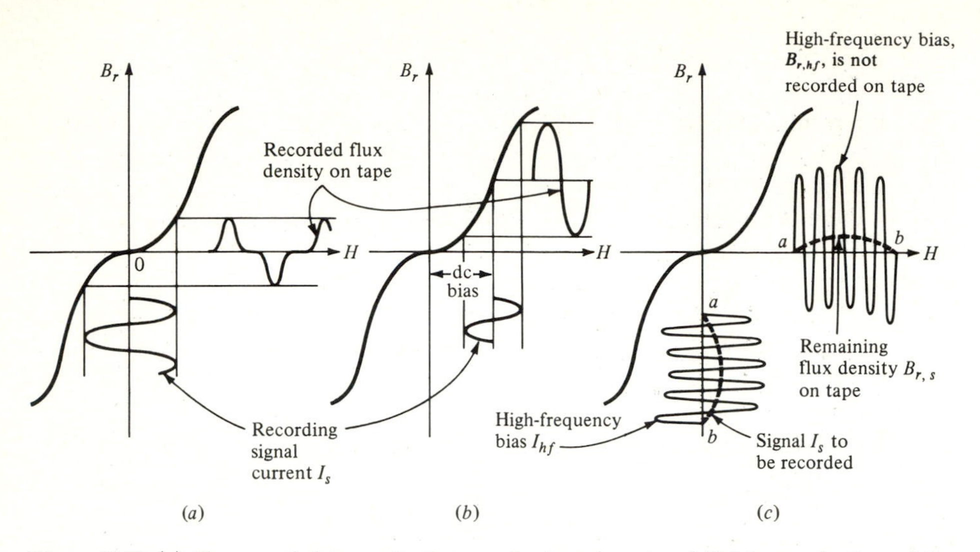

Figure 10.29 (*a*) The recorded magnetization on the tape is not a faithful reproduction of the recording current; (*b*) linear operation is achieved by the use of dc bias; (*c*) linearization by use of a high-frequency bias gives the best results.

superposed on an alternating magnetizing force whose amplitude and frequency (in the range of 50 to 150 kHz) are both considerably greater than those of the signal. The linearity between tape magnetization and applied magnetizing force H, and hence recording current, which the high-frequency biasing technique achieves is shown in Fig. 10.29*c*. The signal current I_s to be recorded is added to the high-frequency biasing current I_{hf} such that the current in the recording head I_{rh} is given by

$$I_{rh} = I_{hf} + I_s \tag{10.67}$$

since there is no mixing or modulation of one signal by the other. The recorded or remaining flux density on the tape should therefore be given by $B_r = B_{r,hf} + B_{r,s}$, as shown in Fig. 10.29*c*. However, due to the very high frequency of the bias, this portion of the signal is not recorded on the tape; this leaves the tape with

$$B_r = B_{r,s} \tag{10.68}$$

where $B_{r,s}$ is the residual flux density due to the signal and is linearly proportional to the recording current.

To understand why $B_{r,hf}$ is not recorded on the tape, we first observe from Fig. 10.27 that recording occurs at the trailing edge of the recording-head gap. The final magnetization in the tape is determined by the last flux encountered by the tape. The amplitude of the flux decreases with distance from the gap. Thus, if the gap flux does not change orientation in an element of tape as the tape advances, this element of tape will acquire a permanent magnetization. On the other hand, the effect of a rapidly reversing gap flux on an element of tape, as the element moves away from the gap, will be to demagnetize that section of tape and leave it with zero magnetization. This is because the gradually decreasing gap flux, which is a result of tape motion, will force each tape element to undergo a number of complete but decreasing hysteresis loops, thus effectively demagnetizing the tape (note that this is the same phenomenon discussed in Sec. 10.6 under demagnetization). If there is no recording current, I_{hf} alone will leave the tape unmagnetized. With a recording current I_s present, the high-frequency biasing flux is not symmetric about zero anymore but rather about a positive or negative flux value determined by the recording current. As the hysteresis loops decrease, they collapse about a point which is different from zero and corresponds to the recording current. The tape thus acquires a magnetization that is linearly proportional to the recording current I_s.

We can represent this more precisely by using the example of a sinusoidal recording current:

$$I_s = I_0 \sin \omega t$$

where $\omega = 2\pi f$ and f is the frequency to be recorded. Since H is proportional to I_s, and residual flux density B_r is proportional to H, we can write for the residual flux on the tape

$$\phi_r = kI_0 \sin \omega t \tag{10.69}$$

where k is a constant of proportionality. Changing from time t to linear coordinate x along the tape, and from frequency f to wavelength λ on the tape, we obtain for the above ϕ_r:

$$\phi_r = kI_0 \sin \frac{2\pi x}{\lambda} \tag{10.70}$$

where $t = x/v$
$\lambda = v/f$ from (10.65)
$ft = x/\lambda$

This is the expression for residual flux variation along the direction of the tape of a sinusoidal signal applied to the recording head.

Reproducing the Recorded Signal

The recorded signal which forms a series of permanent magnets on the tape can now be reproduced electrically by moving the tape past a gap of a playback head, which is usually identical to a recording head. As shown in Fig. 10.27, the field of the recorded magnets projects past the surface of the iron-oxide coating. A portion of the fringing field as it moves past the playback head gap is intercepted by the gap and continues to flow around the core because of the low-reluctance path provided by the core. According to Faraday's law,

$$\mathscr{V} = -N \frac{d\phi}{dt} \tag{10.71}$$

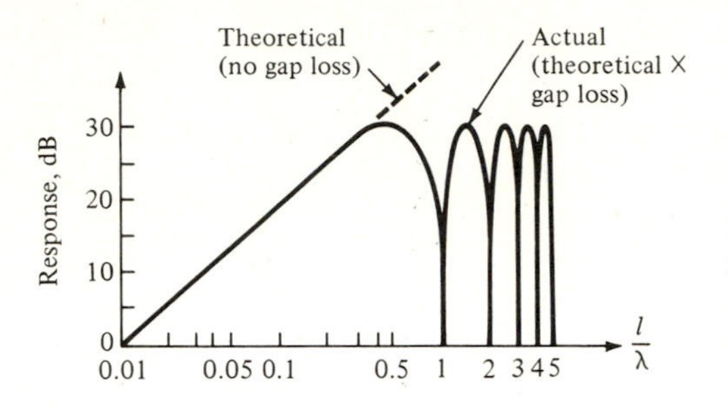

Figure 10.30 Theoretical and actual frequency response curves of a playback head as a function of the ratio of the gap length l to the wavelength λ. The abscissa is also proportional to frequency f since $l/\lambda = fl/v$.

A voltage, which is a faithful reproduction of the original signal, is induced in the playback coil and can now be amplified. If a sinusoidal signal (10.69) was recorded, the induced voltage is

$$\mathscr{V} = k_1 N I_0 \omega \cos \omega t \tag{10.72}$$

where k_1 is a constant. Equation (10.72) shows that the output voltage increases with frequency $f = \omega/2\pi$, a result not obvious at first sight. The frequency dependence arises because the voltage from the playback head is proportional to the rate of change of flux and not to flux itself; i.e., the recovered signal is actually the derivative of the recorded signal rather than the signal itself. The increase in voltage with frequency, or the slope of the playback voltage curve on a logarithmic plot such as Fig. 10.30, is 6 dB per octave or 20 dB per decade. If a compensating network with a $1/f$ response is inserted between the playback head and the following amplifier, this characteristic can be equalized and thus presents no problem.

The Gap Loss in Playback

The true output from a playback head is not the theoretical response which increases linearly with frequency f. It is the theoretical response multiplied by a gap loss G which severely limits the high-frequency response of a playback head. The gap loss G is given by

$$G = k_2 \frac{\sin (\pi l/\lambda)}{\pi l/\lambda} \tag{10.73}$$

The output voltage (10.72) under actual conditions is then given by

$$\mathscr{V}_{\text{out}} = \mathscr{V} G = k_3 \sin \frac{\pi l}{\lambda} \tag{10.74}$$

where l is the gap length, that is, the distance between adjacent surfaces of the north and south pole pieces of a magnetic head, λ is the recorded wavelength on the tape which, from (10.65), is related to signal frequency f by $\lambda f = v$, and k_2, k_3

are constants. The previous expression given in decibels is

$$\mathscr{V}_{\text{out}} = 20 \log \left(\sin \frac{\pi l}{\lambda} \right) \tag{10.75}$$

and is plotted in Fig. 10.30 as the actual curve. Note that for low frequencies, when $l/\lambda \ll 1$, the sine term can be approximated by $\sin(\pi l/\lambda) \cong \pi l/\lambda$ and $\mathscr{V}_{\text{out}}$ is again proportional to the theoretical curve which increases with frequency. The degradation of the high frequencies is such that the output voltage drops to zero whenever the gap l is an integral multiple of the recorded wavelength. Therefore, the shortest measurable wavelength that is recorded on the tape must be many times greater than the gap length; that is, $\lambda \gg l$. For example, a typical gap length for playback heads is 0.2 mil, which means that the gap loss becomes important when the recorded wavelength is 1 mil or less. At a tape speed of 7.5 in/s, a frequency of 7500 Hz will be recorded as a 1-mil wavelength [$f = v/\lambda = (7.5 \text{ in/s})/10^{-3} \text{ in} = 7500 \text{ Hz}$]. Therefore, upon playback, recorded frequencies on the tape which are higher than 7500 Hz will begin to suffer a decrease in output voltage, as represented by Fig. 10.30. The only practical way to increase frequency response is to record and playback at a higher speed v.

The loss due to the finite gap length is of little importance in recording. This is because the recording process takes place from the trailing edge of the gap, as shown in Fig. 10.27. It is this edge, rather than the gap, that is important in the magnetic recording head. During playback, on the other hand, the length of the gap determines the magnetomotive force which leads to the induced output voltage in the playback coil. As long as the gap is much smaller than the length of the recorded bar magnets, the mmf in the gap induced by the fringing field varies in proportion to the strength of the fringing magnetic field as the tape moves past. But when the gap length is on the order of a wavelength, the output voltage drops sharply because a bucking field is induced in the core of the playback head; i.e., both pole faces of the playback head are exposed to either a north–north or south–south pole combination of the tape. If a north pole is next to one pole face, while another north pole is next to the other pole face of the playback head, the first will induce a counterclockwise flux, while the second will induce a clockwise flux in the core. The result is zero flux and zero output voltage from the playback head. This condition is seen to occur for $l/\lambda = 1, 2, 3, \ldots$ in Fig. 10.30. On the other hand, we see that maximum voltage is induced in the playback coil when opposite poles of tape magnetization lie close to, but on opposite sides of, the playback head gap. In Fig. 10.30 this occurs for $l/\lambda = \frac{1}{2}, \frac{3}{2}, \frac{5}{2}, \ldots$.

Another way of looking at the gap loss when the gap length approaches the length of the recorded bar magnets is in terms of a scanning slit. If a narrow vertical slit is moved along a sine wave, we see the amplitude of the sine wave as a point along the slit which varies sinusoidally. As the slit becomes wider with respect to the wavelength of the sine wave, we begin to see an average. For example, for $l = \lambda/2$, the slit averages a half cycle to give a value $2/\pi$ times the peak of the sine wave. For a slit length larger than $\lambda/2$, the response falls rapidly, since the averaging begins to include the opposite polarity of the second half of the wave. Finally, when the slit is as long as the wavelength, $l = \lambda$, the average value seen through it is zero, and therefore the response is zero, as shown by the zeros in Fig. 10.30. This completes the explanation of the loss due to the finite gap length of the playback head. There are other losses that limit the high-frequency response, but the gap loss is the most important one.

PROBLEMS

10.1 A transformer has a core of soft magnetic material. The hysteresis loop is shown in Fig. 9.28. If the volume of the core is 200 cm^3 and the transformer is to be operated at 60 Hz, calculate (approximately) the hysteresis loss for the transformer.

10.2 The hysteresis loop of an iron specimen is plotted to such a scale that 1 cm equals 10^{-2} T and 100 ampere-turns/m. The area of the loop is 15 cm^2. Calculate the energy loss in joules per cubic meter per cycle.

10.3 A certain transformer has a hysteresis loss of 200 W. Doubling the applied voltage to this transformer will result in practically doubling the maximum flux density. If the frequency of the applied voltage remains the same, what will be the new hysteresis loss?

10.4 The eddy-current loss in a transformer that is operated at 25 Hz and at a maximum flux density of 1.1 T is 3 W. Operating the transformer at a frequency of 60 Hz,

(*a*) What is the new eddy-current loss if the maximum flux density is not changed?

(*b*) Reducing the flux density in the same proportion that the frequency is raised, what is the eddy-current loss?

10.5 For a magnetic flux density, which is a sinusoidal function of time, show that the eddy-current loss is given by Eq. (10.11).

10.6 A transformer has a total core loss of 1500 W at a frequency of 35 Hz. The frequency is now changed to 60 Hz, and the applied voltage is adjusted so the flux density in the transformer core is kept constant. The new core loss is measured and found to be 3000 W.

(*a*) Determine the hysteresis loss at 60 Hz.

(*b*) Determine the eddy-current loss at 60 Hz.

10.7 A magnetic shield in the form of a spherical shell of Mu metal has inner and outer radii of 10 and 11 cm, respectively. Calculate the interior magnetic field when the exterior field is uniform, has a value of 1.5 T, and does not vary with time.

10.8 The depth of penetration in a metal is defined as that depth to which the electric or magnetic field falls off to $1/e$ of its value at the surface, where $e = 2.718$. An attenuation of 100 dB requires how many skin depths? What is the length of this in copper at 1 MHz?

10.9 Time-varying magnetic fields are shielded from the interior of a conductor because they produce eddy currents on the surface of the conductor that decay upon penetration. The skin depth $\delta = 1/(\pi f \mu \sigma)^{1/2}$ is a measure of this decay. More specifically, it is the distance at which the current has decreased to $1/e$ of its value at the surface. Show that time-varying electric fields inside a conductor are also limited to the skin depth δ.

10.10 Show that eddy-current power loss is proportional to the square of the relative speed of a conducting object through the magnetic field which it passes.

10.11 An iron ring of mean circumference 20 cm is wound with a toroidal coil of 200 turns. Determine the flux density in the ring when the current in the winding is 0.05, 0.1, and 0.5 A. What value of current gives the highest induced magnetization?

10.12 A ring of soft ferromagnetic material (shown in Fig. 10.8) has an air gap of 1 cm. If the mean length of the ring is 20 cm, its permeability μ is $2000\mu_0$, and the 1000-turn winding carries a current of 2 A, neglecting fringing, calculate

(*a*) The flux density in the air gap.

(*b*) The flux density B and field strength H in the iron.

10.13 In the first example of Sec. 10.4, the flux density in the toroid was calculated as $B = 0.57$ T. Find the magnetization M in the iron toroid. Compare the relative contributions of the magnetic material and that of the current-carrying winding to the total B in the iron.

10.14 In Example 1 of Sec. 10.5, the flux density B in the toroid with a gap was calculated to be $B = 0.0013$ T.

(*a*) Find the magnetization M in the iron toroid.

(*b*) By what factor does the gap decrease the induced magnetization M when compared to the case of no gap considered in the previous problem?

(*c*) Compare this ratio to the ratio of reluctances of the two problems.

10.15 Derive (10.23*b*) from (10.23*a*).

10.16 Find the flux density B in the gap of the electromagnet shown in Fig. 10.8 for the following parameters: $l_g = 0.1$ mm, $l_i + l_g = 10$ cm, $NI = 10$, core of commercial iron (BH curve is given by Fig. 9.31). This is the same problem as that considered in the first example of Sec. 10.5 except for a much smaller gap length. Since the gap length is small, the term l_i/μ_r with respect to l_g in (10.23*a*) can not be neglected, which makes this a more difficult problem as the nonlinearity of the magnetization curve enters the problem through μ_r. Hence, this ceases to be a linear problem, and μ, which before could be taken as a constant, is now a function of H; that is, $\mu = \mu(H)$, with a complicated variation as shown in Fig. 9.31.

Hint: A solution to this problem can be obtained rather easily by trial and error.

(*a*) Guess at H (a good starting point is a value less than $H = NI/l$) and note to what value of μ and B this corresponds to on Fig. 9.31.

(*b*) Using this value of μ, calculate B from (10.23*a*).

(*c*) If the calculated B disagrees with the value of B noted on the BH curve in step (*a*), repeat the calculation until an acceptable agreement is obtained.

10.17 A magnetic circuit core, shown in the accompanying figure, is constructed of commercial iron (see Fig. 9.31 for a BH curve). The cross section of the core is uniform and measures 4×4 cm, which gives for the mean path length the value $l = 60$ cm. Compute the number of ampere-turns (NI) required to produce a flux of 1.6×10^{-3} Wb in the circuit.

(*a*) Calculate the value of NI by ignoring flux fringing in the gap.

(*b*) Calculate NI by compensating for flux fringing.

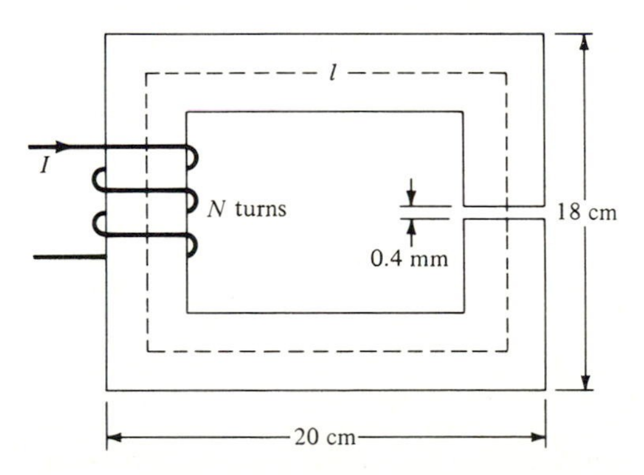

10.18 In the previous problem we calculated NI given flux ϕ in the circuit. The opposite problem of finding flux given NI is more difficult because of the nonlinearity of the BH curve of the core material.

But an iterative procedure gives results very quickly. Calculate the magnetic flux in the gap of the magnetic circuit of the previous problem for an mmf of $NI = 250$.

Hint: In the trial-and-error procedure assume a flux, then use (10.23) and the *BH* curve (Fig. 9.31) to see how close you have come to *NI* of 250. Repeat the procedure until acceptable agreement is reached.

10.19 A magnetic core of square cross section is shown. Find the air-gap flux density when the total number of ampere-turns is 200. Assume the soft ferromagnetic material of the core has a constant relative permeability $\mu_r = \mu/\mu_0 = 4000$. The dimensions of the core are $l_a = l_b = 1$ m, $l_c = 0.34$ m, $l_g = 0.76$ mm, $A = 7.9 \times 10^{-3}$ m^2. Compensate for gap fringing.

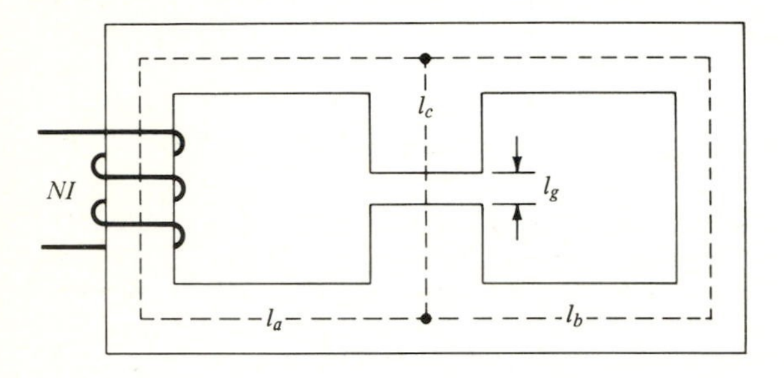

10.20 It is desired to produce an air-gap flux density of 0.2 T in the magnetic circuit of the previous problem. Find the mmf, that is, the total number of ampere-turns required.

10.21 Find the flux ϕ_b that flows through the outside path l_b of the magnetic circuit of Prob. 10.19.

10.22 A magnetic circuit has a nonuniform iron core with windings as shown. Each winding has an mmf of $NI = 400$ ampere-turns. Find the flux in the core if the thickness of the core is 2 cm. The *BH* curve of iron is given in Fig. 9.31.

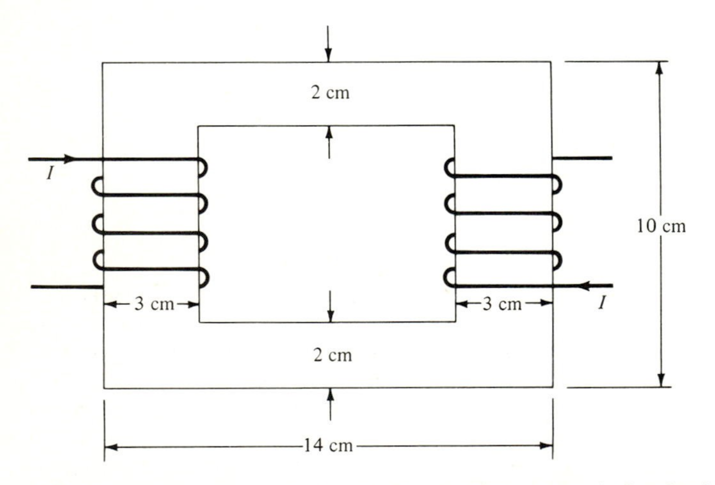

10.23 The permanent magnet shown in Fig. 10.9 has a magnetic flux density of $B_g = 10^{-2}$ T. For the dimension of $l_g = 3$ mm and $l_i = 30$ cm, determine the surface charge density of magnetic poles on the gap faces.

10.24 Calculate the approximate dimensions of a 36 percent cobalt-steel permanent magnet, which would produce a flux density of 4 T in the air gap between pole faces 2.5 by 2.5 cm, separated by 0.25 cm.

10.25 Design a permanent magnet which would set up a flux density of 3600 G in an air gap 0.07 in long and 1.2 in^2 in cross section. Choose any one of the materials in Fig. 10.13.

10.26 Determine the flux in the gap of a magnet if the specifications of the magnet are the same as in the example of Sec. 10.6, except for the use of Alnico XII material in place of Alnico V. Is this as good a magnet as that of the example?

10.27 The construction of an electromagnet is shown in Fig. 10.16*b*. The lower horseshoe is attracted to the upper, when a current I flows in the winding of N turns of the upper horseshoe. If the cross-sectional area is A, the total length of the iron in both parts is l with a separation distance x between the halves, and the permeability of the iron is μ, calculate the force of attraction between the two halves of the magnet.

10.28 A relay is shown in Fig. 10.16*a*. The soft-iron core has a relative permeability $\mu_r = 1000$ and has a winding of 100 turns. The total iron path is 20 cm, and the cross section of horseshoe and armature is 1 cm^2. To start the armature moving requires a force of 50 N when the gap length x is 0.2 cm. Ignore the gap at the pivot between horseshoe and armature.

(*a*) What is the current required to start the armature when the gap length is $x = 0.2$ cm?

(*b*) What is the magnetic pull on the armature when the gap length has decreased to 0.02 cm, for the current found in part (*a*)?

10.29 A horseshoe electromagnet of square pole faces has a winding of N turns. A flat soft-iron armature can rotate about an axis such that it covers both pole faces when $\theta = 0°$. Neglect the reluctance of the electromagnet and the armature. If the armature is held open at a small angle θ when current I flows in the winding,

(*a*) Show that the magnetic flux lines in the air gaps form circles coaxial with the pivot.

(*b*) Calculate the torque about the pivot tending to return the armature to its closed position.

(*c*) Calculate the torque for $a = 1$ cm, $b = c = 5$ cm, $\theta = 1°$, $N = 10$, and $I = 100$ A.

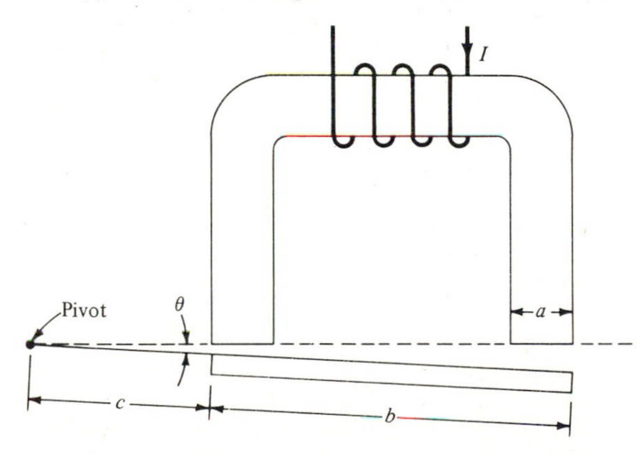

10.30 Audio amplifiers use transformer coupling between the final amplifier stage and a speaker in order to avoid having the dc current of the final stage flow through the speaker coil and to match impedances. We wish to connect an 8-Ω speaker to a vacuum tube which for effective service requires a load resistance of 8000 Ω. Calculate the turn-ratio of the output transformer.

10.31 In a 60-Hz transformer core, the maximum flux density is 1.5 T. What area of core is needed to produce 2 V per turn of winding?

10.32 Design a transformer (Fig. 10.18) to supply 6.3 V at 10 A from a 115-V 60-Hz supply. The core is commercial iron (Fig. 9.31) and is to have a cross section of 8 cm^2. The maximum core flux is to be 1.2 Wb/m^2. Calculate

(*a*) The number of turns in the primary and secondary.

(*b*) The full-load current in the primary.

(*c*) The length of the magnetic circuit if the maximum magnetizing current is to be 10 percent of the full-load current.

10.33 Find the self-inductance L of the magnetic circuit shown in Fig. 10.6. Evaluate L if $\mu = 1000\mu_0$, $N_1 = 100$, $l = 10$ cm, and $A = 10$ cm^2.

10.34 Find the self-inductance L of the magnetic circuit shown in Fig. 10.8. Evaluate L if $\mu = 1000\mu_0$, $N_1 = 100$, $l = 10$ cm, $l_g = 0.1$ cm, and $A = 10$ cm^2.

10.35 Find the self-inductance L and mutual inductance M for the transformer of Fig. 10.18. Evaluate L_{11}, L_{22}, and M if $\mu_r = 1000$, $N_1 = 100$, $N_2 = 1000$, $l = 10$ cm, and $A = 10$ cm^2. Find the coefficient of coupling.

10.36 A two-window transformer is shown in Fig. 10.7.

(*a*) Calculate the self-inductance and mutual inductance in terms of the branch reluctances $\mathcal{R}_1$, $\mathcal{R}_2$, and $\mathcal{R}_3$.

(*b*) Using your answers, verify that if the reluctance $\mathcal{R}_3$ of the center branch becomes indefinitely large, the transformer reverts to one of the type illustrated in Fig. 10.18.

10.37 In the two-window transformer of Fig. 10.7, 40 percent of the magnetic flux due to current in the primary N_1 is shorted through the center branch. For no load on the secondary, find the ratio of secondary to primary voltage.

10.38 Two coplanar and concentric circular rings of radii a and b are situated in air. If one of the rings in much larger, that is, $a \gg b$,

(*a*) Determine approximately the mutual inductance between the rings.

(*b*) Calculate M for $a = 10$ cm and $b = 0.5$ cm.

10.39 Two coaxial rings in air, radii a and b, are separated by a large distance z; that is, $z \gg a, b$.

(*a*) Determine approximately the mutual inductance between the rings.

(*b*) Calculate M for $a = b = 5$ cm and $z = 50$ cm.

10.40 Determine approximately the mutual inductance between two coaxial rings of radii a and b, for any spacing z between the rings, and for the case when one of the rings is much smaller than the other, that is, $a \gg b$.

10.41 An iron-core inductor is used as a filter element (choke coil) in a dc power supply to smooth ripples on the dc current from the power supply. This is an example of an inductor which has superposed dc and ac current flowing through it. If the ac variation ΔI of the dc current I_0 is on the order of the size of the minor loops of Fig. 10.25*b*,

(*a*) Sketch the variation of normalized incremental inductance $L(I_0)/L(0)$ versus dc current I_0 using the slopes of the minor loops of Fig. 10.25*b*.

(*b*) From the shape of the incremental inductance curve, what do you conclude about the filtering properties?

Hint: Refer to the next problem.

10.42 To illustrate the filtering properties of an inductor, consider the following dc power supply consisting of a half-wave rectifier and a filter consisting of a single inductor L. The purpose of the filter is to make the dc voltage appearing across the load resistor R as smooth as possible. If we define the *ripple voltage* as the voltage of the first harmonic (in the case of a half-wave rectifier, the first harmonic is 60 Hz) appearing across R,

(*a*) Derive the ripple factor F for this filter, if F is defined as the ratio of ripple voltage with L present to ripple voltage with $L = 0$.

(*b*) Find F when $R = 100\ \Omega$ and $L = 1$ H.

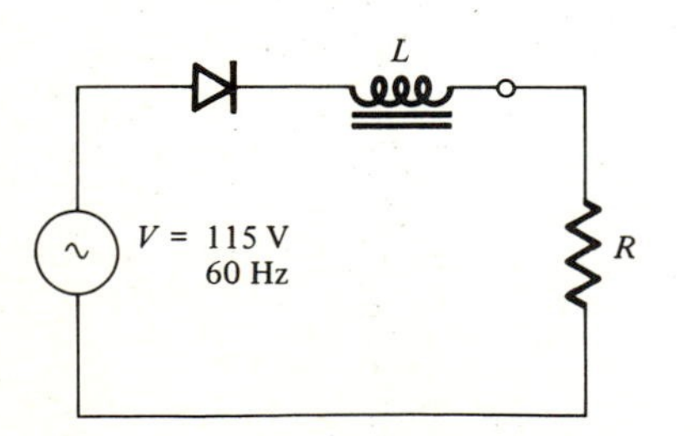

10.43 Discuss the advantages and disadvantages of making the gap in a magnetic recording head larger.

CHAPTER

ELEVEN

MAXWELL'S EQUATIONS

Just as the theory of evolution cannot be credited only to Darwin,† the four equations which are known today as Maxwell's equations cannot be solely attributed to Maxwell. They existed as experimental laws long before him, in fragmentary form as Coulomb's, Gauss', Ampère's, and Faraday's laws. These relations were collected by Maxwell and extended by the introduction of the concept of displacement current and the notion of field. This is the great contribution made by Maxwell in 1862. The experimental laws were combined and a set of partial differential equations postulated which were to apply to all macroscopic (dimensions above atomic size) electromagnetic phenomena. The new theory was introduced in the great work of Maxwell (*Treatise on Electricity and Magnetism*, 1873), where the concept of an electromagnetic field is used throughout the book.

Maxwell's equations as such cannot be derived since they represent mathematical expressions of certain experimental results. Their ultimate justification—as with all experimental laws—is that they have been used and continue to be used to explain and predict electromagnetic phenomena. As with Darwin's theory, the new electromagnetic theory raised doubts and criticism at the time of its publication, even though, looking back, its development undoubtedly must have been the high point of mathematical physics in that century. Maxwell died in 1879, only

† The theory of unending evolution presented in the *Origin of Species* did not originate with Charles Darwin. The idea is older than Aristotle and Lucretius, and much of it was known to Darwin's grandfather, Erasmus Darwin. However, he did collect more indisputable evidence, introduced natural selection, and unified it into a theory.

nine years before Hertz' experiments, which confirmed beyond any question the validity of the new theory.†

Maxwell's equations will be numbered for convenience. It is not implied that such numbering is essential or historic. It is strictly a preference, indicating the ease of development of these equations. Other books might choose not to number Maxwell's equations, or if they do, will probably number them differently from this text.

11.1 FARADAY'S LAW AND MAXWELL'S FIRST EQUATION

The experimental work of Michael Faraday (London, 1831) showed that a changing magnetic field which links a wire loop induces a voltage (emf) in the loop. The induced emf is proportional to the rate of change of the magnetic flux through the loop. The magnetic flux can change with time in several ways. The loop can be fixed in space, while the magnetic field changes with time, as, for example, when it is produced by an alternating current or when a permanent magnet is moved back and forth through the loop. The wire loop can also be moving or changing its shape while in a static magnetic field. The polarity of the induced voltage is given by Lenz' law: It causes a current in the wire loop which produces a magnetic field opposing the change in flux. Combining Lenz' law, which determines the sign in (11.1), with Faraday's experimental results, Faraday's law can be written in the form

$$\boxed{\mathscr{V} = -\frac{d\phi}{dt}} \tag{11.1}$$

where $\mathscr{V}$ is the emf voltage induced in the loop and ϕ is the flux linking the loop (if the loop has N turns, the right side of the above equation should be multiplied by N). If we denote the circuit of the wire loop, shown in Fig. 11.1, by l, the magnetic flux through such a circuit is given by $\phi = \iint_{A(l)} \mathbf{B} \cdot \mathbf{dA}$; that is, ϕ is given by integrating the normal components of the magnetic flux density $\mathbf{B}$ over any surface A, not necessarily a plane, which has the circuit l as a boundary.

It was shown (Sec. 3.1) that emf voltage is related to the work done in moving

† An interesting paper on this subject is by J. Blanchard, Hertz, the Discoverer of Electric Waves, *Proc. IRE*, vol. 26, pp. 505–515, May 1938. An excellent review of the history and development of electromagnetic theory is E. Whittaker, "History of the Theories of Aether and Electricity," Thomas Nelson and Sons, Ltd., London, 1951. Other tutorial papers are C. T. Tai, On the Presentation of Maxwell's Theory, *Proc. IEEE*, vol. 60, pp. 936–945, August 1972; H. A. Haus and J. R. Melcher, Electric and Magnetic Fields, *Proc. IEEE*, vol. 59, pp. 887–895, June 1971; S. A. Schelkunoff, On Teaching the Undergraduate Electromagnetic Theory, *IEEE Trans. Educ.*, vol. E-15, pp. 15–25, February 1972.

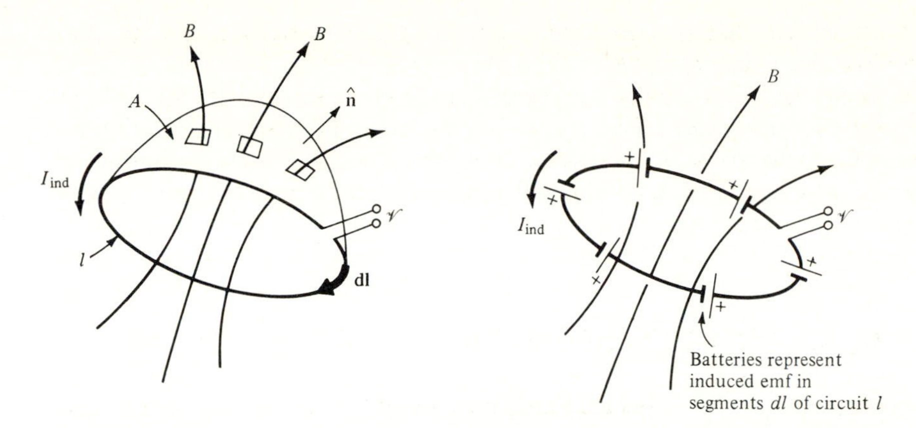

Figure 11.1 Contour l coincides with a wire loop in which a voltage $\mathscr{V}$ is induced by the changing magnetic field B. We can cut the wire thus introducing a small gap and bring $\mathscr{V}$ out to the terminals. The polarity shown is for an increasing B; a decreasing B will have opposite polarity.

a charge around a closed path.† Therefore, if an emf voltage is produced in the circuit l of Fig. 11.1, a force must exist on the electric charges which moves them around the circuit l. This force must be an electric field **E** which is tangential to the circuit l (electric field is force per unit charge). The work per unit charge due to **E**, when added around the circuit l, must be equal to the emf induced in the circuit; that is, $\mathscr{V} = \oint_l \mathbf{E} \cdot \mathbf{dl}$. With these preliminaries, we can write Eq. (11.1) as

$$\boxed{\oint_{l(A)} \mathbf{E} \cdot \mathbf{dl} = -\frac{\partial}{\partial t} \iint_{A(l)} \mathbf{B} \cdot \mathbf{dA}} \tag{11.2}$$

† Any device which is capable of maintaining a potential difference while a *steady* current flows contains an agent we call a *source of emf*, or, loosely speaking, a voltage source. The strength of emf is denoted by the symbol $\mathscr{V}$ and is the work per unit charge, $\mathscr{V} = dW/dq$, done by the emf on the charges which pass through it (the symbol V is reserved for potential difference and voltage drop). The energy of emf is made available when a conductor is connected across the terminals of the emf device, thus forming a closed circuit. The emf maintains an electric field **E** in the conductor which gives rise to steady current in the conductor. Emf voltage can be defined as $\mathscr{V} = \oint \mathbf{E} \cdot \mathbf{dl}$, where **E** is the nonconservative field due to emf. An alternative definition of $\mathscr{V}$ is that it is equal to the potential difference which appears across the open-circuited terminals of an emf device. An emf can exist across a short gap, such as the terminals of a battery, or it can exist distributed in a closed circuit, such as when a magnetic flux through the closed circuit is changing. Devices which produce emf convert some form of energy (chemical, mechanical, heat, solar, etc.) into electric energy. Examples are batteries, generators, thermocouples, solar cells, etc. The time rate of energy conversion, which is the electric power developed by the emf, is given by $P = dW/dt = \mathscr{V}\, dq/dt = \mathscr{V}I$. A charged capacitor is not a source of emf because it cannot sustain a steady current; that is, $\mathscr{V} = \oint \mathbf{E} \cdot \mathbf{dl} = 0$ because, in this case, **E** at all points in the closed circuit is a conservative field. For additional details see Secs. 3.1 to 3.3.

which is Maxwell's first equation in integral form. For circuits that do not move, i.e. are fixed in space, the total time derivative in (11.1) is written as a partial derivative (allowing the partial to be taken inside the integration).†

Figure 11.1 illustrates Faraday's law. The sense of the path l and the surface normal $\hat{\mathbf{n}}$ are related by the right-hand rule, as shown in Fig. A1.2. Figure 11.1 shows a bowl-shaped surface A, bounded by contour l and threaded by an increasing magnetic **B** field. Since each segment dl of the wire loop contributes to the induced emf voltage $\mathscr{V}$, we can think of the induced current as being produced—at any instant of time—by a series of batteries which are distributed along the loop and have the polarities shown.

An important observation can now be made: The above relation is still valid when the conducting wire loop is removed. The changing magnetic field will induce an electric field, regardless of the presence or absence of the wire loop. This illustrates that Eq. (11.2) expresses a relationship between the fields themselves which does not depend on the presence of a conducting path‡ (i.e., independent of the resistance of the path).

We can easily transform (11.2) into a differential expression which is independent of any path and valid at a point in space, by applying Stokes' theorem [Eq. (A1.10) or (8.16)] to the left side of (11.2). Thus

$$\iint_{A(l)} \nabla \times \mathbf{E} \cdot \mathbf{dA} = -\frac{\partial}{\partial t} \iint_{A(l)} \mathbf{B} \cdot \mathbf{dA} \qquad (11.3)$$

Since this statement must be true for any surface, the integrands may be equated, giving

$$\boxed{\nabla \times \mathbf{E} = -\frac{\partial B}{\partial t}} \qquad (11.4)$$

This is Maxwell's first equation, relating the magnetic and electric fields at a point in space.

11.2 GAUSS' LAW AND MAXWELL'S SECOND EQUATION

Gauss' law (covered in Secs. 1.10 and 1.11)

$$\boxed{\oiint_{A(v)} \mathbf{D} \cdot \mathbf{dA} = Q} \qquad (11.5)$$

† For a discussion of the term $d/dt \iint \mathbf{B} \cdot \mathbf{dA}$, for fixed and moving circuits, see Eqs. (12.13) to (12.18).

‡ The betatron, an excellent illustration of Faraday's law for a path of infinite resistance, accelerates charged particles in vacuum in a circular path by means of an electric field which is induced by a changing magnetic field. (D. W. Kerst and R. Serber, Electronic Orbits in the Induction Accelerator, *Phys. Rev.*, vol. 60, p. 53, 1941.)

gives the most important property of electric flux density **D**, namely, that flux out of a closed surface equals the charge enclosed. For a point charge this expression can be easily verified by integrating over a spherical surface.

If a charge Q is distributed continuously throughout a volume v, Gauss' law is

$$\boxed{\oiint_{A(v)} \mathbf{D} \cdot \mathbf{dA} = \iiint_{v(A)} \rho \, dv} \tag{11.6}$$

where ρ is the volume charge density (coulombs per cubic meter) and $v(A)$ is the volume bounded by A. Applying the divergence theorem (1.94) to the left side of the above expression, it can be written as

$$\iiint_{v(A)} \nabla \cdot \mathbf{D} \, dV = \iiint_{v(A)} \rho \, dV \tag{11.7}$$

Since the above statement must be true for any volume, the integrands can be equated to give

$$\boxed{\nabla \cdot \mathbf{D} = \rho} \tag{11.8}$$

This is a differential statement of Gauss' law which expresses the relationship between the electric flux density and charge density at a point. It is also the second Maxwell's equation.

11.3 AMPÈRE'S LAW AND MAXWELL'S THIRD EQUATION

Ampère's Law

Faraday's experimental law, which showed that a time-varying magnetic field produces an electric field, was the basis for Maxwell's first equation. The converse of this, namely, that a time-varying electric field gives rise to a magnetic field, will be expressed by Maxwell's third equation. This is truly Maxwell's equation since he discovered it (1862) while trying to unify the separate theories of electricity and magnetism. Even here Maxwell cannot be given all the credit, as the third equation is really a generalization of Ampère's law (7.18), discovered in 1820, which states that a line integral of magnetic field taken about any given closed path must equal the current enclosed by that path; that is,

$$\oint \mathbf{H} \cdot \mathbf{dl} = I \tag{11.9}$$

where H is the magnetic field strength. For distributed currents, Ampère's circuit law can be written as

$$\oint_{l(A)} \mathbf{H} \cdot \mathbf{dl} = \iint_{A(l)} \mathbf{J} \cdot \mathbf{dA} \tag{11.10}$$

which, by use of Stokes' theorem (8.16), can be converted to the differential form

$$\nabla \times \mathbf{H} = \mathbf{J} \tag{11.11}$$

where $\mathbf{J}$ can be a distributed conduction ($\mathbf{J} = \sigma\mathbf{E}$) or convection current ($\mathbf{J} = \rho\mathbf{v}$).

Displacement Current

The implication of Ampère's law is that a magnetic field can only be produced by a motion of charges. Maxwell showed that there is another current, the displacement current, which can produce a time-varying magnetic field (but not a steady one) just as effectively as a flow of charges can. The introduction of the new current allowed Maxwell to unify the separate laws of electricity and magnetism into an electromagnetic theory. It also led him to predict that electromagnetic waves which can propagate energy must exist and that light is an electromagnetic wave. Let us see how the need for a displacement current arises.

In Sec. 2.3 we derived the continuity equation

$$\nabla \cdot \mathbf{J} + \frac{\partial \rho}{\partial t} = 0 \tag{11.12}$$

and observed that it must be a very general relation because it was derived from the principle of conservation of charge. We can now reason as follows: If we take the divergence of both sides of (11.11), we obtain

$$\begin{aligned} \nabla \cdot \nabla \times \mathbf{H} &= \nabla \cdot \mathbf{J} \\ 0 &= \nabla \cdot \mathbf{J} \end{aligned} \tag{11.13}$$

where (8.26) was used which states that the divergence of any curl is identically zero. Since in the general case the divergence of $\mathbf{J}$ cannot be equal to zero, it follows that Ampère's law in the form $\nabla \times \mathbf{H} = \mathbf{J}$ is not very general. This can be expected as (11.11) is a relationship valid for static fields. Can we make Ampère's law sufficiently general to include the time-varying fields by adding a missing term? As the divergence operation in (11.13) removes the magnetic field part of Ampère's law, the missing term must be a current density, which we will call J_D. We now have $\nabla \times \mathbf{H} = \mathbf{J} + \mathbf{J}_D$ which yields $0 = \nabla \cdot \mathbf{J} + \nabla \cdot \mathbf{J}_D$ in place of (11.13). Comparing this to the continuity equation, we obtain $0 = -\partial\rho/\partial t + \nabla \cdot \mathbf{J}_D$ as the relationship for the new term. Since charge density ρ is related by (11.8) as $\nabla \cdot \mathbf{D} = \rho$, we can substitute for ρ and identify J_D as $J_D = \partial D/\partial t$. Hence the generalized Ampère's law which now satisfies the continuity equation is given by

$$\boxed{\begin{aligned} \nabla \times \mathbf{H} &= \mathbf{J} + \mathbf{J}_D \\ &= \mathbf{J} + \frac{\partial \mathbf{D}}{\partial t} \end{aligned}} \tag{11.14}$$

and in integral form

$$\oint_{l(A)} \mathbf{H} \cdot \mathbf{dl} = \iint_{A(l)} \mathbf{J} \cdot \mathbf{dA} + \iint \frac{\partial \mathbf{D}}{\partial t} \cdot \mathbf{dA} \tag{11.15}$$

This is Maxwell's third equation and is most closely associated with Maxwell. We observe that the new current is a rate of change of electric field. The faster the E field changes, the larger the current. In the static case ($\partial/\partial t = 0$) the displacement current vanishes, and for low frequencies it is not very important since it is significantly smaller than common current densities in conductors. Even at a frequency of 10^6 Hz, a relatively large electric field of 10^4 V/m would only yield a displacement current density of 5.5×10^{-5} A/cm^2. Of course in devices such as capacitors which involve large areas and large *E fields, the total displacement current* I_D can be very large.

The need for displacement current becomes apparent when current flow through a capacitor is considered. Ampère's law, for the case of Fig. 11.2*a*, is $\oint_{l(A)} \mathbf{H} \cdot \mathbf{dl} = I$, since area A is pierced by the wire which carries current I. On the other hand, for the case of Fig. 11.2*b*, area A is not pierced and $\oint_{l(A')} \mathbf{H} \cdot \mathbf{dl} = 0$, unless displacement current is assumed to exist within the capacitor; then $\oint_{l(A')} \mathbf{H} \cdot \mathbf{dl} = \iint_{A'(l)} \partial \mathbf{D}/\partial t \cdot \mathbf{dA}$. This is the argument, attributed to Maxwell: The two areas, A and A', are bounded by the same path l; hence the contour integrals are equal; that is, $\oint_{l(A)} = \oint_{l(A')}$. Therefore, current in the wire must equal the total displacement current in the capacitor, or $I = \iint \partial \mathbf{D}/\partial t \cdot \mathbf{dA}$ (see following example). It appears that here is one of those rare cases where purely mathematical reasoning has preceded and guided the way for experiment.

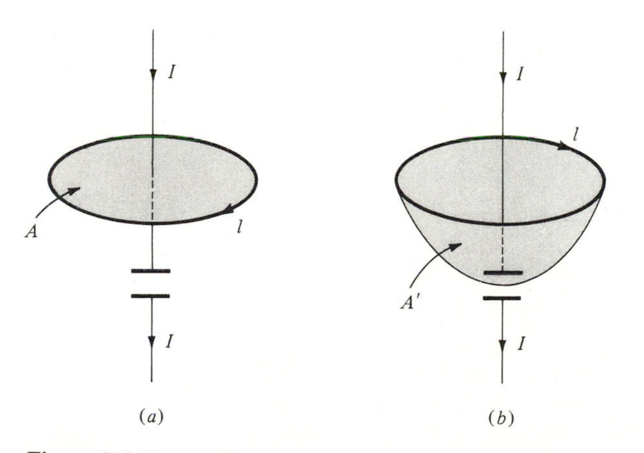

Figure 11.2 Two surfaces of integration. (*a*) Surface A intersects the wire; (*b*) surface A' passes through the capacitor and thus does not intercept current I.

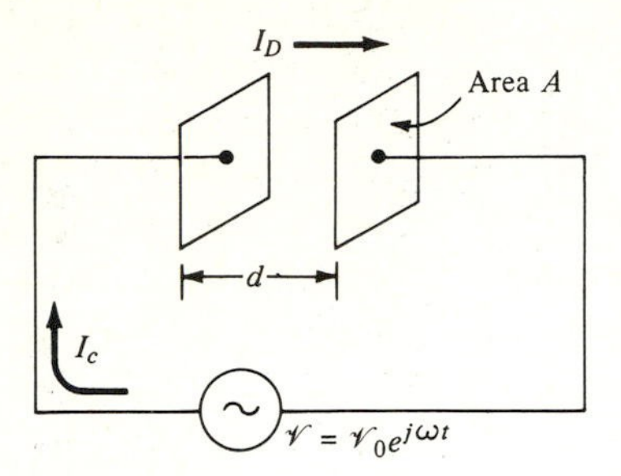

Figure 11.3 An ac circuit containing a parallel-plate capacitor of capacitance $C = \varepsilon_0 A/d$. For simplicity, lead resistance is neglected.

Example Figure 11.3 shows a circuit in which an alternating current I of frequency $f = \omega/2\pi$ flows through the capacitor C. It shows that the conduction current I_C in the connecting wires continues through the capacitor as a displacement current I_D.

Circuit theory tells us that the relationship between capacitor voltage V and charging current I_C is given by (5.2) as

$$I_C = C\frac{dV}{dt} = Cj\omega V_0 e^{j\omega t} = j\omega C V$$

The displacement current I_D through the capacitor is

$$I_D = J_D A = \frac{\partial D}{\partial t} A$$

Since the electric field inside the capacitor is $E = V/d$, we have that

$$I_D = \varepsilon_0 \frac{\partial E}{\partial t} A = \frac{\varepsilon_0 A}{d}\frac{\partial V}{\partial t} = j\omega C V$$

where the parallel-plate capacity from (5.12) is given by $C = \varepsilon_0 A/d$. Therefore, we have shown that

$$I_C = I_D$$

so that the displacement current completes the circuit of the charging current which flows in the connecting leads.

11.4 MAGNETIC FLUX AND MAXWELL'S FOURTH EQUATION

Gauss' law or, equivalently, Maxwell's second equation relates the electric field to its sources, which are electric charges. It was shown in Sec. 7.2 that a similar relationship exists for the magnetic field, except that the magnetic field has no primary source, since no true magnetic pole or charge has ever been isolated. Magnetic poles occur in nature always as equal and opposite pairs. The sources for steady magnetic field are currents or moving charges. In view of this, one can conclude that electric charge is the primary source for electric and magnetic fields. Hence magnetism is a by-product of electricity; it exists only as a result of the motion of electrically charged particles. Nevertheless, theorists have failed to find any good reason why magnetic charge should not exist, although experimentalists have yet to find any sign of the particle. Perhaps the greatest indication for the existence of magnetic charge (sometimes called magnetic monopoles) is the symmetry in Maxwell's equations.† An electric particle gives rise to an electric field and when in motion will produce a magnetic field. There should be magnetic particles that give rise to magnetic fields when stationary and electric fields when in motion. Should magnetic charge ever be discovered, no new equations will be required to describe the effects of magnetic monopoles. Maxwell's equations provide a natural slot for magnetic charge; for example, $\nabla \cdot \mathbf{B} = \rho_m$, where ρ_m is the magnetic charge density. With the addition of magnetic current density J_m, Maxwell's equations would be symmetric.

Since no magnetic charges exist, the magnetic B field cannot terminate on charges and therefore must be solenoidal; i.e., magnetic flux lines must exist as closed lines. All lines entering a closed surface must also leave it. Mathematically, this can be expressed as

$$\boxed{\oiint \mathbf{B} \cdot \mathbf{dA} = 0} \tag{11.16}$$

or as

$$\boxed{\nabla \cdot \mathbf{B} = 0} \tag{11.17}$$

(remember that divergence is a measure of scalar sources). The two statements given above are the integral and differential expressions of Maxwell's fourth equation. Nothing more than the short paragraph describing (11.16) and (11.17) need be said, since, after all, (11.16) is a statement of experimental fact and, as such, cannot be derived.‡

It is interesting to observe that the fourth Maxwell's equation is needed because the remaining equations allow magnetic charge. For example, if we consider Maxwell's first equation

$$\nabla \times \mathbf{E} = -\frac{\partial B}{\partial t} \tag{11.4}$$

and take the divergence of both sides, we obtain

$$\frac{\partial}{\partial t} \nabla \cdot \mathbf{B} = 0 \tag{11.18}$$

† An interesting sketch of the search for magnetic charge appears in K. W. Ford, Magnetic Monopoles, *Sci. Am.*, vol. 209, no. 6, pp. 122–131, December 1963.

‡ For some interesting comments on why $\nabla \cdot \mathbf{B} = 0$, see M. W. Leen, I. Sugai, and P. A. Clavier, Physical Basis for Electromagnetic Theory, *Proc. IRE*, vol. 50, p. 90, January 1962.

A solution to (11.18) is $\nabla \cdot \mathbf{B} =$ constant, implying the existence of magnetic charge. Hence (11.17) is explicitly needed to state that the constant is zero.†

11.5 SUMMARY OF MAXWELL'S EQUATIONS

Maxwell's equations in differential and integral form can be summarized as follows:

$$\nabla \times \mathbf{E} = -\frac{\partial \mathbf{B}}{\partial t} \quad (11.4) \qquad \oint_{l(A)} \mathbf{E} \cdot \mathbf{dl} = -\iint_{A(l)} \frac{\partial \mathbf{B}}{\partial t} \cdot \mathbf{dA} \quad (11.2)$$

$$\nabla \cdot \mathbf{D} = \rho \quad (11.8) \qquad \iint_{A(v)} \mathbf{D} \cdot \mathbf{dA} = \iiint_{v(A)} \rho \, dv \quad (11.6)$$

$$\nabla \times \mathbf{H} = \mathbf{J} + \frac{\partial \mathbf{D}}{\partial t} \quad (11.14) \qquad \oint_{l(A)} \mathbf{H} \cdot \mathbf{dl} = \iint_{A(l)} \mathbf{J} \cdot \mathbf{dA} + \iint_{A(l)} \frac{\partial \mathbf{D}}{\partial t} \cdot \mathbf{dA} \quad (11.15)$$

$$\nabla \cdot \mathbf{B} = 0 \quad (11.17) \qquad \oiint \mathbf{B} \cdot \mathbf{dA} = 0 \quad (11.16)$$

These equations are indefinite unless the constitutive relations are known. For simple media which is linear, isotropic, and homogeneous, the constitutive relations are $\mathbf{D} = \varepsilon\mathbf{E}$, $\mathbf{B} = \mu\mathbf{H}$, and $\mathbf{J} = \sigma\mathbf{E}$, with ε, μ, and σ constants. Maxwell's equations are linear equation which relate the electromagnetic field to its sources ρ and J. Linearity implies that the principle of superposition applies: If each source produces its own field independently, the resulting field due to many sources is the sum of the fields produced independently by each source.

These four relationships together with the Lorentz force law are the fundamental laws of classical electromagnetic theory. They govern all electromagnetic phenomena in media‡ which are stationary with respect to the coordinate system used; they are valid for linear and nonlinear, isotropic and nonisotropic, homogeneous and nonhomogeneous media in the range of frequencies from zero to the highest microwave frequencies, including many phenomena at light frequencies. However, they are macroscopic laws which relate the space and time-average field quantities. As such, they should be applied to regions or volumes

† The Biot-Savart law (6.9), deduced from experiments with steady currents, implies that $\nabla \cdot \mathbf{B} = 0$ (see Prob. 7.1), but only because it is implicit in (6.9) that $\mathbf{B}$ lines exist as closed lines about currents. It is therefore not surprising that $\nabla \cdot \mathbf{B} = 0$ follows from (6.9), because a statement that $\mathbf{B}$ lines close upon themselves is $\nabla \cdot \mathbf{B} = 0$. Biot-Savart law does not preclude, however, the existence of other types of sources for the magnetic field, such as displacement current or magnetic charge.

‡ For a discussion of moving media, see Chap. 12.

whose dimensions are larger than atomic dimensions. Similarly, time intervals of observation should be long enough to allow for an averaging of atomic fluctuations.

Obvious mathematical restrictions on the integral equations are that the field vectors be integrable and the time derivatives of electric and magnetic flux exist. In most physical situations these restrictions are easily satisfied. If the fields are continuous and differentiable, the integral equations can be converted into the set of partial differential equations. This conversion is accomplished by applying the integral equations to infinitesimal regions or by the use of Stokes' and divergence theorem. Differential statements are point relations by nature and thus apply to continuous media. Information about discontinuous media, such as when boundaries are involved, must be obtained from the integral statement of Maxwell's equations. As such, the integral form of Maxwell's equations is a more fundamental one.

With Maxwell's equations one should associate, for completeness, the continuity equation [even though it is contained in Maxwell's equations (11.8) and (11.14)] and Lorentz' force law for a point charge q:

$$\nabla \cdot \mathbf{J} = -\frac{\partial \rho}{\partial t} \tag{11.12}$$

$$\mathbf{F} = q(\mathbf{E} + \mathbf{v} \times \mathbf{B}) \tag{6.16}$$

Sinusoidal Time Variation

Without any assumptions, Maxwell's equations are independent. However, for the common case in which the sources produce fields which vary sinusoidally with time (ac steady-state case) and for the assumption of the continuity equation, the second and fourth Maxwell's equations can be derived from the first and third. For the ac case it is easiest to write the fields in phasor notation in which the time dependence is specified as $\mathbf{E}(x, y, z, t) = \mathbf{E}(x, y, z)e^{j\omega t}$. The multipliers of $e^{j\omega t}$ are called phasors (see footnote p. 478). The operator $\partial/\partial t$ then becomes simply $j\omega$; for example, $\partial\mathbf{E}/\partial t = j\omega\mathbf{E}$. Maxwell's equations for the ac case can then be written as

$$\nabla \times \mathbf{E} = -j\omega\mathbf{B} \qquad \oint \mathbf{E} \cdot \mathbf{dl} = -j\omega \iint \mathbf{B} \cdot \mathbf{dA} \tag{11.19}$$

$$\nabla \times \mathbf{H} = \mathbf{J} + j\omega\mathbf{D} \qquad \oint \mathbf{H} \cdot \mathbf{dl} = \iint \mathbf{J} \cdot \mathbf{dA} + j\omega \iint \mathbf{D} \cdot \mathbf{dA} \tag{11.20}$$

and the continuity equation as

$$\nabla \cdot \mathbf{J} = -j\omega\rho \qquad \oiint \mathbf{J} \cdot \mathbf{dA} = -j\omega \iiint \rho \, dv \tag{11.21}$$

These are the single-frequency statements of Maxwell's equations.

At times we will find it convenient to lump the conduction current $\mathbf{J} = \sigma\mathbf{E}$ with the displacement current $j\omega\varepsilon\mathbf{E}$ to form an equivalent complex displacement current. This is possible only for the single-frequency case and can be done as follows: Using (11.20), we can write

$$\nabla \times \mathbf{H} = \sigma\mathbf{E} + j\omega\varepsilon\mathbf{E}$$

$$= j\omega\varepsilon\left(1 + \frac{\sigma}{j\omega}\right)\mathbf{E}$$

$$= j\omega\varepsilon^{\star}\mathbf{E} \tag{11.22}$$

where $\varepsilon^{\star}$ is now the generalized complex permittivity:

$$\varepsilon^{\star} = \varepsilon\left(1 - \frac{j\sigma}{\omega\varepsilon}\right) \tag{11.23}$$

Thus for the single-frequency case the introduction of the complex permittivity results in a simplification of the third Maxwell's equation.

We can use the complex permittivity to find whether a medium at a given frequency ω acts as a dielectric or a conductor. If conduction current in a medium dominates ($J \gg \partial D/\partial t$), it can be classified as a conducting medium, and then $\sigma/\omega\varepsilon \gg 1$. On the other hand, if displacement current dominates ($\partial D/\partial t \gg J$), we have $\sigma/\omega\varepsilon \ll 1$, and the medium acts as a dielectric. Thus the frequency ω and the ratio σ/ε characterize the medium; when $\omega \cong \sigma/\varepsilon$, the medium has properties of both a conductor and a dielectric. For reference, a table of frequencies at which $\sigma = \omega\varepsilon$ is given in Sec. 13.6. Notice that we already developed in statics [Eqs. (2.36) and (2.37)] a criterion in terms of rearrangement time ($T = \varepsilon/\sigma$) for characterizing media as conductors or dielectrics. Here, using Maxwell's equation, we can develop the same criterion and bring in frequency of the signal more directly.

11.6 MAXWELL'S EQUATIONS FOR MATERIAL MEDIA

We have shown in Sec. 4.8 that the polarization **P** and in Sec. 9.1 that the magnetization **M** can act as sources for the electric and magnetic fields. To show this explicitly, we use the generalized expressions for **D** and **H**:

$$\mathbf{D} = \varepsilon_0\mathbf{E} + \mathbf{P} \tag{4.23}$$

$$\mathbf{H} = \frac{\mathbf{B}}{\mu_0} - \mathbf{M} \tag{9.10}$$

which are valid in arbitrary media (the restriction of linearity, homogeneity, and isotropy is removed). When these are substituted in Maxwell's equations, we obtain

$$\nabla \times \mathbf{E} = -\frac{\partial \mathbf{B}}{\partial t} \tag{11.4}$$

$$\nabla \cdot \mathbf{E} = \frac{1}{\varepsilon_0}(\rho - \nabla \cdot \mathbf{P}) \tag{11.24}$$

$$\nabla \times \mathbf{B} = \mu_0\left(\mathbf{J} + \nabla \times \mathbf{M} + \varepsilon_0\frac{\partial \mathbf{E}}{\partial t} + \frac{\partial \mathbf{P}}{\partial t}\right) \tag{11.25}$$

$$\nabla \cdot \mathbf{B} = 0 \tag{11.17}$$

which is the set of Maxwell's equations that shows the contribution of the medium explicitly. Note that the above set is in terms of the **E** and **B** fields. A similar set in integral form can be written. Thus, as shown before, $\nabla \cdot \mathbf{P}$ acts as a charge density, and $\nabla \times \mathbf{M}$ and $\partial \mathbf{P}/\partial t$ as current densities.

11.7 POTENTIALS FOR TIME-VARYING FIELDS

In Secs. 1.8 and 8.5 we showed that the potential (scalar) for the electrostatic field **E** was ϕ, and that **E** was related to ϕ by $\mathbf{E} = -\nabla\phi$. In Sec. 8.6 we showed that the potential (vector) for the magnetostatic field **B** was **A**, and that **B** was related to **A** by $\mathbf{B} = \nabla \times \mathbf{A}$. We suspect that for time-varying fields, ϕ and **A** are related. Let us use Maxwell's equations to find the generalized potentials.

As $\nabla \cdot \mathbf{B} = 0$ is also true for time-varying fields, we expect

$$\boxed{\mathbf{B} = \nabla \times \mathbf{A}} \tag{11.26}$$

to hold in general. Substituting this in Maxwell's equation (11.4), we obtain

$$\nabla \times \left(\mathbf{E} + \frac{\partial \mathbf{A}}{\partial t}\right) = 0 \tag{11.27}$$

which has a general solution $(\mathbf{E} + \partial \mathbf{A}/\partial t) = -\nabla\phi$. The electric field, therefore, is related to the scalar potential ϕ and the vector potential **A** as

$$\boxed{\mathbf{E} = -\nabla\phi - \frac{\partial \mathbf{A}}{\partial t}} \tag{11.28}$$

Equations (11.26) and (11.28) give the correct potentials for the electromagnetic field. Whereas in statics the potentials satisfied a Poisson-type equation, for the time-dependent case we will see that potentials satisfy wave equations. We will assume linear, isotropic media for which $\mathbf{D} = \varepsilon\mathbf{E}$ and $\mathbf{B} = \mu\mathbf{H}$.

Substituting **E** in (11.8), we obtain

$$\nabla^2\phi + \frac{\partial}{\partial t}(\nabla \cdot \mathbf{A}) = -\frac{\rho}{\varepsilon} \tag{11.29}$$

Substituting **E** and **B** in (11.14), we obtain

$$\nabla^2\mathbf{A} - \nabla(\nabla \cdot \mathbf{A}) = -\mu\mathbf{J} + \mu\varepsilon\nabla\left(\frac{\partial\phi}{\partial t}\right) + \mu\varepsilon\frac{\partial^2\mathbf{A}}{\partial t^2} \tag{11.30}$$

where the identity $\nabla \times \nabla \times \mathbf{A} = \nabla(\nabla \cdot \mathbf{A}) - \nabla^2\mathbf{A}$ was used. These are complicated equations which can be simplified by noting that $\nabla \cdot \mathbf{A}$ has not been specified. In the static case we chose to specify $\nabla \cdot \mathbf{A} = 0$, as given by (8.35). If we now choose

$$\nabla \cdot \mathbf{A} = -\mu\varepsilon\frac{\partial\phi}{\partial t} \tag{11.31}$$

which is known as the gauge condition or Lorentz condition, the set of Eqs. (11.29) to (11.30) uncouples, and we obtain

$$\nabla^2\phi - \mu\varepsilon\frac{\partial^2\phi}{\partial t^2} = -\frac{\rho}{\varepsilon} \tag{11.32}$$

$$\nabla^2\mathbf{A} - \mu\varepsilon\frac{\partial^2\mathbf{A}}{\partial t} = -\mu\mathbf{J} \tag{11.33}$$

Thus the potentials ϕ and $\mathbf{A}$ satisfy wave equations and are defined in terms of their sources ρ and $\mathbf{J}$, respectively. Solutions to these types of equations will be covered in Chap. 13. It can be shown that the solutions are

$$\phi(\mathbf{r}, t) = \iiint_{v'} \frac{\rho(\mathbf{r}', t - R/v)}{4\pi\varepsilon R}\,dv' \tag{11.34}$$

$$\mathbf{A}(\mathbf{r}, t) = \iiint_{v'} \frac{\mu\mathbf{J}(\mathbf{r}', t - R/v)}{4\pi R}\,dv' \tag{11.35}$$

where $v = 1/\sqrt{\mu\varepsilon}$ and $R = |\mathbf{r} - \mathbf{r}'|$ and is the distance between source point $\mathbf{r}'$ and observation point $\mathbf{r}$. These equations show that the potentials are given in terms of their sources, integrated over the volume v' in which the sources exist. The observation point $\mathbf{r}$ can be inside or outside the volume v'. These solutions are similar to the corresponding static ones, (1.16) and (8.41), except for the following differences. An element of charge $\rho\,dv$ contributes to the potential in the same way as in statics, except that the contribution of the element at time $t - R/v$ will be felt at the observation point at the later time t. That is, it takes a time R/v for the contribution to be felt at $\mathbf{r}$, where $v = 1/\sqrt{\mu\varepsilon}$ and is the velocity with which the disturbance travels the distance R. The time $t - R/v$ is called the *retarded time*, for obvious reasons.

The procedure for solving for the electromagnetic fields $\mathbf{E}$ and $\mathbf{B}$ is the same as in the electrostatic case. That is, first we solve for ϕ and $\mathbf{A}$, and then we obtain $\mathbf{E}$ and $\mathbf{B}$ from (11.28) and (11.26), respectively. Notice that in all cases the static forms of the equations are obtained by setting $\partial/\partial t = 0$. A final observation to make is that $\mathbf{A}$ (and in turn $\mathbf{B}$) is related only to current $\mathbf{J}$, ϕ only to charges, but E is related to both charges and currents.

11.8 POTENTIALS IN THE PRESENCE OF POLARIZABLE MEDIA

We have shown that a convenient way to calculate the electric and magnetic fields is to first calculate the potentials from the given currents and charges. The potentials which are

$$\phi = \iiint \frac{\rho'\,dv'}{4\pi\varepsilon_0 R} \qquad \mathbf{A} = \iiint \frac{\mu_0\mathbf{J}'\,dv'}{R}$$

apply to a wide range of problems if the charge and current densities are interpreted to be *total* quantities given by

$$\rho_t = \rho_{\text{free}} - \nabla \cdot \mathbf{P} \tag{11.36}$$

$$\mathbf{J}_t = \sigma\mathbf{E} + \rho\mathbf{v} + \frac{\partial \varepsilon_0 E}{\partial t} + \frac{\partial \mathbf{P}}{\partial t} + \nabla \times \mathbf{M} \tag{11.37}$$

which are obtained from (11.24) and (11.25). Thus in a problem where only free charge is present, $\rho_t = \rho_{\text{free}}$. On the other hand, if no free charge is present, as in the case of a ferroelectric material, the static electric field that such a permanently polarized material will produce can be obtained by letting $\rho_t = \rho_{\text{bound}} = \nabla \cdot \mathbf{P}$. We also see that such a material will not produce a magnetic field, as the only term that could produce it is $\partial\mathbf{P}/\partial t$ which for permanent polarization is zero.

In the total current density J_t the first term is the ordinary conduction current (e.g., conducting wire carrying a current), the second term is a convection current (e.g., a column of moving charges), the third term is the displacement current (the equivalent current of an electromagnetic wave, or the current that flows through a capacitor), the fourth term is a current due to polarization $\mathbf{P}$ that changes with time (considered in Sec. 4.8), and the last term gives the equivalent current due to magnetization $\mathbf{M}$. Thus, if a permanent magnet is the only source of magnetic field, an equivalent current is present in the magnet. The current density of this equivalent current is $\mathbf{J}_t = \nabla \times \mathbf{M}$ and is seen to be the spatial rate of change of $\mathbf{M}$ at right angles to the magnetization $\mathbf{M}$ of the magnet. Substituting $\mathbf{J}_t$ in the vector potential $\mathbf{A}$, the magnetic field of the magnet can be calculated from $\mathbf{B} = \nabla \times \mathbf{A}$.

The point we have tried to make here is that each term in the expressions of ρ_t and $\mathbf{J}_t$ corresponds to a physical situation, implying that each term alone can be the source of the field. For example, if we use $\mathbf{J}_t = \varepsilon_0\, \partial\mathbf{E}/\partial t$, then the magnetic field $\mathbf{B}$ is related by $\nabla \times \mathbf{B} = \nabla \times \nabla \times \mathbf{A} = \mu_0 \varepsilon_0\, \partial\mathbf{E}/\partial t$, which is one of the Maxwell's equations $\nabla \times \mathbf{H} = \varepsilon_0\, \partial\mathbf{E}/\partial t$, as expected.

Example Derive the equations for the potentials, showing the contributions of the medium explicitly.

Using the set of Maxwell's equations stated in Sec. 11.6 and following the procedure for the derivation of (11.32) and (11.33), we obtain

$$\nabla^2\phi - \mu_0\varepsilon_0 \frac{\partial^2 \phi}{\partial t^2} = \frac{-\rho + \nabla \cdot \mathbf{P}}{\varepsilon_0} \tag{11.38}$$

$$\nabla^2\mathbf{A} - \mu_0\varepsilon_0 \frac{\partial^2 \mathbf{A}}{\partial t^2} = -\mu_0 \left(\mathbf{J} + \frac{\partial \mathbf{P}}{\partial t} + \nabla \times \mathbf{M}\right) \tag{11.39}$$

where ρ = free-charge density
$\nabla \cdot \mathbf{P}$ = bound-charge density
$\mathbf{J}$ = current density due to free-charge flow
$\partial\mathbf{P}/\partial t$ = polarization current density
$\nabla \times \mathbf{M}$ = magnetization current density

Note that only charges and equivalent charges contribute to the scalar potential ϕ, whereas only currents and equivalent currents contribute to the vector potential $\mathbf{A}$.

PROBLEMS

11.1 A copper wire 5 mm in diameter carries a conduction current of 1 A at 60 Hz. Find the displacement current in the wire.

11.2 Show that $\partial\mathbf{P}/\partial t$ is a current density.

11.3 Prove that $\oiint (\mathbf{J} + \partial\mathbf{D}/\partial t) \cdot \mathbf{dA} = 0$ for any closed surface.

11.4 The betatron consists of a powerful electromagnet and an evacuated doughnut-shaped tube. Charged particles are accelerated inside the tube by a circular electric field E_ϕ induced by the H_z field which increases linearly with time t. Find the induced electric field when H_z is circularly symmetric ($\partial/\partial\phi = 0$) and is given by

(a) $H_z = t/r^2$.

(b) $H_z = H_0 e^{bt}$.

The r, ϕ, and z coordinates are those of a cylindrical coordinate system.

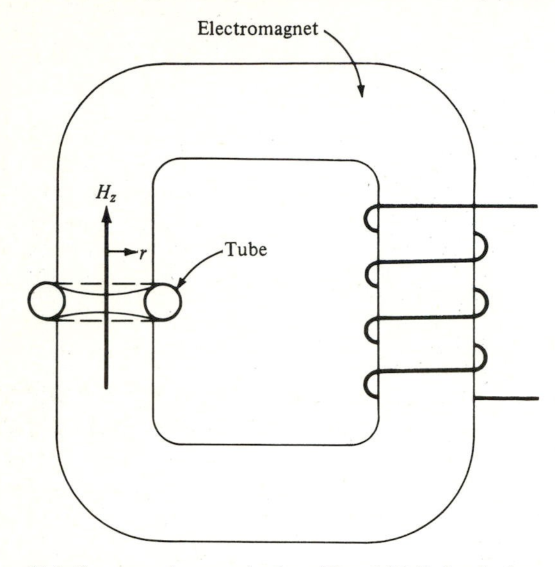

11.5 Compare the magnitudes of **J** and $\partial\mathbf{D}/\partial t$ for the frequencies 60 Hz, 1 kHz, 1 MHz, and 1 GHz for the materials copper ($\sigma = 5.75 \times 10^7$, $\varepsilon = \varepsilon_0$), lead ($\sigma = 0.5 \times 10^7$, $\varepsilon = \varepsilon_0$), seawater ($\sigma \cong 4$, $\varepsilon = 81\varepsilon_0$), and earth ($\sigma \cong 10^{-3}$, $\varepsilon = 10\varepsilon_0$).

11.6 Show that for the interior of a conducting material, the assumption that $\nabla \cdot \mathbf{J} = 0$ can be made for frequencies whose period is as small as 10^{-18} s.

Hint: Since $\nabla \cdot \mathbf{J} = -\partial\rho/\partial t$, examine the limit $\partial\rho/\partial t$ as $\varepsilon/\sigma \to 0$.

11.7 Referring to Fig. 11.2, areas A and A' form a closed surface. Using $I = dQ/dt$ and Gauss' law $Q = \oiint_{A+A'} \mathbf{D} \cdot \mathbf{dA}$, show that the displacement current density through the capacitor must be $\partial D/\partial t$.

11.8 Referring to Fig. 11.2, if the capacitor is a parallel-plate capacitor C with plate spacing d and plate area A, and the applied voltage in the circuit is $V_0 \sin \omega t$, show that $I_{\text{displ}} = I_{\text{wire}} = \omega C V_0 \cos \omega t$.

11.9 Calculate the displacement current at $t = 0$ which flows between the plates of a 20-pF capacitor, for a voltage of $10 \sin \omega t$ applied across the plates. Use $f = 60$ Hz, 1 MHz, 1 GHz.

11.10 Find the total displacement current flowing through the dielectric of a coaxial cable of radii a and b and length l if a voltage source $V_0 \cos \omega t$ is connected between the conducting cylinders. Compare the displacement current to the charging current for the coaxial capacitor.

11.11 A parallel-plate capacitor with circular plates of radius a is being charged at a constant rate. Find the magnitude of the magnetic field **B** at any distance r from the center of the capacitor plates, assuming the E field is confined entirely to the region between the plates. Sketch B versus r.

11.12 A parallel-plate capacitor with circular plates of radius a is connected to an ac generator, which results in the charge Q on the plates varying as $Q = Q_0 \cos \omega t$.

(a) Find the magnetic field at any point between the plates, neglecting fringing.

(b) Calculate the value of peak magnetic field at the edge of the capacitor when $a = 10$ cm, plate separation is 1 cm, and generator voltage is $100 \sin 10^{10}\, t$.

11.13 A sinusoidally varying field of angular frequency ω is applied to a conductor. Show that the ratio of the magnitudes of the conduction and displacement currents is $\sigma/\omega\varepsilon$.

11.14 Determine the electric charge distribution if

(*a*) $\mathbf{D} = y\hat{\mathbf{y}}$.
(*b*) $\mathbf{D} = 3\hat{\mathbf{r}}$.
(*c*) $\mathbf{D} = \hat{\mathbf{r}}/r^2$.

11.15 State Maxwell's differential equations for static electric fields which are created by charges on conducting bodies.

11.16 State Maxwell's differential equations for static magnetic fields produced by steady currents. Assume a medium of permeability μ.

11.17 State Maxwell's equations for a region of empty free space.

11.18 State Maxwell's equations and the scalar and vector potential for a region inside a permanent magnet.

11.19 State Maxwell's equations and the scalar and vector potential for a region inside a permanent ferroelectromagnet.

11.20 Starting with Maxwell's equations, derive the continuity equation.

11.21 Assume the time variation for all fields is e^{st}. State Maxwell's equations in differential form, independent of time.

11.22 Show that in a source-free region ($\rho = 0$, $\mathbf{J} = 0$), a uniform field $\mathbf{B} = \hat{\mathbf{x}}B_0 \sin \omega t$ cannot satisfy Maxwell's equations.

11.23 Write each of the vector differential Maxwell's equations as a set of scalar equations in terms of

(*a*) Rectangular components and coordinates.
(*b*) Cylindrical components and coordinates.
(*c*) Spherical components and coordinates.

11.24 Prove that once a static magnetic field is established inside a perfect conductor, it is "frozen" inside it and will never collapse.

11.25 Prove that $\iint_{A(l)} (\partial \mathbf{B}/\partial t) \cdot \mathbf{dA} = \oint_{l(A)} (\partial \mathbf{A}/\partial t) \cdot \mathbf{dl}$.

11.26 Show that in a charge-free and current-free media, the two divergence equations may be derived from the two curl equations for sinusoidally varying fields.

11.27 Show that for the ac case, with the continuity equation assumed, the two divergence equations may be derived from the two curl equations for regions with finite ρ and $\mathbf{J}$. This fact has made it quite common to refer to the two curl equations alone as Maxwell's equations.

11.28 Prove that the two divergence Maxwell's equations can or cannot be obtained from the continuity equation and the two curl Maxwell's equations

(*a*) Strictly on mathematical grounds.

(*b*) Using the physical fact that magnetic fields from current-carrying wires are divergence-free (see Prob. 7.1).

(*c*) Assuming that $\nabla \cdot B = 0$ everywhere, at least for an instant of time.

(*d*) Postulating that no magnetic charge exists.

CHAPTER

TWELVE

RELATIVITY AND MAXWELL'S EQUATIONS

12.1 INVARIANCE OF MAXWELL'S EQUATIONS

Up to this point we followed a historical development of Maxwell's equations. We showed that a series of careful studies of electrostatic and magnetostatic phenomena gave us various laws that eventually resulted in Maxwell's four equations for the electromagnetic field. Even though these four equations appear as a unit, we are only too aware that they represent generalizations of Gauss', Ampère's, and Faraday's laws. It is actually surprising that these laws existed for a long time as independent laws, without the unity that we associate with them now. Perhaps it was that the laws of electricity and magnetism, now the static case of Maxwell's equations, separate into two distinct pairs ($\nabla \times \mathbf{E} = 0$, $\nabla \cdot \mathbf{D} = \rho$ and $\nabla \times \mathbf{H} = \mathbf{J}$, $\nabla \cdot \mathbf{B} = 0$) which show no apparent connection between the two static fields. Forty years after Maxwell's equations were introduced, a connection between the two static fields was formulated in terms of the special theory of relativity.

Maxwell's equations have a special position in physics since Einstein's special theory of relativity had its origin in electromagnetics. The Lorentz transformation of special relativity was deduced by requiring invariance of Maxwell's equations. Thus under the Lorentz transformation, Maxwell's equations remain invariant; i.e., they remain in the same form when this transformation is applied to them. This means that if we denote a stationary system by unprimed coordinates and a moving system by primed coordinates, Maxwell's equations in these two coordinate systems must have the same form; that is,

$$
\begin{aligned}
\nabla \times \mathbf{E} &= -\frac{\partial \mathbf{B}}{\partial t} & \nabla \times \mathbf{E}' &= -\frac{\partial \mathbf{B}'}{\partial t'} \\
\nabla \times \mathbf{H} &= \mathbf{J} + \frac{\partial \mathbf{D}}{\partial t} & \nabla \times \mathbf{H}' &= \mathbf{J}' + \frac{\partial \mathbf{D}'}{\partial t'} \\
\nabla \cdot \mathbf{D} &= \rho & \nabla \cdot \mathbf{D}' &= \rho' \\
\nabla \cdot \mathbf{B} &= 0 & \nabla \cdot \mathbf{B}' &= 0
\end{aligned}
\tag{12.1}
$$

Such an invariance was verified experimentally, beyond dispute. For example, if Maxwell's equations were to change form for different moving systems, this would mean that electrical, as well as optical, phenomena on a moving airship would be different from those on a stationary airship. Thus one could use the different phenomena to determine the speed of the airship, and, perhaps what is more startling, the absolute speed of an airship could be determined by making electrical measurements at different velocities and angles. That this could not be done was known in the days of Newton. The principle of relativity existed then which required all physical laws to be invariant under a galilean transformation. Galilean relativity simply stated is as follows: All experiments performed on a stationary ship and a ship moving with constant velocity give the same results. After Maxwell's equations were formulated, it was found that they did change form under a galilean transformation; hence, they were not invariant under a galilean transformation. Lorentz was the first one to observe that a transformation, now known as the Lorentz transformation, left Maxwell's equations unchanged, as shown in Eq. (12.1). Thus the special theory of relativity was born and was formulated in general by Einstein.

The main purpose of this chapter is to show that special relativity provides a link between the laws of electricity and magnetism. We can show that electric and magnetic fields appear and vanish depending on the motion of the observer. Furthermore, only one experimentally derived law, namely Coulomb's law, will be needed to deduce Maxwell's equations when the transformations of special relativity are applied to it. Special relativity provides us, therefore, with an alternative way to formulate Maxwell's equations. In this approach, all electromagnetics follow naturally from electrostatics.

12.2 SOME DIFFICULTIES

Let us show that induced emf due to transformer action (the coil is stationary; magnetic field through the coil varies with time) and motion-induced emf (flux-cutting case found in generators where a coil moves through a steady magnetic field) are distinct phenomena; that is, neither can be deduced from the other. In the general case both transformer and motional emf's must be taken into consideration; here again, the use of transformations to a moving coordinate system would eliminate much confusion.†

In the last chapter we stated Faraday's experimental law as

$$\mathscr{V} = -\frac{d\phi}{dt} \tag{12.2}$$

and for a stationary circuit which has a time-varying magnetic field passing through it, we derived

$$\nabla \times \mathbf{E} = -\frac{\partial \mathbf{B}}{\partial t} \tag{12.3}$$

as the relation between the electric and magnetic fields. If we integrate the left-hand side of this expression over the area of a circuit and then apply Stokes' theorem to convert the area integral to a line integral over the contour of the circuit, the resulting expression,

$$\boxed{\mathscr{V}_t = \oint \mathbf{E} \cdot \mathbf{dl} = -\iint \frac{\partial B}{\partial t} \cdot \mathbf{dA}} \tag{12.4}$$

represents the induced emf in the circuit due to transformer action.‡ Nothing was said about the situation when the circuit moves. We did, however, introduce the Lorentz force

$$\mathbf{F} = q(\mathbf{E} + \mathbf{v} \times \mathbf{B}) \tag{12.5}$$

This is the total force which acts on a charge in the presence of an electric and magnetic field. The first term in this expression is Coulomb's force on a charge. The second term was added to the Coulomb's force since an additional force acts on a charge which moves with velocity v through a magnetic $\mathbf{B}$ field.

The second term in Lorentz' force was a by-product of our study of forces between two current-carrying wires. It was obtained as follows: Equation (6.1) states that

$$d\mathbf{F} = \mathbf{I} \times \mathbf{B}\, dl \tag{12.6}$$

which is the force on a current element $\mathbf{I}\, dl$ immersed in a magnetic field, as illustrated in Fig. 12.1. Since the current in the wire is made up of charges (conduction electrons) which drift with a velocity $\mathbf{v}$, the current density at any point in the wire is given by $\mathbf{J} = \rho\mathbf{v}$, where ρ is the charge density of the conduction electrons. By letting $\mathbf{I}\, dl = \mathbf{J}\, dA\, dl$, the force which acts on a small charge $q = \rho\, dA\, dl$ within the wire can be stated as

$$\mathbf{F} = q\mathbf{v} \times \mathbf{B} \tag{12.7}$$

where dl and dA are the length and cross section of an element of the wire, respectively.

† P. Moon and D. E. Spencer, Some Electromagnetic Paradoxes, *J. Franklin Just.*, vol. 260, p. 373, 1955. J. D. Edwards, Trouble with Flux, *IEEE Student Journal*, pp. 29–34, November 1965. C. T. Tai, On the Presentation of Maxwell's Theory, *Proc. IEEE*, vol. 60, pp. 936–945, August 1972.

‡ For a stationary circuit with a steady magnetic field through it, (12.3) gives $\nabla \times \mathbf{E} = 0$. Since this is true for any shaped circuit, we conclude that $\mathbf{E} = 0$. The induced emf is therefore zero.

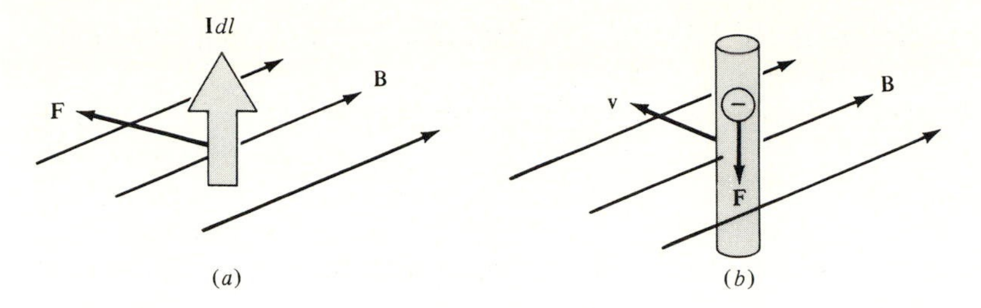

Figure 12.1 (*a*) The force on a current element **I** dl in a magnetic field. Here **I** and **B** are given with **F** as the result. (*b*) A piece of wire is moved with velocity **v** through a magnetic field. Here **B** and **v** are given with the force **F** on the electrons as a result.

We can now use (12.7) to derive the motional emf induced in a conducting wire which moves through a magnetic field (i.e., it cuts flux lines). This is depicted in Fig. 12.1*b*. Since the wire has an abundance of electrons which are free to move in the wire as the wire itself moves through B, the electrons will respond to the force expressed by (12.7) and move along the wire. The force per unit charge which acts on the electrons is equivalent to an induced electric field $\mathbf{E} = \mathbf{F}/q$ along the wire. The motional emf induced in the wire of length L is therefore

$$\boxed{\mathcal{V}_m = \int_0^L \mathbf{E} \cdot \mathbf{dl} = \int_0^L \mathbf{v} \times \mathbf{B} \cdot \mathbf{dl}} \tag{12.8}$$

If the wire forms a closed circuit, the above integral is replaced by a closed-line integral.

Motional emf cannot be derived from the relation (12.3). In the general case of a moving circuit in a time-varying field the total induced emf will be a combination of (12.4) and (12.8).

12.3 GENERAL CASE: MOTIONAL AND TRANSFORMER EMF

It is important to observe that the general case can be derived from Faraday's law (12.2) *or* from Lorentz' force law (12.5). In Faraday's law, taking the total time derivative will properly account for motional and transformer emf. Similarly, in Lorentz' force law, the first term will give transformer emf, whereas the second one will give the motional emf.

Lorentz' Force

The general induction can be derived from Lorentz' force law as follows: The total voltage induced in a closed circuit as a result of the forces which act on the conduction electrons can be written from (12.5) as

$$\boxed{\mathscr{V} = \oint \mathbf{E}' \cdot d\mathbf{l} = \oint (\mathbf{E} + \mathbf{v} \times \mathbf{B}) \cdot d\mathbf{l}} \tag{12.9}$$

where $\mathbf{E}'$ is the total electric field which acts on the conduction electrons. From Sec. 11.7 we can relate the electric field to a vector and scalar potential as

$$\mathbf{E} = -\frac{\partial \mathbf{A}}{\partial t} - \nabla\phi \tag{12.10}$$

If an electrostatic field is not present, then $\mathbf{E} = -\partial\mathbf{A}/\partial t$, and we can write (12.9) as (see Prob. 12.1)

$$\mathscr{V} = -\oint \frac{\partial \mathbf{A}}{\partial t} \cdot d\mathbf{l} + \oint \mathbf{v} \times \mathbf{B} \cdot d\mathbf{l} \tag{12.11}$$

For a closed circuit we can apply Stokes' theorem (A1.10), and since the magnetic field is related to the vector potential by $\mathbf{B} = \nabla \times \mathbf{A}$, we obtain for the total induced voltage due to transformer and motional induction

$$\boxed{\mathscr{V} = \oint \mathbf{E}' \cdot d\mathbf{l} = -\iint \frac{\partial \mathbf{B}}{\partial t} \cdot d\mathbf{A} + \oint \mathbf{v} \times \mathbf{B} \cdot d\mathbf{l}} \tag{12.12}$$

Using (12.4) and (12.8), we observe that $\mathscr{V}$ has two components: $\mathscr{V} = \mathscr{V}_t + \mathscr{V}_m$. Therefore the total emf in a loop of wire can be obtained by using the total force which acts on the conduction electrons.

Faraday's Law

The same result as (12.12) can be obtained from Faraday's law (12.2) by separating the total time derivative into effects due to time change and due to motional change. This essentially amounts to performing a galilean transformation on (12.2). If the total flux ϕ is a function of space and time, we can write Faraday's law as

$$\mathscr{V} = -\frac{d}{dt}\phi(x, y, z, t) = -\frac{d}{dt}\iint_{A(l,t)} \mathbf{B} \cdot d\mathbf{A} \tag{12.13}$$

where the flux density is $B(x, y, z, t)$ and the surface $A(t)$ is bounded by the curve $l(t)$. The total time derivative can be expanded as follows:

$$\frac{d}{dt}\phi = \frac{\partial\phi}{\partial t} + \frac{dx}{dt}\frac{\partial\phi}{\partial x} + \frac{dy}{dt}\frac{\partial\phi}{\partial y} + \frac{dz}{dt}\frac{\partial\phi}{\partial z} \tag{12.14}$$

Hence the total time differentiation operator can be expressed as

$$\frac{d}{dt} = \frac{\partial}{\partial t} + \mathbf{v} \cdot \nabla \tag{12.15}$$

This operator is known as the *convective derivative* and separates into temporal and spatial components. Using identity (A1.22), we obtain

$$\nabla \times (\mathbf{v} \times \mathbf{B}) = \mathbf{v}\nabla \cdot \mathbf{B} - \mathbf{B}\nabla \cdot \mathbf{v} + (\mathbf{B} \cdot \nabla)\mathbf{v} - (\mathbf{v} \cdot \nabla)\mathbf{B} \tag{12.16}$$

Since $\nabla \cdot \mathbf{B} = 0$, and for simplicity, let us assume a constant velocity $\mathbf{v}$. We see that the first three terms on the right-hand side of (12.16) vanish. (See Prob. 12.3 for the general case.) Substituting this result in (12.13), we have

$$\mathscr{V} = -\iint \frac{\partial \mathbf{B}}{\partial t} \cdot d\mathbf{A} + \iint \nabla \times (\mathbf{v} \times \mathbf{B}) \cdot d\mathbf{A} \tag{12.17}$$

Applying Stokes' theorem, we can finally write

$$\mathscr{V} = -\iint \frac{\partial \mathbf{B}}{\partial t} \cdot d\mathbf{A} + \oint (\mathbf{v} \times \mathbf{B}) \cdot d\mathbf{l} \tag{12.18}$$

which is the same result as (12.12), obtained by using the Lorentz force. Therefore, the total induced emf in a loop of wire is equal to the total time derivative of magnetic flux linking the loop.

Example It is instructive to obtain the motional emf shown in Fig. 12.1*b* from Faraday's law. In Fig. 12.2 we have a steady magnetic field through a loop (formed by the U-shaped wire and movable crossbar) in which the flux is changed by varying the area of the circuit. Applying (12.2) to the area Lx of the circuit, we obtain for the induced emf

$$\mathscr{V}_m = -\frac{d(BLx)}{dt} = -BL\frac{dx}{dt} = -BLv \tag{12.19}$$

Therefore Faraday's law (12.2) gives the same result as (12.8). The induced emf will result in an induced current I which circulates around the loop. By cutting the loop anywhere, thus creating a short gap, the induced emf will appear across the gap. This problem is similar to that shown in Fig. 6.37.

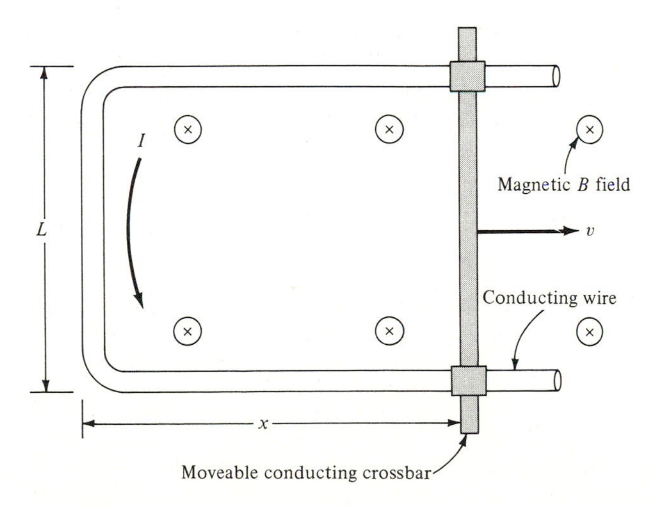

Figure 12.2 The induction of motional emf by varying the area of the loop, where v is the velocity of the crossbar.

12.4 BEWARE OF MOVING FLUX LINES

In the preceding pages we have shown that the total emf voltage induced in a coil can be obtained either by applying Faraday's law or by using the total force on a charge responsible for the induction of emf. However, it is not always that simple. For example, if a charge q moves parallel to a long, straight stationary wire that carries a steady current I, we know from the magnetic force law (12.7) that the charge will curve toward the wire (assuming I and motion of q are in same direction). The B field through which q moves is that of the current-carrying wire (see Fig. 6.19). The difficulty comes when we consider a different situation in which the current-carrying wire is moved along its axial direction past a stationary charge q. Will the charge remain at rest, or will it move toward the wire? That is, will an observer riding along with the moving charge (as in the first case) and an observer moving with the wire (as in the second case) see different things? From experience we know that the same effect will be observed in both cases: The charge moves toward the wire. This is also what the newtonian principle of relativity would tell us. The second case, however, is the baffling one, since to an observer stationary with respect to the charge, the wire that moves past must appear charged; i.e., it now has an electric field which attracts the charge. This is hard to understand at first, since we know that a current-carrying wire appears uncharged to an observer stationary with respect to the wire. One might be quick to point out that the magnetic lines of the moving wire also move past the stationary charge; hence (12.7) is still applicable, with $\mathbf{v}$ the velocity of the moving flux lines. This is erroneous and in general doesn't make sense since lines of $\mathbf{E}$ and $\mathbf{B}$ are not physical entities. They don't exist and are only used to help visualize the field at a point in space. For example, if the moving wire were hidden from view, one could not establish by measuring the force on the stationary charge whether a wire carrying a current moved past or whether the stationary charge was attracted by some hidden charge. Anyhow, it makes no sense to say that if a current element or magnet is moved, $\mathbf{B}$ lines move with the current element or magnet. One can only say that at any given point the magnetic field changes when the source of the field is moved; the magnetic field might even vanish, and an electric field might appear, as discussed above.† The force on a charge, given by the Lorentz force, is determined by the values of $\mathbf{E}$ and $\mathbf{B}$ at the charge, whether the sources of the flux lines are moving or not. A B field will produce a force on the charge only if the charge is moving at some velocity $\mathbf{v}$. Examples such as these show that electricity and magnetism are not independent, but are somehow related.

The above-posed problem of charge and current-carrying wire can be treated after the introduction of relativity. For the time being we can explain the problem

† In the second case in which the current-carrying wire is moved in the direction of its axis past a point, the magnetic field at that point does not even change with time because the wire is assumed to be long. Only when the ends of the wire move past that point does the field change.

as follows: In the first case, when the wire is stationary and the charge q moves parallel to the wire, the force on the charge is part of Lorentz' force, $\mathbf{F} = q\mathbf{v} \times \mathbf{B}$, where $\mathbf{v}$ is the velocity of the charge with respect to the wire. The magnetic field $\mathbf{B}$ of the wire is produced by the motion of the conduction electrons which have a density ρ_- in the wire, while the positive atomic charges of density ρ_+ remain fixed inside the wire. The wire is uncharged, because $\rho_- = \rho_+$ at every point in the wire. No electric field appears from an uncharged wire carrying a current since the moving conduction electrons stay neutralized as they move through the lattice structure of positive atoms. In the second case, when the wire moves and the charge is stationary, we see that both ρ_- and ρ_+ have velocities with respect to the stationary charge. However, since the velocities of ρ_- and ρ_+ are different, we can show that a relativistic transformation of ρ_- and ρ_+ to the coordinate system of the stationary charge will yield different magnitudes for ρ_- and ρ_+. Hence, for this case $\rho_- \neq \rho_+$, and a net charge appears on the wire which "creates" an electric field and hence attracts the stationary charge.

12.5 FROM THE GALILEAN TO THE LORENTZ TRANSFORMATION

Let us review briefly the events that shaped the present Maxwell-Lorentz-Einstein picture of electrodynamics. It is well known that all laws of motion, whether they be Newton's laws or Maxwell's equations for the motion of an electromagnetic field, must be associated with a frame of reference.† At the same time, a physical process cannot depend on a coordinate frame the observer may have chosen from which to observe it. Therefore, physical laws must be written in such a way that their form is preserved for different coordinate systems. The transformation that will leave a physical law invariant for different frames which move relative to one another becomes then of fundamental importance. That transformation in the days of Newton was the galilean transformation.

$$x' = x - u_0 t \qquad y' = y \qquad z' = z \qquad t' = t \tag{12.20}$$

which is here expressed for two inertial frames, the primed and the unprimed, moving along the x axis at velocity u_0 relative to one another. For a long time it had been known that Newton's laws of mechanics are invariant under a galilean transformation. Nevertheless, Newton believed that his laws were related to an absolute frame of reference. This notion, however, was purely metaphysical, since he himself formulated the principle of relativity in mechanics according to which it is impossible to detect a uniform rectilinear motion of a body, or a frame of reference, relative to this absolute system. For example, accelerations and forces are identical in two frames of reference, since from (12.20)

$$\frac{d^2x'}{dt'^2} = \frac{d^2x}{dt^2}$$

Therefore, $F' = ma'$ transforms under the galilean transformation into $F = ma$; that is, it is invariant under this transformation. The reason for invariance is that velocity does not appear in the equations of motion.

† Frames of reference in which free bodies move with constant velocity unless acted upon by external forces are called *inertial frames of reference*; i.e., inertial frames are those in which the law of inertia holds. All inertial frames move with a constant velocity relative to one another.

If the galilean transformation is applied to Maxwell's equations, they assume different forms in different coordinate systems, because the equations contain the velocity of propagation of electromagnetic waves. This velocity, when added vectorially, has different values in different coordinate systems that move with respect to each other. The difference in velocity had several implications; one, it would have been possible to detect an absolute frame of reference by electromagnetic or optical means. It was assumed at that time that light needed a medium, called the ether, through which to propagate. The fact that Maxwell's equations were not invariant under a galilean transformation implied that there was a preferred reference system in which the ether was at rest. The velocity of light in that absolute-rest frame would be equal to c, and presumably in another reference frame the velocity of light would not be equal to c. That the velocity of light should have different values in two coordinate systems moving relative to each other under the galilean transformation can be easily seen from (12.20). If the velocity of light in the (unprimed) S frame is given by dx/dt, then in the S' frame which moves relative to S at u_0, the velocity of light is

$$\frac{dx'}{dt'} = \frac{dx}{dt} - u_0$$

However, Fizeau's experiment, the Michelson-Morley experiment, and many others which tried to examine the accuracy of Maxwell's equations in moving reference frames showed beyond doubt that the speed of light is the same in any direction in all inertial coordinate systems. It was then necessary to conclude that Maxwell's equations must have the same form in all inertial reference systems. In 1904 Lorentz, trying to account for this conclusion, found a new transformation under which Maxwell's equations remained in the same form in different coordinate systems (or inertial frames) moving relative to each other, and thus laid the groundwork for special relativity. It is now known as the Lorentz transformation

$$x' = \frac{x - u_0 t}{\sqrt{1 - u_0^2/c^2}} \qquad y' = y \qquad z' = z \qquad t' = \frac{t - xu_0/c^2}{\sqrt{1 - u_0^2/c^2}} \tag{12.21}$$

Shortly thereafter Poincare showed that all equations of electrodynamics are invariant in form under a Lorentz transformation. A simple check shows that $dx'/dt' = dx/dt = c$ for the Lorentz transformation. It was Einstein in 1905, however, who generalized these ideas to what is presently known as special relativity. He showed that the Lorentz transformation can be obtained by adopting Newton's principle of relativity (his first postulate: All physical laws are the same in all inertial systems) and assuming that the free-space light velocity is a universal constant independent of the motion of the source (his second postulate). Even though special relativity had its origin in electromagnetism, in Einstein's approach one does not have to depend on electrodynamics explicitly since Maxwell's equations are just one of the physical laws covered by the first postulate. Thus, special relativity has far-reaching results. It renders meaningless the question of detecting motion relative to an ether; hence the ether concept itself becomes meaningless. Our classical concept of time as a variable independent of the spatial coordinates and of relative motion is destroyed by the second postulate and must be replaced by a complicated and interwoven space-time concept. Simultaneity of events had to be revised. Events which are simultaneous in one coordinate system are not necessarily simultaneous in another coordinate system moving relative to the first. These ideas were so startling that they were vehemently resisted by many (and perhaps still are). Many ingenious attempts were made to invent theories which would avoid the second postulate. However, so many experiments have been performed by now which verify these predictions, that only theories consistent with special relativity need be considered.

The classical laws of electrodynamics which were not invariant under the galilean transformation are invariant under the Lorentz transformation. Newton's equations of mechanics, however, which were invariant under the galilean transformation are now found to be not invariant under the Lorentz transformation. This meant that either they had to be discarded or had to be replaced by relativistic equations in which the mass m of a moving body was related to its velocity v and its rest mass m_0 by the relationship

$$m = \frac{m_0}{\sqrt{1 - v^2/c^2}} \tag{12.22}$$

Fortunately this was the only correction needed in the laws of mechanics. At low velocities, where $v^2/c^2 \approx 0$, the relativistic equations reduce to Newton's laws of motion. The necessity of relativistic corrections in the laws of motion was soon verified experimentally. For example, electrons in a particle accelerator can be accelerated close to the velocity of light such that their mass will increase by factors of thousands. Equation (12.22), when expanded, will lead to a term m_0c^2, which was identified by Einstein, Lewis, and others as its *rest energy* $E = m_0 c^2$. This prediction of the rather large rest energy which a mass possesses was then verified in the fission experiment and nuclear devices. Perhaps the most simple and elegant illustration of $E = m_0 c^2$ is given when an electron and its antiparticle, the positron, both of mass m_0, meet. They annihilate each other and give off two gamma rays. The energy in the gamma rays is identical to the rest energy of the two particles before they collided; that is, $2m_0 c^2$.

12.6 MAXWELL'S EQUATIONS FROM COULOMB'S LAW AND SPECIAL RELATIVITY

The development of Maxwell's equations in this book is historical, for it is based on a sequence of experimentally based postulates. Had special relativity existed before Oersted discovered magnetism (1820), the laws of magnetism as well as Maxwell's equations could have been predicted by applying special relativity to the laws of electrostatics. At the present time we can say that there is an alternative approach to Maxwell's equations which is based on the two postulates of special relativity and the experimentally based law of Coulomb. As this approach is more sophisticated and as it does not uncover any new relations, we will confine ourselves to the main points of this approach without going into the details.†

Assume a static electric field $E'(x', y', z')$ is established in the S' frame by a charge distribution $\rho'(x', y', z')$ which is stationary in the S' frame. The equations for the electrostatic field in S' are

$$\nabla' \cdot \mathbf{E}' = \frac{\rho'}{\varepsilon_0} \tag{12.23}$$

$$\nabla' \times \mathbf{E}' = 0 \tag{12.24}$$

In the S' frame, the force on a small test charge q', which can be either stationary or moving in S', is given by

$$F' = q'E' \tag{12.25}$$

The test charge q' experiences only an electrostatic force. Any B field created by the steady motion of q' does not produce a magnetic force on the stationary charge distribution ρ' (see related discussion in Sec. 12.4).

† For details the student is referred to a short paper by D. A. Driscoll, Maxwell's Equations—Derived from Coulomb's Law and Special Relativity, *IEEE Student Journal*, pp. 31–35, May 1968, and a more complete paper by R. S. Elliott, Relativity and Electricity, *IEEE Spectrum*, pp. 140–152, March 1966, as well as to some texts: E. G. Cullwick, "Electromagnetism and Relativity, with Particular Reference to Moving Media and Electromagnetic Induction," John Wiley & Sons, Inc., New York, 1959; R. S. Elliott, "Electromagnetics," McGraw-Hill Book Company, New York, 1966; J. A. Kong, "Theory of Electromagnetic Waves," John Wiley & Sons, Inc., New York, 1975.

Suppose the S' frame moves with an x-directed velocity u_0 relative to a stationary S frame. Assume our laboratory in which we make observations is stationary in the S frame. The force exerted on a test charge as observed in the S frame can be found now by applying the Lorentz transformation (12.21) to (12.25), with the result that

$$\mathbf{F} = q[E'_x\hat{\mathbf{x}} + \beta(E'_y\hat{\mathbf{y}} + E'_z\hat{\mathbf{z}})] + q\mathbf{v} \times \left(\frac{\beta u_0 \hat{\mathbf{x}}}{c^2} \times \mathbf{E}'\right) \tag{12.26}$$

where $\mathbf{v}$ is the velocity of the test charge q as measured in the S frame and $\beta = (1 - u_0^2/c^2)^{-1/2}$.

For the case when the test charge is stationary in the S frame, the force acting on the test charge is due to an electric field $\mathbf{E}$ alone; that is,

$$\mathbf{F} = q\mathbf{E} = q[E'_x\hat{\mathbf{x}} + \beta(E'_y\hat{\mathbf{y}} + E'_z\hat{\mathbf{z}})] \tag{12.27}$$

As the relative velocities are usually much smaller than the velocity of light, $\beta \cong 1$ and (12.27) reduces to $\mathbf{F} = q\mathbf{E} \cong q\mathbf{E}'$.

The additional force term in (12.26) which appears when the test charge is moving is identified with a magnetic field B; that is,

$$\boxed{\mathbf{B} = \frac{\beta u_0 \hat{\mathbf{x}}}{c^2} \times \mathbf{E}'} \tag{12.28}$$

This is the same magnetic field that appears throughout the book. For example, for nonrelativistic velocities, when $\beta \cong 1$, (12.28) is the same as (6.10), which gives the magnetic field produced by a moving charge. The difference is that in the historical development, the Biot-Savart law (6.10) had to be postulated, whereas in the relativistic development, the magnetic field is a natural consequence of electrostatics in moving coordinates. A relativistic transformation of Coulomb's law gives us, therefore, Lorentz' force law

$$\mathbf{F} = q(\mathbf{E} + \mathbf{v} \times \mathbf{B}) \tag{12.29}$$

Maxwell's equations can now be obtained by transforming the laws of electrostatics, (12.23) and (12.24), to the S frame. Such a transformation involves operating with the divergence and curl on the derived $\mathbf{B}$ of (12.28).

Example Calculate the magnetic force between two charges q_1 and q_2 moving with the same velocity v along a parallel path. Assume $v \ll c$, such that $\beta \cong 1$. The geometry is that shown in Fig. 6.10.

The magnetic force is given by the second term in (12.26). Letting $v = u_0$, $q = q_1$, $E' = q_2\hat{\mathbf{R}}/4\pi\varepsilon_0 R^2$, where R is the distance between q_1 and q_2, we obtain the attractive force as

$$F_m = \frac{q_1 q_2 v^2}{4\pi\varepsilon_0 c^2 R^2} = \frac{\mu_0 q_1 q_2 v^2}{4\pi R^2} \tag{12.30}$$

which is the same as (6.14), which was derived from the postulated Biot-Savart law.

Conclusion

A relativistic approach to the derivation of Maxwell's equations is an approach which emphasizes from the start that electric and magnetic fields have no independent existence. A pure electric field **E** in one reference frame S transforms into an electric field **E**′ and magnetic field **B**′ in another frame S' that is moving relative to the first. However, for $u_0 < c$, no Lorentz frame S' exists that would transform a pure **E** in S to a pure **B**′ in S'. **E** and **B** are therefore completely interrelated, and one should speak properly of an electromagnetic field instead of **E** and **B** separately.

PROBLEMS

12.1 Show that the result (12.12) is still correct, even if electrostatic fields are present. Discuss the relationship of electrostatic fields to emf.

Hint: See Secs. 3.1 and 3.2 and footnote on page 443.

12.2 If the crossbar in Fig. 12.2 oscillates in the x direction with the velocity

$$\mathbf{v}(t) = \hat{\mathbf{x}} v_m \cos \omega t$$

and the magnetic field also varies with time as

$$\mathbf{B} = \hat{\mathbf{z}} B_m \cos \omega t$$

obtain an expression for the induced emf in the circuit. The unit vector $\hat{\mathbf{z}}$ is out of the page.

12.3 Show that for a continuous vector field **c** with continuous derivatives, we have the identity

$$\frac{d}{dt} \iint \mathbf{c} \cdot d\mathbf{A} = \iint \left[\frac{\partial \mathbf{c}}{\partial t} + \mathbf{v} \nabla \cdot \mathbf{c} + \nabla \times (\mathbf{c} \times \mathbf{v}) \right] \cdot d\mathbf{A}$$

Hint: If $\mathbf{r} = \hat{\mathbf{x}}x + \hat{\mathbf{y}}y + \hat{\mathbf{z}}z$ is the radius vector drawn from the origin to the point (x, y, z), what is $\nabla \cdot \mathbf{r}$? Since $d\mathbf{r}/dt = \mathbf{v}$, what is $\nabla \cdot \mathbf{v}$? Use these results with identity (12.16) to show the above identity.

12.4 An open circuit is formed by a straight wire of length L.

(*a*) Starting with the Lorentz force (12.5) and following a procedure similar to Eqs. (12.9) to (12.12), show that the induced emf is $\mathscr{V} = vBL$ if the wire moves with velocity v at right angles to the B field (assume the B field and wire axis are normal to each other, as shown in Fig. 12.1*b*).

(*b*) Starting with Faraday's law (12.2) and for the same conditions of motion through B as given in (*a*), can you obtain the same result for induced emf? Notice that the method outlined in Eqs. (12.13) to (12.18) depends on the presence of a closed circuit as does the method shown in the example following (12.18).

12.5 If the sliding bar in Fig. 12.2 is replaced by a sliding strip which stays in contact with the stationary U-shaped part as the conducting strip slides with velocity v, show that the induced emf is given by $\mathscr{V} = vBL$. Calculate the induced voltage for $v = 10$ m/s, $H = 1$ A/m, $L = 10$ cm.

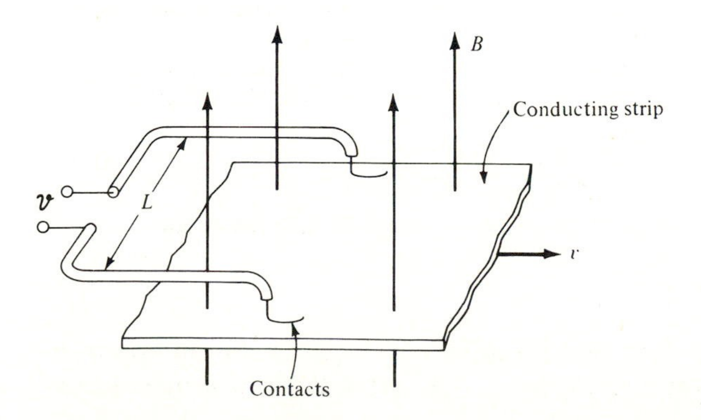

12.6 Show that the answer to Prob. 6.27 is $\mathscr{V} = \omega B l^2/2$.

12.7 A conducting single-spoke wheel is free to rotate about its axis in a magnetic field B which is parallel to the axis.

(*a*) If the spoke length is l and the wheel rotates with angular velocity ω rad/s, find the induced emf voltage $\mathscr{V}$ between axis and rim. Calculate $\mathscr{V}$ for 500 r/min, $l = 20$ cm, $B = 0.5$ T.

(*b*) If a battery is connected to the circuit, a current I will flow through the spoke and rim. Find the torque on the wheel.

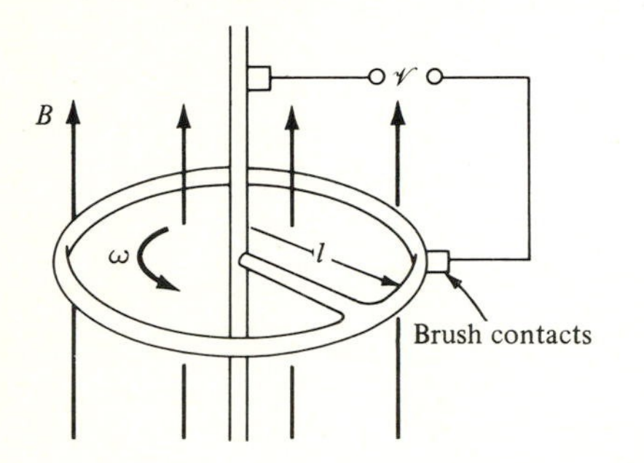

12.8 (*a*) Do the Faraday disc-generator Prob. 6.28.

(*b*) In the figure for Prob. 6.28 the circuit containing the voltmeter is in a plane normal to the disc. If this circuit is positioned so it is in the plane of the disc, will the results be different?

12.9 Do Prob. 6.30.

12.10 Show that the galilean transformation (12.20) leaves Newton's laws invariant.

12.11 Show that the galilean transformation can be obtained from the Lorentz transformation in the limit $u_0/c \to 0$.

12.12 Show that the inverse Lorentz transformation is

$$x = \beta(x' + u_0 t') \qquad y = y' \qquad z = z' \qquad t = \beta\left(t' + \frac{u_0 x'}{c^2}\right)$$

CHAPTER

THIRTEEN

APPLICATIONS OF MAXWELL'S EQUATIONS: EM WAVES AND PROPAGATION OF ENERGY

13.1 THE WAVE EQUATION

Maxwell's equations, given in Sec. 11.5 and restated here,

$$\nabla \times \mathbf{E} = -\frac{\partial \mathbf{B}}{\partial t} \tag{13.1}$$

$$\nabla \cdot \mathbf{D} = \rho \tag{13.2}$$

$$\nabla \times \mathbf{H} = \mathbf{J} + \frac{\partial \mathbf{D}}{\partial t} \tag{13.3}$$

$$\nabla \cdot \mathbf{B} = 0 \tag{13.4}$$

describe all behavior of the electric and magnetic field vectors **E** and **H**. The vectors **D** and **B** are related by the constitutive relations $\mathbf{D} = \varepsilon\mathbf{E}$ and $\mathbf{B} = \mu\mathbf{H}$. The four equations are a formidable set of coupled partial differential equations in **E** and **H** which cannot be solved in general. Is it possible to describe at least some behavior of electromagnetic fields by a simpler set of equations? We have already seen that in the static case this is possible.† Any effort to simplify the above equations should start by trying to uncouple the set, i.e., to obtain differential equations in **E** or **H** alone. Let us take the curl of (13.1) and substitute (13.3) for

† Maxwell's equations for the static case are a set of uncoupled differential equations: $\nabla \times \mathbf{E} = 0$, $\nabla \cdot \mathbf{D} = \rho$ for **E** and $\nabla \times \mathbf{H} = \mathbf{J}$, $\nabla \cdot \mathbf{B} = 0$ for **H**.

curl **H.** This gives us

$$\nabla \times \nabla \times \mathbf{E} = -\mu \frac{\partial \mathbf{J}}{\partial t} - \mu\varepsilon \frac{\partial^2 \mathbf{E}}{\partial t^2} \tag{13.5}$$

Using identity (A1.24), $\nabla \times \nabla \times \mathbf{E} = \nabla(\nabla \cdot \mathbf{E}) - \nabla^2 \mathbf{E}$, and the Maxwell equation, $\nabla \cdot \mathbf{E} = \rho/\varepsilon$:

$$\boxed{\nabla^2 \mathbf{E} - \mu\varepsilon \frac{\partial^2 \mathbf{E}}{\partial t^2} = \mu \frac{\partial \mathbf{J}}{\partial t} + \nabla \frac{\rho}{\varepsilon}} \tag{13.6}$$

A similar procedure for **H** gives us

$$\boxed{\nabla^2 \mathbf{H} - \mu\varepsilon \frac{\partial^2 \mathbf{H}}{\partial t^2} = \nabla \times \mathbf{J}} \tag{13.7}$$

An equation whose left-hand side has the form of (13.6) or (13.7) is called a *wave equation*, because the solution to such equations give propagating waves. The right-hand side expresses merely the sources for the wave fields **E** and **H**, which are currents and charges for the **E** field, and currents only for the **H** field. The wave equation is a commonly occurring equation in many branches of engineering, physics, etc., as wave phenomena are present in many fields.

As a result of the uncoupling process we have found that the **E** and **H** fields satisfy wave equations which are simpler and can be more readily solved than Maxwell's equations. But since these equations were obtained by differentiating, we have lost some information in going from Maxwell's equations to the wave equation. Hence there is no way to reconstruct Maxwell's equations from the wave equation, which therefore describes only a subset of possible electromagnetic (EM) field behavior. For example, solutions to the wave equation contain unknown coefficients. These coefficients can be determined from boundary conditions which in turn are derivable only from Maxwell's equations. Thus, to obtain a complete solution to the wave equation, additional information must be drawn directly from Maxwell's equations.

13.2 WAVE EQUATION FOR FREE SPACE (LOSSLESS CASE)

To learn the nature of waves, we can consider a large, empty volume of space. As no sources can be present in empty space: $\mathbf{J} = \rho = 0$. Fields in such a region must satisfy the wave equation

$$\boxed{\nabla^2 \mathbf{E} - \mu\varepsilon \frac{\partial^2 \mathbf{E}}{\partial t^2} = 0} \tag{13.8}$$

The solution to any differential equation consists of two parts. One is the solution to the homogeneous equation, obtained with all source terms assumed zero; the other is the particular solution which gives the response of a system to a particular forcing function (source). It is the solution to the homogeneous equation which is characteristic of the system (in our case, an electromagnetic field in free space), for it contains terms relating only to the system. Thus it is the solution to (13.8) which characterizes our system.

We should now note that if it were not for the displacement current $\partial \mathbf{D}/\partial t$, electromagnetic waves would not be possible. Maxwell's equations leading to (13.8) are

$$\nabla \times \mathbf{E} = -\frac{\partial \mathbf{B}}{\partial t} \tag{13.9}$$

$$\nabla \times \mathbf{H} = \frac{\partial \mathbf{D}}{\partial t} \tag{13.10}$$

A wave is possible because a time-changing magnetic field generates a time-varying electric field, which in turn generates a varying magnetic field, and so on. This coupling of the two fields is illustrated in Fig. 13.1. Note that the right-hand rule gives the sense of the **B** field surrounding $\partial \mathbf{D}/\partial t$, because **J** and $\partial \mathbf{D}/\partial t$ in (13.3) have the same direction, and the right-hand rule was developed for current **J**. On the other hand, the sense of the surrounding **E** field generated by $\partial \mathbf{B}/\partial t$ is opposite to that of the right-hand rule, reflecting the minus sign as given by Lenz' law [see Eq. (11.1)]. The different senses of the surrounding fields in Fig. 13.1 are represented in (13.9) and (13.10) by the opposite signs.

The propagating characteristics of the EM wave are contained in the solution to (13.8). To obtain the characteristics, we do not need the most general case, represented by (13.8). It is sufficient to assume propagation in one direction, which we will choose as the z direction, and only one component of the **E** field, say $\mathbf{E} = \hat{\mathbf{x}}E_x$. Equation (13.8) then simplifies to

$$\frac{\partial^2 E_x}{\partial z^2} - \mu\varepsilon \frac{\partial^2 E_x}{\partial t^2} = 0 \tag{13.11}$$

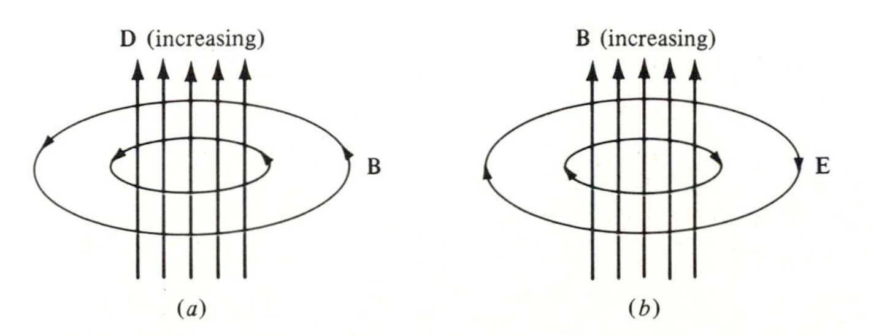

Figure 13.1 (*a*) An increasing electric flux generates a surrounding magnetic field; (*b*) a surrounding electric field is created by an increasing magnetic flux.

which is known as a one-dimensional scalar wave equation ($E_y = E_z = 0$ and $\partial/\partial x = \partial/\partial y = 0$). From a purely dimensional analysis of (13.11), we observe that the constant $\mu\varepsilon$ must be related to a velocity v which we will shortly show to be the velocity of the wave:

$$v = \frac{1}{\sqrt{\mu\varepsilon}} = \frac{1}{\sqrt{\mu_r \varepsilon_r}} \frac{1}{\sqrt{\mu_0 \varepsilon_0}} = \frac{c}{\sqrt{\mu_r \varepsilon_r}} = \frac{c}{n} \tag{13.12}$$

where $\mu_r = \mu/\mu_0$ is relative permeability
$\varepsilon_r = \varepsilon/\varepsilon_0$ is relative permittivity
$n = \sqrt{\mu_r \varepsilon_r}$ is index of refraction of medium in which wave travels
$c = 3 \times 10^8$ m/s is velocity of light in vacuum

We can now rewrite (13.11) as

$$\boxed{\frac{\partial^2 E_x}{\partial z^2} - \frac{1}{v^2}\frac{\partial^2 E_x}{\partial t^2} = 0} \tag{13.13}$$

By inspection, this has a general solution which has the form $g(z \pm vt)$ or $g(t \pm z/v)$:

$$\boxed{E_x(z, t) = g_1(z - vt) + g_2(z + vt)} \tag{13.14}$$

where g is an arbitrary function representing the shape (sinusoidal, square wave, pulse, etc.) of the wave excited by a remote transmitting source. A second-order differential equation such as (13.13) has two independent solutions. We have written the two solutions explicitly as g_1 and g_2, noting that for a linear system the sum of two individual solutions is also a solution (an example of linear superposition). That (13.14) is a solution can be easily checked by substituting (13.14) into (13.13).

Solution (13.14) represents two traveling waves, one traveling along the positive z direction, the other along the negative z direction. The best way to study a wave is to pick a fixed point on a wave and follow that point as the wave progresses. For solution $g_1(z - vt)$ we can do this by keeping the argument $z - vt$ constant, which will correspond to a fixed point on g_1. As time t increases, z must increase, for the argument to stay constant. The rate at which z must increase to hold the argument and in turn the value of g_1 is

$$\frac{d}{dt}(z - vt = \text{constant})$$

$$\frac{dz}{dt} - v = 0 \tag{13.15}$$

which gives us the velocity of the fixed point as $dz/dt = v$. The wave therefore travels along the $+z$ direction at a velocity† $v = 1/\sqrt{\mu\varepsilon}$. As g represents E, v is the

† The fact that in vacuum $v = 1/\sqrt{\mu_0 \varepsilon_0} = c$ led Maxwell to state that light is just one form of electromagnetic radiation.

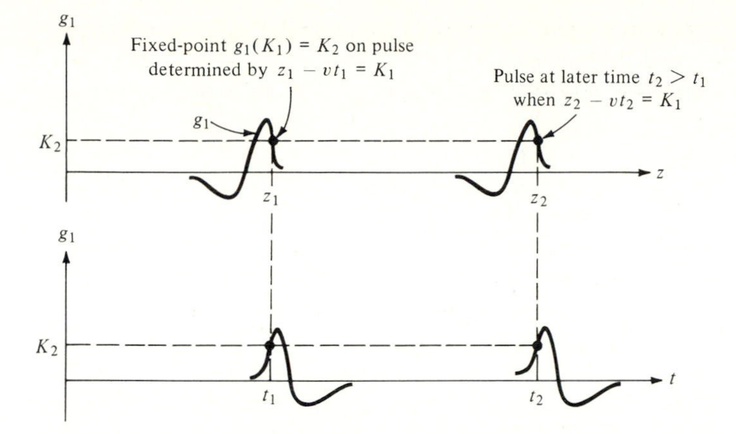

Figure 13.2 An EM pulse is shown at two positions. The pulse shape in the z direction is given by g_1. In the x and y directions, the pulse is assumed uniform (an infinite plane pulse). Constant K_1 specifies the argument for which the value of g_1 is the constant K_2. As K_1 is varied, the pulse shape g_1 is traced out along the z axis (t held fixed) and along the t axis (z held fixed). The fixed point on the pulse has progressed to position 2 in the time interval $t_2 - t_1 = (z_2 - z_1)/v$.

velocity with which a fixed value of E advances along the z axis. Figure 13.2 shows an EM wave pulse traveling along the positive z axis.

It is important to note that the variables z and t in the argument of the waveshape g_1 must occur in the combination $z - vt$. Only then can we speak of having a wave which travels at a constant velocity with unchanged shape. By similar reasoning, we can say that $g_2(z + vt)$ represents a wave traveling with velocity v in the $-z$ direction. Because z and t in $z - vt$ have opposite signs for a forward-traveling wave, the pulses along the z and t axes appear reversed in Fig. 13.2. Sketches for a backward-traveling wave $g_2(z + vt)$ would show them to be facing in the same direction on the z and t axis, respectively.

We have shown that the solution to (13.8) leads to waves that can exist in free space. Even though the electric and magnetic fields of the waves start out on sources, they detach themselves from charges and currents and move through empty space as independent entities, unattached to any electric charge or current. This is the implication of the solution to the free-space wave equation.

13.3 WAVE EQUATION IN MATERIAL MEDIA (LOSSY CASE)

Waves which propagate in a medium other than vacuum will experience some loss due to absorption in the medium. If the propagation is in the atmosphere, the absorption loss is small. If on the other hand, it is in a conducting medium such as salt water, ionized gas, or a metallic medium, the losses can be very high. The ohmic loss can be accounted for by substituting the conduction current density $\mathbf{J} = \sigma\mathbf{E}$ in the wave equations (13.6) and (13.7). Thus the wave equation for regions free of sources are

$$\nabla^2 \mathbf{E} - \mu\varepsilon \frac{\partial^2 \mathbf{E}}{\partial t^2} - \sigma\mu \frac{\partial \mathbf{E}}{\partial t} = 0 \tag{13.16}$$

and

$$\nabla^2 \mathbf{H} - \mu\varepsilon \frac{\partial^2 \mathbf{H}}{\partial t^2} - \sigma\mu \frac{\partial \mathbf{H}}{\partial t} = 0 \tag{13.17}$$

The effect of the new loss term ($\sigma\mu\ \partial\mathbf{E}/\partial t$ and $\sigma\mu\ \partial\mathbf{H}/\partial t$) in the above equations is to attenuate a propagating wave, because energy is drained from the wave to provide for the ohmic heating losses in the medium. When this term is small, for example, when the absorbing medium is a low-loss dielectric ($\sigma \approx 0$), a wave will undergo a small (exponential) decrease in amplitude as it propagates in the medium (see Fig. 13.5). On the other hand, when the conduction loss is large ($\sigma \gg 1$), the exponential decrease of a wave will be so rapid that one can hardly speak of propagation (see Fig. 13.7). It is more of a diffusion into the medium. Whether there is primarily diffusion or propagation depends on the relative sizes of the last two terms in (13.16), which can be identified with displacement current and conduction current as follows:

$$\nabla^2 \mathbf{E} - \mu \frac{\partial}{\partial t}\left(\underbrace{\frac{\partial \mathbf{D}}{\partial t}}_{\text{displacement current}} + \underbrace{\mathbf{J}}_{\text{conduction current}}\right) = 0 \tag{13.18}$$

In a medium, if displacement current dominates and conduction current is so negligible that it can be ignored ($\partial\mathbf{D}/\partial t \gg \mathbf{J}$), Maxwell's equation tells us that the magnetic field is produced by displacement current and that propagation in such a medium is characterized by the lossless wave equation; that is,

$$\boxed{\nabla \times \mathbf{H} = \frac{\partial \mathbf{D}}{\partial t} \quad \text{and} \quad \nabla^2 \mathbf{E} - \mu\varepsilon \frac{\partial^2 \mathbf{E}}{\partial t^2} = 0} \tag{13.19}$$

whereas if conduction current dominates ($\mathbf{J} \gg \partial\mathbf{D}/\partial t$), the magnetic field is produced by conduction current and propagation is characterized by a diffusion equation, which is similar to the equation that governs diffusion of heat or gases; that is,

$$\boxed{\nabla \times \mathbf{H} = \mathbf{J} \quad \text{and} \quad \nabla^2 \mathbf{E} - \sigma\mu \frac{\partial \mathbf{E}}{\partial t} = 0} \tag{13.20}$$

This last equation is also known as the eddy-current equation, because it is of the same form as the equation for the current density $\mathbf{J}$. Since $\mathbf{J} = \sigma\mathbf{E}$, we have $\nabla^2\mathbf{J} - \sigma\mu\ \partial\mathbf{J}/\partial t = 0$ as the equation which governs the current behavior in a conducting medium.†

† An exactly similar equation also holds for **H**.

We have shown that in a medium in which conduction current dominates (conductors), the EM fields obey a diffusion equation, whereas in a medium in which displacement current dominates (dielectrics), the EM fields obey a wave equation. The implications of the two types of equations, (13.19) and (13.20), will be explored in the next sections.

13.4 SINUSOIDAL PLANE WAVES

Plane waves are waves that vary only in the direction of propagation and are uniform in planes normal to the direction of propagation. In (13.11) we considered such a wave. It propagated in the z direction. The **E** field had only an E_x component which has the same value at every point in a plane parallel to the xy plane.

It appears that as solutions to the general vector wave equation (13.8) are hopelessly complicated, plane waves are introduced primarily to make the mathematics simpler. Fortunately, this is not the case. It is well known in more advanced studies of EM fields that an arbitrary field or wave can always be represented as a spectrum of plane waves.† Therefore, plane waves can be considered as the building blocks in more complicated waves. Even of more importance is that the fields radiated by any transmitting antenna look like plane waves at distances far from the source. This is depicted in Fig. 13.3, where over a finite area ΔA, which is normal to the propagating direction z, the E and H fields are approximately planar. The farther one gets from the antenna, the better the approximation is. The fact that plane waves are simple and obey a scalar wave equation is a welcome mathematical convenience.

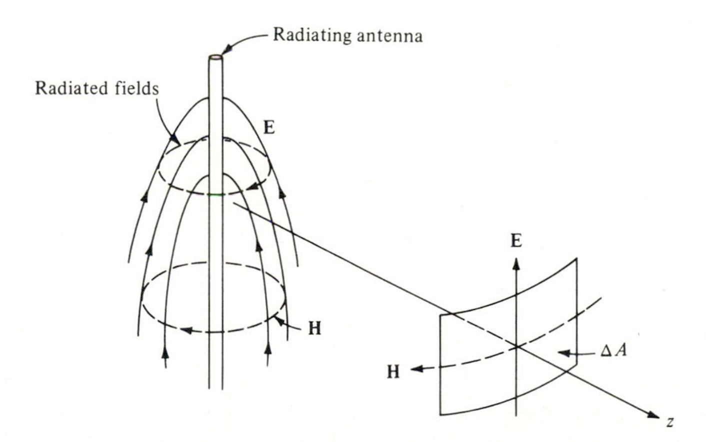

Figure 13.3 A vertical tower antenna radiates a field which spreads out in a radial direction from the antenna. Far from the antenna the field in area ΔA is a plane wave.

† P. C. Clemmow, "The Plane Wave Spectrum Representation of Electromagnetic Fields," Pergamon Press, Oxford, 1966.

Let us now choose the time behavior of the fields as sinusoidal; i.e., the fields oscillate at a single frequency $f = \omega/2\pi$ Hz. Again, the motivation for this is not just to consider simpler fields but is based upon two reasons. One is that many transmitting sources (radio, microwave, and optical) operate at such a narrow band of frequencies that the single-frequency approximation applies. The other is that any periodic wave can be represented as a Fourier series of sinusoids and any nonperiodic wave such as a pulse can be represented as a continuous spectrum of harmonics by the Fourier integral. For example, the pulse shown in Fig. 13.2 can be constructed from an infinite set of sinusoidally varying plane waves. These plane waves interfere constructively at the location of the pulse in such a way as to yield the pulse shape and interfere destructively every place else to give zero. Thus the general time case can be reduced to a problem involving sinusoids, which we will now proceed to develop.

Using the phasor notation,† the sinusoidal time variation of an E field polarized in the x direction can be represented by

$$\mathbf{E}(z, t) = E_x(z, t)\hat{\mathbf{x}} = E_x(z)e^{j\omega t}\hat{\mathbf{x}} \tag{13.21}$$

Substituting into the source-free wave equation (13.16), which is applicable to the lossy case ($\sigma \neq 0$) as well as the lossless case ($\sigma = 0$), we obtain

$$\frac{\partial^2 E_x(z)}{\partial z^2} + \omega^2 \mu\varepsilon \left(1 - \frac{j\sigma}{\omega\varepsilon}\right) E_x(z) = 0 \tag{13.22}$$

where the common factor $\hat{\mathbf{x}}e^{j\omega t}$ has been deleted. This is a relatively simple wave equation since it depends only on a single space variable. Equation (13.22) determines the space behavior of a uniform plane wave which varies sinusoidally with time. Using the complex permittivity $\varepsilon^\star$, defined in (11.23) as

$$\varepsilon^\star = \varepsilon \left(1 - \frac{j\sigma}{\omega\varepsilon}\right) \tag{13.23}$$

we can write (13.22) in the form of a lossless wave equation

$$\boxed{\frac{\partial^2 E_x}{\partial z^2} + \omega^2 \mu\varepsilon^\star E_x = 0} \tag{13.24}$$

which is better known as the equation of *simple harmonic motion* and has the solution

$$E_x(z) = E_0^i e^{-j\beta^\star z} + E_0^r e^{j\beta^\star z} \tag{13.25}$$

† Sinusoidal time variation can be represented by the real part (Re) of an exponential; that is, $\cos \omega t = \text{Re}\, e^{j\omega t} = \text{Re}\,(\cos \omega t + j \sin \omega t) = \cos \omega t$. As superposition applies in a linear system (Maxwell's equations and the wave equation are linear in media for which μ, ε, σ are constants), we can drop the operator Re and simply work with $e^{j\omega t}$. After a solution to a problem (using $e^{j\omega t}$) has been worked out, to give it physical meaning, we take the real part which is then referred to as the instantaneous solution.

where $\beta^\star$ is a complex phase-propagation constant† given by $\beta^{\star 2} = \omega^2 \mu \varepsilon^\star$. In general $\beta^\star$ will have real and imaginary parts which are given by $\beta^\star = \omega\sqrt{\mu\varepsilon^\star} = \beta - j\alpha$. E_0^i and E_0^r are the amplitudes of the forward (incident) and backward (reflected) traveling waves, respectively. If we assume there are no reflections ($E_0^r = 0$), we have propagation in one direction only. Putting back the time dependence, the single-frequency uniform-plane wave solution to (13.24) is‡

$$\boxed{E_x(z, t) = E_0 e^{j(\omega t - \beta^\star z)} = E_0 e^{-\alpha z} e^{j(\omega t - \beta z)}} \tag{13.26}$$

where $\omega t - \beta z$ is the phase of the wave.

It is interesting to observe that by introducing the complex permittivity $\varepsilon^\star$, the wave equation (13.24) and its solution (13.26) give the correct behavior of plane waves in dielectric as well as conducting media simply by letting $\varepsilon^\star \to \varepsilon$ and $\varepsilon^\star \to -j\sigma/\omega$, respectively. Thus for a dielectric medium, in which displacement current dominates and conductive current is negligible $[J/(\partial D/\partial t) = \sigma/\omega\varepsilon \ll 1]$, we have

$$\boxed{\frac{\partial^2 E_x}{\partial z^2} + \beta^2 E_x = 0 \qquad E_x = E_0 e^{j(\omega t - \beta z)}} \tag{13.27}$$

where $\beta^2 = \omega^2 \mu\varepsilon$ and the approximation that $\sigma/\omega\varepsilon = 0$ is used. In highly conducting media ($\sigma/\omega\varepsilon \gg 1$), we let $\varepsilon^\star \to -j\sigma/\omega$ and obtain for (13.24) and (13.26)

$$\boxed{\frac{\partial^2 E_x}{\partial z^2} - j\omega\sigma\mu E_x = 0 \qquad E_x = E_0 e^{-z/\delta} e^{j(\omega t - z/\delta)}} \tag{13.28}$$

where $\beta^\star = (1 - j)/\delta$, and where $\delta = (\omega\mu\sigma/2)^{-1/2} = (\pi f \mu\sigma)^{-1/2}$ and is known as the skin depth or depth of penetration of a wave in a conducting medium. Thus if the wave has an amplitude E_0 at some point in the conducting medium, in a distance equal to $z = \delta$, the amplitude of that wave will have decreased by a factor of $1/e$. Since δ can be very small for good conductors even at low frequencies, the wave decreases exponentially very rapidly as it propagates into the medium (see values for δ on page 400 and Table 13.1). Such rapid decrease is more characteristic of diffusion than of propagation. This is as expected, for (13.28) is really a diffusion equation; it is the time-independent form of (13.20) which is a diffusion equation. What is surprising is that for harmonic time variation we can obtain the solution to a diffusion equation from a solution to a wave equation. But note that even

† Other books define a complex propagation constant $\gamma = \alpha + j\beta$ by letting $\omega^2\mu\varepsilon^\star = -\gamma^2$. The proper relationship between $\beta^\star$ and γ is $\gamma = j\beta^\star$ or $\beta^\star = \beta - j\alpha$. The term α is known as the *attenuation constant* and β as the *phase constant* or phase-propagation constant. Note that phase and phase constant have meaning only in reference to sinusoidally varying waves (single-frequency waves).

‡ Note that this is a phasor expression. To convert this to a physically meaningful expression, one takes the real part of Eq. (13.26), called the instantaneous value $E_x(z, t)_{\text{inst}} = \text{Re}\ (13.26) = E_0 e^{-\alpha z} \cos(\omega t - \beta z)$.

though the diffusion part in (13.28), which is $e^{-z/\delta}$, heavily dominates the solution, a traveling-wave part is present in the solution. We will elaborate on (13.27) and (13.28) in the following two sections.

The Transverse Nature of Plane Waves

Maxwell's equation (13.2) for free space (or any other homogeneous and isotropic medium for which $\rho = 0$) is

$$\nabla \cdot \mathbf{E} = \frac{\partial E_x}{\partial x} + \frac{\partial E_y}{\partial y} + \frac{\partial E_z}{\partial z} = 0 \tag{13.29}$$

If we apply this statement to plane waves for which there is no variation of the field with x or y, (13.29) reduces to

$$\frac{\partial E_z}{\partial z} = 0 \tag{13.30}$$

A solution to this equation is that $E_z =$ constant. Therefore, E_z cannot have any variations with x, y, or z. Such a solution cannot be a wave. Since an exactly similar argument holds for **H**, we conclude that $H_z = E_z = 0$ for a wave that travels in the z direction. An EM wave which has only components transverse to the direction of propagation is called a *TEM wave*, an abbreviation for transverse electric and magnetic.

Relation between Electric and Magnetic Fields in a Plane Wave

Starting with (13.17) and using a similar procedure that was followed for E, we can obtain H as

$$H(z, t) = H_0 e^{j(\omega t - \beta^\star z)} \tag{13.31}$$

For a wave traveling in the z direction, H can be H_x or H_y but not H_z, just as in the case of E. For a relationship between E and H, we must go back to Maxwell's equations. Thus, for sinusoidal time variation and for a plane wave which has only an E_x component, (13.1) gives

$$\nabla \times E_x \hat{\mathbf{x}} = -j\omega\mu \mathbf{H} \tag{13.32}$$

which in rectangular coordinates simplifies to

$$\hat{\mathbf{y}} \frac{\partial E_x}{\partial z} = -j\omega\mu \mathbf{H} \qquad \text{or} \qquad \frac{\partial E_x}{\partial z} = -j\omega\mu H_y \tag{13.33}$$

since $\partial/\partial y = \partial/\partial x = 0$ for a z-directed plane wave. This determines that a plane wave which has an E_x component can have only an H_y component of magnetic field. Substituting for E_x from (13.26) and differentiating, we obtain for the above equation

$$-j\beta^\star E_0 e^{j(\omega t - \beta^\star z)} = -j\omega\mu H_y \tag{13.34}$$

or

$$H_y = \frac{\beta^\star}{\omega\mu} E_x = \frac{\beta^\star}{\omega\mu} E_0 e^{j(\omega t - \beta^\star z)} \tag{13.35}$$

Using (13.31), we find that the amplitude H_0 of the magnetic field is related to that of the electric field by $H_0 = E_0(\beta^\star/\omega\mu)$. We can now make the important observation that E and H are at right angles to each other in a plane wave, and furthermore, that the direction of propagation, the direction of the H field, and the direction of the E field are mutually orthogonal to each other. It is common to write the above result in a form often called Ohm's law for a plane wave:

$$\boxed{E_x = \eta^\star H_y} \tag{13.36}$$

where
$$\eta^\star = \frac{\omega\mu}{\beta^\star} \tag{13.36a}$$

and is called the complex characteristic, intrinsic, or wave impedance of the medium. The units of η are volt per ampere or ohm. For vacuum $\eta^\star$ is real and is $\eta_0 = \omega\mu_0/\omega\sqrt{\mu_0\varepsilon_0} = \sqrt{\mu_0/\varepsilon_0} = 377\ \Omega$.

If the electric field in the plane wave had only an E_y component, the analogous relationship to (13.36) would be $E_y = -\eta^\star H_x$. We can now generalize as follows: If the direction of propagation is given by the $\hat{\mathbf{z}}$ vector, Ohm's law for plane waves is given by

$$\boxed{\hat{\mathbf{z}} \times \mathbf{E} = \eta^\star \mathbf{H} \qquad \text{or} \qquad \hat{\mathbf{z}} \times \mathbf{H} = -\frac{\mathbf{E}}{\eta^\star}} \tag{13.37}$$

13.5 PLANE WAVES IN INSULATING OR DIELECTRIC MEDIA

This is the case of propagation of plane waves in vacuum, air, or any other dielectric medium which has practically no loss. The displacement current dominates, and the plane wave solution that applies is (13.27), with the other constants being

$$\varepsilon^\star = \varepsilon = \varepsilon_r\varepsilon_0$$

$$\beta^\star = \beta = \omega\sqrt{\mu\varepsilon} = \omega\sqrt{\mu_0\varepsilon_0}\sqrt{\varepsilon_r} = \beta_0\sqrt{\varepsilon_r}$$

$$\eta^\star = \eta = \frac{E_x}{H_y} = \sqrt{\frac{\mu}{\varepsilon}} = \frac{1}{\sqrt{\varepsilon_r}}\sqrt{\frac{\mu_0}{\varepsilon_0}} = \frac{\eta_0}{\sqrt{\varepsilon_r}} = \frac{120\pi}{\sqrt{\varepsilon_r}} \tag{13.38}$$

where it was assumed that the approximation $\sigma = 0$ is valid and the permeability μ of the medium (except for ferromagnetic medium) is that of vacuum. For vacuum $\varepsilon_0 = 8.85 \times 10^{-12}\ \text{Fm}^{-1}$, $\mu_0 = 4\pi \times 10^{-7}\ \text{Hm}^{-1}$, $\eta_0 = 377\ \Omega \cong 120\pi\ \Omega$. The electric and magnetic fields are then given by

$$\boxed{\begin{aligned} E_x &= E_0\, e^{j(\omega t - \beta z)} \\ H_y &= \sqrt{\frac{\varepsilon}{\mu}}\, E_0\, e^{j(\omega t - \beta z)} \end{aligned}} \tag{13.39}$$

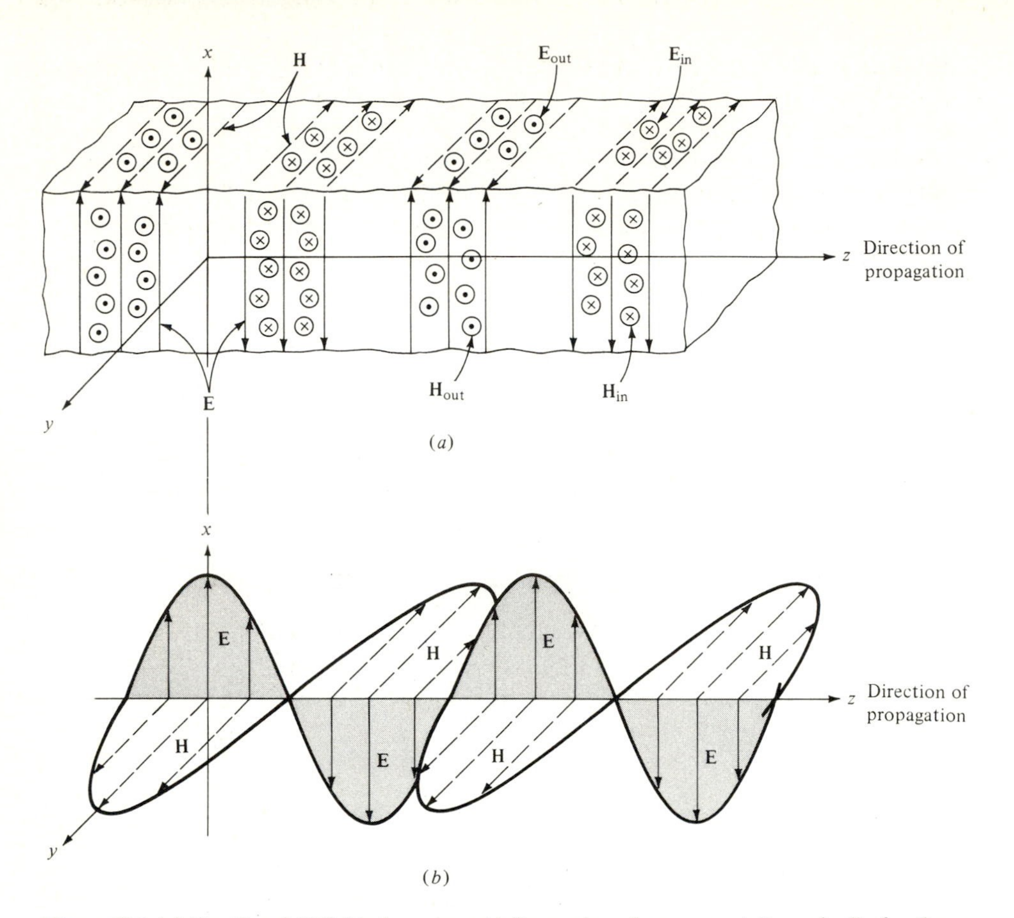

Figure 13.4 (*a*) The **E** and **H** fields in a sinusoidally varying plane wave. A "snapshot" of a three-dimensional section of a plane wave showing the relationship between the **E** and **H** fields. (*b*) An alternative representation of a plane wave showing the sinusoidal nature of the fields and the orthogonality between **E, H,** and the direction of propagation.

and their variation along the direction of propagation is shown in Fig. 13.4. The sinusoidal variation which is shown in the figure is obtained by taking the real part of (13.39) in the usual manner when phasor notation is employed; i.e., the instantaneous values are given by Re $E_0 e^{j(\omega t - \beta z)} = E_0 \cos(\omega t - \beta z)$. This figure suggests that once the wave is set in motion, it continues in space unattenuated. The E and H fields are interdependent and should not be thought of as independent sets of waves, but as different aspects of the same phenomenon.

A single-frequency plane wave is thus characterized by its polarization (direction in which the **E** vector points), its amplitude E_0, and its phase $\psi = \omega t - \beta z$, as, for example,

$$\mathbf{E} = \hat{\mathbf{x}} E_0 e^{j\psi} \tag{13.40}$$

All three can be measured, and all three can be used to impose information on the plane wave by modulating polarization, amplitude, or phase. But the wave nature resides strictly in the phase term. Thus at a fixed point along the z axis, an observer could measure a phase change that increases linearly with time, $\psi(t) \propto \omega t$, as the wave moves past. Similarly, if we could freeze time, we would see a phase change $\psi(z) \propto \beta z$ along the axis of propagation. Hence ω is a temporal phase-shift constant (phase shift in radians per unit time), and β is a spatial phase-shift constant (phase shift in radians per unit distance). A *period* is defined as the time T during which a wave undergoes a phase shift of 2π:

$$\omega T = 2\pi \qquad \text{or} \qquad T = \frac{2\pi}{\omega} \tag{13.41}$$

A *wavelength* is defined as the distance λ during which a wave undergoes a phase shift of 2π:

$$\beta\lambda = 2\pi \qquad \text{or} \qquad \lambda = \frac{2\pi}{\beta} \tag{13.42}$$

The wavelength λ plays the same role in the space domain as period T plays in the time domain. The relationship between the phase-propagation constant β and velocity v of the wave is given by the solution to the wave equation (13.25) as

$$\beta = \omega\sqrt{\mu\varepsilon} = \frac{\omega}{v} \tag{13.43}$$

For sinusoidal waves, the velocity v is called the phase velocity. It is the velocity with which a given value of E or H advances along the z axis. Since in a sinusoidal wave a given value of E or H is specified by the value of the phase angle ψ, the velocity of the wave is appropriately referred to as the phase velocity. In other words, an observer moving with velocity v alongside the wave observes a constant phase ψ.

A medium in which the phase velocity remains constant as the frequency of the wave is varied is referred to as a *nondispersive medium*. As $v = 1/\sqrt{\mu\varepsilon}$, a nondispersive medium must have μ and ε which are not functions of frequency; vacuum is an example.

Wave Propagation in a Dielectric with Small Losses

Wave propagation when displacement current dominates but a small amount of energy is extracted from the wave because the medium is absorbent represents a practical situation. Then

$$\varepsilon^{\star} = \varepsilon\left(1 - j\frac{\sigma}{\omega\varepsilon}\right) \tag{13.44}$$

where $\sigma/\omega\varepsilon \ll 1$, that is, small but not zero. The complex phase-propagation constant is

$$\beta^\star = \omega\sqrt{\mu\varepsilon^\star} = \omega\sqrt{\mu\varepsilon}\sqrt{1 - j\frac{\sigma}{\omega\varepsilon}}$$

$$\approx \beta\left(1 - j\frac{\sigma}{2\omega\varepsilon}\right)$$

$$= \beta - j\frac{\sigma}{2}\sqrt{\frac{\mu}{\varepsilon}}$$

$$\boxed{= \beta - j\alpha} \tag{13.45}$$

where the binomial approximation $(1 \pm \Delta)^{1/2} \approx 1 \pm \dfrac{\Delta}{2}$ for $\Delta \ll 1$ was used. Hence the electric field is

$$E_x = E_0 e^{j(\omega t - \beta^\star z)} = E_0 e^{-\alpha z} e^{j(\omega t - \beta z)} \tag{13.46}$$

where α is the attenuation coefficient $\alpha = (\sigma/2)\sqrt{\mu/\varepsilon}$ measured in nepers per meter (Np/m). The exponent of e is then in the dimensionless units of neper. The electric field (as well as the magnetic field) now experiences a small exponential attenuation. Small because the decrease in a distance of one wavelength is small; that is,

$$\alpha z\bigg|_{z=\lambda} = \left(\frac{\sigma}{2}\sqrt{\frac{\mu}{\varepsilon}}\right)\left(\frac{2\pi}{\omega\sqrt{\mu\varepsilon}}\right) = \frac{\pi\sigma}{\omega\varepsilon} \ll 1 \tag{13.47}$$

Figure 13.5 shows the instantaneous values of the E field with a small attenuation modulating an otherwise sinusoidal spatial variation which has a wavelength $\lambda = 2\pi/\beta$.

The intrinsic or characteristic impedance of the medium which has a finite conductivity σ is

$$\eta^\star = \sqrt{\frac{\mu}{\varepsilon^\star}} = \sqrt{\frac{\mu}{\varepsilon}}\,\frac{1}{\sqrt{1 - j\dfrac{\sigma}{\omega\varepsilon}}} = \eta\left(1 + j\frac{\sigma}{2\omega\varepsilon}\right) \tag{13.48}$$

Hence, the loss adds a small reactive component to the intrinsic impedance, which for most practical purposes can be ignored; that is, $\eta^\star \cong \eta = \sqrt{\mu/\varepsilon}$.

Nomenclature Used in Reference Books

There are two types of loss mechanisms which attenuate a wave. The first (already considered) arises when the dielectric is slightly conducting. The second arises when energy is dissipated in the course of the polarization process even though the conductivity of the dielectric is zero (dipoles experience friction as they

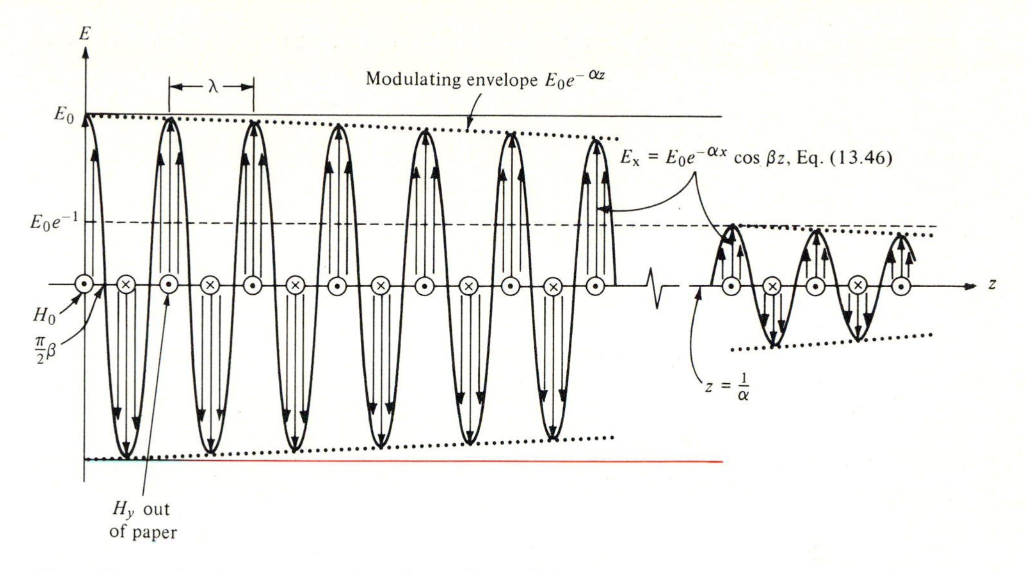

Figure 13.5 A slightly absorbing dielectric medium will impose a small exponential attenuation on the propagating fields. In practical situations the distance $z = 1/\alpha$ at which the field has decayed to E_0e^{-1} is usually very large.

flip back and forth in a sinusoidal field thus extracting energy from the field.† As both loss mechanisms generate heat, each can be represented by a conductivity σ. The complex permittivity Eq. (13.23) can now be generalized to reflect conduction and polarization losses as

$$\boxed{\varepsilon^\star = \varepsilon' - j\varepsilon'' - j\frac{\sigma}{\omega}} \tag{13.49}$$

where $\varepsilon'/\varepsilon_0$ is the dielectric constant of the material and the total effective conductivity is

$$\sigma_{\text{eff}} = \sigma + \omega\varepsilon'' \tag{13.50}$$

The ratio of conduction current to displacement current in the lossy dielectric is called the *loss tangent* or *dissipation factor*:

$$\boxed{\tan\phi = \frac{\sigma_{\text{eff}}}{\omega\varepsilon'} = \frac{\sigma + \omega\varepsilon''}{\omega\varepsilon'}} \tag{13.51}$$

Values of dielectric losses are tabulated in reference books under a variety of names, such as loss tangent, dissipation factor, power factor. The loss tangent

† Because of such friction (polarization damping forces), the polarization vector **P** will lag behind the applied **E** field. The difference in time phase between **P** and **E** is accounted for by a permittivity with an imaginary part; that is $\varepsilon = \varepsilon' - j\varepsilon''$.

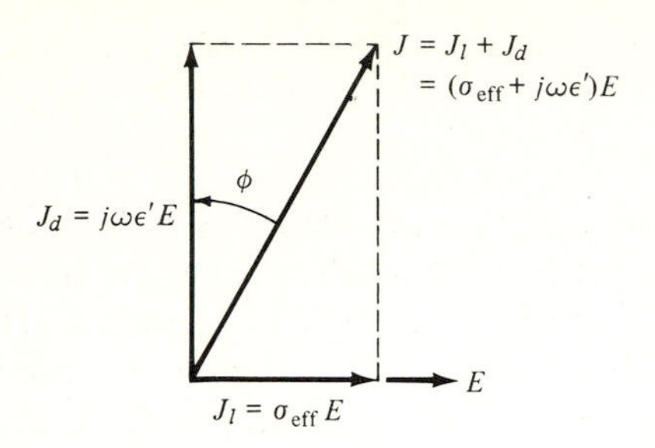

Figure 13.6 The loss angle ϕ. Power factor is $\sin\phi = \cos(\pi/2 - \phi)$, where $\pi/2 - \phi$ is the angle by which **J** leads **E**. The loss tangent is $\tan\phi = J_{\text{loss}}/J_{\text{displ.}} = \sigma_{\text{eff}}/\omega\varepsilon'$.

relates to the power factor which is defined as $\sin\phi$. These relationships are illustrated in Fig. 13.6. Since the losses in most dielectrics are small, we see that loss tangent = dissipation factor $\approx$ power factor $\approx \phi$. The loss tangent (13.51) includes conduction and polarization losses. At microwave frequencies, because of large values of ω, losses due to polarization damping forces dominate ($\omega\varepsilon'' \gg \sigma$) and $\tan\phi \approx \varepsilon''/\varepsilon'$.

Example What is the loss per kilometer for a plane wave propagating in dry earth? The frequency is 1 MHz.

At this frequency, dry soil has a conductivity of $\sigma = 10^{-5}$ S/m and a relative permittivity of $\varepsilon_r = 3$. Hence, $\sigma/\omega\varepsilon \cong 0.06 \ll 1$, which means that displacement current dominates and the effect of the conductivity is to attenuate the propagating wave. The value of the attenuation coefficient, using (13.45), is given as

$$\alpha = \frac{\sigma}{2}\sqrt{\frac{\mu}{\varepsilon}} = \beta\left(\frac{\sigma}{2\omega\varepsilon}\right) = 3.6 \times 10^{-2}(0.03) = 1.1 \times 10^{-3}\ \text{Np/m}$$

where $\beta = \omega/v = 2\pi f/(v_0/\sqrt{\varepsilon_r}) = 2\pi \times 10^6/(3 \times 10^8/\sqrt{3}) = 3.6 \times 10^{-2}$ rad/m. In 1 km of propagation the amplitude will have decreased from one to

$$e^{-(1.1 \times 10^{-3})(10^3)} = e^{-1.1} = 0.33$$

or by 20 log (0.33) = 9.5 dB, which for many applications is a tolerable loss.

Example Calculate the loss per kilometer for a plane wave propagating in distilled water at a frequency of 25 GHz.

The dissipation factor and dielectric constant ε_r at this frequency are given as 0.3 and 34, respectively. Since the dissipation factor is equal to $\sigma_{\text{eff}}/\omega\varepsilon = \varepsilon''/\varepsilon'$, we have for the attenuation coefficient, using (13.45)

$$\alpha = \beta\frac{\text{dissipation factor}}{2} = \sqrt{\varepsilon_r}\beta_0\frac{0.3}{2} = (\sqrt{34})(524)(0.15) = 460\ \text{Np/m}$$

where $\beta_0 = \omega/v_0 = 2\pi f/v_0 = 2\pi(2.5 \times 10^{10})/3 \times 10^8 = 524$ rad/m. In 1 km of propagation the amplitude will have decreased from one to

$$e^{-(460)(10^3)} = e^{-4.6 \times 10^5} \cong 0$$

or by 4×10^6 dB. Clearly, communication is not possible. Even for a distance of 1 cm, the loss is $20 \log e^{-4.6} = 40$ dB, a very large value. Hence, communication (or radar) which uses such high-frequency microwaves is not possible. Other means of communication, which employ acoustic waves (sonar) or very low-frequency radio waves (see example in next section) must be used. The case for seawater which has higher conductivities than distilled water is even worse. The extreme rapid attenuation of high-frequency waves in water explains why the presence of water in the atmosphere (rain, fog) causes severe attenuation of such waves.

We might point out, that for dissipation factors ($\sigma/\omega\varepsilon$) larger than 0.1, the two-term binomial approximation for the attenuation coefficient α given by (13.45) is not sufficiently accurate. Additional terms in the binomial approximation must be carried, that is, $(1 \pm \Delta)^n = 1 \pm n\Delta + [n(n-1)/2]\,\Delta^2 \pm \cdots$. However, even when $\sigma/\omega\varepsilon = 0.3$ was used in (13.45), as was the case in this example, the error is small.

13.6 PLANE WAVES IN CONDUCTING MEDIA

In conducting media the conduction current dominates the displacement current

$$\frac{J}{\partial D/\partial t} = \frac{\sigma}{\omega\varepsilon} \gg 1 \tag{13.52}$$

to such an extent that we ignore the displacement current completely and substitute for $\varepsilon^\star = \varepsilon(1 - j\sigma/\omega\varepsilon)$ simply $\varepsilon^\star = -j\sigma/\omega$. The wave equation and its solution† for this case is (13.28), with the other constants being

$$\beta^\star = \omega\sqrt{\mu\varepsilon^\star} \cong \omega\sqrt{\mu\left(-\frac{j\sigma}{\omega}\right)} = (1-j)\sqrt{\frac{\omega\mu\sigma}{2}} = \frac{1-j}{\delta} = \beta - j\alpha \tag{13.53}$$

where the phase-propagation constant is $\beta = 1/\delta$, the attenuation constant is $\alpha = 1/\delta$,

$$\delta = \sqrt{\frac{2}{\omega\mu\sigma}} = \sqrt{\frac{1}{\pi f\mu\sigma}} \tag{13.54}$$

$\sqrt{-j} = e^{-j\pi/4} = (1-j)/\sqrt{2}$, and the intrinsic impedance of the conducting medium is

$$\eta^\star = \frac{E_x}{H_y} = \sqrt{\frac{\mu}{\varepsilon^\star}} \cong \sqrt{\frac{j\omega\mu}{\sigma}} = (1+j)\sqrt{\frac{\omega\mu}{2\sigma}} \tag{13.55}$$

We still have wave propagation in the conducting medium, since solution (13.28) contains the term $e^{j(\omega t - z/\delta)}$, which is a traveling wave whose phase constant is

† The instantaneous values of the fields are obtained by taking the real part of the phasor expression (13.28):

$$E_x = E_0 e^{-z/\delta} \cos(\omega t - z/\delta)$$

$$H_y = E_0(\sigma/\omega\mu)^{1/2} e^{-z/\delta} \cos(\omega t - z/\delta - \pi/4)$$

and are plotted in Fig. 13.7.

given by $\beta = 1/\delta$; that is,

$$\beta = \frac{\omega}{v} = \frac{2\pi}{\lambda} = \frac{1}{\delta} \tag{13.56}$$

In good conductors σ is very large, which implies that both the attenuation constant α and phase-propagation constant β are also large and the velocity v, wavelength λ, and skin depth δ are very small. This means that an existing wave in a conducting medium is rapidly attenuated and has a large phase shift per unit length. Phase velocity v is given by

$$v = \frac{\omega}{\beta} = \omega\delta = \sqrt{\frac{2\omega}{\mu\sigma}} \tag{13.57}$$

and is small when σ is large. The wavelength undergoes a great contraction as a wave goes from free space into a conducting medium. The effective wavelength in the conducting medium is given by

$$\lambda = 2\pi\delta \tag{13.58}$$

For example, in seawater $\varepsilon_r = \varepsilon/\varepsilon_0 = 81$, and $\sigma = 4$ S/m. A 10^4-Hz wave, which in free space has a wavelength $\lambda = v/f = (3 \times 10^8 \text{ m/s})/10^4 \text{ Hz} = 30$ km, upon penetrating into seawater remains a 10^4-Hz wave but with a wavelength of $\lambda = 2\pi\delta = 2\pi/\sqrt{\pi f \mu\sigma} = 2\pi/\sqrt{\pi \times 10^4 \times 4\pi \times 10^{-7} \times 4} =$ 16 m. The velocity of this wave in free space is the velocity of light, but in the conducting medium it slows to 1.6×10^5 m/s.

For copper, even at the low frequency of 60 Hz, the changes are even more dramatic. A 60-Hz wave in free space has a wavelength of $\lambda = 5 \times 10^3$ km (3110 mi) and a velocity of $v = 3 \times 10^8$ m/s. Upon penetrating into copper ($\sigma = 5.8 \times 10^7$ S/m), it becomes a wave of $\lambda = 5.36$ cm and $v = 3.2$ m/s, or about 7.2 mi/h.

It should again be pointed out that we are concerned with the characteristics of the wave in the medium and not with how the wave was established in the medium. We imply from the above discussion that when a wave impinges from free space (or some other dielectric which readily supports a wave for large distances) onto a metallic block, a small part of the wave will penetrate into the metal. Most of the wave will bounce back, i.e., will be reflected from the metal. The reason why a small portion of the wave does penetrate into the metal is that the metal in question is an imperfect conductor ($\sigma \neq \infty$), which allows the existence of a rapidly attenuated wave (13.28). If it were a perfect conductor ($\sigma = \infty$), no field, and therefore no waves, could exist inside as $\delta = 0$. Note also that $v = \lambda = 0$ for $\sigma = \infty$, which is another way of stating that a wave cannot exist.

Depth of Penetration or Skin Depth

We have seen from the discussion following (13.28) that the skin depth δ

$$\boxed{\delta = \frac{1}{\sqrt{\pi f \mu\sigma}}} \tag{13.59}$$

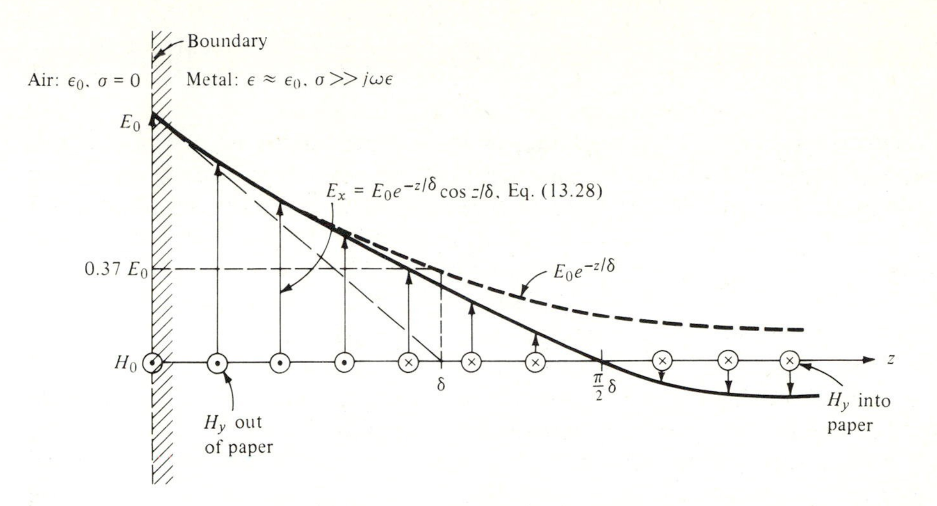

Figure 13.7 A wave on the surface of a metal has field components E_0 and H_0. As the wave penetrates into the conducting material, the fields decrease exponentially. The conduction current variation is that of the electric field because $J = \sigma E$.

denotes the distance of propagation in which E, H, and J have decreased by the factor $1/e$ or by approximately 63 percent. Since the distance δ for conductors is small (for copper see page 400), a field incident from free space onto a block of metal will exist in the metal for only a short distance measured from the surface, as shown in Fig. 13.7. The origin of the term skin depth lies in the fact that time-varying fields and current densities exist in only a thin layer on the surface of good conductors. For example, a signal which is propagating on a coaxial transmission line travels with the velocity of light in the region surrounding the inner conductor. The current in the wire may be found from the line integral of the magnetic field which exists outside the inner wire, using Ampère's law $\oint \mathbf{H} \cdot \mathbf{dl} = I$. If we have a microwave signal of 10 GHz and the wire is copper, δ is 6.6×10^{-4} mm. Hence, the current flows in the outermost layer of the wire. Most of the interior of the inner wire does not contribute to current flow. Since only the surface coating is important at microwave frequencies, guiding structures can be made of poorly conducting material, provided the surface is plated with silver, copper, or any other good conducting material which will hold the I^2R losses to tolerable limits (see Prob. 13.24).

Skin depth was also discussed in the section on boundary conditions (Sec. 8.8) and in connection with magnetic shielding (Sec. 10.2), where it was shown that the induced eddy currents and the accompanying magnetic field penetrate a conducting material for a distance of only a few skin-depth layers δ. Hence, an effective magnetic shield for time-varying fields can be made from nonmagnetic but conducting material such as aluminum, provided the thickness of the shield exceeds several skin depths. In this chapter we have shown rigorously that the penetration of an electromagnetic field into a conducting material is attenuated by the exponential factor $e^{-z/\delta}$, where z is the perpendicular distance into the material.

Skin depth versus frequency is given in Table 13.1. The third column gives the frequency at which the conduction current is equal to the displacement current. Hence below that frequency the material in question acts as a conductor, and above that frequency it acts as a dielectric. The crossed-out entries for δ denote that at these frequencies the displacement current dominates, and therefore skin depth δ as given by $1/\sqrt{\pi f \mu \sigma}$ loses its meaning. At these frequencies the material acts as a dielectric and (13.46) gives the behavior of waves in it. The finite conductivity in this case imposes an attenuation on the waves which is given by $\alpha = (\sigma/2)(\mu/\varepsilon)^{1/2}$, and an equivalent skin depth which can be obtained from it ($\delta = 1/\alpha$). The difference between the equivalent δ and δ is as follows: In the case of a dielectric the wave undergoes many cycles of oscillation in a distance of $\delta = 1/\alpha = 2/(\eta\sigma)$ (see Fig. 13.5), whereas in the case of a conductor the wave has completed only one-sixth of a cycle in a distance of $\delta = 1/(\pi f \mu \sigma)^{1/2}$ (see Fig. 13.7).

Note that rearrangement time $T = \varepsilon/\sigma$, developed in statics in (2.36) and (2.37), served to distinguish between conductors and dielectrics in the same manner as the factor $\omega\varepsilon/\sigma$ does. A brief discussion of this is found following Eq. (11.23).

The skin depth δ is dependent on permeability μ but not on permittivity ε. The reason for this is that the induced eddy currents are proportional to dB/dt and not to dH/dt.

Example: Submarine communication† Seawater is a poor medium for communications because it is a good conductor. If a sufficiently high frequency is used so that displacement current greatly exceeds conduction current, the attenuation constant α, from (13.45), is

$$\alpha = \frac{1}{\delta} = \frac{\sigma}{2}\sqrt{\frac{\mu}{\varepsilon}} \tag{13.60}$$

For seawater ($\sigma = 4$ S/m, $\varepsilon = 81\varepsilon_0$), this gives an α of 84 Np/m. The equivalent skin depth δ is, therefore, only 1.19 cm, and the penetration at such frequencies is, for all practical purposes, nonexistent. For this reason, frequencies which are low enough so that seawater acts as a conducting medium must be used in any attempt at communication. At low frequencies, where displacement current can be neglected, the frequency-dependent attenuation constant is given by

$$\alpha = \frac{1}{\delta} = \sqrt{\pi f \mu \sigma} \tag{13.61}$$

which at 10 kHz gives an α of 0.4 Np/m and a skin depth of $\delta = 2.5$ m. Because attenuation for one skin depth is 8.686 dB a communication path cannot be many skin depths long; at 10 kHz, the attenuation is 87 dB for a distance of only 25 m. The size of underwater antennas is small when compared to the size of free-space antennas. For example, a 10-kHz wave which has a wavelength λ of 30 km in air, has a wavelength of only $\lambda = 2\pi\delta = 15.7$ m in seawater. Hence, in seawater a $\lambda/2$ dipole has dimensions of 7.8 m and can be easily mounted on a submarine.

13.7 FLOW OF ENERGY AND THE POYNTING VECTOR

We have shown that the energy density of a static electric field is $\frac{1}{2}\varepsilon E^2$, and that of a static magnetic field is $\frac{1}{2}\mu H^2$. When the fields vary with time, these energy

† R. K. Moore, Radio Communication in the Sea, *IEEE Spectrum*, November 1967.

Table 13.1 Skin depth of conducting materials

Material†	Conductivity σ, S/m	Relative permittivity $\varepsilon_r = \varepsilon/\varepsilon_0$, $\varepsilon_0 = 8.854 \times 10^{-12}$	Frequency at which $\sigma = \omega\varepsilon$	Skin depth $\delta = 1/\sqrt{\pi f \mu \sigma}$, m	Skin depth δ at frequency f			
					60 Hz	10 kHz	1 MHz	10 GHz
Silver	6.2×10^7	1	1.1×10^{18} Hz	$0.064/\sqrt{f}$	8.2 mm	0.64 mm	0.064 mm	0.64 μm
Aluminum	3.7×10^7	1	6.7×10^{17} Hz	$0.083/\sqrt{f}$	10.7 mm	0.83 mm	0.083 mm	0.83 μm
Solder	0.7×10^7	1	1.2×10^{17} Hz	$0.185/\sqrt{f}$	24 mm	1.85 mm	0.185 mm	1.85 μm
Seawater	4	81	0.9 GHz	$250/\sqrt{f}$	32 m	2.5 m	0.25 m	2.5 mm
Fresh water	10^{-3}	81	0.2 MHz	$\frac{1.6 \times 10^4}{\sqrt{f}}$	2.1 km	160 m	16 m	0.16 m
Wet earth	10^{-3}	10	1.8 MHz	1.6×10^4	2.1 km	160 m	16 m	0.16 m
Dry earth	10^{-5}	3	60 kHz	$\frac{1.6 \times 10^5}{\sqrt{f}}$	21 km	1.6 km	160 m	1.6 m

† Since these are nonmagnetic materials, $\mu \approx \mu_0 = 4\pi \times 10^{-7}$ H/m.

densities will also vary with time. But more important is the fact that the time-varying electromagnetic field can carry energy. Thus, as a wave propagates, it transfers energy. If we consider a volume, a wave incident on it can transfer energy to the volume (receiver case). If, on the other hand, a radiating source is inside the volume, the electromagnetic field produced by the source can transport energy out of the volume (transmitter case).

To put energy transport on a formal basis, consider the identity

$$\nabla \cdot (\mathbf{E} \times \mathbf{H}) = \mathbf{H} \cdot \nabla \times \mathbf{E} - \mathbf{E} \cdot \nabla \times \mathbf{H} \tag{13.62}$$

Using Maxwell's equations $\nabla \times \mathbf{E} = -\partial \mathbf{B}/\partial t$ and $\nabla \times \mathbf{H} = \mathbf{J} + \partial \mathbf{D}/\partial t$, the above identity becomes

$$\nabla \cdot (\mathbf{E} \times \mathbf{H}) = -\mathbf{H} \cdot \frac{\partial \mathbf{B}}{\partial t} - \mathbf{E} \cdot \frac{\partial \mathbf{D}}{\partial t} - \mathbf{E} \cdot \mathbf{J} \tag{13.63}$$

If we integrate this expression over the volume $v(A)$ which is bounded by surface A, we obtain

$$\boxed{\oint\!\!\oint_{A(v)} (\mathbf{E} \times \mathbf{H}) \cdot \mathbf{dA} = -\frac{\partial}{\partial t} \iiint_{v(A)} \left(\frac{\mathbf{H} \cdot \mathbf{B}}{2} + \frac{\mathbf{E} \cdot \mathbf{D}}{2} \right) dv - \iiint_{v(A)} \mathbf{E} \cdot \mathbf{J}\, dv} \tag{13.64}$$

where the left side was converted to a closed-surface integration by Gauss' divergence theorem (1.94). For constant ε and μ we also have that $\partial(\mathbf{D} \cdot \mathbf{E})/\partial t = 2\mathbf{E} \cdot \partial \mathbf{D}/\partial t$ and $\partial(\mathbf{H} \cdot \mathbf{B})/\partial t = 2\mathbf{H} \cdot \partial \mathbf{B}/\partial t$.

The first term of (13.64) represents a flow of power through the surface A which bounds volume v. Recall that E is in volts per meter and H in amperes per meter, which makes EH a quantity in watts per square meter. The vector $\mathbf{E} \times \mathbf{H}$ represents the amount of energy per unit time crossing a unit area on the surface A. This power density is called the *Poynting vector* $\mathscr{P}$

$$\boxed{\mathscr{P} = \mathbf{E} \times \mathbf{H} \qquad \text{W/m}^2} \tag{13.65}$$

Such a vector is perpendicular to the plane determined by **E** and **H** and is in the direction of energy flow.† The vector $\mathscr{P}$ forms a triad with the vectors **E** and **H**, as shown in Fig. 13.8. Thus, a negative value for the Poynting vector integral ($\oint\!\!\oint \mathscr{P} \cdot \hat{\mathbf{n}}\, dA = -|\oint\!\!\oint \mathscr{P} \cdot \mathbf{dA}|$) represents inward energy flow through the surface A (note that the direction of the surface element **dA** is represented by the outward normal $\hat{\mathbf{n}}$; that is, $\mathbf{dA} = \hat{\mathbf{n}}\, dA$). We denote this case as the receiver case. Similarly, a positive value for the integral ($\oint\!\!\oint \mathscr{P} \cdot \mathbf{dA} = +|\oint\!\!\oint \mathscr{P} \cdot \mathbf{dA}|$ because $\hat{\mathbf{n}}$ and $\mathscr{P}$ are mostly in same direction) represents outward flow through the surface; this is the transmitter case. Both cases are illustrated in Fig. 13.8.

The second term in (13.64) represents the time rate of increase of energy storage in the magnetic and electric fields inside the volume v.

† For a plane wave (13.36), the Poynting's vector (13.65) becomes $\mathscr{P} = \mathbf{E} \times \mathbf{H} = E_x H_y \hat{\mathbf{z}} = (E_x^2/\eta)\hat{\mathbf{z}}$. In free space the z-directed power flow is then given by $\mathscr{P}_z = E_x^2/120\pi$ W/m².

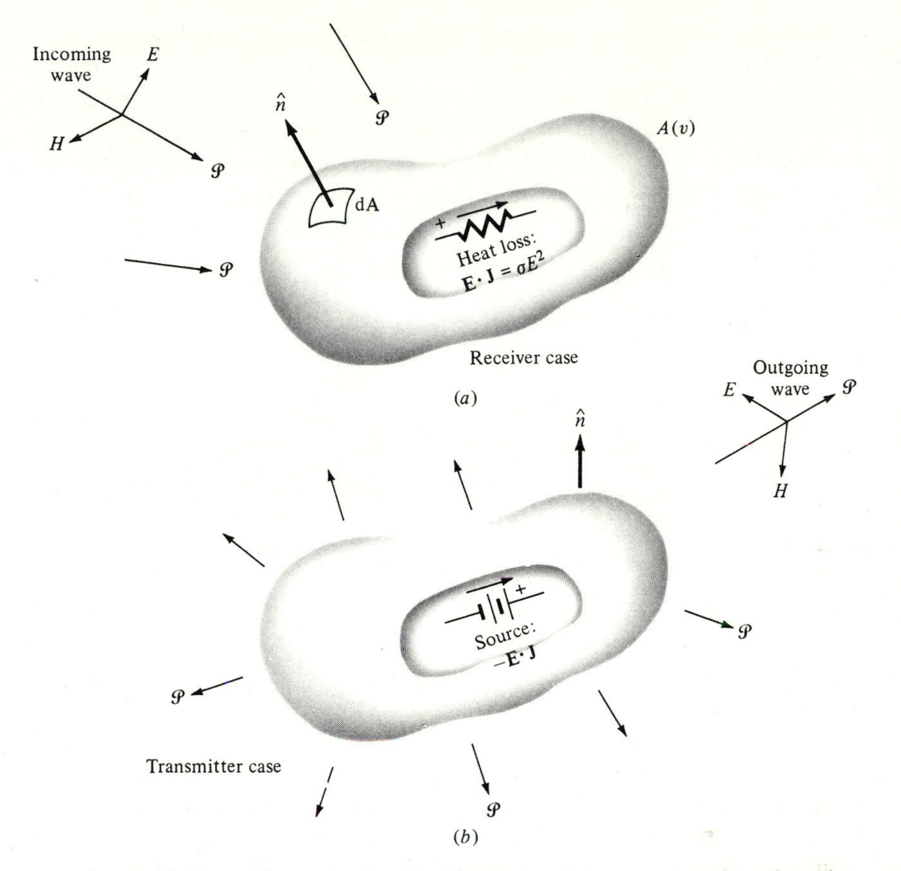

Figure 13.8 (*a*) The receiver case: An incoming wave transports energy into the volume, some of which is dissipated in the I^2R losses of the receiver. (*b*) The transmitter case: The energy of a source, represented by a battery, is transported out of volume v by the Poynting's vector term $\oiint \mathbf{E} \times \mathbf{H} \cdot \mathbf{dA}$.

For the case when all sources are outside the volume v, the third term is an ohmic loss term. The **E** and **J** are in the same direction, and it represents the rate of energy dissipated in heat (the usual I^2R loss). The source term is then the Poynting's vector term as it accounts for the energy inflow. For this situation we can state (13.64) explicitly as

Receiver case:

$$\left| \oiint \mathcal{P} \cdot \mathbf{dA} \right| = \frac{\partial}{\partial t} \iiint \left(\frac{\mathbf{H} \cdot \mathbf{B}}{2} + \frac{\mathbf{E} \cdot \mathbf{D}}{2} \right) dv + \iiint \sigma \mathbf{E}^2 \, dv \quad (13.66)$$

↑		↑		↑
Source term; inflow of energy due to sources outside v	=	rate of increase of electric and magnetic energy in v	+	energy dissipated in heat

where the substitution $\mathbf{J} = \sigma \mathbf{E}$ was made.

On the other hand when the sources are located inside the volume v, the third term in (13.64) must contain the source term. The $\mathbf{E}$ and $\mathbf{J}$ when representing sources are in opposite directions. Hence, the third term in (13.64), when due to sources, changes sign. For simplicity let us assume that ohmic losses inside volume v are zero. The equation of energy balance per unit time, for sources inside volume v is then given by

Transmitter case:

$$\left|\iiint \mathbf{E}\cdot\mathbf{J}\,dv\right| = \frac{\partial}{\partial t}\iiint\left(\frac{\mathbf{H}\cdot\mathbf{B}}{2}+\frac{\mathbf{E}\cdot\mathbf{D}}{2}\right)dv + \left|\oiint \mathscr{P}\cdot d\mathbf{A}\right| \tag{13.67}$$

Source term = rate of increase of stored energy in v + outflow of energy from volume v

Example: Consider the case of a round wire carrying current I_z, as shown in Fig. 13.9. If the resistance per unit length of wire is R', show by using Poynting's vector that the power dissipated is I_z^2R' watts per unit length.

From Ohm's law, the electric field in the wire is

$$E_z = I_zR' \qquad \text{V/m}$$

The magnetic field at the surface of the wire is given from (7.5) as

$$H_\phi = \frac{I_z}{2\pi r} \qquad \text{A/m}$$

Poynting's vector is therefore radially inward:

$$\mathscr{P} = \mathbf{E}\times\mathbf{H} = E_zH_\phi\hat{\mathbf{r}} = \frac{I_z^2R'}{2\pi r}\hat{\mathbf{r}} = \mathscr{P}_r\hat{\mathbf{r}} \qquad \text{W/m}^2$$

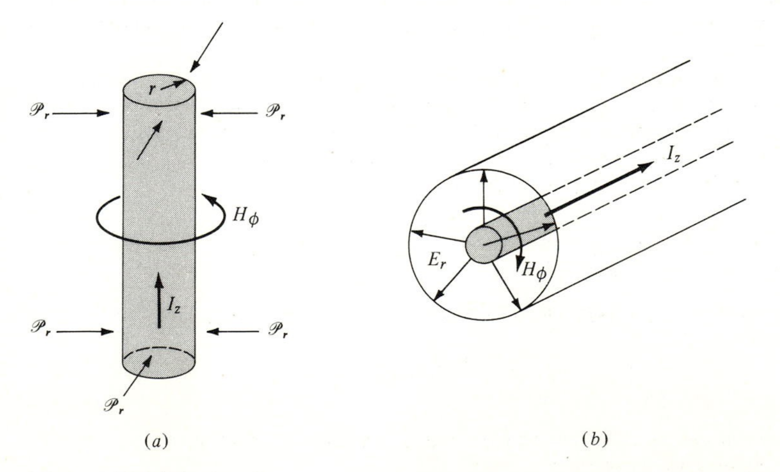

Figure 13.9 (a) A wire of radius r and length l carrying current I_z, dissipates I_z^2R' watts per unit length in heat. (b) In a coaxial transmission line there are I_z^2R losses in the center as well as the outer conductor. The energy transmitted along the line is by the transverse field, E_r and H_ϕ.

and when integrated over a cylindrical surface of the wire gives

$$\oint\!\!\oint \mathscr{P} \cdot d\mathbf{A} = \iint \mathscr{P}_r \, dA = \mathscr{P}_r(2\pi r l) = I_z^2 R' l = I_z^2 R \qquad \text{W}$$

where $2\pi rl$ is the surface area of the wire and the resistance of a wire of length l is $R = R'l$. Therefore, the energy per unit time entering the wire from the outside field is the power $I_z^2 R'$ in watts per unit length which is dissipated in heat inside the resistor. This is a remarkable result and shows that power dissipated in a wire or resistor can be obtained by computing the energy entering it from the electromagnetic field. In the Poynting vector point of view the dissipated energy does not enter through the connecting wires but from the space around the wires. In this view, a source such as a battery sets up the electric and magnetic fields. The energy flows in this field and enters the wire through the cylindrical surface.

If the wire is part of a coaxial cable, as shown in Fig. 13.9*b*, electromagnetic energy can be transported in the direction of the axis of the cable. A source sets up a field in the nonconducting space between the inner and outer conductors with E_z, E_r, and H_ϕ components. The E_r and H_ϕ components give a Poynting vector in the axial direction of the cable; that is, $\mathscr{P}_z = E_r H_\phi$. The energy that is transferred down the cable to a load is then given by

$$\iint \mathscr{P} \cdot d\mathbf{A} = \iint \mathscr{P}_z \, dA = \iint E_r H_\phi \, dA$$

and flows in the nonconducting space between the inner and outer conductors. The integration of $\mathscr{P}_z$ over the cross section gives the power flow. The integration of $\mathscr{P}_r$ (which is equal to $E_z H_\phi$) over a cylindrical surface of length l of a cable gives the power loss in length l of the cable. The radial energy that flows into the wire and is dissipated in heat is the penalty we have to pay for using an imperfect conductor in our transmission line.

Example: Use the Poynting's vector method to show that the energy stored in a capacitor is $W = \frac{1}{2}VQ$ [see Eq. (5.8)].

This is one more example illustrating that any problem involving the flow of electric or magnetic energy can be solved by applying the Poynting vector. The only restriction to this method is that the electric and magnetic fields must be related. If they arise from independent sources, meaningless results can be obtained, as, for example, a Poynting's vector obtained for the case of a static point charge Q located at the center of a small loop of wire carrying direct current I.

Figure 13.10 shows a parallel-plate capacitor of capacitance $C = \varepsilon A/d$ being charged by current I flowing in the connecting wires. If we neglect fringing, the electric field exists only inside the capacitor. The field is uniform and is given by E_z. The magnetic field created by I in the wire or, equivalently, by the displacement current $(\partial D/\partial t)A$ inside the capacitor can be found from Maxwell's equation (see example in Sec. 11.3)

$$\oint \mathbf{H} \cdot d\mathbf{l} = \iint \frac{\partial \mathbf{D}}{\partial t} \cdot d\mathbf{A} \qquad \text{inside capacitor}$$

Using the symmetry of the problem, the above expression gives for the magnetic field

$$H_\phi = \frac{\varepsilon r}{2} \frac{\partial E_z}{\partial t}$$

Poynting's vector inside the capacitor therefore has an r component only: $\mathbf{E} \times \mathbf{H} = E_z H_\phi \hat{\mathbf{r}} = \mathscr{P}_r \hat{\mathbf{r}}$. The total power flow across the gap surface is

$$\oint\!\!\oint \mathscr{P}_r \, dA = E_z \frac{\varepsilon r}{2} \frac{\partial E_z}{\partial t} 2\pi r d = Ad \frac{\partial}{\partial t}\left(\frac{1}{2} \varepsilon E_z^2\right)$$

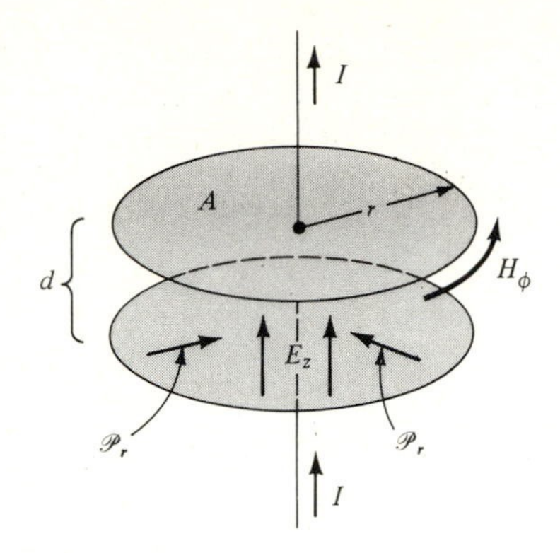

Figure 13.10 A parallel-plate capacitor being charged.

This is the rate of energy flow into the volume of the capacitor due to the charging current, recalling that $\frac{1}{2}\varepsilon E^2$ is the energy density of an electric E field and Ad is the volume of the capacitor. Comparing this to (13.66), we see that $\oiint \mathscr{P} \cdot \mathbf{dA}$ equals the rate at which the stored electrostatic energy increases.†

Since energy is equal to the time integral of power, we have for energy W stored in the capacitor,

$$W = \int \left(\oiint \mathscr{P} \cdot \mathbf{dA} \right) dt = Ad\tfrac{1}{2}\varepsilon E_z^2 = \tfrac{1}{2}(E_z d)(D_z A) = \tfrac{1}{2}VQ$$

where the voltage V across the capacitor is $V = E_z d$ and the charge Q on the plates is $Q = \rho_s A = D_z A$. Note that the boundary condition for normal components of E field on a conducting surface is $D = \rho_s$, where ρ_s is the surface charge density.

13.8 POYNTING'S VECTOR FOR SINUSOIDAL TIME VARIATION

Sinusoidal time variation at a fixed frequency ω is of practical interest, as many of our sources generate sinusoidal outputs. Even if the output is not sinusoidal, it can be represented as a summation of sinusoidal components of different amplitudes, phases, and frequencies. In (13.65) we expressed the instantaneous Poynting vector $\mathscr{P}(t)$. Usually, the average energy flow per unit time is of more practical importance. For periodic time variation average energy flow can be expressed as instantaneous energy flow averaged over a cycle. The time-averaged Poynting's vector then becomes

$$\mathscr{P}_{\text{ave}} = \frac{1}{T}\int_0^T \mathscr{P}(t)\, dt \tag{13.68}$$

† In Chap. 5, Eq. (5.8), an alternative view was used. It was shown that energy storage in a capacitor increases because charge on the capacitor plates increases as a result of current flow.

where $T = 1/f = 2\pi/\omega$ is the period of the signal. For sinusoidal time variation it is convenient to use the complex notation (phasor method) and represent the instantaneous values of **E** and **H** as the real parts of complex exponentials $\mathbf{E}'e^{j\omega t}$ and $\mathbf{H}'e^{j\omega t}$, respectively, where $\mathbf{E}'$ and $\mathbf{H}'$ are complex vector functions of position. We can write then

$$\mathbf{E}_{\text{inst}} = \text{Re}\,(\mathbf{E}'e^{j\omega t}) = \text{Re}\,[(\mathbf{E}_r + j\mathbf{E}_i)(\cos\omega t + j\sin\omega t)]$$
$$= \mathbf{E}_r\cos\omega t - \mathbf{E}_i\sin\omega t \tag{13.69}$$

where the real part of $\mathbf{E}'$ is $\mathbf{E}_r$ and the imaginary part is $\mathbf{E}_i$, with similar notation for $\mathbf{H}'$. Poynting's vector representing the power flow at instant t can now be expressed as†

$$\boxed{\begin{aligned}\mathscr{P}(t) &= \mathbf{E}_{\text{inst}} \times \mathbf{H}_{\text{inst}} \\ &= (\mathbf{E}_r \times \mathbf{H}_r)\cos^2\omega t + (\mathbf{E}_i \times \mathbf{H}_i)\sin^2\omega t \\ &\quad - [(\mathbf{E}_r \times \mathbf{H}_i) + (\mathbf{E}_i \times \mathbf{H}_r)]\sin\omega t\cos\omega t\end{aligned}} \tag{13.70}$$

If we time average the instantaneous Poynting's vector according to (13.68), we obtain

$$\boxed{\mathscr{P}_{\text{ave}} = (\mathbf{E} \times \mathbf{H})_{\text{ave}} = \tfrac{1}{2}(\mathbf{E}_r \times \mathbf{H}_r + \mathbf{E}_i \times \mathbf{H}_i)} \tag{13.71}$$

because the time average of $\cos^2\omega t$ or $\sin^2\omega t$ is one-half, and the $\sin\omega t\cos\omega t$ term averages to zero. This is the desired result and gives the average power flow in steady-state ac problems. It can be written in a more convenient form using the complex notation as

$$\boxed{\mathscr{P}_{\text{ave}} = (\mathbf{E} \times \mathbf{H})_{\text{ave}} = \tfrac{1}{2}\,\text{Re}\,(\mathbf{E} \times \mathbf{H}^*)} \tag{13.72}$$

where the asterisk represents a complex conjugate; that is, $\mathbf{E}'^* = (\mathbf{E}_r + j\mathbf{E}_i)^* = \mathbf{E}_r - j\mathbf{E}_i$. In deriving (13.72), we note that $\mathbf{E} \times \mathbf{H}^* = (\mathbf{E}'e^{j\omega t}) \times (\mathbf{H}'^*e^{-j\omega t}) = \mathbf{E}' \times \mathbf{H}'^* = (\mathbf{E}_r + j\mathbf{E}_i) \times (\mathbf{H}_r - j\mathbf{H}_i) = (\mathbf{E}_r \times \mathbf{H}_r + \mathbf{E}_i \times \mathbf{H}_i) + j(\mathbf{E}_i \times \mathbf{H}_r - \mathbf{E}_r \times \mathbf{H}_i)$. This representation is similar to that used in circuits and transmission lines where the use of complex exponentials for voltage V and current I gives for the average power: $\text{Power}_{\text{ave}} = \frac{1}{2}\,\text{Re}\,VI^*$.

Example: Calculate the instantaneous and average Poynting vector for the plane wave given by (13.39).

Rewriting (13.39) in the notation of this section, we have

$$\mathbf{E}(z, t) = \mathbf{E}'(z)e^{j\omega t} = \hat{\mathbf{x}}E_x(z)e^{j\omega t} = \hat{\mathbf{x}}E_0e^{-j\beta z}e^{j\omega t} \tag{13.73}$$

$$\mathbf{H}(z, t) = \mathbf{H}'e^{j\omega t} = \hat{\mathbf{y}}H_y(z)e^{j\omega t} = \hat{\mathbf{y}}H_0e^{-j\beta z}e^{j\omega t} \tag{13.74}$$

† When expressing the instantaneous Poynting vector which gives real power flow at an instant of time, the instantaneous values of E and H must be used. However, it is possible to construct an expression for instantaneous power flow $\mathscr{P}(t)$ in terms of phasor expressions (see Prob. 13.31).

The instantaneous Poynting vector is given by (13.70), with $E_r = E_0 \cos \beta z$, $E_i = -E_0 \sin \beta z$, and $H_r = H_0 \cos \beta z$, $H_i = -H_0 \sin \beta z$,

$$\begin{aligned}\mathscr{P}_z(z, t) &= E_0 H_0(\cos^2 \beta z \cos^2 \omega t + \sin^2 \beta z \sin^2 \omega t \\ &\quad + 2 \cos \beta z \sin \beta z \sin \omega t \cos \omega t) \\ &= E_0 H_0 \cos^2 (\omega t - \beta z) = \tfrac{1}{2} E_0 H_0 [1 + \cos 2 (\omega t - \beta z)]\end{aligned} \tag{13.75}$$

The average value of Poynting's vector is given by (13.71) as

$$\begin{aligned}\mathscr{P}_{\text{ave}} &= \tfrac{1}{2}(\hat{\mathbf{x}} E_0 \cos \beta z \times \hat{\mathbf{y}} H_0 \cos \beta z + \hat{\mathbf{x}} E_0 \sin \beta z \times \hat{\mathbf{y}} H_0 \sin \beta z) \\ &= \frac{1}{2} E_0 H_0 \hat{\mathbf{z}} = \frac{1}{2} \frac{E_0^2}{\eta} \hat{\mathbf{z}}\end{aligned} \tag{13.76}$$

or by (13.72) as

$$\mathscr{P}_{\text{ave}} = \tfrac{1}{2} \text{ Re } (\hat{\mathbf{x}} E_0 e^{j(\omega t - \beta z)} \times \hat{\mathbf{y}} H_0^* e^{-j(\omega t - \beta z)}) = \tfrac{1}{2} \text{ Re } (E_0 H_0 \hat{\mathbf{z}}) = \tfrac{1}{2} E_0 H_0 \hat{\mathbf{z}}$$

where the amplitudes of the plane wave are assumed to be real; that is, $E_0 = E_0^*$ and $H_0 = H_0^*$. On the other hand, if E_0 or H_0 is complex, either because of a complex intrinsic impedance η (note that $E_0 = \eta H_0$), as in the case of a lossy dielectric, or because E and H have a specified phase shift with respect to each other, the above expressions for $\mathscr{P}(t)$ and $\mathscr{P}_{\text{ave}}$ readily take care of this.†

The relationship between instantaneous and average power flow is easily seen if we plot (13.75). Without losing generality, we choose $z = 0$ as the point to examine $\mathscr{P}_{\text{inst}}$. Expression (13.75) then simplifies to

$$\mathscr{P}(t) = \mathscr{P}_{\text{ave}}(1 + \cos 2\omega t) \tag{13.77}$$

and is plotted in Fig. 13.11. Note that $\mathscr{P}_{\text{inst}}$ is a pulsating quantity, with two pulses per period T. An identical figure would be obtained if we freeze time, say at $t = 0$, and plot the spatial distribution of $\mathscr{P}_{\text{inst}}$. The horizontal axis would be the z axis with two pulses per wavelength λ.

Relation between Energy Density and Power Flow in a Plane Wave—Energy Velocity

In an electromagnetic wave the energy stored in the electric field at any instant is equal to the energy stored in the magnetic field:

$$\frac{1}{2} \mu H^2 = \frac{1}{2} \frac{\mu E^2}{\eta^2} = \frac{1}{2} \varepsilon E^2 \tag{13.78}$$

where $E = \eta H$ and $\eta = \sqrt{\mu/\varepsilon}$. The total energy density in a wave is therefore εE^2 or μH^2. In other words, the energy carried by a wave is shared on the average equally by the electric and magnetic fields.

If the fields vary sinusoidally with time, we should be careful to differentiate between instantaneous, peak, and average values. The instantaneous *total* energy density in a sinusoidal plane wave is $\varepsilon E_0^2 \cos^2 (\omega t - \beta z) = \mu H_0^2 \cos^2 (\omega t - \beta z)$. The peak total energy density is therefore $\varepsilon E_0^2 = \mu H_0^2$, and the average total energy

† For example, if there is an additional time-phase angle between the electric and magnetic fields; that is, $E_0 \to E_0$, but $H_0 \to H_0 e^{j\delta}$, then (13.76) becomes $\mathscr{P}_{\text{ave}} = \tfrac{1}{2} E_0 H_0 \cos \delta \, \hat{\mathbf{z}}$. See Prob. 13.29.

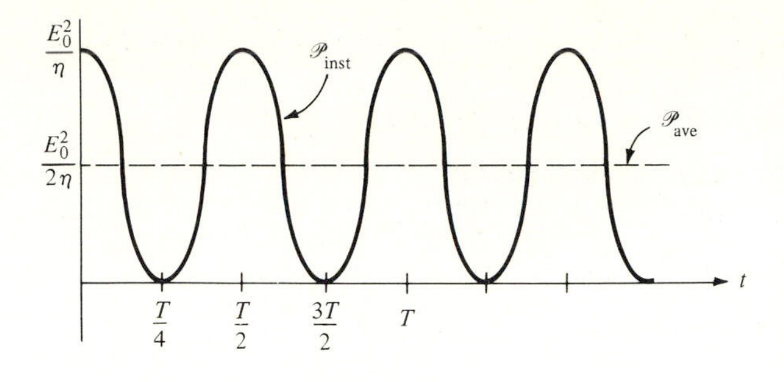

Figure 13.11 Plot of the instantaneous Poynting's vector $\mathscr{P}(t)$ and average $\mathscr{P}_{\text{ave}}$ for the spatial position $z = 0$. The traveling nature of the wave can be displayed by superimposing on the figure a plot for a different position, say $z = \lambda/8$.

density is $\frac{1}{2}\varepsilon E_0^2 = \frac{1}{2}\mu H_0^2$, because the average of $\cos^2(\omega t - \beta z)$ is one-half (see also Prob. 13.32).

We now observe that if we divide power flow by energy density, a quantity with dimensions of velocity, which is called the *energy velocity* v_e, is obtained:

$$\boxed{v_e = \frac{\text{Poynting's vector}}{\text{energy density}}} \tag{13.79}$$

For example, for the sinusoidal wave we have, from (13.76),

$$\mathscr{P}_{\text{ave}} = \frac{1}{2}\frac{E_0^2}{\eta} = v\left(\frac{1}{2}\varepsilon E_0^2\right) = v(\text{average total energy density}) \tag{13.80}$$

where $\eta = (\mu/\varepsilon)^{1/2} = \varepsilon v$. Thus, for a nondispersive medium the phase velocity, group velocity,† and energy velocity are the same. For a situation such as propagation through lossy media where group velocity often loses its meaning, the energy velocity v_e can always be associated with velocity of energy flow. The energy velocity can be an instantaneous or an average one depending on the use of instantaneous or average values in the right-hand side of (13.79).

Example: What are the magnitudes of the electric and magnetic fields in a free-space microwave beam which has a power density $\mathscr{P}_{\text{ave}} = 100 \text{ W/m}^2$?

The average total energy density from (13.80) is

$$\frac{1}{2}\varepsilon E_0^2 = \frac{\mathscr{P}_{\text{ave}}}{v} = \frac{100 \text{ W/m}}{3 \times 10^8 \text{ m/s}} = 3.3 \times 10^{-7} \text{ J/m}^3$$

The peak value for the electric field is therefore

$$E = \sqrt{\frac{2}{\varepsilon}\frac{\mathscr{P}_{\text{ave}}}{v}} = \sqrt{\frac{(2)(3.3 \times 10^{-7})}{8.85 \times 10^{-12}}} = 270 \text{ V/m}$$

† *Phase velocity* refers to the velocity with which a plane of constant phase in a sinusoidally varying wave travels [see Eq. (13.43)]. *Group velocity* refers to the velocity with which information which is modulated on a sinusoidally varying wave travels. As a modulated wave is composed of sinusoidal waves of many frequencies, phase velocity for a modulated wave has meaning only if all component waves travel at the same velocity. A medium in which this is possible is called *nondispersive.*

For the magnetic field

$$B = \mu H = \mu \frac{E}{\eta} = \frac{E}{v} = \frac{270}{3 \times 10^8} = 9 \times 10^{-7}\ \text{T}$$

The electric field is seen to be much stronger than the magnetic field which is only 9×10^{-3} G, a magnitude difficult to detect. For comparison, earth's magnetic field is about $\frac{1}{2}$ G. If the root-mean-square values for the field are desired, the above number should be divided by $\sqrt{2}$. We should also observe that at no time was the frequency ω of the wave considered in the calculations. Thus we conclude that values for E and B would be the same in any 100-W/m^2 beam, whether it be a light, microwave, or radio beam.

Complex Poynting Vector Theorem

We can formulate Poynting's theorem for time-averaged quantities by using the complex Poynting vector $\mathbf{E} \times \mathbf{H}^*$ in (13.63). In such a formulation the balance between real power flow is more obvious than in the more general formulation of (13.64). Thus

$$\nabla \cdot (\mathbf{E} \times \mathbf{H}^*) = -j\omega \mathbf{B} \cdot \mathbf{H}^* + j\omega \mathbf{D}^* \cdot \mathbf{E} - \mathbf{E} \cdot \mathbf{J}^* \tag{13.81}$$

where Maxwell's equations $\nabla \times \mathbf{E} = -j\omega\mathbf{B}$ and $\nabla \times \mathbf{H}^* = -j\omega\mathbf{D}^* + \mathbf{J}^*$ have been used and * denotes complex conjugate. Integrating the above equation throughout volume $v(A)$ and applying the divergence theorem (1.94) to the left-hand side gives

$$\oint\!\!\oint \mathbf{E} \times \mathbf{H}^* \cdot \mathbf{dA} = -j\omega \iiint (\mathbf{B} \cdot \mathbf{H}^* - \mathbf{E} \cdot \mathbf{D}^*)\, dv - \iiint \mathbf{E} \cdot \mathbf{J}^*\, dv \tag{13.82}$$

which is the complex Poynting vector theorem. If we now divide by 2 and take the real part of (13.82), assuming the volume integrands are real we obtain

$$\boxed{\frac{1}{2} \text{Re} \oint\!\!\oint_A \mathbf{E} \times \mathbf{H}^* \cdot \mathbf{dA} = -\frac{1}{2} \iiint_v \mathbf{E} \cdot \mathbf{J}^*\, dv} \tag{13.83}$$

which allows the left-hand side to be identified with (13.72). This equation expresses the balance between real power flow. Note that the energy-storage terms are absent because, for sinusoidal time dependence, there can be no average increase in energy stored.

In the receiver case, when net power flows across surface A into region v, (13.83) can be written as

$$\oint\!\!\oint \mathscr{P}_{\text{ave}} \cdot (-\mathbf{dA}) = \frac{1}{2} \iiint_v \sigma |\mathbf{E}|^2\, dA \tag{13.84}$$

For inflow of power, the left-hand side is positive because $-\mathbf{dA}$ represents an elemental vector area directed into volume v. The right-hand side expresses the power dissipated in heat by the conduction currents $\mathbf{J} = \sigma\mathbf{E}$ that are induced in v by the incoming electromagnetic field.

If the dielectric that occupies volume v has an additional loss due to the polarization process, additional energy will be extracted from the incoming wave to overcome the polarization damping forces.† To account for this, we use the complex permittivity $\varepsilon^\star$, which was defined in (13.49) as $\varepsilon^\star = \varepsilon' - j\varepsilon''$. Substituting $\varepsilon^\star$ in $\mathbf{D}^*$ of (13.82) gives for $j\omega\mathbf{E}\cdot\mathbf{D}^* = j\omega\mathbf{E}\cdot(\varepsilon' + j\varepsilon'')\mathbf{E}^* = j\omega\varepsilon'|E|^2 - \omega\varepsilon''|E|^2$. With the additional real term, (13.84) becomes

$$-\oiint \mathscr{P}_{\text{ave}} \cdot \mathbf{dA} = \frac{1}{2}\iint (\sigma + \omega\varepsilon'')|\mathbf{E}|^2\, dv \tag{13.85}$$

This energy-balance equation shows that the incoming power is dissipated in heat generated by the induced conduction currents and in heat generated by the dielectric losses. From (13.85) we note that the additional dielectric loss can be represented by an equivalent conductivity σ_d, where $\sigma_d = \omega\varepsilon''$. Any losses due to magnetization damping forces, if present in volume v, can be similarly accounted for by using the complex permeability $\mu^\star = \mu' - j\mu''$ in the $\mathbf{B}$ term of (13.82). The additional magnetization loss would then show up as an equivalent conductivity‡ $\sigma_m = \omega\mu''/\eta^2$ in (13.85). All three losses can be accounted for by replacing σ in (13.84) by an effective conductivity $\sigma_{\text{eff}} = \sigma + \omega\varepsilon'' + \omega\mu''/\eta^2$.

The transmitter case, in which a radiating source exists inside volume v and transports energy out of v, is given from (13.83) as

$$\oiint \mathscr{P}_{\text{ave}} \cdot \mathbf{dA} = \frac{1}{2}\iiint \mathbf{E}\cdot\mathbf{J}_s^*\, dv \tag{13.86}$$

where the current of the internal source is represented in (13.83) by $\mathbf{J}^* = -\mathbf{J}^*_{\text{source}}$.

The remaining imaginary part of (13.82) which is

$$\text{Im}\frac{1}{2}\oiint \mathbf{E}\times\mathbf{H}^*\cdot\mathbf{dA} = -2\omega\iiint\left(\frac{\mu}{4}\mathbf{H}\cdot\mathbf{H}^* - \frac{\varepsilon}{4}\mathbf{E}\cdot\mathbf{E}^*\right)dv \tag{13.87}$$

shows that the imaginary part of the complex rate of energy flow across surface A is equal to 2ω times the difference in the time-averaged values of the magnetic and electric energies stored. Note that the time-averaged stored electric energy density (see Prob. 13.32) is given by $\frac{1}{2}(\frac{1}{2}\varepsilon\mathbf{E}\cdot\mathbf{E}^*) = \varepsilon|\mathbf{E}|^2/4$. Since the real part of (13.82) relates to actual power transfer, the expression (13.83) is useful in practical situations; (13.87) merely balances the reactive powers.

Example: The average electric power consumption in the United States is 2×10^{11} W. If this amount of power were to be carried by an electromagnetic wave, what would the size of the beam be, assuming that an antenna exists that could collimate such a beam?

Let us assume that the electric field in such a beam is only limited by the dielectric breakdown strength of air, which is 3×10^6 V/m. Using this value, the power density of such a beam would be

$$\mathscr{P} = EH = \frac{E^2}{\eta} = \frac{(3\times 10^6\ \text{V/m})^2}{120\pi\ \Omega} = 2.4 \times 10^{10}\ \text{W/m}^2$$

† This loss in a dielectric is related to the friction which the induced dipoles experience as they are flipping back and forth in response to the sinusoidal time variation of the inducing field.

‡ The energy density of the magnetization loss is $\omega\mu''|H|^2$. To obtain σ_m, we have assumed that the relationship between E and H is that of plane waves; that is, $E = \eta H$.

If the cross-sectional area of the beam is A, we have

$$2 \times 10^{11} = \mathscr{P}A = \frac{E^2}{\eta}A \qquad \text{W}$$

Solving for A, we obtain the beam cross section as

$$A = \frac{2 \times 10^{11}}{2.4 \times 10^{10}} = 8.4 \text{ m}^2$$

which is surprisingly small for a beam that carries this amount of energy. Notice that at no point did the frequency of the wave enter the discussion, as the above formulas are valid at any frequency. Presumably any frequency could be used. The magnitudes of E are the same if we use a 2×10^{11} W radio beam, a 2×10^{11} W microwave beam, or a 2×10^{11} W light beam. Of course, at present very large obstacles exist in the design of such a system, even when small amounts of power are involved. For example, what kind of an antenna could focus such a beam, what materials that could handle such enormous energy densities could be used in the construction of the antenna, how would a receiving antenna convert the beam energy into usable energy, etc?

For comparison, we can state that the power density of all electromagnetic radiation from the sun at the earth's surface is 1400 W/m². To collect 2×10^{11} W would then take an area of $A = 2 \times 10^{11}/1.4 \times 10^3 = 1.4 \times 10^8$ m², which is a square of 12 km by 12 km, not at all a large area considering that all electric power consumed in the United States could be produced there.

13.9 FORCE OF AN ELECTROMAGNETIC WAVE AND RADIATION PRESSURE

In the previous two sections we have shown that an electromagnetic wave can carry energy. It can also exert a force in the direction of propagation. The *radiation pressure*† is the force per unit area of a wave on a material upon which it impinges.

In general, the rate of doing work, or the power $P = \Delta W/\Delta t$, is given by

$$P = Fv \qquad \text{W} \tag{13.88}$$

where F is the force doing the work, $\Delta W = F\,\Delta l$, and v is the speed given by $v = \Delta l/\Delta t = (\varepsilon_0 \mu_0)^{-1/2}$. For a sinusoidally varying plane wave, the time-average Poynting vector is given by (13.80) as

$$\frac{P}{\Delta A} = \mathscr{P}_{\text{ave}} = v w_{\text{ave}} \equiv \frac{Fv}{\Delta A} \qquad \text{W/m}^2 \tag{13.89}$$

where the average total energy density of the wave is $w_{\text{ave}} = \frac{1}{2}\varepsilon_0 E^2$ J/m³. Using pressure p which is defined as force per unit area, $p = F/\Delta A$, we have for the radiation pressure

$$\boxed{\frac{F}{\Delta A} = p = \frac{\mathscr{P}_{\text{ave}}}{v} = w_{\text{ave}} \qquad \text{N/m}^2} \tag{13.90}$$

† A. Ashkin, The Pressure of Laser Light, *Sci. Am.*, pp. 62–72, February 1972.

The radiation pressure of the wave is therefore equal to the average total energy density in the wave and acts in the direction of travel of the wave.

The radiation force of the wave is transferred to an object upon which it impinges. Hence, if an object completely absorbs an incoming wave, a force equal to the power P of the wave divided by the velocity of the wave, $F = P/v$, acts on it. On the other hand, if a body, such as a metallic object, completely reflects a wave, the force acting on the body is twice that for absorption, or $F = 2P/v$ (once for absorption and once for reemission of the wave from the metallic object). This implies that when a transmitter emits a wave of power P in a certain direction, the transmitter itself must recoil in the opposite direction because the emission is equivalent to a force $F = P/v$ acting on the transmitter.

Example: A plane wave, with electric field amplitude E_0, is normally incident upon a perfectly conducting plane surface. (*a*) Find the average pressure exerted on the conducting surface. (*b*) Calculate the average pressure when the power in the plane wave is 10 W/m^2.

SOLUTION
(*a*) $p = 2\mathscr{P}_{\text{ave}}/v = 2(\frac{1}{2}E_0^2/\eta)/(\varepsilon_0 \mu_0)^{-1/2} = \varepsilon_0 E_0^2 \quad \text{N/m}^2$
(*b*) $p = 2\mathscr{P}_{\text{ave}}/v = 2(10)/3 \times 10^8 = 6.7 \times 10^{-8} \quad \text{N/m}^2$
which is a very small pressure.

Sometimes it is more desirable to use momentum M in place of force F. As $F = \Delta M/\Delta t$, and as $F = P/v = (\Delta W/\Delta t)/v$, we have that the momentum of an electromagnetic wave is $M = W/v$, where W is the electromagnetic energy $W = \int w_{\text{ave}}\, dv_{ol}$.

Example: A 100-W collimated beam of electromagnetic radiation is turned on for 2 s. (*a*) Determine the momentum of the electromagnetic pulse. (*b*) Determine the recoil force on the source emitting the beam.

SOLUTION
(*a*) $M = W/v = Pt/v = (100)(2)/(3 \times 10^8) = 6.7 \times 10^{-7}$ kg · m/s
(*b*) $F = P/v = (100)/(3 \times 10^8) = 3.3 \times 10^{-7}$ N
Notice that the recoil force is independent of the frequency of the source. A radio or light source, as long as the power output is the same, will experience the same recoil.

Total Electromagnetic Force on a Conducting Plane

In Secs. 2.12 and 5.15 we showed that an **E** field, whether it points toward or away from a conducting surface, produces a force which pulls on the surface. On the other hand, we have just shown that a wave which impinges on a surface will push on it. We can generalize† and say that a normal field (**E** or **B**) on a surface will always produce a pull, whereas a tangential field (**E** or **B**) will always produce a push. On a perfectly conducting surface, only a normal electric field $E_n = \hat{\mathbf{n}} \cdot \mathbf{E}$

† C. C. Johnson, "Field and Wave Electrodynamics," sec. 1.19, McGraw-Hill Book Company, New York, 1965.

and a tangential magnetic field $\mathbf{H}_t = \hat{\mathbf{n}} \times \mathbf{H}$ can exist. Hence, we can write for the total force per unit area, or pressure, on the conducting surface

$$\mathbf{p} = \frac{\mathbf{F}}{\Delta A} = \left(\frac{\varepsilon_0 E_n^2}{2} - \frac{\mu_0 H_t^2}{2}\right)\hat{\mathbf{n}} \tag{13.91}$$

where $\hat{\mathbf{n}}$ is a unit normal to the conducting surface. The minus sign denotes a push on the surface. The above expression is an instantaneous pressure. For sinusoidally varying fields, H_t is given by $H_t = H_0 \cos(\omega t - \beta z)$ and the normal electric field by $E_n = E_0 \cos(\omega t - \beta z)$, where H_0 and E_0 are the amplitudes. The time-averaged pressure is then

$$\mathbf{p}_{\text{ave}} = \frac{1}{2}\left(\frac{\varepsilon_0 E_0^2}{2} - \frac{\mu_0 H_0^2}{2}\right)\hat{\mathbf{n}} \tag{13.92}$$

because the time average of $\cos^2(\omega t - \beta z)$ is equal to one-half.

The magnetic force term in (13.91) is derived from (6.5), which gives the magnetic force on a current density $\mathbf{J}$ immersed in a magnetic field $\mathbf{B}$ as $\Delta \mathbf{F}_m / \Delta v_{ol} = \mathbf{J} \times \mathbf{B}$. Since a current in a conductor flows only on the surface in a sheet of (skin) depth Δn, we have $\Delta \mathbf{F}_m / \Delta A = \mathbf{J}\, \Delta n \times \mathbf{B} = \mathbf{K} \times \mathbf{B}$, where K is the sheet current discussed in (8.75), and an elemental volume is given as $\Delta v_{ol} = \Delta A\, \Delta n$. Boundary condition (8.93) relates sheet current to the magnetic field on the conducting surface as $\mathbf{K} = \hat{\mathbf{n}} \times \mathbf{H}$. Substituting this in the magnetic force expression, expanding the vector triple product, and multiplying the resulting expression by one-half (because the magnetic field acting on the sheet current exists on only one side of the current), we obtain

$$\frac{\Delta \mathbf{F}_m}{\Delta A} = \frac{1}{2}\mathbf{K} \times \mathbf{B} = \frac{1}{2}(\hat{\mathbf{n}} \times \mathbf{H}) \times \mathbf{B} = -\frac{1}{2}\mu_0 H^2 \hat{\mathbf{n}} \tag{13.93}$$

which is the magnetic term in (13.91). Note that the vector triple product results in a term $B_n = \hat{\mathbf{n}} \cdot \mathbf{B}$ which is a normal magnetic field and is equal to zero on a conducting surface.

The explanation of the magnetic radiation pressure is that the electrons carrying the induced surface current are pushed by the Lorentz force in the direction of propagation of the wave. The electrons, as they collide with atoms in the material, in turn push on the conducting material, giving rise to radiation pressure. Although, in general, radiation pressure is small, it can lead to large-scale effects, such as in comets. Comet tails are forced to point away from the sun by the sun's radiation pressure.

Example: Calculate the average pressure which a sinusoidal plane wave with a Poynting vector $\mathscr{P}_{\text{ave}} = 10\ \text{W/m}^2$ produces when it impinges normally on a large plane conducting surface.

SOLUTION The electric force term in (13.92) is zero, because for a normally incident wave $E_n = 0$. This leaves the magnetic "push" term; that is, $p_{\text{ave}} = p_{m,\text{ave}} = \mu_0 H_0^2/4$. Poynting's vector is related to the incident electric and magnetic fields as $\mathscr{P}_{\text{ave}} = E_0^2/2\eta = \eta H_0^2/2$. However, on the surface of a conductor, the total magnetic field due to a normally incident wave of amplitude H_0 is $H = 2H_0$ [see

Eq. (14.8) or Prob. 14.3]. This gives

$$p_{m,\text{ave}} = \frac{\mu_0(2H_0)^2}{4} = \frac{2\mathscr{P}_{\text{ave}}}{v} = \frac{2(10)}{3 \times 10^8} = 6.7 \times 10^{-8}\ \text{N/m}^2$$

which agrees with the result in the example of the previous section.

PROBLEMS

13.1 The wave equations for the **E** and **H** fields as given by (13.6) and (13.7) were derived from Maxwell's equations. Did we lose information in going from Maxwell's equations to the wave equations? Explain.

13.2 Explain why the source term [the right-hand term of (13.7)] of the wave equation for the **H** field contains only currents and not charges.

13.3 Why are the wave equations for the **E** and **H** fields the same in free space?

13.4 Show that $g_1(z - vt)$, $g_2(z + vt)$, and $g_1 + g_2$ are solutions to the wave equation (13.13).

13.5 (*a*) Show that only functions which have an argument $A(z \pm vt)$ or $A(t \pm z/v)$ are solutions to the wave equation. A is an arbitrary function.

(*b*) Show that $e^{j\omega(t - bz)}$, $\sin \omega(t - bz)$, and $(t - bz) \sin \omega(t - bz)$, where b and ω are constants are solutions of the wave equation.

13.6 For an x-axis that increases to the right, show that $A \sin (\omega t - kx)$ is a wave traveling to the right.

13.7 A traveling wave is described by the equation $E(z, t) = 0.3 \cos (2z + 20t)$. Find

(*a*) The wave speed.
(*b*) The wavelength λ.
(*c*) The frequency.
(*d*) The amplitude.

13.8 A traveling wave is given by the equation $E(z, t) = A \exp (-at^2 - bz^2 - 2\sqrt{ab}\, zt)$.

(*a*) In what direction is the wave traveling?
(*b*) What is the speed of the wave?

13.9 A pressure wave in water is described by the wave equation $\partial^2 p/\partial z^2 = \rho k(\partial^2 p/\partial t^2)$, where the pressure in newtons per square meter is p, the mass density ρ is 10^3 kg/m^3, and the compressibility constant k is equal to 4.8×10^{-10} m^2/N. Find the velocity of propagation of the pressure wave.

13.10 A plane wave propagates in a certain medium with an electric field given by $\mathbf{E}(z, t) = 5\hat{\mathbf{y}} \cos (10^9 t + 30z)$. Find

(*a*) The amplitude of the E field.
(*b*) The angular frequency ω of the wave.
(*c*) The phase-propagation constant β.
(*d*) The phase velocity.
(*e*) The direction of wave travel.
(*f*) The dielectric constant of the medium, assuming it to be nonmagnetic; that is, $\mu \cong \mu_0$.
(*g*) The magnetic field $\mathbf{H}(z, t)$.

13.11 Seawater at a frequency of $f = 4 \times 10^8$ Hz has the following constants: $\mu_r = \mu/\mu_0 \cong 1$, $\varepsilon_r = \varepsilon/\varepsilon_0 \cong 81$, and $\sigma \cong 4.4$.

(*a*) What is the ratio of conduction current to displacement current at this frequency; that is, $J/(\partial D/\partial t) = \sigma/\omega\varepsilon$?

(*b*) If one had to approximate seawater at this frequency by a conducting or dielectric medium, which one would be more accurate?

(*c*) Find the attenuation constant α for a plane wave that propagates in seawater. The attenuation constant α is given by $\beta^\star = \beta - j\alpha$, where $\beta^{\star 2} = \omega^2\mu\varepsilon^\star$, as given in (13.25).

(*d*) If we approximate seawater by a conducting medium, the attenuation constant α is then given by (13.28) or (13.53). How does this value compare with the value for α obtained in part *c*?

(*e*) If we were to use the expression for attenuation constant α given by (13.45), what value for α would we obtain? Why is it inappropriate to use this expression for this case?

13.12 At the frequencies $f = 1$ kHz, 10 MHz, 1 GHz, are the following materials conducting or dielectric media?

(*a*) Seawater.
(*b*) Fresh water.
(*c*) Wet earth.
(*d*) Dry earth.

13.13 Determine the loss per kilometer for a plane wave propagating in wet earth at a frequency $f = 0.5$ MHz. The parameters of wet earth at this frequency are $\sigma = 10^{-3}$, $\mu_r = 1$, and $\varepsilon_r = 10$.

13.14 Determine the loss per kilometer for a plane wave propagating in dry earth at a frequency $f = 0.5$ MHz. The parameters of dry earth at this frequency are $\sigma = 10^{-5}$, $\mu_r = 1$, and $\varepsilon_r = 3$.

13.15 Freshly fallen snow has a loss tangent of 0.02 and a dielectric constant of $\varepsilon_r = 1.2$ at the frequency $f = 1$ MHz. Calculate the loss per kilometer for a plane wave propagating in freshly fallen snow at 1 MHz.

13.16 A certain nonmagnetic material has a dielectric constant of $\varepsilon_r = 9$ and a dissipation factor of 0.1 which is assumed to be constant with frequency. Calculate the attenuation in decibels for a wave after it propagates 200 m in the material at the frequencies

(*a*) $\omega = 10^4$.
(*b*) $\omega = 10^6$.
(*c*) $\omega = 10^8$.

Note that decibel attenuation is defined as $20 \log E(z = 200 \text{ m})/E(z = 0) = 20 \log e^{-\alpha z}$.

13.17 A 1-kHz plane wave is partially transmitted from air into seawater.

(*a*) Find the wavelength λ of this wave in air and in water.
(*b*) Find the velocity v in air and in water.
(*c*) What is the frequency of this wave in water?

13.18 A submarine with its antenna just below the water surface receives a 1-kHz signal which registers 20 db above noise level.

(*a*) How far down can the submarine dive before it loses the signal in the noise?
(*b*) How long is its $\lambda/2$ dipole antenna?

13.19 Calculate the attenuation in decibels for a distance of five skin depths.

13.20 The expression for skin depth is given by $\delta = 1/\sqrt{\pi f \mu \sigma}$. Why is δ dependent upon permeability μ but not on permittivity ε?

13.21 If the magnitude of H in a plane wave is 10^{-3} A/m, what is the magnitude of E for a plane wave in free space?

13.22 A uniform plane wave is propagating through a nonmagnetic medium. Determine the relative dielectric constant of the medium if

(*a*) The intrinsic impedance is 200 Ω.
(*b*) The wavelength at 10 GHz is 1.5 cm.

13.23 Determine the wave impedance (E/H) for silver.

13.24 One of the many reasons that skin depth δ is valuable for engineering purposes is that as the frequency increases and the current is forced to flow in a smaller and smaller layer on the surface of a conductor, the high-frequency resistance of a conductor can be obtained by assuming that the entire current flow is uniformly distributed over a depth δ. This approximation is valid for curved conductors such as wires, as long as the radius a of the wire is larger than the skin depth δ; that is, $a \gg \delta$. Since the dc resistance of a piece of wire of length L and cross section πa^2 is given from (2.16) or (3.34*c*) as $R_{dc} = L/(\sigma \pi a^2)$, the ac resistance is therefore $R_{ac} = L/(\sigma 2\pi a \delta)$, where the cross section through which the current flows is given by the circumference $2\pi a$ and thickness δ. The increase of the resistance of a wire as the frequency is increased from dc is therefore

$$R_{ac} = R_{dc}\frac{a}{2\delta} = \frac{R_{dc}a\sqrt{\pi f\mu\sigma}}{2}$$

Determine the factor by which the resistance of a copper wire 3 mm in radius increases as the frequency changes from dc to

(*a*) 100 KHz.
(*b*) 10 MHz.
(*c*) 10 GHz.

13.25 A plane wave propagates in free space with a peak electric field of $E = 10$ V/m. Find

(*a*) The peak Poynting vector.
(*b*) The average Poynting vector.
(*c*) The peak value of the magnetic H field.

13.26 The electric field of a plane wave is given by $E_x = E_0 \cos(\omega t - \beta z)$. Find the instantaneous Poynting vector.

13.27 The electric field of a plane wave is given by $E_x = E_0 e^{j(\omega t - \beta z)}$. Find the instantaneous Poynting vector. What is its peak value when $E_0 = 10$ V/m.

13.28 A plane wave in free space of frequency 10 MHz has an average Poynting vector of 2 W/m^2. Find

(*a*) The wavelength and velocity of the wave.
(*b*) The peak electric field E and magnetic field H.

13.29 In a lossy dielectric the electric and magnetic fields are not in time phase because the intrinsic impedance is complex; that is, $E_x/H_y = \eta^* = \sqrt{\mu/\varepsilon^*}$, as given by (13.48). If the electric field of a plane wave is given by

$$E_x(z, t) = E_0 e^{j(\omega t - \beta^* z)} = E_0 e^{-\alpha z} e^{j(\omega t - \beta z)}$$

show that

(*a*) The magnetic field is

$$H_y(z, t) = \frac{E_0}{\eta^*} e^{(j\omega t - \beta^* z)} = \frac{E_0}{|\eta^*|} e^{j(\omega t - \beta^* z - \phi)}$$

where η^* is expressed in polar form as $\eta^* = |\eta^*|e^{j\phi}$.

(*b*) Show that the average power flow of the plane wave is along the z axis and is given by Poynting's vector in the form

$$\mathscr{P}_{z,\text{ave}} = \frac{1}{2}\frac{E_0^2}{|\eta^*|} e^{-2\alpha z} \cos\phi$$

(*c*) Derive the above Poynting vector from (13.72) and (13.76).

13.30 What is the relationship between **E** and **B** in a plane wave? What can you say about the relative magnitudes of E and B in a plane wave?

13.31 If **E**′ and **H**′ are the complex multipliers of $e^{j\omega t}$, show that the instantaneous Poynting vector is given by

$$\mathscr{P}(t) = \tfrac{1}{2}\,\text{Re}\,[(\mathbf{E}' \times \mathbf{H}'^*) + (\mathbf{E}'e^{j\omega t}) \times (\mathbf{H}'e^{j\omega t})]$$

13.32 Referring to (13.76), show that the time-average stored electric and magnetic energy densities are given by $(\frac{1}{2}\varepsilon E^2)_{\text{ave}} = \frac{1}{2}\varepsilon\frac{1}{2}\mathbf{E}\cdot\mathbf{E}^* = (\varepsilon/4)|E|^2$ and by $(\frac{1}{2}\mu H^2)_{\text{ave}} = (\mu/4)|H|^2$, respectively.

Hint: Use a procedure similar to that in the derivation of (13.71).

13.33 The power density of all electromagnetic radiation from the sun at the earth's surface is 1400 W/m^2.

(*a*) Calculate the rms electric field E at the earth, assuming that all the sunlight is concentrated at a single frequency.

(*b*) Assuming the sun radiates isotropically, what is the power output of the sun? The sun-earth distance is 1.49×10^8 km.

(*c*) Calculate the total power received by the earth. The radius of the earth is 6.37×10^3 km.

13.34 Assume that a 1-W 5-GHz transmitter located on the moon radiates isotropically. The earth-moon distance is 3.8×10^5 km. Find

(*a*) The rms electric and magnetic fields at the earth.

(*b*) The average Poynting vector at the earth.

(*c*) The average energy density.

(*d*) The time it takes a signal to reach earth.

13.35 Discuss the interpretation of the Poynting vector, for the case of a static point charge Q located at the center of a small loop of conducting wire carrying a direct current I. Is power radiated by this arrangement?

13.36 The Poynting vector due to all electromagnetic radiation from the sun at the earth's surface is 1.4 kW/m^2. Find the radiation pressure of the sun's radiation on an object on the earth's surface if

(*a*) The object is an absorber.

(*b*) The object is a metallic reflector.

13.37 A collimated beam of cross section 10 cm^2 is switched on for 10 s. If the electric field amplitude of the beam wave is 10 V/m, calculate the force and momentum of the pulsed beam.

13.38 The magnetic pull of a magnet is given by (10.35). Rewriting (13.91) to apply to the case of a magnet, show that the magnetic pull per pole is given by $F/\Delta A = \frac{1}{2}\mu_0 H^2$.

13.39 A 3-kW radiation beam is used to accelerate a spaceship that weighs 10,000 kg. If the beam is sufficiently well focused so that beam spreading can be ignored, the approximation that it radiates in one direction can be made. Calculate the change in speed which the radiation imparts to the ship in one week.

CHAPTER

FOURTEEN

APPLICATION OF MAXWELL'S EQUATIONS: REFLECTION OF EM WAVES

14.1 REFLECTION OF ELECTROMAGNETIC WAVES

When a wave in one medium impinges upon a second medium with different permittivity ε, permeability μ, or conductivity σ, the wave will in general be partially transmitted into the second medium and be partially reflected from it. Two waves will then be present in the first medium, one traveling forward (toward the second medium), the other backward. The existence of the reflected wave was already predicted in the solution to the wave equation (13.24), which being a second-order differential equation, has two solutions given by (13.25) as

$$E_x(z, t) = E_0^i e^{j(\omega t - \beta z)} + E_0^r e^{j(\omega t + \beta z)} \tag{14.1}$$

where E_0^i is the amplitude of the forward, or incident, wave and E_0^r is the amplitude of the reflected wave. In this section we will be concerned with finding the amplitude E_0^r of the reflected wave. We will see that E_0^r is characterized by the differences in ε, μ, and σ between the first and second media. Clearly, if there is no difference in the three parameters, the two media are electrically the same; the incident wave continues to see the same media, and no reflection occurs. The assumption will also be made that a boundary between two media is sharp or sudden. For practical purposes we will call any transition region (between two media) that is small with respect to wavelength λ a *sharp boundary*.

From a different point of view we can say that the reflected wave arises because boundary conditions at the junction between two different media cannot be satisfied by the incident wave alone. Using solution (14.1), the boundary condition on the tangential components of **E** and **H**, which are summarized in Sec. 8.9, can be satisfied. We can now define a *reflection coefficient* Γ as the ratio of reflected to incident amplitude; that is, $\Gamma = E_0^r/E_0^i$.

14.2 REFLECTION OF PLANE WAVES FROM PERFECT CONDUCTOR—NORMAL INCIDENCE

Let us assume the conducting medium occupies the half space $z > 0$, as shown in Fig. 14.1. A plane wave incident from free space onto the boundary $z = 0$ will be reflected. Thus, in the free-space region $z < 0$, we have waves traveling in both directions and the total field there is

$$E_x = E_0^i e^{j(\omega t - \beta z)} + E_0^r e^{j(\omega t + \beta z)} \tag{14.2}$$

According to (8.92), the total tangential electric field at the surface of a conductor must be zero; that is, for all values of time t,

$$E_x \Big|_{z=0} = 0 = (E_0^i + E_0^r)e^{j\omega t} \tag{14.3}$$

which implies that at the boundary the amplitude of the reflected wave E_0^r is equal and opposite to that of the incident wave, or $E_0^r = -E_0^i$. The reflection coefficient is therefore $\Gamma = E_0^r/E_0^i = -1$ which gives for the total field in the free-space region

$$E_x = E_0^i(e^{-j\beta z} - e^{j\beta z})e^{j\omega t}$$

$$\boxed{= -2jE_0^i \sin \beta z e^{j\omega t}} \tag{14.4}$$

where $e^{\pm jx} = \cos x \pm j \sin x$ was used. The instantaneous value of the total field is the real part of (14.4), or

$$\boxed{E_{x,\text{inst}} = 2E_0^i \sin \beta z \sin \omega t} \tag{14.5}$$

and is plotted in Fig. 14.1*a* and *d*. This shows that for complete reflection ($E_0^r = -E_0^i$), the incident and reflected wave combine to produce a pure standing wave. It is not a traveling wave because neither (14.4) nor (14.5) can be written as a function which has an argument ($\omega t \pm \beta z$), characteristic of a traveling wave. The total electric field varies sinusoidally in the direction normal to the conducting surface. It always has a zero (called a node) at the surface and at distances $z = -n\pi/\beta = -n\lambda/2$, where $n = 0, 1, 2, \ldots$. At these distances the incident and reflected waves are 180° out of phase with each other at all times.† On the other

† At the boundary and at all nodes the electric field of the incident wave is generally not zero; its variation with time is given by the real part of (14.2) as $E_0^i \cos \omega t$. Similarly for the reflected wave. It is only the sum of incident and reflected fields which add to zero at the boundary and all nodes.

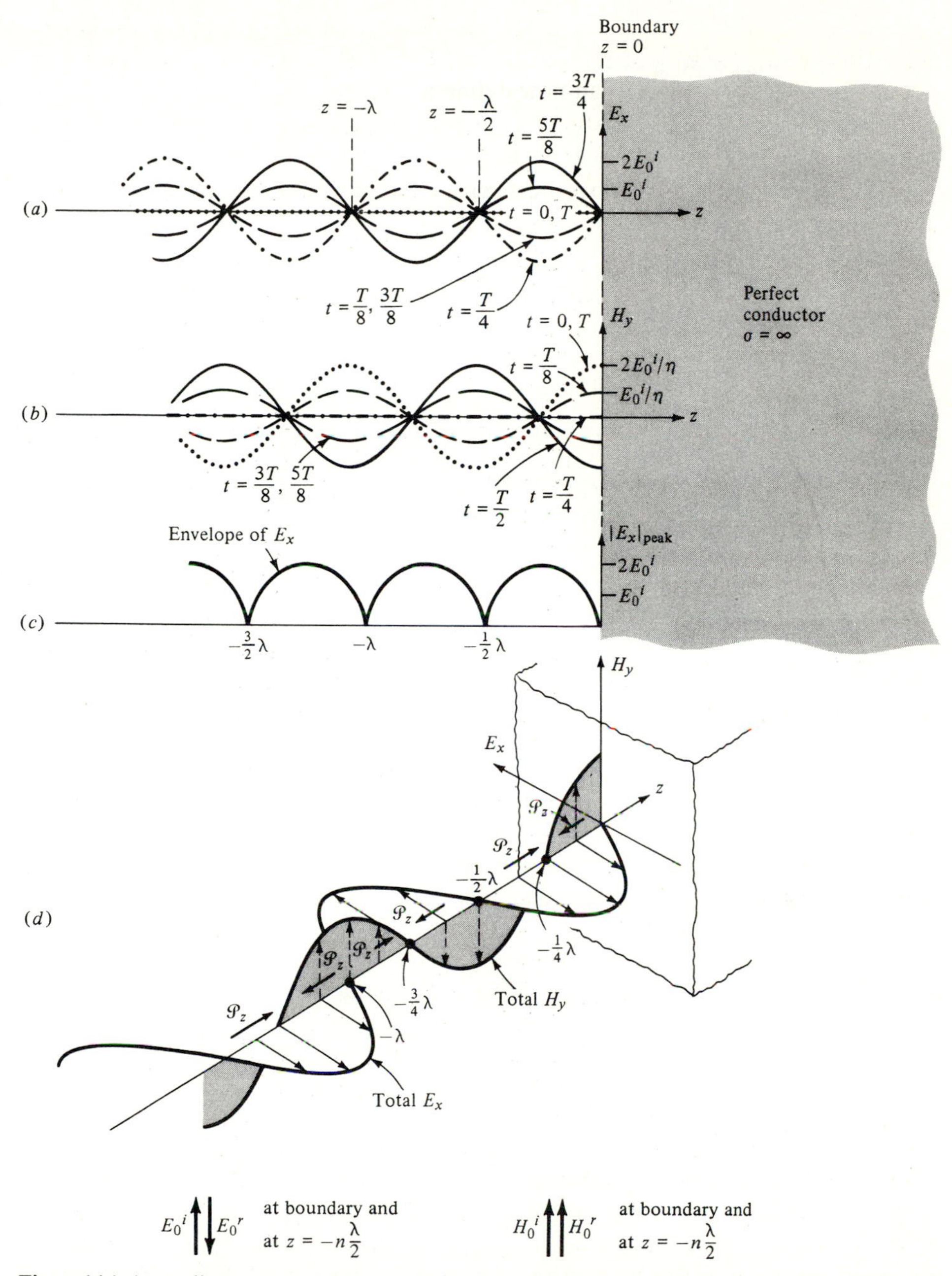

Figure 14.1 A standing wave pattern created when two oppositely traveling waves interfere. (*a*) Total electric field (14.5) is shown. (*b*) Total magnetic field (14.8) is shown. (*c*) The usual representation of a standing wave is by the absolute peak value of the electric field (called an envelope); an effective-reading field strength meter would record such a variation when moved along the z axis. (*d*) The fields in a standing wave at time $t = T/8$. The energy [represented by P_z (14.10)] surges back and forth along the z axis indicating that net energy flow along this axis is zero.

hand, at distances $z = -(2n + 1)\lambda/4$, the incident and reflected fields are in phase and produce a maximum in the total electric field. It is important to realize that the positions of zero field, maximum field, as well as any others do not travel in the z direction but are stationary with respect to z. However, all field points on a standing wave bob up and down according to sin ωt, as shown in Fig. 14.1*a*.

The total magnetic field in the free-space region $z < 0$ is similarly composed of an incident and a reflected component:

$$H_y = H_0^i e^{j(\omega t - \beta z)} + H_0^r e^{j(\omega t + \beta z)}$$

$$= \frac{E_0^i}{\eta} e^{j(\omega t - \beta z)} - \frac{E_0^r}{\eta} e^{j(\omega t + \beta z)} \tag{14.6}$$

where the relationship between the magnetic and electric field in an incident and reflected wave is given by (13.37); that is, $E_x^i = \eta H_y^i$ and $E_x^r = -\eta H_y^r$. Substituting $E_0^r = -E_0^i$ in the above equation, we obtain

$$H_y = \frac{E_0^i}{\eta}(e^{-j\beta z} + e^{j\beta z})e^{j\omega t}$$

$$\boxed{= \frac{2E_0^i}{\eta} \cos \beta z e^{j\omega t}} \tag{14.7}$$

The instantaneous value of the total magnetic field is then

$$\boxed{H_{y,\text{inst}} = \frac{2E_0^i}{\eta} \cos \beta z \cos \omega t} \tag{14.8}$$

and is plotted in Fig. 14.1*b* and *d*. Magnetic field is seen to have a maximum at the conductor surface and at the distances $z = -n\pi/\beta = -n\lambda/2$, which are the same distances at which the total electric field is zero. Similarly, zeros of magnetic field and maxima of electric field occur at $z = -(2n + 1)\pi/2\beta = -(2n + 1)\lambda/4$. At the surface of the conductor the reflected magnetic field (unlike the reflected electric field) is in the same direction as the incident field, which makes the total magnetic field at the surface $H_y = 2H_0^i$.

Poynting's Vector for a Standing Wave

It is obvious that a pure standing wave cannot transport energy. The incident wave carries energy in one direction, and the reflected wave carries it in exactly the opposite direction. The net energy transport for perfect reflection is therefore zero. This can be seen from (13.72) which gives

$$\mathscr{P}_{\text{ave}} = \tfrac{1}{2} \text{Re}\,(\mathbf{E} \times \mathbf{H}^*) = \tfrac{1}{2} \text{Re}\,(-2jE_0^i \sin \beta z e^{j\omega t})\left(\frac{2E_0^i}{\eta} \cos \beta z e^{-j\omega t}\right)\hat{\mathbf{z}} = 0 \tag{14.9}$$

because $\mathbf{E} \times \mathbf{H}^*$ for a standing wave is imaginary. Even though total magnetic and electric fields for the standing wave are still mutually perpendicular in space, as shown in Fig. 14.1*d*, the electric field is multiplied by $j(=e^{j\pi/2})$, denoting that the fields are now in time quadrature. The instantaneous expressions (14.5) and (14.8) display time quadrature explicitly because $E_x \propto \sin \omega t$ and $H_y \propto \cos \omega t$. The instantaneous Poynting's vector is given by

$$\mathscr{P}_z(t) = E_x H_y = 4\sqrt{\frac{\varepsilon}{\mu}}(E_0^i)^2 \sin \beta z \cos \beta z \sin \omega t \cos \omega t$$

$$= \sqrt{\frac{\varepsilon}{\mu}}(E_0^i)^2 \sin 2\beta z \sin 2\omega t \qquad \text{W/m}^2 \tag{14.10}$$

which averages to zero, because the average of a sinusoid is zero. Figure 14.1*d* shows the flow of $\mathscr{P}_z(t)$ for the time $t = T/8$. At this time and at distance $z = -\lambda/8$, $\mathscr{P}_z$ as given by (14.10) reaches the maximum value of $\sqrt{\varepsilon/\mu}\,(E_0^i)^2$. The power flow surges back and forth along the z axis, denoting that half the time power flows from source to conductor, and the other half, power is returned from conductor to source.

Referring to Fig. 14.1, we observe that at two times during each cycle, all the energy is in the magnetic field ($z = 0, -\lambda/2, \ldots$), 90° later, all the energy is stored in the electric field ($z = -\lambda/4, -3\lambda/4, \ldots$). It is instructive to examine the instantaneous energy densities separately in a pure standing wave ($E_0^r = -E_0^i$). The electric energy density w_E from (5.46) is

$$w_E = \tfrac{1}{2}\varepsilon E_x^2 = 2\varepsilon(E_0^i)^2 \sin^2 \beta z \sin^2 \omega t \qquad \text{J/m}^3 \tag{14.11}$$

and the magnetic energy density w_M from (7.45) is

$$w_M = \tfrac{1}{2}\mu H_y^2 = 2\varepsilon(E_0^i)^2 \cos^2 \beta z \cos^2 \omega t \qquad \text{J/m}^3 \tag{14.12}$$

These energy densities are plotted in Fig. 14.2 for three times, $t = 0$, $T/8$, and $T/4$, corresponding to those in Fig. 14.1. The energy thus oscillates back and forth, being completely in the magnetic field at $t = 0$, completely in the electric field a quarter of a cycle (spatial $\lambda/4$ or temporal $T/4$) later, and so forth. In between these times the energy is flowing either from the magnetic field to the electric field or vice versa.

The average energy velocity of a standing wave is zero, denoting that no energy is transported. Thus from (13.79) and (14.9) we have

$$v_e = \frac{P_{\text{ave}}}{\text{average energy density}} = \frac{0}{\varepsilon(E_0^i)^2} = 0 \tag{14.13}$$

The instantaneous energy velocity, on the other hand, is given by

$$v_{e,\text{inst}} = \frac{P_{\text{inst}}}{\omega_E + \omega_M} = \frac{1}{\sqrt{\mu\varepsilon}}\,\frac{\sin 2\beta z \sin 2\omega t}{1 + \cos 2\beta z \cos 2\omega t} \tag{14.13a}$$

and is seen to fluctuate between negative and positive values, which can be related to flow from the magnetic to electric energy and vice versa.

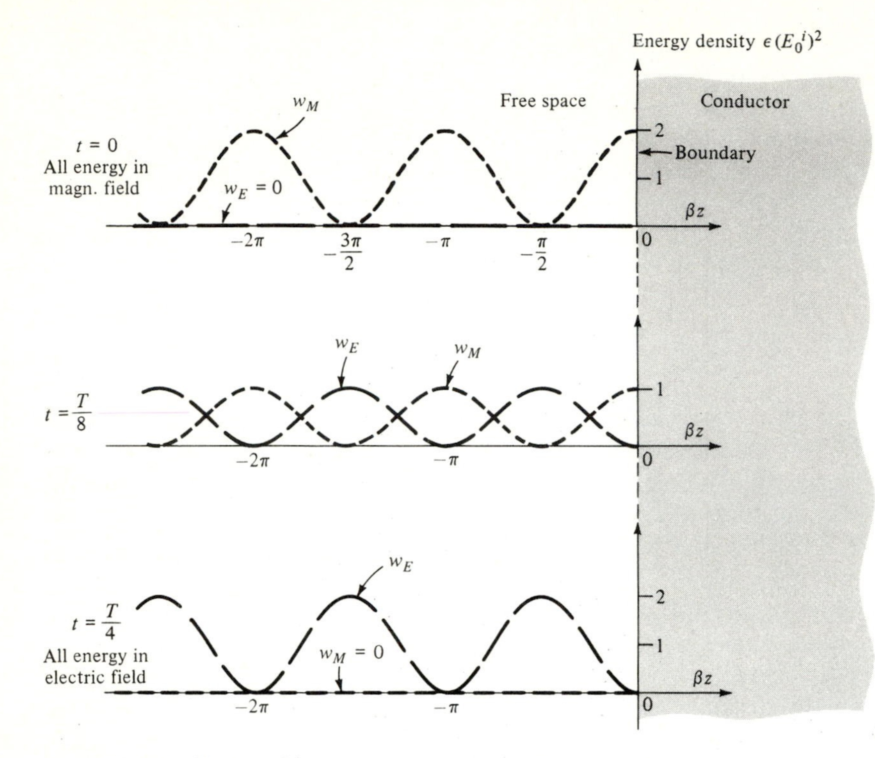

Figure 14.2 Oscillation of the total energy between the magnetic and electric fields in a standing wave. The same pictures are obtained in the time domain, simply by changing the horizontal axis from βz to ωt.

The relationship between the incident Poynting's vector $\mathscr{P}_z^i$ and reflected Poynting's vector $\mathscr{P}_z^r$ at the conducting surface is shown in Fig. 14.3. Since we know that energy cannot penetrate to the inside of a perfect conductor, the energy carried by the incident wave $\mathscr{P}_z^i$ must be reflected and carried away by $\mathscr{P}_z^r$ in the opposite direction. Thus, once it is established (from boundary conditions) that the reflected electric field amplitude at the surface is equal and opposite to the incident amplitude ($E_0^r = -E_0^i$), the reflected Poynting's vector can then be used to determine that the reflected magnetic amplitude at the surface is equal and in the same direction as the incident amplitude; that is, $H_0^r = H_0^i$. Recall that $\mathscr{P} = \mathbf{E} \times \mathbf{H}$, which implies that vectors **E**, **H**, and $\mathscr{P}$ form a positive triad (corresponding to the unit vectors $\hat{\mathbf{x}}$, $\hat{\mathbf{y}}$, $\hat{\mathbf{z}}$ in an x-y-z coordinate system). Therefore, knowing the direction of two legs ($\mathscr{P}_z^r$ and E^r) in the reflected wave triad determines the direction of the third leg H^r.

Power Loss in the Surface of a Conductor

The average Poynting's vector at the surface of a conductor is given by use of (14.7) and (13.55) as

$$\mathscr{P}_{z,\text{ave}} = \frac{1}{2}\text{Re } E_x H_y^* = \frac{1}{2}\text{Re } \eta^\star\,|H_y|^2 = \frac{1}{2}\text{Re }(1+j)\sqrt{\frac{\mu\omega}{2\sigma}}\left|\frac{2E_0^i}{\eta}\right|^2$$

$$= \sqrt{\frac{2\omega}{\sigma\mu}}\,\varepsilon(E_0^i)^2 = \sqrt{\frac{2\omega}{\sigma\mu}}\,\mu(H_0^i)^2 \tag{14.14a}$$

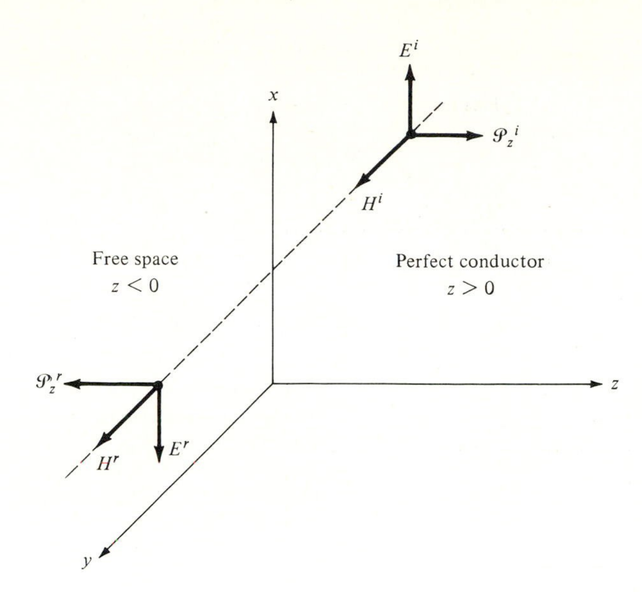

Figure 14.3 Relationship between incident Poynting's vector $\mathcal{P}_z^i$ and reflected $\mathcal{P}_z^r$ at a perfectly conducting surface (xy plane). The incident wave travels in the positive z direction.

Equation (14.14*a*) gives the total power which flows from the field into the conductor and is dissipated in the form of heat. An alternative expression for the power loss in terms of surface current $\mathbf{K} = \hat{\mathbf{n}} \times \mathbf{H}$ [Eq. (8.93)] and intrinsic resistivity of the conductor $R_s = \sqrt{\omega\mu/2\sigma}$ can be given as

$$\boxed{\mathcal{P}_{\text{ave}} = \tfrac{1}{2} R_s |\mathrm{K}|^2 \qquad \text{W/m}^2} \tag{14.14b}$$

which is in the familiar RI^2 form. It is useful for the calculation of power loss in the walls of transmission lines, wave guides, cavities, etc. Note that the pressure on the conductor is $p = \mathcal{P}_{\text{ave}}/v = \varepsilon(E_0^i)^2 = \mu(H_0^i)^2$, where the velocity in the conducting medium is given by (13.57) as $v = \sqrt{2\omega/\sigma\mu}$. This is in agreement with the expression for pressure developed in Eqs. (13.90) to (13.92).

Example: A plane wave is incident normally on a large sheet of copper. If the frequency and peak electric field strength E_0^i of the incident wave is 100 MHz and 1 V/m, respectively, find the power absorbed per unit area by the copper sheet.

For copper, $\sigma = 5.8 \times 10^7$ S/m, $\mu = \mu_0$, and $\varepsilon = \varepsilon_0$. Power absorbed in the sheet is

$$\mathcal{P}_{\text{ave}} = \sqrt{\frac{(2)(2\pi \times 10^8)}{(5.8 \times 10^7)(4\pi \times 10^{-7})}}\, 8.85 \times 10^{-12}(1)^2 = 3.67 \times 10^{-8} \qquad \text{W/m}^2$$

For comparison, the incident power density is $(E_0^i)^2/2\eta = 1/(2)(377) = 1.33 \cdot 10^{-3}$ W/m^2.

14.3 PLANE WAVE PROPAGATING IN ANY DIRECTION

Up to this point we have talked about plane waves that travel parallel to the z axis, such as (14.1). Now we need to express a plane wave traveling in any direction. Let the direction of propagation be given by the unit vector $\hat{\boldsymbol{\beta}}$, as shown in

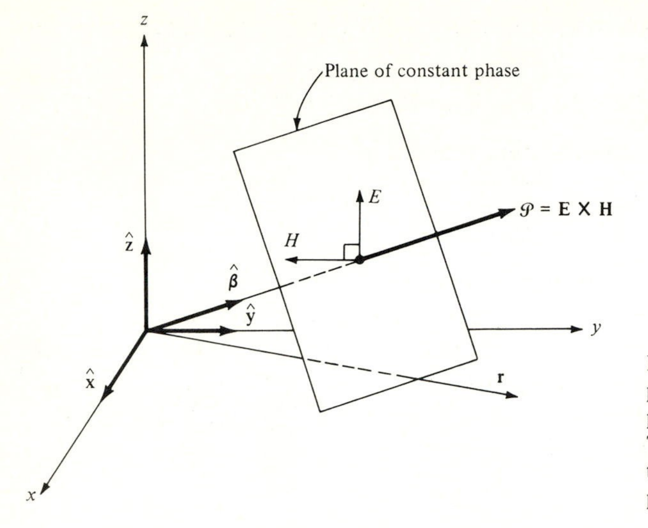

Figure 14.4 The plane of constant phase is shown for a plane wave propagating in the $\hat{\boldsymbol{\beta}}$ direction. The propagation vector $\boldsymbol{\beta} = \hat{\boldsymbol{\beta}}\beta$ thus gives both the direction and phase shift per unit length.

Fig. 14.4. Then the electric field of a linearly polarized plane wave is given by

$$\boxed{\mathbf{E} = \mathbf{E}_0 e^{j(\omega t - \boldsymbol{\beta} \cdot \mathbf{r})}} \tag{14.15}$$

where the polarization direction is that of the E field, namely, $\hat{\mathbf{E}}_0$, and the direction of propagation is given by the unit vector $\hat{\boldsymbol{\beta}}$ where $\boldsymbol{\beta} = \hat{\boldsymbol{\beta}}\beta$. The planes of constant phase are shown in Fig. 14.4 and are obtained by setting the phase in (14.15) equal to a constant for any time t; that is,

$$\boldsymbol{\beta} \cdot \mathbf{r} = \text{constant} \tag{14.16}$$

which is the equation of a plane in spherical coordinates whose normal is $\hat{\boldsymbol{\beta}}$. In the more familiar rectangular coordinate system, the equation for the equiphase planes becomes

$$\boldsymbol{\beta} \cdot \mathbf{r} = \beta_x x + \beta_y y + \beta_z z = \beta(\cos \theta_x x + \cos \theta_y y + \cos \theta_z z) = \text{constant} \tag{14.16a}$$

where the radial vector $\mathbf{r}$ is equal to $\mathbf{r} = x\hat{\mathbf{x}} + y\hat{\mathbf{y}} + z\hat{\mathbf{z}}$, $\boldsymbol{\beta} = \beta_x \hat{\mathbf{x}} + \beta_y \hat{\mathbf{y}} + \beta_z \hat{\mathbf{z}}$, and θ_x, θ_y, and θ_z are the angles vector $\boldsymbol{\beta}$ makes with the x, y, and z axes, respectively.

In (13.29) and (13.30) we showed that in a plane wave the $\mathbf{E}$ vector is perpendicular to the direction of propagation. We can show this in general by using one of the Maxwell's equations for free space, $\nabla \cdot \mathbf{E} = 0$. Since $\mathbf{E}_0$ is a constant vector, using (A1.18) gives

$$\nabla \cdot \mathbf{E} = \mathbf{E}_0 \cdot \nabla e^{-j\beta \hat{\boldsymbol{\beta}} \cdot \mathbf{r}} = (\mathbf{E}_0 \cdot \hat{\boldsymbol{\beta}}) j\beta e^{-j\beta(\hat{\boldsymbol{\beta}} \cdot \mathbf{r})} = 0 \tag{14.17}$$

The term in parentheses must be zero since the exponential term and the $j\beta$ term are not zero in general. Thus,

$$\boxed{\mathbf{E}_0 \cdot \hat{\boldsymbol{\beta}} = 0} \tag{14.17a}$$

which shows that in a plane wave the $\mathbf{E}$ vector is normal to the direction of propagation.

The magnetic field vector can be obtained by using Maxwell's curl equation (11.19), which gives

$$-j\omega\mu\mathbf{H} = \nabla \times \mathbf{E} = \nabla \times (\mathbf{E}_0 e^{-j\boldsymbol{\beta}\cdot \mathbf{z}})e^{j\omega t} \tag{14.18}$$

Using identity (A1.19), the curl equation reduces to

$$\boxed{\mathbf{H} = (\hat{\boldsymbol{\beta}} \times \mathbf{E})\sqrt{\frac{\varepsilon}{\mu}}} \tag{14.19}$$

which implies that **H** is perpendicular to both $\hat{\boldsymbol{\beta}}$ and **E**, as shown in Fig. 14.4. The above equation also implies that E and H are related by the wave impedance $\eta = \sqrt{\mu/\varepsilon}$. This had previously been concluded in connection with (13.37) but is developed here more rigorously. As shown in Fig. 14.4, Poynting's vector $\mathbf{E} \times \mathbf{H}$ coincides with the wave direction $\hat{\boldsymbol{\beta}}$.

14.4 REFLECTION BY A PERFECT CONDUCTOR—INCIDENCE AT ANY ANGLE

Arbitrary incidence on a perfect conductor can be conveniently split into two cases. One, in which the polarization is with the **E** field in the plane of incidence, and the other, in which the **E** field is normal to the plane of incidence. The general case can be considered a superposition of these two. The plane of incidence is the plane formed by the normal to the reflecting surface and the direction $\hat{\boldsymbol{\beta}}$ of incidence.

Case 1: Electric Field Parallel to Plane of Incidence

An incident wave, polarized with the **E** field in the plane of incidence, impinges on a perfectly conducting surface at an angle θ_i, as shown in Fig. 14.5. The direction of incidence is given by the unit vector $\hat{\boldsymbol{\beta}}_i$ which is equal to $\hat{\boldsymbol{\beta}}_i = \hat{\mathbf{E}} \times \hat{\mathbf{H}}$, as Poynting's vector gives the direction of propagation of a wave. For the direction of the **E** field chosen as shown, the **H** field is out of the paper; that is, $\mathbf{H} = H_y\hat{\mathbf{y}}$. From (14.19) we also find that $E^i/H_y^i = E^r/H_y^r = \eta$ where the superscripts i and r denote incident and reflected, respectively. The total electric field in the free-space region can now be written as

$$\mathbf{E}(x, z) = \mathbf{E}^i(x, z) + \mathbf{E}^r(x, z) = \mathbf{E}_0^i e^{j(\omega t - \boldsymbol{\beta}_i \cdot \mathbf{r})} + \mathbf{E}_0^r e^{j(\omega t - \boldsymbol{\beta}_r \cdot \mathbf{r})} \tag{14.20}$$

where the time dependence was included to show explicitly that the incident and reflected waves are traveling in the directions $\hat{\boldsymbol{\beta}}_i$ and $\hat{\boldsymbol{\beta}}_r$, respectively. The spatial phase factor has components along the x and z axis which are

$$\boldsymbol{\beta}_i \cdot \mathbf{r} = \beta(x \sin\theta_i + z \cos\theta_i)$$

$$\boldsymbol{\beta}_r \cdot \mathbf{r} = \beta(x \sin\theta_r - z \cos\theta_r) \tag{14.21}$$

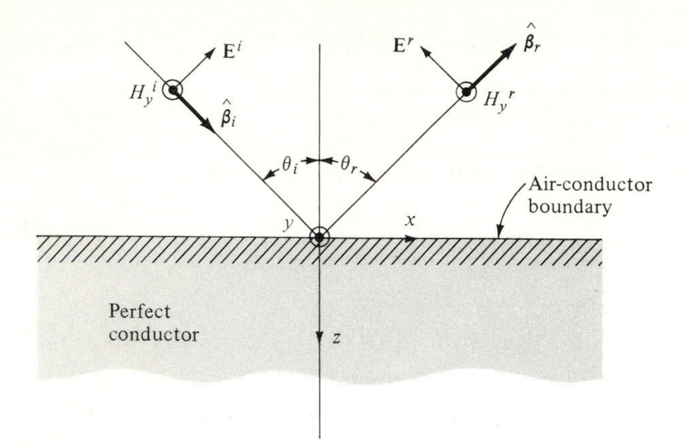

Figure 14.5 A plane wave, polarized with the **E** field in the plane of incidence, impinges on a perfect conductor. Half a cycle later ($T/2$ or $\lambda/2$) the directions of E and H would be reversed but with the direction of power flow ($\mathbf{E} \times \mathbf{H}$) remaining the same. A dot (cross) represents a direction out of (into) paper.

The total electric field is seen to have x and z components:

$$\mathbf{E}(x, z) = E_x(x, z)\hat{\mathbf{x}} + E_z(x, z)\hat{\mathbf{z}} \tag{14.22}$$

where

$$E_x = E_x^i + E_x^r = E_0^i \cos\theta_i e^{-j\boldsymbol{\beta}_i\cdot\mathbf{r}} - E_0^r \cos\theta_r e^{-j\boldsymbol{\beta}_r\cdot\mathbf{r}} \tag{14.23}$$

and

$$E_z = E_z^i + E_z^r = -E_0^i \sin\theta_i e^{-j\boldsymbol{\beta}_i\cdot\mathbf{r}} - E_0^r \sin\theta_r e^{-j\boldsymbol{\beta}_r\cdot\mathbf{r}} \tag{14.24}$$

where for convenience the common factor $e^{j\omega t}$ was omitted.

The total magnetic field is given by

$$\mathbf{H}(x, z) = H_y(x, z)\hat{\mathbf{y}} = [H_y^i(x, z) + H_y^r(x, z)]\hat{\mathbf{y}}$$
$$= [H_0^i e^{-j\boldsymbol{\beta}_i\cdot\mathbf{r}} + H_0^r e^{-j\boldsymbol{\beta}_r\cdot\mathbf{r}}]\hat{\mathbf{y}} \tag{14.25}$$

where H_0^i and H_0^r are the amplitudes of the incident and reflected magnetic fields, respectively.

The relationship between the incident and reflected amplitudes is provided by the boundary conditions, which for a perfect conductor, state that total tangential E field at the surface must be zero. Total field means the combination of incident and reflected fields. Thus, using (14.23),

$$E_x(x, 0) = E_0^i \cos\theta_i e^{-j\beta x \sin\theta_i} - E_0^r \cos\theta_r e^{-j\beta x \sin\theta_r} = 0 \tag{14.26}$$

In order for this equation to be satisfied for all x's, the phase terms, which are the exponents, must be equal to each other. Equating, we obtain that

$$\boxed{\theta_i = \theta_r} \tag{14.27}$$

The angle of reflection is thus seen to be equal to the angle of incidence (this is also known as Snell's law of reflection). Substituting (14.27) into (14.26), we obtain

$$\boxed{E_0^r = E_0^i} \tag{14.28}$$

With these results the total field in the free-space region can then be written as follows:

$$\mathbf{E} = E_x\hat{\mathbf{x}} + E_z\hat{\mathbf{z}} = 2E_0^i[-j\cos\theta_i \sin(\beta z\cos\theta_i)\hat{\mathbf{x}} - \sin\theta_i \cos(\beta z\cos\theta_i)\hat{\mathbf{z}}]\cdot e^{-j\beta x\sin\theta_i} \tag{14.29}$$

$$\mathbf{H} = H_y\hat{\mathbf{y}} = \hat{\mathbf{y}}\frac{2E_0^i}{\eta}\cos(\beta z\cos\theta_i)e^{-j\beta x\sin\theta_i} \tag{14.30}$$

Putting back the time factor $e^{j\omega t}$ in the above two equations, we observe that in the x direction, the total field acts like a traveling wave with a phase constant $\beta_x = \beta\sin\theta_i$, but in the z direction, the field acts like a standing wave. This is not unexpected since the incident and reflected waves travel in the same direction along the x axis but in opposite directions along the z axis. The nature of these waves is further clarified by considering power flow parallel and perpendicular to the surface.

Average power flow parallel to the conducting surface is given by the x component of the average Poynting's vector, which is

$$\mathscr{P}_{x,\text{ave}} = \frac{1}{2}\operatorname{Re} E_z H_y^* = \frac{2(E_0^i)^2}{\eta}\sin\theta_i\cos^2(\beta z\cos\theta_i) \tag{14.31}$$

Thus, for glancing incidence ($\theta_i \to 90°$), $\mathscr{P}_x = 2(E_0^i)^2/\eta$, and the power flow is maximum as expected. On the other hand, for normal incidence ($\theta_i = 0°$), power flow in the x direction is zero; that is, $\mathscr{P}_{x,\text{ave}} = 0$.

It is interesting to observe that the phase velocity in the x direction is given by

$$v_x = \frac{\omega}{\beta_x} = \frac{\omega}{\beta\sin\theta_i} = \frac{v}{\sin\theta_i} = \frac{1}{\sqrt{\mu\varepsilon}\sin\theta_i} \tag{14.32}$$

which is obtained from (14.29) or (14.30) by observing the motion of a plane of constant phase, that is, by letting $(\omega t - \beta x\sin\theta_i) = \text{constant}$ and differentiating this expression with respect to time t as outlined in (13.15). Thus, for glancing incidence the phase velocity v_x approaches the velocity v of a plane wave in free space; that is, $v = 1/\sqrt{\mu\varepsilon}$. But as normal incidence ($\theta_i \to 0°$) is approached, the phase velocity in the x direction goes to infinity; that is, $v_x \to \infty$. We observe that for all angles of incidence the phase velocity in the x direction is larger than the velocity of light $1/\sqrt{\mu\varepsilon}$, as shown in Fig. 14.6. No fundamental principles are violated by a phase velocity which is greater than the velocity of light, as the phase velocity in a given direction is merely the velocity of progression of a surface of constant phase. For example, as shown in Fig. 14.6, the phase velocity of the incident plane wave with direction $\boldsymbol{\beta}_i$ is at the velocity of light $v = 1/\sqrt{\mu\varepsilon}$. In the x direction, the phase velocity of the plane wave is greater than the velocity of light. The energy velocity in the x direction, however, is always less than the velocity of light; from (13.79) we obtain

$$v_{e,x} = \frac{\mathscr{P}_{x,\text{ave}}}{\frac{1}{2}(\frac{1}{2}\varepsilon|E_z^i|^2 + \frac{1}{2}\mu|H_y|^2)} = \frac{2\sin\theta_i}{\sqrt{\mu\varepsilon}(1 + \sin^2\theta_i)} \tag{14.33}$$

where (14.29), (14.30), and (14.31) were used.

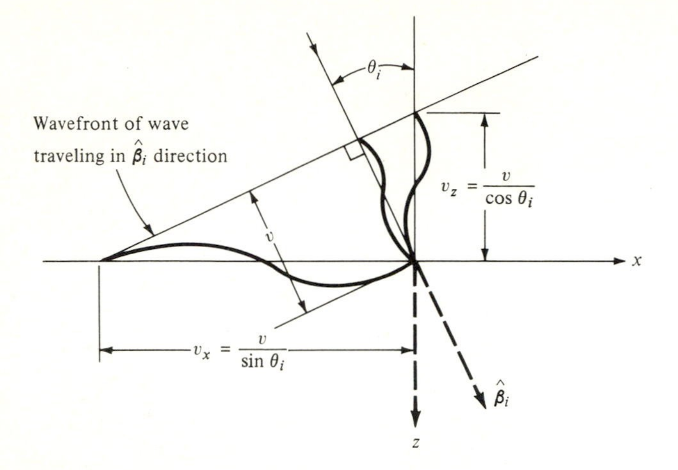

Figure 14.6 The phase velocity of a wave in a direction along its normal $\hat{\boldsymbol{\beta}}_i$ is v. Constant phase planes in the x and z directions advance with velocities $v/\sin\theta_i$ and $v/\cos\theta_i$, respectively.

Average power flow perpendicular to the conducting surface is zero, since the average Poynting's vector in that direction is zero; that is,

$$\boxed{\mathscr{P}_{z,\text{ave}} = \tfrac{1}{2}\,\text{Re}\,E_x H_y^* = 0} \tag{14.34}$$

because E_x and H_y are 90° out of time phase (E_x is multiplied by $j = e^{j\pi/2}$, but H_y is not). A standing-wave pattern is observed in the z direction.† The zeros (nodes) for the E_x field are given by $\sin(\beta z \cos\theta_i) = 0$; they occur at the conducting plane and at distances given by

$$\boxed{z = n\frac{\lambda}{2\cos\theta_i} \qquad n = 0, 1, 2, \ldots} \tag{14.35}$$

from the conducting plane. The zeros of the magnetic field H_y and of E_z are given by $\cos(\beta z \cos\theta_i) = 0$. The standing-wave pattern is similar to that shown in Fig. 14.1 except for the standing-wave zeros which occur at distances larger than multiples of $\lambda/2$. For normal incidence, when $\theta_i = 0$, the positions of the zeros are the same as those shown in Fig. 14.1.

Case 2: Electric Field Normal to Plane of Incidence

This polarization is depicted in Fig. 14.7. For this case the entire E field is in the y direction (out of the paper) and is given by $\mathbf{E} = E_y\hat{\mathbf{y}}$. Since it is easier to apply boundary conditions to the electric field, we will start with the electric field. In the

† The E_x component being parallel to the conducting surface acts like the total electric field for the case of normal incidence considered in Sec. 14.2.

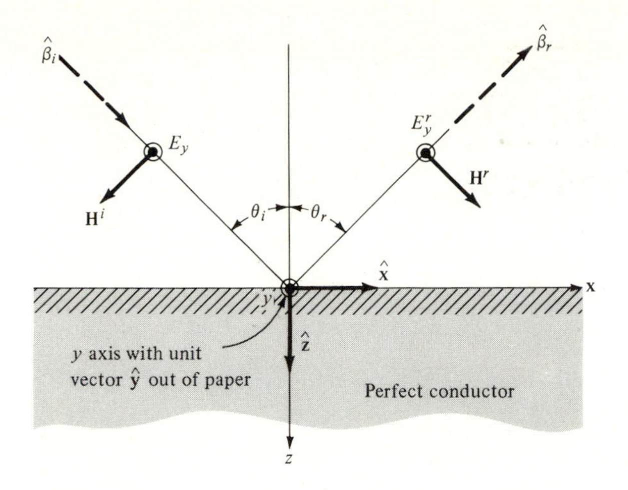

Figure 14.7 Plane wave polarized with the E field out of the page and incident at angle θ_i on a perfectly conducting surface.

free-space region ($z < 0$) the total E field is given by the combination of incident and reflected fields:

$$E_y(x, z) = E_0^i e^{-j\boldsymbol{\beta}_i \cdot \mathbf{r}} + E_0^r e^{-j\boldsymbol{\beta}_r \cdot \mathbf{r}} \tag{14.36}$$

where the phase terms are given by (14.21). The boundary condition on the E field at the conducting surface is $E_y(x, 0) = 0$, which implies that $E_0^r = -E_0^i$ and, as expected, $\theta_r = \theta_i$. The reflected E field is, therefore, opposite to that shown in Fig. 14.7 (it is into the paper). The total E field is then, from (14.36),

$$\boxed{E_y(x, z) = -2jE_0^i \sin(\beta z \cos\theta_i) e^{-j\beta x \sin\theta_i}} \tag{14.37}$$

It is expected that the reflected magnetic field in the figure must also be reversed since Poynting's vector $\mathbf{E}_0^r \times \mathbf{H}_0^r$ must be in the direction of the reflected wave which is given by the unit vector $\hat{\boldsymbol{\beta}}_r$.

To obtain the **H** field, we can use (14.19), which is $\eta\mathbf{H} = \hat{\boldsymbol{\beta}} \times \mathbf{E}$. The total **H** field in the free-space region by use of (14.36) and the fact that $E_0^r = -E_0^i$ is then

$$\eta\mathbf{H} = \eta(\mathbf{H}^i + \mathbf{H}^r) = \hat{\boldsymbol{\beta}}_i \times \hat{\mathbf{y}} E_0^i e^{-j\boldsymbol{\beta}_i \cdot \mathbf{r}} - \hat{\boldsymbol{\beta}}_r \times \hat{\mathbf{y}} E_0^i e^{-j\boldsymbol{\beta}_r \cdot \mathbf{r}} \tag{14.38}$$

where, from Fig. 14.7, the direction vectors of the incident and reflected waves are

$$\hat{\boldsymbol{\beta}}_{i,r} = \hat{\mathbf{x}} \sin\theta_i \pm \hat{\mathbf{z}} \cos\theta_i \tag{14.39}$$

and

$$\hat{\boldsymbol{\beta}}_{i,r} \times \hat{\mathbf{y}} = \hat{\mathbf{z}} \sin\theta_i \mp \hat{\mathbf{y}} \cos\theta_i \tag{14.40}$$

where $\hat{\mathbf{x}}, \hat{\mathbf{y}}, \hat{\mathbf{z}}$ are the unit vectors along the x, y, z axes. The components of the total magnetic field are therefore

$$\boxed{\eta H_x = -2E_0^i \cos\theta_i \cos(\beta z \cos\theta_i) e^{-j\beta x \sin\theta_i}} \tag{14.41}$$

$$\boxed{\eta H_z = -2jE_0^i \sin\theta_i \sin(\beta z \cos\theta_i) e^{-j\beta x \sin\theta_i}} \tag{14.42}$$

Again, as expected, we have a standing-wave distribution in the z direction, because the reflected and incident waves travel in opposite directions along the z axis. Zeros of the tangential E field (E_y), zeros of the normal H field (H_z), and maxima of H_x occur at the conducting plane and at parallel planes which are spaced at distances nz, where z is given by (14.35) as $z = \lambda/2 \cos \theta_i$. In the x direction both the incident and reflected waves progress to the right with the same velocity $v_x = \omega/\beta_x = \omega/\beta \sin \theta_i = v/\sin \theta_i$. Hence, in this direction the wave behaves as does a traveling wave.

Example: Find the induced surface current **K** in the perfectly conducting surface for the geometry shown in Fig. 14.7.

The sheet current K (in amperes per meter) is determined by the total tangential magnetic field at the surface. From boundary condition (8.93) we have

$$\mathbf{K} = \hat{\mathbf{n}} \times \mathbf{H} = \hat{\mathbf{n}} \times (\mathbf{H}^i + \mathbf{H}^r) \tag{14.43}$$

where the normal $\hat{\mathbf{n}}$ to the surface for the geometry of Fig. 14.7 is $\hat{\mathbf{n}} = -\hat{\mathbf{z}}$. By symmetry considerations or by direct calculation we can show that (see also Fig. 14.3 and Prob. 14.3)

$$\hat{\mathbf{n}} \times \mathbf{H}^i = \hat{\mathbf{n}} \times \mathbf{H}^r \tag{14.44}$$

Equation (14.43) then becomes

$$\mathbf{K} = 2(\hat{\mathbf{n}} \times \mathbf{H}^i) = 2\left(\frac{\varepsilon}{\mu}\right)^{1/2} [\hat{\mathbf{n}} \times (\hat{\boldsymbol{\beta}}_i \times \mathbf{E}^i)] \tag{14.45}$$

or alternatively

$$\mathbf{K} = 2(\hat{\mathbf{n}} \times \mathbf{H}^r) = 2\left(\frac{\varepsilon}{\mu}\right)^{1/2} [\hat{\mathbf{n}} \times (\hat{\boldsymbol{\beta}}_r \times \mathbf{E}^r)] \tag{14.46}$$

where (14.19) was used to express the magnetic field in terms of the electric field.

For the geometry of Fig. 14.7 and by use of Eqs. (14.38) to (14.40) we have

$$\mathbf{K} = 2(-\hat{\mathbf{z}} \times \mathbf{H}^i)\Big|_{z=0} = \hat{\mathbf{y}} \frac{2E_0^i}{\eta} \cos \theta_i e^{-j\beta x \sin \theta_i} \tag{14.47}$$

The instantaneous value of the sheet current **K** can be obtained by multiplying by $e^{j\omega t}$ and taking the real part. Carrying out this operation, we obtain

$$K_y = \frac{2E_0^i \cos \theta_i}{\eta} \cos (\omega t - \beta x \sin \theta_i) \tag{14.48}$$

For normal incidence $\theta_i = 0$; the instantaneous induced surface current density is

$$K_y = \frac{2E_0^i}{\eta} \cos \omega t \tag{14.49}$$

which agrees with (14.8).

Note that (14.47) could have also been obtained by using in (14.43) the total magnetic field **H** as given by (14.41) and (14.42).

14.5 REFLECTION BY A DIELECTRIC—NORMAL INCIDENCE

A plane electromagnetic wave that is incident normally on a dielectric medium will be partially reflected from and partially transmitted into the dielectric. The reflected and transmitted waves are shown in Fig. 14.8. In region 1 the total

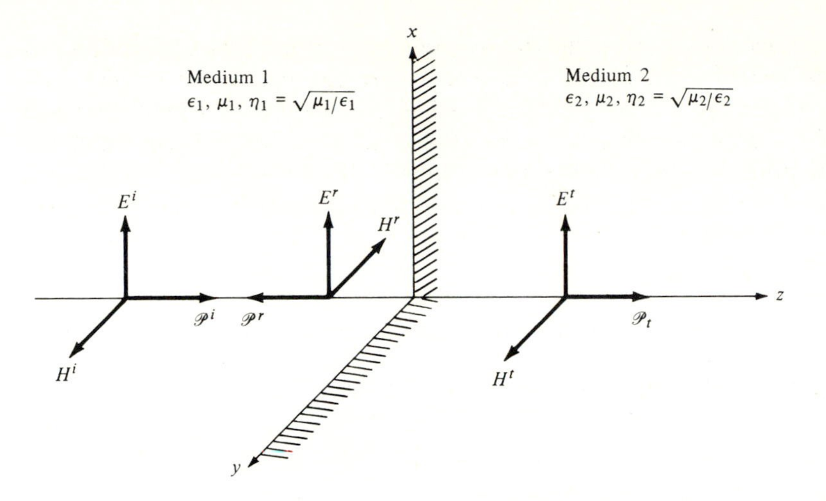

Figure 14.8 The boundary between two dielectric media is the xy plane. Medium 1 contains the incident and reflected waves, medium 2 the transmitted wave.

electric field is composed of the incident and reflected fields, or

$$
\begin{aligned}
E_x(z) &= E_0^i e^{-j\beta z} + E_0^r e^{j\beta z} \qquad z \le 0 \\
&= E_0^i e^{-j\beta z}\left(1 + \frac{E_0^r}{E_0^i} e^{j2\beta z}\right) \\
&= E_0^i e^{-j\beta z}[1 + \Gamma(z)]
\end{aligned}
\tag{14.50}
$$

where for convenience the $e^{j\omega t}$ factor has been omitted. We see that in the region $z \le 0$, the total field can be represented by the incident field multiplied by a factor $1 + \Gamma(z)$, where $\Gamma(z)$ is a reflection coefficient anywhere along the negative z axis defined by

$$
\boxed{\Gamma(z) = \frac{E_x^r(z)}{E_x^i(z)} = \frac{E_0^r}{E_0^i} e^{j2\beta z} = \Gamma e^{j2\beta z}}
\tag{14.51}
$$

where Γ is the reflection coefficient evaluated at the boundary:

$$
\boxed{\Gamma = \Gamma(0) = \frac{E_0^r}{E_0^i}}
\tag{14.52}
$$

Γ thus gives the relationship between the reflected and incident amplitudes at the boundary. Γ may be complex since a phase shift can be introduced by the reflection process. In the case when medium 2 is a perfect conductor, we have seen from (14.3) that $\Gamma = -1$, which implies that at the boundary a phase shift of 180° exists between the incident and the reflected electric field.

The discontinuity which the incident wave experiences arises from the fact that the wave initially travels in a medium with impedance $\eta_1 = \sqrt{\mu_1/\varepsilon_1}$ and then strikes a medium with impedance $\eta_2 = \sqrt{\mu_2/\varepsilon_2}$. The relationship between the electric and magnetic fields in the respective waves is given by (14.19) as $E_x^i = \eta_1 H_y^i$, $E_x^r = -\eta_1 H_y^r$, and $E_x^t = \eta_2 H_y^t$, where the unit vector $\hat{\boldsymbol{\beta}}$ coincides with the direction of power flow as given by Poynting's vector $\mathscr{P} = \mathbf{E} \times \mathbf{H}$. We can now find the reflection coefficient Γ and transmission coefficient τ from the boundary conditions. From (8.84) and (8.85), the tangential components of E and H must be continuous across the boundary ($z = 0$), which requires that

$$E_0^i + E_0^r = E_0^t$$

$$H_0^i + H_0^r = H_0^t \tag{14.53}$$

Dividing the top equation by E_0^i and the bottom equation by H_0^i, we obtain, respectively,

$$\boxed{1 + \Gamma = \tau} \tag{14.54}$$

$$\boxed{1 - \Gamma = \frac{\eta_1}{\eta_2}\tau} \tag{14.55}$$

where the transmission coefficient τ is defined as $\tau = E_0^t/E_0^i$. Solving these two simultaneous equations, we obtain

$$\boxed{\Gamma = \frac{\eta_2 - \eta_1}{\eta_2 + \eta_1}} \tag{14.56}$$

and the transmission coefficient

$$\boxed{\tau = \frac{2\eta_2}{\eta_2 + \eta_1}} \tag{14.57}$$

Thus, we see that no reflection will occur when the impedances of the media are the same; that is, $\eta_2 = \eta_1$. We can then speak of a matched condition. For example, for certain ferromagnetic materials the ratio μ_2 to ε_2 can be made the same as that of free space. Then, even though the ferromagnetic material is physically discontinuous, an incident wave will enter such a material without reflection; that is, $\tau = 1$ because $\Gamma = 0$.

A common situation, especially at high frequencies, is that the permeabilities of dielectrics do not differ much from free space, so that $\mu_2 \cong \mu_1 \cong \mu_0$. The expression for the reflection coefficient then simplifies to

$$\Gamma = \frac{\sqrt{\varepsilon_1} - \sqrt{\varepsilon_2}}{\sqrt{\varepsilon_1} + \sqrt{\varepsilon_2}} \tag{14.58}$$

This expression can be generalized to make it applicable to conductive or lossy media. Using the generalized complex permittivity $\varepsilon^\star$ which was introduced in (11.23) or (13.23) as $\varepsilon^\star = \varepsilon - j\sigma/\omega$, the reflection coefficient can be written as

$$\Gamma = \frac{\sqrt{\varepsilon_1^\star} - \sqrt{\varepsilon_2^\star}}{\sqrt{\varepsilon_1^\star} + \sqrt{\varepsilon_2^\star}} \tag{14.58a}$$

Example: Find the reflection coefficient for a 1-MHz electromagnetic wave in air which is incident normally on a copper medium in the form of a large sheet.

For medium 1, which is air, we have $\sigma_1 = 0$ and $\varepsilon_1 = \varepsilon_0 = 8.85$ pF/m. For medium 2, which is copper, we have $\sigma_2 = 5.8 \times 10^7$ S/m, $\varepsilon_2 = \varepsilon_0$, and $\mu_2 = \mu_0 = 4\pi \times 10^{-7}$ H/m.

For copper at this frequency, $\sigma/\omega\varepsilon \gg 1$. Hence, $\varepsilon^\star \cong -j\sigma/\omega$. The reflection coefficient Γ can then be written as

$$\Gamma = \frac{\sqrt{\varepsilon_0} - \sqrt{\varepsilon_2^\star}}{\sqrt{\varepsilon_0} + \sqrt{\varepsilon_2^\star}} = \frac{\sqrt{j\varepsilon_0\omega/\sigma} - 1}{\sqrt{j\varepsilon_0\omega/\sigma} + 1} \cong -0.9999996$$

where $\sqrt{j\varepsilon_0\omega/\sigma} = (1 + j)\sqrt{\varepsilon_0\omega/2\sigma}$

$$= (1 + j)\sqrt{(8.85 \times 10^{-12})(2\pi \times 10^6)/(2)(5.8 \times 10^7)} \cong (1 + j)2.19 \times 10^{-7}.$$

We therefore see that the reflection coefficient for copper differs negligibly from that for a perfect conductor, for which Γ is -1. For most practical purposes we can consider copper to be a perfect reflector at this frequency.

Because $\Gamma \cong -1$, very little power flows into the metal. This can also be seen from the ratio of E to H, which for a metal is $E/H = \eta^\star \cong \sqrt{\mu_0/(-j\sigma/\omega)} = \sqrt{(\mu_0/\varepsilon_0)(j\omega\varepsilon_0/\sigma)}$. Hence, the ratio of E to H is much smaller than the free-space value $\sqrt{\mu_0/\varepsilon_0} \cong 120\pi\ \Omega$. Because of the smallness of this ratio, we can almost neglect E within the metal. Poynting's vector at the surface is, therefore, very small, which in turn means that only a small amount of energy flows into the metal.

Example: Find the energy relationship at the boundary between two dielectrics.

Multiplying corresponding sides of (14.54) and (14.55), we obtain

$$\boxed{1 = \Gamma^2 + \frac{\eta_1}{\eta_2}\tau^2} \tag{14.59}$$

which expresses conservation of energy per unit time at the boundary. Thus, for unit incident power on the boundary, the reflected power is equal to the reflection coefficient Γ squared, and the transmitted power is equal to the third term involving the transmission coefficient τ.

Relationship between Reflection Coefficient Γ and Standing-Wave Ratio (SWR)

In Sec. 14.2 it was shown that a totally reflected wave, combines with the incident wave to form a standing wave. No traveling-wave component remains where there is perfect reflection; hence, no energy can be transported, and Poynting's vector is zero. For partial reflection, on the other hand, the reflected wave is smaller in amplitude than the incident wave. The reflected wave now combines with an equally strong part of the incident wave to form a standing wave. The remainder of the incident wave is a traveling wave which carries energy, has a finite Poynting's vector, and continues into medium 2. The total field in region 1, therefore, consists of a traveling-wave part and a standing-wave part. This can be shown by rearranging (14.50) as

$$\begin{aligned} E_x(z) &= E_0^i e^{-j\beta z} + E_0^r e^{j\beta z} \\ &= (1+\Gamma)E_0^i e^{-j\beta z} + \Gamma(2jE_0^i \sin \beta z) \\ &= (1+\Gamma)(\text{traveling wave}) + \Gamma(\text{standing wave}) \end{aligned} \tag{14.60}$$

where $1 + \Gamma = \tau$. Thus, when there is no reflection, $\Gamma = 0$, the transmission coefficient becomes $\tau = 1$, and only the traveling wave remains in regions 1 and 2. When region 2 is a perfect conductor, $\Gamma = -1$, the transmission coefficient is $\tau = 0$, and only the standing wave remains in region 1. This corresponds to a case covered previously in Eq. (14.4). For cases between no reflection and perfect reflection, the reflection coefficient Γ gives the strength of the standing wave present in region 1, and $1 + \Gamma$ gives the strength of the traveling wave in region 1 which continues into region 2. Thus, besides the usual interpretation for reflection coefficient Γ and transmission coefficient τ, one can use Γ as a measure of the amplitude of a standing wave in region 1, and τ as a measure of a traveling wave in regions 1 and 2.

For graphical representation the instantaneous values of expression (14.60) are needed. Multiplying by $e^{j\omega t}$ and taking the real part, we obtain the instantaneous total field in region 1 as

$$E_{x,\text{inst}} = \tau E_0^i \cos(\omega t - \beta z) - \Gamma 2E_0^i \sin \beta z \sin \omega t \tag{14.61}$$

For the case when $\Gamma = -1$, the above equation is equal to that of (14.5).

Figure 14.9 shows the standing-wave envelopes of (14.61) for three values of reflection coefficient.† The envelope gives the maximum values of the time oscilla-

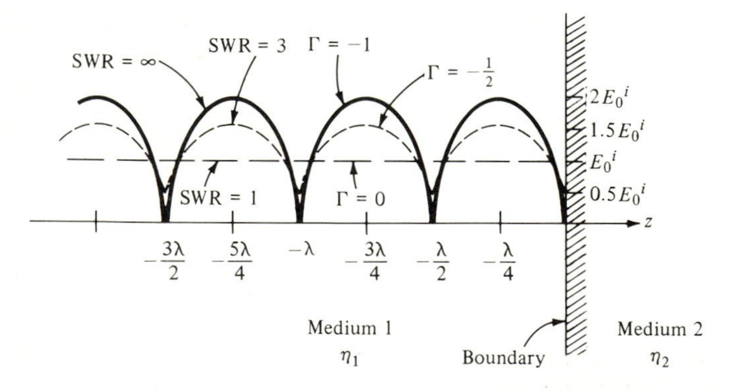

Figure 14.9 Standing-wave envelopes for the matched case ($\Gamma = 0$ when $\eta_2 = \eta_1$), for reflection at the boundary of two dielectrics ($\Gamma = -\frac{1}{2}$ when $\eta_1 = 3\eta_2$), and for reflection at a perfect conductor ($\Gamma = -1$ when $\eta_2 = 0$).

† Envelope of $E_x = |E_x|_{\text{peak}} = E_0^i \sqrt{(1+\Gamma)^2 \cos^2 \beta z + (1-\Gamma)^2 \sin^2 \beta z}$. Note that the envelope is not a sine curve; the maxima are broad with relatively sharp minima.

tions of the total field. Figure 14.1 shows in detail the time variations of the standing wave for $\Gamma = -1$. The reason why standing-wave envelopes are of interest is that field-strength meters, when moved along the z direction, would record such a variation. For example, on transmission lines the standing-wave pattern is measured by inserting a movable probe in a narrow slot cut along the transmission line. The probe is connected to a crystal which rectifies the induced current in the probe; the resultant current is then read by a dc meter.

Ordinarily, we do not need to know the exact shape of the standing-wave pattern. It is sufficient to know the maximum and minimum values. We characterize a standing wave by a factor called the *standing-wave ratio* (SWR), which is obtained by dividing the maximum value of the standing-wave envelope by the minimum value; that is, $\text{SWR} = E_{max}/E_{min}$. Since the maximum value is the result of constructive interference between the incident and reflected waves and the minimum the result of destructive interference, the standing-wave ratio is given by

$$\boxed{\text{SWR} = \frac{E_{max}}{E_{min}} = \frac{|E_0^i| + |E_0^r|}{|E_0^i| - |E_0^r|} = \frac{1 + |\Gamma|}{1 - |\Gamma|}} \tag{14.62}$$

where the definition of the reflection coefficient Γ given in (14.52) was used. SWR can now be used to express the magnitude of the reflection coefficient Γ as

$$\boxed{|\Gamma| = \frac{\text{SWR} - 1}{\text{SWR} + 1}} \tag{14.63}$$

Thus, a simple measurement of SWR gives us $|\Gamma|$. SWR can have values between one and infinity. Referring to Fig. 14.9, we observe that for the matched case, when there is no reflection, $\Gamma = 0$ and $\text{SWR} = 1$. For perfect reflection, such as in the case of a perfect conductor, $\Gamma = -1$ and $\text{SWR} = \infty$. For the case when media 1 and 2 are perfect dielectrics† for which η is real, we obtain by use of (14.56) that

$$\text{SWR} = \begin{cases} \dfrac{\eta_2}{\eta_1} & \text{if } \eta_2 > \eta_1 \\[2ex] \dfrac{\eta_1}{\eta_2} & \text{if } \eta_1 > \eta_2 \end{cases} \tag{14.64}$$

This is an unexpectedly simple result. The SWR is seen to be the ratio of the impedances of the two media. Furthermore, for perfect dielectrics, the reflection coefficient Γ must be real, which makes the reflected field ($E_0^r = \Gamma E_0^i$) at the boundary either in phase or out of phase with the incident field. This in turn implies that the boundary must be a position of a maximum or minimum. If $\eta_2 > \eta_1$, Γ is positive, reflected and incident waves add, and the boundary coincides with a maximum of electric field (minimum of magnetic field). On the other

† A perfect dielectric is one which has no losses.

hand, the boundary $z = 0$ coincides with a minimum of electric field (maximum of magnetic field) when $\eta_1 > \eta_2$ and Γ is negative. The limit of the latter case, as $\eta_2 \to 0$, corresponds to reflection from a perfect conductor, illustrated in Fig. 14.1.

Reflection phenomena at the boundary between two media are summarized in Table 14.1.

Example: The formula for the reflection coefficient $\Gamma = (\eta_2 - \eta_1)/(\eta_2 + \eta_1)$ shows that a wave can enter a second dielectric without reflecting from it only if the impedances of dielectric 1 and 2 are matched; that is, $\eta_2 = \eta_1$. Since dielectrics are nonferromagnetic ($\mu_2 = \mu_1 = \mu_0$), the condition $\eta_2 = \eta_1$ implies that $\varepsilon_2 = \varepsilon_1$. Therefore, reflection will be zero when the two media are either identical or, if different, have the same permittivities. In general, different materials have different permittivities.

Recently a new type of absorber material has become available. It is a mixture of ferrite (high μ) and barium titanate (high ε) materials, such that $\eta_2 = \sqrt{\mu/\varepsilon} \cong \sqrt{\mu_0/\varepsilon_0} = 377\ \Omega$. Hence, even though the material presents a physical discontinuity, a wave from free space can enter this material without reflection, because an impedance match exists. If such a material could be made sufficiently absorbing, energy entering the material would be continuously absorbed without reflection. These materials could then be used to coat highly reflecting objects and render them invisible. Such materials have been developed and over a limited bandwidth reduce the reflectivity of an object.

14.6 REFLECTION WITH SEVERAL DIELECTRICS PRESENT

Figure 14.10 shows a boundary between two dielectric media, an incident wave (E^i, H^i), a reflected wave (E^r, H^r), and a transmitted wave (E^t, H^t). A wave impedance at any distance l from the boundary plane is defined as

$$Z(z) = \frac{E_x^{\text{total}}(z)}{H_y^{\text{total}}(z)} \tag{14.65}$$

If only a single forward-traveling wave exists in medium 1, this ratio reduces to $Z(z) = \eta_1$, which is the intrinsic, or characteristic, impedance of the medium.

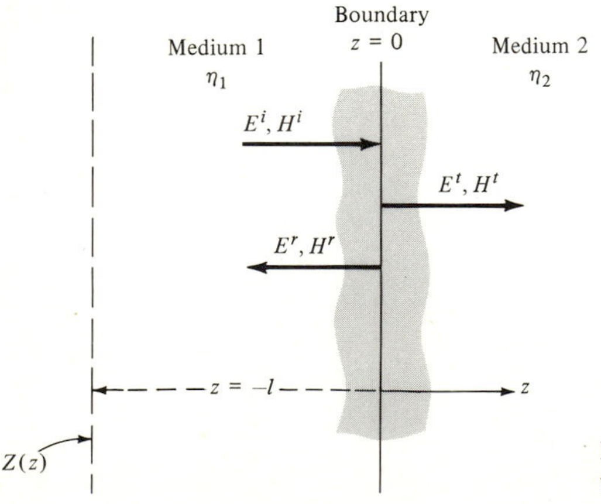

Figure 14.10 A boundary at $z = 0$ separates two dielectric media.

Table 14.1 Plane wave reflection characteristics at boundary between two media

Impedance of medium 1 $\eta_1 = \sqrt{\mu_1/\varepsilon_1}$	Impedance of medium 2 $\eta_2 = \sqrt{\mu_2/\varepsilon_2}$	Reflection coefficient Γ	Transmission coefficient τ	Standing-wave ratio SWR
Generalized lossy media† $\eta_1^\star = \sqrt{\mu_1^\star/\varepsilon_1^\star}$ $\mu_1^\star = \mu_1' - j\mu_1''$ $\varepsilon_1^\star = \varepsilon_1 - j\sigma_1/\omega = \varepsilon_1' - j\varepsilon_1'' - j\sigma_1/\omega$	Generalized lossy media† $\eta_2^\star = \sqrt{\mu_2^\star/\varepsilon_2^\star}$ $\mu_2^\star = \mu' - j\mu''$ $\varepsilon_2^\star = \varepsilon_2 - j\sigma_2/\omega = \varepsilon_2' - j\varepsilon_2'' - j\sigma_2/\omega$	$\Gamma = \dfrac{\eta_2^\star - \eta_1^\star}{\eta_2^\star + \eta_1^\star}$	$\tau = \dfrac{2\eta_2^\star}{\eta_2^\star + \eta_1^\star}$	$\text{SWR} = \dfrac{1+\lvert\Gamma\rvert}{1-\lvert\Gamma\rvert}$
Free space $\eta_0 = 376.7\ \Omega \cong 120\pi\ \Omega$ $\varepsilon_0 = 8.854 \times 10^{-12}$ F/m $\mu_0 = 4\pi \times 10^{-7}$ H/m $\sigma = 0$	Dielectric with small conductive losses $\eta_2^\star \cong \eta_2\left(1 + j\dfrac{\sigma}{2\omega\varepsilon_2}\right) \cong \eta_2[1 + j\sigma/(2\omega\varepsilon_2)]$ $\varepsilon_2^\star = \varepsilon_2\left(1 - j\dfrac{\sigma}{\omega\varepsilon_2}\right)$ $\mu_2^\star = \mu_0$, $\varepsilon_r = \varepsilon_2/\varepsilon_0$ $\sigma/\omega\varepsilon_2 \ll 1$	$\dfrac{\eta_2^\star - \eta_0}{\eta_2^\star + \eta_0} = \dfrac{1 - \sqrt{\varepsilon_r} + j\dfrac{\sigma}{2\omega\varepsilon_2}}{1 + \sqrt{\varepsilon_r} + j\dfrac{\sigma}{2\omega\varepsilon_2}}$	$\dfrac{2\left(1 + j\dfrac{\sigma}{2\omega\varepsilon_2}\right)}{1 + \sqrt{\varepsilon_r} + j\dfrac{\sigma}{2\omega\varepsilon_2}}$	$\cong \sqrt{\varepsilon_r}$
Free space	Perfect dielectric $\eta_2 = \sqrt{\mu_0/\varepsilon_2}$ $\varepsilon_2^\star = \varepsilon_2$ $\varepsilon_r = \varepsilon_2/\varepsilon_0$ $\mu_2^\star = \mu_0$ $\sigma = 0$	$\dfrac{1 - \sqrt{\varepsilon_r}}{1 + \sqrt{\varepsilon_r}}$	$\dfrac{2}{1 + \sqrt{\varepsilon_r}}$	$\sqrt{\varepsilon_r}$
Perfect dielectric $\eta_1 = \sqrt{\mu_0/\varepsilon_1}$ $\varepsilon_1, \mu_0, \sigma = 0$	Perfect dielectric $\eta_2 = \sqrt{\mu_0/\varepsilon_2}$ $\varepsilon_2, \mu_0, \sigma = 0$	$\dfrac{\sqrt{\varepsilon_1} - \sqrt{\varepsilon_2}}{\sqrt{\varepsilon_1} + \sqrt{\varepsilon_2}}$	$\dfrac{2\sqrt{\varepsilon_1}}{\sqrt{\varepsilon_1} + \sqrt{\varepsilon_2}}$	$\sqrt{\varepsilon_1/\varepsilon_2}$ if $\varepsilon_1 > \varepsilon_2$ $\sqrt{\varepsilon_2/\varepsilon_1}$ if $\varepsilon_1 < \varepsilon_2$
Free space	Good conductor $\eta_2 \cong \sqrt{\mu_0/(-j\sigma/\omega)} = (1 + j)\sqrt{\mu\omega/2\sigma}$ $\varepsilon_0, \mu_0, \sigma/\omega\varepsilon_0 \gg 1$	$\dfrac{\sqrt{j\varepsilon_0\omega/\sigma} - 1}{\sqrt{j\varepsilon_0\omega/\sigma} + 1} \cong -1 + \sqrt{\dfrac{\varepsilon_0\omega}{2\sigma}}$	$\dfrac{2}{1 + \sqrt{\sigma/(j\omega\varepsilon_0)}}$	$\cong \sqrt{\dfrac{2\sigma}{\varepsilon_0\omega}}$
Free space	Perfect conductor $\eta_2 = \sqrt{j\mu\omega/\sigma} = 0$ $\varepsilon_0, \mu_0, \sigma = \infty$	-1	0	∞
η_1	$\eta_2 = \eta_1$	0	1	1

† See Eqs. (13.49) and (13.85).

When medium 2 differs from 1, reflected waves are present, which means that the wave impedance in medium 1 now differs from the constant value η_1. The wave impedance is found by writing the total fields in medium 1 as

$$E_x(z) = E_0^i e^{-j\beta z} + E_0^r e^{j\beta z} = E_0^i(e^{-j\beta z} + \Gamma e^{j\beta z}) \tag{14.66}$$

and

$$H_y(z) = H_0^i e^{-j\beta z} + H_0^r e^{j\beta z} = \frac{E_0^i}{\eta_1}(e^{-j\beta z} - \Gamma e^{j\beta z}) \tag{14.67}$$

The minus Γ in $H_y(z)$ results from the relationship (14.19). The wave impedance at the distance $z = -l$ from the boundary is therefore

$$Z(z)\Big|_{z=-l} = \eta_1 \left(\frac{e^{j\beta l} + \Gamma e^{-j\beta l}}{e^{j\beta l} - \Gamma e^{-j\beta l}}\right) = \eta_1 \left(\frac{\eta_2 \cos \beta l + j\eta_1 \sin \beta l}{\eta_1 \cos \beta l + j\eta_2 \sin \beta l}\right) \tag{14.68}$$

where the expression for reflection coefficient $\Gamma = (\eta_2 - \eta_1)/(\eta_2 + \eta_1)$ was used. As a check, we observe that (14.68) gives $Z(-l) = \eta_1$ when $\eta_2 = \eta_1$. The above expression can be thought of as an impedance Z in medium 1 that a wave sees when medium 1 is terminated or loaded with medium 2. $Z(z)$ varies about η_1 and is a function of l and the type of terminating medium. For example, when medium 2 is a perfect conductor, $\eta_2 = 0$ and $\Gamma = -1$, which gives $Z = j\eta \tan \beta l$ for the wave impedance.

Expression (14.68) will be helpful when working with reflection problems involving cascaded media, as shown in Fig. 14.11. Here we have three distinct media separated by two parallel boundaries. When media 1 and 3 are free space and 2 is a dielectric, the resultant configuration can be used to study reflection and transmission by dielectric panels. For example, radomes,† which are dome-shaped structures constructed from dielectric panels and are used to enclose microwave and radar antennas, are a particularly interesting application.

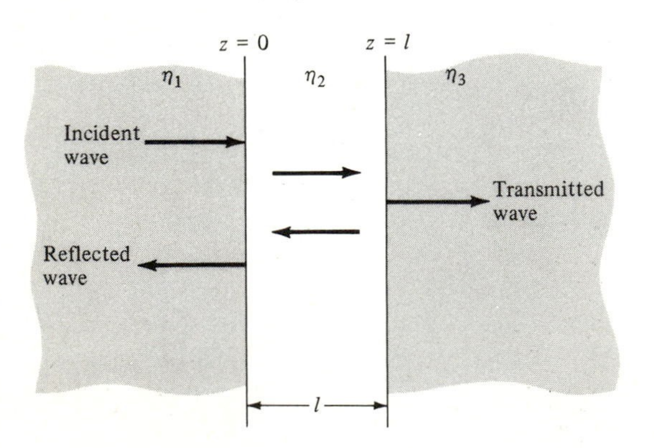

Figure 14.11 Three dielectric media separated by boundaries at $z = 0$ and $z = l$.

† S. A. Silver, "Microwave Antenna Theory and Design," chap. 14, McGraw-Hill Book Company, New York, 1949.

To find the expressions for the reflected and transmitted waves, for the configuration shown in Fig. 14.11, we reason as follows: Medium 3 has a single wave traveling to the right; the wave impedance is therefore $Z_3 = \eta_3$. Medium 2 has waves bouncing back and forth which can be lumped into one wave traveling to the left and one to the right. The wave impedance in medium 2 is then given by (14.68). Thus just to the right of boundary $z = 0$, that is, at $z = 0^+$, we have

$$Z_2(0^+) = \eta_2 \left(\frac{\eta_3 \cos \beta_2 l + j\eta_2 \sin \beta_2 l}{\eta_2 \cos \beta_2 l + j\eta_3 \sin \beta_2 l} \right) \tag{14.69}$$

which now becomes the load impedance for medium 1. Therefore, the reflection coefficient at $z = 0$ for an incident wave in medium 1 is given by

$$\Gamma = \frac{E_0^r}{E_0^i} = -\frac{H_0^r}{H_0^i} = \frac{Z_2(0^+) - \eta_1}{Z_2(0^+) + \eta_1} \tag{14.70}$$

where (14.56) was used as the basis for the reflection coefficient. If reflection coefficients are introduced at the first and second boundary, the above formula can be rearranged in the useful form

$$\boxed{\Gamma = \frac{\Gamma_{12} + \Gamma_{23} e^{-2j\beta_2 l}}{1 + \Gamma_{12}\Gamma_{23} e^{-2j\beta_2 l}}} \tag{14.71}$$

where $\Gamma_{12} = (\eta_2 - \eta_1)/(\eta_2 + \eta_1)$ and $\Gamma_{23} = (\eta_3 - \eta_2)/(\eta_3 + \eta_2)$. Thus for a unity amplitude incident wave, Γ is the amplitude of the reflected wave in medium 1. The reflected power is given by $|\Gamma|^2$.

A reader familiar with transmission line practice will note the similarity between (14.56) and the reflection coefficient for a transmission line of characteristic impedance Z_0 which is terminated by a load impedance Z_L; that is, $\Gamma = (Z_L - Z_0)/(Z_L + Z_0)$. The input impedance of a transmission line of length l which is terminated by a load impedance Z_L is similarly given by (14.68) by substituting Z_0 for η_1 and Z_L for η_2. As a matter of fact, the equations for transmission-line propagation and those of plane-wave propagation in infinite media are completely analogous.†

Let us consider some special cases of practical importance.

Case 1: Quarter-Wave Matching Section

Referring to Fig. 14.11, suppose we desire to eliminate the reflected wave in medium 1. This can be done by making l a quarter wavelength in medium 2, that is, $\beta_2 l = \pi/2$, and by choosing η_2 to be the geometric mean of the intrinsic impedances on both sides, that is, $\eta_2 = \sqrt{\eta_1 \eta_3}$. With this choice of parameters, (14.69) becomes

$$Z_2(0^+) = \sqrt{\eta_1 \eta_3} \left(\frac{0 + j\sqrt{\eta_1 \eta_3}}{0 + j\eta_3} \right) = \eta_1 \tag{14.72}$$

† E. C. Jordan and K. G. Balmain, "Electromagnetic Waves and Radiating Systems," Prentice Hall Inc., Englewood Cliffs, N.Y., sec. 5.15, 1968.

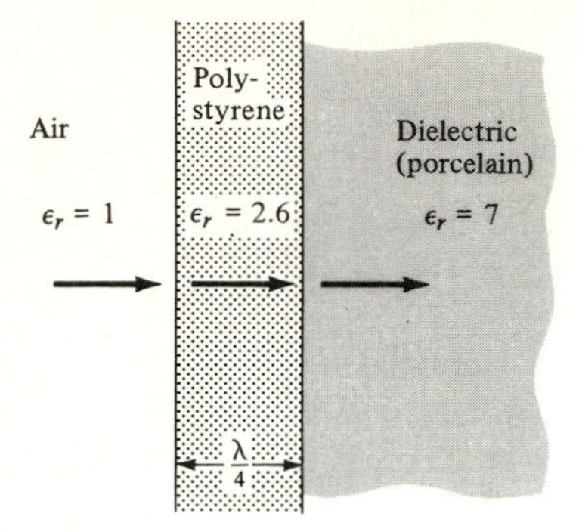

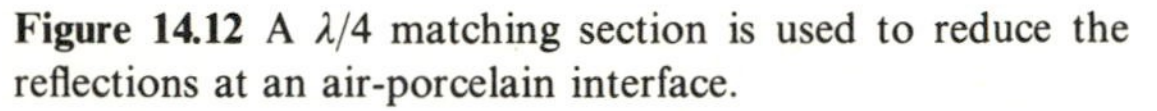

Figure 14.12 A $\lambda/4$ matching section is used to reduce the reflections at an air-porcelain interface.

which when substituted in (14.70) gives

$$\Gamma = \frac{\eta_1 - \eta_1}{\eta_1 + \eta_1} = 0 \tag{14.73}$$

Hence the reflected wave is effectively eliminated. Such matching is perfect only at the frequency given by $\omega = (2n + 1)\pi/(2l\sqrt{\mu_2 \varepsilon_2})$, where $n = 0, 1, 2, \ldots$. Quarter-wave transformers are useful at practically all frequencies. Even at light frequencies, a $\lambda/4$ coating on optical lenses is used to reduce the amount of reflected light.†

Example: It is desired to reduce the reflections at an air-porcelain interface by the use of a $\lambda/4$ plate. Find the thickness of the plate and a suitable material at a frequency of 10 GHz.

Figure 14.12 shows the geometry. From Table 1.1 or 2.3 we find that the dielectric constant of porcelain is 7. Since for air $\eta_1 = \sqrt{\mu_0/\varepsilon_0} = 377\ \Omega$, and for porcelain $\eta_3 = \sqrt{\mu_0/\varepsilon_r \varepsilon_0} = 142\ \Omega$, the condition for a $\lambda/4$ plate gives us $\eta_2 = \sqrt{\eta_1 \eta_3} = 231\ \Omega$ as the intrinsic impedance of the matching $\lambda/4$ plate. Assuming that no ferromagnetic material is to be used in the matching, so that $\mu_2 = \mu_0$, we obtain $\varepsilon_r = (\eta_1/\eta_2)^2 = 2.65$ as the dielectric constant of the matching section. From Table 2.3, an appropriate material for the $\lambda/4$ plate would be polystyrene. At 10 GHz, the thickness of the plate is $l = \pi/2\beta_2 = \pi/2\omega\sqrt{\mu_0 \varepsilon_2} = \pi/2\omega\sqrt{\varepsilon_r}\sqrt{\mu_0 \varepsilon_0} = (\pi \times 3 \times 10^8)/2(2\pi \times 10^{10})\sqrt{2.65} = 0.46 \times 10^{-2}\ \text{m} \cong 0.5\ \text{cm}$.

Case 2: Half-Wave Matching Section

Again we seek to eliminate the reflected wave in medium 1, but with the restriction that medium 1 be the same as 3; that is, $\eta_1 = \eta_3$. In other words, the dielectric panel of thickness l is to be a transparent window, usually referred to as a dielectric window. A radome (a dome-shaped shell used to enclose microwave antennas) is a special case of a dielectric window.

We can eliminate reflections in medium 1 by requiring that the dielectric slab be a half wavelength in thickness, measured in the material of the slab; that is, $\beta_2 l = n\pi$, where $n = 1, 2, 3, \ldots$. With these conditions, (14.69) gives

$$Z_2(0^+) = \eta_2 \left[\frac{\eta_3(\pm 1) + j0}{\eta_2(\pm 1) + j0}\right] = \eta_1 \tag{14.74}$$

† J. M. Stone, "Radiation and Optics," chap. 16, McGraw-Hill Book Company, New York, 1963.

and the reflection coefficient (14.70) becomes

$$\Gamma = 0 \tag{14.75}$$

Example: A radome† is to be constructed from polystyrene ($\varepsilon_r = 2.54$). How thick must the radome walls be at an operating frequency of 10 GHz?

The first resonance, which also gives the thinnest sheet, is at

$$l = \frac{\lambda_2}{2} = \frac{v_2}{2f} = \frac{3 \times 10^8}{2\sqrt{2.54} \times 10^{10}} = 9.4 \times 10^{-3} \text{ m} \cong 1 \text{ cm}$$

Thus at this thickness of radome material, all the power of the enclosed antenna is transmitted to the outside and none is reflected.

Example: Plot a graph, showing the frequency dependence of a dielectric panel (nonferromagnetic: $\mu_2 = \mu_0$) which is immersed in air; that is, $\eta_1 = \eta_3 = \sqrt{\mu_0/\varepsilon_0} = 377\ \Omega$.

Using (14.71) and setting $\Gamma_{12} = -\Gamma_{23} = (\eta_2 - \eta_1)/(\eta_2 + \eta_1) = (1 - \sqrt{\varepsilon_r})/(1 + \sqrt{\varepsilon_r})$, where ε_r is the dielectric constant of the panel, we obtain for the reflection coefficient from the panel

$$\Gamma = \Gamma_{12}\left(\frac{1 - \cos 2\beta_2 l + j \sin 2\beta_2 l}{1 - \Gamma_{12}^2 \cos 2\beta_2 l + j\Gamma_{12}^2 \sin 2\beta_2 l}\right) \tag{14.76}$$

The reflected power relative to unit amplitude of the incident wave is, therefore, given by

$$|\Gamma|^2 = \frac{4\Gamma_{12}^2 \sin^2 \beta_2 l}{1 + \Gamma_{12}^4 - 2\Gamma_{12}^2 \cos 2\beta_2 l} \tag{14.77}$$

The value of $|\Gamma|^2$ is zero when $\sin \beta_2 l = 0$; that is, the panel or film is an integral number of half-wavelengths thick. When the panel is an odd number of quarter-wavelengths thick, a maximum value for the reflected power is obtained, which is $|\Gamma|^2 = 4\Gamma_{12}^2/(1 + \Gamma_{12}^2)^2$. Figure 14.13 is a graph of reflected power $|\Gamma|^2$ versus phase $\beta_2 l$ in the panel for several values of Γ_{12}. The nulls at phase $\beta_2 l = n\pi$, where the values $n = 0, 1, 2, \ldots$ are exhibited as well as the in-between maxima.

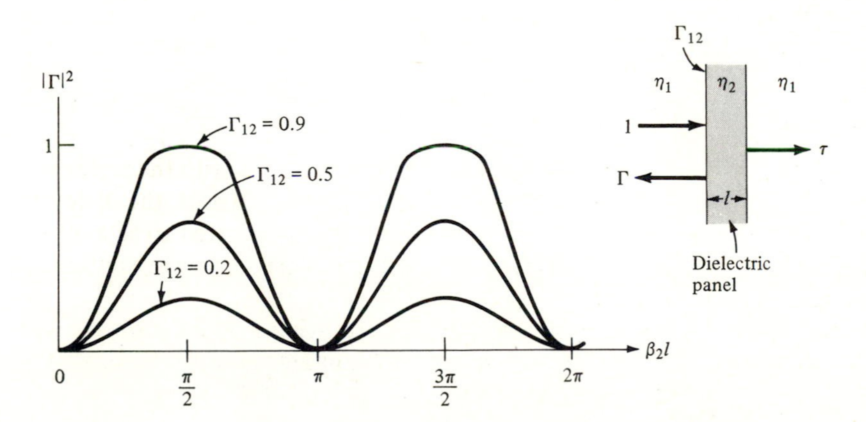

Figure 14.13 Frequency dependence of reflected power from a dielectric panel of thickness l.

† H. Jasik, Editor, "Antenna Engineering Handbook," McGraw-Hill Book Company, New York, 1961, Chap. 32.

Note that we can obtain the transmission coefficient τ for the panel from conservation of energy (assuming the panel is lossless), which can be stated as

$$|\Gamma|^2 + |\tau|^2 = 1 \tag{14.78}$$

This gives for the power transmission coefficient

$$|\tau|^2 = \frac{(1 - \Gamma_{12}^2)^2}{(1 - \Gamma_{12}^2)^2 + 4\Gamma_{12}^2 \sin^2 \beta_2 l} \tag{14.79}$$

14.7 REFLECTION BY A DIELECTRIC—INCIDENCE AT ANY ANGLE

A plane wave incident at angle θ on a boundary between two dielectrics ε_1 and ε_2 will be partially transmitted into and partially reflected from the second dielectric. The transmitted wave is refracted into the second medium; that is, its direction is different from that of the incident wave. The angles of reflection θ_r and refraction θ_t can be related to the angle of incidence θ with the aid of Fig. 14.14.

Consider first the incident and reflected rays. Since both are in the same medium, their velocities will be the same and the distance CA covered by the incident ray is equal to the distance $0B$ covered by the reflected ray. Therefore

$$CA = 0B$$

$$0A \sin \theta = 0A \sin \theta_r$$

$$\sin \theta = \sin \theta_r$$

$$\theta = \theta_r \tag{14.80}$$

As expected—and as already shown in (14.27)—the angle of reflection is equal to the angle of incidence.

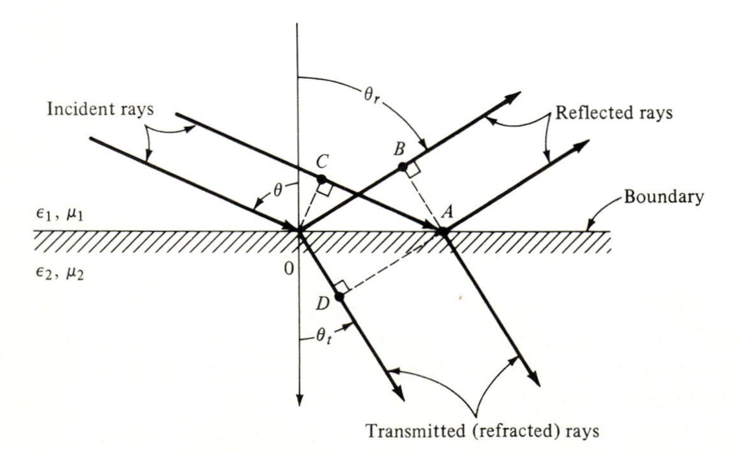

Figure 14.14 Construction for determining Snell's law of refraction.

To find the relationship between the incident and refracted directions, we observe that it takes the incident ray the same time to cover distance CA as it takes the refracted ray to cover distance $0D$; that is,

$$\frac{CA}{v_1} = \frac{0D}{v_2} \tag{14.81}$$

where the velocity in medium 1 is $v_1 = 1/\sqrt{\mu_1 \varepsilon_1}$ and in medium 2 the velocity is $v_2 = 1/\sqrt{\mu_2 \varepsilon_2}$ (note that for most dielectrics $\mu_2 = \mu_1 = \mu_0$; hence $v_1 = 1/\sqrt{\mu_0 \varepsilon_1}$ and $v_2 = 1/\sqrt{\mu_0 \varepsilon_2}$). From Fig. 14.14 we observe that $CA = 0A \sin\theta$ and $0D = 0A \sin\theta_t$, so that

$$\boxed{\frac{\sin\theta}{\sin\theta_t} = \frac{v_1}{v_2} = \sqrt{\frac{\mu_2\varepsilon_2}{\mu_1\varepsilon_1}} = \sqrt{\frac{\varepsilon_2}{\varepsilon_1}}\Bigg|_{\mu_1=\mu_2=\mu_0}} \tag{14.82}$$

This equation is *Snell's law of refraction*. The *index of refraction*, *n*, for a medium is defined as the ratio of phase velocity in the medium to velocity of light in free space. The ratio v_1/v_2 is therefore a ratio of index of refractions. If medium 1 is free space, v_1/v_2 is equal to $n_2 = \sqrt{\varepsilon_r}$, where the relative permittivity (dielectric constant) of medium 2 is $\varepsilon_r = \varepsilon_2/\varepsilon_0$.

The directions of the reflected and refracted waves are now determined. We must now find the amplitudes of these waves. As in the case of arbitrary polarization, Sec. 14.4, we can resolve the electric field of the incident wave into parallel and perpendicular components. Arbitrary polarization is therefore a superposition of two cases: One in which **E** is parallel to the plane of incidence, and one in which **E** is perpendicular. Recall that the plane of incidence is formed by the normal to the reflecting surface and the direction of incidence.

Example: A wave is incident at an angle of 30° from air onto glass ($\mu_2 = \mu_0; \varepsilon_2 = 6\varepsilon_0$). Calculate the angle of refraction θ_t for the transmitted wave.

From (14.82) we have

$$\sin\theta_t = \sin\theta \sqrt{\frac{\varepsilon_1}{\varepsilon_2}} = \sin(30^\circ)\sqrt{\frac{1}{6}} = 0.204$$

$$\theta_t = 11.8^\circ$$

Case 1: Parallel Polarization (E in Plane of Incidence)

Figure 14.15 shows a parallel polarized wave incident at angle θ on dielectric medium 2. The wave reflects at the same angle and refracts into medium 2 at angle θ_t. To determine the amount of the wave reflected and the amount transmitted, we will follow the procedure outlined in Sec. 14.4.

For this polarization the total H field is parallel to the boundary surface. As the tangential component of **H** is continuous across the boundary, because surface

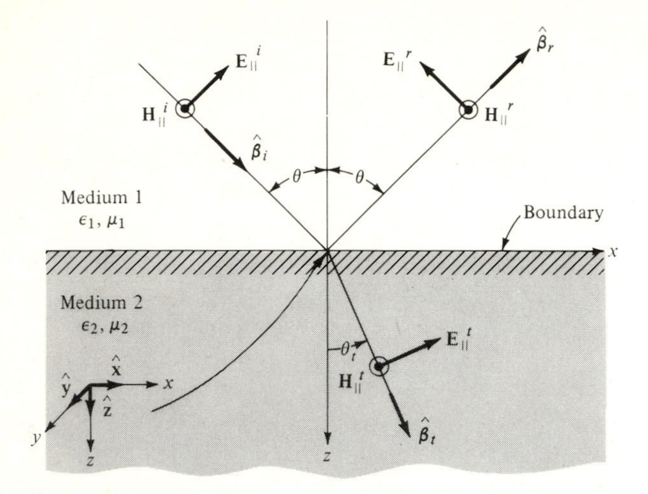

Figure 14.15 Reflection and refraction of a parallel-polarized wave. Medium 1 and 2 are lossless, nonferromagnetic dielectrics for which $\mu_2 = \mu_1 = \mu_0 = 4\pi \times 10^{-7}$ H/m. This is not overly restrictive as most dielectrics fall into this category.

currents cannot flow in a dielectric,† we have that

$$H_{\|}^{i} + H_{\|}^{r} = H_{\|}^{t} \qquad \text{at } z = 0 \tag{14.83}$$

where H_0 is the magnitude of the traveling magnetic field $\mathbf{H}_{\|} = \hat{\mathbf{y}} H_{\|} = \hat{\mathbf{y}} H_0 e^{-j\boldsymbol{\beta}\cdot\mathbf{r}}$ and $\hat{\mathbf{y}}$ is the unit vector in the y direction. Since E and H are related by $\eta = \sqrt{\mu/\varepsilon}$, the above equation can be written as

$$\sqrt{\frac{\varepsilon_1}{\mu_1}}\, E_{\|}^{i} + \sqrt{\frac{\varepsilon_1}{\mu_1}}\, E_{\|}^{r} = \sqrt{\frac{\varepsilon_2}{\mu_2}}\, E_{\|}^{t} \tag{14.84}$$

which, considering that $\mu_2 = \mu_1 = \mu_0$, becomes

$$E_{\|}^{i} + E_{\|}^{r} = \sqrt{\frac{\varepsilon_2}{\varepsilon_1}}\, E_{\|}^{t} \tag{14.85}$$

The tangential components of E must also be continuous across the boundary [see Eq. (8.84)]. Using Fig. 14.15, we have that

$$E_{\|}^{i} \cos\theta - E_{\|}^{r} \cos\theta = E_{\|}^{t} \cos\theta_t \qquad \text{at } z = 0 \tag{14.86}$$

If we solve the two simultaneous equations (14.85) and (14.86), treating the reflected field $E_{\|}^{r}$ and transmitted field $E_{\|}^{t}$ as unknowns, with $E_{\|}^{i}$ as the source field, we obtain for the reflected field

$$E_{\|}^{r} = E_{\|}^{i} \left(\frac{\sqrt{\varepsilon_2/\varepsilon_1}\cos\theta - \cos\theta_t}{\sqrt{\varepsilon_2/\varepsilon_1}\cos\theta + \cos\theta_t} \right) \tag{14.87}$$

† As shown in Secs. 8.8 and 8.9, surface current can flow only on a conductor.

and for the transmitted field

$$E_{\parallel}^{t} = E_{\parallel}^{i}\left(\frac{2\cos\theta}{\sqrt{\varepsilon_2/\varepsilon_1}\cos\theta + \cos\theta_t}\right) \tag{14.88}$$

We can now define a reflection coefficient $\Gamma_{\parallel}$ at $z = 0$ for the parallel polarized case as

$$\Gamma_{\parallel} = -\frac{E_{\parallel}^{r}}{E_{\parallel}^{i}} = \frac{\eta_2\cos\theta_t - \eta_1\cos\theta}{\eta_2\cos\theta_t + \eta_1\cos\theta} = -\left.\frac{\sqrt{\varepsilon_2/\varepsilon_1}\cos\theta - \cos\theta_t}{\sqrt{\varepsilon_2/\varepsilon_1}\cos\theta + \cos\theta_t}\right|_{\mu_2=\mu_1} \tag{14.89}$$

and a transmission coefficient as†

$$\tau_{\parallel} = \frac{E_{\parallel}^{t}}{E_{\parallel}^{i}} = \frac{2\eta_2\cos\theta}{\eta_2\cos\theta_t + \eta_1\cos\theta} = \left.\frac{2\cos\theta}{\sqrt{\varepsilon_2/\varepsilon_1}\cos\theta + \cos\theta_t}\right|_{\mu_2=\mu_1} \tag{14.90}$$

The minus sign in $\Gamma_{\parallel}$ was introduced so that $\Gamma_{\parallel}$ would conform with the earlier definition of the reflection coefficient Γ given by (14.56). For example, for normal incidence ($\theta = \theta_t = 0°$), $\Gamma_{\parallel} \to \Gamma$ and $\tau_{\parallel} \to \tau$, where Γ and τ are given by (14.56) and (14.57), respectively.‡ Also, when medium 2 is a perfect conductor§, $\eta_2 = 0$ and $\Gamma_{\parallel} = -1$, which again agrees with (14.3). The variation of $\Gamma_{\parallel}$ with angle is shown in Fig. 14.19.

The relationship between reflection and transmission coefficients, using (14.89) and (14.90), is

$$1 + \Gamma_{\parallel} = \frac{\cos\theta_t}{\cos\theta}\tau_{\parallel} \tag{14.91}$$

$$1 - \Gamma_{\parallel} = \frac{\eta_1}{\eta_2}\tau_{\parallel} \tag{14.92}$$

Combining the above two equations, we obtain

$$1 = \Gamma_{\parallel}^2 + \frac{\eta_1\cos\theta_t}{\eta_2\cos\theta}\tau_{\parallel}^2 \tag{14.93}$$

which expresses the conservation of energy at the boundary. Note that (14.59), which expresses conservation of energy for normal incidence, is a special case of the above equation.

The complete field in both regions can now be expressed. In medium 1 we have, for the total field expressed in terms of the incident and reflected waves,

$$\mathbf{E}_{\parallel}^{\text{total}} = \mathbf{E}_{\parallel}^{i} + \mathbf{E}_{\parallel}^{r} = \mathbf{E}_0^{i} e^{-j\boldsymbol{\beta}_i\cdot\mathbf{r}} + \mathbf{E}_0^{r} e^{-j\boldsymbol{\beta}_r\cdot\mathbf{r}} \tag{14.94}$$

† The expressions for $\Gamma_{\parallel}$ or $\tau_{\parallel}$ can be stated solely in terms of the incidence angle θ by substituting Snell's law (14.82) in Eqs. (14.89) and (14.90).

‡ Note that in Fig. 14.8, E^i and E^r are in the same direction, whereas in Fig. 14.15, $E_{\parallel}^i$ and $E_{\parallel}^r$ for the case of normal incidence are in opposite directions. Hence the minus sign in (14.89).

§ One can obtain results for a perfect conductor from those of a dielectric by letting ε of the dielectric go to infinity ($\varepsilon \to \infty$). Equivalently, one can introduce the generalized complex permittivity $\varepsilon \to \varepsilon^{\star} = \varepsilon - j\sigma/\omega$, and for a perfect conductor let $\sigma \to \infty$. Either way, we obtain $\eta = 0$ for a perfect conductor.

where from Fig. 14.15 $\mathbf{E}_0^i = (\hat{\mathbf{x}} \cos\theta - \hat{\mathbf{z}} \sin\theta)E_0^i$, $\mathbf{E}_0^r = -(\hat{\mathbf{x}} \cos\theta + \hat{\mathbf{z}} \cos\theta)E_0^r$, and $\boldsymbol{\beta}_i$ and $\boldsymbol{\beta}_r$ are given by Eq. (14.21). The reflection coefficient can be related to the scalar amplitudes E_0^i and E_0^r as follows:

$$\Gamma_{\parallel} = -\frac{E_{\parallel}^r}{E_{\parallel}^i}\bigg|_{z=0} = -\frac{E_0^r e^{-j\boldsymbol{\beta}_r \cdot \mathbf{r}}}{E_0^i e^{-j\boldsymbol{\beta}_i \cdot \mathbf{r}}}\bigg|_{z=0} = -\frac{E_0^r}{E_0^i} \tag{14.95}$$

This allows us to write the field in region 1 in terms of total x and y components as

$$\begin{aligned} \mathbf{E}_{\parallel}^{\text{total}} &= E_x\hat{\mathbf{x}} + E_z\hat{\mathbf{z}} \\ &= \cos\theta\, E_0^i(e^{-j\boldsymbol{\beta}_i \cdot \mathbf{r}} + \Gamma_{\parallel} e^{-j\boldsymbol{\beta}_r \cdot \mathbf{r}})\hat{\mathbf{x}} \\ &\quad + \sin\theta\, E_0^i(-e^{-j\boldsymbol{\beta}_i \cdot \mathbf{r}} + \Gamma_{\parallel} e^{-j\boldsymbol{\beta}_r \cdot \mathbf{r}})\hat{\mathbf{z}} \end{aligned} \tag{14.96}$$

If we substitute for β_i and β_r, the components of the total electric field become

$$\boxed{E_x = \cos\theta\, E_0^i e^{-j\beta x \sin\theta}(e^{-j\beta z \cos\theta} + \Gamma_{\parallel} e^{j\beta z \cos\theta})} \tag{14.97}$$

$$\boxed{E_z = \underbrace{\sin\theta\, E_0^i e^{-j\beta x \sin\theta}}_{\text{traveling-wave part}}\underbrace{(-e^{-j\beta z \cos\theta} + \Gamma_{\parallel} e^{j\beta z \cos\theta})}_{\text{standing-wave part}}} \tag{14.98}$$

where $E_0^i = \eta_1 H_0^i$ and $E_0^r = \eta_1 H_0^r$.

As in the case of a perfect conductor (Sec. 14.4), we observe that in the x direction we have a traveling-wave field and in the z direction (which is normal to the boundary), a standing-wave field. The difference is that the minima of the standing-wave field do not reach zero as in the case of a perfectly conducting boundary (for which $\Gamma = -1$). Hence, the standing-wave ratio SWR, which is given by (14.62), is a finite number larger than 1 but not equal to infinity as it would be in the case of a perfect conductor. The fact that the reflection coefficient $\Gamma_{\parallel}$ in the standing-wave part is not equal to unity suggests that we can decompose this part—as was done in (14.60)—into a traveling wave and a standing wave in the z direction. In the case of (14.97), the standing-wave part is equal to $[(1 + \Gamma_{\parallel})e^{-j\beta z \cos\theta} + \Gamma_{\parallel} 2j \sin(\beta z \cos\theta)]$. The energy which flows in the z direction from medium 1 into medium 2 is carried by the traveling wave. In passing we should mention that the perfectly conducting case [see Eq. (14.29)] can be obtained from the above equations simply by substituting $\Gamma_{\parallel} = -1$.

The magnetic field in region 1 is similarly given by

$$\boxed{\begin{aligned} \mathbf{H}_{\parallel}^{\text{total}} &= \mathbf{H}_{\parallel}^i + \mathbf{H}_{\parallel}^r \\ &= \mathbf{y} H_0^i e^{-j\beta x \sin\theta}(e^{-j\beta z \cos\theta} - \Gamma_{\parallel} e^{j\beta z \cos\theta}) \end{aligned}} \tag{14.99}$$

where

$$\mathbf{H}_{\parallel}^i = \hat{\mathbf{y}} H_0^i e^{-j\boldsymbol{\beta}_i \cdot \mathbf{r}} = \hat{\mathbf{y}}\frac{E_0^i}{\eta_1} e^{-j\boldsymbol{\beta}_i \cdot \mathbf{r}} \tag{14.100}$$

and

$$\begin{aligned} \mathbf{H}_{\parallel}^r &= \hat{\mathbf{y}} H_0^r e^{-j\boldsymbol{\beta}_r \cdot \mathbf{r}} = \hat{\mathbf{y}}\frac{E_0^r}{\eta_1} e^{-j\boldsymbol{\beta}_r \cdot \mathbf{r}} \\ &= -\mathbf{y}\Gamma_{\parallel}\frac{E_0^i}{\eta_1} e^{-j\boldsymbol{\beta}_r \cdot \mathbf{r}} \end{aligned} \tag{14.101}$$

The transmitted or refracted fields in medium 2 are

$$\begin{aligned}\mathbf{E}_{\|}^{t} &= \mathbf{E}_0^t e^{-j\boldsymbol{\beta}_t\cdot\mathbf{r}} = (\hat{\mathbf{x}}\cos\theta_t - \hat{\mathbf{z}}\sin\theta_t)E_0^t e^{-j\boldsymbol{\beta}_t\cdot\mathbf{r}} \\ &= (\hat{\mathbf{x}}\cos\theta_t - \hat{\mathbf{z}}\sin\theta_t)\tau_{\|} E_0^i e^{-j\boldsymbol{\beta}_t\cdot\mathbf{r}}\end{aligned} \tag{14.102}$$

and

$$\mathbf{H}_{\|}^{t} = \mathbf{H}_0^t e^{-j\boldsymbol{\beta}_t\cdot\mathbf{r}} = \hat{\mathbf{y}} H_0^t e^{-j\boldsymbol{\beta}_t\cdot\mathbf{r}} = \hat{\mathbf{y}}\frac{\tau_{\|} E_0^i}{\eta_2} e^{-j\boldsymbol{\beta}_t\cdot\mathbf{r}}$$

where

$$\begin{aligned}\tau_{\|} &= \left.\frac{E_{\|}^t}{E_{\|}^i}\right|_{z=0} = \left.\frac{E_0^t e^{-j\boldsymbol{\beta}_t\cdot\mathbf{r}}}{E_0^i e^{-j\boldsymbol{\beta}_i\cdot\mathbf{r}}}\right|_{z=0} \\ &= \frac{E_0^t e^{-j\beta_2 x\sin\theta_t}}{E_0^i e^{-j\beta_1 x\sin\theta}} = \frac{E_0^t}{E_0^i} = \frac{\eta_2 H_0^t}{\eta_1 H_0^i}\end{aligned} \tag{14.103}$$

and $\boldsymbol{\beta}_t\cdot\mathbf{r} = \beta_2(x\sin\theta_t + z\cos\theta_t)$. The above expression for transmission coefficient $\tau_{\|}$ was reduced to a simple form by use of Snell's law (14.82).

Alternative Derivation

The above reflection and transmission coefficients, $\Gamma_{\|}$ and $\tau_{\|}$, can also be derived by applying the principle of conservation of energy at the boundary. Thus, since the power transmitted per unit area in a wave is equal to E^2/η, the power in the incident wave striking the boundary surface is $(E_{\|}^i)^2\cos\theta/\eta_1$, that reflected is $(E_{\|}^r)^2\cos\theta/\eta_1$, and that transmitted is $(E_{\|}^t)^2\cos\theta_t/\eta_2$. At the boundary we must have then

$$(E_{\|}^i)^2\frac{\cos\theta}{\eta_1} = (E_{\|}^r)^2\frac{\cos\theta}{\eta_1} + (E_{\|}^t)^2\frac{\cos\theta_t}{\eta_2} \tag{14.104}$$

This equation is equivalent to (14.93). If the boundary condition (14.86) is substituted in the above equation, we can then solve for $\Gamma_{\|}$ and $\tau_{\|}$. Thus the reflection and transmission coefficients can be derived either by conservation of energy and the boundary condition on the electric fields or by matching both electric and magnetic fields at the boundary [Eqs. (14.83) and (14.86)].

Brewster's Angle (Polarizing Angle)

An examination of (14.87) or (14.89) shows the possibility for no reflected wave at some angle θ of incidence. Let us call this angle *Brewster's angle* θ_B. To find angle θ_B for which $\Gamma_{\|} = 0$, we set the numerator of these expressions equal to zero; that is,

$$\sqrt{\frac{\varepsilon_2}{\varepsilon_1}}\cos\theta_B = \cos\theta_t \tag{14.105}$$

Making use of Snell's law (14.82), we can solve for Brewster's angle and obtain

$$\theta_B = \sin^{-1}\sqrt{\frac{\varepsilon_2}{\varepsilon_1+\varepsilon_2}} = \tan^{-1}\sqrt{\frac{\varepsilon_2}{\varepsilon_1}} \tag{14.106}$$

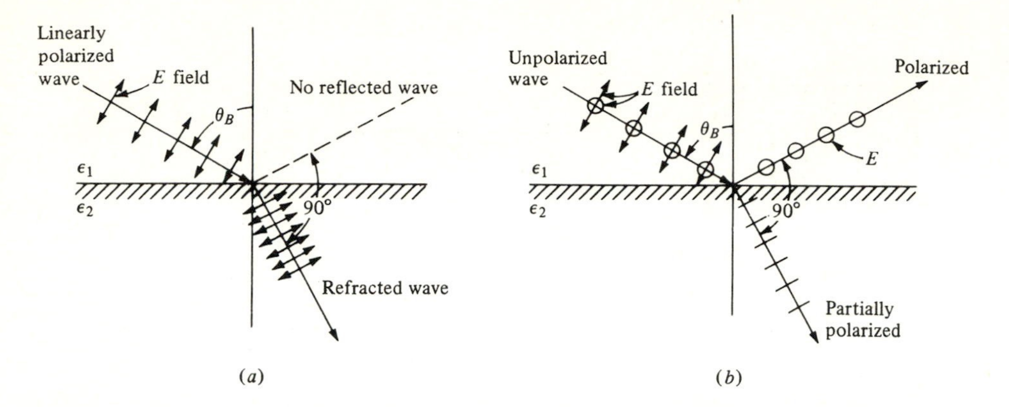

Figure 14.16 (*a*) A linearly parallel-polarized wave, incident at θ_B, has no reflected wave; (*b*) an unpolarized wave, such as ordinary light, has a polarized reflected wave and a partially polarized transmitted wave. Note that for incidence at θ_B the angle between the reflected and refracted rays is 90°.

This angle is indicated in Fig. 14.19. Note that a Brewster's angle exists for either $\varepsilon_1 > \varepsilon_2$ or $\varepsilon_1 < \varepsilon_2$. Thus, if a parallel-polarized wave strikes the boundary at Brewster's angle θ_B, there will be no reflected wave, and the incident wave will be totally transmitted into medium 2. If the incident wave is not entirely parallel polarized, that is, if it has a parallel as well as a perpendicular component, only the perpendicularly polarized component will be reflected. For incidence at θ_B the reflected wave is therefore always linearly polarized, with the E field normal to the plane of incidence. Even a circularly polarized wave incident at θ_B becomes linearly polarized upon reflection. Hence, θ_B is also known as the *polarizing angle.* As a matter of fact, polarized light was first discovered by this phenomenon, and today Brewster's angle is the basis for many practical applications. Figure 14.16 illustrates this effect for a polarized and an unpolarized wave. The refracted wave in Fig. 14.16*b* is partially polarized because in it the parallel-polarized component predominates since no portion of it was reflected. We should note at this time that polarization perpendicular to the plane of incidence has no corresponding Brewster's angle (see Fig. 14.19).

Example. A gas laser, shown in Fig. 14.17, has two quartz windows which are oriented at Brewster's angle θ_B. The purpose of these windows is to reduce reflection losses for radiation which is polarized with an **E** vector in the plane of the paper. This causes the output radiation to be polarized as shown, because the perpendicular polarization (**E** vector out of paper) has reflection losses at the windows which does not allow the oscillations between the mirrors of the external optical resonator to build up sufficiently to overcome the threshold. Such windows cause the laser output to be almost completely linearly polarized. What is Brewster's angle if the index of refraction n for quartz is $n = 1.5$ ($n = \sqrt{\varepsilon_r}$)?

$$\theta_B = \tan^{-1}\sqrt{\varepsilon_r} = \tan^{-1} n = \tan^{-1} 1.5 = 56.5°$$

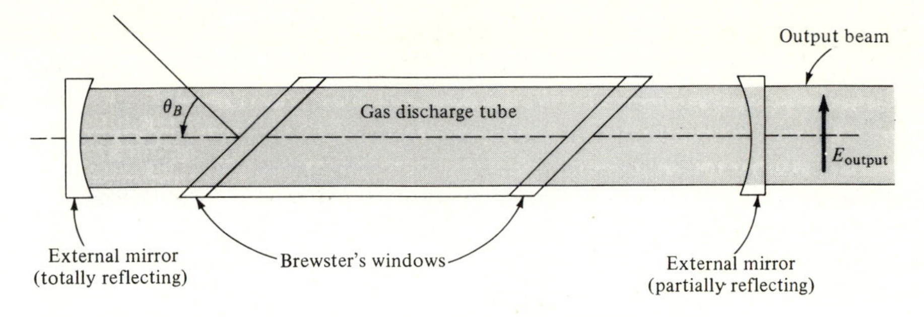

Figure 14.17 A typical gas laser with quartz windows at Brewster's angle.

Case 2: Perpendicular Polarization (E Normal to Plane of Incidence)

This case of polarization is shown in Fig. 14.18. Continuity of the tangential components of **H** at the boundary is expressed by

$$H_{\perp}^{i} \cos\theta - H_{\perp}^{r} \cos\theta = H_{\perp}^{t} \cos\theta_t \tag{14.107}$$

Because E and H are related by η as $E = \eta H$, (14.107) can also be written as

$$E_{\perp}^{i} \cos\theta - E_{\perp}^{r} \cos\theta = \frac{\eta_1}{\eta_2} E_{\perp}^{t} \cos\theta_t \qquad \text{at } z = 0 \tag{14.108}$$

Similarly, continuity of tangential components of E is expressed by

$$E_{\perp}^{i} + E_{\perp}^{r} = E_{\perp}^{t} \qquad \text{at } z = 0 \tag{14.109}$$

Solving these two simultaneous equations for the reflected field $E_{\perp}^{r}$ and refracted field $E_{\perp}^{t}$ gives for the reflection coefficient (at $z = 0$)

$$\Gamma_{\perp} = \frac{E_{\perp}^{r}}{E_{\perp}^{i}} = \frac{\eta_2 \cos\theta - \eta_1 \cos\theta_t}{\eta_2 \cos\theta + \eta_1 \cos\theta_t} = \left.\frac{\cos\theta - \sqrt{\varepsilon_2/\varepsilon_1}\cos\theta_t}{\cos\theta + \sqrt{\varepsilon_2/\varepsilon_1}\cos\theta_t}\right|_{\mu_2=\mu_1} \tag{14.110}$$

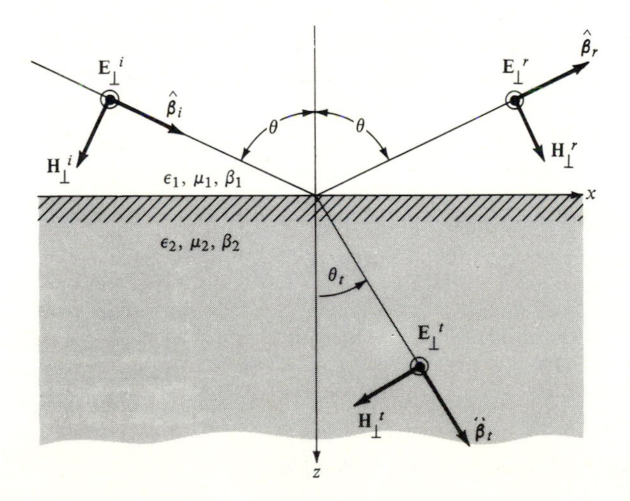

Figure 14.18 Reflection and refraction of a perpendicularly polarized wave.

and for the transmission coefficient (at $z = 0$)

$$\tau_\perp = \frac{E_\perp^t}{E_\perp^i} = \frac{2\eta_2 \cos\theta}{\eta_2 \cos\theta + \eta_1 \cos\theta_t} = \frac{2\cos\theta}{\cos\theta + \sqrt{\varepsilon_2/\varepsilon_1}\cos\theta_t}\bigg|_{\mu_2=\mu_1} \tag{14.111}$$

where the last form in each of the above two equations is valid for nonferromagnetic dielectrics, for which $\mu_2 = \mu_1 = \mu_0 = 4\pi \times 10^{-7}$ H/m. The above equations can be stated solely in terms of the incidence angle θ by making use of Snell's law (14.82); that is, $\sqrt{\varepsilon_2}\cos\theta_t = \sqrt{\varepsilon_2(1-\sin^2\theta_t)} = \sqrt{\varepsilon_2 - \varepsilon_1 \sin^2\theta}$.

The relationship between the reflection and transmission coefficients for perpendicular polarization is, from (14.109), seen to be

$$1 + \Gamma_\perp = \tau_\perp \tag{14.112}$$

which leads to an energy-balance equation, similar to Eq. (14.93). Also, if medium 2 is a perfect conductor, $\eta_2 = 0$ and $\Gamma_\perp = -1$, and $\tau_\perp = 0$ as expected.

The total electric field in region 1, which is entirely in the positive y direction, is now given by

$$\begin{aligned}\mathbf{E}_\perp^{\text{total}} &= \mathbf{E}_\perp^i + \mathbf{E}_\perp^r = \hat{\mathbf{y}}E_0^i e^{-j\boldsymbol{\beta}_i\cdot\mathbf{r}} + \hat{\mathbf{y}}E_0^i\Gamma_\perp e^{-j\boldsymbol{\beta}_r\cdot\mathbf{r}} \\ &= \hat{\mathbf{y}}E_0^i e^{-j\beta x\sin\theta}(e^{-j\beta z\cos\theta} + \Gamma_\perp e^{j\beta z\cos\theta})\end{aligned} \tag{14.113}$$

where $\Gamma_\perp = (E_\perp^r/E_\perp^i)|_{z=0} = E_0^r/E_0^i$ and the phase factors for the incident and reflected waves are $\boldsymbol{\beta}_{i,r}\cdot\mathbf{r} = \beta x\sin\theta \pm \beta z\cos\theta$. The total magnetic field is given by

$$\begin{aligned}\mathbf{H}_\perp^{\text{total}} = \mathbf{H}_\perp^i + \mathbf{H}_\perp^r &= -\hat{\mathbf{x}}\frac{E_0^i}{\eta_1}\cos\theta(e^{-j\boldsymbol{\beta}_i\cdot\mathbf{r}} - \Gamma_\perp e^{-j\boldsymbol{\beta}_r\cdot\mathbf{r}}) \\ &\quad + \hat{\mathbf{z}}\frac{E_0^i}{\eta_1}\sin\theta(e^{-j\boldsymbol{\beta}_i\cdot\mathbf{r}} + \Gamma_\perp e^{-j\boldsymbol{\beta}_r\cdot\mathbf{r}})\end{aligned} \tag{14.114}$$

Just as for parallel polarization, the field has a traveling-wave character in the x direction and a standing-wave character in the z direction.

As a check on the equations, the following limiting cases can be considered. Substituting $\eta_2 = \eta_1$ or $\varepsilon_2 = \varepsilon_1$, the total fields should reduce to the incident wave. Letting the incidence angle be zero, we should get the same results as those in Sec. 14.5. For the case when medium 2 is a perfect conductor, $\eta_2 = 0$ or $\varepsilon_2 = \infty$, the results should reduce to those given in Sec. 14.4.

In medium 2 the fields are

$$\mathbf{E}_\perp^t = \hat{\mathbf{y}}E_0^t e^{-j\boldsymbol{\beta}_t\cdot\mathbf{r}} = \hat{\mathbf{y}}\tau_\perp E_0^i e^{-j\beta_2(x\sin\theta_t + z\cos\theta_t)} \tag{14.115}$$

$$\mathbf{H}_\perp^t = \frac{E_0^i}{\eta_2}\tau_\perp(-\,\hat{\mathbf{x}}\cos\theta_t + \,\hat{\mathbf{z}}\sin\theta_t)e^{-j\beta_2(x\sin\theta_t + z\cos\theta_t)} \tag{14.116}$$

where $\tau_\perp = (E_\perp^t/E_\perp^i)|_{z=0} = E_0^t/E_0^i$ and $H_0^t = \eta_2 E_0^t = \eta_2\tau_\perp E_0^i$.

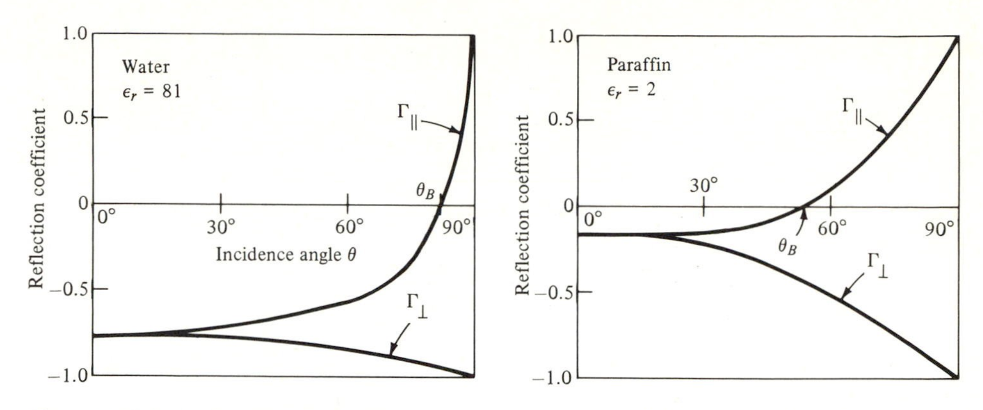

Figure 14.19 Reflection coefficient versus incidence angle θ for parallel and perpendicularly polarized waves incident from air ($\varepsilon_r = 1$) onto water ($\varepsilon_r = \varepsilon_2/\varepsilon_0 = 81$) and paraffin ($\varepsilon_r = \varepsilon_2/\varepsilon_0 = 2$).

Figure 14.19 shows the reflection coefficient for both polarizations for two materials. We observe from these figures that for perpendicular polarization there is no angle similar to Brewster's angle θ_B for which the reflected wave vanishes. The reason for this is that the numerator of Eq. (14.110) can never be zero for a nonferromagnetic dielectric; that is, $\cos\theta \neq \sqrt{\varepsilon_2/\varepsilon_1}\,\cos\theta_t$ when $\varepsilon_2 > \varepsilon_1$ because Snell's law requires that $\theta_t < \theta$ for $\varepsilon_2 > \varepsilon_1$. However, if we consider a material with the same permittivities, $\varepsilon_2 = \varepsilon_1$, but different permeabilities, $\mu_1 \neq \mu_2$, then the numerator of (14.110) can in effect be zero, because $\eta_2\cos\theta - \eta_1\cos\theta_t = \sqrt{\mu_2}\,\cos\theta - \sqrt{\mu_1}\,\cos\theta_t = 0$ exists. Such a material would then have a Brewster's angle for perpendicular polarization, that is, an incident angle giving no reflections.

14.8 TOTAL REFLECTION

We have seen that for common dielectrics only the parallel-polarized waves have an angle θ_B for which there is no reflection. There is a second phenomenon which exists for both polarizations. Total reflection can take place at the interface between two dielectric media, for a wave that is passing from a medium with a larger permittivity to a medium with a smaller permittivity. Referring to Fig. 14.20*a* and assuming that medium 1 and medium 2 are such that $\varepsilon_1 > \varepsilon_2$, we see that a wave incident at an angle θ_1 will pass into medium 2 at a larger angle; that is, $\theta_2 > \theta_1$. This is given by Snell's law (14.82) which states that $\sin\theta_1 = \sqrt{\varepsilon_2/\varepsilon_1}\,\sin\theta_2$. If the angle of incidence θ_1 is now increased until $\theta_2 = 90°$, as shown in Fig. 14.20*b*, we see that for still further increases of θ_1, the incident wave must be totally reflected back into medium 2. We will call the angle in medium 1 when $\theta_2 = 90°$, the *critical angle* θ_c; that is,

$$\boxed{\theta_c = \sin^{-1}\sqrt{\frac{\varepsilon_2}{\varepsilon_1}}} \tag{14.117}$$

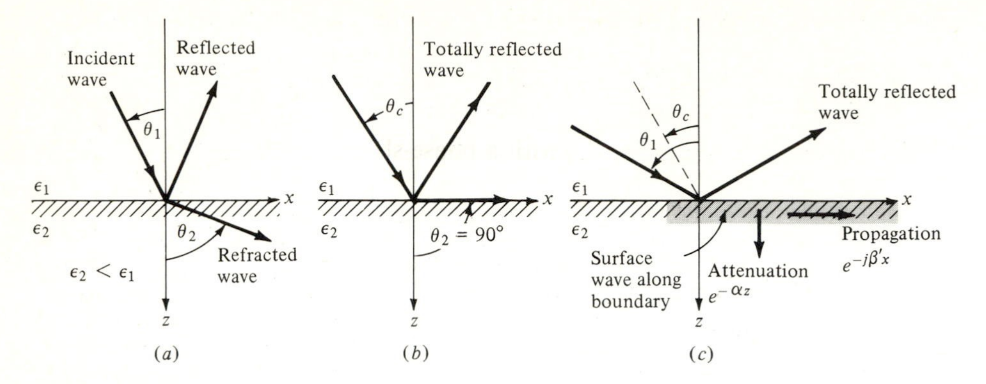

Figure 14.20 (*a*) If medium 1 has a permittivity larger than medium 2 ($\varepsilon_1 > \varepsilon_2$), a wave incident from medium 1 will refract into medium 2 at a larger angle; (*b*) because θ_2 cannot exceed 90°, the incident wave is completely reflected back into medium 1 when $\theta_1 \geqslant \theta_c$; (*c*) surface wave along boundary, when $\theta_1 > \theta_c$.

Whenever $\theta_1 > \theta_c$, Snell's law states that $\sin\theta_2 = \sqrt{\varepsilon_1/\varepsilon_2}\,\sin\theta_1 > 1$. The only solution to $\sin\theta_2 > 1$ is for θ_2 to be an imaginary angle; that is, $\theta_2 = \pi/2 + j\theta_2'$. We do not have to pursue the meaning of the imaginary angle any further, since for physical interpretation of total reflection we do not need θ_2 but need $\sin\theta_2$ and $\cos\theta_2$, and both can be related to the real angle θ_1 by Snell's law. Thus, for $\theta_1 > \theta_c$,

$$\sin\theta_2 = \sqrt{\frac{\varepsilon_1}{\varepsilon_2}}\sin\theta_1$$

$$\cos\theta_2 = \sqrt{1 - \sin^2\theta_2} = \sqrt{1 - \frac{\varepsilon_1}{\varepsilon_2}\sin^2\theta_1}$$

$$= \pm j\sqrt{\frac{\varepsilon_1}{\varepsilon_2}\sin^2\theta_1 - 1} \tag{14.118}$$

The physical interpretation of the imaginary angle θ_2 and the imaginary $\cos\theta_2$ will lead to a surface wave along the boundary between the two media. Before proceeding with this, let us first derive the reflection coefficients for total reflection.

A more direct way to see that total reflection takes place is to consider $\Gamma_\perp$ and $\Gamma_\parallel$ for $\theta_1 > \theta_c$. Both coefficients reduce to the form $(a + jb)/(a - jb) = 1e^{j\phi}$, where $\phi = 2\tan^{-1} b/a$, which shows the unit magnitude of both reflection coefficients. For example, for parallel polarization, after substituting $\cos\theta_2$ as given by (14.118), Eq. (14.89) becomes (note that $\theta_t \equiv \theta_2$)

$$\Gamma_\parallel = -\frac{\sqrt{\varepsilon_2/\varepsilon_1}\cos\theta_1 \mp j\sqrt{\varepsilon_1/\varepsilon_2\sin^2\theta_1 - 1}}{\sqrt{\varepsilon_2/\varepsilon_1}\cos\theta_1 \pm j\sqrt{\varepsilon_1/\varepsilon_2\sin^2\theta_1 - 1}} = 1e^{j\phi_\parallel} \tag{14.119}$$

For perpendicular polarization, the reflection coefficient (14.110) becomes

$$\Gamma_{\perp} = \frac{\cos\theta_1 \mp j\sqrt{\sin^2\theta_1 - \varepsilon_2/\varepsilon_1}}{\cos\theta_1 \pm j\sqrt{\sin^2\theta_1 - \varepsilon_2/\varepsilon_1}} = 1e^{j\phi_{\perp}} \qquad (14.120)$$

Thus the wave is totally reflected, with a phase-shift angle $\phi_{\parallel}$ or $\phi_{\perp}$.

Example: The basis for optical fibers is total internal reflection. An optical fiber has a center core to which the propagating wave is confined. The index of refraction of the core in a typical fiber is $n = \sqrt{\varepsilon_r} = 1.45$. The core is surrounded by a cladding which has an index of refraction of $n = 1.44$. What is the angle θ of reflection for waves in the center core?

Using (14.117), $\theta_c = \sin^{-1} 1.44/1.45 = \sin^{-1} .993 \cong 83°$. As $\theta \geq \theta_c$ for propagation, the rays of the incident wave are practically parallel to the axis of the fiber. Nevertheless, for a fiber of 50 μm in diameter, the light rays are reflected many thousand times per meter of fiber length. The number of reflections is inversely proportional to the diameter of the fiber.

Surface Waves

Although, under total reflection conditions, there is no energy transfer into medium 2, a field does exist in medium 2; otherwise boundary conditions which require continuity of the tangential fields across the boundary would be violated. As an example, let us use the case of perpendicular polarization to see what the field in medium 2 is. After substituting for $\sin\theta_2$ and $\cos\theta_2$ from (14.118), the transmitted electric field (14.115) is found to be

$$\boxed{\begin{aligned} E_y &= \tau_{\perp} E_0^i e^{-j\beta_2(x\sin\theta_2 + z\cos\theta_2)} \\ &= \tau_{\perp} E_0^i e^{-\alpha z} e^{-j\beta' x} \end{aligned}} \qquad (14.121)$$

where

$$\tau_{\perp} = \frac{2\cos\theta_1}{\cos\theta_1 + j\sqrt{\sin^2\theta_1 - \varepsilon_2/\varepsilon_1}} \qquad (14.122)$$

$$\alpha = \beta_2 \sqrt{\frac{\varepsilon_1}{\varepsilon_2}\sin^2\theta_1 - 1} \qquad (14.123)$$

$$\beta' = \beta_2 \sqrt{\frac{\varepsilon_1}{\varepsilon_2}}\sin\theta_1 \qquad \beta_2 = \omega\sqrt{\mu_2\varepsilon_2} = 2\pi/\lambda_2 \qquad (14.124)$$

A field therefore does exist in medium 2 and is shown in Fig. 14.20*c*. The factor $e^{-j\beta' x}$ represents a wave which propagates along the x axis (parallel to the boundary surface) with an apparent phase velocity $v' = \omega/\beta' = \omega/\beta_2\sqrt{\varepsilon_1/\varepsilon_2}\sin\theta_1$, which is less than the phase velocity ω/β_2 of an ordinary plane wave in medium 2. The factor $e^{-\alpha z}$ indicates that the amplitude of this wave† is exponentially damped in a direction normal to the boundary.

† We have chosen the + sign in (14.118) in order to obtain an exponentially decreasing field; the − sign would give an exponentially increasing field which violates the conservation of energy principle.

The exponential damping is very rapid, and in a distance of a few wavelengths the field is, for practical purposes, unobservable. Such a wave is therefore tightly bound to the surface. Waves of this type which propagate at reduced velocity parallel to a surface and are attenuated normally to a surface are called *surface waves*.†

It can be easily shown that Poynting's vector for a surface wave in a direction normal to the surface is zero. Therefore, no transfer of energy to medium 2 takes place, and all the energy which is incident on the boundary reappears in the reflected wave in medium 1. Summarizing, we can say that total reflection takes place when a wave is incident at an angle exceeding the critical angle θ_c and is accompanied by a surface wave in the medium with the smaller permittivity.

Example: (*a*) Calculate the critical angle θ_c for a water-air interface. (*b*) Calculate the attenuation in a direction normal to the boundary for a surface wave in air when a wave of amplitude E_0^i in water is incident at 30° onto the water-air boundary. State the attenuation in decibels for a distance of λ from the surface.

The properties of water are $\varepsilon_r = \varepsilon_1/\varepsilon_0 = 81$ and $\mu_r = 1$, and of air $\varepsilon_r = \varepsilon_2/\varepsilon_0 = 1$ and $\mu_r = 1$. The critical angle, using (14.117), is therefore

$$\theta_c = \sin^{-1}\left(\tfrac{1}{81}\right)^{1/2} = 6.4^\circ$$

The strength of the surface wave at the surface is given by Eq. (14.121) as

$$|E_y|_{z=0} = |\tau_\perp E_0^i| = 1.55E_0^i$$

where
$$\tau_\perp = \frac{2\cos\theta_1}{\cos\theta_1 + j\sqrt{\sin^2\theta_1 - \varepsilon_2/\varepsilon_1}} = \frac{2\cos 30^\circ}{\cos 30^\circ + j\sqrt{\sin^2 30^\circ - 1/81}}$$

$$= 1.55e^{-j39^\circ}$$

The surface wave attenuates exponentially with distance from the surface as

$$|E_y| = 1.55E_0^i e^{-\alpha z}$$

The decibel attenuation at a distance of $z = \lambda$ is, therefore,

$$\text{dB} = 20\log\frac{E_y(z=\lambda)}{E_y(z=0)} = 20\log e^{-\alpha\lambda} = 20\log e^{-27.8}$$

$$= 20(-27.8)\log e = -240$$

where
$$\alpha = \frac{2\pi}{\lambda}\sqrt{\frac{\varepsilon_1}{\varepsilon_2}\sin^2\theta_1 - 1} = \frac{2\pi}{\lambda}\sqrt{81(\tfrac{1}{2})^2 - 1} = \frac{27.8}{\lambda}$$

This is a very rapid attenuation which shows that the surface wave exists in only a very small layer near the boundary surface.

Polarization

As the phases for the parallel and perpendicularly polarized components of a reflected wave are different ($\phi_\parallel \neq \phi_\perp$ in general), a linearly polarized incident

† A tutorial treatment of surface waves is given by S. A. Schelkunoff, Surface Waves, *IEEE Trans. Antennas and Propagation*, vol. AP-7, December, 1959.

wave which has both polarization components will become an elliptically polarized wave upon total reflection. We can see this as follows: Combining two electric fields (E_x and E_y) which are at right angles in space to each other and which have the same magnitude, frequency, and phase, the result will be a linear plane polarized wave with the resultant **E** field at a 45° angle to both the original fields, as shown in Fig. 14.21. The tip of the **E** vector, if we view the wave in the direction of propagation, will follow a line, hence the name linear polarization. Such a wave can be produced by two crossed dipole antennas fed by the same transmission line. On the other hand if the two crossed E components have different phases, the tip of the resultant **E** vector will trace out an ellipse. A special case, that of circular polarization, occurs when the two crossed components are of equal magnitude but have a phase difference of 90°. As in the case of linear polarization when the **E** and **H** fields are at right angles to each other, so too in elliptical polarization, the ellipse representing **E** has its major axis perpendicular to that of the ellipse representing **H.** The **E** field rotation in space of an elliptically polarized wave can be easily observed with ordinary instruments.

Polarization can be described mathematically as follows: Assume that a plane wave [such as (13.39)] which is traveling in the z direction has x and y components:

$$E_x = E_1 \cos \omega t \tag{14.125}$$

$$E_y = E_2 \cos (\omega t + \zeta) \tag{14.126}$$

where for convenience we have let $z = 0$. The time-phase angle between the two components is ζ. The above two equations are the parametric equations for an ellipse with the major axis tilted with respect to the x axis. The tip of the **E** vector $\mathbf{E} = E_x \hat{\mathbf{x}} + E_y \hat{\mathbf{y}}$ rotates as a function of time and traces out an ellipse called the *polarization ellipse.* The axial ratio (AR), which is the ratio of major to minor axes, is given by AR $= E_2/E_1$.

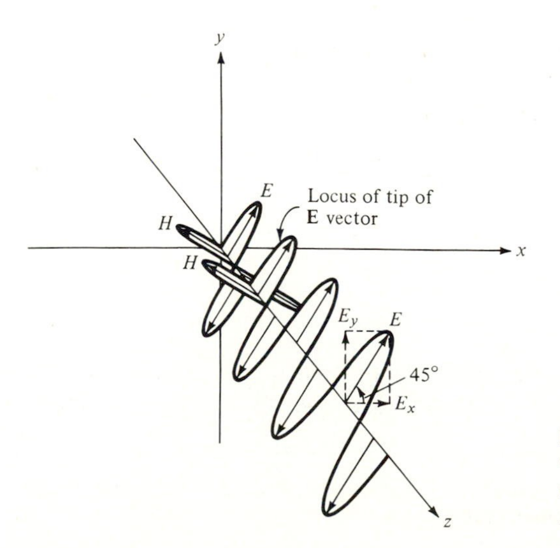

Figure 14.21 A linearly polarized plane wave in which the **E** field oscillates along a line which makes a 45° angle with the x axis.

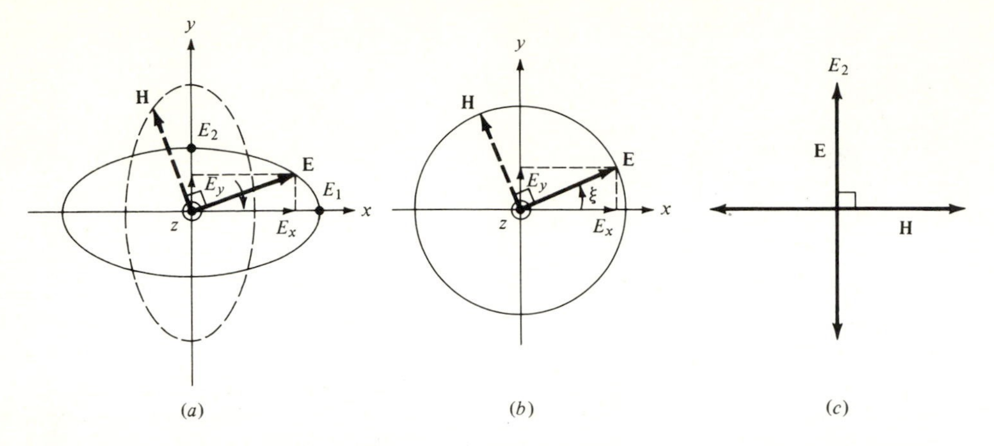

Figure 14.22 Elliptic polarization (*a*), circular polarization (b), and linear polarization (*c*) as viewed in the plane $z = 0$. The z axis, which is the direction of wave propagation, is out of the paper.

To illustrate elliptic polarization, let us choose $\zeta = 90°$ and $E_1 \neq E_2$. Then $E_x = E_1 \cos \omega t$ and $E_y = -E_2 \sin \omega t$, which gives the equation of an ellipse

$$\frac{E_x^2}{E_1^2} + \frac{E_y^2}{E_2^2} = 1 \tag{14.127}$$

and is shown in Fig. 14.22*a*. As the positive z axis is out of the paper, the wave is viewed as approaching the reader. The rotation of the **E** vector is therefore clockwise.

Circular polarization is obtained when the magnitudes of the E_x and E_y components are equal; that is, $E_1 = E_2$ and $\zeta = 90°$, which gives an axial ratio AR = 1. The ellipse reduces now to a circle, as (14.127) yields

$$E_x^2 + E_y^2 = E_1^2 \tag{14.128}$$

which is the equation of a circle. Figure 14.22*b* shows the circular polarization which is followed by the tip of the **E** vector. The **E** vector is constant in magnitude and rotates with angular velocity ω. The rotation can be shown by expressing the instantaneous angle ξ between the **E** vector and the x axis as

$$\xi = \tan^{-1} \frac{E_y}{E_x} = \tan^{-1} \frac{-E_1 \sin \omega t}{E_1 \cos \omega t} = -\omega t \tag{14.129}$$

The **E** vector rotates with angular velocity $d\xi/dt = -\omega$. In the time of one period T or, equivalently, in the distance of one wavelength λ, the **E** vector completes one revolution. If the wave is out of the paper, the minus sign implies that **E** rotates clockwise. Had we chosen $\zeta = -90°$, then $\xi = \omega t$ and the **E** vector would rotate counterclockwise. Figure 14.23 shows the counterclockwise rotation of the electric field in a circularly polarized wave as it progresses along the z axis.

Linear polarization in the y direction is obtained when $E_1 = 0$, and in the x direction when $E_2 = 0$. Linear polarization at an angle of 45° to the x axis results when $E_1 = E_2$ with $\zeta = 0$, as shown in Fig. 14.21. Figures 14.22*c* and 13.4 show vertical linear polarization.

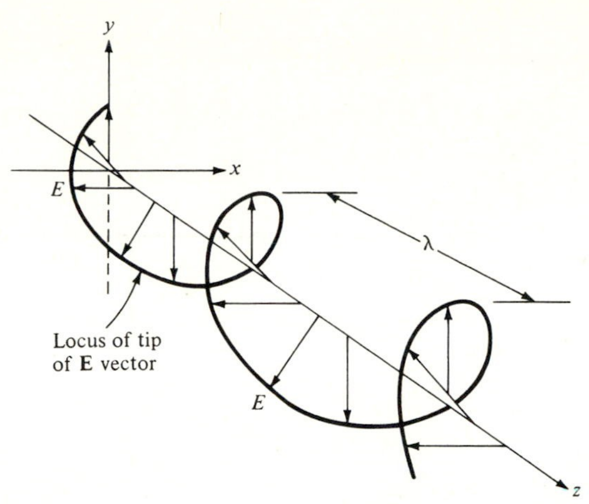

Figure 14.23 A circularly polarized wave in which the **E** field rotates as the wave advances.

PROBLEMS

14.1 Show that the average Poynting vector for a standing wave is zero. Use the instantaneous values (14.5) and (14.8).

14.2 Using the eddy-current point of view (or Lenz' law), show that the reflected magnetic field from a conducting surface [Fig. 14.1 and Eq. (14.6)] is in phase with the incident magnetic field at the reflecting surface. Assume incidence perpendicular to the surface.

14.3 An incident wave impinges normally on a large, plane, perfectly conducting body, as shown in the figure below. The strength and direction of the power flow is given by Poynting's vector, $\mathscr{P} = \mathbf{E} \times \mathbf{H}$.

(*a*) Show that the reflected component of the electric field on the surface must be equal and opposite to the incident electric field; that is, $E^i + E^r = 0$. Similarly, show that on the surface $\mathbf{H}_{\text{total}} = 2\mathbf{H}^i$. Hence, the surface current flowing is $\mathbf{K} = \hat{\mathbf{n}} \times 2\mathbf{H}^i$ A/m.

(*b*) When K is taken as twice the incident magnetic field, the induced current is usually referred to as the geometric optics current. Such a current serves as a good approximation to currents on conducting surfaces which are curved, as long as the radius of curvature is large with respect to wavelength. Show that for an incident wave at an arbitrary angle to the surface, the magnetic field on the surface is given by $\mathbf{H}_t = \mathbf{t}2H^i$, where the incident magnetic field is $\mathbf{H}^i = \hat{\mathbf{a}}H^i$. The direction of the incident magnetic field is given by the unit vector $\hat{\mathbf{a}}$, which is perpendicular to the wave travel. The unit vector $\mathbf{t}$ is tangent to the surface, which gives the direction of the total surface field: $\mathbf{t} = \hat{\mathbf{a}} - (\hat{\mathbf{a}} \cdot \hat{\mathbf{n}})\hat{\mathbf{n}}$. The induced surface current is then $\mathbf{K} = \hat{\mathbf{n}} \times \mathbf{H}_t$.

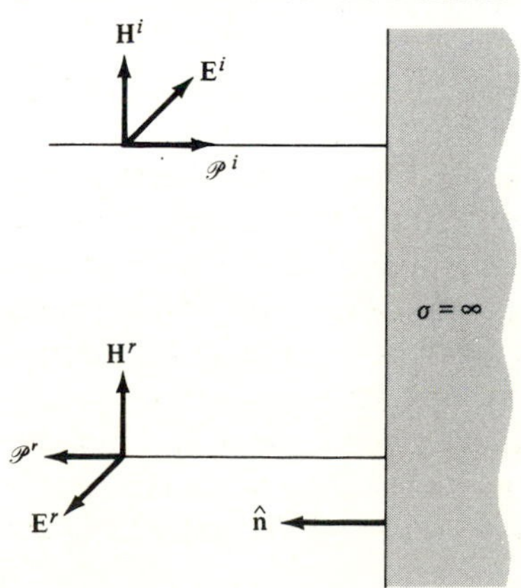

14.4 Show that the div and curl operations on the electric field of a plane wave reduce to the algebraic operations

$$\nabla \cdot \mathbf{E} = -j\boldsymbol{\beta} \cdot \mathbf{E} \qquad \nabla \times \mathbf{E} = -j\boldsymbol{\beta} \times \mathbf{E}$$

where $\mathbf{E} = \mathbf{E}_0 e^{j(\omega t - \boldsymbol{\beta} \cdot \mathbf{r})}$ and $\mathbf{E}_0$ is a constant vector giving the direction and magnitude of the electric field.

14.5 Referring to the result of the previous problem, we observe that a plane wave is a spectral component in the spatial domain, in the same sense that a sinusoid is a spectral component in the time domain. Hence, one can expect that an arbitrary wave can be decomposed in a spectrum of plane waves which propagate in different directions. Show the similarities between plane waves in the spatial domain and sinusoids ($e^{j\omega t}$) in the time domain.

Hint: Note that in the case of spectral components, differential operations convert to algebraic operations.

14.6 Using Fig. 14.5 and Eq. (14.31), we observe that at glancing incidence a plane wave transfers power in the x direction given by $\mathscr{P}_x = 2(E_0^i)^2/\eta$. Why is this four times larger than the power transfer for a plane wave in free space which is $\mathscr{P} = (E_0)^2/2\eta$, as given by (13.76)?

14.7 A sinusoidal plane wave with Poynting vector $\mathscr{P}_i$ is normally incident on the surface of the sea from above the water. Calculate the reflected and transmitted Poynting vectors, $\mathscr{P}_r$ and $\mathscr{P}_t$, respectively, in terms of $\mathscr{P}_i$. The parameters of seawater are $\sigma = 4$, $\varepsilon_r = 81$, and $\mu_r = 1$. Consider (a) $f = 1$MHz (b) $f = 10$GHz.

14.8 A 2-GHz plane wave in air with a peak electric field of 2 V/m is incident normally on a large copper sheet. Find the average power absorbed by the sheet per square meter of area.

14.9 A 1-GHz 1-kW microwave beam is normally incident upon a 1-m^2 sheet of copper foil of 10-μm thickness. Calculate the power

(*a*) Which is reflected from the foil.
(*b*) Which is transmitted into the foil.
(*c*) Which reemerges on the other side of the foil.

14.10 A plane wave in free space is incident normally on a large plane dielectric slab of dielectric constant 5. The amplitude of the electric field in the incident wave is 20 V/m. What is the amplitude inside the dielectric?

14.11 A plane wave in free space is incident normally on a large conducting slab. If the incident wave is $E_x = 100 \cos (10^9 t - \beta z)$ V/m, find the total fields (electric and magnetic) in free space. Assume the plane surface of the conductor is at $z = 0$.

14.12 A 500-MHz plane wave in free space is incident normally on a large plane dielectric slab ($z > 0$) of dielectric constant 3. If $\mathbf{E}^i = 10e^{-j\beta z}\,\hat{\mathbf{x}}$ is the incident wave, find $\mathbf{E}^r$, $\mathbf{E}^i_{\text{dielectric}}$, and $\mathscr{P}^t_{z,\text{ave,dielect.}}/\mathscr{P}^i_{z,\text{ave,free space}}$.

14.13 The electric field of a plane wave which propagates in the $-z$ direction in free space is given by $\mathbf{E} = 2\hat{\mathbf{x}} + 3\hat{\mathbf{y}}$ V/m. Find the magnetic field strength $\mathbf{H}$.

14.14 The magnetic field of a plane wave which is incident on a large, plane, perfect conductor at angle $\theta_i = 45°$ (use Fig. 14.5) is $\mathbf{H} = H_0\hat{\mathbf{y}} \cos (10^6\pi t - \boldsymbol{\beta}_i \cdot \mathbf{r})$. Calculate the density of average power flow in the x direction.

14.15 Find the peak value of an induced surface current when a plane wave is incident at an angle on a large, plane, perfectly conducting sheet. The surface of the sheet is located at $z = 0$ and $\mathbf{E}^i = 10\hat{\mathbf{y}} \cos (10^{10}t - \beta x/\sqrt{2} - \beta z/\sqrt{2})$ V/m.

14.16 Calculate the reflection coefficient for a 10-GHz plane wave in air which is incident normally on a large brass sheet.

14.17 Verify Eq. (14.44).

14.18 Find the induced surface current $\mathbf{K}$ (in amperes per meter) for the polarization, incidence, and geometry shown in Fig. 14.5, and compare to the result given in Eq. (14.48).

14.19 Find the percentage of power transmitted from air into polystyrene for a normally incident wave. Assume the interface is plane and large.

14.20 Starting with the first statement in (14.60), show the second and third statements.

14.21 Equation (14.60) gives the total electric field when reflection is present. Show that the total magnetic field is given by

$$H_y = \frac{(1+\Gamma)E_0^i}{\eta} e^{-j\beta z} - \frac{2\Gamma E_0^i}{\eta} \cos \beta z$$

Hence the traveling-wave part of the magnetic field can be obtained from the traveling-wave part of the electric field by dividing by the intrinsic impedance η, but such a simple operation is not valid for the standing-wave part. Explain.

14.22 Using (14.60) and the results of the previous problem, show that the average Poynting's vector (13.71) for the case of reflection gives

$$\mathscr{P}_{\text{ave}} = (1-\Gamma^2)\frac{(E_0^i)^2}{2\eta_1} = \frac{\eta_1}{\eta_2}\tau^2\frac{(E_0^i)^2}{2\eta_1} = \frac{(\tau E_0^i)^2}{2\eta_2}$$

where (14.59) was used. Notice that this is the power that flows in medium 2, as τE_0^i is the electric field that is transmitted into medium 2. Thus the division of power at a discontinuity is as expressed by (14.59): Incident power $\{\frac{1}{2}[(E_0^i)^2/\eta_1]\}$ minus reflected power $\{\frac{1}{2}[(\Gamma E_0^i)^2/\eta_1]\}$ equals transmitted power $\{\frac{1}{2}[(\tau E_0^i)^2/\eta_2]\}$.

14.23 We have learned that a traveling wave can carry energy ($\mathscr{P}_{\text{ave}} \neq 0$) but that a standing wave cannot ($\mathscr{P}_{\text{ave}} = 0$ for a standing wave). For the general case which includes reflection, the electric field can be divided into a traveling- and standing-wave part as shown in (14.60). If we use the traveling-wave part to calculate Poynting's vector, we obtain

$$\mathscr{P}_{\text{ave}} = (1+\Gamma)^2\frac{(E_0^i)^2}{2\eta_1}$$

which gives the wrong result by comparison with the correct result given in Prob. 14.22. Why does this give the wrong result? Explain.

14.24 Repeat Prob. 14.22, but for the Poynting vector expression use (13.72).

14.25 For the case of reflection, obtain the Poynting vector as given in Prob. 14.22, but start with the electric field (14.61).

14.26 Verify Eq. (14.55).

14.27 In Sec. 14.5 the reflection and transmission coefficients were derived. Obtain the reflection and transmission coefficients for the reflected and transmitted magnetic fields, H_y^r and H_y^t, respectively.

14.28 Find the thickness and dielectric constant of a quarter-wave matching plate at 1 GHz such that reflections at an air-quartz surface ($\varepsilon_r = 5$) are absent.

14.29 What is the polarizing angle for an air-water ($\varepsilon_r = 81$) interface at which plane waves pass from
(*a*) air into water
(*b*) water into air.

14.30 Calculate the percentage of incident energy transmitted through a dielectric panel ($\varepsilon_r = 2.5$) at 10 GHz. The thickness of the panel is 1 cm.

14.31 A perpendicularly polarized plane wave is incident on an air-mica ($\varepsilon_r = 6$) interface. Calculate the transmission coefficient for an angle of incidence of 60°.

14.32 In Sec. 14.6, following Eq. (14.75) we found the wall thickness of a radome at which it will transmit without reflection all the antenna power to the outside. If the thickness of the radome material is reduced to 0.5 cm, what is the percentage of the incident power that will be reflected by the radome walls?

14.33 Derive (14.71) from (14.70).

14.34 Given (14.96) to (14.98), use Maxwell's equations to find the magnetic field.

14.35 Derive (14.99).

14.36 For parallel polarization, derive the analogous field to (14.121) which exists in the less dense medium.

14.37 For a surface wave such as (14.121), find the Poynting vector normal to and tangential to the surface. What conclusions can you make?

CHAPTER
FIFTEEN

TRANSMISSION LINES

GUIDE TO THE CHAPTER

A treatment of transmission lines, if it is to be of practical importance in the areas of computers and microwaves, must cover the topics of transients (Sec. 15.3 to 15.5), lossless steady-state lines (Sec. 15.6 to 15.9), and lossy lines (Sec. 15.10). A detailed development of the three topics would require the space of several chapters. The following plan was adopted for treatment in a single chapter:

1. The circuits approach, in which the unknowns are the voltage V and current I along the line, is used to derive the transmission-line equations. This is a relatively short procedure (Sec. 15.2) when compared to that of the more general fields approach in which the unknowns are the electric and magnetic field. A property of two-conductor lines (coaxial, twinlead, stripline) is that the electric and magnetic fields on the line are transverse to the direction of wave propagation. Such transverse electromagnetic fields are known as TEM modes. For TEM modes the scalar quantities V and I are uniquely related to the vector **E** and **H** fields of the transmission line. Hence for two-conductor lines, the simpler circuits approach is adequate. On the other hand, wave guides which are hollow, single-conductor structures and dielectric rod guides have electromagnetic fields with components in the direction of wave propagation. Such field configurations (known as *higher-order modes*) can be analyzed only in terms of the fields approach.
2. Graphical methods such as Smith charts and bounce diagrams are omitted. Before the advent of pocket calculators, Smith charts were the only convenient means for transmission-line calculations—although it can be argued that Smith charts are also helpful in understanding steady-state situations on transmission lines. Today, calculations with pocket calculators are more convenient and more accurate than with Smith charts. Bounce diagrams aid in visualizing multiple reflections on transmission lines. Both graphical methods are omitted from this chapter in the interest of brevity.

15.1 INTRODUCTION

The principal use of transmission lines is in the transmission of signals and power. Transmission lines are large in one dimension and small in the other two. At

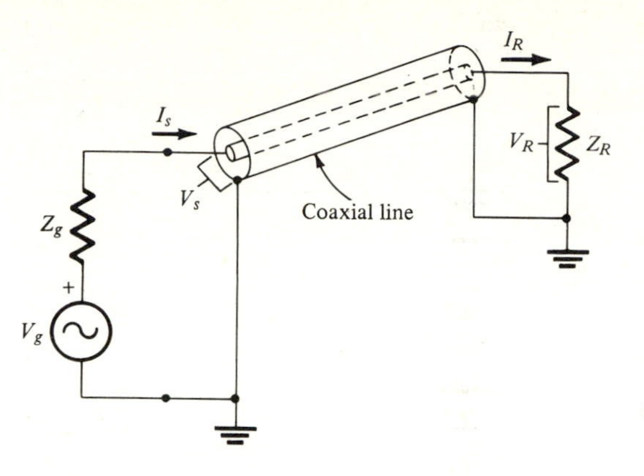

Figure 15.1 A coaxial transmission line is connected to a source at the sending end and to a load impedance Z_R at the receiving end.

frequencies used for power transmission, the transverse dimensions are very small compared with wavelength λ; for example, for a frequency of 60 Hz, the wavelength is 5000 km and the transverse dimensions are normally of the order of a meter or less. Even the longitudinal dimension is only a fraction of a wavelength, as the length of power-transmission lines rarely exceeds 500 km. At higher frequencies the length of transmission lines can be many wavelengths, with the cross-sectional dimension still remaining a fraction of a wavelength. For example, at 1 GHz, which is the upper-frequency limit for most practical lines such as the coaxial and twin lead types, the cross-sectional dimension is of the order of 0.03λ. Above 1 GHz, the losses in ordinary transmission lines become prohibitive. In the frequency range of 1 to 100 GHz, wave guides which are not as lossy are used. The cross-sectional dimensions of wave guides are on the order of λ. At 10 GHz, which is a typical frequency in the operation of wave guides, a 10-m-long wave guide has several hundred wavelengths. Above 100 GHz, wave guides become impractical as their small cross-sectional dimensions require very precise manufacturing techniques. Recently optical wave guides have become available which have losses (2 dB/km) far below those of ordinary coaxial lines ($\approx$300 dB/km) or wave guides ($\approx$30 dB/km). They operate at frequencies in the range of 500 THz. Their cross-sectional dimensions are large in terms of wavelength, usually exceeding 100λ (typical fiber diameters are 75 to 100 μm). Of course any practical length of fiber optic is many million wavelengths.

A typical setup is shown in Fig. 15.1. A coaxial transmission line is connected to a signal generator whose internal impedance and voltage are Z_g and V_g, respectively. At the sending end the current is I_s and the voltage across the line is V_s. At the receiving end the transmission line is terminated by a load impedance Z_R. A signal traveling on the line will cause current I_R and voltage V_R as shown.

15.2 THE UNIFORM TRANSMISSION LINE

A parallel-wire transmission line, as shown in Fig. 15.2, has series inductance and resistance as well as shunt capacitance and conductance distributed uniformly

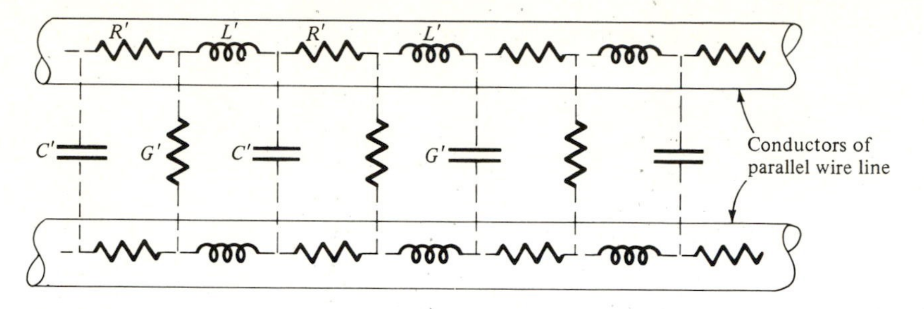

Figure 15.2 Parallel-wire transmission line with distributed parameters shown. It should be emphasized that the primed quantities are per unit length. For example, the series resistance of an elemental section of transmission line is $R = R'\,dz$.

along the line. A transmission line is therefore a distributed parameter network in which the quantities R', L', G', and C' are given per unit length of line.† The conventional lumped-circuit theory can be applied to an infinitesimal length dz of transmission line, as shown in Fig. 15.3. Such a short section can be treated as a four-terminal network with lumped series resistance $R'\,dz$, lumped series inductance $L'\,dz$, lumped shunt conductance $G'\,dz$, and lumped shunt capacitance $C'\,dz$. The result of such an analysis will reveal that waves of voltage V and current I travel along the line. The voltage and current waves are intimately related to the waves of **E** and **H** in the space between the conductors of the transmission line. Because of the nature of one-dimensional wave propagation the scalar quantities V and I uniquely determine the vector **E** and **H** fields of the transmission line. Hence it is simpler to describe the electromagnetic fields on the transmission line in terms of voltage, current, and impedance.

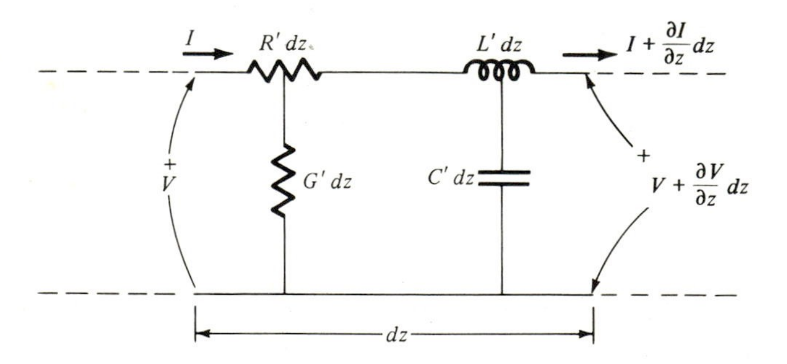

Figure 15.3 The equivalent circuit for an infinitesimal length dz of transmission line.

† We will use two parallel wires to represent any two-conductor transmission line, such as the coaxial line, stripline, twin line, etc. In Secs. 5.5, 5.6, and 7.7 the expressions for capacitance and inductance per unit length for a parallel-wire and coaxial transmission line are derived. In Probs. 15.1 and 15.2 the stripline is considered.

Kirchhoff's voltage law applied to the network of Fig. 15.3 gives

$$-V + (R'\,dz)I + (L'\,dz)\frac{\partial I}{\partial t} + V + \frac{\partial V}{\partial z}\,dz = 0 \tag{15.1}$$

Kirchhoff's current law results in

$$I - (G'\,dz)V - (C'\,dz)\frac{\partial V}{\partial t} - \left(I + \frac{\partial I}{\partial z}\,dz\right) = 0 \tag{15.2}$$

Canceling the common factor dz, we obtain the following equations:

$$-\frac{\partial V}{\partial z} = L'\frac{\partial I}{\partial t} + R'I \tag{15.3}$$

$$-\frac{\partial I}{\partial z} = C'\frac{\partial V}{\partial t} + G'V \tag{15.4}$$

which form the basis for the study of transmission lines from a (distributed) circuit point of view.

The Lossless Line

The important features of propagation and reflection of signals can be better illustrated if we study the ideal transmission line. Thus let us ignore I^2R losses and insulation leakage losses G^2V by neglecting R' and G'. Fortunately, in most practical situations line losses are truly negligible, which makes the lossless case of great practical significance. The equations of a lossless line are†

$$-\frac{\partial V}{\partial z} = L'\frac{\partial I}{\partial t} \tag{15.5}$$

$$-\frac{\partial I}{\partial z} = C'\frac{\partial V}{\partial t} \tag{15.6}$$

By differentiating one equation with respect to time and the other with respect to distance, substituting, and rearranging, we find that

$$\frac{\partial^2 V}{\partial z^2} - L'C'\frac{\partial^2 V}{\partial t^2} = 0 \tag{15.7}$$

$$\frac{\partial^2 I}{\partial z^2} - L'C'\frac{\partial^2 I}{\partial t^2} = 0 \tag{15.8}$$

† Note that these equations are equivalent to Maxwell's equations (13.9) and (13.10), which for the case of z-directed plane waves with E_x and H_y components reduce to $-\partial E_x/\partial z = \mu\,\partial H_y/\partial t$ and $-\partial H_y/\partial z = \varepsilon\,\partial E_x/\partial t$.

These are one-dimensional wave equations. Therefore, we have voltage and current waves on the line which travel with velocity $v = 1/\sqrt{L'C'}$. The solution† for V and I anywhere on the line is a combination of incident and reflected waves:

$$V = V^i\left(t - \frac{z}{v}\right) + V^r\left(t + \frac{z}{v}\right) \tag{15.9}$$

$$I = I^i\left(t - \frac{z}{v}\right) + I^r\left(t + \frac{z}{v}\right) \tag{15.10}$$

The current I can be expressed in terms of the voltages in the incident and reflected waves by substituting (15.9) in (15.5) and integrating with respect to time t. This gives

$$I = \frac{1}{Z_0}\left[V^i\left(t - \frac{z}{v}\right) - V^r\left(t + \frac{z}{v}\right)\right] \tag{15.11}$$

where

$$Z_0 = L'v = \sqrt{\frac{L'}{C'}} \tag{15.12}$$

and is known as the *characteristic impedance* of the line (actually as L' and C' are real, the proper term should be characteristic resistance). Z_0 is the ratio of voltage to current of an individual wave on the line. That is, for the incident wave $Z_0 = V^i/I^i$, and for the reflected wave $Z_0 = -V^r/I^r$. It is incorrect to say that $Z_0 = V/I$, as can be easily seen by taking the ratio of (15.9) to (15.11). The minus sign in Ohm's law for the reflected wave, $Z_0 = -V^r/I^r$, denotes that in Fig. 15.3 current is in the negative z direction.

The question of what characteristic impedance is can now be answered as follows: It is the ratio of voltage to current on an infinite transmission line.

Example: Infinite line with dc voltage input Figure 15.4 shows a battery voltage V_b suddenly applied to an infinite line. As a result a step voltage and step current advance along the line with velocity $v = 1/\sqrt{L'C'}$. In other words at any time t after the switch is closed, Eq. (15.9) implies that voltage $V = V^i = V_b$, and (15.11) implies that a current $I = I^i = V_b/Z_0$ exists on the line up to the point $z = vt$, and zero voltage and current beyond.

The outflow of energy from the battery goes into charging the capacitance and establishing a current in the inductance of the line. The energy ΔW_C stored in the capacitance of a line of length dz is equal to the energy ΔW_L in the inductance; that is,

$$\begin{aligned}\Delta W_C &= \tfrac{1}{2}(C'\,dz)V_b^2 = \tfrac{1}{2}(C'\,dz)(Z_0 I)^2 \\ &= \tfrac{1}{2}(C'\,dz)\left(\frac{L'}{C'}I^2\right) = \tfrac{1}{2}(L'\,dz)I^2 = \Delta W_L\end{aligned} \tag{15.13}$$

† The case of one-dimensional wave propagation in free space is analogous to the propagation of voltage and current waves on a transmission line. Therefore the solution to (15.7) and (15.8) is obtained by the same procedure that was used in (13.11) to (13.15).

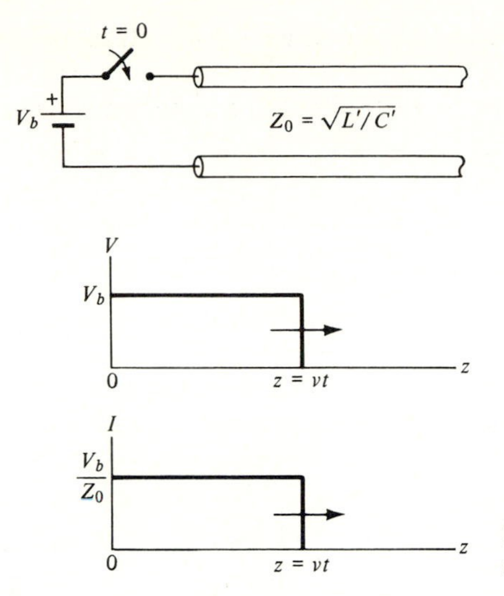

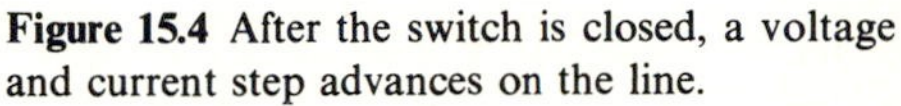
Figure 15.4 After the switch is closed, a voltage and current step advances on the line.

where the units of C' and L' are given as coulomb per meter and henry per meter, respectively. The total energy in an infinitesimal section of line is given by

$$\Delta W = \Delta W_C + \Delta W_L = C'V_b^2\, dz \tag{15.14}$$

which implies that the total energy which leaves the battery in time t is

$$W = C'V_b^2 z = C'V_b^2 vt = \frac{V_b^2}{Z_0} t \tag{15.15}$$

That is, the battery supplies power continuously at the rate $P = dW/dt = V_b^2/Z_0$. As the line is assumed to be infinite, no reflected wave exists.

15.3 REFLECTION AND TRANSMISSION AT A DISCONTINUITY AT THE END OF A LINE

Assume a finite length of line is terminated by a load resistance R_R, as shown in Fig. 15.5. The subscript R denotes receiving end. The voltage V_R at the load is now the sum of an incident and reflected voltage†, as given by (15.9),

$$V_R = V^i + V^r \tag{15.16}$$

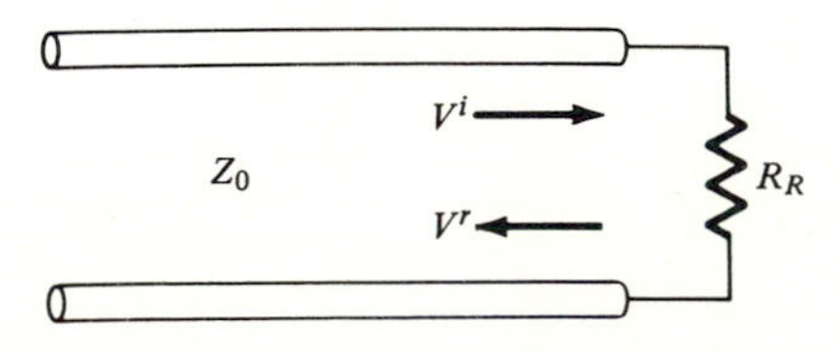

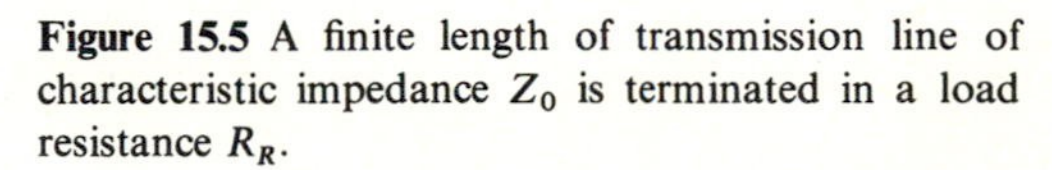
Figure 15.5 A finite length of transmission line of characteristic impedance Z_0 is terminated in a load resistance R_R.

† For convenience, we will refer to the voltage in the incident and reflected waves as incident and reflected voltage.

The total current which flows in R_R is similarly given as

$$I_R = I^i + I^r = \frac{1}{Z_0}(V^i - V^r) \tag{15.17}$$

At the termination therefore

$$R_R = \frac{V_R}{I_R} = \frac{V^i + V^r}{I^i + I^r} = \frac{V^i + V^r}{(1/Z_0)(V^i - V^r)} \tag{15.18}$$

The reflection coefficient Γ is defined as the ratio of reflected voltage to incident voltage. Using the above equation, we can solve for Γ:

$$\boxed{\Gamma = \frac{V^r}{V^i} = \frac{R_R - Z_0}{R_R + Z_0}} \tag{15.19}$$

Note that Γ in terms of the currents in the incident and reflected waves is $\Gamma = -I^r/I^i$. The transmission coefficient τ is defined as the ratio of the voltage in the load to the incident voltage and is given by

$$\boxed{\tau = \frac{V_R}{V^i} = 1 + \Gamma = \frac{2R_R}{R_R + Z_0}} \tag{15.20}$$

The reflection coefficient Γ is an important parameter in transmission line practice because it expresses the relative strength of the reflected wave which is generated by discontinuities on the line. It can have values between $-1 \leq \Gamma \leq 1$, where the extreme values are due to a shorted transmission line ($\Gamma = -1$ for $R_R = 0$) and to an open-circuited line ($\Gamma = 1$ for $R_R = \infty$).

A special case occurs between the two extremes: When the line is terminated with a load resistance $R_R = Z_0$, the line is matched ($\Gamma = 0$) as no reflected voltage is generated by the load, and all power ($V^i I^i$) in the incident wave is absorbed in R_R. Therefore a second answer to the question of what a characteristic impedance is can be given as follows: It is that resistance which when connected to a *finite* length of transmission line, will not result in a reflected wave.

Example: The short-circuited line A transmission line of length l is shorted at the receiving end, as shown in Fig. 15.6. At time $t = 0$, the switch is closed, connecting a battery of voltage V_b to the

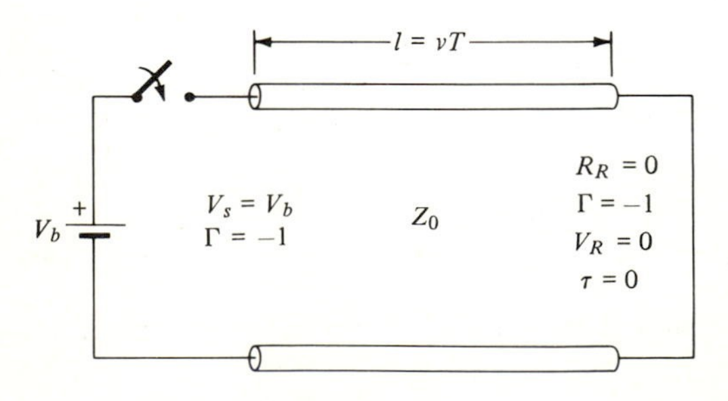

Figure 15.6 A shorted transmission line to which a voltage V_b is suddenly connected. T is the time it takes a wave traveling with velocity v to reach the receiving end.

sending end of the transmission line. (*a*) Determine the voltage and current in the incident and reflected waves. (*b*) Determine the voltage and current at the sending and receiving ends.

SOLUTION (*a*) We readily observe that infinite current must eventually flow if the battery remains connected. The manner in which the current builds, however, is interesting. For times $0 \le t < T$, a wave advances on the line which is identical to that on the infinite line shown in Fig. 15.4. At time $t = T$, the incident wave reaches the short which results in the launching of a reflected wave, since the total voltage at the short must be zero; that is, $V_R = V^i + V^r = 0$, therefore $V^r = -V^i$, or $\Gamma = V^r/V^i = -1$. On the other hand the current in the reflected wave is $I_R = -\Gamma I^i = I^i$, and therefore it adds to the current in the incident wave. Using (5.10), we see that the total current in R_R is $I_R = I^i + I^r = V_b^i/Z_0 - V_b^r/Z_0 = 2V^i/Z_0$. The reflected wave (for $T \le t < 2T$) leaves behind zero voltage and $2V_b/Z_0$ current on the transmission line. At time $t = 2T$, the reflected wave reaches the battery. An ideal battery, which is assumed here, has zero internal resistance and therefore has a reflection coefficient of $\Gamma = -1$. The wave therefore reflects from the battery as it reflects from a short. The wave continues to bounce back and forth on the line as the current builds, and the voltage on the line fluctuates between zero and the battery voltage V_b. Figure 15.7 shows the waves at three instances of time. A check on the above figures can be made since they must be consistent with the end conditions of the line. As $R_R = 0$, the voltage at $z = l$ divided by current at $z = l$ should give zero.

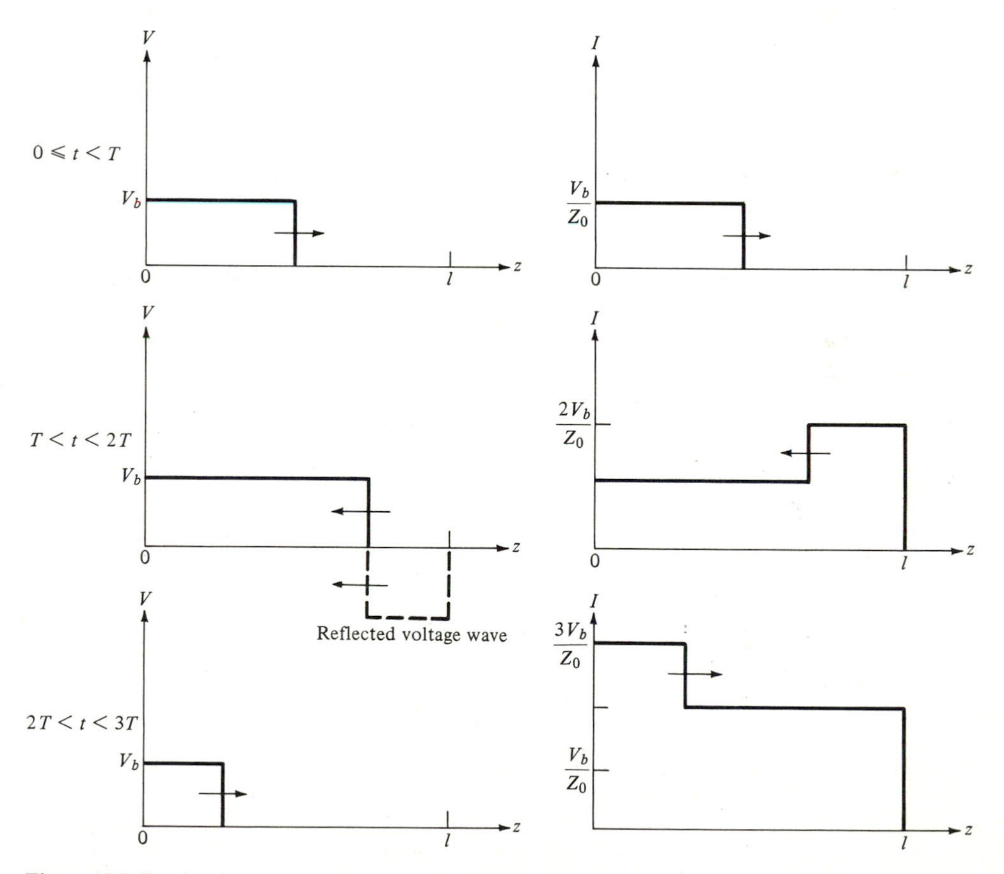

Figure 15.7 Total voltages and currents on the shorted line. Current in the incident and reflected waves is related by $I^r = -\Gamma I^i$.

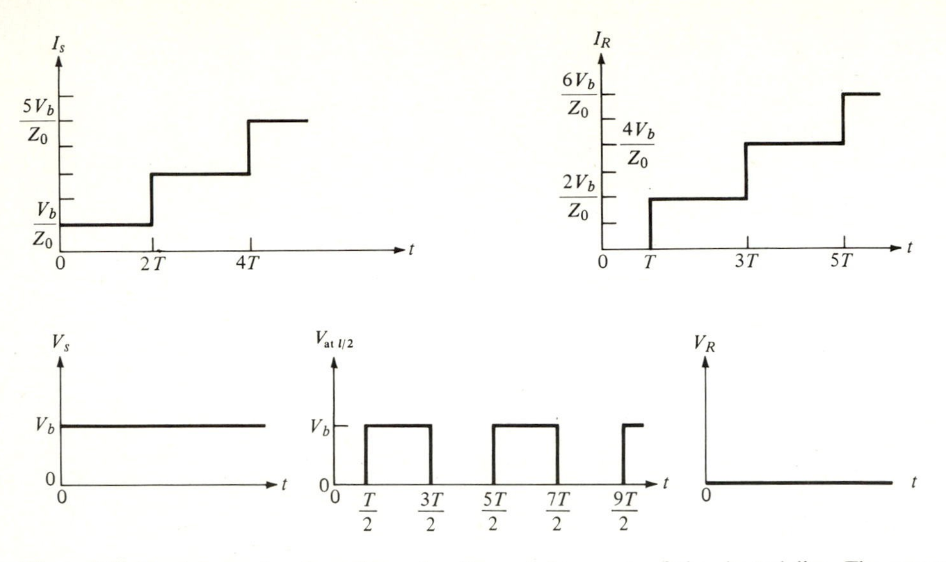

Figure 15.8 Receiving-end and sending-end voltages and currents of the shorted line. The center figure shows the pulsating voltage at the midpoint on the line.

Similarly the voltage at the sending end should be V_b. (b) The total voltages and currents versus time at the sending and receiving ends are shown in Fig. 15.8.

Example: The open-circuited line Figure 15.9 shows a transmission line which is open at the receiving end. At the sending end a battery in series with a resistance R is connected at $t = 0$ to the line. Show the variation in current and voltage.

SOLUTION After the switch is thrown and for $0 < t < 2T$, the input impedance of the line is the characteristic impedance Z_0. During this time interval the sending-end voltage is therefore

$$V_s = I_s Z_0 = \frac{V_b}{3Z_0 + Z_0} Z_0 = \frac{V_b}{4} \tag{15.21}$$

which gives rise to a wave of voltage V_s which travels toward the open end where it is reflected with the reflection coefficient

$$\Gamma = \lim_{R_R \to \infty} \frac{R_R - Z_0}{R_R + Z_0} = \lim_{R_R \to \infty} \frac{1 - Z_0/R_R}{1 + Z_0/R_R} = 1$$

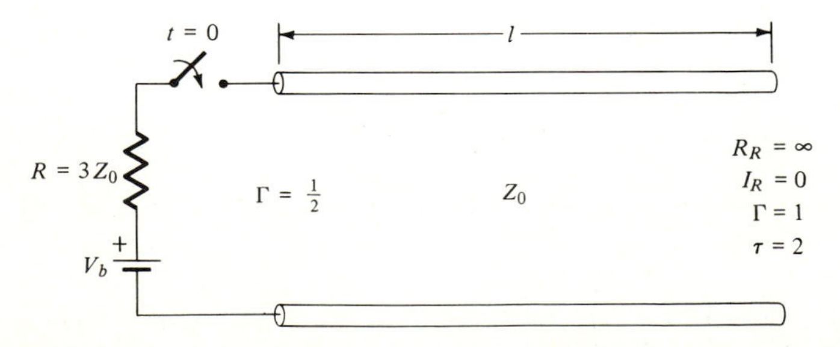

Figure 15.9 An open-circuited transmission line suddenly connected to a battery and resistance R.

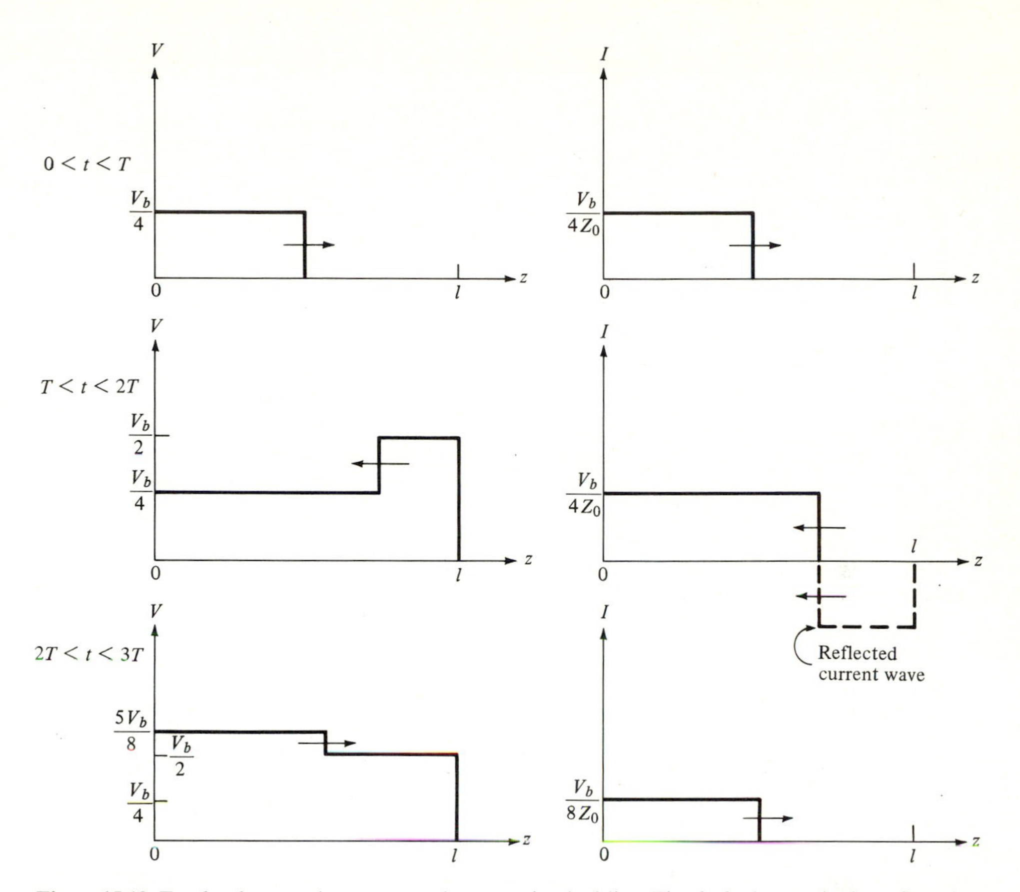

Figure 15.10 Total voltage and current on the open-circuited line. The dashed curve is the reflected current wave. Note that the relation between current in the incident and reflected waves is $I^r = -\Gamma I^i$.

and travels back toward the sending end where it is again reflected with the coefficient $\Gamma_s = (3Z_0 - Z_0)/(3Z_0 + Z_0) = \frac{1}{2}$. Figure 15.10 shows the wave at three instances of time on the line.

The voltage on the line increases in steps until it reaches the battery voltage V_b, while the current on the line goes to zero in steps, as shown in Fig. 15.11. We can consider the arrangement of Fig. 15.9 as one in which the transmission line is charged to voltage V_b. This problem is similar to that of charging a capacitor from a battery through a resistance (see Fig. 5.2). As a matter of fact, treating the open-circuited line as a capacitor with capacitance $C = C'l$, where C' is the capacitance per unit length of the line, the time constant of the equivalent charging circuit of Fig. 15.9 is

$$RC = (3Z_0)(C'l) = \left(3\sqrt{\frac{L'}{C'}}\right)(C'l) = 3\sqrt{L'C'}\,l = \frac{3l}{v} = 3T \tag{15.22}$$

where T is the time it takes the wave to travel a distance l. Using this time constant in (5.7), we can write for capacitor voltage,

$$V_s = V_b(1 - e^{-t/3T}) \tag{15.23}$$

which is plotted as the dashed curve in Fig. 15.11. The charging current is similarly given by use of Eq. (5.6) and is shown dashed in Fig. 15.11. This figure displays the different nature of the responses in a lumped and distributed network.

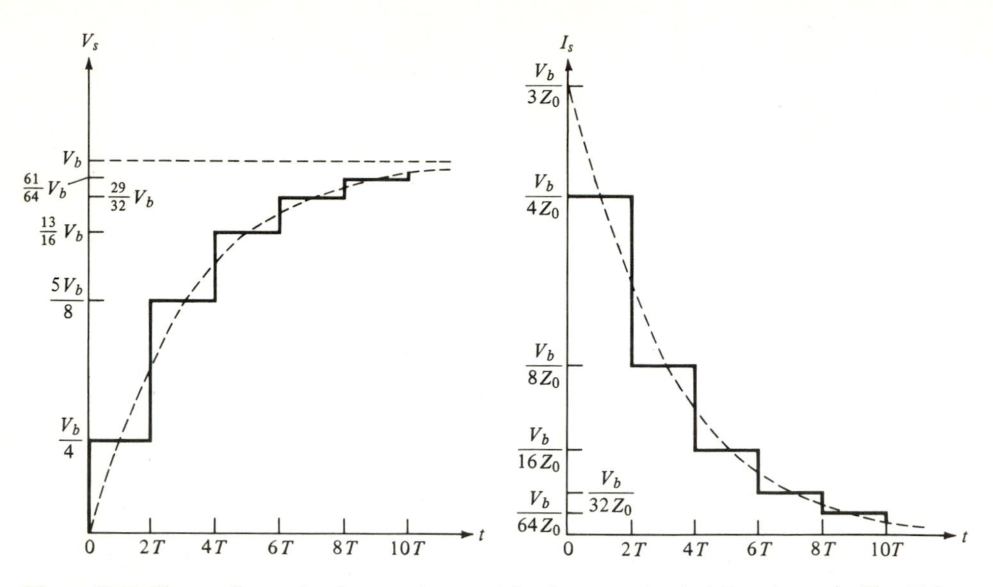

Figure 15.11 The sending-end voltage and current for the open-circuited line shown in Fig. 15.9. The dashed curve is the charging voltage and current for the equivalent lumped *RC* circuit.

Example: Discharging a charged line Figure 15.12 shows a transmission line which is charged to voltage V_b. At time $t = 0$, a resistance R is connected to the line. Show the variation in current and voltage during discharge of the line.

SOLUTION The closing of the switch disturbs the equilibrium condition of the line. The transient gives rise to a wave which travels down the line. To calculate the magnitude of the transient, we use the equivalent circuit, Fig. 15.12*b*, which gives

$$V_s = -I_s R = \frac{V_b}{Z_0 + R} R \tag{15.24}$$

for the voltage imposed across the sending end of the line. Therefore as V_s is different from V_b, a wave is launched which leaves behind a voltage V_s. A transient can only be avoided if $V_s = V_b$, which is possible only for $R = \infty$. For finite R, $V_s < V_b$, and the launched wave whose amplitude is V^i wipes out part of the initial voltage V_b on the line. To find V^i we observe that the voltage across R is the sum of the initial

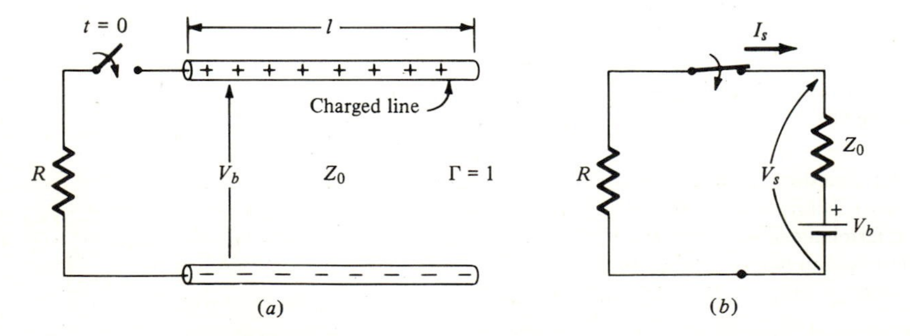

Figure 15.12 (*a*) An open-circuited transmission line initially charged to voltage V_b; (*b*) equivalent sending-end circuit at the time the switch is closed.

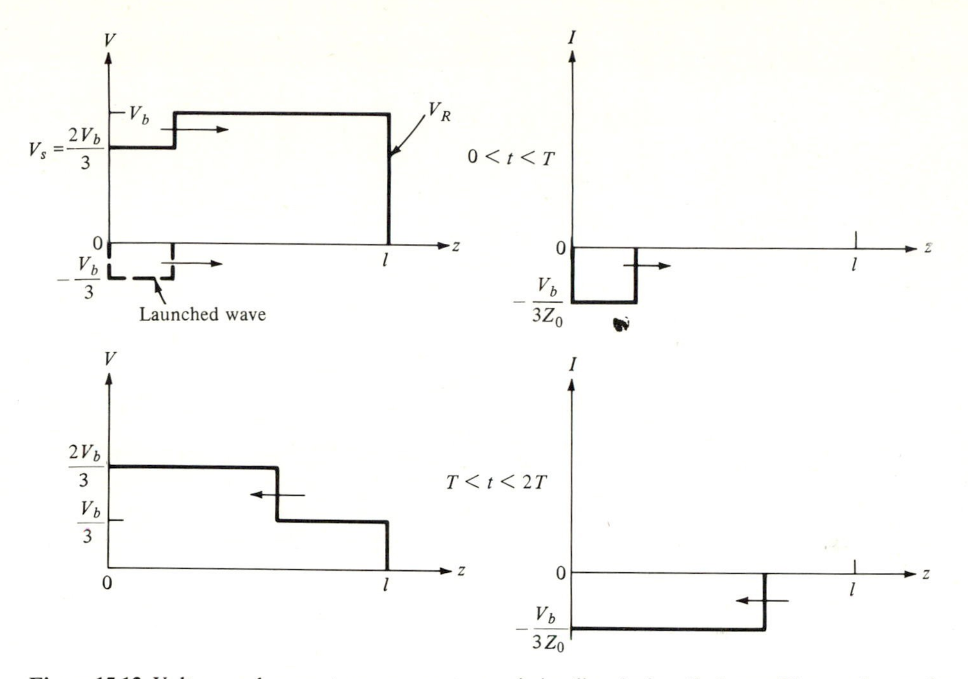

Figure 15.13 Voltage and current waves on a transmission line during discharge. The graphs are for the case $R = 2Z_0$. The dashed graph is the incident wave launched at $t = 0$.

voltage V_b and the voltage V^i of the launched wave:

$$V_s = V_b + V^i \tag{15.25}$$

Substituting (15.24), we can solve for V^i:

$$V^i = -V_s\frac{Z_0}{R} = -V_b\frac{Z_0}{Z_0 + R} = I_s Z_0 \tag{15.26}$$

When this wave reaches the open end, it reflects with a coefficient $\Gamma = 1$. Upon reaching R, it will reflect again with a coefficient $\Gamma = (R - Z_0)/(R + Z_0)$. These reflections will continue until the voltage on the line reaches zero.

Figure 15.13 shows the voltage and current waves at two instances of time, for the case $R = 2Z_0$. The sending-end voltages and currents for three different values of R are shown in Fig. 15.14. The three cases correspond to overdamped, critically damped, and underdamped circuits. The case of $R = Z_0$ is of special interest: A wave of voltage $-V_b/2$ is launched. It reflects at the open end and leaves behind zero voltage on the line as it moves toward R. When the reflected wave reaches R, it is absorbed because $R = Z_0$. No reflection takes place after $t = 2T$, and the line can be considered discharged.

Charged lines can be used in the generation of pulses, in which the pulse width is controlled by the length of the line.

Example: Transmission line with pulsed input voltage A coaxial transmission line shown in Fig. 15.15*a* has a characteristic impedance of 50 Ω, velocity of propagation of 200 m/μs, length of 400 m, and is terminated in a load resistance $R_R = 16.7$ Ω. At the sending end the line is connected to a pulse generator which has an internal resistance of 150 Ω and produces a 40-V 1-μs pulse. Find the sending-end voltage and current waveforms.

Solution At $t = 0$, the pulse is unaffected by the receiving end and begins to travel down the line as if the line were infinitely long. Initially the 40-V pulse divides between the generator resistance R_g

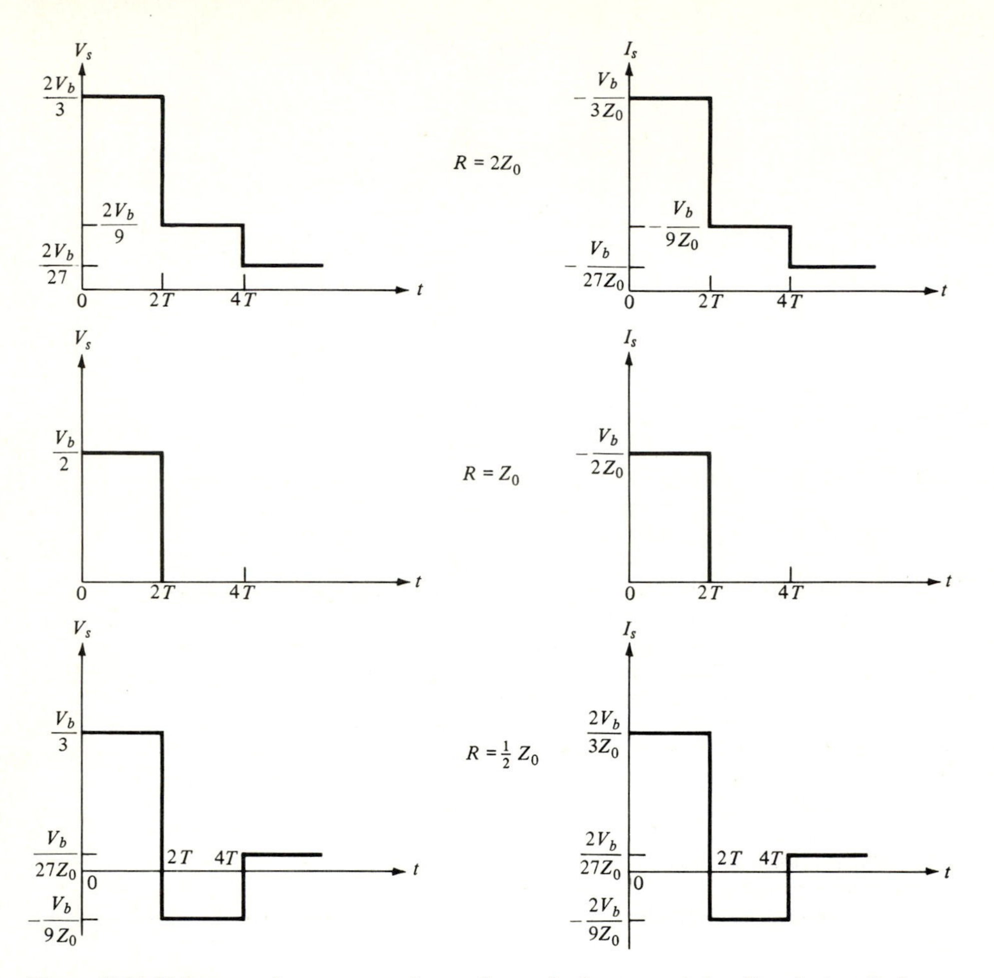

Figure 15.14 Voltages and currents at the sending end of a transmission line during discharge. The three cases correspond to $R = 2Z_0$, Z_0, and $Z_0/2$.

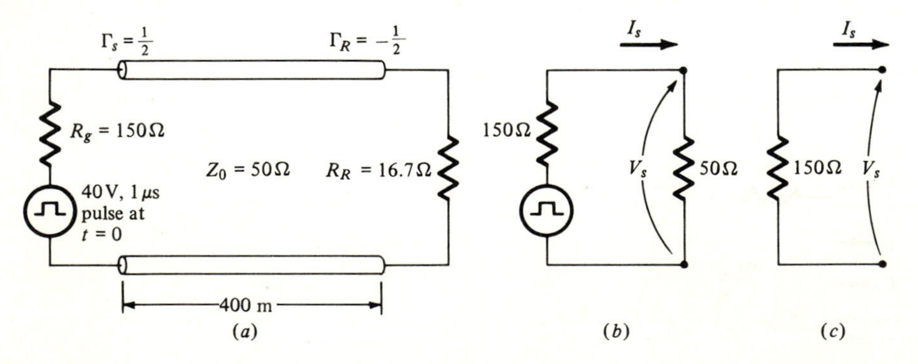

Figure 15.15 (*a*) A pulse generator connected to a transmission line; (*b*) the equivalent sending-end circuit at $t = 0$; (*c*) the equivalent sending-end circuit at the time the reflected pulse arrives ($t = 4\mu s$).

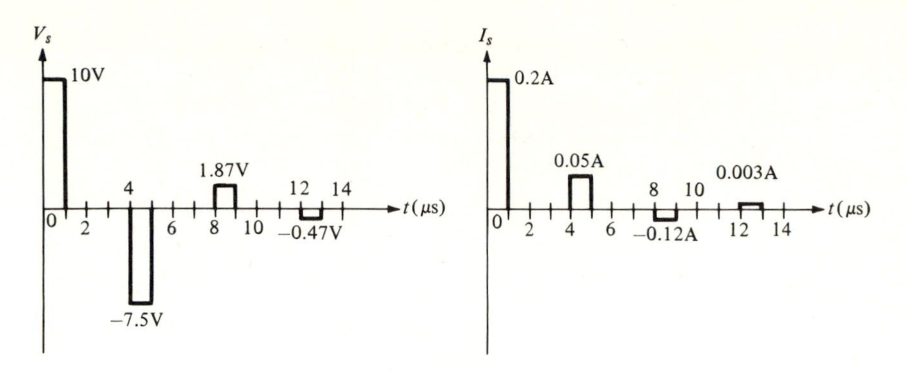

Figure 15.16 Sending-end voltage and current waveforms.

and the characteristic impedance Z_0 of the line to give a sending-end pulse of (see Fig. 15.15*b*)

$$V_s = \frac{40\ \text{V}}{R_g + Z_0} Z_0 = \frac{40\ \text{V}}{150 + 50} 50 = 10\ \text{V} \tag{15.27}$$

The 10-V pulse travels to the load in 2 μs and is partly absorbed and partly reflected with the coefficient $\Gamma_R = -\frac{1}{2}$. The reflected pulse, which is -5 V, travels back toward the generator in 2 μs and is partly reflected by the mismatched resistance R_g. The reflection coefficient at the generator is $\Gamma_s = \frac{1}{2}$, which results in a reflected pulse of -2.5 V. From this point on, the process repeats itself, with the pulse inverted and halved at the load and halved at the generator.

When the reflected pulse arrives at the sending end, it leaves a medium with a characteristic impedance Z_0 and encounters the resistance R_g. Reflection will result unless $R_g = Z_0$. The total voltage at the sending end is the sum of voltages in the incident and reflected waves. Thus

$$V_s = V^i + V^r = V^i(1 + \Gamma_s) \tag{15.28}$$

The equivalent circuit for this time is given by Fig. 15.15*c*. Using this circuit, the sending-end current is $I_s = -V_s/R_g$. For example, at $t = 4\ \mu$s, $V_s = -5(1 + \frac{1}{2}) = -7.5$ V and $I_s = 7.5/150 = 0.05$ A. The current is positive as it has the positive direction shown in Fig. 15.15*c*. V_s and I_s versus time are shown in Fig. 15.16.

15.4 REFLECTION AND TRANSMISSION AT A DISCONTINUITY ON THE LINE

A discontinuity can exist anywhere along the transmission line. Figure 15.17*a* presents a general case. It is composed of three separate discontinuities: (1) A

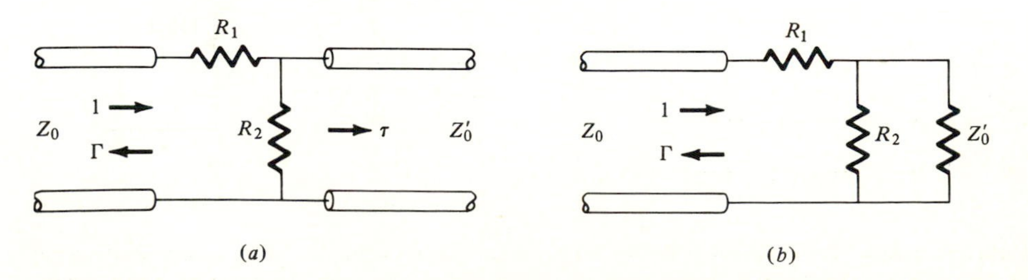

Figure 15.17 (*a*) General discontinuity on a line; (*b*) equivalent circuit of the discontinuity.

transmission line of characteristic impedance Z_0 interrupted by a lumped series resistance R_1 ($R_2 = \infty$, $Z_0 = Z_0'$); (2) A transmission line shunted by a lumped resistance R_2 ($R_1 = 0$, $Z_0 = Z_0'$); (3) Two transmission lines of characteristic impedance Z_0 and Z_0' connected to each other ($R_1 = 0$, $R_2 = \infty$). It is much more likely that one of the separate discontinuities will be encountered in practice. However, as the general case is not difficult to solve, it will be presented. A voltage wave of unit amplitude is assumed to be incident on the discontinuity. The reflected wave will then have amplitude Γ and the transmitted wave amplitude τ as shown. The equivalent circuit for the incident wave is shown in Fig. 15.17*b*. Hence the equivalent resistance of the discontinuity is given by

$$R_{eq} = R_1 + \frac{R_2 Z_0'}{R_2 + Z_0'} \tag{15.29}$$

The reflection coefficient, which is defined as $\Gamma = V^r/V^i$, is therefore, using (15.19),

$$\Gamma = \frac{R_{eq} - Z_0}{R_{eq} + Z_0} = \frac{R_1 + R_2 Z_0'/(R_2 + Z_0') - Z_0}{R_1 + R_2 Z_0'/(R_2 + Z_0') + Z_0} \tag{15.30}$$

The transmission coefficient τ gives that portion of the incident wave which continues on the second transmission line. Hence τ is the ratio of the voltage across the parallel combination of R_2 and Z_0' to the voltage in the incident wave, or

$$\tau = (1 + \Gamma)\frac{R_2 Z_0'/(R_2 + Z_0')}{R_1 + R_2 Z_0'/(R_2 + Z_0')} = \frac{2R_2 Z_0'}{(R_1 + Z_0)(R_2 + Z_0') + R_2 Z_0'} \tag{15.31}$$

Note that the amplitude of the transmitted wave, $1 + \Gamma$, divides between R_1 and the second transmission line, which explains the presence of the factor multiplying $1 + \Gamma$ in Eq. (15.31).

Example: A transmission line of characteristic impedance Z_0 is shunted at some point on the line by a resistance R, as shown in Fig. 15.18. Find the power transmitted past the discontinuity, if the incident voltage $V^i = 10$ V and $Z_0 = R = 300\ \Omega$.

SOLUTION Using (15.30) and (15.31), the reflection and transmission coefficients are

$$\Gamma = \frac{-Z_0}{2R + Z_0} \qquad \tau = 1 + \Gamma = \frac{2R}{2R + Z_0} \tag{15.32}$$

Applying conservation of energy at the discontinuity, we can state that power in the incident wave must equal the sum of power in the reflected and transmitted waves and power absorbed in R; that is,

$$P^i = P^r + P^t + P_R$$

$$\frac{(V^i)^2}{Z_0} = \frac{(\Gamma V^i)^2}{Z_0} + \frac{(\tau V^i)^2}{Z_0} + \frac{(\tau V^i)^2}{R} \tag{15.33}$$

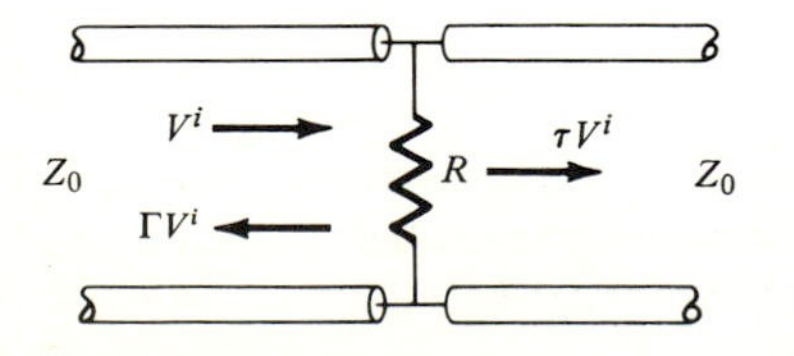

Figure 15.18 A transmission line shunted by resistance R.

Thus the transmitted power is

$$P^t = \frac{(V^i)^2}{Z_0}\left(1 - \Gamma^2 - \frac{\tau^2 Z_0}{R}\right) \tag{15.34}$$

Using the given values, we find that the incident power is $P^i = (V^i)^2/Z_0 = 0.33$ W and the transmitted power is $P^t = \tau^2 P^i = (2/3)^2\, 0.33 = 0.148$ W. Therefore only 44 percent of the incident power gets past the discontinuity.

15.5 TRANSMISSION LINE WITH CAPACITIVE TERMINATION

Reflection at an inductive or capacitive termination, such as shown in Fig. 15.19, is more difficult to analyze because now the voltage-current relationship across the termination is a function of time. This makes the reflection coefficient a function of time also.

Consider the capacitor C in Fig. 15.19 to be initially uncharged; that is, $V_C(0) = (1/C)\int_{-\infty}^{0} I\, dt = 0$. The total voltage V across the termination is the sum of incident and reflected voltages and is related to total current I by

$$V = V^i + V^r = RI + \frac{1}{C}\int_0^t I\, dt \tag{15.35}$$

The total current at $z = l$ is similarly given by

$$I = I^i + I^r = \frac{V^i}{Z_0} - \frac{V^r}{Z_0} \tag{15.36}$$

Differentiating (15.35) with respect to time and noting that $dV^i/dt = 0$, we can obtain a differential equation for the reflected voltage as

$$\frac{dV^r}{dt} + \frac{V^r}{C(Z_0 + R)} = \frac{V^i}{C(Z_0 + R)} \tag{15.37}$$

whose solution is

$$V^r = V^i\left[1 - \frac{2Z_0}{Z_0 + R}\, e^{-t/C(Z_0 + R)}\right] = V^i\Gamma(t) \tag{15.38}$$

where $V^i = V_b$ and $t = 0$ is the time at which the pulse arrives at the termination (switch is closed at $t = -T$, where $T = l/v$). As the capacitor is assumed to be initially uncharged, it acts initially (at $t = 0$) as a short circuit, and the reflection of

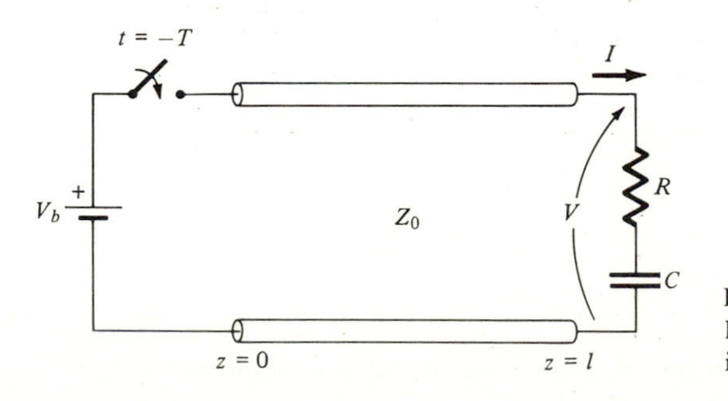

Figure 15.19 A capacitive termination of lumped resistance R and lumped capacitance C in series.

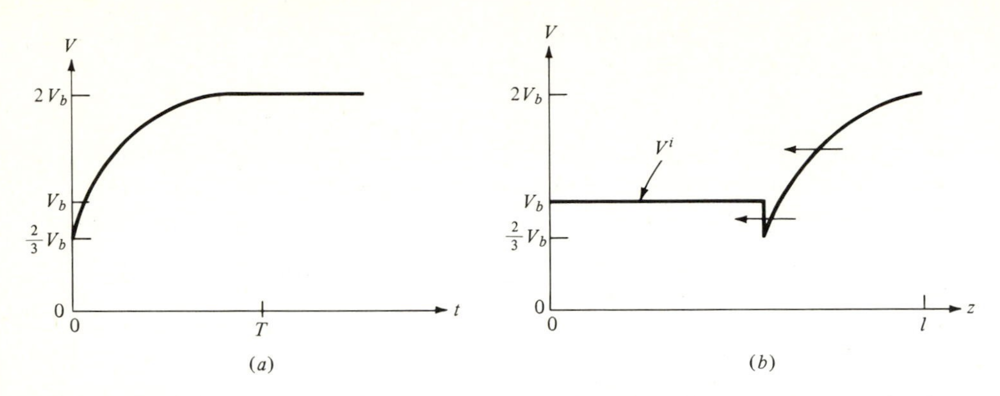

Figure 15.20 (*a*) Total voltage at the receiving end for $R = Z_0/2$; (*b*) total voltage on the line for $0 < t < T$.

the wave is caused by resistance R. The initial condition which was used to determine the unknown constants in solution (15.38) is therefore $V^r = V^i(R - Z_0)/(R + Z_0) = V^i\Gamma(0)$. The above solution can be checked by using the limiting cases: $R = \infty$ gives $\Gamma = 1$; the capacitor short-circuits by setting $C = \infty$ which gives $\Gamma = (R - Z_0)/(R + Z_0)$; $R = 0$ gives $\Gamma(t) = 1 - 2e^{-t/CZ_0}$.

Figure 15.20 gives the waveform of the total voltage at the receiving end and the total voltage on the line sometime after reflection. C was chosen to be

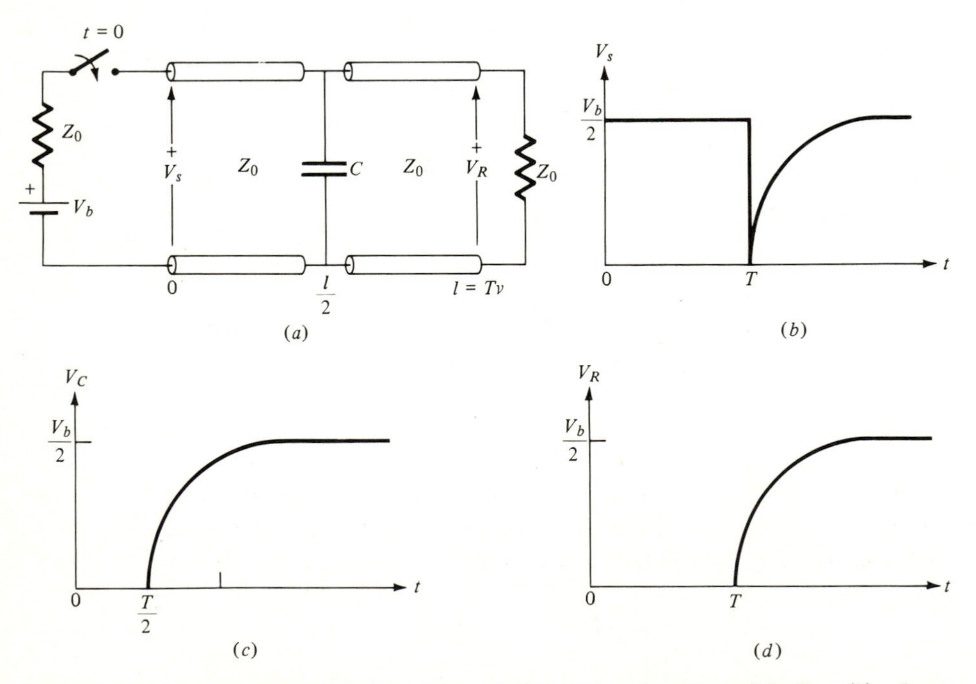

Figure 15.21 (*a*) A transmission line shunted by capacitance C at the center of the line; (*b*) voltage variation at the sending end, (*c*) across C, and (*d*) at the receiving end.

sufficiently small such that the time constant $(Z_0 + R)C \ll T$. The capacitor can therefore be considered fully charged in a time less than T and thereafter acts as an open circuit.

Example: A transmission line shown in Fig. 15.21 has a discontinuity in the middle of the line in the form of a lumped capacitance C which is initially uncharged. Find the voltage waveform at the sending end, receiving end, and across C.

SOLUTION At $t = 0$, a voltage step of magnitude $V_s = V_b/2$ starts at the sending end. At $t \geq T/2$, the voltage step reflects from C with a reflection coefficient $\Gamma(t) = -e^{-2(t-T/2)/CZ_0}$, as given by (15.38). The transmitted part of the voltage step is determined by the transmission coefficient $\tau(t) = 1 + \Gamma(t)$ and is absorbed by the matched termination Z_0. To show the complete variation in the waveform, we choose C such that the charging time is small compared to T; that is, $Z_0 C \ll T$. The capacitor, once fully charged, acts as an open circuit and does not disturb the line thereafter. The voltage anywhere on the transmission line then settles down to the value $V_b/2$.

15.6 LOSSLESS TRANSMISSION LINE WITH SINUSOIDAL VOLTAGE SOURCES

No restriction on the time variation of the voltages which were applied to the lines was made in the previous sections. As most practical generators have sinusoidally varying output voltages, we consider a voltage of frequency $f = \omega/2\pi$:

$$V(0, t) = V \cos \omega t = \text{Re } Ve^{j\omega t} \tag{15.39}$$

applied to the sending end at $z = 0$. For convenience, the phasor notation in which sinusoids are represented by exponentials is used (see footnote page 478). The applied voltage causes a traveling wave on the transmission line which is

$$V(z, t) = \text{Re } V^i e^{j\omega(t - z/v)} \tag{15.40}$$

where ω gives the phase shift per unit time and ω/v gives the phase shift per unit length. Phase velocity on the line is given by (15.8) as $v = 1/\sqrt{L'C'}$. The phase-propagation constant β and wavelength λ are then given by $\omega/v = \beta = 2\pi/\lambda$. When reflection is present, the total voltage and current on the line are made up of oppositely traveling waves which, using (15.9) and (15.11), are given by

$$V(z, t) = \text{Re } (V^i e^{j(\omega t - \beta z)} + V^r e^{j(\omega t + \beta z)}) \tag{15.41}$$

$$I(z, t) = \text{Re} \frac{1}{Z_0} (V^i e^{j(\omega t - \beta z)} - V^r e^{j(\omega t + \beta z)}) \tag{15.42}$$

where $V^i/I^i = Z_0$ and $V^r/I^r = -Z_0$. As voltages and currents everywhere along the line oscillate at the frequency of the applied voltage, the factor $e^{j\omega t}$ can be omitted. The phasor voltages and currents are then written in the simple form

$$V(z) = V^i e^{-j\beta z} + V^r e^{j\beta z} \tag{15.43}$$

$$I(z) = \frac{1}{Z_0} (V^i e^{-j\beta z} - V^r e^{j\beta z}) \tag{15.44}$$

remembering that the actual quantities are the instantaneous voltages and currents which are obtained by remultiplying by $e^{j\omega t}$ and taking the real part (Re) of the resulting expressions.

In practice the most useful measurements are those with respect to the receiving end of the line. Therefore the origin $z = 0$ is taken at the receiving end, as shown in Fig. 15.22. The reflection coefficient Γ is the ratio of reflected to incident voltage at the point of reflection. It is determined by setting $z = 0$ in Eqs. (15.43) and (15.44) and solving for Γ, which gives

$$\boxed{\Gamma = \frac{V^r}{V^i} = \frac{Z_R - Z_0}{Z_R + Z_0}} \tag{15.45}$$

where the load impedance at the receiving end is $Z_R = V_R/I_R = V(0)/I(0) = Z_0(V^i + V^r)/(V^i - V^r)$. Γ is in agreement with that of (15.19) but is now valid for a complex load impedance. Problems involving reactive loads (Γ a complex constant) can now be handled as easily as those with resistive loads (Γ a real constant). The impedance at an arbitrary position $z = -d$ on the line can be found by dividing (15.43) by (15.44):

$$\boxed{Z(d) = \frac{V(d)}{I(d)} = Z_0 \frac{1 + \Gamma e^{-j2\beta d}}{1 - \Gamma e^{-j2\beta d}} = Z_0 \frac{Z_R \cos \beta d + jZ_0 \sin \beta d}{Z_0 \cos \beta d + jZ_R \sin \beta d}} \tag{15.46}$$

where $e^{\pm jx} = \cos x \pm j \sin x$ was used, and

$$V(d) = V^i e^{j\beta d}(1 + \Gamma e^{-j2\beta d}) = \frac{V_R}{1 + \Gamma} e^{j\beta d}(1 + \Gamma e^{-j2\beta d}) \tag{15.47}$$

$$I(d) = \frac{V^i}{Z_0} e^{j\beta d}(1 - \Gamma e^{-j2\beta d}) = \frac{V_R}{Z_0(1 + \Gamma)} e^{j\beta d}(1 - \Gamma e^{-j2\beta d}) \tag{15.48}$$

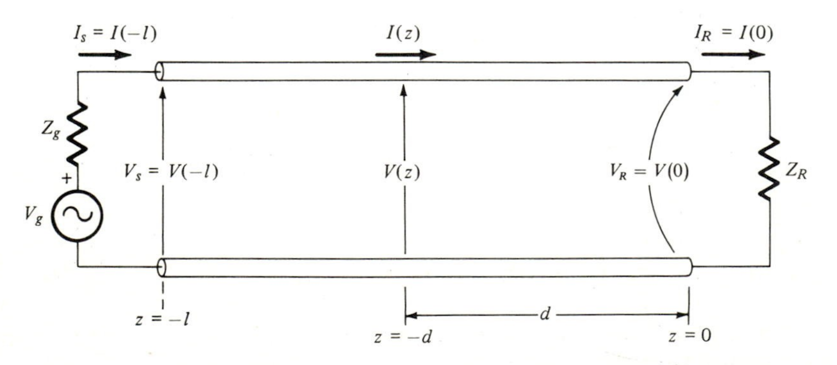

Figure 15.22 A transmission line terminated by an arbitrary load impedance Z_R.

Note that d is a positive distance measured from the receiving end. Expression (15.46) shows that the impedance of the line is periodic (with period $\beta d = \pi$) and varies between maximum and minimum values about Z_0. Multiplying numerator and denominator of (15.46) by I_R/Z_0, voltage and current anywhere on the line are determined as

$$V(d) = V_R \cos \beta d + jI_R Z_0 \sin \beta d \tag{15.49}$$

$$I(d) = I_R \cos \beta d + j\frac{V_R}{Z_0} \sin \beta d \tag{15.50†}$$

Sending-end V_s and I_s are given in terms of receiving-end V_R and I_R by substituting $d = l$ in the above equations, where $V(d = l) = V_s$ and $I(d = l) = I_R$.

Example: An RG-58C/U coaxial cable has a characteristic impedance of 50 Ω and a velocity of propagation of 200 m/μs. Find the input impedance of a 50-m cable which is terminated by a load of 25 Ω. The operating frequency is 27 MHz.

SOLUTION

$$\beta d = \frac{\omega}{v} d = \frac{2\pi(27 \times 10^6)}{200 \times 10^6} 50 = 42.41$$

Using (15.46)

$$Z(d) = 50\frac{25 + j50 \tan 42.41}{50 + j25 \tan 42.41} = 100\ \Omega$$

Two cases are of special interest:

1. A line short-circuited at the receiving end has $Z_R = 0$, and therefore $V_R = 0$. Substituting $Z_R = 0$ in (15.46), the input impedance is

$$Z_{sc} = jZ_0 \tan \beta d \tag{15.53}$$

As no power can be dissipated in a short, the input impedance of a short-circuited line is purely reactive. Figure 15.23*a* shows the variation of reactance with distance d from the short. We observe that for $d < \lambda/4$, the short-circuited line is inductive, and for $\lambda/4 < d < \lambda/2$, it is capacitive. For $d = \lambda/4$, the input

† Frequently an expression in terms of sending-end quantities V_s, I_s, and Z_s is needed. Setting $V(z = 0) = I_s Z_s$ and $I(z = 0) = I_s$ in (15.43) and (15.44) and solving for V^i and V^r, we obtain

$$V(z) = V_s \cos \beta z - jI_s Z_0 \sin \beta z \tag{15.51}$$

$$I(z) = I_s \cos \beta z - j\frac{V_s}{Z_0} \sin \beta z \tag{15.52}$$

where $Z_s = V_s/I_s$ and z is a positive distance measured from the sending end which in this case is at $z = 0$. Sending-end and receiving-end quantities can be related by setting $z = l$, where $V_R = V(l)$ and $I_R = I(l)$; thus $V_R = V_S \cos \beta l - jI_S Z_0 \sin \beta l$, and $I_R = I_s \cos \beta l - j(V_s/Z_0) \sin \beta l$.

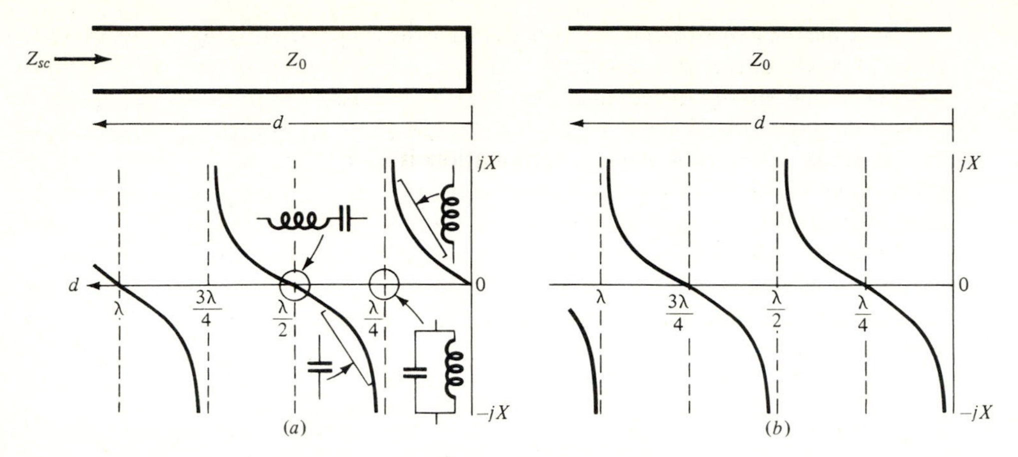

Figure 15.23 The input impedance of a short-circuited transmission line (*a*) and an open-circuited line (*b*) is purely reactive. Theoretically any value of reactance can be obtained by varying *d*.

impedance is theoretically infinite, and the line acts as an antiresonant circuit comparable to a lumped parallel *LC* circuit. For $d = \lambda/2$, the input impedance is zero, and the line acts as a resonant circuit in the sense of a lumped series *LC* circuit. This pattern repeats itself with a periodicity of $\lambda/2$. Figure 14.1 shows the voltage and current variation on a short-circuited line (let $E_x \to V, H_y \to I$).

2. A line which is open-circuited at the receiving end has $Z_R = \infty$, and therefore $I_R = 0$. Substituting $Z_R = \infty$ in (15.46) gives for input impedance

$$\boxed{Z_{oc} = -jZ_0 \cot \beta d} \tag{15.54}$$

An open-circuited line is capacitive for $d < \lambda/4$, is inductive for $\lambda/4 < d < \lambda/2$, and so on. A shorted line which is inductive is therefore equivalent to an open-circuited line which is longer by $\lambda/4$. Figure 15.23*b* shows the input impedance of an open-circuited line.

Example: A lossless transmission line for which $Z_0 = 400\ \Omega$ is a quarter wavelength long. The sending end is connected to a generator which has an internal impedance $Z_g = 50\ \Omega$ and a voltage $V_g = 10$ V. What is the voltage at the open-circuited receiving end?

SOLUTION Using either (15.50) or (15.51) and the substitution $\beta d = \pi/2$, results in

$$V_R = -jI_S Z_0$$

I_S can be found from the equivalent sending-end circuit. Using (15.54), we obtain $Z_s = 0$, which gives $I_S = V_g/Z_g = 0.2$ A. The receiving-end voltage is therefore

$$V_R = -j(0.2)(400) = -j80 \text{ V}$$

In terms of instantaneous values, $V_g(t) = \sqrt{2}\,(10 \cos \omega t)$ and $V_R(t) = \sqrt{2}\,(80 \sin \omega t)$. The open-circuited $\lambda/4$ line acts therefore as a voltage step-up device. Note that the voltage step-up V_R/V_S is infinite, as V_S for an open-circuited quarter-wave lossless line is zero.

As a short or open circuit cannot accept power, complete reflection must take place at such terminations; that is, $|\Gamma| = 1$. For the same reason, complete reflection also takes place at a reactive termination, $\pm jX_R$, which gives a reflection coefficient $|\Gamma| = |(\pm jX_R - Z_0)/(\pm jX_R + Z_0)| = 1$. This is not unexpected since any purely reactive termination is equivalent to a finite length of short-circuited or open-circuited line. Hence a load which is an inductor or capacitor can be replaced by a section of short-circuited or open-circuited line.

In conclusion, a third answer to the question of what a characteristic impedance is can be stated as follows: Z_0 is the square root of the product of input impedances of a finite-length transmission line with alternate short-circuit and open-circuit termination; that is,

$$\boxed{Z_0 = \sqrt{Z_{sc} Z_{oc}}} \tag{15.55}$$

15.7 REFLECTION AND VOLTAGE STANDING-WAVE RATIO *S*

As the oppositely traveling waves of (15.43) undergo different phase changes, there will be positions of voltage maxima and minima on the line, corresponding to positions of constructive and destructive interference. The *standing-wave ratio S* is defined as the ratio of maximum voltage magnitude to minimum voltage magnitude,

$$\boxed{S = \frac{V_{\max}}{V_{\min}} = \frac{|V^i| + |V^r|}{|V^i| - |V^r|} = \frac{1 + |\Gamma|}{1 - |\Gamma|}} \tag{15.56}$$

where $\Gamma = V^r/V^i$. Depending on the strength of the reflected wave, the values of S may range from one to infinity. Note that S can be expressed just as well in terms of current; that is, $S = I_{\max}/I_{\min}$, where $I_{\max} = (|V^i| + |V^r|)/Z_0$ and $I_{\min} = (|V^i| - |V^r|)/Z_0$. The standing-wave ratio is related to the magnitude of the reflection coefficient as

$$\boxed{|\Gamma| = \frac{S - 1}{S + 1}} \tag{15.57}$$

S is easily measured by moving a pick-up probe along an open line or, for a coaxial line, by cutting a thin axial slot in the outer cylinder along which the pick-up probe can be moved. The probe is connected to a detector which reads the absolute values of the variation in voltage along the line. For example, if the line is terminated in a pure resistance R_R, the meter of the detector would read, using (15.49),

$$\begin{aligned} |V(d)| &= V_R\sqrt{\cos^2 \beta d + (Z_0/R_R)^2 \sin^2 \beta d} \\ &= V^i\sqrt{(1 + \Gamma)^2 \cos^2 \beta d + (1 - \Gamma)^2 \sin^2 \beta d} \end{aligned} \tag{15.58}$$

where $V_R = I_R R_R$
$V_R = V^i + V^r$
$I_R = (V^i - V^r)/Z_0$

As (15.58) is periodic with period $\beta d = \pi$, the distance between minima on the line is $d = \lambda/2$.

A position where incident and reflected waves interfere to give a maximum in voltage, is a position of minimum in current. This result is a direct consequence of the minus sign in (15.44). At this position on the line the impedance is a maximum and is purely resistive:

$$\boxed{R_{\max} = \frac{V_{\max}}{I_{\min}} = Z_0 \frac{|V^i| + |V^r|}{|V^i| - |V^r|} = Z_0 S} \tag{15.59}$$

Similarly, $V_{\min}$, $I_{\max}$, and $R_{\min}$ occur together, a quarter wavelength from $V_{\max}$, $I_{\min}$, and $R_{\max}$:

$$\boxed{R_{\min} = \frac{V_{\min}}{I_{\max}} = Z_0 \frac{|V^i| - |V^r|}{|V^i| + |V^r|} = \frac{Z_0}{S}} \tag{15.60}$$

$R_{\min}$ is again purely resistive. Recall that for a lossless line, Z_0 is a pure resistance.

The magnitude distributions of voltage and current for five different terminations are plotted in Fig. 15.24 using Eq. (15.58). The variations are similar to those of the fields in plane waves shown in Figs. 14.1 and 14.9. The following observations can now be made: For purely resistive terminations, the receiving end is a position of an impedance maximum or minimum. This follows either from (15.59) and (15.60) or directly from a calculation of S, which for pure resistance termination gives

$$S = \frac{1 + |\Gamma|}{1 - |\Gamma|} = \frac{1 + (R_R - Z_0)/(R_R + Z_0)}{1 - (R_R - Z_0)/(R_R + Z_0)}$$

which simplifies to

$$\boxed{S = \begin{cases} \dfrac{R_R}{Z_0} & \text{for } R_R > Z_0 \\[2ex] \dfrac{Z_0}{R_R} & \text{for } R_R < Z_0 \end{cases}} \tag{15.61}$$

The value of an unknown impedance at the receiving end can now be easily determined from relative measurements in voltage or current along the line, provided the receiving end is a position of maximum or minimum. If this is the case the unknown termination is purely resistive and has a value $R_R = Z_0(V_{\max}/V_{\min})$ or $R_R = Z_0(V_{\min}/V_{\max})$, where Z_0 can be calculated from the dimensions of the line or obtained directly from measurement as $Z_0 = V_{\min}/I_{\min} = V_{\max}/I_{\max}$.

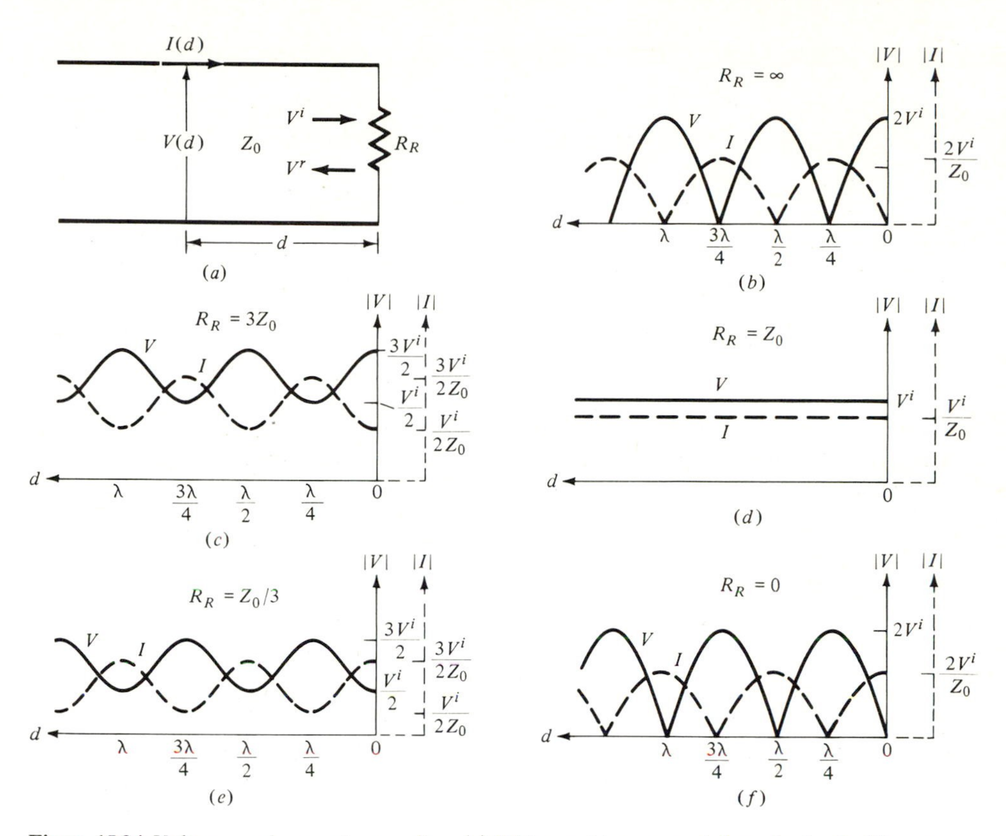

Figure 15.24 Voltages and currents on a line. (*a*) Distance d is measured from the load; (*b*) open-circuited line for which $R_R = \infty$, $\Gamma = 1$, $S = \infty$; (*c*) termination $R_R = 3Z_0$ for which $\Gamma = \frac{1}{2}$, $S = 3$; (*d*) matched line for which $R_R = Z_0$, $\Gamma = 0$, $S = 1$; (*e*) termination $R_R = Z_0/3$ for which $\Gamma = -\frac{1}{2}$, $S = 3$; (*f*) short-circuited line for which $R_R = 0$, $\Gamma = -1$, $S = \infty$.

Determining an Unknown Impedance

Suppose the transmission line is terminated by an unknown impedance. If the maximum or minimum voltage is at a position other than the receiving end, the unknown impedance Z_R is complex and has a resistive as well as reactive part. To determine Z_R, the reflection coefficient Γ must be determined first. Z_R is then given by (15.45) as

$$\boxed{Z_R = Z_0 \frac{1+\Gamma}{1-\Gamma} = Z_0 \frac{1+|\Gamma|e^{j\theta}}{1-|\Gamma|e^{j\theta}}} \tag{15.62}$$

The absolute value of Γ is determined by a measurement of the standing-wave ratio S of the line which is terminated in Z_R; that is, $|\Gamma| = (S-1)/(S+1)$. To determine the phase angle θ of Γ, the distance from Z_R to the first minimum of voltage is needed.† A voltage minimum occurs when the phase $2\beta d$ of $\Gamma e^{-j2\beta d}$ in (15.47) is 180°. That is,

$$\Gamma e^{-j2\beta d} = |\Gamma| e^{j(\theta - 2\beta d)} = |\Gamma| e^{-j\pi}$$

which gives

$$\theta = -\pi + 2\beta d_{\min} \tag{15.63}$$

The reflection coefficient due to an unknown termination is therefore

$$\boxed{\Gamma = |\Gamma| e^{j\theta} = \frac{S-1}{S+1} e^{j(-\pi + 2\beta d_{\min})}} \tag{15.64}$$

Figure 15.25 shows two cases of reactive terminations.

Example: A transmission line of characteristic impedance 300 Ω is terminated in an unknown load Z_R. The measured standing-wave ratio is found to be $S = 3$, the distance between minima is 50 cm, and the distance from Z_R to the first minimum is 30 cm. Find Z_R.

The real and imaginary components of Z_R, from Eq. (15.62), can be expressed directly in terms of S and $d_{\min}$ as

$$R = \frac{SZ_0}{S^2 \cos^2 \beta d_{\min} + \sin^2 \beta d_{\min}} \tag{15.65}$$

$$X_R = \frac{-Z_0(S^2 - 1)\cos \beta d_{\min} \sin \beta d_{\min}}{S^2 \cos^2 \beta d_{\min} + \sin^2 \beta d_{\min}} \tag{15.66}$$

where $Z_R = R_R + jX_R$. The wavelength is $\lambda = 100$ cm, and $\beta d_{\min} = (2\pi/\lambda)d_{\min} = (2\pi/100)30 = 1.88$, which gives $Z_R = 510.22 + j399.87$ Ω.

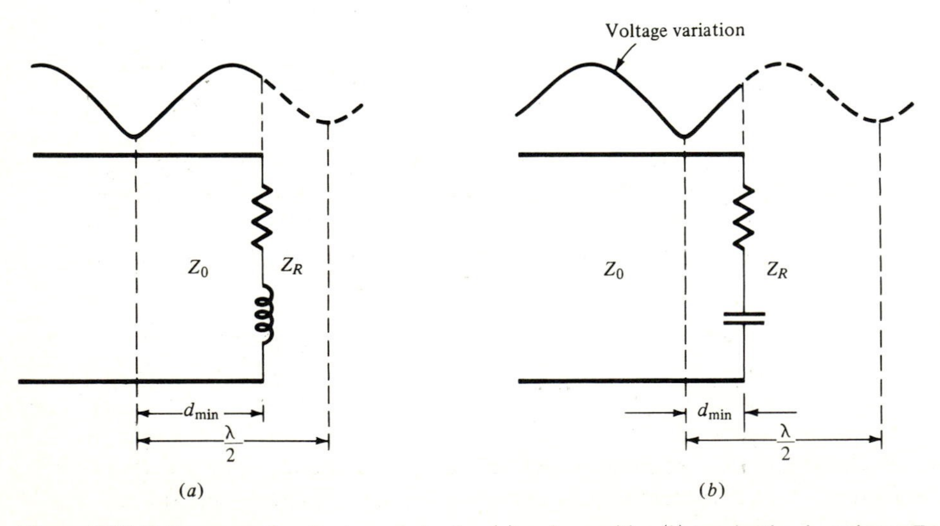

Figure 15.25 Voltage variation due to an inductive (*a*) and capacitive (*b*) terminating impedance Z_R.

† The minimum, rather than the maximum, is chosen because the minimum is sharper.

Power Transfer by a Line

The average power transmitted along a line and delivered to the load is $P_R = \frac{1}{2}\,\mathrm{Re}\, VI^*$, where V and I are total voltage and total current and the asterisk denotes a complex conjugate. For a lossless line for which Z_0 is real, using (15.47) and (15.48) results in

$$P_R = \frac{1}{2}\mathrm{Re}\,\frac{|V^i|^2}{Z_0}(1 + \Gamma e^{-j2\beta d})(1 - \Gamma^* e^{j2\beta d})$$

$$= \frac{1}{2}\mathrm{Re}\,\frac{|V^i|^2}{Z_0}(1 - |\Gamma|^2 + \Gamma e^{-j2\beta d} - \Gamma^* e^{j2\beta d})$$

$$\boxed{= \frac{1}{2}\frac{|V^i|^2}{Z_0}(1 - |\Gamma|^2)} \tag{15.67}$$

as $\Gamma e^{-j2\beta d} - \Gamma^* e^{j2\beta d}$ is purely imaginary. The first term is the power in the incident wave, $P^i = |V^i|^2/2Z_0$; the second term represents power in the reflected wave. Maximum power is delivered to the load when the load is matched to the line ($Z_R = Z_0$; $\Gamma = 0$) because then all the power in the incident wave is absorbed by the load. To deliver the same power to the load when reflections are present requires more power in the incident wave; the higher voltages which come with the increased power could cause insulation breakdown on the line. Optimum operating conditions therefore require a value of standing-wave ratio as close to one as possible. Expressing (15.67) as

$$\boxed{\frac{P_R}{P^i} = 1 - |\Gamma|^2 = \frac{4S}{(1+S)^2}} \tag{15.68}$$

gives the ratio of power P_R which is absorbed by the load to power P^i which would reach the load if the system (transmission line and load) were matched. A matched system has therefore the greatest power capability.

Power capability, based on insulation breakdown ($V_{\mathrm{max}} < V_{\mathrm{dielectric\ breakdown}}$), can be expressed using (15.67) and (15.61) as

$$P_R = \frac{1}{2}\frac{|V^i|^2}{Z_0}(1 - |\Gamma|)(1 + |\Gamma|) = \frac{1}{2}\frac{|V^i|^2}{Z_0}\frac{(1 + |\Gamma|)^2}{S} = \frac{1}{2}\frac{V_{\mathrm{max}}^2}{Z_0 S} \tag{15.69}$$

which shows that for the same power delivered to the load, V_{max} must increase by the factor $\sqrt{S}$ as S increases. Any deviation from the matched condition ($S = 1$) increases the voltage.

For example, a 50-Ω line terminated by a load of 50 Ω transmits a power of 450 W which results in a voltage of $\sqrt{(450)(50)} = 150$ V(rms) on the line. The same line with the load changed to 500 Ω would have a maximum voltage of $150(1 + |\Gamma|) = 150(1 + 9/11) = 273$ V(rms).† Thus a change in the standing-wave ratio from $S = 1$ to $S = 10$ practically doubles the maximum voltage on the line. If the matched line operates near maximum rated voltage, the unmatched line can be expected to breakdown. Furthermore, should a maximum voltage position coincide with the sending end, damage could be easily done to the solid-state elements in the transmitter which are particularly sensitive to voltage overloads.

15.8 IMPEDANCE MATCHING AND TRANSFORMATION

In general it is desirable (1) to match the load impedance to the line as this eliminates reflected waves and (2) to match the generator impedance to the line as this results in maximum power transfer from generator to line.

Impedance Matching Using Stub Lines

A complex load impedance can be matched to a lossless line by adding the reactance of a shorted stub either in series or in shunt to the line. A shorted shunt stub is shown in Fig. 15.26.

A study of Eq. (15.46) shows that the impedance $Z(d)$ of a line, which is terminated in a complex load Z_R, has at some distance d_1 on the line a real part which is Z_0; that is, $Z(d_1) = Z_0 + jX$. If at this distance the line is broken and a reactance $-jX$ is added in series with the line, the combination of line and reactance will have an impedance $(Z_0 + jX) - jX = Z_0$. The line appears to be terminated at d_1 in its characteristic impedance and is therefore matched. Between sending end and d_1 the line is smooth, with $S = 1$.

In practice, shunt stubs are used for matching since the discontinuity created by breaking the line disturbs the fields of the line. To determine the shunt stub position, it is easiest if the calculations are done on an admittance basis. The admittance $Y(d)$ of the line can be obtained by inverting (15.46):

$$Y(d) = Y_0 \frac{Y_R + jY_0 \tan \beta d}{Y_0 + jY_R \tan \beta d} \tag{15.70}$$

where $Y_R = 1/Z_R$ and $Y_0 = 1/Z_0$. Assume the standing-wave ratio S is caused by a purely resistive load $R_R > Z_0$, then $S = R_R/Z_0$ as stated by (15.61), the load is at a position of a voltage maximum, and the distance from the load to a point d_1 on the line where the real part of (15.70) is equal to Y_0 is given by

$$\boxed{\tan \beta d_1 = \sqrt{S}} \tag{15.71}$$

† The voltage is even higher if the transmitter coupling is readjusted so that 450 W is delivered to the 500-Ω load. Then $\frac{1}{2}V_{\max}^2/(Z_0 S) = \frac{1}{2}V_i^2/Z_0$, and at positions of voltage maxima: $V_{\max} = V_i\sqrt{S} = 150\sqrt{10} = 474.3$ V(rms).

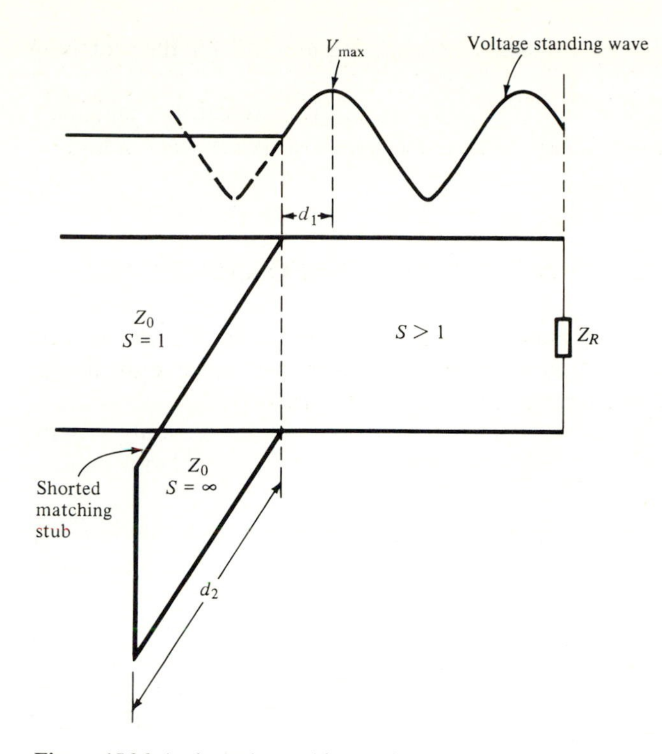

Figure 15.26 A shorted matching stub of a length d_2 is placed across the line at a distance d_1 (toward sending end) from any voltage maximum.

A shorted stub has a purely imaginary admittance which, from (15.53), is equal to $Y_{sc} = -jY_0 \cot \beta d_2$. If the stub is connected at d_1, the imaginary part of (15.70) will be canceled if the stub length d_2 is adjusted such that

$$\boxed{\tan \beta d_2 = \frac{\sqrt{S}}{S-1}} \tag{15.72}$$

which leaves the line with a termination equal to Y_0 at d_1. The line is therefore matched since an incident wave is not reflected at the position of the stub even though reflected waves do exist between the stub and the load.

In the case of a complex load impedance, the line is matched when the position of the stub is at a distance d_1 measured from any voltage maximum toward the sending end, as shown in Fig. 15.26. The matched condition of the line is indicated by the nonvarying voltage to the left of the stub position. Therefore, any load impedance (complex or real, known or unknown) can be matched to the line if the standing-wave ratio S and the position of V_{max} which the load produces can be measured or calculated. The design equations† for the matching stub system are (15.71) and (15.72).

Theoretically, an open-circuited stub could be used to cancel the imaginary part of $Y(d)$. In practice, however, short-circuited stubs are used because a short is less sensitive to external influences and radiates less than the open-circuit end of a line.

Any load impedance (except a purely reactive one) can be matched with a single shorted stub. However, the position of a stub on the line (especially on a coaxial line) cannot be easily changed should a changing load require rematching. When single-stub matching is inconvenient, double-stub matching, in which two stubs whose length may be varied, can be used. The location of the stubs on the line is arbitrary but remains fixed during the matching procedure. Matching is accomplished by varying the length of the two stubs. A double-stub system, although it can match a broad range of impedances, cannot match all impedances.

Example: Design a single-stub system to match a load admittance of $Y_R = 0.0275 + j0.0175$ to a line of characteristic impedance of 100 Ω.

SOLUTION The reflection coefficient Γ and distance $d_{\min}$, Eq. (15.63), from the load to the first voltage minimum is needed first.

$$\Gamma = \frac{Z_R - Z_0}{Z_R + Z_0} = \frac{Y_0 - Y_R}{Y_0 + Y_R} = \frac{1 - Y_R/Y_0}{1 + Y_R/Y_0} = \frac{1 - (2.75 + j1.75)}{1 + (2.75 + j1.75)} = 0.598e^{-j2.793}$$

$$S = \frac{1 + |\Gamma|}{1 - |\Gamma|} = \frac{1 + 0.598}{1 - 0.598} = 3.976$$

$$d_{\min} = \frac{\theta + \pi}{2\beta} = \frac{-2.793 + \pi}{4\pi}\lambda = 0.0277\lambda$$

Using (15.71) and (15.72),

$$d_1 = \frac{\lambda}{2\pi}\arctan\sqrt{3.976} = 0.176\lambda$$

$$d_2 = \frac{\lambda}{2\pi}\arctan\frac{\sqrt{3.976}}{3.976 - 1} = 0.0939\lambda$$

The stub which has the same characteristic impedance as the line, has a length of 0.0939λ and is to be placed across the line at a distance $d_1 = 0.176\lambda$ (toward the generator) from any voltage

† A different matching stub system results if the design is with respect to a voltage minimum. As S at a voltage minimum is given by $S = Z_0/R$, which is the inverse of the S used to derive (15.71) to (15.72), we can simply invert S in (15.71) to (15.72) and obtain

$$\tan\beta d_1 = \sqrt{\frac{1}{S}} \qquad \tan\beta d_2 = \frac{\sqrt{S}}{1 - S} \tag{15.73}$$

where d_1 is measured (toward the sending end) from any voltage minimum. As $S > 1$, (15.73) gives a negative length d_2 for the shorting stub. This can be changed to a positive length by adding $\lambda/2$ to d_2 (Fig. 15.23 shows that impedance is unchanged at intervals of $\lambda/2$), thus $\tan(\beta d_2 - \pi) = \sqrt{S}/(1 - S)$ can be used. One of the designs will give a stub location which is nearer to the load.

maximum. The distance of the stub position from the load would be $d = d_{min} + \lambda/4 + d_1 = 0.454\lambda$.

An alternative stub system, using (15.73), is

$$d_1 = \frac{\lambda}{2\pi} \arctan \sqrt{\frac{1}{3.976}} = 0.0740\lambda$$

$$d_2 = \frac{\lambda}{2\pi}\left(\arctan \frac{\sqrt{3.976}}{1 - 3.976} + \pi\right) = 0.406\lambda$$

The position of this stub from the load is $d = d_{min} + d_1 = 0.102\lambda$. As this position is closer to the load than the previous one, it is preferred since more of the line operates under matched conditions.

Half-Wave Transformer

A study of Eq. (15.46) shows that impedances are repeated at intervals of $\lambda/2$. The input impedance of a line of half wavelength which is terminated in Z_R is

$$Z\left(\frac{\lambda}{2}\right) = Z_0 \frac{Z_R \cos \pi + jZ_0 \sin \pi}{Z_0 \cos \pi + jZ_R \sin \pi} = Z_R \tag{15.74}$$

A half-wavelength line is therefore a 1 to 1 transformer. It is useful in connecting load to source when they are not adjacent to each other.

Example: Maximum power transfer results when load resistance matches generator resistance. For maximum power transfer a 50-Ω generator should be connected to a 50-Ω load by a line of 50-Ω characteristic impedance. If the available line has $Z_0 = 300$ Ω and $v_p = 2 \times 10^8$ m/s, find the length of the line for maximum power transfer at 27 MHz.

If the length l is chosen to be an integral number of half wavelength, the 50-Ω load impedance will be imaged at the sending end. Therefore, $l = n(\lambda/2) = n[(2 \times 10^8)/2 \times 27 \times 10^6] = n3.7$ m, where $n = 1, 2, \ldots$. The 300-Ω line will operate with a standing wave ratio of $S = Z_0/R_R = 300/50 = 6$. As the half-wave transformer is not dependent on Z_0, a line of any characteristic impedance will give a perfect match as long as the length restriction of $n(\lambda/2)$ is adhered to. Only the standing-wave ratio will increase as the characteristic impedance deviates from R_R. If the parameters (R', G') of the line are such that the lossless line approximation is valid, the increased losses of operating the line at a high standing-wave ratio S will be insignificant as long as insulation breakdown due to the increased voltage does not occur. The 300-Ω line operates at a voltage which is larger than the voltage on a 50-Ω line by a factor of $V_{max}/V^i = \sqrt{S} = \sqrt{6} = 2.45$ [see Eq. (15.69)]. Figure 15.27 shows the voltage distribution on a line of length $l = 7.4$ m (one wavelength).

Quarter-Wave Transformer

A quarter-wavelength line is an impedance inverter, as its input impedance [using Eq. (15.46)]

$$Z_s = Z\left(\frac{\lambda}{4}\right) = Z_0 \frac{Z_R \cos \pi/2 + jZ_0 \sin \pi/2}{Z_0 \cos \pi/2 + jZ_R \sin \pi/2} = \frac{Z_0^2}{Z_R} \tag{15.75}$$

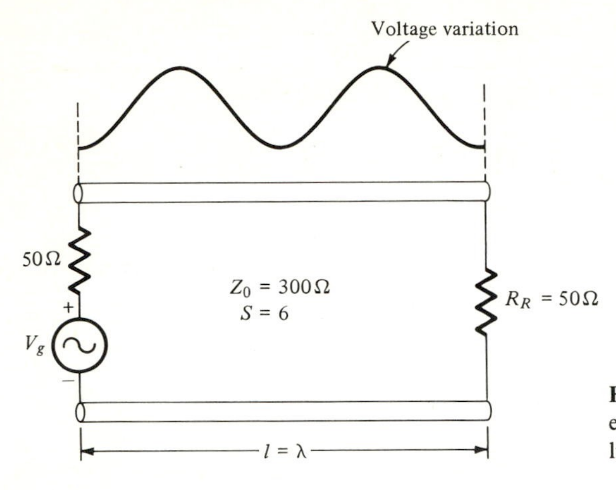

Figure 15.27 A perfectly matched system even though standing waves exist on the line.

varies inversely with Z_R. It can be considered to be a transformer for matching a load of Z_R ohms to a generator of Z_s ohms. The matching section must have a characteristic impedance which is the geometric mean of the generator and load impedances; that is,

$$Z_0 = \sqrt{Z_s Z_R} \tag{15.76}$$

Quarter-wave sections are frequently used to match (1) two transmission lines of different characteristic impedances, and (2) a transmission line to a resistive load, such as an antenna.

Example: Design a quarter-wavelength section to match a $\lambda/2$ dipole whose antenna resistance is 75 Ω to a transmission line of $Z_0 = 50$ Ω.

Using (15.75), the quarter-wave transformer must have a characteristic impedance of $Z_0 = \sqrt{(50)(75)} = 61.3$ Ω. The matching system is shown in Fig. 15.28*a*.

Example: Design a quarter-wavelength section to match a complex load Z_R to a line whose characteristic impedance is Z_0 (note that for lossless lines, Z_0 is real).

SOLUTION A quarter-wave section transforms a complex impedance into another complex impedance. Therefore Z_R must first be converted into a pure resistance by adding an appropriate length of line to Z_R. If the length is chosen to be $d_{\min}$, (15.63), which is the distance to the first voltage minimum, then the input impedance is purely resistive and has a value $R_{\min} = Z_0/S$ according to (15.60). As neither S nor $d_{\min}$ is given, both must be calculated by first calculating the reflection coefficient $\Gamma = (Z_R - Z_0)/(Z_R + Z_0) = |\Gamma| e^{j\theta}$. Standing-wave ratio S and distance $d_{\min}$ are then determined by $S = (1 + |\Gamma|)/(1 - |\Gamma|)$ and $d_{\min} = (\theta + \pi)/2\beta$. The characteristic impedance of the quarter-wave transformer is then given by (15.75) as $\sqrt{(Z_0/S)Z_0} = Z_0/\sqrt{S}$. The design is shown in Fig. 15.28*b*. The choice of position of a voltage maximum would have resulted in a quarter-wave section whose characteristic impedance is $Z_0\sqrt{S}$.

The frequency sensitivity of quarter-wave transformers is a disadvantage. Only at the frequency called the *center frequency* at which the length of the transformer section is exactly $\lambda/4$, will the standing-wave ratio S on the main line

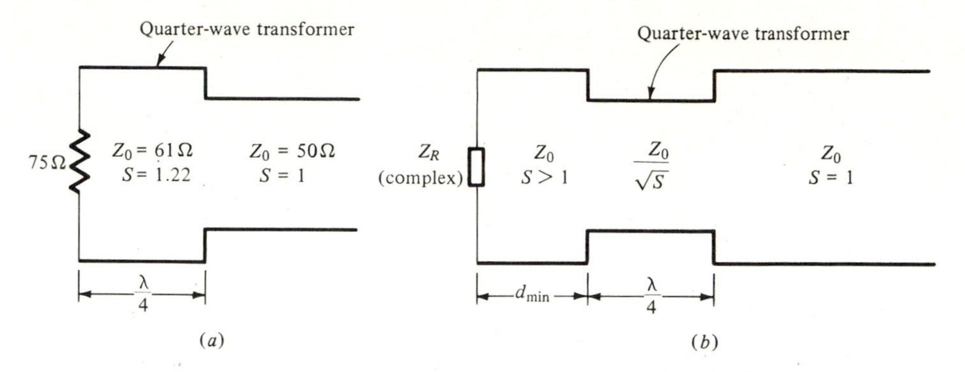

Figure 15.28 Matching a resistive load (*a*) and a complex impedance load (*b*) by use of a quarter-wave section.

be unity. At frequencies above (below) the center frequency, the transformer section will appear longer (shorter) than $\lambda/4$, with the result that S will be greater than unity. The *bandwidth* of the quarter-wave transformer is defined as the range of frequencies for which S has a value lower than some number, usually 1.5. Frequency sensitivity can be decreased by cascading quarter-wave sections, which provides a smoother transition between the two impedances which have to be matched. For example, when a single $\lambda/4$ section is used to match a load of 400 Ω to a line of 50 Ω, the bandwidth at a center frequency of 300 MHz is found to be 64 MHz. The bandwidth with two $\lambda/4$ sections in series is approximately doubled. After four sections, little improvement is obtained by cascading more sections.†

15.9 EFFECT OF STANDING WAVES ON POWER DELIVERED TO LOAD

In the transfer of power from source to load, an unwarranted degree of importance is frequently attached to the subject of standing waves. In many practical situations, the difference between a line operating at large values of standing-wave ratio S and a flat line ($S = 1$) is negligible. We have already shown in Fig. 15.27 that a lossless line can operate at a high value of S and still deliver maximum power to the load. (The condition for maximum power transfer is that load impedance and source impedance be complex conjugates of each other.)

Consider a line which transfers power from a transmitter to an antenna, as shown in Fig. 15.29. Let us assume that the power levels involved are such that the voltage on the transmission line, even at high values of S, is well below the rated voltage of the line.

† S. Guccione, Nomograms Speed Design of $\lambda/4$ Transformers, *Microwaves*, August 1975.

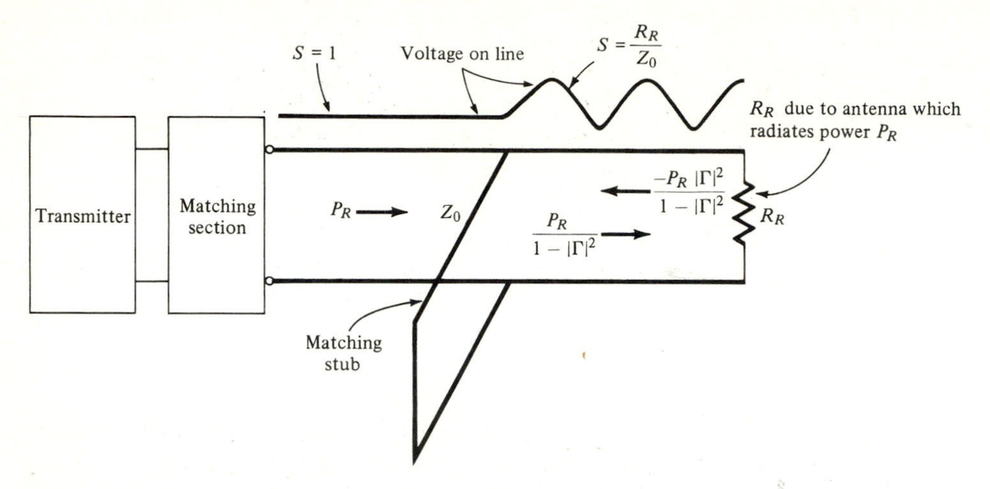

Figure 15.29 A typical transmission system consisting of a transmitter, transmission line with matching stub, and an antenna. Power P_R is supplied by the transmitter, delivered to the antenna, and radiated by it. Between matching stub and antenna, we have $P^i - P^r = P^i - |\Gamma|^2 P^i \equiv P_R$, which gives for incident power $P^i = P_R/(1 - |\Gamma|^2)$ and reflected power $P^r = |\Gamma|^2 P^i$.

Case 1: $R_R = Z_0$ (Matching Stub Absent)

The load (representing an antenna) is matched to the line. As no reflected waves exist, standing waves on the line are absent ($S = 1$). When the transmitter is properly loaded its maximum output is P_m watts. The purpose of the matching network is to transform the line impedance Z_0 to the proper impedance for maximum transmitter output. As the line is assumed to be lossless, power $P_R[= \frac{1}{2}(|V_R|^2/R_R)]$ absorbed at the receiving end must be equal to the power taken from the transmitter. For a properly loaded transmitter, $P_R = P_m$. The power flowing in the incident wave is therefore $P_{m^i} = \frac{1}{2}(|V_m^i|^2/Z_0)$, where V_m^i is the voltage of the incident wave under matched conditions. For example, when $P_m = 100$ W, $V_m^i = V_R = 70.7$ V(rms) for a line with a characteristic impedance of 50 Ω and a load $R_R = 50$ Ω.

Case 2: $R_R \neq Z_0$ (Matching Stub Absent)

A new higher resistance load is connected to the line without altering the setting of the matching network. This is the typical mismatched case in which standing waves are present on the line. The power absorbed by the load is given by (15.67) as

$$P_R = \frac{1}{2}\frac{|V_R|^2}{R_R} = \frac{1}{2}\frac{|V_m^i|^2}{Z_0}(1 - |\Gamma|^2)$$

and is less than that in case 1. The power not absorbed is reflected by the mismatched load, travels back to the input end, and reduces the power taken from the transmitter. The transmitter, therefore, has to deliver only power P_R which is less than the power P_m under matched conditions.

For example, for $R_R = 500\ \Omega$, $S = R_R/Z_0 = 500/50 = 10$, $|\Gamma| = (S-1)/(S+1) = 9/11 = 0.82$, incident power $P_m = 100$ W, absorbed power $P_R = 100[1 - (9/11)^2] = 33$ W, and reflected power equals 67 W. Transmitter output is therefore $100 - 67 = 33$ W. The maximum voltage on the line is $V_m^i(1 + |\Gamma|) = 70.7(1 + 9/11) = 128.6$ V(rms); receiving-end voltage $V_R = V^i + V^r = \sqrt{2R_R P_R} = \sqrt{(500)(33)} = 128.6$ V(rms).

The conclusion is that the reflected wave changes the impedance of the input of the line [see Eq. (15.46)] to a value different from 50 Ω. The conditions for maximum transmitter output therefore no longer exist. However, the matching network can be readjusted for maximum output; the actual power put into the line is then the same as that for a matched load, that is, P_m. As power on a lossless line is conserved, the power P_R absorbed by the load is equal to P_m, or

$$\frac{1}{2}\frac{|V^i|^2}{Z_0}(1 - |\Gamma|^2) = \frac{1}{2}\frac{|V_m^i|^2}{Z_0} \tag{15.77}$$

where V^i is the increased voltage in the incident wave after readjustment. The new values on the line are $V^i = V_m^i/\sqrt{1 - |\Gamma|^2}$ and $V_{max} = V^i(1 + |\Gamma|) = V_m^i\sqrt{S}$, where V_m^i is the incident voltage on the matched line.

If the transmitter puts 100 W into the line, $P_R = 100$ W, $V^i = 70.7/\sqrt{1 - (9/11)^2} = 123$ V(rms), $V_{max} = 70.7\sqrt{10} = 223.6$ V(rms), and $V_R = \sqrt{2R_R P_R} = \sqrt{500 \times 100} = 223.6$ V(rms). Even though the transmitter output is only 100 W, the power in the incident wave is $\frac{1}{2}(|V^i|^2/Z_0) = \frac{1}{2}(|\sqrt{2} \times 123|^2/50) = 302.5$ W and in the reflected wave $(9/11)^2\ 302.5 = 202.5$ W. Thus if transmitter and transmission line can tolerate the excessive voltages of a mismatched line, the two antenna systems, one matched and the other mismatched, will be equally effective, as both radiate 100 W of power.

Case 3: $R_R \neq Z_0$ (Matching Stub Present)

This is the usual arrangement when antenna resistance differs from the characteristic impedance of the line. The ideal position for the matching stub is at the antenna, as then the entire length of the line is matched. In practice, a position which is a fraction of a wavelength from the antenna can be found. However, if the power levels involved are such as to permit operation with standing waves on the line, then the position of the stub is not important, and the matching stub itself might not be necessary if a matching network between transmitter and the line is present because a matching network can perform the function of a stub located at the input of the transmission line. The important criterion is the radiated power which is independent of the standing-wave ratio S on a lossless line and depends only on the actual power put into a line which feeds an antenna. (For a lossless line, input power equals output power.)

In case the line is lossy, the effect of standing waves is to increase the losses on the line. For lines with low losses, the increased loss due to standing waves may still be inconsequential in comparison with the power delivered to the load. For example, RG-58/U, which is a general-purpose, small size, flexible coaxial cable, has a loss when matched of 1 dB per 15 m at 30 MHz. The additional loss due to mismatch is 0.2 dB for $S = 2$, 1 dB for $S = 5$, and 3 dB for $S = 10$.

15.10 LINES WITH LOSS

The analysis of the previous sections was based on the assumption that there is no power loss in the line itself. Every line consumes some power, partly because of the resistance of the wires (I^2R loss) and partly because the dielectric, which separates the parallel conductors, is not a perfect insulator (V^2G loss). For lines with small but finite losses, the effect of losses on characteristic impedance, input impedance, voltage on the line, etc. is usually negligible so that the lossless line analysis is valid. However, for special conditions, losses and the attenuation of signals which they cause cannot be ignored. These conditions are (1) when the signals are to be transmitted for great distances along the line, (2) the line is to be used at very high frequencies, as line losses increase with frequency, (3) the line carries large amounts of power which creates heat that must be dissipated, and (4) when transmission lines are used as circuit elements and resonant circuits, where the assumption of zero dissipation leads to circuits which have infinite Q values.

The ac steady-state analysis of transmission lines with losses is exactly the same as for the lossless line, except that impedance per unit length $j\omega L'$ is replaced by $Z' = R' + j\omega L'$, and admittance per unit length $j\omega C'$ is replaced by $Y' = G' + j\omega C'$. This substitution is a direct consequence of replacing the lossless line equations (15.5) and (15.6) by the lossy line equation (15.3) and (15.4). The characteristic impedance for the lossy line is then complex

$$Z_0 = \sqrt{\frac{Z'}{Y'}} = \sqrt{\frac{R' + j\omega L'}{G' + j\omega C'}} \tag{15.78}$$

and reduces to a pure resistance $\sqrt{L'/C'}$ in the lossless case ($R' = G' = 0$) and in the high-frequency case ($R' \ll \omega L'$, $G' \ll \omega C'$), which explains why high-frequency lines are referred to as lossless lines. The complex Z_0 indicates that in an individual traveling wave, voltage and current are not in phase.

The phase-propagation constant of a single traveling wave is given by Eq. (15.40) as $j\beta = j\omega/v = j\omega\sqrt{L'C'} = \sqrt{(j\omega L')(j\omega C')}$, and for the lossy line it is replaced by a propagation constant

$$\gamma = \alpha + j\beta = \sqrt{Z'Y'} = \sqrt{(R' + j\omega L')(G' + j\omega C')} \tag{15.79}$$

where α is called the attenuation constant, as it gives the rate of exponential attenuation of a traveling wave. The voltage and power in an incident wave on an infinite line are now, using Eqs. (15.43) and (15.44)

$$V(z) = V^i e^{-\gamma z} = V^i e^{-\alpha z - j\beta z} \qquad P(z) = \tfrac{1}{2}\,\mathrm{Re}\,VI^* = \tfrac{1}{2}\,\mathrm{Re}\,V^i I^{i*} e^{-2\alpha z} \tag{15.80}$$

DC Line

The characteristics of a dc line are determined by the limit as $\omega \to 0$, which gives $\beta = 0$, $\alpha = \sqrt{R'G'}$, and a voltage on the line of $V(z) = V^i \exp(-\sqrt{R'G'}\,z)$. There is no phase shift, as the wavelength at zero frequency is infinite, leaving only exponential attenuation caused by conductor (I^2R) and insulation (V^2G) losses. The leakage conductance between the inner and outer conductor of a coaxial line was determined in Eq. (3.36d) as $G' = G/l = 2\pi\sigma/\ln(b/a)$.

For RG58/CU, a commonly used coaxial line, we have $G' = 2\pi(10^{-12})/\ln(1.47/0.45) = 5.3 \cdot 10^{-12}$ S/m, where the conductivity of the dielectric is $\approx 10^{-12}$ S/m, and 1.47 and 0.45 are the outer and inner radius in millimeters. The series resistance for a copper wire of 0.45 mm radius is $R' = 1/(\sigma\pi(.45)^2) = 0.027\ \Omega$/m, which gives $\alpha = \sqrt{R'G'} = \sqrt{(0.027)(5.3 \cdot 10^{-12})} = 3.8 \cdot 10^{-7}$ Np/m. A voltage is then attenuated by 8.7 dB in the distance of $1/\alpha = 2632$ km. Hence attenuation for dc lines is rather small.

Low-Frequency Line (RC Cable)

To reduce space requirements and cost, a telephone cable is made simply by twisting a well-insulated pair of wires. Such construction results in very small values of G' and L' such that at audio frequencies, the approximation $Z' \cong R'$, $Y' \cong j\omega C'$ is valid. When series resistance and shunt capacitance dominate, the RC cable is characterized by, respectively,

$$Z_0 \cong \sqrt{\frac{R'}{j\omega C'}} = (1-j)\sqrt{\frac{R'}{\omega C'}} \tag{15.81}$$

$$\gamma = \alpha + j\beta \cong (1+j)\sqrt{\frac{\omega R'C'}{2}} = \frac{R'}{|Z_0|}e^{j\pi/4} \tag{15.82}$$

The RC cable has substantial distortion as higher frequencies (in a signal which travels on the transmission line) are attenuated more and travel faster: ($v_p = \omega/\beta = \sqrt{2\omega/R'C'}$). For a line to be distortion free, attenuation α and velocity v_p must be independent of frequency. Distortion is decreased by increasing the inductance on the line (inductive loading). Typical values for telephone lines are given in Prob. 15.42 to 15.44.

High-Frequency Line

At higher than audio frequencies coaxial, twin-lead, and strip lines are used, as the twisted type presents too much loss. At high frequencies reactances dominate ($R' \ll j\omega L'$, $G' \ll j\omega C'$), such that

$$Z_0 \cong \sqrt{\frac{R' + j\omega L'}{j\omega C'}} = \sqrt{\frac{L'}{C'}}\sqrt{1 + \frac{R'}{j\omega L'}} \cong \sqrt{\frac{L'}{C'}}\left(1 - j\frac{R'}{2\omega L'}\right) \cong \sqrt{\frac{L'}{C'}} \tag{15.83}$$

$$\gamma = \alpha + j\beta = \sqrt{(R' + j\omega L')j\omega C'} \cong j\omega\sqrt{L'C'}(1 + R'/j2\omega L') = \frac{R'}{2Z_0} + j\omega\sqrt{L'C'} \tag{15.84}$$

where insulation losses in comparison to wire losses are negligible ($G' \approx 0$). Since $R'/2\omega L'$ is small with respect to unity, the characteristic impedance has a small reactive part. But for most practical purposes Z_0 can be assumed to be purely resistive and equal to $Z_0 \cong \sqrt{L'/C'}$. For γ, on the other hand, $R'/2\omega L'$ cannot be ignored as without it there would be no attenuation. The attenuation term, in the form of the attenuation coefficient α, can become substantial at high frequencies as R' increases (due to skin effect) with frequency f as $\sqrt{f}$. The increase in R' is stated in Prob. 13.24 where it is shown that

$$\frac{R'_{ac}}{R'_{dc}} = \frac{\text{area}_{dc}}{\text{area}_{ac}} = \frac{a}{2}\sqrt{\pi f \mu \sigma} \tag{15.85}$$

for a wire of radius a and conductivity σ. Thus for a coaxial line the ac resistance is given by $R'_{ac} = 4.16 \cdot 10^{-8}\sqrt{f}\,(1/a + 1/b)\,\Omega/\text{m}$, where a and b are the inner and outer radii of the line, $\mu = 4\pi \times 10^{-7}$, and $\sigma = 5.8 \times 10^7$. As most of the resistance is due to the inner wire, the attenuation coefficient for a coaxial line is

$$\alpha = 2.08 \cdot 10^{-8}\sqrt{f}\,\frac{1}{Z_0 a} \tag{15.86}$$

For example, for the RG 58/CU line ($Z_0 = 50\ \Omega$) at 1 GHz, $\alpha = 0.039$ Np/m, which gives an attenuation of 8.68 dB per 34.2 m of line.

Input Impedance of a Line With Losses

Similar to Eq. (15.46), we now have

$$Z(d) = Z_0\frac{1 + \Gamma e^{-2\gamma d}}{1 - \Gamma e^{-2\gamma d}} = Z_0\frac{Z_R\cosh\gamma d + Z_0\sinh\gamma d}{Z_0\cosh\gamma d + Z_R\sinh\gamma d} \tag{15.87}$$

for the input impedance to a line of length d that is terminated by a load impedance Z_R. Whereas in the lossless case the input impedance for a short-circuited line of length $d = \lambda/4$ is given by (15.53) as infinite, with losses taken into consideration, the input impedance turns out to be large but not infinite, i.e., for $Z_R = 0$, $d = \lambda/4$

$$Z_{sc} = Z_0 \tanh (\alpha + j\beta)d = Z_0 \tanh \left(\alpha \frac{\lambda}{4} + j\frac{\pi}{2}\right) \cong \frac{4Z_0}{\alpha\lambda} = \frac{8Z_0^2}{R'\lambda}$$

where (15.84) was used to express the last term.

For certain conditions a lossy line will appear to be matched irrespective of the termination. If $2\alpha d$ in the $\Gamma e^{-2\gamma d}$ term of (15.87) is large, the term can be neglected in comparison to unity and we obtain $Z(d) \cong Z_0$. Therefore, if a line is sufficiently lossy such that little of the incident signal returns to the sending end after reflection, the line appears to be matched, i.e., its input impedance appears to be Z_0. The following is a fourth way to answer the question of what characteristic impedance is: It is the input impedance of a finite line which is lossy ($\alpha d \gg 1$).

PROBLEMS

15.1 A stripline is a parallel two-conductor transmission line in which each conductor is a strip, as shown in the figure. Ignoring fringing and leakage flux, show that the parameters of a stripline are $L' = \mu a/b$ H/m, $C' = \varepsilon b/a = \varepsilon_r \varepsilon_0 b/a$ F/m, and $Z_0 = \sqrt{L'/C'} = \sqrt{\mu/\varepsilon}\, a/b = (\eta_0/\sqrt{\varepsilon_r})a/b$ Ω, where $\eta_0 = \sqrt{\mu_0/\varepsilon_0} = 377$ Ω.

Hint: To find L', treat the stripline as a strip cut from two infinite parallel current sheets shown in Fig. 7.12. To find C', treat the stripline as a parallel-plate capacitor.

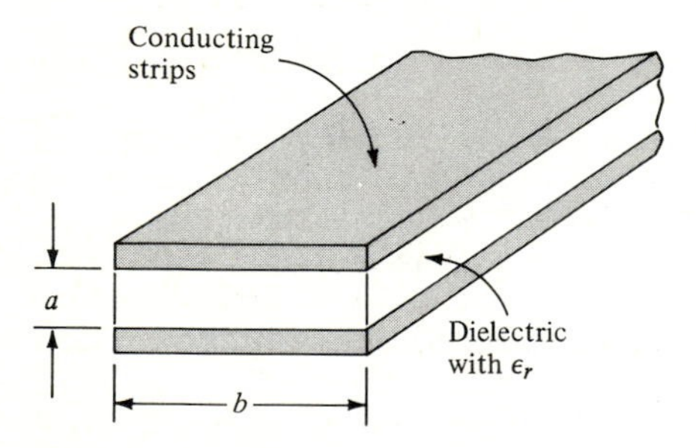

15.2 At frequencies above 1 GHz, microstrip transmission lines are frequently used. The width of the ground plane is at least several times the width b of the strip, and the thickness a of the dielectric is small as compared to b. Ignoring fringing and leakage flux, show that the characteristic impedance is given by $Z_0 = (\eta_0/\sqrt{\varepsilon_r})(a/b)$, where η_0 is the free-space impedance (377 Ω) and ε_r is the dielectric constant.

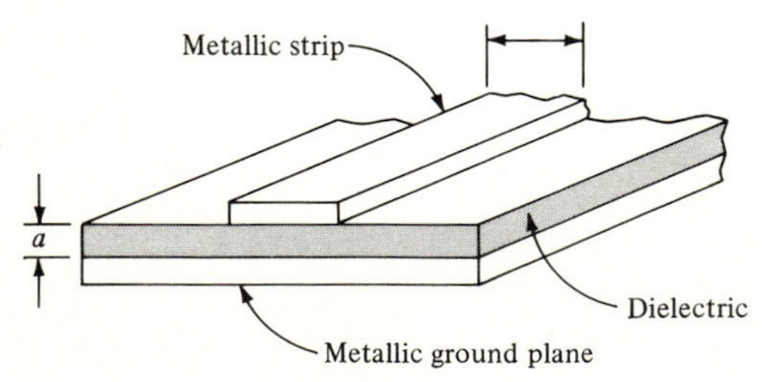

15.3 Show that $e^{j\omega(t-bz)}$, $\sin \omega(t - bz)$, and $(t - bz) \sin \omega(t - bz)$ (where b and ω are constants) are solutions of the wave equation.

15.4 For a terminating resistance R_R and for arbitrary time functions calculate the fraction of the incident power reflected and the fraction of the incident power transmitted to R_R.

15.5 Find the reflection and transmission coefficients for $R_R = 0$, $Z_0/3$, Z_0, $3Z_0$, and ∞.

15.6 Figure 15.13 gives the total voltages and currents during discharge of a transmission line for the case $R = 2Z_0$ of Fig. 15.14.

(*a*) Sketch figures similar to Fig. 15.13 for the case of $R = Z_0$ of Fig. 15.14.

(*b*) Sketch figures similar to Fig. 15.13 for the case of $R = Z_0/2$ of Fig. 15.14.

15.7 The open-circuited line shown is being charged by a battery of voltage V_b in series with $R = Z_0/2$. Sketch the sending-end current and voltage versus time.

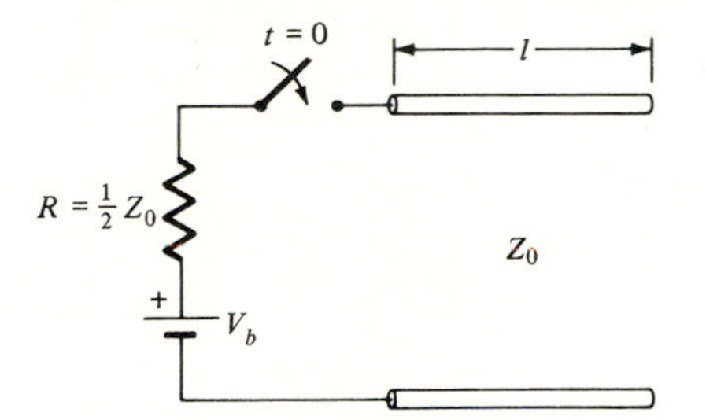

15.8 Using the transmission line of Fig. 15.15, assume the input pulse is 6 μs long (the reflected pulse arrives at the receiving end before the input pulse has ended). Plot a graph, similar to Fig. 15.16, for the sending-end voltages and currents.

15.9 Using the transmission line shown in Fig. 15.15*a* in which the termination R_R is replaced by a short-circuit, sketch and label the voltage and current waveforms as a function of time at the sending end, for an input pulse of 32 V and 1 μs.

15.10 Repeat the above problem for a 6 μs pulse.

15.11 A transmission line of characteristic impedance $Z_0 = 200\ \Omega$ and of length l is terminated by a load resistance of 800 Ω. At $z = l/3$, it is shunted by a resistance of 300 Ω. The line is connected to a battery of 6 V which has an internal resistance of 400 Ω. $T = l/v$, where v is the phase velocity.

(*a*) At $t = 0$, what is the voltage and current in the voltage and current waves which start to travel down the line?

(*b*) At $t = T/3$, when the wave reaches the 300-Ω shunt, what is the total voltage and current in the shunt?

(*c*) At $t = T$, when the wave reaches the 800-Ω termination, what is the total voltage and current in the termination?

(*d*) At $t = T$, what is the power absorbed by the 800-Ω load?

(*e*) After the transients have died down ($t \to \infty$), what is the steady-state voltage at the sending end and at the receiving end?

(*f*) Plot the sending-end voltage versus time.

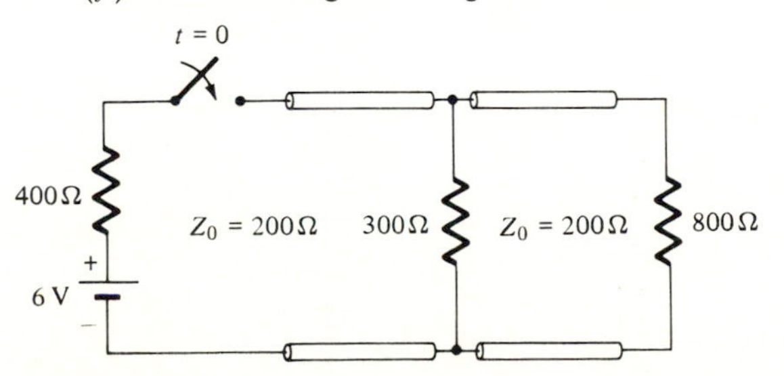

15.12 A 74-Ω line is driven and terminated by 100-Ω lines as shown. If a step pulse with rise time T_1 is applied as shown, plot the waveform at the sending end and label the values of the steps.

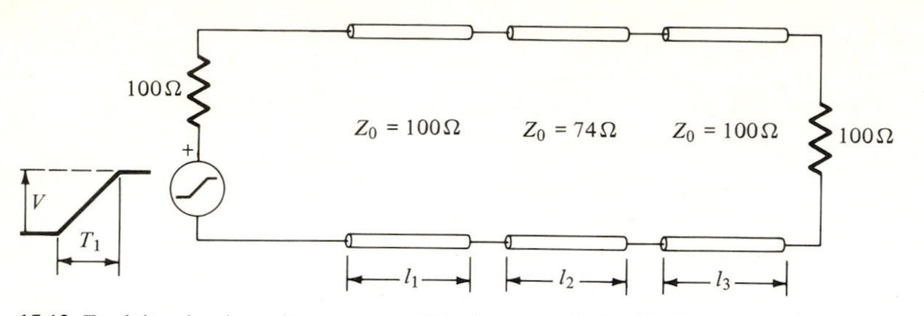

15.13 Explain why the voltage across C for the case of Fig. 15.20 rises to twice the voltage incident on the capacitor, whereas for the case of Fig. 15.21c it rises only to the value of the voltage incident on the capacitor.

15.14 A transmission line is terminated by a resistance R and capacitance C in series, as shown in Fig. 15.19. Find the voltage V_c across the capacitance and sketch it for the case of $R = Z_0$.

15.15 A transmission line of length l and characteristic impedance Z_0 is terminated in a capacitance C. Capacitor C is to be charged by connecting a battery of voltage V_b to the other end of the transmission line. The battery has an internal resistance Z_0. Find the charging voltage $V_c(t)$.

15.16 For Fig. 15.21 sketch the current at the sending end, receiving end, and across C.

15.17 A transmission line of length l and characteristic impedance Z_0 is terminated in an inductor of inductance L. The other end of the line is suddenly connected to a battery with internal resistance Z_0. Find the voltage and current waveforms at the receiving end.

15.18 A coaxial cable of type RG-58A/U has a velocity of propagation of 2×10^8 m/s and a capacitance of 100 pF/m. A 40-m-long cable is terminated by a resistance of 25 Ω.

(a) Find the characteristic impedance of the cable.

(b) At an operating frequency of 100 MHz, find the input impedance.

15.19 A lossless transmission line has a characteristic impedance of 55 Ω and is 2.25 m long. It is terminated by $Z_R = 115 + j75$ Ω and is driven by a generator whose frequency, internal impedance, and voltage are 150 MHz, 40 Ω, and 100 V(rms), respectively.

(a) Find the sending-end impedance of the line, given that an infinite line would have a phase change of 180° every 0.95 m of line.

(b) Find the power dissipated by Z_R.

15.20 A transmission line of 400-Ω characteristic impedance is terminated by an inductance whose impedance is $Z_R = j600$ Ω. Find the length (in wavelength) of a short-circuited line of the same characteristic impedance which is equivalent to Z_R and could be used to replace Z_R.

15.21 A quarter-wave lossless line has a characteristic impedance of 300 Ω. It is terminated by a resistance of 600 Ω. If the sending-end voltage is 100 V(rms), find the magnitude of the receiving-end voltage.

15.22 A high-frequency oscillator is made by connecting a shorted length of a 300-Ω characteristic-impedance transmission line between the grid and plate of a tube. What should the length of the line be to tune to 500 MHz if the capacitance between grid and plate is 3 pF and $v_p = 3 \times 10^8$ m/s?

15.23 Find the voltage at the open-circuited receiving end of a transmission line which has a characteristic impedance of 50 Ω, is 5/16 wavelength long, and is connected at the sending end to a generator with internal impedance of 20 Ω and voltage of 10 V.

15.24 Derive the set of equations (15.49) and (15.50) for voltage and current on the line.

15.25 Equations (15.49) and (15.50) give the voltage and current on the line in terms of receiving-end voltage and current. Sometimes it is convenient to state voltage and current in terms of voltage V^i of the incident wave. Derive

$$V(d) = V^i[(1 + \Gamma) \cos \beta d + j(1 - \Gamma) \sin \beta d]$$

$$I(d) = \frac{V^i}{Z_0}[(1 - \Gamma) \cos \beta d + j(1 + \Gamma) \sin \beta d]$$

15.26 Show that V and I on the line can be written as

$$V(d) = \frac{V_R(Z_R + Z_0)}{2Z_R} e^{j\beta d}(1 + \Gamma e^{-j2\beta d})$$

$$I(d) = \frac{I_R(Z_R + Z_0)}{2Z_0} e^{j\beta d}(1 - \Gamma e^{-j2\beta d})$$

where $V_R = Z_R I_R$ and d is a positive distance measured from the receiving end.

15.27 It is desired to express voltage and current on the line as measured from the sending end. Use (15.43) and (15.44) where sending end and receiving end are at $z = 0$ and $z = l$, respectively, to show that

$$V(z) = V_R^i(1 + \Gamma e^{-j2\beta(l-z)})e^{j\beta(l-z)}$$

$$I(z) = \frac{V_R^i}{Z_0}(1 - \Gamma e^{-j2\beta(l-z)})e^{j\beta(l-z)}$$

where $V_R^i = V^i e^{-j\beta l}$ is the value of the incident wave at the receiving end.

15.28 Show that $Z_0 = V_{\min}/I_{\min} = V_{\max}/I_{\max}$.

15.29 Find the load impedance on a line for which $Z_0 = 50\ \Omega$, on which the standing-wave ratio is 4, and on which a minimum of current occurs at the load.

15.30 Derive (15.65) and (15.66).

15.31 Show that an unknown load impedance Z_R is determined by

$$Z_R = Z_0 \frac{1 - jS \tan \beta d_{\min}}{S - j \tan \beta d_{\min}}$$

where S is the standing-wave ratio on the line and $d_{\min}$ is the distance from the load to the first voltage minimum.

15.32 A transmission line of characteristic impedance of 75 Ω and phase velocity of 200 m/μs is terminated in an unknown load Z_R. Find Z_R if the line frequency is 1 GHz, standing-wave ratio on the line is $S = 4$, and the distance from Z_R to the first voltage minimum is 10 cm.

15.33 A lossless line of characteristic impedance 300 Ω is terminated in an unknown load which produces a standing-wave ratio of 3 on the line, with the first voltage minimum 11 cm from the load. Short-circuiting the load, the minimum moves to a position 20 cm from the load. Find the value of load impedance.

15.34 Express the ratio of power reflected to power absorbed for a transmission line which is delivering power to a load and calculate it for standing-wave ratios of

(*a*) $S = 2$.
(*b*) $S = 5$.
(*c*) $S = 10$.

15.35 A transmission line operates with a standing-wave ratio of 5. The characteristic impedance of the line is 300 Ω, and the maximum voltage on the line is 150 V(rms). Find the power which is delivered to the load.

15.36 Show that for the same power flowing in a line with standing waves as in a line which is matched (also referred to as a smooth or flat line), the voltage increase in the unmatched line is given by $V_{\max} = V_{\text{flat}}\sqrt{S}$. Calculate the voltage increase when $S = 6$.

15.37 Prove that a purely reactive load cannot be matched to a lossless line.

15.38 The voltage standing-wave ratio on a transmission line is 2.1. If the first two voltage minima occur 1.25 and 2.77 m from the load, find the length and location of a single shorted shunt stub which will match the line.

15.39 Design a single-stub system to match a load impedance of $25 - j25\ \Omega$ to a line of characteristic impedance of 50 Ω. If there is more than one stub position, use the one closest to the load. Give values for stub length and distance from the load for operating frequencies of (take $v_p = 2 \times 10^8$ m/s)

(a) 27 MHz.
(b) 150 MHz.
(c) 500 MHz.

15.40 Design a quarter-wave transformer system to match a line of characteristic impedance of 300 Ω to a load of impedance

(a) 100 Ω.
(b) $100 + j100$ Ω.

15.41 A 10-V battery is placed across the sending end of a cable which has $R' = 53$ Ω/km and $G' = 0.93 \cdot 10^{-6}$ S/km. Calculate α and Z_0 for the line. If a 100-km-long cable is terminated in its characteristic impedance, compute the ratio of receiving-end voltage to sending-end voltage.

15.42 A telephone cable has the parameters $R' = 37$ Ω/km, $L' = 0.6$ mH/km, $C' = 0.04$ μF/km, and $G' = 1$ μS/km. Determine α, Z_0, and the neper loss per kilometer. Express the loss in decibels per kilometer. Assume $f = 1$ kHz.

15.43 An open-wire transmission line is constructed of two parallel wires spaced 3 cm apart, each of 1.29 mm radius. At 1 kHz it has the following constants: $R' = 2$ Ω/km, $L' = 1.4$ mH/km, $C' = 8.8$ nF/km, and G' is negligible. At 100 MHz it has these constants: $R' = 0.6 \cdot 10^3$ Ω/km, $L' = 1.3$ mH/km, $C' = 8.8$ nF/km, and G' is negligible. Calculate α, Z_0, and the neper loss per kilometer (in decibels) for each of the two frequencies, and then compare the two.

15.44 A standard open-wire telephone line consists of copper wires of 0.1 in diameter and spaced 12 in apart. The distributed constants of such a line are $R' = 10$ Ω/mi, $L' = 3.7$ mH/mi, $C' = 8.4$ nF/mi, and $G' = 0.8$ μS/mi. At a frequency of 795.8 Hz ($\omega = 5000$) and for a 300-mi length of line, calculate the voltage, current, and power at the receiving end if a generator of 600-Ω internal resistance, and generated voltage of 2 V, is connected at the sending end. Assume line is terminated in Z_0.

15.45 Show that for an open-circuited transmission line of length $d = \lambda/2$, the input impedance is given by $Z_{oc} = 2Z_0/\alpha\lambda = 4Z_0^2/(R'\lambda)$.

15.46 In the example following Eq. (15.54) it was stated that the voltage stepup V_R/V_S for a quarter-wave lossless line is infinite. Show that by taking losses into consideration, the voltage stepup is large, but finite.

APPENDIX

THE MATHEMATICS OF CURL. VECTOR IDENTITIES

A1.1 CURL AND THE CIRCULATION OF A VECTOR FIELD†

Let us consider a turbulent fluid field in which **F** is the velocity at any point in the moving fluid. An appropriate measure of the fluid's behavior might be its rotation. Rotation must have a continuous component of the fluid velocity along a circular path l. This being the case, we see that circulation $\oint_l \mathbf{F} \cdot \mathbf{dl} \neq 0$, as shown for two-dimensional turbulence in Fig. A1.1. The scalar value that is obtained for circulation can be used as a measure of rotation for two-dimensional flow because the direction of rotation in two-dimensional flow is normal to the plane of flow. For three-dimensional flow, however, the scalar value of circulation alone is limited as a measure of rotation because we need to know the direction of rotation as well as the magnitude.

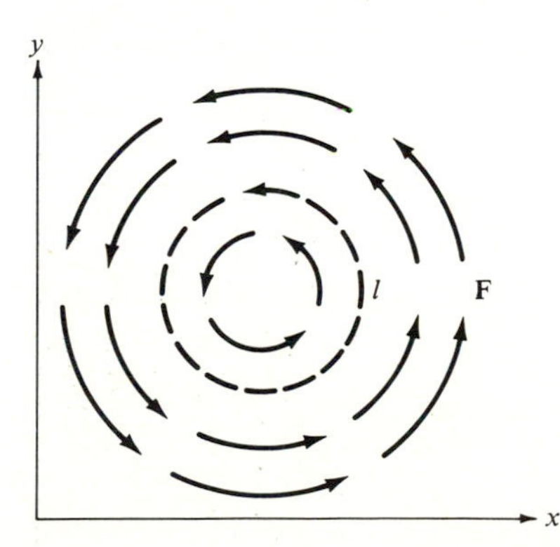

Figure A1.1 Circulation of a turbulent field **F** around a curve l.

† See Sec. 8.1 and R. B. McQuistan, "Scalar and Vector Fields: A Physical Interpretation," John Wiley & Sons, Inc., New York, 1965.

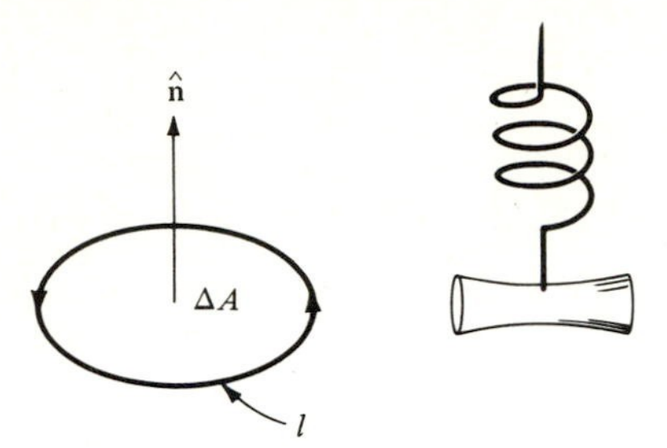

Figure A1.2 The direction that a right-handed corkscrew advances is the direction of an element of area bounded by a contour.

To obtain a better indication of the turbulence, let us derive a new vector operator, the curl, which will give the direction and magnitude of rotation in a fluid. Consider a plane loop l which encloses a small area ΔA. The direction of the normal $\hat{\mathbf{n}}$ of ΔA will be specified with respect to the direction of contour l by the right-handed "corkscrew rule," as shown in Fig. A1.2. If we form the ratio $C/\Delta A$, where C denotes the circulation of $\mathbf{F}$ about the path l, we will designate by (curl $F)_n$ the limit of $C/\Delta A$ as the area goes to zero:

$$(\text{curl } \mathbf{F})_n = \lim_{\Delta A_n \to 0} \frac{\oint_l \mathbf{F} \cdot \mathbf{dl}}{\Delta A_n} \tag{A1.1}$$

where the subscript n on ΔA denotes that the normal of ΔA is given by $\hat{\mathbf{n}}$.

In Eq. (A1.1) the normal $\hat{\mathbf{n}}$ to area ΔA_n may or may not be in the direction of fluid rotation. For example, for whirlpool turbulence in a river, circulation calculated for a contour oriented parallel to the water surface would give a maximum value, whereas a contour oriented perpendicular to the surface would yield zero circulation. It can be seen that a vector exists which is denoted by curl $\mathbf{F}$ such that $(\text{curl } \mathbf{F})_n = \hat{\mathbf{n}} \cdot \text{curl } \mathbf{F}$ is the component of curl $\mathbf{F}$ in the direction $\hat{\mathbf{n}}$. For a small path of integration, (A1.1) can then be written as

$$\oint_l \mathbf{F} \cdot \mathbf{dl} = (\text{curl } \mathbf{F}) \cdot \Delta\mathbf{A} \tag{A1.2}$$

where $\Delta\mathbf{A} = \hat{n}\, \Delta A$. The curl of a vector $\mathbf{F}$ is therefore a vector whose magnitude is the circulation per unit area of an infinitesimally small loop so oriented that the circulation is a maximum.

The definition of curl, (A1.1), is not dependent on a coordinate system. However, when working in a particular coordinate system, we can find a simpler operation than that given in (A1.1) for calculating the curl of a vector function. To illustrate, let us choose the rectangular coordinate system. To compute the z component of the curl, we start with the incremental area $\Delta A = \Delta x\, \Delta y$ in the xy plane, as shown in Fig. A1.3. The line integral of the circulation about ΔA, starting the integration

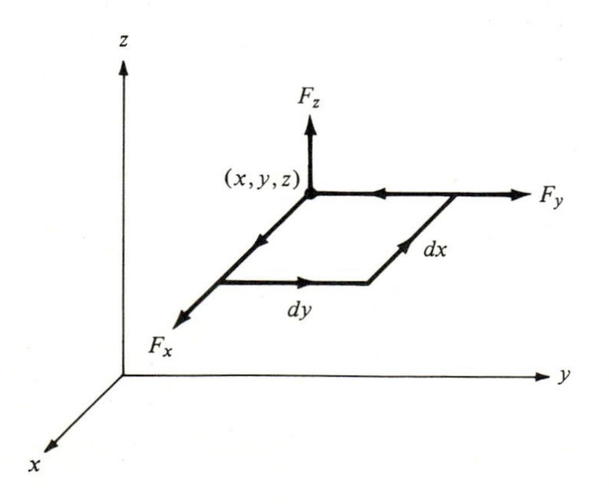

Figure A1.3 An element of area in the xy plane. The arrows on the contour bounding the area show the right-handed sense of integration with respect to the z axis.

at point (x, y, z) and following the arrows, is then

$$\oint \mathbf{F} \cdot \mathbf{dl} = F_x\Big|_y dx + F_y\Big|_{x+dx} dy - F_x\Big|_{y+dy} dx - F_y\Big|_x dy \tag{A1.3}$$

In the above expression we have assumed that a function at point b can be obtained if the function is known at point a, simply by $\phi(b) = \phi(a) + \partial\phi/\partial x|_a \, dx$, where $dx = b - a$, provided points a and b are separated by a small distance. To be more formal, we are expanding the function ϕ in a Taylor series about point a; all terms but the first two in the expansion are neglected, since, if dx is small, higher powers of dx will be negligible. Returning to (A1.3), we can say that

$$F_x\Big|_{y+dy} = F_x\Big|_y + \frac{\partial F_x}{\partial y}\Big|_y dy$$

$$F_y\Big|_{x+dx} = F_y\Big|_x + \frac{\partial F_y}{\partial x}\Big|_x dx \tag{A1.4}$$

Substituting the above expressions, (A1.3) becomes

$$\oint \mathbf{F} \cdot \mathbf{dl} = \left(\frac{\partial F_y}{\partial x} - \frac{\partial F_x}{\partial y}\right) dx\, dy \tag{A1.5}$$

Using the definition of curl, (A1.1), we finally obtain

$$(\text{curl } \mathbf{F})_z = \frac{\oint \mathbf{F} \cdot \mathbf{dl}}{dx\, dy} = \frac{\partial F_y}{\partial x} - \frac{\partial F_x}{\partial y} \tag{A1.6}$$

Similarly, by taking incremental areas in the yz plane and xz plane, respectively, we would obtain

$$(\text{curl } \mathbf{F})_x = \frac{\partial F_z}{\partial y} - \frac{\partial F_y}{\partial z}$$

$$(\text{curl } \mathbf{F})_y = \frac{\partial F_x}{\partial z} - \frac{\partial F_z}{\partial x} \tag{A1.7}$$

When these components are multiplied by the corresponding unit vectors and added, a vector is formed which represents the curl operation on $\mathbf{F}$; that is,

$$\text{curl } \mathbf{F} = \hat{\mathbf{x}}\left(\frac{\partial F_z}{\partial y} - \frac{\partial F_y}{\partial z}\right) + \hat{\mathbf{y}}\left(\frac{\partial F_x}{\partial z} - \frac{\partial F_z}{\partial x}\right) + \hat{\mathbf{z}}\left(\frac{\partial F_y}{\partial x} - \frac{\partial F_x}{\partial y}\right) \tag{A1.8}$$

The above form shows that the curl operation is equivalent to a sequence of derivative operations. The curl operation can be expressed in terms of the del operator $\nabla = \hat{\mathbf{x}}\, \partial/\partial x + \hat{\mathbf{y}}\, \partial/\partial y + \hat{\mathbf{z}}\, \partial/\partial z$ as curl $\mathbf{F} = \nabla \times \mathbf{F} = (\hat{\mathbf{x}}\, \partial/\partial x + \hat{\mathbf{y}}\, \partial/\partial y + \hat{\mathbf{z}}\, \partial/\partial z) \times (F_x\hat{\mathbf{z}} + F_y\hat{\mathbf{y}} + F_z\hat{\mathbf{z}})$. The sequence with which the del operator components operate on the components of $\mathbf{F}$ can be obtained by treating the operator ∇ as a vector and using the rules for cross-product multiplication. As this is a lengthy operation, we can use a memory aid in the form of the rules for expanding a determinant. Thus

$$\text{curl } \mathbf{F} = \nabla \times \mathbf{F} = \begin{vmatrix} \hat{\mathbf{x}} & \hat{\mathbf{y}} & \hat{\mathbf{z}} \\ \dfrac{\partial}{\partial x} & \dfrac{\partial}{\partial y} & \dfrac{\partial}{\partial z} \\ F_x & F_y & F_z \end{vmatrix} \tag{A1.9}$$

where the expansion of the determinant must be about the first row, otherwise (A1.9) will not reduce to (A1.8). In coordinates other than rectangular, the curl operation in terms of derivatives is given on the inside of the back cover.

We have now obtained a measure of rotation which is valid at a point. This should be clear since derivatives, in terms of which the curl operator is defined, are point relations which are evaluated at a

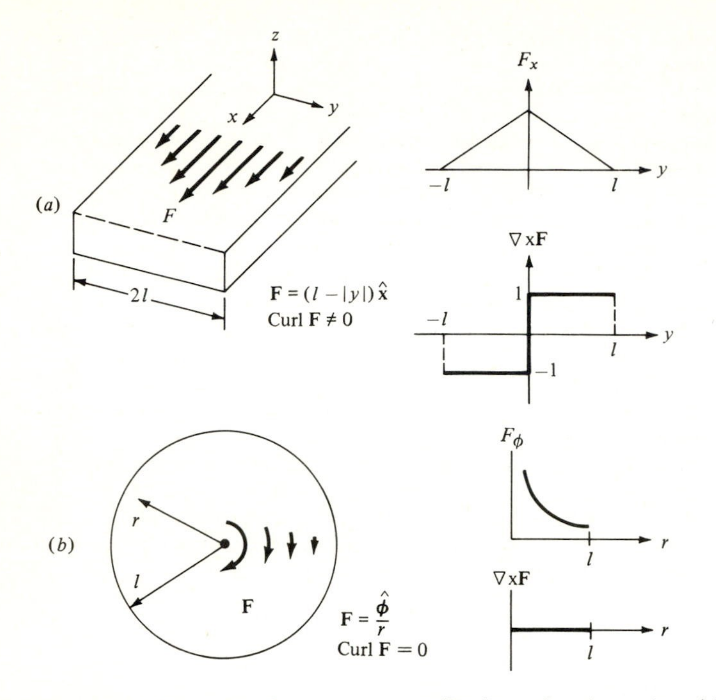

Figure A1.4 Examples of apparent anomalies in curl and rotation. (*a*) Straight-line motion for which the curl is not zero (for example, water flow in a channel); (*b*) rotary motion for which the curl is zero (approximated by water rotation in a long cylinder due to a spinning axial rod).

point. It should be noted that circular flow by itself does not guarantee nonzero curl, nor is circular flow required for nonzero curl. These apparent anomalies are illustrated in Fig. A1.4, where **F** represents the velocity of the fluid.

The mathematical definition of curl is seen to be straightforward, but its visualization may still present problems. Whenever possible, representing the field by a fluid flow helps, since the more familiar behavior of turbulent eddies can then be associated with curl. A device which will literally measure the curl of a turbulent fluid is the paddle wheel shown in Fig. A1.5, also referred to as a *curl meter*. If this device is immersed in a turbulent fluid, it will spin clockwise or counterclockwise,

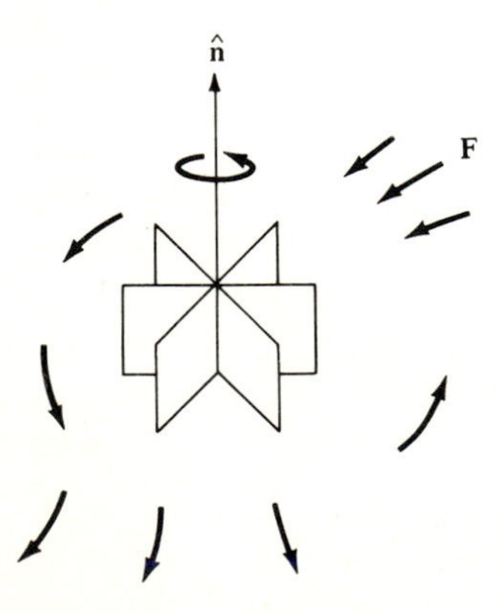

Figure A1.5 Paddle wheel for measuring the curl of a turbulent fluid. **F** represents the velocity at any point in the fluid. The rate of rotation of the paddle wheel is a measure of the component of curl **F** in the direction $\hat{\mathbf{n}}$. In other words, the paddle wheel measures $(\text{curl } \mathbf{F})_n = \hat{\mathbf{n}} \cdot \text{curl } \mathbf{F}$, where $\hat{\mathbf{n}}$ is the direction of the paddle-wheel axis.

corresponding to negative and positive values of curl. If the velocity is the same on both sides of the immersion point, there will be no tendency for the wheel to rotate since the forces on the paddles are balanced. This corresponds to $\oint_l \mathbf{F} \cdot d\mathbf{l} = 0$, or zero curl, where F is the fluid velocity in the neighborhood of the immersion point. If the velocity is greater on one side than on the other so that $\oint_l \mathbf{F} \cdot d\mathbf{l} \neq 0$, the wheel will turn with a rate of rotation proportional to $(\text{curl } \mathbf{F})_n = \hat{\mathbf{n}} \cdot \mathbf{F}$. The direction of the paddle-wheel axis is $\hat{\mathbf{n}}$, and the relationship between curl and circulation is given by (A1.1). Furthermore, when the direction of the curl vector at a particular point in the fluid is unknown, the orientation of the paddle can be changed until a position is found where the rotation of the paddle wheel is a maximum. In this position the direction of the paddle-wheel axis corresponds to the direction of the curl **F** vector, while the maximum rotation is a measure of the magnitude of the curl.

As with any other measuring device, the paddle wheel must be small enough so as not to significantly disturb the flow. Ideally, the device should have zero dimension, for then it would not disturb the field. Of course, such a device could not intercept any energy and thus could not function as a meter, since any measuring device must extract some energy from the field and convert that to an indication. The best, or smallest, measuring apparatus that could be made to work would still be limited by Heisenberg's uncertainty principle.†

The curl of a field was originally called by Maxwell and others the rotation of the field since at that time an electromagnetic field was usually visualized in terms of a mechanical model.‡ In German literature, the name rotation or "rot" is still being used for curl. However, the name curl should not be associated with curvature of the field lines, for a field consisting of closed circles may have zero curl nearly everywhere, and a straight line field may have a finite curl (see Fig. A1.4).

A1.2 STOKES' THEOREM

Stokes' theorem, which is analogous to the divergence theorem, relates a surface integral to a closed-line integral about the contour along the edge of this surface (see Sec. 8.3). It should be noted that a closed curve does not define a surface uniquely, since the same curve could bound many surfaces, two-dimensional and three-dimensional ones. Stokes' theorem states that the line integral of a vector **F** around a closed line l in space is equal to the surface integral of the curl of **F** over the surface A which is bounded by l:

$$\oint_{l(A)} \mathbf{F} \cdot d\mathbf{l} = \iint_{A(l)} \nabla \times \mathbf{F} \cdot d\mathbf{A} \tag{A1.10}$$

[$l(A)$ is the curve l which bounds the area A; $A(l)$ is the area A bounded by the curve l.] To prove the theorem, let us divide the surface A—any surface, not necessarily a plane—into elements of area $\Delta\mathbf{A}_1$, $\Delta\mathbf{A}_2, \ldots, \Delta\mathbf{A}_n$ as shown in Fig. A1.6. For any one of the small areas, using (A1.2),

$$\oint_{l_i} \mathbf{F} \cdot d\mathbf{l}_i = (\nabla \times \mathbf{F}) \cdot \Delta\mathbf{A}_i \tag{A1.11}$$

However, since each interior edge is shared by two elemental areas, the line integral along the common edge will cancel since the circulations in the two adjacent areas are equal and opposite. When all the small contour integrals are added, only those having an edge on the curve l will remain. Thus we can write

$$\oint_l \mathbf{F} \cdot d\mathbf{l} = \sum_i^N \oint_{l_i} \mathbf{F} \cdot d\mathbf{l}_i$$

† G. Gamow, The Principle of Uncertainty, *Sci. Am.*, pp. 51–57, September 1958.
‡ F. J. Dyson, Innovation in Physics, *Sci. Am.*, pp. 74–82, September 1958.

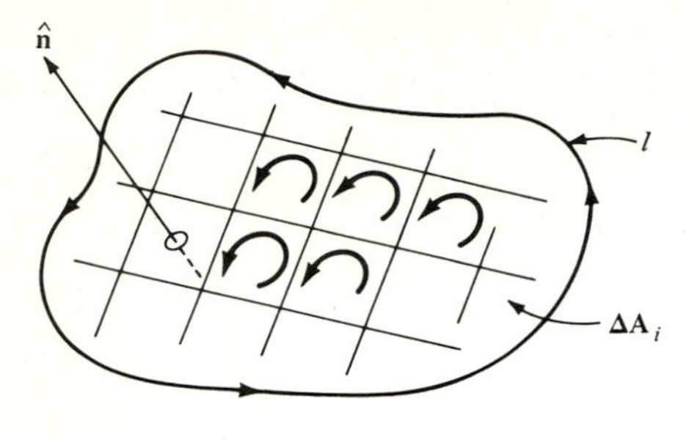

Figure A1.6 An arbitrary surface bounded by l, and its division into small area elements. A field that has a curl is assumed to permeate the surface.

Substituting (A1.11) in the right side of the above equation, we obtain

$$\oint_l \mathbf{F} \cdot \mathbf{dl} = \sum_i^N (\nabla \times \mathbf{F}) \cdot \Delta\mathbf{A}_i$$

In the limit as $N \to \infty$ and $\Delta A \to 0$, the summation becomes an integral; that is,

$$\oint_l \mathbf{F} \cdot \mathbf{dl} = \lim_{N \to \infty} \sum_i^N \cdots = \iint_A (\nabla \times \mathbf{F}) \cdot \mathbf{dA} \tag{A1.12}$$

A1.3 VECTOR IDENTITIES

A list of vector identities is presented below. In the following identities, symbols like **F** and ϕ are vector and scalar functions of space; that is, $\mathbf{F}(x, y, z)$ and $\phi(x, y, z)$.

$$\mathbf{A} \times (\mathbf{B} \times \mathbf{C}) = (\mathbf{A} \cdot \mathbf{C})\mathbf{B} - (\mathbf{A} \cdot \mathbf{B})\mathbf{C} \tag{A1.13}$$

$$\mathbf{A} \cdot \mathbf{B} \times \mathbf{C} = \mathbf{B} \cdot \mathbf{C} \times \mathbf{A} = \mathbf{C} \cdot \mathbf{A} \times \mathbf{B} \tag{A1.14}$$

$$\begin{aligned}(\mathbf{A} \times \mathbf{B}) \cdot (\mathbf{C} \times \mathbf{D}) &= \mathbf{A} \cdot \mathbf{B} \times (\mathbf{C} \times \mathbf{D}) \\ &= \mathbf{A} \cdot [(\mathbf{B} \cdot \mathbf{D})\mathbf{C} - (\mathbf{B} \cdot \mathbf{C})\mathbf{D}] \\ &= (\mathbf{A} \cdot \mathbf{C})(\mathbf{B} \cdot \mathbf{D}) - (\mathbf{A} \cdot \mathbf{D})(\mathbf{B} \cdot \mathbf{C})\end{aligned} \tag{A1.15}$$

$$(\mathbf{A} \times \mathbf{B}) \times (\mathbf{C} \times \mathbf{D}) = [(\mathbf{A} \times \mathbf{B}) \cdot \mathbf{D}]\mathbf{C} - [(\mathbf{A} \times \mathbf{B}) \cdot \mathbf{C}]\mathbf{D} \tag{A1.16}$$

$$\nabla(\phi + \psi) = \nabla\phi + \nabla\psi \tag{A1.17}$$

$$\nabla \cdot (\phi\mathbf{F}) = \phi\nabla \cdot \mathbf{F} + \mathbf{F} \cdot \nabla\phi \tag{A1.18}$$

$$\nabla \times (\phi\mathbf{F}) = \phi\nabla \times \mathbf{F} + \nabla\phi \times \mathbf{F} \tag{A1.19}$$

$$\nabla(\phi\psi) = \phi\nabla\psi + \psi\nabla\phi \tag{A1.20}$$

$$\nabla \cdot (\mathbf{F} \times \mathbf{G}) = \mathbf{G} \cdot \nabla \times \mathbf{F} - \mathbf{F} \cdot \nabla \times \mathbf{G} \tag{A1.21}$$

$$\begin{aligned}\nabla \times (\mathbf{F} \times \mathbf{G}) &= \mathbf{F}\nabla \cdot \mathbf{G} - \mathbf{G}\nabla \cdot \mathbf{F} \\ &\quad + (\mathbf{G} \cdot \nabla)\mathbf{F} - (\mathbf{F} \cdot \nabla)\mathbf{G}\end{aligned} \tag{A1.22}$$

$$\nabla(\mathbf{F}\cdot\mathbf{G}) = (\mathbf{F}\cdot\nabla)\mathbf{G} + (\mathbf{G}\cdot\nabla)\mathbf{F} + \mathbf{F}\times(\nabla\times\mathbf{G}) + \mathbf{G}\times(\nabla\times\mathbf{F}) \tag{A1.23}$$

$$\nabla\times\nabla\times\mathbf{F} = \nabla\nabla\cdot\mathbf{F} - \nabla^2\mathbf{F} \tag{A1.24}$$

$$\nabla\cdot(\phi\nabla\psi) = \nabla\phi\cdot\nabla\psi + \phi\nabla^2\psi \tag{A1.25}$$

$$\nabla\times\nabla\phi = 0 \tag{A1.26}$$

$$\nabla\cdot\nabla\times\mathbf{F} = 0 \tag{A1.27}$$

In the above identities the operator $\mathbf{F}\cdot\nabla$ is defined as

$$\mathbf{F}\cdot\nabla = F_x\frac{\partial}{\partial x} + F_y\frac{\partial}{\partial y} + F_z\frac{\partial}{\partial z} \tag{A1.28}$$

Besides Gauss' and Stokes' theorems,

$$\oiint_{A(V)} \mathbf{F}\cdot\mathbf{dA} = \iiint_{V(A)} \nabla\cdot\mathbf{F}\,dV \tag{A1.29}$$

$$\oint_{l(A)} \mathbf{F}\cdot\mathbf{dl} = \iint_{A(l)} \nabla\times\mathbf{F}\cdot\mathbf{dA} \tag{A1.30}$$

the following theorems can be derived but are of lesser importance in electromagnetic field problems than Gauss' and Stokes' theorems:

$$\oint_{l(A)} \mathbf{F}\times\mathbf{dl} = \iint_{A(l)} [(\nabla\cdot\mathbf{F})\hat{\mathbf{n}} - \nabla(\mathbf{F}\cdot\hat{\mathbf{n}})]\,dA \tag{A1.31}$$

$$\oiint_{A(V)} \hat{\mathbf{n}}\times\mathbf{F}\,dA = \iiint_{V(A)} \nabla\times\mathbf{F}\,dV \tag{A1.32}$$

$$\oiint_{A(V)} \hat{\mathbf{n}}\phi\,dA = \iiint_{V(A)} \nabla\phi\,dV \tag{A1.33}$$

$$\oint_{l(A)} \phi\,\mathbf{dl} = \iint_{A(l)} \hat{\mathbf{n}}\times\nabla\phi\,dA \tag{A1.34}$$

In the above identities $\hat{\mathbf{n}}$ is a unit vector normal to the surface element dA of the surface A (note that $\mathbf{dA}$ can be written as $\hat{\mathbf{n}}\,dA$). The directions of $\hat{\mathbf{n}}$ and the sense of the contour l are related by the "corkscrew rule" (Fig. A1.2); that is, the sense in which the contour l is described is that of rotation of a right-handed corkscrew advancing in the direction of $\hat{\mathbf{n}}$.

Index

Index

Rectangular Coordinates

$$\nabla f = \hat{\mathbf{x}}\,\frac{\partial f}{\partial x} + \hat{\mathbf{y}}\,\frac{\partial f}{\partial y} + \hat{\mathbf{z}}\,\frac{\partial f}{\partial z}$$

$$\nabla \cdot \mathbf{F} = \frac{\partial F_x}{\partial x} + \frac{\partial F_y}{\partial y} + \frac{\partial F_z}{\partial z}$$

$$\nabla \times \mathbf{F} = \hat{\mathbf{x}}\left(\frac{\partial F_z}{\partial y} - \frac{\partial F_y}{\partial z}\right) + \hat{\mathbf{y}}\left(\frac{\partial F_x}{\partial z} - \frac{\partial F_z}{\partial x}\right) + \hat{\mathbf{z}}\left(\frac{\partial F_y}{\partial x} - \frac{\partial F_x}{\partial y}\right)$$

$$\nabla^2 f = \frac{\partial^2 f}{\partial x^2} + \frac{\partial^2 f}{\partial y^2} + \frac{\partial^2 f}{\partial z^2}$$

$$\nabla^2 \mathbf{F} = \hat{\mathbf{x}}\,\nabla^2 F_x + \hat{\mathbf{y}}\,\nabla^2 F_y + \hat{\mathbf{z}}\,\nabla^2 F_z$$

Cylindrical Coordinates

$$\nabla f = \hat{\boldsymbol{\rho}}\,\frac{\partial f}{\partial \rho} + \hat{\boldsymbol{\phi}}\,\frac{1}{\rho}\frac{\partial f}{\partial \rho} + \hat{\mathbf{z}}\,\frac{\partial f}{\partial z}$$

$$\nabla \cdot \mathbf{F} = \frac{1}{\rho}\frac{\partial}{\partial \rho}(\rho F_\rho) + \frac{1}{\rho}\frac{\partial F_\phi}{\partial \phi} + \frac{\partial F_z}{\partial z}$$

$$\nabla \times \mathbf{F} = \hat{\boldsymbol{\rho}}\left(\frac{1}{\rho}\frac{\partial F_z}{\partial \phi} - \frac{\partial F_\phi}{\partial z}\right) + \hat{\boldsymbol{\phi}}\left(\frac{\partial F_r}{\partial z} - \frac{\partial F_z}{\partial \rho}\right) + \hat{\mathbf{z}}\,\frac{1}{\rho}\left(\frac{\partial}{\partial \rho}[\rho F_\phi] - \frac{\partial F_r}{\partial \phi}\right)$$

$$\nabla^2 f = \frac{1}{\rho}\frac{\partial}{\partial \rho}\left(\rho\,\frac{\partial f}{\partial \rho}\right) + \frac{1}{\rho^2}\frac{\partial^2 f}{\partial \phi^2} + \frac{\partial^2 f}{\partial z^2}$$

$$\nabla^2 \mathbf{F} = \nabla(\nabla \cdot \mathbf{F}) - \nabla \times \nabla \times \mathbf{F}$$

Spherical Coordinates

$$\nabla f = \hat{\mathbf{r}}\,\frac{\partial f}{\partial r} + \hat{\boldsymbol{\theta}}\,\frac{1}{r}\frac{\partial f}{\partial \theta} + \hat{\boldsymbol{\phi}}\,\frac{1}{r\sin\theta}\frac{\partial f}{\partial \phi}$$

$$\nabla \cdot \mathbf{F} = \frac{1}{r^2}\frac{\partial}{\partial r}(r^2 F_r) + \frac{1}{r\sin\theta}\frac{\partial}{\partial \theta}(F_\theta \sin\theta) + \frac{1}{r\sin\theta}\frac{\partial F_\phi}{\partial \phi}$$

$$\nabla \times \mathbf{F} = \hat{\mathbf{r}}\,\frac{1}{r\sin\theta}\left(\frac{\partial}{\partial \theta}(F_\phi \sin\theta) - \frac{\partial F_\theta}{\partial \phi}\right) + \hat{\boldsymbol{\theta}}\,\frac{1}{r}\left(\frac{1}{\sin\theta}\frac{\partial F_r}{\partial \phi} - \frac{\partial}{\partial r}(rF_\phi)\right)$$

$$+ \hat{\boldsymbol{\phi}}\,\frac{1}{r}\left(\frac{\partial}{\partial r}(rF_\theta) - \frac{\partial F_r}{\partial \theta}\right)$$

$$\nabla^2 f = \frac{1}{r^2}\frac{\partial}{\partial r}\left(r^2\,\frac{\partial f^2}{\partial r}\right) + \frac{1}{r^2\sin\theta}\frac{\partial}{\partial \theta}\left(\sin\theta\,\frac{\partial f}{\partial \theta}\right) + \frac{1}{r^2\sin^2\theta}\frac{\partial^2 f}{\partial \phi^2}$$

$$\nabla^2 \mathbf{F} = \nabla(\nabla \cdot \mathbf{F}) - \nabla \times \nabla \times \mathbf{F}$$